HIGH-VELOCITY IMPACT PHENOMENA

CONTRIBUTORS

A. J. CABLE

W. J. CARTER

J. K. DIENES

J. N. FRITZ

JOHN W. GEHRING, JR.

C. M. GLASS

NORRIS L. HICKERSON

R. G. MCQUEEN

S. P. MARSH

R. B. POND

WILLIAM J. RAE

T. D. RINEY

J. W. TAYLOR

J. M. WALSH

High-Velocity Impact Phenomena

Edited by RAY KINSLOW

DEPARTMENT OF ENGINEERING SCIENCE
TENNESSEE TECHNOLOGICAL UNIVERSITY
COOKEVILLE, TENNESSEE

1970

ACADEMIC PRESS New York and London

ACADEMIC PRESS, INC.
111 Fifth Avenue, New York, New York 10003

United Kingdom Edition published by
ACADEMIC PRESS, INC. (LONDON) LTD.
Berkeley Square House, London W1X 6BA

LIBRARY OF CONGRESS CATALOG CARD NUMBER: 71-91425

PRINTED IN THE UNITED STATES OF AMERICA

LIST OF CONTRIBUTORS

Numbers in parentheses indicate the pages on which the authors' contributions begin.

A. J. CABLE (1), Aerophysics Branch, Von Karman Gas Dynamics Facility, Aro, Inc., Arnold Engineering Development Center, Arnold Air Force Station, Tennessee

W. J. CARTER (293), Los Alamos Scientific Laboratory, University of California, Los Alamos, New Mexico

J. K. DIENES* (45), Gulf General Atomic, La Jolla, California

J. N. FRITZ (293), Los Alamos Scientific Laboratory, University of California, Los Alamos, New Mexico

JOHN W. GEHRING, JR. (105, 463), Impact Physics Laboratory, General Motors Corporation and AC Electronic Defense Research Laboratories, Santa Barbara, California

C. M. GLASS (419), Terminal Ballistics Division, U. S. Army Ballistics Research Laboratory, Aberdeen, Maryland

NORRIS L. HICKERSON (23), Tennessee Technological University, Cookeville, Tennessee

R. G. MCQUEEN (293), Los Alamos Scientific Laboratory, University of California, Los Alamos, New Mexico

S. P. MARSH (293), Los Alamos Scientific Laboratory, University of California, Los Alamos, New Mexico

R. B. POND (419), The Johns Hopkins University, Baltimore, Maryland

* Present address: Systems, Science and Software, La Jolla, California.

WILLIAM J. RAE (213), Cornell Aeronautical Laboratory, Inc., Buffalo, New York

T. D. RINEY* (157), Space Sciences Laboratory, General Electric Company, King of Prussia, Pennsylvania

J. W. TAYLOR (293), Los Alamos Scientific Laboratory, University of California, Los Alamos, New Mexico

J. M. WALSH* (45), Gulf General Atomic, La Jolla, California

* Present address: Systems, Science and Software, La Jolla, California.

CONTENTS

IV. Theory of Impact on Thin Targets and Shields and Correlation with Experiment

JOHN W. GEHRING, JR.

V. Numerical Evaluation of Hypervelocity Impact Phenomena

T. D. RINEY

VI. Analytical Studies of Impact-Generated Shock Propagation: Survey and New Results

WILLIAM J. RAE

VII. The Equation of State of Solids from Shock Wave Studies

R. G. MCQUEEN, S. P. MARSH, J. W. TAYLOR, J. N. FRITZ, AND W. J. CARTER

VIII. Metallurgical Observations and Energy Partitioning

R. B. Pond and C. M. Glass

IX. Engineering Considerations in Hypervelocity Impact

John W. Gehring, Jr.

PREFACE

Terminal ballistics has been the subject of formal study for more than two centuries. The incentive for such a study was largely motivated by warfare with its need for the development of faster projectiles and stronger armor. Current interest in high-speed impact is due largely to the need for information concerning the meteoroid hazard to space vehicles. The speeds of this interplanetary debris exceed by an order of magnitude or more the impact velocities common to traditional ballistics, and the mechanics of the impact process are quite different.

During the past decade considerable effort has been expended toward the generation of experimental data as well as the formulation of realistic theories to explain the dynamic mechanical response of materials and the many phenomena that are involved.

The reaction of materials to the extremely high pressures resulting from hypervelocity impact is of great fundamental importance to the physicist. The nature of the strong shock waves generated by such impact makes the subject one of interest to the seismologist. These extremely high pressures and shock waves can be used to harden and form metals, bond dissimilar materials, and be utilized by the engineer in many other ways in addition to the problem of spacecraft design.

The book covers a wide range of pertinent topics dealing with impact phenomena. Scientists and engineers not already familiar with the dynamics of impulsive loading may acquire a background and develop guidelines for further studies and investigation. The specialist in any aspect of the field may have his viewpoint broadened and learn more of other phases of the problem. The book should be of especial value to the design engineer as it includes discussions of practical aspects associated with specific problems such as spacecraft shielding, reentry vehicles, space radiators, solar panels, and pressure vessels; information concerning the use of laminates, composites, fillers, and honeycombs;

and data relating to specific engineering materials such as the Hugoniot coefficients and isentropic values of fifty-five substances given in Appendix E.

The authors of the various chapters were chosen both for their outstanding research dealing with the topic assigned and for their ability to present and explain the subject in a clear and comprehensible manner.

CHAPTER I

HYPERVELOCITY ACCELERATORS

A. J. CABLE

AEROPHYSICS BRANCH, VON KARMAN GAS DYNAMICS FACILITY, ARO, INC.
ARNOLD ENGINEERING DEVELOPMENT CENTER
ARNOLD AIR FORCE STATION, TENNESSEE

I. Introduction

In the study of hypervelocity impact phenomena, it is first necessary to develop methods of accelerating projectiles of interest to hypervelocities. The devices used to perform this task, hypervelocity accelerators, fall basically into two broad divisions: gun accelerators and explosive accelerators.

Gun accelerators include all accelerators in which the projectile is guided during acceleration, while in explosive accelerators the projectile is essentially unguided.

II. Gun Accelerators

A. Two-Stage Light-Gas Guns

The most commonly used gun for hypervelocity acceleration is the two-stage light-gas gun first developed at the New Mexico School of Mines (NMSM) in 1948 by Crozier and Hume (*1*) and since followed by many others (*2–19*). Figure 1 is a diagrammatic sketch of such a launcher. The combustion of a propellant composition in the powder chamber bursts a diaphragm and drives the piston along the pump tube, compressing a light gas (usually hydrogen). A diaphragm separating the launch tube from the pump tube bursts at a predetermined pressure and allows the compressed light gas to propel the projectile along the launch tube. In some versions, the propellant in the powder chamber is surrounded by a light gas (hydrogen or helium). In another version, the powder chamber is part of the pump tube, and the diaphragm and light gas are not used.

There are also variations in the shape of the launch-tube–pump-tube change of area. In some launchers, the change is abrupt, as in the Arnold Engineering Development Center (AEDC) launchers (*7–10, 12–14*). In others, a taper is used to encourage piston extrusion, as in the "accelerated reservoir guns" of the Ames Research Center, National Aeronautics and Space Administration (NASA Ames) (*11*), and General Motors Defense Research Laboratories (GMDRL) (*16*). Figure 2 is a photograph of a 0.5-in.-caliber, two-stage light-gas gun at the AEDC, which is typical of the smaller-caliber guns used in impact work. Figure 3 is a plot of launch velocity versus total launch weight for two-stage light-gas guns taken from Lukasiewicz (*20*). This shows the velocities and launched weights which could be obtained in late 1966. These range from 0.044 g (10^{-4} lb) launched at 37,000 ft/sec (11.3 km/sec) to 2500 g (5.5 lb) at 15,500 ft/sec (4.7 km/sec).

The total in-gun weight includes the weight of the model being tested and also that of the sabot, which adapts the model to the launcher bore and carries it undamaged to the muzzle. At the muzzle, the sabot is separated from the projectile either mechanically or aerodynamically. In most cases the sabot is manufactured of a plastic material of which polycarbonate is the most used. The sabots vary from a simple cylinder carrying the model in a dimple in the front face to multipiece sabots interlocked together enclosing the model completely. Many simple accelerators dispense with sabots, but the projectile can possibly then be damaged during launch, which reduces the accuracy of the data obtained.

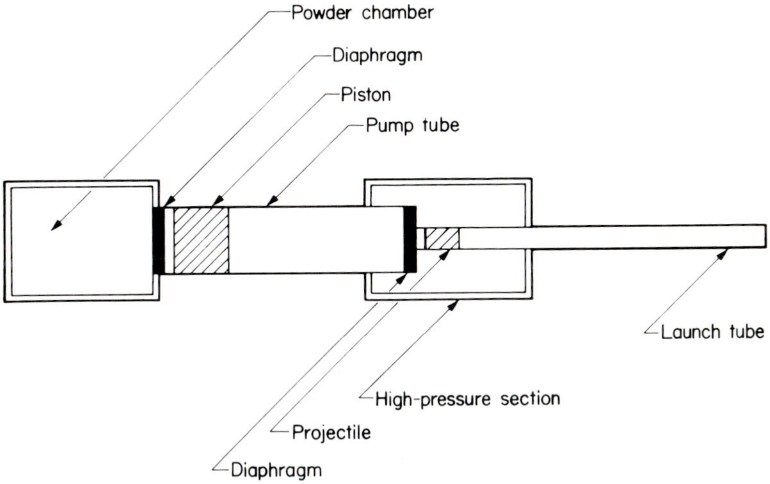

FIG. 1. Schematic arrangement of two-stage light-gas gun.

FIG. 2. AEDC VKF hypervelocity impact range S-1.

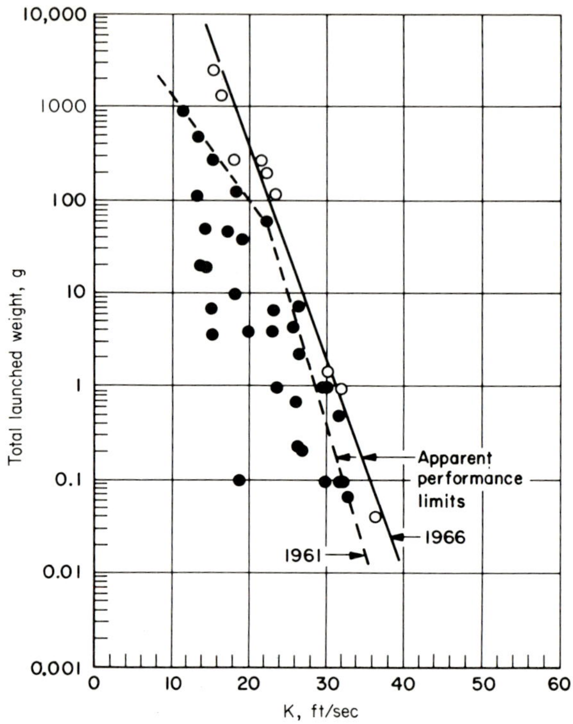

FIG. 3. Maximum launch velocities and weights for two-stage light-gas gun. (○) 1963–66 data (*20*).

1. *Theory of Two-Stage Light-Gas Guns*

The theory of the operation of light-gas guns is now well established, starting from the early work of Stephenson at the AEDC (*8*, *12*). In this theory the projectile in the launch tube was assumed to be driven by a gas expanding from a reservoir at a constant pressure corresponding to the peak pressure in the high-pressure section, empirical factors being used to relate the piston motion to the projectile motion. This technique, after preliminary firings had been made, gave accurate predictions of launch velocity. Then Wilenius at the Canadian Armament Research and Development Establishment (CARDE) (*21*), based on the work of Stanyukovitch (*22*), and, independently, Smith of the Royal Armament Research and Development Establishment (RARDE) (*23*) developed theories of the "constant base pressure" light-gas gun. In this gun the projectile is assumed to be accelerated by a constant base pressure, which not only simplifies the theory but also minimizes the maximum acceleration experienced by the projectile. Unfortunately, it is not possible to obtain this in practice, although a maximum acceleration of only 24% greater than the constant base pressure value has been reported by Cable (*24*). Other contributions to the theory of constant base pressure launchers have been made by Curtis of GMDRL (*25*) and Wilenius of CARDE (*26*).

In 1963, Piacesi at the Naval Ordnance Laboratory (NOL) (*27*) and Baer at the Ballistic Research Laboratories (BRL) (*28*) independently developed performance theories using the von Neumann–Richtmeyer "*q*" method (*29*) for which the assumption of constant base pressure was no longer needed. This launcher theory was capable of predicting both projectile acceleration and launch velocity, as demonstrated by DeWitt (*30*).

2. *Augmentation of Two-Stage Light-Gas Guns*

Due to the present limitation of the light-gas gun to velocities of around 32,000–37,000 ft/sec (10–11 km/sec) with 0.5-in.- to 0.22-in.-caliber guns, attempts have been made to augment their velocities by various techniques. The driving gas has been heated using an electrical discharge of energy in the pump tube by Eckerman of Avco (*17*) and Swift of the Naval Research Laboratory (NRL) (*19*), and also by electrical discharge in the launch tube by Howell of the Denver Research Institute (DRI) (*31*). Another technique has been the "traveling reservoir" (*32*), in which a portion of the piston is carried by its own momentum along the launch tube, further propelling the projectile. In the "muzzle squirter" (*33*), the momentum of the sabot causes it to extrude through a taper at the muzzle, enhancing the projectile velocity.

Another similar technique to the "muzzle squirter" has been developed by Cooley, Exotech, Inc. (*34*), and is described by him as a "shock plasma acceleration technique" (SPLAT). In this technique the "booster," consisting of a tube of plastic (nylon, polypropylene, or polycarbonate), is located inside the gun bore near the muzzle (Fig. 4). Its inner diameter is somewhat smaller than the caliber of the gun though larger than the sphere carried on the sabot. When the sabot impacts the booster, the booster material is shocked and converges toward the center of the gun, producing a "jet" which accelerates the projectile. A 220 Swift gun has been boosted in velocity from 9000 ft/sec (2.7 km/sec) to 15,000 ft/sec (4.6 km/sec). For one shot with a 0.50-in.-caliber, two-stage light-gas gun, the velocity was augmented from 21,000 ft/sec (6.4 km/sec) to 24,600 ft/sec (7.5 km/sec), and the sphere model was preceded by the plastic. This has also occurred in the "muzzle squirter" when attempts were made to launch metal spheres.

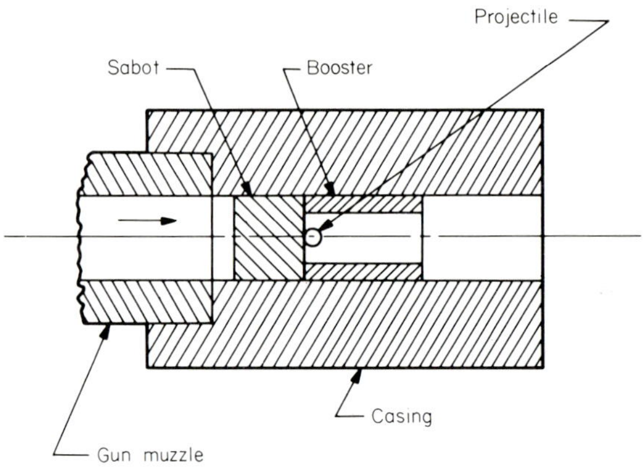

FIG. 4. Cross section of a "SPLAT" velocity augmenter.

These techniques have all augmented the muzzle velocity, when the unaugmented velocity has been low, but none have reliably produced velocities greater than those obtained by the highest-velocity unaugmented two-stage light-gas guns.

B. Electromagnetic Guns

A number of electromagnetic guns (*35–41*) have been developed but have not been as widely used as the light-gas guns. The induction solenoid gun uses the linear induction motor principle. The projectile which is a conductor is contained in a long coil and is inherently centered in the coil (Fig. 5). By varying either the frequency of the current applied to the windings or the spacing of the windings, the projectile is accelerated by the interaction of the eddy currents induced in it with the moving magnetic flux. The fundamental difficulty with this gun is to provide polyphase pulse power of the order of a million kilowatts and frequencies varying from a few hertz to several hundred kilohertz. A four-coil gun of this type was built by Salisbury, Zenith Radio Research Corporation (*36*), and each coil was separately pulsed. This gun produced a velocity of 1150 ft/sec (0.35 km/sec) with a 2.4-g projectile. Attempts to produce higher velocities caused collapse of the coils and of the projectiles. Harris, University of Utah (*37*), produced velocities of 660 ft/sec (0.20 km/sec) with a 4.5-g projectile with a similar device.

Another type of electromagnetic gun is the parallel rail or dc gun (Armour Research Foundation) (*35*), in which the projectile is in sliding contact with two rails; and when the current flows in the rails, the

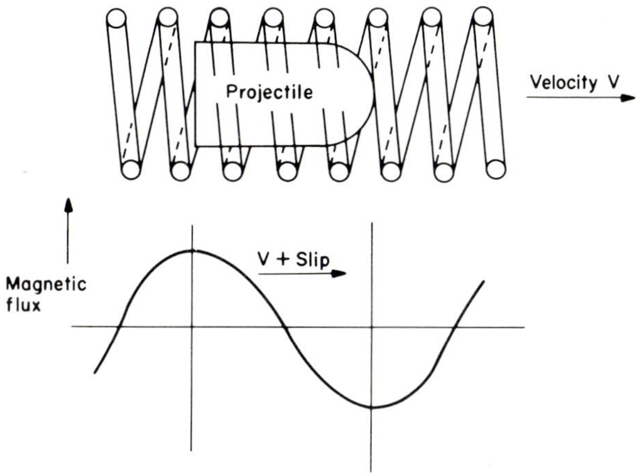

FIG. 5. Principle of operation of induction solenoid gun.

projectile is accelerated along the rails (Fig. 6). To reduce excessive power requirements, these electromagnetic guns must be contained in evacuated chambers, and then problems arise due to arcing of the contacts. MB Associates (*38*) describe a rail gun which has launched 0.0024-g nylon spheres to 19,700 ft/sec (6.0 km/sec). A further modification of the rail gun has been made by Rosen, North American Aviation (*39*), in an attempt to overcome the velocity limitation due to the Joule heating of the projectile, by the addition of an externally generated magnetic field. This rail gun has produced a velocity of 9500 ft/sec (2.9 km/sec) with a 0.0052-g aluminum model, which is considerably lower than that obtained by the rail gun without the externally generated magnetic field. Chapman *et al.* (MB Associates) (*40*) have produced a velocity of 31,000 ft/sec (9.5 km/sec) with 0.010-g projectiles, which, however, broke up, by explosively compressing the magnetic field to produce very high field strengths (2 MG at 2.4×10^6 psi). A similar technique has been reported by Chapman *et al.* (*41*) where a "coherent luminous cloud" was projected at 34,000 ft/sec (10.5 km/sec).

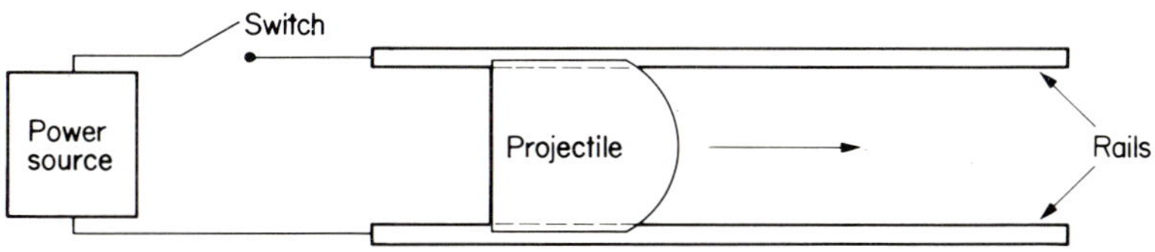

FIG. 6. Parallel-rail electric gun.

C. Explosively Driven Guns

A number of guns have been built where additional explosive energy is added to the projectile during acceleration along the gun barrel. In the "traveling charge gun" of BRL (*42*), the projectile carries its own propellant and accelerates like a rocket guided in a long tube. Its two main advantages are: (1) the projectile is subjected to a low, more nearly constant acceleration, eliminating the possibility of projectile failure; and (2) theoretically, except for relativistic considerations, there is no upper limit to the velocity obtainable. The disadvantages are: (1) very long barrel lengths are required; (2) very high-burning-rate propellants are required with burning rates three orders of magnitude greater than conventional propellants; and (3) with the high burning rates needed, the possibility of detonation of the propellant can occur. To date, because of the very high structural weight required to contain the high combustion pressures developed, it has not proved possible to obtain the high mass ratio (ratio of total to structural weight) required to obtain hypervelocities, despite the obvious attractions of this type of gun.

Frankford Arsenal (*43*) describes a launcher where the projectile is successively accelerated by sequential impulses from high-explosive stages. The projectile passes through each explosive stage, which is initiated after some suitable time interval. Velocities of up to 8600 ft/sec (2.6 km/sec) were obtained with 7.4-g projectiles and two explosive stages, but difficulties with the control of the initiation of the explosives were experienced.

Lithium hydride has been exploded to augment the velocity of a light-gas gun, as it provides a gas of molecular weight similar to that of helium. However, in the experiments described by Bergson *et al.*, General Electric Company (GE) (*44*), a velocity of only 1200 ft/sec (0.37 km/sec) was obtained with a 4.5-g projectile.

A more promising hypervelocity acceleration technique has been described by Physics International (PI) and Stanford Research Institute (SRI) (*45–47*). A high-explosive charge surrounding a metal liner is initiated from the end to produce a shaped-charge effect; however, the cone angle of the collapsing liner is too low for the metal to form a jet, but jetting of the gas (hydrogen or helium) contained in the liner does occur and gas velocities of 130,000 ft/sec (140 km/sec) have been reported. Figure 7, based on work reported by PI (*46*), shows the method of operation of such a gun with a collapsing conical liner. With a cylindrical liner, a velocity 26,000 ft/sec (8.0 km/sec) has been obtained for 0.1-g projectiles, and one of 19,000 ft/sec (5.8 km/sec) has been obtained with a 102-g projectile (*46*).

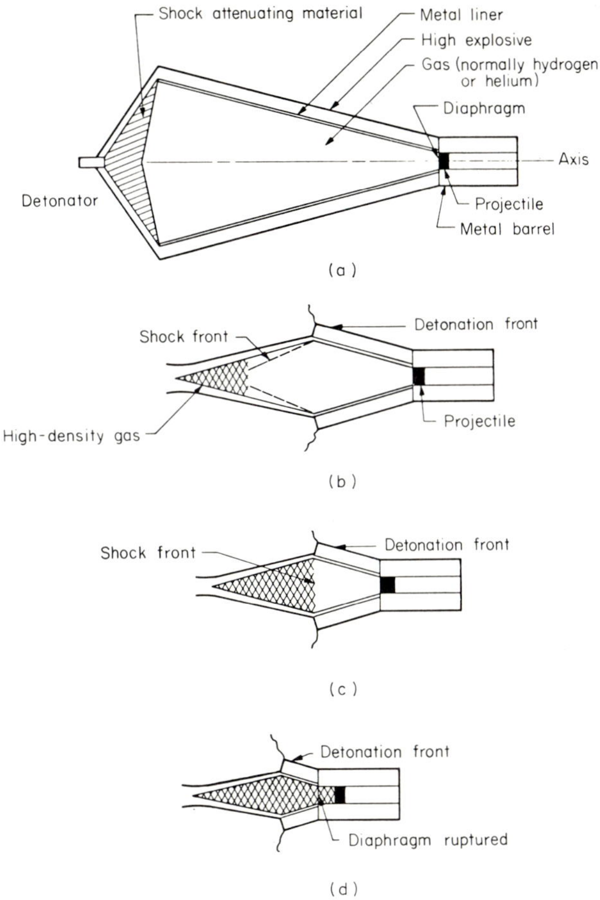

FIG. 7. Collapsing liner shaped-charge gun. (a) Schematic; (b) position of liner T_1 μsec after detonation; (c) position of liner T_2 μsec after detonation; (d) position of liner T_3 μsec after detonation.

D. The Exploding-Wire Accelerator

The exploding wire accelerator uses the energy of a fine wire exploded by the rapid discharge of a high-voltage capacitor through it to accelerate the projectile. An exploding-wire accelerator is described by Scherrer, Technical Operations, Inc. (*48*), in which the pressure vessel containing the exploding wire is water in a suitable container, as shown in Fig. 8. The water is prevented from entering the evacuated launch tube by a thin plastic diaphragm. The discharge of 500 J of energy through a

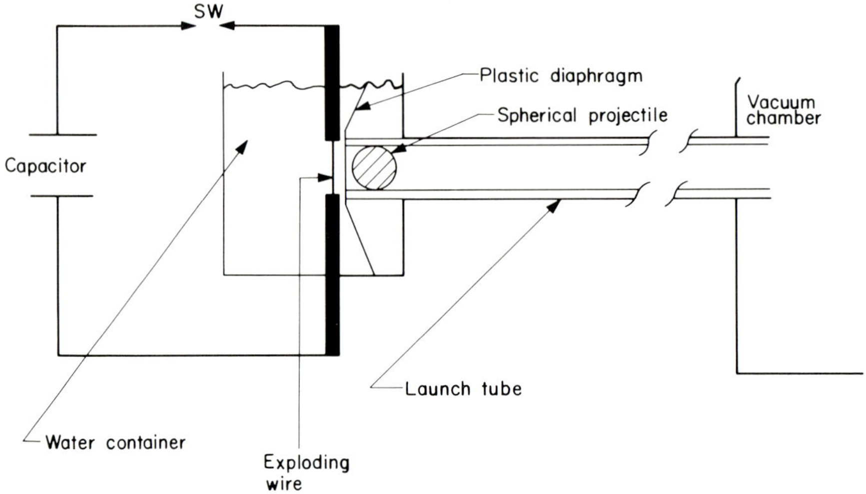

FIG. 8. Exploding-wire hypervelocity accelerator.

0.00025-in. aluminum foil launched a 0.128-g projectile at 8200 ft/sec (2.5 km/sec).

Another version, the exploding foil backstrap accelerator, is described by Clark, Air Force Materials Laboratory (AFML) (*49*), and a sketch of it is shown as Fig. 9. In this accelerator, the current flowing in the

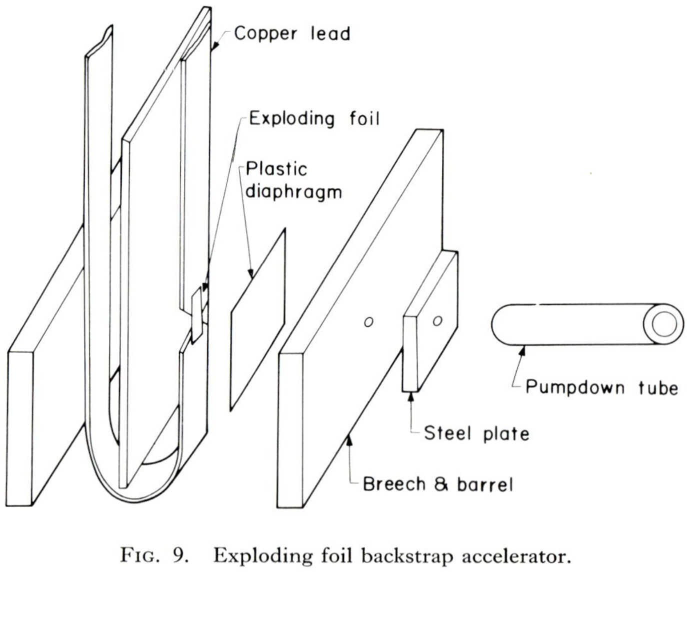

FIG. 9. Exploding foil backstrap accelerator.

backstrap creates a magnetic field which repels the plasma created by the exploding foil, thus accelerating the plasma and the projectile which is a disk from the plastic diaphragm. The mass of the projectile has been measured by measuring the momentum of a light-weight target consisting of thin aluminum sheets spaced by thicker sheets of foamed plastic. A maximum velocity of 35,000 ft/sec (10.7 km/sec) with a mass of 0.0008 g has been recorded with this accelerator. The disadvantages of this type of accelerator are the difficulty in determining projectile mass and shape and its limitation to launching thin plastic disks.

III. Explosive Accelerators

The other major hypervelocity acceleration device accelerates the projectile by explosive means, and the only guidance is that obtained by the positioning of the projectile in the explosive device. These devices will be discussed under three subdivisions: high-explosive and shaped-charge accelerators, plasma drag accelerators, electrostatic accelerators.

A. High-Explosive and Shaped-Charge Accelerators

Poulter (SRI) (*50*) and Willig (GE) (*51*) describe methods of accelerating projectiles by laying them on the surface or embedding them in the surface of a high-explosive charge, as shown in Fig. 10. To prevent blast damage to the target, baffles can be used or the end of the explosive charge can be shaped. Projectile damage can occur with this method of launch and damage can be reduced by embedding the spherical projectile in the explosive or having inert material between the projectile and explosive charge. The metal plate projectiles suffer from the problem of buckling. A plate velocity of as high as 30,000 ft/sec (9.1 km/sec) is reported by Christian (*52*). Clark *et al.* (*53*) and Kineke (*54*) describe air cavity high-explosive charge projectors (Fig. 11), which have been used to launch disks. At Picatinny Arsenal (*53*), velocities of up to 15,600 ft/sec (4.75 km/sec) have been obtained with 8-g aluminum disks, and at the BRL (*54*), 16,400 ft/sec (5.01 km/sec) with 0.18-g projectiles.

A plate acceleration technique known as the "target plate accelerator," in which the target plate is projected at a stationary projectile, is described by Kreyenhagen *et al.* (Aerojet-General Corporation) (*55*). The plate is laid on an explosive block initiated by a line wave generator, and the plate accelerates by folding off the explosive block, as is demonstrated in Fig. 12. An aluminum plate has been launched at 11,500 ft/sec (3.5 km/sec) using this method. It is useful where the projectile material is too fragile to withstand launching accelerations; but since the target

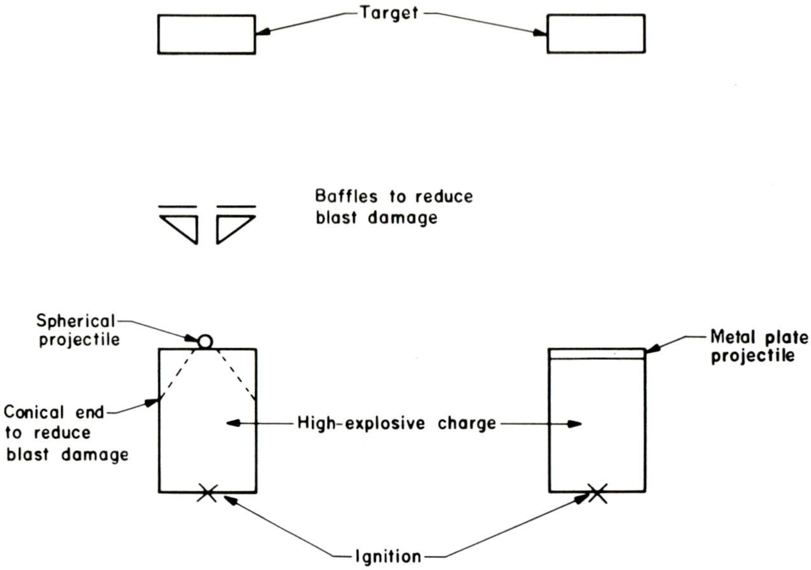

FIG. 10. Types of high-explosive accelerators.

plate cannot be recovered, the diameter of the hole produced by impact is obtained by x-ray photography, and the target is limited to thin plates which can be completely penetrated.

One of the most-used explosive projectors is the shaped-charge explosive projector (Fig. 13). The projectiles are formed by the jetting

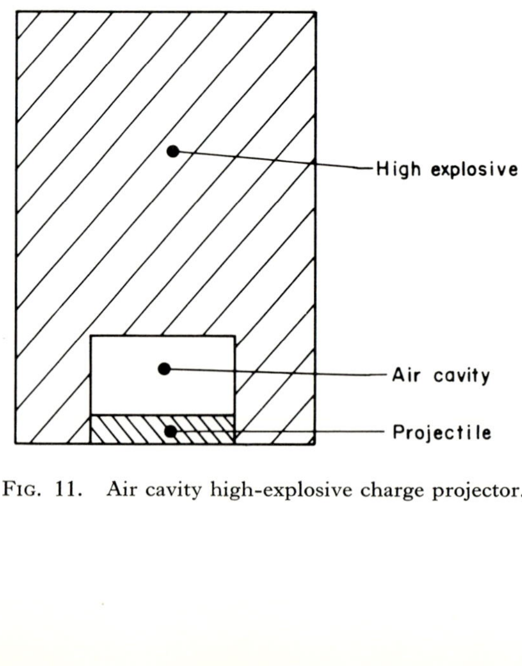

FIG. 11. Air cavity high-explosive charge projector.

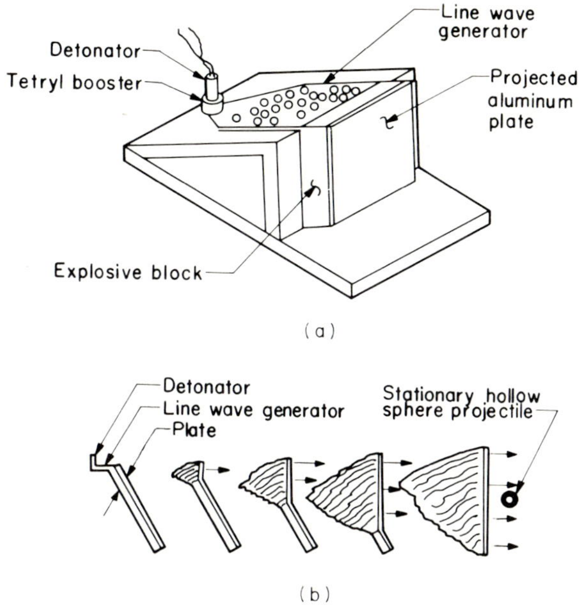

FIG. 12. Target plate accelerator. (a) Plate acceleration technique; (b) folding off of accelerating plate.

of the conical liner. The jet, which is formed by the shaped-charge process, contains a velocity gradient; and thus the jet stretches as it flies and breaks up into individual projectiles, which are dispersed laterally by the eccentric initiation to produce separated impacts on the target. Since the initial breakup of the jet occurs in a more or less random

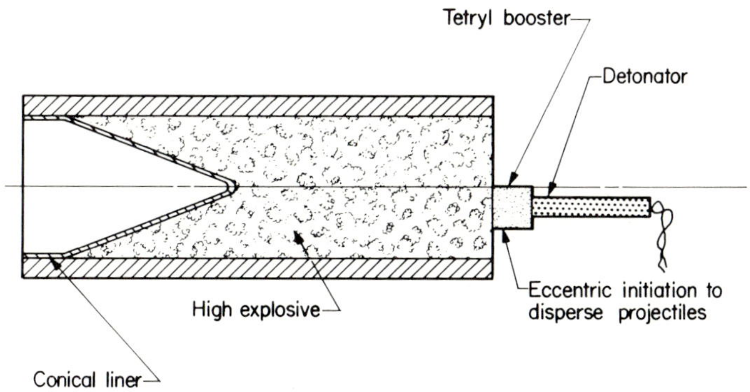

FIG. 13. Shaped-charge explosive projector.

manner, the velocity, mass, shape, size, and orientation of the individual fragments must be determined before impact. These measurements are most usually made by the use of orthogonal x-ray shadow graphs. The liner material determines the fragment material, and the velocity is varied by changing the cone angle of the shaped charge, the velocity increasing with reduction of cone angle. Usually, such explosives as Composition B, HMX 9404, ERS-9, and Kel-F are used for these projectors. Kreyenhagen (*55*) quotes velocities of up to 33,000 ft/sec (10 km/sec) from a shaped charge of a similar type to that illustrated.

In the inhibited shaped-charge projector (Fig. 14), a plastic inhibitor is used to cut off all fragments of the jet except for the first and largest fragment of the jet. A semicircular deflector charge is used to deflect the

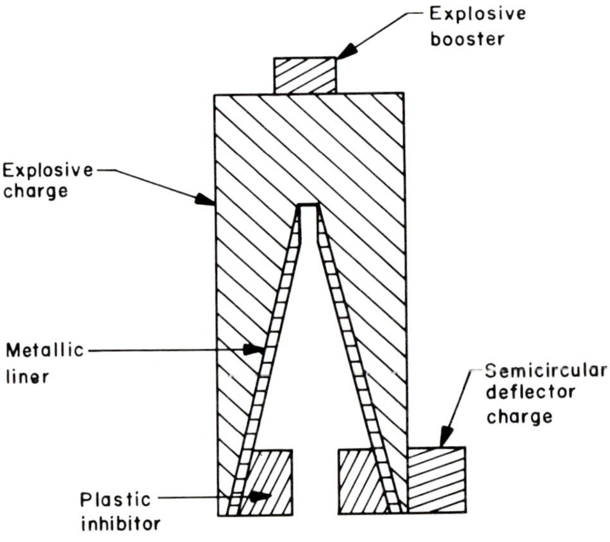

FIG. 14. Inhibited shaped-charge projector.

latter parts of the jet and the inhibitor so that these items do not impact the target. The inhibited shaped charge is described in detail by Gehring and Wenzel (GMDRL) (*56*, *57*). Wenzel reports velocities of up to 54,100 ft/sec (16.5 km/sec) for beryllium projectiles weighing 0.08 g.

The shaped-charge accelerator is capable of producing very high velocities with metallic projectiles. Its main disadvantages are associated with the uncertainty of determining accurately the mass and shape of the projectile being launched.

B. Plasma Drag Accelerators

Projectiles can be accelerated to hypervelocities by their drag in a high-speed plasma. One version of this type of accelerator is shown in Fig. 15. The discharge of a large high-voltage capacity bank causes the lithium wire to explode into a high-speed plasma, breaking the diaphragm holding the models and carrying them along the launch tube, accelerating them to hypervelocities. As many as 10,000 small (50-μ, 0.0019-in.-diam) glass spheres are launched at a time. Baffles and an electromagnetic shutter eliminate the projectiles not directed along the axis and also the slower ones before a few strike the target. Detailed descriptions of the development of such a launcher are given by Scully and Posever (NAA) (*58–60*). The maximum velocity obtained with this accelerator has been 65,600 ft/sec (20 km/sec) with a glass sphere weighing 1.46×10^{-7} g.

This accelerator is limited to launching materials which are transparent to the radiation from the very high plasma temperature, thus preventing their ablation, and to spheres because of the optical technique used to determine, in flight, the integrity of the model. The particular version described by Scully and Posever (*58–60*) appears also to be limited to 50-μ-(0.0019-in.) diam spheres.

Another plasma drag accelerator is described by Bloxsom (*61*) and Carlson (General Dynamics/Astronautics) (*62*), and a diagram is shown in Fig. 16. With this version, the capacitor bank is discharged into an arc chamber containing hydrogen, and the hydrogen plasma expanding into an expansion cone accelerates the projectiles. Carlson (*62*) reports

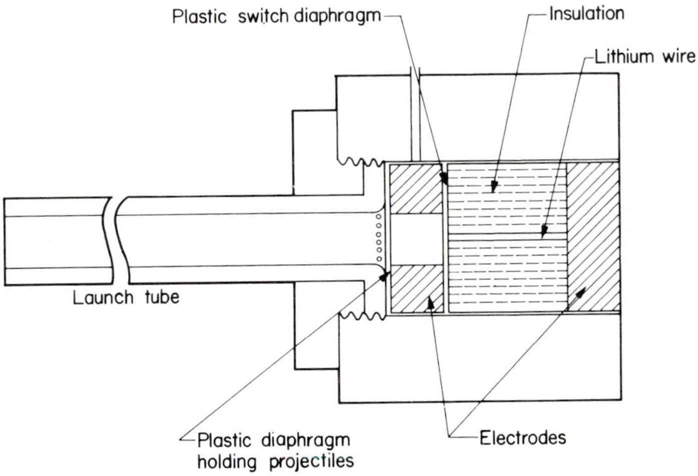

FIG. 15. Plasma drag accelerator (57).

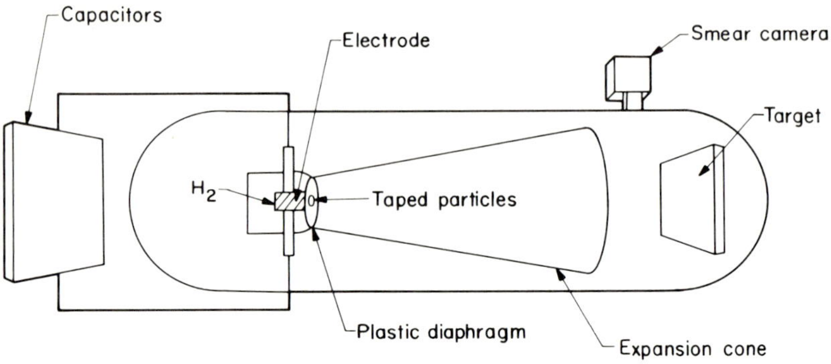

FIG. 16. Plasma drag accelerator (*59*).

that a velocity of 99,000 ft/sec (30 km/sec) has been achieved with this accelerator launching 0.0014-in.-diam glass spheres weighing 5.9×10^{-8} g.

C. Electrostatic Accelerators

An electrostatic accelerator, shown in Fig. 17 and described by Friichtenicht, TRW Systems, Inc. (*63*), uses the 100-kV electrostatic potential to accelerate very small particles (1.5-μ, 6×10^{-5}-in.-diam) of iron carbonyl to velocities of the order of 5600 ft/sec (1.7 km/sec). Large numbers of particles are impacted on the target, and crater parameters are correlated to the impacting particle parameters by statistical methods. The detector is used to determine the velocity and charge on the particle and, with two orthogonal sets of plates in it, gives a measure of the particle trajectory.

A later version of this accelerator (*64*) uses a 2-MV van de Graaff machine to provide the accelerating potential. Velocities of as high as 92,000 ft/sec (28 km/sec) with very small (0.2-μ, 8×10^{-6}-in.) iron particles are given by Friichtenicht *et al.* (*64*). As the craters produced by these very small particles are also very small, specialized techniques must be used to determine the crater parameters.

IV. Summary

All hypervelocity accelerators used for impact studies in the United States, Canada, and the United Kingdom as of June 1964 are listed by Battelle Memorial Institute (*65*). The list includes 151 operational accelerators, of which 83 (55%) are two-stage light-gas guns. The

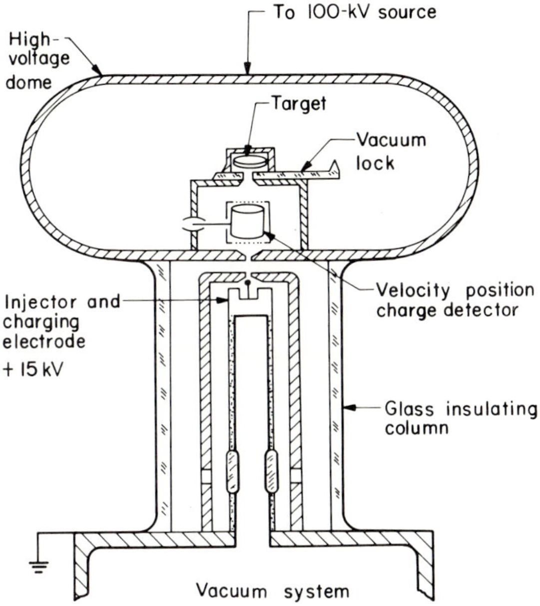

FIG. 17. Electrostatic accelerator.

light-gas gun, although limited in velocity to about 33,000 ft/sec (10 km/sec), is not as limited in the size, weight, material, and shape of projectile it can launch as are the other types of hypervelocity accelerators. Its main advantage is that the projectile weight and dimensions can be accurately determined and the launch velocity accurately predicted.

Attempts to augment the light-gas gun velocity by the addition of electrical or explosive energy has only proved successful with relatively low-performance guns and adds little or nothing to the velocity of the high-performance guns.

The shaped-charge explosive accelerator is capable of producing higher velocities (16.5 km/sec, 54,000 ft/sec) than the light-gas gun, but is limited in the shape and material of the projectile which it can launch. Since the mass and shape of projectile have to be determined by indirect methods after launch, the accuracy of these measurements is less than that for the light-gas gun projectile.

The highest launch velocities, 99,000 ft/sec (30 km/sec) and 66,000 ft/sec (20 km/sec), have been obtained with the plasma drag accelerators, and the projectiles are limited to very small glassy spheres weighing approximately 10^{-7} g, whose integrity is not without doubt at the highest velocities.

The electrostatic accelerator can also produce very high launch velocities, 92,000 ft/sec (28 km/sec), but the projectile size launched is very small (8×10^{-6} in.) so that determination of projectile and crater parameters is difficult and more liable to error than with, say, the light-gas gun.

REFERENCES

1. W. D. Crozier and W. Hume, *J. Appl. Phys.* **28**, 892 (1957).
2. A. C. Charters, B. P. Denardo, and V. J. Rossow, Development of a high-velocity free flight launcher—the Ames light-gas gun, NACA RM A55G11 (Dec. 1955).
3. A. E. Siegel and Z. I. Slawsky, A hypervelocity gun using a shock-compressed steam heated propellant, NavOrd Rept. 4345 (July 1956).
4. A. C. Charters, B. P. Denardo, and V. J. Rossow, Development of a piston-compressor type light gas gun for launching of free flight models at high velocity, NACA TN 4143 (November 1957).
5. F. Smith, The development of a hypersonic launcher, ARDE Rept. (B) 18/58 (1958).
6. C. Bioletti and B. Cunningham, A high velocity gun employing a shock-compressed light gas. NASA TN D-307 (Feb. 1960).
7. M. E. Lord, Performance of a 40-mm combustion-heated light-gas gun launcher, AEDC-TN-60-176 (Oct. 1960).
8. W. B. Stephenson and D. E. Anderson, Design of a large two-stage light-gas model launcher, AEDC-TR-61-6 (June 1961).
9. W. B. Stephenson, The performance of a small two-stage light-gas gun used for impact testing, AEDC-TN-61-166 (Jan. 1962).
10. W. B. Stephenson and D. E. Anderson, Performance of two-stage light-gas model launchers, Ann. Meeting IAS, 30th, New York (Jan. 1962).
11. J. S. Curtis, An accelerated reservoir light-gas gun, NASA TN D-1144 (Feb. 1962).
12. W. B. Stephenson and R. E. Knapp, Performance of a two-stage launcher using hydrogen, AEDC-TDR-62-32 (March 1962).
13. D. E. Anderson and M. D. Prince, Design of light-gas model launchers for hypervelocity research, *Symp. Hypervelocity Techniques, 2nd, Denver* (1962).
14. D. E. Anderson, Design of light-gas model launchers for hypervelocity research, AEDC-TDR-62-97 (May 1962).
15. H. Carter, D. A. Oakes, R. Piacesi, and B. M. Shepard, The design and testing of the naval ordnance laboratory's two-stage gun, NOLTR 62-112 (June 1962).
16. A. C. Charters and J. S. Curtis, High velocity guns for free flight ranges, TM 62/207, GM Defense Res. Lab. (1962).
17. J. Eckerman and W. L. McKay, Performance of a 3-stage arc heated light gas gun, *Proc. Symp. Hypervelocity Impact, 6th, Cleveland, Ohio* **1**, **247** (Aug. 1963).
18. A. J. Cable, The performance of the RARDE $\frac{1}{4}$-in. calibre hypervelocity launcher, RARDE Memorandum 14/64, (April 1964); Proc. Meeting AGARD-NATO Specialists, "The Fluid Dynamic Aspects of Space Flight," Marseilles (April 20–24, 1964); AGARDograph 187.
19. H. F. Swift, A study of electrically augmented gas guns. *Proc. Symp. Hypervelocity Impact, 7th, Tampa, Florida* (November 1964).

20. J. Lukasiewicz, Constant acceleration flows and applications to high speed guns, AEDC-TR-66-181 (Nov. 1966); *AIAA J.* **5**, No. **11**, 1955 (1967).
21. G. P. T. Wilenius, M. Cloutier, and P. L. Cowan, A theoretical analysis of a constant base pressure light gas gun, CARDE TM 703/62 (1962).
22. K. P. Stanyukovich, "Unsteady Motion of a Continuous Media." Pergamon Press, Oxford, 1960.
23. F. Smith, Theory of two-stage hypervelocity launcher to give constant driving pressure at the model. RARDE Rept. (B) 5/63 (1963); also *J. Fluid Mech.* **17**, 113 (1963).
24. A. J. Cable, J. R. DeWitt, and M. D. Prince, Launcher optimization studies, AEDC-TR-65-204 (Oct. 1965).
25. J. S. Curtis, An analysis of the interior ballistics of the constant base pressure gun, Symp. Hypervelocity Techniques, 3rd, Denver, Colorado (March 1964).
26. G. P. T. Wilenius, P. L. Cowan, and M. Cloutier, The constant base pressure light gas gun. Symp. Hypervelocity Techniques, 3rd, Denver, Colorado (March 1964).
27. R. Piacesi, D. F. Gates, and A. E. Siegel, Computer analysis of two-stage hypervelocity model launchers, NOL TR 62-87 (Feb. 1963).
28. P. G. Baer and H. C. Smith, Experimental and theoretical studies on the interior ballistics of light gas guns. Symp. Hypervelocity Impact, 6th, Cleveland, Ohio (Aug. 1963).
29. J. von Neumann and R. D. Richtmeyer, *J. Appl. Phys.* **21**, 232 (1932).
30. J. R. DeWitt and A. J. Cable, A comparison of experimental and theoretical launcher kinematics. AEDC-TR-65-203 (Oct. 1965).
31. W. G. Howell and J. P. Kottenstette, Electrical augmentation of a light gas gun, AEDC-TR-65-32 (Jan. 1965).
32. J. P. Kottenstette and W. G. Howell, The travelling reservoir light gas gun and development and analysis on a third stage accelerator for a light gas gun, Symp. Hypervelocity Impact, 7th, Tampa, Florida (Nov. 1964).
33. W. G. Howell, R. F. Recht, and T. W. Ipson, Hypervelocity augmentation techniques, Symp. Hypervelocity Impact, 6th, Cleveland, Ohio (April 1963).
34. W. C. Cooley, A shock plasma acceleration technique for velocity augmentation, Symp. Hypervelocity Techniques, 7th, Tampa, Florida (Nov. 1964).
35. K. W. Miller and R. M. Bergslein, Study of electromagnetic gun, Symp. Hypervelocity Impact Effects, 2nd, Washington, D.C. (May 1957).
36. W. W. Salisbury, A. four-coil induction accelerator, Symp. Hypervelocity Impact Effects, 2nd, Washington, D.C. (May 1957).
37. L. D. Harris, A coil type of electromagnetic accelerator, Symp. Hypervelocity Impact Effects, 2nd, Washington, D.C. (May 1957).
38. D. E. Brast and D. R. Sawle, Study of a rail-type MHD hypervelocity accelerator, Symp. Hypervelocity Impact, 7th, Tampa, Florida (Nov. 1964).
39. F. D. Rosen, A magnetically augmented rail gun, Symp. Hypervelocity Impact, 7th, Tampa, Florida (Nov. 1964).
40. R. L. Chapman, D. E. Harms, and G. P. Sorenson, The magnetohydrodynamic hypervelocity gun, Symp. Hypervelocity Impact, 6th, Cleveland, Ohio (April 1963).
41. R. L. Chapman, M. J. Morley, D. Bernstein, and R. C. Goettelman, The magnetic gradient hypervelocity accelerator, Symp. Hypervelocity Impact, 7th, Tampa, Florida (Nov. 1964).
42. P. G. Baer, The travelling charge gun as a hypervelocity launching device, Symp. Hypervelocity Impact, 4th, Eglin AFB, Florida (April 1960).
43. W. E. Fogg and C. W. Fleischer, Acceleration of projectiles with the sequenced high

explosive impulse launcher, Symp. Hypervelocity Impact, 4th, Eglin AFB, Florida (April 1960).
44. M. H. Bengson, T. K. Slawecki, and F. J. Willig, A multistage hypervelocity projector, Symp. Hypervelocity Impact, 4th, Eglin AFB, Florida (April 1960).
45. C. S. Godfrey, Gas ejectors for hypervelocity projection, Symp. Hypervelocity Impact, 7th, Tampa, Florida (Nov. 1964).
46. E. T. Moore, D. Mumma, C. S. Godfrey, and D. Bernstein, Explosive gas guns for hypervelocity acceleration. Symp. Hypervelocity Techniques, 4th, Tullahoma, Tennessee (November 1965).
47. J. K. Crosby, S. P. Gill, and G. R. Fowles, Explosive compression and acceleration of helium for hypervelocity applications, Symp. Hypervelocity Techniques, 4th, Tullahoma, Tennessee (Nov. 1965).
48. V. E. Scherrer and P. I. Richards, An exploding wire hypervelocity projector, Symp. Hypervelocity Impact, 4th, Eglin AFB, Florida (April 1960).
49. W. H. Clark, A. K. Hopkins, J. E. Myrberg, R. W. O'Niel, V. E. Scherrer, and M. C. Stevens, AFML exploding foil gun, development, diagnostics, data, Symp. Hypervelocity Impact, 7th, Tampa, Florida (Nov. 1964).
50. T. C. Poulter, Acceleration of small particles with high explosives, Symp. Hypervelocity Impact Effects, 2nd, Washington, D.C. (May 1957).
51. F. J. Willig, Acceleration of masses to hypervelocities by explosive means, Symp. Hypervelocity Impact Effects, 2nd, Washington, D.C. (May 1957).
52. R. H. Christian, R. E. Duff, and F. L. Yarger, Equation of state of gases by shock wave measurements. II. The dissociation energy of nitrogen, *J. Chem. Phys.* **23**, 11, 2045 (1955).
53. E. N. Clark, A. Mackenzie, F. H. Schmitt, and I. L. Kintish, Studies of hypervelocity impact on lead, Symp. Hypervelocity Impact, 4th, Eglin AFB, Florida (April 1960).
54. J. H. Kineke, Jr., An experimental study of crater formation in metallic targets, Symp. Hypervelocity Impact, 4th, Eglin AFB, Florida (April 1960).
55. K. N. Kreyenhagen, J. E. Ferguson, R. R. Randall, and J. P. Joyce, Special explosive projectors. I. Shaped charge accelerator. II. Target plate accelerator, Symp. Hypervelocity Impact, 6th, Cleveland, Ohio (Aug. 1963).
56. J. W. Gehring, Jr., D. R. Christman, and A. R. McMillan, Hypervelocity impact studies concerning the meteoroid hazard to aerospace materials and structures, Conf. AIAA Structures and Materials, 5th, Palm Springs, California (April 1964).
57. A. B. Wenzel and J. W. Gehring, Jr., Techniques for launching 0.01 to 25-gram discrete projectiles at velocities up to 54,100 ft/sec, Symp. Hypervelocity Techniques, 4th, Tullahoma, Tennessee (Nov. 1965).
58. C. N. Scully and P. L. Cowan, Hypervelocity gun for micrometeorite impact simulation employing capacitor discharge in a condensed phase, Symp. Hypervelocity Impact, 4th, Tampa, Florida (April 1960).
59. F. C. Posever, F. L. Rish, and C. N. Scully, An evaluation of impact effects on micrometeoroid shielding configurations for velocities up to 60,000 ft/sec, Ann. Conf. AIAA Structures and Materials, 5th, Palm Springs, California (April 1964).
60. C. N. Scully, E. A. Escallier, F. D. Rosen, and J. D. O'Keefe, Electrothermal gun for hypervelocity ballistics research, Symp. Hypervelocity Impact, 7th, Tampa, Florida (Nov. 1964).
61. D. Bloxsom, Experimental production of hypervelocity pellets by means of condenser discharges into hydrogen, Symp. Hypervelocity Impact, 4th, Eglin AFB, Florida (April 1960).
62. R. E. Carlson and J. A. Fager, Effects of hypervelocity impact on semi-infinite plate

and spacecraft hulls from 18,000 fps to 100,000 fps, Conf. AIAA Structures and Materials, 6th, Palm Springs, California (April 1965).
63. J. F. Friichtenicht and B. Hamermesh, Ballistic impacts by microscopic projectiles, Symp. Hypervelocity Impact, 4th, Eglin AFB, Florida (April 1960).
64. J. F. Friichtenicht, J. C. Slattery, and D. O. Hansen, Electrostatic accelerators—experimental techniques, Symp. Hypervelocity Impact, 7th, Tampa, Florida (Nov. 1964).
65. J. L. Harp, D. N. Gideon, and R. A. Stein, Hypervelocity-projector facilities, Battelle Memorial Inst. BAT-197-21-4 (June 1964).

CHAPTER II

STRESS WAVE PROPAGATION IN SOLIDS

NORRIS L. HICKERSON

TENNESSEE TECHNOLOGICAL UNIVERSITY
COOKEVILLE, TENNESSEE

I. Introduction

The hypervelocity impact of a projectile with a solid target results in an extremely complex phenomenon. A complete description of this behavior would involve consideration of all phases of continuum mechanics theory. In the initial, high-pressure phases of the impact, the material behaves essentially as an inviscid, compressible fluid since the pressures are high with respect to the maximum shear stresses that can be developed within the material. A crater forms which expands rapidly for a time and a shock wave emanates from its surface. A stage of plastic deformation follows which apparently decays rapidly into a spherical elastic wave which continues through the target (*1*). As pointed out by Rae (*2*), a complete theory for the description of the hypervelocity impact phenomena would involve not only the above phases but also other situations such as melting and resolidification, vaporization and condensation, and the kinetics of phase change.

Several models have been proposed for the various stages of the hypervelocity impact problem. The inviscid, compressible fluid model has been applied by Bjork (*3–5*), Walsh *et al.* (*6–8*), and Riney and Heyda (*9–11*). A numerical method for the solution of problems of the type of the hydrodynamic phase of the hypervelocity impact problem has been presented by Baron *et al.* (*12*). The propagation of spherical elastic-plastic waves has been considered by several investigators, among them Friedman *et al.* (*13*) and, more recently, Garg (*14*). The linear-elastic model has been applied by Kinslow (*15*), among others.

This chapter will deal with the classical theory of stress wave propagation in a solid continuous medium since most of the models attempting to predict the response of a solid target to the impact of a high-velocity projectile make use of this theory. First, the theory of stress wave propagation in an infinite, elastic medium will be discussed. Then the reflection of stress waves from surfaces involving various boundary conditions will be treated. The remaining section of the chapter discusses wave propagation in laminated materials.

II. Stress Wave Propagation in an Infinite, Elastic Medium

The theory of stress wave propagation in elastic solids has a relatively long history. The problem was treated in the nineteenth century by

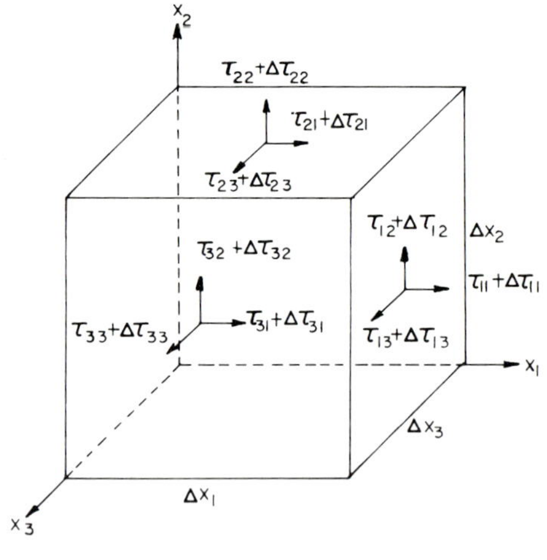

FIG. 1. Stresses on an element.

several of the outstanding contributors to the theory of mechanics and mathematical physics of that time, among them Rayleigh (*16*) and Poisson (*17*). In the early part of this century, elastic wave propagation was discussed by Love (*18*) and Lamb (*19*), among others. More recently, applications in electronics have generated increased interest in the study of high-frequency elastic waves. This aspect of the problem has been developed extensively by Mindlin (*20*).

A. Equations of Motion

The equations of motion for an elastic medium may be obtained from a physical point of view by applying Newton's second law of motion to the stresses on a small finite element of the medium. Assuming the values of τ in Fig. 1 to be the average values of stress on the faces of an element in the form of a rectangular parallelepiped, the resultant force in the x_1 direction is

$$\begin{aligned}&-\tau_{11}(\Delta x_1\, \Delta x_2\, \Delta x_3) + (\tau_{11} + \Delta\tau_{11})(\Delta x_1\, \Delta x_2\, \Delta x_3)\\&-\tau_{12}(\Delta x_1\, \Delta x_2\, \Delta x_3) + (\tau_{12} + \Delta\tau_{12})(\Delta x_1\, \Delta x_2\, \Delta x_3)\\&-\tau_{13}(\Delta x_1\, \Delta x_2\, \Delta x_3) + (\tau_{13} + \Delta\tau_{13})(\Delta x_1\, \Delta x_2\, \Delta x_3)\end{aligned}$$

From Newton's second law of motion, neglecting body forces, this quantity is equal to $\rho(\Delta x_1\, \Delta x_2\, \Delta x_3)\, \ddot{u}_1$, where ρ is the mass density of the material and $\ddot{u}_1$ represents the acceleration in the x_1 direction. Dividing through by the volume of the element, $\Delta x_1\, \Delta x_2\, \Delta x_3$, and letting Δx_1, Δx_2, and Δx_3 approach zero, the average stresses become stresses at a point and the equation becomes

$$\partial\tau_{11}/\partial x_1 + \partial\tau_{12}/\partial x_2 + \partial\tau_{13}/\partial x_3 = \rho\ddot{u}_1 \tag{1}$$

Similarly, by considering the forces in the x_2 and x_3 directions, the equations

$$\partial\tau_{21}/\partial x_1 + \partial\tau_{22}/\partial x_2 + \partial\tau_{23}/\partial x_3 = \rho\ddot{u}_2 \tag{2}$$

and

$$\partial\tau_{31}/\partial x_1 + \partial\tau_{32}/\partial x_2 + \partial\tau_{33}/\partial x_3 = \rho\ddot{u}_3 \tag{3}$$

may be obtained.

These equations may be expressed in more compact form as

$$\tau_{ij,j} = \rho\ddot{u}_i \tag{4}$$

making use of standard index notation.

For an isotropic medium, stress may be related to strain using Lamé's constants, λ and μ, giving

$$\tau_{ij} = \lambda\delta_{ij}\epsilon_{kk} + 2\mu\epsilon_{ij} \tag{5}$$

where δ_{ij} is the Kronecker delta. The equations of motion then become

$$(\lambda\delta_{ij}\epsilon_{kk} + 2\mu\epsilon_{ij})_{,j} = \rho\ddot{u}_i \tag{6}$$

Combining these equations with the strain-displacement equations

$$\epsilon_{ij} = \tfrac{1}{2}(u_{i,j} + u_{j,i})$$

gives the displacement equations of motion for an isotropic, elastic medium

$$(\lambda + \mu)\, u_{i,ij} + \mu u_{j,ii} = \rho\ddot{u}_j \tag{7}$$

or alternately

$$(\lambda + \mu)\, \nabla\nabla \cdot \mathbf{u} + \mu\, \nabla^2\mathbf{u} = \rho\ddot{\mathbf{u}} \tag{8}$$

In studying wave propagation, it is convenient to define $\nabla \cdot \mathbf{u} = \epsilon_{11} + \epsilon_{22} + \epsilon_{33}$ as the dilatation, Δ. This term represents the volume change of a unit cube. Another useful definition is that of the rotation, $\omega = \frac{1}{2}\nabla x\mathbf{u}$.

Obtaining the dot product of the operator ∇ with both sides of Eq. (8) gives the equation

$$(\lambda + 2\mu)\, \nabla^2\Delta = \rho\ddot{\Delta} \tag{9}$$

while performing the curl operation on both sides of Eq. (8), thus eliminating the dilatation, yields

$$\mu\, \nabla^2\boldsymbol{\omega} = \rho\ddot{\boldsymbol{\omega}} \tag{10}$$

As pointed out by Love (*18*), Eqs. (9) and (10) are of the form

$$c^2\, \nabla^2\phi = \ddot{\phi} \tag{11}$$

the hyperbolic partial differential equation known as the wave equation. In this equation c represents the velocity of wave propagation.

From Eq. (9) it may be seen that dilatational waves propagate through the medium at a velocity of $[(\lambda + 2\mu)/\rho]^{1/2}$. Likewise, from Eq. (10), the distortional waves propagate at a velocity of $(\mu/\rho)^{1/2}$. The dilatational waves and the distortional waves are often referred to as irrotational waves and equivoluminal waves, respectively. As shown by Kolsky (*21*), in the case of dilatational waves the motion is along the direction of propagation and the motion of distortional waves is paralled to the wave front.

B. Properties of the Wave Equation

It is useful to study the properties of the wave equation when the solution is a function of only one coordinate variable and time. The resulting equation thus simplified is

$$c^2\phi_{\cdot xx} = \phi_{\cdot tt}$$

for $\phi = \phi(x, t)$. If the transformation

$$\xi = x - ct$$

$$\eta = x + ct$$

is introduced, the resulting equation expressed in the new coordinate system is

$$\phi_{\cdot \xi\eta} = 0$$

This equation may be integrated directly to give

$$\phi_{(\xi,\eta)} = f_1(\xi) + f_2(\eta)$$

where f_1 is a function of ξ alone and f_2 is a function of η only. The solution, when transformed back into the original coordinates, has the form

$$\phi_{(x,t)} = F_1(x - ct) + F_2(x - ct) \tag{12}$$

Love (*18*) and Kolsky (*21*) point out that this solution represents plane waves propagated with velocity c. The terms F_1 and F_2 are arbitrary functions which may be determined from the initial conditions, $\phi(x, 0)$ and $\phi_{,t}(x, 0)$. If $\phi(x, 0) = A(x)$ and $\phi_{,t}(x, 0) = B(x)$ then the solution of Eq. (12) may be expressed in terms of these initial conditions as

$$\phi(x, t) = \tfrac{1}{2}[A(x - ct) + A(x + ct)] + \tfrac{1}{2}c \int_{x-ct}^{x+ct} B(\alpha)\, d\alpha \tag{13}$$

This form of the solution is known as D'Alembert's formula. The implications of D'Alembert's formula may be represented as shown in Fig. 2 for the case of $\phi_{,t}(x, 0) = 0$. An initial value of $\phi(x, 0) = A(x)$ results in a plane wave specified by $\frac{1}{2}A(x - ct)$ traveling along the positive x axis with velocity c, and one given by $\frac{1}{2}A(x + ct)$ traveling in the opposite direction.

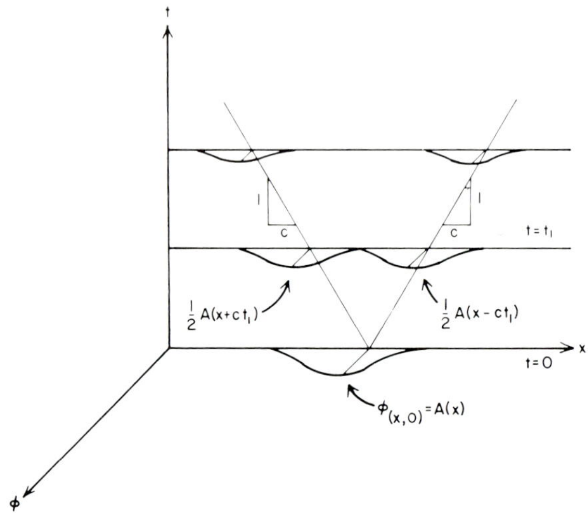

FIG. 2. Propagation of an initial disturbance.

The family of lines $x + ct = \text{const}$ and $x - ct = \text{const}$ are the characteristic curves of the equation $c^2\phi_{,xx} = \phi_{,tt}$. These characteristic curves determine the wave propagation behavior in the medium. As shown in Fig. 2, the characteristics give the "direction" of wave propagation in the *xt* plane.

The significance of the characteristics may be further explored by considering the solution of the equation at a point $P_{(x,\,t)}$ (*22*). From

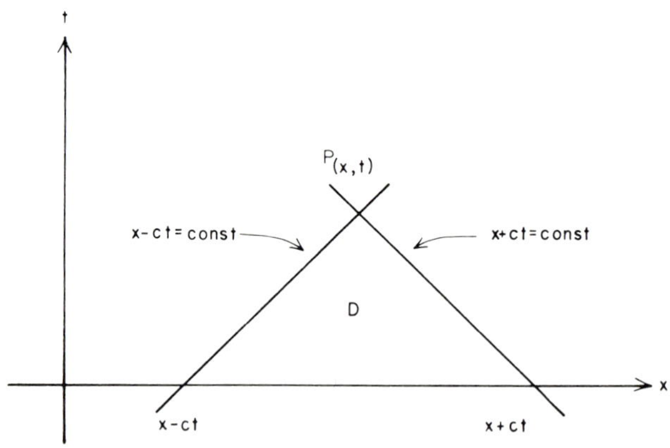

FIG. 3. Domain of dependence of solution at a point $P_{(x,t)}$.

Eq. (13) it is seen that $\phi_{(x,t)}$ depends on the initial data between the points $x - ct$ and $x + ct$ shown on the x axis in Fig. 3. The domain D, bounded by the pair of characteristics passing through $P_{(x,y)}$ and that portion of the initial curve between the points at which these particular characteristics intersect the line $t = 0$, is thus known as the *domain of dependence* of the solution ϕ with respect to point P. The solution at any point inside D is determined by the data on this portion of the initial curve. The *range of influence* of the segment $x_1 \leqslant x \leqslant x_2$ shown in Fig. 4 may be defined as all points x at any t such that

$$(x_1 - ct) \leqslant x \leqslant (x_2 + ct).$$

The solution ϕ at all points in R is influenced, but not necessarily completely determined, by the initial data specified on the segment of the initial line between x_1 and x_2.

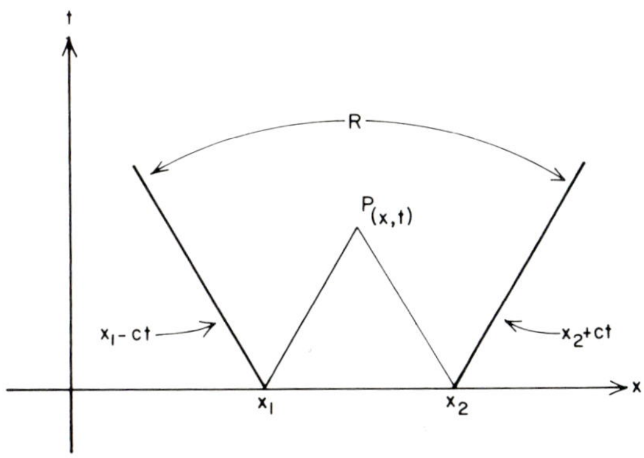

FIG. 4. Range of influence of initial line segment $x_1 \leqslant x \leqslant x_2$.

For cases in which more independent variables than two are involved, the problem of interpreting the characteristics proceeds in a manner similar to that outlined for the one-dimensional wave equation. For example, if $\phi = \phi(x, y, t)$ the wave equation becomes

$$c^2(\phi_{,xx} + \phi_{,yy}) = \phi_{,tt}$$

the equation of cylindrical waves. The characteristics of this partial differential equation are defined by the surface of a cone (*23*). Figure 5 indicates how the notion of the domain of dependence may be extended to this case. The solution at point $P_{(x,y,t)}$ depends on the data on the trace of the characteristic cone on the xy initial plane.

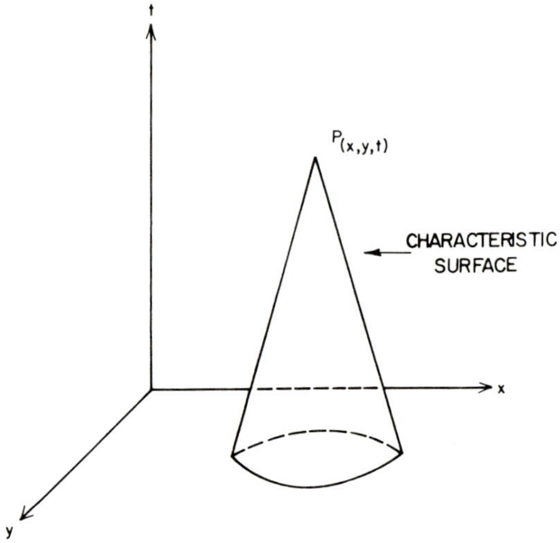

FIG. 5. Domain of dependence for equation of cylindrical waves.

For the case of spherical waves the wave equation takes the form

$$c^2(\phi_{\cdot xx} + \phi_{\cdot yy} + \phi_{,zz}) = \phi_{\cdot tt}\,. \tag{14}$$

Substituting $r^2 = x^2 + y^2 + z^2$, where r is the radius vector from the point in the medium from which the disturbance originates, leads to the following terms :

$$\phi_{\cdot xx} = \frac{x^2}{r^2}\phi_{\cdot rr} + \frac{1}{r}\frac{r^2 - x^2}{r^2}\phi_{\cdot r}$$

$$\phi_{\cdot yy} = \frac{y^2}{r^2}\phi_{\cdot rr} + \frac{1}{r}\frac{r^2 - y^2}{r^2}\phi_{\cdot r}$$

$$\phi_{\cdot zz} = \frac{z^2}{r^2}\phi_{\cdot rr} = \frac{1}{r}\frac{r^2 - y^2}{r^2}\phi_{\cdot r}$$

Equation (14) then becomes

$$c^2(\phi_{\cdot rr} + 2/r\phi_{\cdot r}) = \phi_{\cdot tt} \qquad \text{or} \qquad c^2(r\phi)_{\cdot rr} = (r\phi)_{\cdot tt}$$

This equation has the same form as the one-dimensional wave equation and represents spherical waves caused by a disturbance at a point in the medium. Its solution is analagous to Eq. (12) and may be expressed as

$$\phi_{(r,t)} = \frac{1}{r}[F_1(r - ct) + F_2(r + ct)]$$

III. Stress Wave Propagation in an Elastic Half Space

It was seen that in an infinite elastic medium there were two possible types of wave that can be propagated: dilatational waves and distortional waves. When a plane boundary is introduced, it becomes of interest to study the reflection of these waves from the boundary surface when various conditions are present on the surface.

A third type of wave motion is possible when the medium is bounded by a surface. Plane surface waves which propagate over the boundary and whose effect does not extend but a small distance below the surface may be generated. These waves are known as Rayleigh surface waves after Lord Rayleigh who first investigated them (*24*).

A. Effects of Various Boundary Conditions

The effect of various boundary conditions on the reflection of stress waves has been developed in detail by Mindlin (*25*). It has been previously shown that dilatational waves (pressure, or P waves) propagate with a velocity of $c_P = [(\lambda + \mu/\rho)]^{1/2}$ and distortional waves (shear, or S waves) propagate with a velocity of $c_S = (\mu/\rho)^{1/2}$. The P waves travel with a greater velocity than do S waves and are sometimes referred to as primary waves, while the S waves may be referred to as secondary waves.

For the purpose of discussion, consider waves specified by

$$\phi = A \cos 2\pi/L(\mathbf{n} \cdot \mathbf{r} - ct) \tag{15}$$

where $\mathbf{n}$ is a wave normal in an arbitrary direction and $\mathbf{r}$ is the position vector, $\mathbf{e}_1 x_1 + \mathbf{e}_2 x_2 + \mathbf{e}_3 x_3$. This function represents an infinite train of plane sinusoidal waves. As shown in Fig. 6, A is the amplitude of the waves and L is the wavelength. It is convenient to indicate the wave-number $2\pi/L = \alpha$ and the circular natural frequency $\omega = \alpha c$.

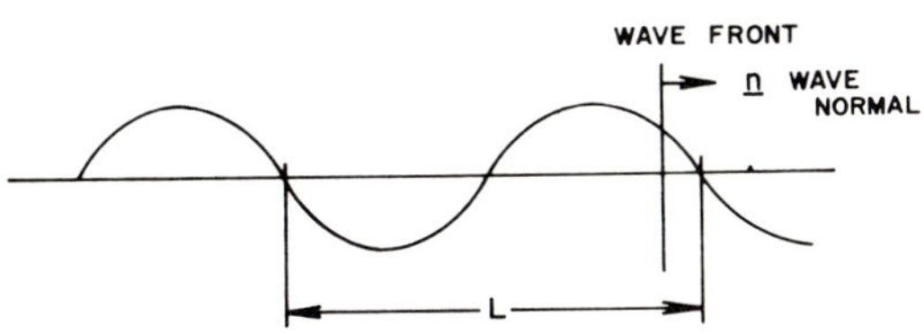

FIG. 6. Infinite train of plane, sinusoidal waves.

1. *P Wave Incident on a Stress-Free Boundary*

Consider dilatational waves, as shown in Fig. 7, incident on the stress-free surface $x_3 = 0$ and with the wave normal making an angle γ with the

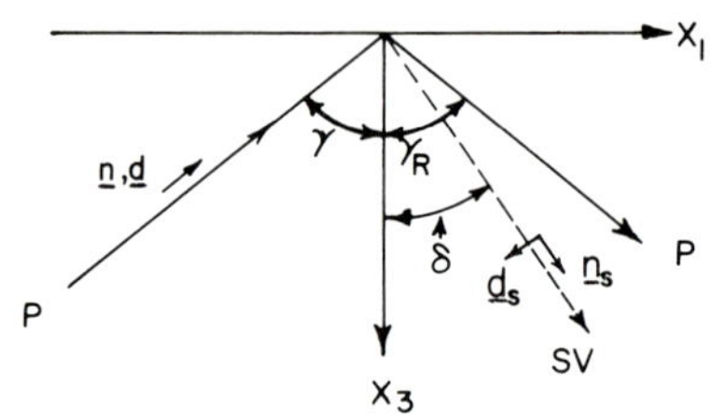

FIG. 7. *P* wave incident on stress-free surface.

x_3 axis. Using the wave form indicated by Eq. (15) the displacements in the direction of the wave normal are

$$\mathbf{u} = A\mathbf{d} \cos \alpha(\mathbf{n} \cdot \mathbf{r} - ct)$$

where **d** is a vector in the direction of the displacement and is equal to **n** in this case. For the direction of the wave normal shown in Fig. 7,

$$\mathbf{n} = \mathbf{e}_1 \sin \gamma - \mathbf{e}_3 \cos \gamma \qquad \text{and} \qquad \mathbf{n} \cdot \mathbf{r} = x_1 \sin \gamma - x_3 \cos \gamma$$

The components of displacement for the specified wave motion are

$$\begin{aligned} u_1 &= A \sin \gamma \exp[i\alpha(x_1 \sin \gamma - x_3 \cos \gamma - ct)] \\ u_2 &= 0 \\ u_3 &= -A \cos \gamma \exp[i\alpha(x_1 \sin \gamma - x_3 \cos \gamma - ct)] \end{aligned} \tag{16}$$

expressing the cosine function in exponential form. Stresses on the surface $x_3 = 0$ then become

$$\begin{aligned} \tau_{31} &= \mu(\partial u_1/\partial x_3 + \partial u_3/\partial x_1) \\ &= -2i\alpha\mu A \sin \gamma \cos \gamma \exp[i\alpha(x_1 \sin \gamma - ct)] \\ \tau_{32} &= \mu(\partial u_2/\partial x_3 + \partial u_3/\partial x_2) \\ &= 0 \\ \tau_{33} &= \lambda\nabla \cdot \mathbf{u} + 2\mu(\partial u_3/\partial x_3) \\ &= i\alpha A(\lambda + 2\mu \cos^2 \gamma) \exp[i\alpha(x_1 \sin \gamma - ct)] \end{aligned} \tag{17}$$

If the amplitude of the reflected wave is denoted by A_R and the angle that it makes with the x_3 axis is γ_R, the vector **n** becomes $\mathbf{e}_1 \sin \gamma_R + \mathbf{e} \cos \gamma_R$ and the corresponding displacements may be written

$$\begin{aligned} u_{1R} &= A_R \sin \gamma_R \exp[i\alpha_R(x_1 \sin \gamma_R - x_3 \cos \gamma_R - ct)] \\ u_{2R} &= 0 \\ u_{3R} &= A_R \cos \gamma_R \exp[i\alpha_R(x_1 \sin \gamma_R - x_3 \cos \gamma_R - ct)] \end{aligned} \tag{18}$$

The stresses on the surface $x_3 = 0$ due to the reflected wave are then

$$\begin{aligned} \tau_{31R} &= 2i\alpha_R \mu A_R \sin\gamma_R \cos\gamma_R \exp[i\alpha_R(x_1 \sin\gamma_R - ct + a)] \\ \tau_{32R} &= 0 \\ \tau_{33R} &= i\alpha_R A_R(\lambda + 2\mu\cos^2\gamma_R) \exp[i\alpha_R(x_1 \sin\gamma_R - ct + a)] \end{aligned} \tag{19}$$

where a represents a possible change of phase upon reflection.

It is seen that the values of stress given by the incident and reflected waves at $x_3 = 0$ cannot be combined to satisfy the condition that both the normal and shearing stress be zero on the boundary. As discussed by Mindlin (*25*) and Kolsky (*21*), the conclusion may be drawn that a P wave cannot reflect as a simple P wave from a stress-free surface. The normal stress for all x_3 and t on the surface $x_3 = 0$ becomes zero if $\alpha_R = \alpha$, $\gamma_R = \gamma$, $A_R = -A$, and $a = 0$, or if $A_R = A$ and $a = \pi$. The shearing stress on the boundary cannot be zero under thees conditions.

It may be shown that the reflection of a P wave from a stress-free surface serves to generate a reflected shear wave with displacements in the x_1x_3 plane (SV waves). As shown in Fig. 7, the displacement vector $\mathbf{d}_s$ is perpendicular to the wave normal in this case. For this situation

$$\mathbf{d}_s = -\mathbf{e}_1 \cos\delta + \mathbf{e}_3 \sin\delta \qquad \text{and} \qquad \mathbf{n} = \mathbf{e}_1 \sin\delta + \mathbf{e}_3 \cos\delta$$

The wave normal makes the angle δ with the x_3 axis as shown. The displacements for this shear wave are

$$\mathbf{u}_S = A_S \mathbf{d}_S \exp[i\alpha_S(\mathbf{n}_S \cdot \mathbf{r} - c_S t)]$$

where the S subscripts denote quantities associated with the shear wave. The displacement components due to the reflected SV wave are

$$\begin{aligned} u_{1S} &= -A_S \cos\delta \exp[i\alpha_S(x_1 \sin\delta + x_3 \cos\delta - c_S t)] \\ u_{2S} &= 0 \\ u_{3S} &= A_S \sin\delta \exp[i\alpha_S(x_1 \sin\delta + x_3 \cos\delta - c_S t)] \end{aligned} \tag{20}$$

The corresponding stresses on the plane $x_3 = 0$ are

$$\begin{aligned} \tau_{31_S} &= -\mu i\alpha_S A_S(\cos^2\delta - \sin^2\delta) \exp[i\alpha_S(x_1 \sin\delta - c_S t)] \\ \tau_{32_S} &= 0 \\ \tau_{33_S} &= 2\mu i\alpha_S A_S \sin\delta \cos\delta \exp[i\alpha_S(x_1 \sin\delta - c_S t)] \end{aligned} \tag{21}$$

Superimposing the stresses on $x_3 = 0$ leads to the relation $\alpha c = \alpha_R c = \alpha_S c_S$. Therefore $\alpha_R = \alpha$ and $\alpha/\alpha_S = c_S/c = (\mu/\rho)^{1/2}/(\lambda + 2\mu/\rho)^{1/2}$ or

$c_S/c = [\mu/(\lambda + 2\mu)]^{1/2}$. In general αc is equal to ω, the circular frequency, so the frequency remains unchanged on reflection.

An examination of the equations for stress also leads to the result that $\alpha \sin \gamma = \alpha_S \sin \delta$. Therefore $\sin \gamma / \sin \delta = \alpha_S/\alpha = c/c_S$. Since c is greater than c_S (dilatational waves travel faster than shear waves), $\sin \gamma$ is greater than $\sin \delta$.

In order to relate the amplitudes of the incident and reflected waves the stresses on the boundary may be summed and equated to zero. Superimposing the shearing stresses on $x_3 = 0$ leads to the equation

$$-2 \sin \gamma \cos \gamma + 2(A_R/A) \sin \gamma \cos \gamma - [(\lambda + 2\mu)/\mu)/\mu]^{1/2}(A_S/A)(\cos^2 \delta - \sin^2 \delta) = 0 \tag{22}$$

while adding the corresponding normal stresses yields

$$(\lambda + 2\mu \cos^2 \gamma)(1 + A_R/A) + 2\mu[(\lambda + 2\mu)/\mu]^{1/2}(A_S/A) \sin \delta \cos \delta = 0 \tag{23}$$

Simplifying Eqs. (22) and (23) and denoting $[(\lambda + 2\mu)/\mu]^{1/2} = m$ gives the following two simultaneous equations in A_R/A and A_S/A:

$$\sin 2\gamma(1 - A_R/A) + (m^{A_S}/A) \cos 2\delta = 0$$

$$m^2 \cos 2\delta(1 + A_R/A) + (m^{A_S}/A) \sin 2\delta = 0$$

Solving these gives

$$A_R/A = \frac{\sin 2\gamma \sin 2\delta - m^2 \cos^2 2\delta}{\sin 2\gamma' \sin 2\delta + m^2 \cos^2 2\delta}$$

$$A_S/A = \frac{2m \sin 2\gamma \sin 2\delta}{\sin 2\gamma \sin 2\delta + k^2 \cos^2 2\delta}$$

It may be observed that there is no reflected shear wave for the case of normal incidence, $\gamma = 0$. For certain materials and certain angles of incidence of the P wave, complete mode conversion is possible; i.e., there is no reflected P wave, only a reflected shear wave.

2. *P and SV Waves Incident on a Surface with Mixed Boundary Conditions*

An examination of Eqs. (16)–(19) developed in Section III.B indicates that a P wave may be reflected as a pure P wave if the boundary conditions are

$$u_3 = 0 \quad \text{and} \quad \tau_{31} = 0 \quad \text{on} \quad x_3 = 0$$

or

$$u_1 = 0 \quad \text{and} \quad \tau_{33} = 0 \quad \text{on} \quad x_3 = 0$$

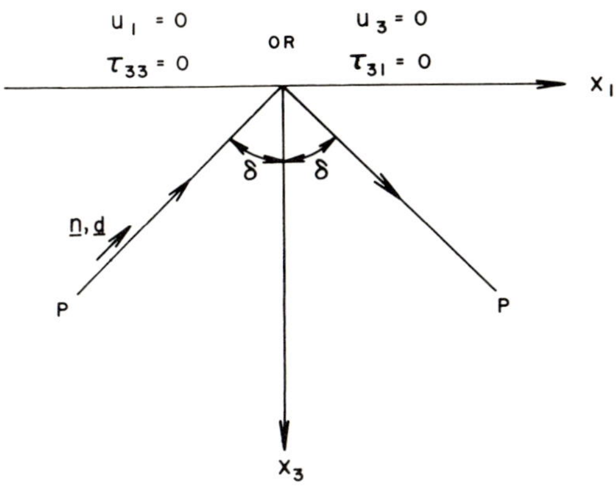

FIG. 8. Reflection of P wave from surface with mixed boundary conditions.

This behavior is illustrated in Fig. 8. For the first set of conditions to be satisfied A must equal A_R, $\alpha = \alpha_R$, and $\gamma = \gamma_R$. Similarly, the second set of conditions are satisfied if $A = -A_R$. An incident SV wave has displacements given by

$$\mathbf{u}_S = A_S \mathbf{d} \exp[i\alpha_S(\mathbf{n}_S \cdot \mathbf{r} - c_S t)]$$

where

$$\mathbf{d}_S = \mathbf{e}_1 \cos\delta + \mathbf{e}_3 \sin\delta \qquad \text{and} \qquad \mathbf{n}_S = \mathbf{e}_1 \sin\delta - \mathbf{e}_3 \cos\delta$$

The displacement components for this wave are then

$$\begin{aligned} u_{1S} &= A_S \cos\delta \exp[i\alpha_S(x_1 \sin\delta - x_3 \cos\delta - c_S t)] \\ u_{2S} &= 0 \\ u_{3S} &= A_S \sin\delta \exp[i\alpha_S(x_1 \sin\delta - x_3 \cos\delta - c_S t)] \end{aligned} \tag{24}$$

The corresponding stresses on $x_3 = 0$ are

$$\begin{aligned} \tau_{31_S} &= \mu i\alpha_S A_S(\cos^2\delta - \sin^2\delta) \exp[i\alpha_S(x_1 \sin\delta - c_S t)] \\ \tau_{32_S} &= 0 \\ \tau_{33_S} &= -2\mu i\alpha_S A_S \sin\delta \cos\delta \exp[i\alpha_S(x_1 \sin\delta - c_S t)] \end{aligned} \tag{25}$$

These may be combined with the reflected shearing displacements given in Eqs. (20) and (21) to satisfy conditions $u_1 = 0$ and $\tau_{33} = 0$ if A_S reflected is the negative of A_S incident and $u_3 = 0$ and $\tau_{31} = 0$ for A_S

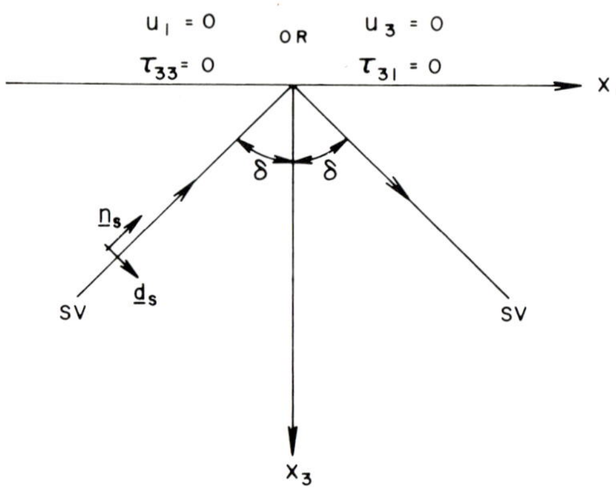

FIG. 9. Reflection of *SV* wave from surface with mixed boundary conditions.

incident equal to A_S reflected. Under these conditions, then, an *SV* wave will reflect from a boundary as a pure *SV* wave as indicated in Fig. 9.

3. *SV Wave Incident on a Stress-Free Surface*

As shown in Fig. 10, an *SV* wave incident on a stress-free surface results in general in a reflected *SV* wave and also a reflected *P* wave. An examination of Eqs. (21) and (25) for reflected and incident *SV* waves, respectively, shows that the condition that all stresses vanish on the surface $x_3 = 0$ cannot be satisfied by these waves alone. The equations for corresponding stress components on $x_3 = 0$ given by Eqs. (19), (21), and (25) may be superimposed and equated to zero to give two simultaneous equations in A_{SR}/A_S and A_R/A_S. Here A_{SR} refers to the amplitude of the reflected *SV* wave, A_R the amplitude of the reflected *P*

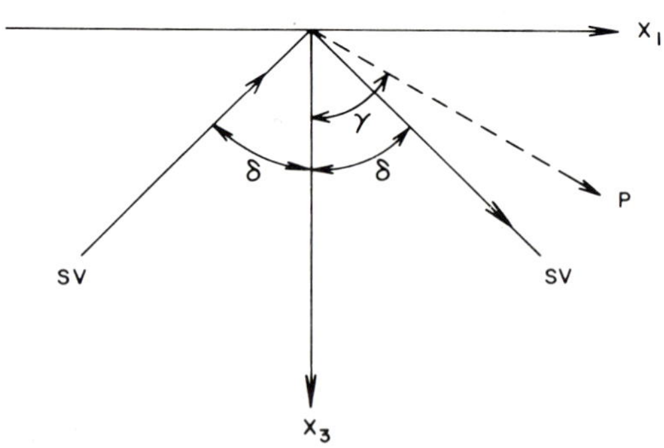

FIG. 10. Reflection of *SV* wave from stress-free surface.

wave, and A_S the amplitude of the incident SV wave. The equations thus obtained are

$$\cos 2\delta(1 + A_{SR}/A_S)\, m - (A_R/A_S) \sin 2\gamma = 0 \tag{26}$$

$$\sin 2\delta(1 - A_{SR}/A_S) + (A_R/A_S)(1/m)(\lambda + 2\mu \cos^2 \gamma)/\mu = 0 \tag{27}$$

which may be solved to yield

$$A_R/A_S = \frac{m \sin 4\delta}{m^2 \cos^2 2\delta + \sin 2\delta \sin 2\gamma} \tag{28}$$

$$A_{SR}/A_S = \frac{\sin 2\delta \cos 2\delta - m^2 \cos^2 2\delta}{m^2 \cos^2 2\delta + \sin 2\delta \sin 2\gamma} \tag{29}$$

From Eq. (28) it is seen that there is no reflected P wave when the incident SV wave makes an angle of 45° with the x_3 axis. From Eq. (29) $A_{SR}/A_S = -1$ for this situation indicating that the amplitude of the reflected SV wave is the negative of the incident SV wave.

4. *Shear Waves with Motion in the x_2x_3 Plane—SH Waves*

Shear waves may be propagated with displacement parallel to the x_2 axis and wave normal in the x_2x_3 plane. The wave normal for the incident SH wave is $\mathbf{n}_S = \mathbf{e}_1 \sin \delta - \mathbf{e}_3 \cos \delta$ and the displacement direction $\mathbf{d} = \mathbf{e}_2$. As in the previous cases the displacement has the general form

$$\mathbf{u} = A\mathbf{d} \exp[i\alpha(\mathbf{n} \cdot \mathbf{r} - vt)]$$

In the case discussed here, the u_1 and u_3 components are zero and

$$u_2 = A_{SH} \exp[i\alpha_{SH}(x_1 \sin \delta - x_3 \cos \delta - c_S t)]$$

The corresponding stresses on the surface $x_3 = 0$ are

$$\tau_{31} = 0$$

$$\tau_{32} = -\mu i \alpha_{SH} A_{SH} \cos \delta \exp[i\alpha_{SH}(x_1 \sin \delta - x_3 \cos \delta - c_S t)]$$

$$\tau_{33} = 0$$

For the reflected wave, $\mathbf{n}_S = \mathbf{e}_1 \sin \delta + \mathbf{e}_3 \cos \delta$ and

$$u_2 = A_{SH} \exp[i\alpha_{SH}(x_1 \sin \delta + x_3 \cos \delta - c_S t)]$$

τ_{31} and τ_{33} are zero for the reflected wave as they were for the incident wave and

$$\tau_{32} = \mu i \alpha_{SH} A_{SH} \exp[i\alpha_{SH}(x_1 \sin \delta - x_3 \cos \delta - c_S t)]$$

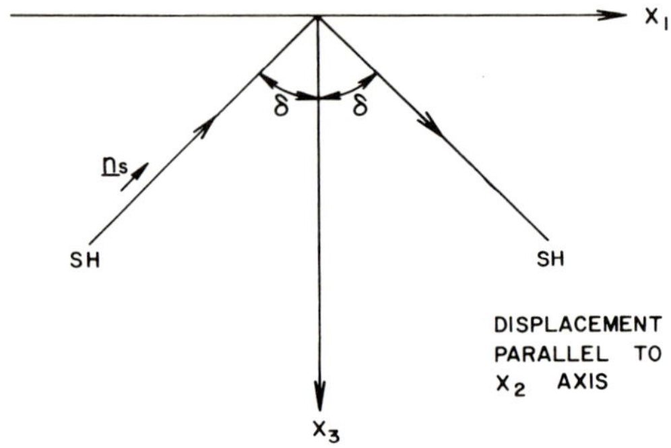

FIG. 11. Reflection of *SH* wave from a stress-free surface.

on $x_3 = 0$. The condition of a stress-free boundary is satisfied for A_{SH} and α_{SH} the same for the incident and reflected *SH* waves. The reflection of these waves is illustrated in Fig. 11.

B. Rayleigh Surface Waves

As mentioned earlier in this section, a form of elastic wave motion is possible in an isotropic half-space with a stress-free boundary surface in which the disturbance extends only a small distance from the boundary. Consider the situation in which plane waves propagate in the x_1 direction, the x_3 axis is directed into the medium, and there is no displacement in the x_2 direction. The equations of motion are satisfied by solutions which decay exponentially with increasing x_3.

It is useful to express the displacements **u** by the Helmholtz resolution (*25*)

$$\mathbf{u} = \nabla F + \nabla \times \mathbf{G}; \qquad \nabla \cdot \mathbf{G} = 0 \tag{30}$$

where F and **G** are potential functions. Recalling the displacement equation of motion

$$\lambda \nabla^2 \mathbf{u} + (\lambda + \mu) \nabla\nabla \cdot \mathbf{u} = \rho \ddot{\mathbf{u}}$$

and substituting the expression for **u** in Eq. (30) gives

$$\nabla[(\lambda + 2\mu)/\rho \, \nabla^2 F - \ddot{F}] + \nabla \times [(\mu/\rho) \, \nabla^2 \mathbf{G} - \ddot{G}] = 0$$

This equation is satisfied by the following relations (21):

$$c \, \nabla^2 F = \ddot{F} \tag{31}$$

$$c_S \, \nabla^2 \mathbf{G} = \ddot{\mathbf{G}} \tag{32}$$

This indicates that F is a potential for dilatational waves and G is a potential for distortional waves. The expansion of Eq. (30) gives

$$\begin{aligned} u_1 &= \partial F/\partial x_1 + \partial G_3/\partial x_2 - \partial G_2/\partial x_3 \\ u_2 &= \partial F/\partial x_2 + \partial G_1/\partial x_3 - \partial G_3/\partial x_1 \\ u_3 &= \partial F/\partial x_3 + \partial G_2/\partial x_1 - \partial G_1/\partial x_2 \end{aligned} \tag{33}$$

In the case under consideration there will be u_1 and u_3 components of displacement but no u_2 component. Equation (32) becomes

$$c_S \nabla^2 G_2 = \ddot{G}_2 \tag{34}$$

and the expansion of Eqs. (33) is reduced to

$$\begin{aligned} u_1 &= \partial F/\partial x_1 - \partial G_2/\partial x_3 \\ u_3 &= \partial F/\partial x_3 + \partial G_2/\partial x_1 \end{aligned} \tag{35}$$

$(G_1 = G_3 = 0)$. Introducing a dilatational potential function

$$F = f_{(x_3)} \exp[i\alpha(x_1 - ct)]$$

and a distortional potential

$$G_2 = g_{(x_3)} \exp[i\alpha(x_1 - ct)]$$

and substituting into Eqs. (31) and (34) yields the following ordinary differential equations which must be satisfied by $f_{(x_3)}$ and $g_{(x_3)}$:

$$\begin{aligned} d^2f/dx_3{}^2 - \alpha^2 f &= -\alpha^2/c^2 f \\ d^2g/dx_3{}^2 - \alpha^2 g &= -\alpha^2/c_S{}^2 g \end{aligned}$$

Defining p^2 as $\alpha^2[1 - (1/c)^2]$ and q^2 as $\alpha^2[1 - (1/c_S)^2]$, the above equations are

$$\begin{aligned} d^2f/d^2x_3 - p^2 f &= 0 \\ d^2g/d^2x_3 - q^2 g &= 0 \end{aligned}$$

The general solution of these equations is

$$f_{(x_3)} = A_1 e^{-px_3} + A_2 e^{px_3}$$

$$g_{(x_3)} = B_1 e^{-qx_3} + B_2 e^{qx_3}$$

Since the wave motion does not increase with increasing x_3, A_2 and B_2 must equal zero. The solution then becomes

$$f_{(x_3)} = A_1 e^{-px_3}$$

$$g_{(x_3)} = B_1 e^{-qx_3}$$

The potential functions F and G_3 are then

$$F = A_1 \exp[-px_3 + i\alpha(x_1 - ct)]$$

$$G_2 = B_1 \exp[-qx_3 + i\alpha(x_1 - c_S t)]$$

From these functions displacements may be derived by Eqs. (30). Stresses may then be found from the displacements. Kolsky (*21*) shows that the surface waves described here travel with a velocity slightly less than that of distortional waves.

IV. Elastic Waves in Laminated Media

There has been a great deal of interest in the use of laminated materials in applications where hypervelocity impact is involved. The multiple reflections involved in the passage of elastic waves through a material with many laminations has been treated by Kinslow (*1*). In the discussion presented here the theory of the refraction and reflection of elastic waves at the boundary between two media will be reviewed.

A wave (either dilatational or distortional) which is incident on the surface separating two separate media resluls in two reflected waves, as shown in the previous examples, and two waves that are refracted into the adjacent media. As Kolsky (*21*) implies there is refraction of waves into the bounding medium even when that medium is air. The examples cited in Section III are then strictly correct only in a vacuum where there can be no refracted waves.

The matching conditions that must be satisfied at the interface between two media may be summarized as

$$\mathbf{u}_a = \mathbf{u}_b$$

$$\boldsymbol{\tau}_a = \boldsymbol{\tau}_b$$

where a and b refer to the two media under consideration. The contributions to the conditions at the interface from the incident wave and each of the reflected and refracted waves are superimposed to obtain relations between the amplitudes of the incident and reflected waves.

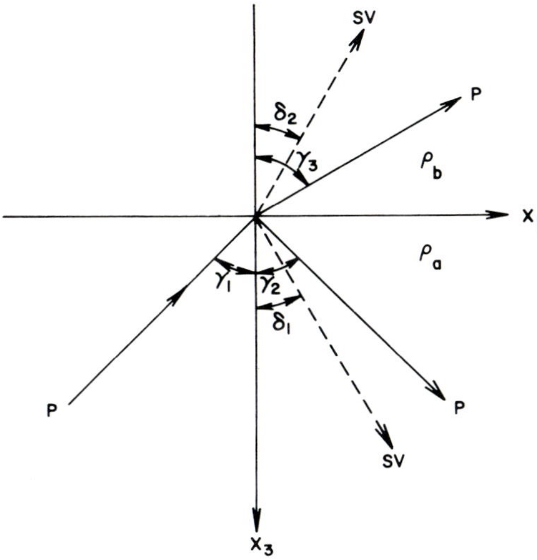

FIG. 12. *P* wave incident on a plane interface between two media.

Consider a plane *P* wave, traveling in the x_1–x_3 plane and incident on the surface $x_3 = 0$ (Fig. 12). In Fig. 12 the incident wave originates in a medium whose density is ρ_a and the refracted waves travel in a medium with density ρ_b. The boundary conditions are satisfied (*21*) if the following relation exists between the angles of incidence, reflection, and refraction, shown in Fig. 12, and the velocities of wave propagation in the two media:

$$\sin \gamma_1/c_1 = \sin \gamma_2/c_1 = \sin \gamma_1/c_2 = \sin \gamma_3/c_3 = \sin \gamma_2/c_4$$

The velocities c_1 and c_3 are the velocities of propagation of dilatational waves in the two media and c_2 and c_4 are the velocities of propagation of distortional waves.

The quantity ρc is in general known as the characteristic impedance of a medium. This quantity is used by Kinslow (*1*) to determine the behavior of pressure pulses which encounter a boundary between two media. If P_0, P_t, and P_r represent the original pressure, the transmitted pressure, and the reflected pressure, respectively, then the following "impedance mismatch" equations may be determined:

$$P_t = [2\rho_b c_3/(\rho_b c_3 + \rho_a c_1)]\, P_0$$

$$P_r = [(\rho_b c_3 + \rho_a c_1)/(\rho_b c_3 + \rho_a c_1)]\, P_0$$

with $P_t/P_r = 2\rho_b c_3/(\rho_b c_3 - \rho_a c_1)$. These relations assume that the characteristic impedance remains constant. From this information an approach may be made to the construction of pressure waves as they progress through a laminated material.

V. Summary

This chapter has attempted to present some of the established results describing the propagation of elastic waves through various media. Most of the theories currently being used to predict the response of materials to the hypervelocity impact of a projectile make use of these results at some stage of the analysis. Of course the problem of predicting the wave propagation involved in the hypervelocity impact problem is much more complicated than that represented by purely elastic response. Plastic waves are also involved in a complete description of the phenomenon. The study of plastic waves from both the analytical and experimental point of view is in a relatively early stage of development compared to the theory of elastic wave propagation. According to Garg (*14*) the propagation of elastic-plastic waves in an infinite region may be discussed in terms of four time-dependent zones. These are as follows: (1) an undisturbed zone, (2) an elastic precursor zone which precedes the plastic zone and terminates at the elastic front, (3) the plastic zone, and (4) an elastic unloading zone. A complete theory for the propagation of waves due to hypervelocity impact must include the reflection and refraction of waves of this type from various boundary surfaces as well as the other types of phenomenon mentioned in Section I.

REFERENCES

1. R. Kinslow, Stress Waves in Laminated Materials, AIAA Paper No. 67-140 (1967).
2. W. J. Rae, A critical review of impact theories, unpublished report presented at Highwater Conf. Meteriod Impact, Highwater Quebec, McGill Univ., sponsor (1966).
3. R. L. Bjork, *Proc. Symp. Hypervelocity Impact, 3rd,* Armour Res. Foundation, Illinois Inst. of Technol. (1958).
4. R. L. Bjork, *J. Geophys. Res.* **66**, 3379 (1961).
5. R. L. Bjork, *Proc. Intern. Astronaut. Congr., 10th* **2**, 505 (1960).
6. J. M. Walsh and J. H. Tillotson, *Proc. Symp. Hypervelocity Impact, 6th* **2**, 59 (1963).
7. J. M. Walsh, W. E. Johnson, J. K. Dienes, J. H. Tillotson, and D. R. Yates, Summary report on the theory of hypervelocity impact, General Atomic Div., General Dynamics Corp., Rept. GA-5119 (1964).
8. J. M. Walsh and W. E. Johnson, *Proc. Symp. Hypervelocity Impact, 7th* **2**, 1 (1965).
9. T. D. Riney, Theoretical Hypervelocity Impact Calculations Using the PICWICK Code, General Electric Co. Rept. R64SD13 (1964).

10. T. D. Riney, *in* "Developments in Mechanics" (S. Ostrach and R. H. Scranlan, eds.), p. 419. Pergamon Press, Oxford, 1965.
11. J. F. Heyda and T. D. Riney, Peak axial pressures in semi-infinite media under hypervelocity impact, General Electric Rept. R645SD87 (1964).
12. M. L. Baron, R. Skalak, and C. P. Lecht, *J. Eng. Mech. Div. Am. Soc. Civil Engrs.* **91**, 129 (1965).
13. M. B. Friedman, H. H. Bleich, and R. Parnes, *J. Eng. Mech. Div. Am. Soc. Civil Engrs.* **91**, 129 (1965).
14. S. K. Garg, *J. Appl. Math. Phys.* **19**, 243 (1968).
15. R. Kinslow, *Proc. Symp. Hypervelocity Impact, 6th, Cleveland, Ohio,* **2**, 273 U.S. Army, U.S. Air Force, U.S. Navy; sponsors (1963).
16. Lord Rayleigh, *London Math. Soc. Proc.* **20**, 225 (1888).
17. S. D. Poisson, *Mem. Acad. Sci. Paris* **8** (Ser. 2), 623 (1829).
18. A. E. H. Love, "A Treatise on the Mathematical Theory of Elasticity," 4th ed., Chapter XIII. Dover, New York, 1927.
19. H. Lamb, *Proc. Roy. Soc. (London)* **93**, 114 (1917).
20. R. D. Mindlin, U.S. Army Signal Corps Engineering Lab. Rept., DA-36-039 SC-56772, Fort Monmouth, New Jersey (1955).
21. H. Kolsky, "Stress Waves in Solids." Dover, New York, 1963.
22. P. R. Garabedian, "Partial Differential Equations," Chapter 4. Wiley, New York, 1964.
23. J. Hadamard, "Lectures on Cauchy's Problem in Linear Partial Differential Equations," Book II, Chapter I. Yale University Press, New Haven, Connecticut, 1923.
24. Lord Rayleigh, *London Math. Soc. Proc.* **17**, 4 (1887).
25. R. D. Mindlin, *Proc. Symp. Naval Structural Mechanics, 1st,* p. 199 (1960).

CHAPTER III

THEORY OF IMPACT: SOME GENERAL PRINCIPLES AND THE METHOD OF EULERIAN CODES

J. K. DIENES* AND J. M. WALSH*

GULF GENERAL ATOMIC, LA JOLLA, CALIFORNIA

* Present address: Systems, Science and Software, La Jolla, California.

This article is primarily a review of research performed by the authors and their colleagues between 1962 and 1967, and most of the material originally appeared in topical reports. The effort was divided into two main parts. In the first, some general principles concerning hypervelocity impact were formulated using a combination of theoretical arguments and hydrodynamic codes. The second part consisted of a series of improvements to the available hydrodynamic codes which made possible more general, accurate, and economical numerical solutions to specific impact problems. A general review of this research is given in Section I.

In Section II the effects of projectile shape, velocity, and density are accounted for by late-stage equivalence, a theory stating the conditions that must be met in order that two projectiles produce the same effects at late times. The ideas of late-stage equivalence, dimensional analysis, self-similarity, and linear scaling are related in Section III, and some self-similar solutions are examined.

The hydrodynamic codes used in developing the ideas mentioned above are the subject of the remainder of the article. The continuous Eulerian code OIL is discussed briefly in Section IV. This code accounts for the main features of the compressible, inviscid flow of a material, including the behavior of both the vapor and condensed phases. However, the terminal stages of crater formation cannot be calculated directly with such a code because material strength effects are not included. The last part, Section V, discusses a method which does allow shear strength to be accounted for in a generalization of the OIL code. Cratering calculations describing the impact of an aluminum sphere onto both soft and hard aluminum are presented and compared with experiment to demonstrate that the code does indeed account for material strength in a reasonable manner. Finally, two calculations are compared which verify that projectiles which are late-stage equivalent do indeed produce similar craters.

I. Introduction

Because of the extreme distortion of both projectile and target following a high-speed impact, it seems essential to consider the equations of motion in very general terms to describe adequately the entire process. To illustrate this, a typical interaction is shown in Fig. 1. The most striking feature is the spreading out of the projectile into a thin crescent as it becomes embedded in the target. Assuming that the sphere becomes uniformly spread over the surface of the crater, and that the crater is hemispherical, the final thickness of the projectile would be less than 1% of the crater radius. Thus the theory must account for very large strains

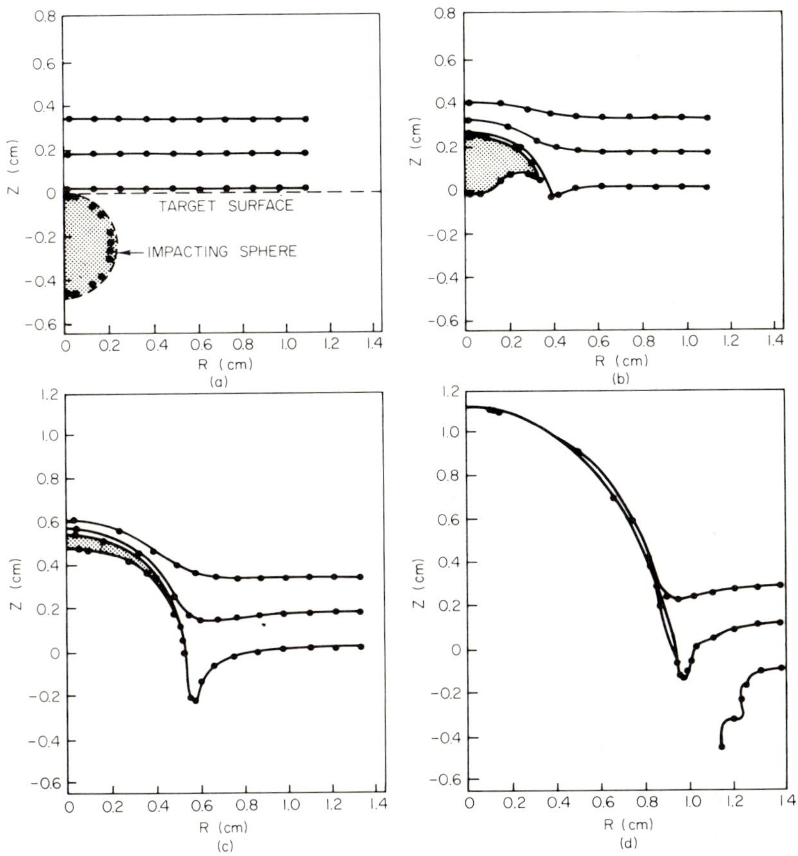

FIG. 1. The deformation of an aluminum sphere striking a hard aluminum target at 7.35 km/sec. At 13.3 μsec the motion is essentially terminated. (a) $t = 0$; (b) $t = 0.69$ μsec; (c) $t = 1.89$ μsec; (d) $t = 13.3$ μsec.

realistically. In addition to the geometric complications, the appropriate system of equations is highly nonlinear as a result of the convective terms, important variations in density, the complexity of an equation of state describing both the vapor and condensed phases of the material, and finally the nonlinear character of the stress–strain relation in the plastic regime. Thus, the various simplifications which suggest themselves, such as assuming spherical symmetry, a rigid projectile, or an ideal gas equation of state, seem to be too restrictive to provide a useful picture of the process.

In view of these theoretical complications there has been considerable emphasis on experiments to provide information about impact. Although much useful information has been obtained in the laboratory, the maximum velocity of carefully controlled projectiles is limited to about 9 km/sec. Even at lower velocities, however, it is possible to extract only limited information about the dynamics of impact because of difficulties in measuring the unusually high pressures and large strains incurred within the projectile and the portion of the target adjacent to it.

Thus there exists a powerful incentive to study impact by means of hydrodynamic codes. This method makes use of general purpose digital computers to integrate numerically the equations of motion, and it is not inhibited by nonlinearities in these equations. The accuracy of the results is limited only by the resolution of the finite-difference approximations and our knowledge of material properties. A collection of articles on various codes is given in the volume edited by Alder *et al.* (*1*).

Codes of the Lagrangian type (*2*) follow the motion of fixed elements of mass, and the finite-difference grid is thought of as fixed in the material. These Lagrangian codes are conceptually straightforward, and easily handle the boundary conditions at free surfaces and contact surfaces between different materials. Their limitation is in the inaccuracy of the numerical approximation when the cells are significantly distorted (strains greater than unity). There is a more specific difficulty when the cells fold over on themselves resulting in negative masses and related pathological behavior.

In Eulerian codes the grid is fixed in space, and the continuum passes through it. The material can be represented either by discrete point masses, as in the particle-in-cell method (*3*), or in a continuous manner, as in the OIL code (*4*). Eulerian codes can handle flows with large distortions in a satisfactory manner, but may be limited by not accurately calculating free surface motion or conditions at material interfaces. The work described in this article has made use of the Eulerian codes which are described in Sections IV and V.

Codes make it possible to obtain quite a complete description of an impact by means of tables or graphs of velocity, pressure, density, and internal energy throughout the flow as a function of time. In a typical problem ten million words of information are printed, and the machine cost may be a thousand dollars.

An important feature of solutions by machine is that one can investigate idealized situations such as the motion of material with a polytropic gas equation of state (discussed in Section III) or the behavior of targets without strength (discussed in Section II). Furthermore, the sensitivity of the results to individual parameter variations (strength, density, sound speed) can be conveniently determined.

The principal limitation of the hydrocode method is that it is difficult to generalize the results to include the effects of several variables entering into a problem. For instance, it is unlikely that the functional dependence on γ appearing in the expression

$$\frac{P_2}{P_1} = \frac{(3\gamma - 1)/(\gamma + 1)(P_1/P_0) - (\gamma - 1)/(\gamma + 1)}{(\gamma - 1)/(\gamma + 1)(P_1/P_0) + 1}$$

for the reflected pressure ratio behind a shock reflected on a rigid wall (*5*, p. 153), could be generated. In some cases this limitation can be partially remedied by dimensional analysis. A second limitation of the method of hydro code solutions is that it does not necessarily provide physical insight, although detailed analysis of a variety of related problems may be instructive. In spite of these limitations, hydro codes do provide a useful tool for investigating impact problems not amenable to experiment. They command particular attention for impact research because of their potential, resulting from rapid progress both in computers and in numerical integration methods.

In the earliest application of hydro codes to impact problems (*6*, *7*, *13*) the particle-in-cell method (PIC) was used to calculate some axisymmetric impacts. (In this method the fluid is approximated by a large number of discrete particles contained in cells which are fixed in space. Each particle moves about independently with the velocity of the cell that contains it.) One result of the early impact calculations (*13*) was to show that two like-material impacts at different velocities lead to the same late-stage flow when the projectiles are characterized by the same value of Lv_0^{α}, where L denotes a characteristic dimension of the projectile, v_0 the impact velocity, and α was determined to be about 0.58. It was also shown that at high velocities the late-stage flow does not depend on the density of the projectile and the details of its shape except through their influence on the mass M (provided neither the density ratio nor the shape is too extreme). These three results can be summarized in the statement that at high velocities and late times projectile effects depend only on a single parameter, $Mv_0^{1.74}$, and not on the details of its shape, density, or velocity. The subject of late-stage equivalence is discussed extensively in Section II, and a related subject, self-similar flows in impact theory, is reviewed in Section III.

The PIC code was successful in calculating flows with large compressions and high internal energy, but at late times, when the pressure is under about a million atmospheres, the results are subject to large fluctuations. These are caused by the discrete change in the number of particles (usually 9–49) in a cell when a particle crosses a cell boundary.

The change in number causes a change in the density of a cell, and leads to the pressure fluctuations. This difficulty led to the development of the OIL code (*4*) which treates the mass in a continuous manner. This code made it possible to follow flows in which the shock pressure is as low as 10,000 atm with good resolution, and calculations with this code provided further information verifying late-stage equivalence (LSE). This code is discussed in Section IV, and it is the basis for the calculations discussed in Sections II and III.

Finally, strength effects were introduced into a modified version of the OIL code known as OIL–RPM (rigid-plastic material). As this name suggests, the rigid-plastic approximation of plasticity theory (*8*) is used to characterize the deformation. This is justified by arguing that the distortions of interest in impact involve strains generally greater than 1%, whereas it is only necessary to account for elastic behavior in shear when the shear deformation is 0.5% or below. In this approximation no elastic distortion is accounted for in the calculation of the shear stresses.

The strength code was used in two calculations which could be compared directly with experimental craters resulting from aluminum-on-aluminum impacts at 7.3 km/sec. The calculations agreed with experiment to within 10% in crater depth for both soft and hard target materials. The agreement is about as good as that between repeated experiments, which is on the order of 10% in crater depth. In addition, the calculated shock pressure was compared with experimental data (*9*). The agreement is within 20% at pressures above 15 kbar, but differences as great as 50% are found in the 1–15-kbar regime. Details are given in Section V.

Calculations were also done with this code which verified predictions of LSE that crater size would remain fixed as projectile velocity is increased and size is decreased in such a way that $Lv_0^{0.58}$ remains fixed.

II. Late-Stage Equivalence: An Asymptotic Theory for Late Times

In the theory of strong explosions it is natural to represent the explosive by a single parameter, its energy, and to assume that the size, the duration of burning, asymmetries, and other details of the energy release do not significantly affect the process at late times. In particular, this is the basis for Taylor's theory (*10*). One might suspect that a similar situation would hold for hypervelocity impact, and this is the basis for early speculations by Stanyukovitch (*11*) and Whipple (*12*). The energy hypothesis is somewhat unsatisfactory, however, if one suspects that the momentum of a projectile may also play a role in determining its ultimate effects. In particular, it seems intuitive that as the momentum of a projectile goes down, while its energy remains constant, the overall effect it produces must somehow go down. In fact, some authors (*6*) have suggested that projectile momentum is the dominant parameter in impact. In the description of impact that follows the projectile is characterized by a single parameter which is neither momentum nor energy, but depends on both these quantities.

A number of hydrodynamic codes are currently in existence which are capable of calculating the hydrodynamic phase of the impact process. One of these, the OIL code, is described in Section IV. It was used to compute a number of impacts, which were systematically compared in four series of calculations. In the first series it was found that the late-stage flow was unchanged when the projectile size L and the velocity v_0 were varied in such a way that $Lv_0{}^{\alpha}$ remained constant. More precisely, it was found that the parameter α in the expression $Lv_0{}^{\alpha}$ could be determined in such a way that at late times the velocity, pressure, and density profiles in both the axial and radial directions were asymptotically the same for pairs of projectiles characterized by the same value of $Lv_0{}^{\alpha}$. The value of α determined was 0.58 ± 0.01 for a wide range of velocities and materials. (If the flow were to depend on projectile energy, α would be $\frac{2}{3}$, and if the projectile momentum were the dominant parameter, α would be $\frac{1}{3}$.) In the second series of calculations comparisons were made which showed the equivalence of interactions between different kinds of materials. In the last two series projectile mass and velocity were held constant while the shape (series three) and the density (series four) were varied. More detailed discussions of the manner in which these calculations and comparisons were made are given in what follows.

A. Late-Stage Equivalence with Velocity Variations

A series of like-material impacts was computed for iron, aluminum, lead, and polyethylene. In each (six-problem) series the impact velocity was varied from $1.24c_0$ to $80c_0$, where c_0 is the material "sound speed" $c_0 = (\partial p/\partial \rho)_s^{1/2}$. Thus the velocities considered range from those attainable experimentally to velocities well beyond that attained by meteors. The initial geometry for each interaction was that of a right-circular cylinder of equal length and diameter interacting with a thick target plate of the same material. Parameters defining the various problems are listed in Table I.

For each material, the problems were planned and analyzed in pairs (1.24 and $2.48c_0$, 4.96 and $9.92c_0$, 40 and $80c_0$). The purpose was to

TABLE I

IMPACT VELOCITIES AND PROBLEM PARAMETERS FOR A SERIES OF CYLINDRICAL PROJECTILE IMPACT CALCULATIONS

Series	Iron (cm/sec)	Lead (cm/sec)	Aluminum (cm/sec)	Polyethylene plastic (cm/sec)	v_0/c_0
A	5×10^5	2.5×10^5	6.5×10^5	3.5×10^5	1.24
B	10^6	5×10^5	1.3×10^6	7×10^5	2.48
A	2×10^6	10^6	2.6×10^6	1.4×10^6	4.96
B	4×10^6	2×10^6	5.2×10^6	2.8×10^6	9.92
A	1.6×10^7	8×10^6	2.1×10^7	1.14×10^7	40
B	3.2×10^7	1.6×10^7	4.2×10^7	2.28×10^7	80

assure comparable space resolution at late times so that comparison of the two flows for late-stage equivalence could be made without possible bias due to unequal space resolution. In each pair, the twofold increase in impact velocity was offset by reducing the projectile dimensions to $\frac{2}{3}$ the former values. Thus the problems were run on a basis intermediate between that given by keeping projectile kinetic energy and projectile momentum constant. More precisely, the indicated change in velocities and dimensions corresponds to $\alpha = 0.585$. The close agreement seen below between the as-run problems (prior to changing the relative size of two problems by scaling) therefore corresponds to values of α which are very near to 0.585.

For the mid- and high-velocity pairs, the detailed comparisons of late stages of the interactions show very good agreement, in both the plots of pressure, density, and velocity profiles from individual cell quantities and in the integrated momentum curves. Typical results of these comparisons are reproduced as Figs. 2–4.

For the lowest velocities considered (5×10^5 to 10^6 cm/sec in iron, 2.5×10^5 to 5×10^5 cm/sec for lead, etc.), late-stage equivalence is less exact. The radial and axial momentum curves are again brought into good agreement by a single value of α, but there persists to late times some tendency for the higher-velocity impact to be associated with a more intense but narrower pulse. Thus, in the iron pair, the shock for the 10^6-cm/sec impact has attenuated to 250 kbar while that for the low-velocity impact is down to 200 kbar at the same radius; for lead, an analogous comparison is 90 and 70 kbar. It seems apparent that the failure to achieve precise equivalence is characteristic of low velocities and would become more pronounced if the velocity were further reduced.

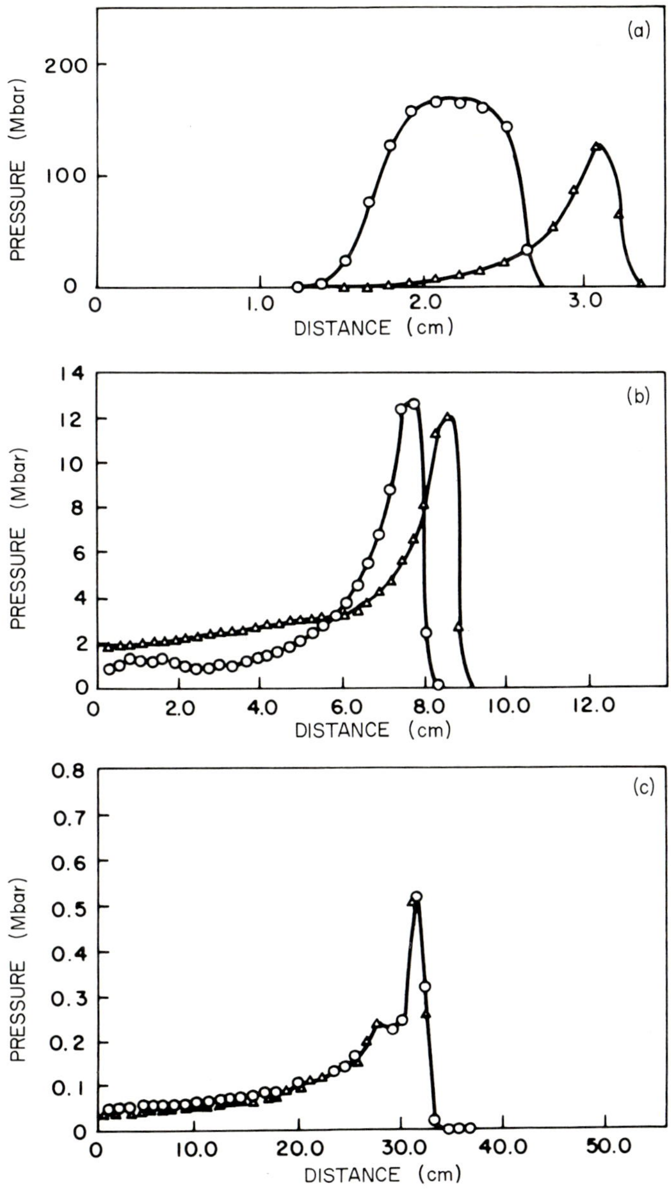

FIG. 2. Convergence to late-stage equivalence of two impact flows. Pressure versus distance is given at three times. Plots are from the 8×10^6-cm/sec and 1.6×10^7-cm/sec lead impacts. Distance is measured along a line 45° from the axis of symmetry, with origin at the center of impact on the target surface. (a) $t \sim 0.4$ μsec; (b) $t \sim 2.6$ μsec; (c) $t \sim 4.6$ μsec.

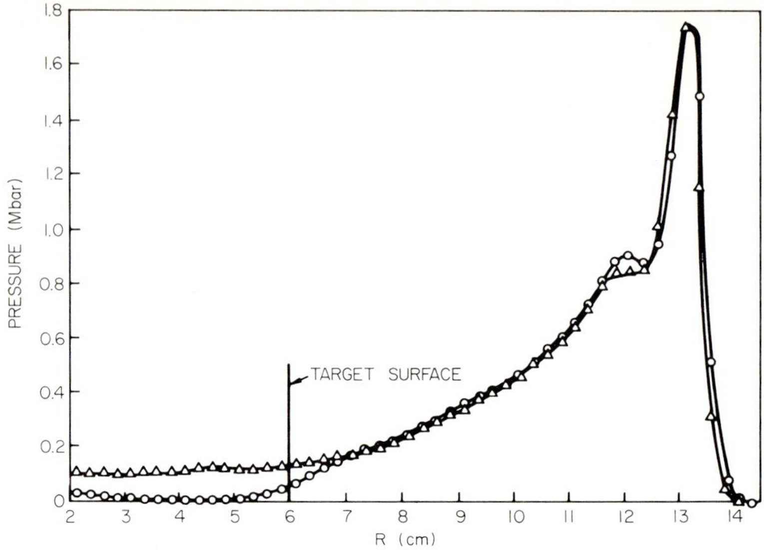

FIG. 3. Comparison of axial pressure profiles for the 2×10^6-cm/sec and 4×10^6-cm/sec iron impacts, showing late-stage agreement. (○) 2×10^6; (△) 4×10^6.

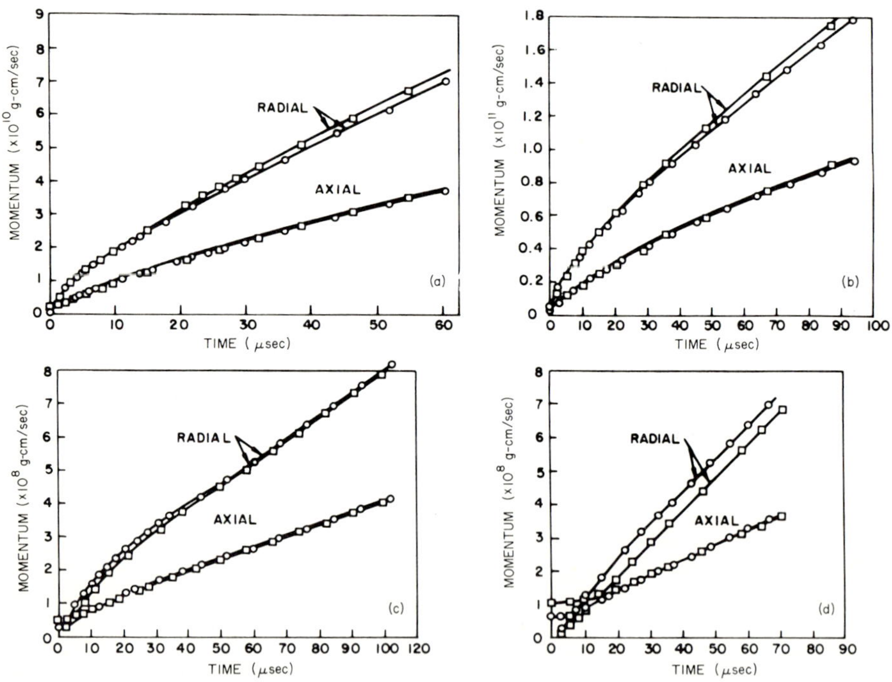

FIG. 4. Comparison of radial and axial momenta; high velocity refers to the two highest velocities given for that material in Table I, midvelocity refers to the third and fourth highest, and low velocity refers to the two lowest velocities; data for the lower-velocity member of each comparison are plotted as circles. (a) High-velocity aluminum; (b) high-velocity iron; (c) midvelocity plastic; (d) low-velocity lead.

Values of α were obtained from the various comparisons by scaling the relative size of the problems until the best agreement was obtained between the two solutions. (The most precise scale factors are those given by requiring that the axial and radial momenta agree.) The value of α is then calculated from $L_1 v_1^{\alpha} = L_2 v_2^{\alpha}$, where L_1 and L_2 are the characteristic linear dimensions after scaling and v_1, v_2 are the impact velocities. The results show a remarkable uniformity in α for all twelve comparisons. Individual values for the eight higher-velocity pairs lie between 0.56 and 0.59 and the low-velocity comparisons indicate values of α very near 0.60. Precisions on the individual determinations are typically 3–4%.

B. Material Equivalence

Comparisons to test the equivalence of iron, lead, aluminum, and plastic interactions are possible from any pair of problems listed in a given line of Table I. Dimensionless plots of the momentum curves and representative pressure profiles were made for each of the possible comparisons and typical results are reproduced as Figs. 5 and 6. The general agreement is seen to be good. It may be noted that the results for polyethylene correlate somewhat less satisfactorily than those for the metals, as would be expected from the fact that the polyethylene equation of state does not conform to those for metals (*13*). Comparable results were obtained for impacts at lower velocities.

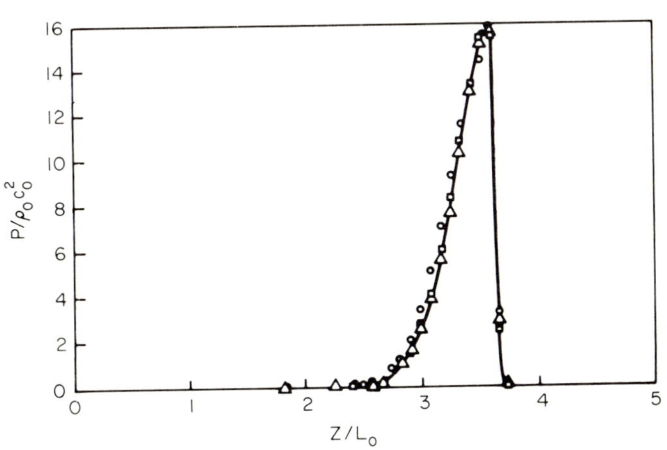

Fig. 5. Comparison of late-stage axial pressure profiles in dimensionless variables for polyethylene, aluminum, and lead impacts of equal v_0/c_0. (○) Polyethylene, 2.8×10^6 cm/sec; (△) aluminum, 5.2×10^6 cm/sec; (□) lead, 2×10^6 cm/sec.

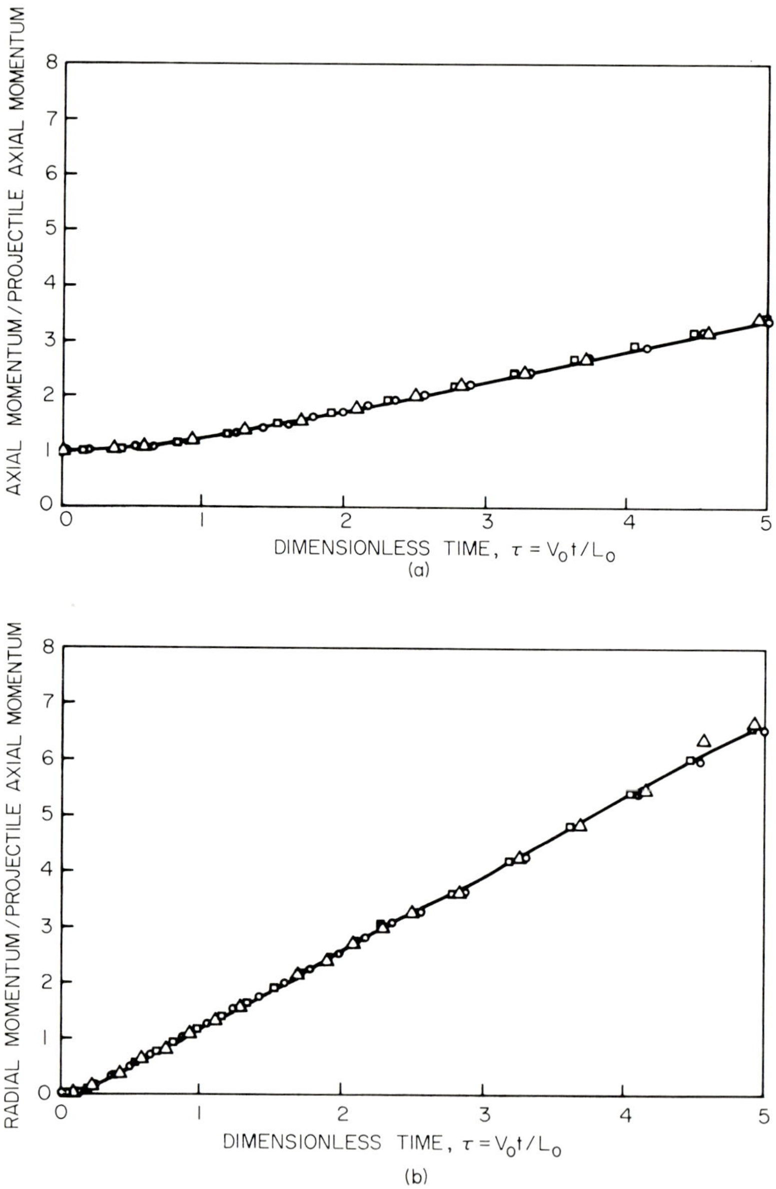

FIG. 6. Comparison of momenta versus time in dimensionless variables for polyethylene, aluminum, and lead impacts of equal v_0/c_0. (○) Polyethylene, 3.5×10^5 cm/sec; (△) aluminum, 6.5×10^5 cm/sec; (□) lead, 2.5×10^5 cm/sec.

C. Projectile Shape Effects

A series of 12 impact interactions was studied to explore effects of projectile shape. Four shapes were considered for each of three impact velocities:

(1) right-circular cylinder, equal length and diameter;
(2) sphere;
(3) rod, three diameters long; and
(4) disk, three thicknesses in width.

The velocities were $1.24c_0$, $4.96c_0$, and $40c_0$ (corresponding to three of the cylindrical projectile impacts listed in Table I) and the projectile masses were the same in all problems. The impacting material was taken to be lead, although the results apply also to other solids by virtue of the material equivalence studies.

In the analysis of the computed results, particular attention was given to determining whether the different projectile shapes lead to a late-stage equivalence and to evaluating the relative effectiveness of the various shapes.

The simplest and most definitive results were obtained in the comparisons of spherical and cylindrical projectiles. For each of the three impact velocities, these projectiles lead to practically identical late-stage flows, both in the total axial and radial momenta and in the comparison of instantaneous pressure, density, and velocity plots. Substantial differences are limited to the very early times in which the projectile is embedding itself in the target plate. Finally, the relative scale of the two flows is such that the projectiles are equivalent on an equal-mass basis.

Late-stage equivalence is somewhat less exact for the disk and rod projectile geometries. At the two higher velocities, however, the flows are in reasonably good general agreement after a shock penetration distance which is large compared to initial projectile dimensions. This is demonstrated by the vector field plots for the $4.96c_0$ (10^6 cm/sec for lead) impact, as seen in in Fig. 7. A close comparison of these results, however, shows that even at late times the disturbance for the rod impact is slightly more elongated in the axial direction than that for either the disk or the sphere. Except for this characteristic of the rod impacts, the late-stage agreement found in the high-velocity comparisons (10^6 and 8×10^6 cm/sec) is quite good. Precise comparisons of the momentum curves and the associated scale factors show that the rod impact is equivalent to that for the sphere or cylinder on a projectile equal-mass basis. The disk is slightly, but perceptibly, less effective. At both velocities, it would require an increase of some 15% in disk mass for this projectile to be as effective as the others.

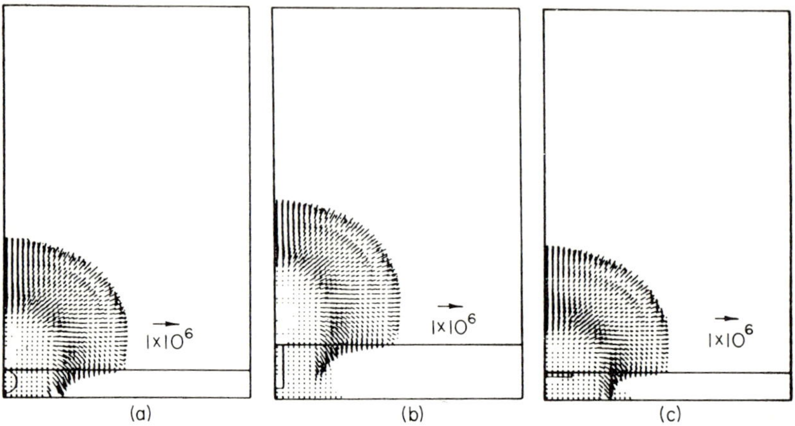

FIG. 7. Comparison of the mass flux distributions for sphere, rod, and disk having equal mass impacting at 10^6 cm/sec. (a) $t = 59.7$ μsec; (b) $t = 64$ μsec; (c) $t = 58$ μsec.

The above rod and disk projectiles were purposely chosen to have moderately high aspect ratios, with the objective of testing the limits of late-stage equivalence. Substantially more extreme aspect ratios will certainly lead to flows which remain qualitatively different throughout the hydrodynamic regime. Such dissimilar flows are, in fact, also evident for the present aspect ratios at the lowest velocity (2.5×10^5) studied. The target plate disturbance due to the rod, for example, was much more concentrated near the axis than that for the disk, and the impact configurations are expected to remain dissimilar indefinitely.

Analytical results can be cited for the limiting extremes of thin-disk and long-rod geometries. For disk projectiles of sufficiently extreme aspect ratio, impact disturbances will become equivalent as essentially plane waves. The slab impact considerations for an ideal gas of $\gamma = 1.5$, discussed in Section III, indicates equivalence on a basis which is approximately midway between that for equal-energy and equal-momentum projectiles. Because of equation-of-state similarities, the result should apply also to solid impact for sufficiently high velocities and sufficiently extreme aspect ratio. It may be that the relative ineffectiveness of the disk with an aspect ratio of three is indicative of the fact that α should decrease to about 0.51 in this limit.

D. DENSITY EFFECT STUDIES

In most applications, the projectile and target are of different materials, and it is desirable to extend present considerations to include such cases. Dimensional considerations suggest that one should introduce at least two new parameters to characterize unlike-material impacts—namely, the density and sound speed of the projectile in addition to those of the target. It seems reasonable to assume, however, that variations in projectile density are more significant than those in projectile sound speed, and the subject of unlike-material impact is usually equated to that of density effect studies.

Several comparisons have been made to evaluate possible effects due to changing projectile density. These include:

1. Three ideal-gas impacts, $\gamma = 1.5$, were computed in which the initial projectile density was $(1.5)^3 = 3.375$ times the target plate density, equal to the target density, and $(1.5)^{-3}$ times the target density. The intial masses of the three projectiles were the same. The problems were run using the continuous Eulerian code. The calculations were carried to a time about 125-fold greater than the time required for the pellet to bury itself in the target, and detailed cross comparisons of the flows were made in order to evaluate late-stage equivalence and the relative effectiveness of the three densities. When consideration is given to the fact that projectile density varies by a factor of 11.4, the flows are in remarkably good agreement. Shock penetration distances at late times agree to 10% with the maximum discrepancy occurring on the axis of symmetry. Axial momenta in the three cases agree within 5%, and the spread in radial momentum is 20% at late times. The direction of indicated discrepancies is such that the high-density projectile is a more effective projectile, but the effect is a negligibly small one.

2. Two calculations were made using special equations of state in order to investigate possible projectile density effects for metal–metal impacts at velocities around 10^6 cm/sec. The target material was normal lead and the projectiles were taken to have a modified lead equation of state for which the densities were, respectivley, $(1.5)^3$ and $(1.5)^{-3}$ times those of normal lead. This procedure was used to provide a comparison in which the only equation-of-state variation was the density ratio for the target and projectile. The calculations were carried out using the particle-in-cell code, and a relatively high mass resolution of 36 particles per cell was used in order to facilitate the comparison of cell quantities. The two problems were run with equal initial projectile masses. The remarks made above for gas impacts apply almost verbatim to the current comparison. Axial momenta agree well and the total spread in radial momentum at late times is only 25%. The effects of the 11.4-fold variation in projectile density, with projectile mass kept constant, are quite small; within 20% in projectile mass, the two projectiles are equivalent on an equal-mass basis.

3. Three impacts were computed in which tungsten, iron, and beryllium projectiles interacted with iron targets at a velocity of 4×10^6 cm/sec. The particle-in-cell code was used; the interactions were again found to be approximately equivalent at late times for equal initial projectile masses.

Substantial premiums on projectile density have occasionally been reported from empirical correlations (*14*, *15*) of experimental impact data. The problem has not been the subject of previous theoretical treatment, however, and the results described above are believed to be sufficient to show that such effects do not persist to velocities of 10^6 cm/sec and greater.

E. Overall Effects of Solid–Solid Impact Based on Late-Stage Equivalence

1. *Penetration Relation*

The close agreement of two impact flows during the hydrodynamic phase of the interaction is useful when one observes that the subsequent strength-dependent phase of the interaction will also be the same. It follows that the craters in the two cases must agree in shape and in size. This result and the equivalence observations described above can be summarized by the penetration relation

$$p/d = K(\rho_p/\rho_t)^{1/3}(v_0/c_0)^{0.58}$$

Here p is the crater depth or some other characteristic crater dimension, d is the diameter of a sphere with the mass of the projectile, ρ_t and c_0 are target density and sound speed, ρ_p and v_0 are projectile density and velocity, and K is a constant to be determined from a single impact experiment. The quantity K is a constant of the target material, depending on its strength.

From the results given in the preceding discussion one can expect the penetration relation to be reliable for solid impacts in which (1) the impact velocity is greater than about twice target sound speed; (2) the projectile aspect ratio is within a factor of 3 of unity; and (3) the projectile density is within a factor of about 3 of the target density. Additional calculations might show that these results are conservative.

2. *Momentum Multiplication*

One important effect in hypervelocity impact is that the final momentum imparted to a target may be many times that of the incident projectile. This enhancement is due to the recoil resulting from ejecta thrown backwards. Material moving backward can be observed in the neighborhood of the crater lip on velocity plots such as in Fig. 7, or the displacement plots in Fig. 1.

This dependence of target momentum upon impact velocity is determined by recalling that the late-stage momentum I depends only on $Mv_0^{3\alpha}$ and not on M and v_0 individually. This is expressed functionally by

$$I = f(Mv_0^{3\alpha})$$

On the other hand, simple linear scaling shows that the impulse must be directly proportional to projectile mass M so that the functional dependence must be

$$I = kMv_0^{3\alpha}$$

Finally, momentum multiplication r_I, defined as the ratio of total target impulse to initial projectile momentum, satisfies

$$r_I = I/I_0 = kMv_0^{3\alpha}/Mv_0 = kv_0^{3\alpha-1}$$

Hence, momentum multiplication increases as the 0.74 power of impact velocity, when the value $\alpha = 0.58$ is taken into account.

3. *Target Hardness Effects*

If two impacts have produced essentially identical flows at some moderate time after impact when the flow is still hydrodynamic, it is to be expected that the flow at subsequent times will also be the same. Hence, characteristics of the target which affect the flow at a very late time or low pressure will not alter the equivalence of the two impacts. Target strength is such a characteristic, and we can conclude that target strength will play the same role for all projectiles with the same value of $Mv_0^{1.74}$. Hence, the effect of target strength will persist at very high velocities.

In experiments at Ballistic Research Laboratories by Kineke and Richards (*16*) velocities of 9.7×10^5 cm/sec and 1.55×10^6 cm/sec were attained for aluminum and beryllium projectiles, respectively, both of which were fired into soft and hard aluminum targets.

The ratio of crater volumes in the two targets (same impacting projectile) was found to be a constant, about 3, for the entire range of impact velocities from a few times 10^5 cm/sec to the maximum values given above. Thus target strength effects continue to influence crater dimensions up to the highest impact velocities attained.

4. *Principal Conclusions*

The principal conclusions from late-stage equivalence are as follows:

(1) Crater dimensions increase as the 0.58 power of the impact velocity.

(2) Changing projectile density at constant projectile mass leaves unchanged the overall effects on the target.

(3) Spherical and unit-aspect-ratio cylindrical projectiles of equal mass have equivalent impact effects except at very early times. Projectile aspect ratios as extreme as 3 are unimportant for impacts at five times target sound speed and greater.

(4) Target strength plays an important role in limiting crater size even for arbitrarily high velocities. The ratio of crater dimensions for hard and soft targets is independent of the impact velocity.

(5) The ratio of total target impulse to initial projectile momentum increases as the 0.74 power of the impact velocity.

Conclusions (1), (4), and (5) are applicable to impact velocities which are greater than about twice sound speed, and (2) pertains to impact velocities greater than about five times target sound speed.

III. Dimensional Analysis, Self-Similarity, and Late-Stage Equivalence

The general mathematical problem of determining the flow produced by a hypervelocity projectile is not tractable analytically. However, important simplifications can be made in the functions defining velocity, pressure, and density profiles by using principles which allow the independent variables to be collapsed into groups. Three stages of simplification are possible in the problem of perpendicular impact, while retaining axisymmetric flow. A fourth simplification could be obtained by assuming spherical symmetry, ultimately reducing the problem to one in ordinary differential equations, but spherical symmetry is not assumed in this article.

Linear scaling is valid so long as dissipative effects are not important. Assuming that is the case, the important material properties are the density (ρ), the sound speed (c), and the Grüneisen ratio (G). The velocity at a radius (r) and a depth (z) below the free surface can be written in the general form

$$u = F(z, r, t, v_0, c_0, \rho_0, G_0, c_1, \rho_1, G_1, L_0)$$

where L_0 denotes a characteristic projectile dimension and v_0 is the impact velocity. Zero subscripts are used to denote projectile properties, and unity subscripts denote properties of the target. Similar functional dependences apply to the pressure, density, and internal energy. Since for many pairs of materials the ratio of sound speeds is near one and the Grüneisen ratio does not vary significantly from 1.5, the velocity expression can be approximately simplified to

$$u = F(z, r, t, v_0, c_0, \rho_0, \rho_1, L_0)$$

Dimensional analysis can be used to further simplify the expression. The nondimensional coordinates are written

$$\zeta_0 = z/L_0, \qquad \xi_0 = r/L_0$$

The other nondimensional variables can be selected as

$$\beta = v_0/c_0, \qquad \nu = \rho_0/\rho_1, \qquad \tau = tv_0/L_0$$

and the velocity can then be written

$$u = v_0 F(\zeta_0, \xi_0, \tau, \beta, \nu)$$

This expression satisfies the requirement of linear scaling since it is left unchanged by the transformation

$$L_0 \to \lambda L_0, \qquad z \to \lambda z, \qquad r \to \lambda r, \qquad t \to \lambda t, \qquad v_0 \to v_0$$

which expresses mathematically the requirement that a projectile B, larger by a factor λ than projectile A, generates a flow field at time λt which can be obtained by stretching the A field at time t by a factor λ. Linear scaling would not apply if viscosity or other dissipative factors were included in the analysis.

As a working hypothesis one can further assume that a single parameter $L_0 v_0^{\alpha}$ characterizes the projectile, where α is a parameter to be determined. (An α of $\frac{1}{3}$ would indicate that momentum, $\theta\rho_0 L_0^3 v_0^2$, is the significant parameter, and an α of $\frac{2}{3}$ indicates that energy, $\theta\rho_0 L_0^3 v_0^2/2$, is the significant quantity. Here θL_0^3 denotes projectile volume.) The length and time variables can then be collapsed into two constellations with the result

$$u = v_0(L_0/z)^{1/\alpha} F(\zeta_0^{\alpha+1}/\tau_0^{\alpha}, \xi_0^{\alpha+1}/\tau_0^{\alpha}, \beta^{\alpha+1}/\tau_0, \nu)$$

The discussion in the previous section on late-stage equivalence has shown that projectile mass is the dominant quantity when projectile density is varied, and consequently ρ_0 must appear only in the combination $\rho_0 L_0{}^3$. Then ν is eliminated if we introduce the parameter $L = L_0(\rho_0/\rho_1)^{1/3} = \nu^{1/3}L_0$ and ζ, ξ, and τ are defined as $\zeta = z/L$, $\xi = r/L$, $\tau = u_0 t/L$. Then

$$u = v_0(L/z)^{1/\alpha} F(\xi^{\alpha+1}/\tau^\alpha, \zeta^{\alpha+1}/\tau^\alpha, \beta^{\alpha+1}/\tau)$$

$$v = v_0(L/z)^{1/\alpha} G(\xi^{\alpha+1}/\tau^\alpha, \zeta^{\alpha+1}/\tau^\alpha, \beta^{\alpha+1}/\tau)$$

$$p = \rho_0 v_0{}^2(L/z)^{2/\alpha} H(\xi^{\alpha+1}/\tau^\alpha, \zeta^{\alpha+1}/\tau^\alpha, \beta^{\alpha+1}/\tau)$$

$$\rho = \rho_0\, J(\xi^{\alpha+1}/\tau^\alpha, \zeta^{\alpha+1}/\tau^\alpha, \beta^{\alpha+1}/\tau)$$

The form of the equations cannot be further simplified for condensed material impacts.

For very high impact velocities the equation of state at early times can be represented by the high-pressure approximation

$$p = GE\rho$$

where E is the internal energy per unit mass and G is Grüneisen's ratio. This is the ideal-gas equation with $G = \gamma - 1$. The equation of state documented by Tillotson (*17*) gives values of G near 0.5 for pressures above 10 Mbars. Under these conditions the small disturbance sound speed c_0 can be dropped from the problem (β is a large parameter) and

$$u = v_0(L/z)^{1/\alpha} F(\xi^{\alpha+1}/\tau^\alpha, \zeta^{\alpha+1}/\tau^\alpha)$$

$$v = v_0(L/z)^{1/\alpha} G(\xi^{\alpha+1}/\tau^\alpha, \zeta^{\alpha+1}/\tau^\alpha)$$

$$p = \rho_0 v_0{}^2(L/z)^{2/\alpha} H(\xi^{\alpha+1}/\tau^\alpha, \zeta^{\alpha+1}/\tau^\alpha)$$

$$\rho = \rho_0\, J(\xi^{\alpha+1}/\tau^\alpha, \zeta^{\alpha+1}/\tau^\alpha)$$

These equations are self-similar. That is, the solution at time λt can be constructed from the solution at time t by a stretching of axes in which

$$r \to \lambda^{\alpha/(\alpha+1)} r, \qquad z \to \lambda^{\alpha/(\alpha+1)} z, \qquad t \to \lambda t$$

$$u \to u/\lambda^{1/(\alpha+1)}, \qquad p \to p/\lambda^{2/(\alpha+1)}, \qquad \rho \to \rho$$

The reduction to two variables leaves a set of partial differential equations and the flow is not amenable to analytical solution without additional approximations. It is possible, however, to draw certain conclusions from the form of the late-stage self-similar solution:

(1) The dimensions of the disturbance increase as $t^{\alpha/(\alpha+1)}$.

(2) Pressures at corresponding points are in the ratio $(\xi_1/\xi_2)^{-2/\alpha}$,

where ξ_1/ξ_2 is the ratio of linear dimensions at the two times. In particular, the shock pressure attenuates as $\xi_2^{-2/\alpha}$. The velocities at corresponding points are in the ratio $(\xi_1/\xi_2)^{-1/\alpha}$, and the velocity of the shock, and the material velocity there, decrease as $\xi_2^{-1/\alpha}$.

(3) The total positive axial and total positive radial momenta increase as $t^{(3\alpha-1)/(\alpha+1)}$. This result follows by noting that distances increase as $t^{\alpha/(\alpha+1)}$, mass engulfed therefore goes as $t^{3\alpha/(\alpha+1)}$, and velocities decrease as $t^{-1/(\alpha+1)}$, so that the momentum within corresponding elements of volume varies as $t^{(3\alpha-1)/(\alpha+1)}$.

The Eulerian code has been used to treat numerically the problem of a right-circular cylinder of unit aspect ratio impacting a semi-infinite target. The value $\frac{3}{2}$ was chosen for γ because solids at extremely high shock pressures ($\gtrsim$1000 Mbar) have Hugoniot curves and expansion isentropes which approximate those of a $\gamma = \frac{3}{2}$ ideal gas. The values $v_0 = 10^6$ cm/sec, $\rho_0 = 1$ g/cm^3, and a projectile diameter of 3.6 cm were used to complete the problem specification.

A point of particular interest concerning the computed solution is whether or not the flow becomes self-similar as indicated by the assumption of late-stage equivalence; after an early phase (two or three times the time required for the initial projectile to move its own length), the flow geometry tends to a configuration which does indeed remain self-similar as it enlarges.

In this late-stage configuration, the shock wave closely approximates a spherical shape, with the (moving) center of the sphere located at about one-third of the shock penetartion distance. The target material which has been engulfed by the shock is largely compressed into a thin shell near the shock front. Progressive stages of this flow can be visualized from the velocity and mass flux plots reproduced in Fig. 8.

Tests of other flow properties characteristic of the similarity solution, and also determinations of α, are possible from plots of the integrated momenta and the shock attenuation. Since the momenta must vary as $t^{(3\alpha-1)/(\alpha+1)}$ if the assumption of late-stage equivalence is valid, one can examine the log–log plots of integrated radial and axial momenta versus time to (a) test if the plots become linear at late times, (b) test whether the late-time slopes of the two momentum plots are in agreement, and (c) determine α from the slopes. The momentum curves are plotted as Fig. 9, where it is seen that the plots are quite linear at late times and that the slope $(3\alpha - 1)/(\alpha + 1)$ is in good agreement with both plots if one chooses $\alpha = 0.59$. (The slope for $\alpha = \frac{1}{3}$ and that for $\alpha = \frac{2}{3}$ are indicated on the graph for comparison.) Figure 10 shows log–log plots of shock penetration and shock strength versus time, both for the axis of

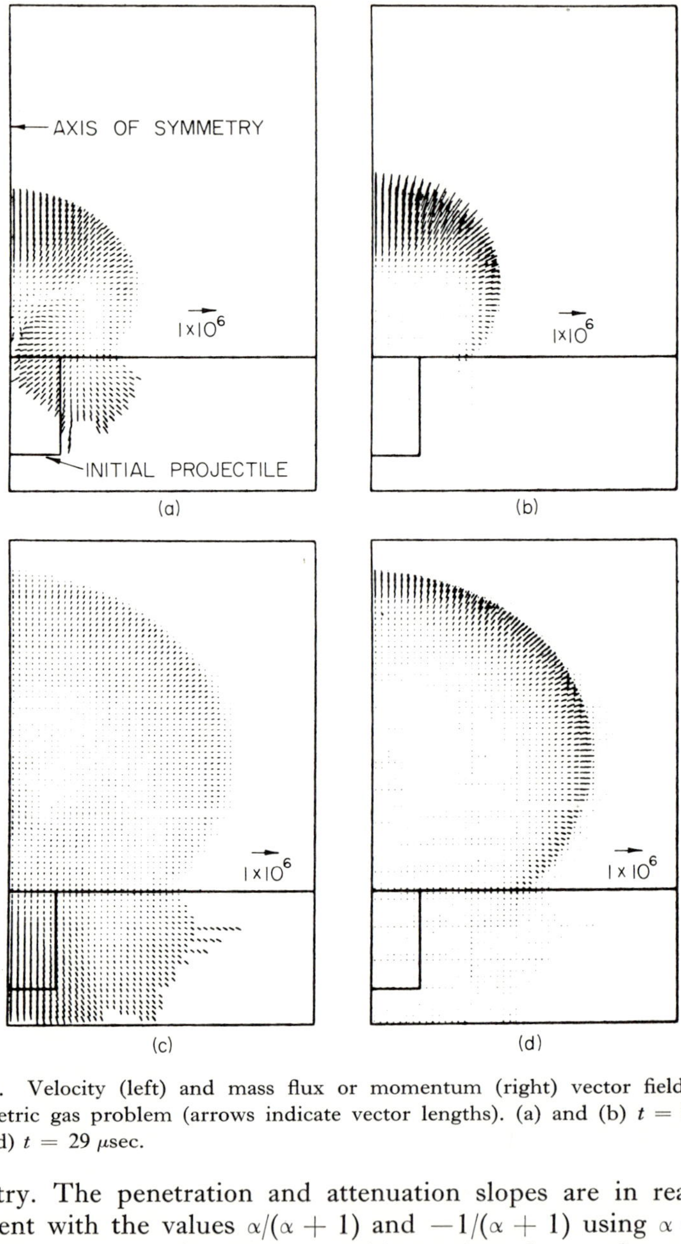

FIG. 8. Velocity (left) and mass flux or momentum (right) vector fields for the axisymmetric gas problem (arrows indicate vector lengths). (a) and (b) $t = 9.75$ μsec; (c) and (d) $t = 29$ μsec.

symmetry. The penetration and attenuation slopes are in reasonable agreement with the values $\alpha/(\alpha + 1)$ and $-1/(\alpha + 1)$ using $\alpha = 0.59$. (A slightly lower α, about 0.55, would correlate the results somewhat better; but the value of α from the momentum curves is considerably more accurate.)

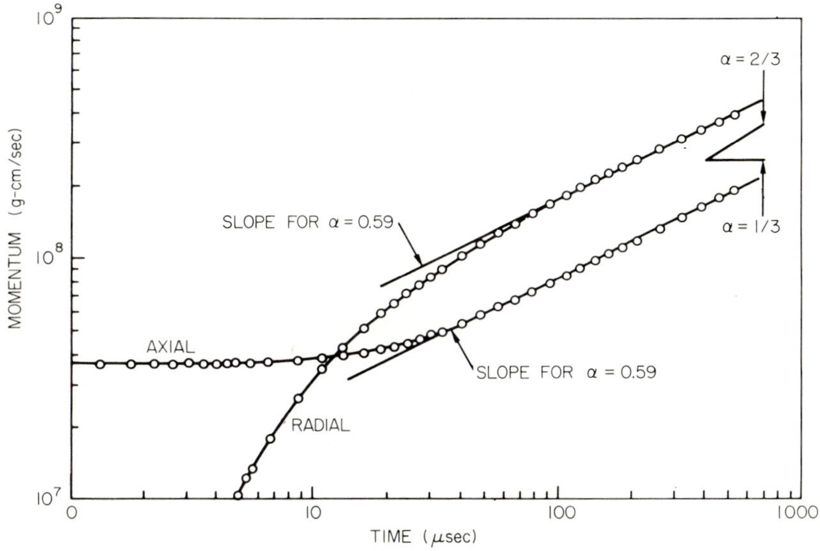

FIG. 9. Radial and axial momenta for the $\gamma = 1.5$ ideal-gas impact; the slopes necessary for late-stage equivalence on an $\alpha = \frac{1}{3}$ and $\alpha = \frac{2}{3}$ basis are indicated at the upper right.

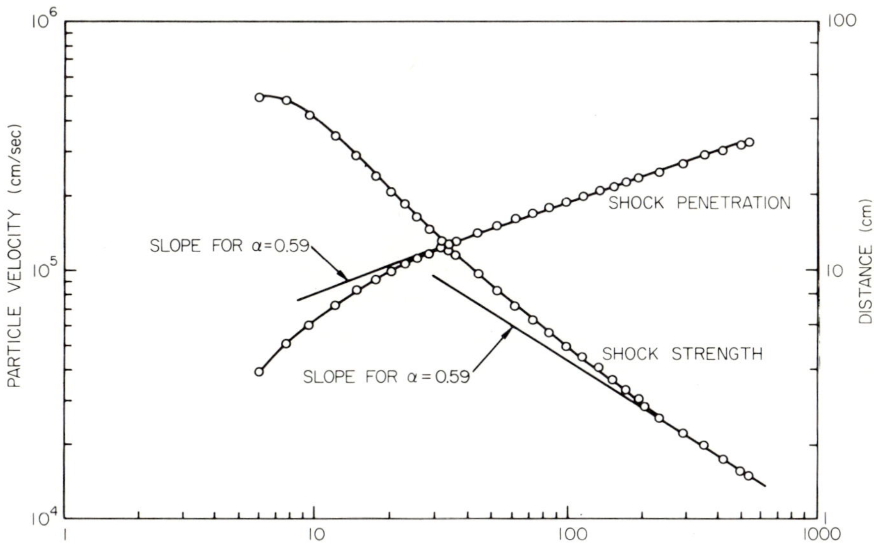

FIG. 10. Shock penetration and shock strength versus time for the $\gamma = 1.5$ impact.

Finally, velocity, density, and pressure profiles at various times can be compared to test whether the flow remains self-similar. The results of a representative comparison—the velocity distributions along 45° for times different by a factor of three—are shown as Fig. 11. It is seen that if the

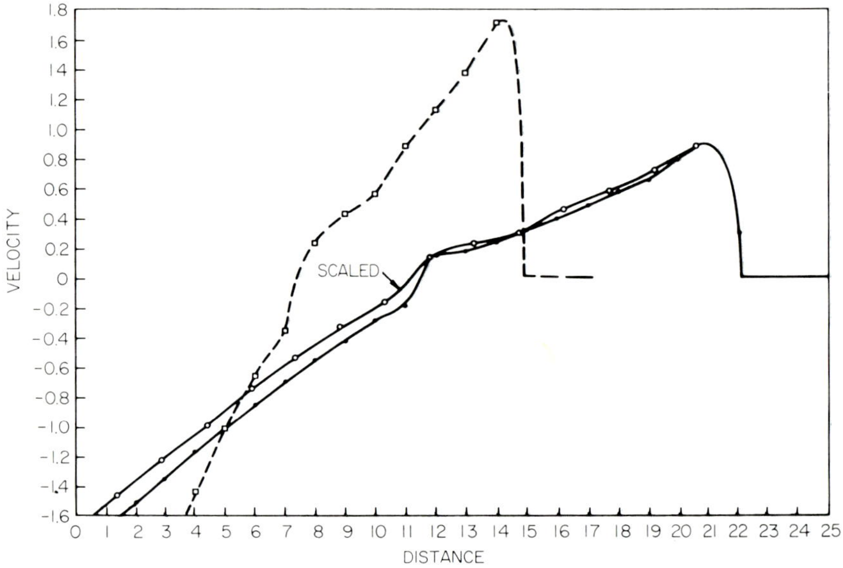

FIG. 11. Velocity versus distance along a line 45° from the axis of symmetry at two times; the scaled curve is determined from the earlier data and the similarity relations.

curve for the early time is scaled to a later time by the similarity relations $Z \sim t^{-\alpha/(\alpha+1)}$ and $v \sim t^{-1/(1+\alpha)}$, the agreement is good. The value $\alpha = 0.59$ was used.

The results of all comparisons are, then, compatible with the assumption that the flow has tended to the self-similar form indicated by late-stage equivalence. It is also of interest to note that the value $\alpha = 0.59$ is in good agreement with the values obtained for condensed materials, despite substantial equation-of-state differences and formal differences in the method of analysis of the numerical results.

Using the assumption that the flow is self-similar, Rae and Kirchner (*18*) have reported some analytical progress on the problem of axisymmetric ideal-gas impact. By considering the flow equations as specialized to the axis of symmetry and by requiring that the solution pass smoothly through a singularity in the differential equations, they find that the exponent N in the similarity variable r/t^N can be evaluated to a good approximation.* Values $N = 0.37$–0.39 are obtained, as compared to $N = 0.40$ for the spherical-blast problem.

A consequence of the late-stage-equivalence interpretation, on the other hand, is a relation between α and the similarity exponent N. Specifically, comparison with the present similarity variable $Z^{\alpha+1}/\tau^{\alpha}$ shows that $N = \alpha/(\alpha + 1)$. Using the present result, $\alpha = 0.59$, gives $N = 0.37$. Rae gives N the value 0.375 for $\gamma = 3$ and the value 0.367 for $\gamma = 2$. Hence the value for $\gamma = \frac{3}{2}$ is found by extrapolation to be 0.363.

The above agreement is, of course, quite gratifying. The numerical integrations and the analytical treatment by Rae and Kirchner (*18*) are based on entirely different approximations, so that the agreement is a significant confirmation of both results.

Late-Stage Equivalence in One-Dimensional Impact

The flow resulting from the impact of a slab on a semiinfinite target has some of the essential features of the axisymmetric impact problem, and yet is amenable to a relatively simple analytical treatment when the slab and target are described by the polytropic equation of state and are initially cold. In this impact problem, the notion of late-stage equivalence serves to interpret a similarity solution which otherwise has rather puzzling features. Specifically, the similarity solution discussed in the remainder of this section has infinite kinetic energy and zero momentum, and consequently special physical ideas are required to provide a useful interpretation to this flow. Although a mathematical solution has been known for some time (*19*), an interpretation which allowed the solution to be connected with a specific impact did not exist until it was realized that impacts are characterized by a parameter of the form Lv_0^{α}, and that α cannot be determined by energy or momentum arguments. It will be shown that the boundary condition that pressure must vanish at the free surface determines α, which, in consequence, may be considered an eigenvalue.

1. *Reduction of the Flow Equations by the Similarity Method*

The Lagrangian equations of motion are used to describe the flow. The independent variables are x, the initial distance of a particle from the free surface, and t, the time from the instant of impact. The dependent

* The principal approximations, in addition to assuming self-similarity, are as follows: The shock wave in the target is a hemisphere and an estimate is made for the magnitude of $d^2p/d\theta^2$ at the singular point. The last approximation, and the fact that N is found to be a weak function of γ, accounts for the small variation in N.

variables are the pressure p, velocity u, and specific volume V. The laws of mass and momentum conservation and the condition that the expansion behind the shock be isentropic provide three equations for these three variables. The equations of motion for a material with initial specific volume V_1 are, then,

$$\dot{V} = V_1 u_x \qquad \text{(conservation of mass)} \tag{1}$$

$$\dot{u} = V_1 p_x \qquad \text{(conservation of momentum)} \tag{2}$$

$$pV^\gamma = f(x) \qquad \text{(isentropic expansion)} \tag{3}$$

The boundary conditions at the shock front are the Rankine–Hugoniot relations (5) between the shock pressure p_s, the material velocity u_s, the shock velocity v_s, and the specific volume V_s behind a strong shock being driven into a material with a perfect-gas equation of state and no internal energy :

$$p_s = [2/(\gamma + 1)](v_s^2/V_1) \tag{4}$$

$$V_s = [(\gamma - 1)/(\gamma + 1)]\, V_1 \tag{5}$$

$$v_s = [(\gamma + 1)/2]\, u_s \tag{6}$$

A separation of the dependent variables into a product of two terms may lead to a reduction of the flow equations to a set of ordinary differential equations. This process is justified if it leads to a solution satisfying the flow equations and boundary conditions given by Eqs. (1)–(6). An additional condition, for the impact problem, but not the blast wave problem, is that the pressure at the free surface must be zero. A natural procedure is to make the variables nondimensional by normalizing to their values at the shock front. Then reduction to a set of ordinary differential equations by a separation of variables occurs if the (nondimensional) dependent variables depend only on a single independent variable η, which in turn may be a function of both x and t. The essential feature of the similarity method is that if the flow depends only on a single dimensional parameter of the impact, Lv_0^α, and not on the slab thickness L or the slab velocity v_0 individually, then the only nondimensional independent variable that can be constructed is $Lv_0^\alpha t^\alpha / x^{\alpha+1}$, or some function of this quantity. A convenient choice is to let the space variable be measured in units of distance to the shock front; that is,

$$\eta = x/x_s(t) \tag{7}$$

where

$$x_s(t) = \theta(Lv_0^\alpha t^\alpha)^{1/(\alpha+1)} \tag{8}$$

is the position of the shock front at time t; θ and α are initially undetermined parameters. The pressure, velocity, and specific volume can be written as the product of a dimensional, time-dependent part and a nondimensional part that depends only on η:

$$p = p_s(t)\,\bar{p}(\eta), \qquad u = u_s(t)\,\bar{u}(\eta), \qquad V = V_s(t)\,\bar{V}(\eta) \tag{9}$$

An additional application of dimensional analysis determines the time dependence of the variables at the shock:

$$p_s = \frac{\theta_1}{V_1}\left(\frac{Lv_0{}^{\alpha}}{x_s(t)}\right)^{2/\alpha}, \qquad u_s = \theta_2\left(\frac{Lv_0{}^{\alpha}}{x_s(t)}\right)^{1/\alpha}, \qquad V_s = \theta_3 V_1 \tag{10}$$

The nondimensional parameters θ_1, θ_2, and θ_3 can be calculated as a function of θ by the use of the shock conditions, Eqs. (4)–(6),

$$\theta_1 = \frac{2}{\gamma+1}\left(\frac{\alpha}{\alpha+1}\right)^2 \theta^{2(1+\alpha)/\alpha}, \quad \theta_2 = \frac{\alpha}{\alpha+1}\frac{2}{\gamma+1}\theta^{(\alpha+1)/\alpha}, \quad \theta_3 = \frac{\gamma-1}{\gamma+1} \tag{11}$$

and the relation between shock position and velocity

$$\dot{x}_s(t) = v_s \tag{12}$$

Substitution of Eqs. (9) into Eqs. (1)–(3) separates the variables and leads to the set of ordinary differential equations below, in which primes denote differentiation with respect to η:

$$\bar{u}' + [(\gamma-1)/2]\,\bar{V}'\eta = 0 \tag{13}$$

$$\eta\bar{u}' + \bar{u}/\alpha = \bar{p}' \tag{14}$$

$$\bar{p}\bar{V}^{\gamma}\eta^{2/\alpha} = 1 \tag{15}$$

with the boundary conditions at the shock front

$$\bar{p}(1) = \bar{u}(1) = \bar{V}(1) = 1 \tag{16}$$

and, at the free surface,

$$\bar{p}(0) = 0 \tag{17}$$

2. *Solution in the Case* $\gamma = \frac{7}{5}$

For general γ, no closed-form solution to Eqs. (13)–(15) with the boundary conditions (16) and (17) is known. However, for $\gamma = \frac{7}{5}$ a remarkably simple solution exists, as first pointed out by Häfele (*19*). If it is assumed that $\bar{p} = \eta$, Eq. (14) gives

$$\bar{u} = \alpha + (1 - \alpha)/\eta^{1/\alpha}$$

However, Eqs. (13) and (15) together result in the solution

$$\bar{u} = 1 - \frac{1}{2} \frac{\alpha + 2}{\alpha - 2/(\gamma - 1)} (1 - \eta^{1-(1+2/\alpha)/\gamma})$$

These expressions for $\bar{u}$ are consistent only if γ and α have the special values $\gamma = \frac{7}{5}$, $\alpha = \frac{3}{2}$. Then

$$\bar{u} = (3 - \eta^{-2/3})/2 \tag{18}$$

The particle position $y(x, t)$ measured from the free surface is obtained by integration:

$$y(x, t) = [(5 - \eta^{-2/3})/4]\, x_s(t)$$

The initial position of a particle at $y = 0$ is found by putting $\eta = 1/5^{3/2} = 0.0896$. The nondimensional material velocity there is $\bar{u} = -1$. Hence only about 9% of the shocked material has moved past the free surface at any time, but the momentum and energy of this material are by no means negligible.

The total momentum of the shocked material is zero and its kinetic energy is infinite. Thus energy or momentum criteria will not serve to connect the similarity solution to a specific impact. This difficulty is resolved by the principle of late-stage equivalence as follows.

The total energy of the flow to the right of x is given by the integral

$$E = \int_x^{x_s(t)} \left(\frac{u^2}{2} + \frac{pV}{\gamma - 1}\right) dx/V \tag{19}$$

and the momentum to the right of x is

$$I = \int_x^{x_s(t)} u\, dx/V \tag{20}$$

Substitution of the $\gamma = \frac{7}{5}$ solution shows that for small x or, what is equivalent, for small η

$$E \sim \tfrac{3}{8} u_s^2 x_s x^{-1/3}/V_1$$

which becomes infinite as x goes to zero, as mentioned above, and

$$I \sim \tfrac{3}{2} u_s x_s x^{1/3}/V_1$$

which vanishes as the cube root of x. Then

$$EI = \tfrac{9}{16}u_s{}^3 x_s{}^2/V_1{}^2$$

With the limits taken in this way, the product is finite and independent of x. If it is hypothesized that EI for the flow is the same as E_0I_0 for the impacting slab, then θ is determined by substitution from Eqs. (8) and (10) as in the equation below:

$$EI = \tfrac{9}{16}u_s{}^3 x_s{}^2/V_1 = \tfrac{9}{16}\theta_2{}^3 v_0{}^3 L^2/V_1{}^2 = E_0I_0 = \frac{L_0v_0}{V_1}\cdot\frac{L_0v_02}{2V_1}$$

Then

$$\theta_2 = (\tfrac{8}{9})^{1/3}$$

Using Eq. (11), the complete set of constants is obtained:

$$\theta_1 = \tfrac{6}{5}(\tfrac{8}{9})^{2/3}, \qquad \theta_2 = (\tfrac{8}{9})^{1/3}, \qquad \theta_3 = \tfrac{1}{6}, \qquad \theta = (\tfrac{64}{9})^{1/3} \tag{21}$$

Although EI is finite in the above calculation, no way has been found of proving analytically that it must be equal to E_0I_0, the energy–momentum product for an impacting slab. Such a proof would require following in some manner the flow from the early stages, where a solution is available because the flows are simple waves and the shock strength is constant, through the later stages, where no analytical solution is available to the finite-slab problem. However, the assumption that EI is equal to E_0I_0 does lead to accurate numerical results for finite-slab impacts, as will be shown in the following paragraph. Thé term EI may be thought of as an invariant integral quantity for the flow. This interpretation is strengthened by considering the energy E_+ and the momentum I_+ (which are not invariant!) of that portion of the material going in the same direction as the original slab. It is found that E_+ decreases as $t^{-1/5}$ and I_+ increases as $t^{1/5}$. However, the product E_+I_+ is constant, and a lengthy but straightforward integration shows that it is equal to 0.5132 of the product taken over the entire flow.

To verify this hypothesis that the EI product for the slab and the asymptotic similarity flow are equal, a finite-slab impact problem was solved numerically be means of the SPUTTER code. The target material was chosen to have a density of 1 g/cm^3, the velocity of the projectile was 10^6 cm/sec, and the mass per unit area of the projectile was taken to be 1.2 g/cm^2. At impact the shock pressure was 0.3 Mbar.

Pressure, velocity, and density data obtained from the computer solution are presented in Figs. 12–14 and are compared with the analytical

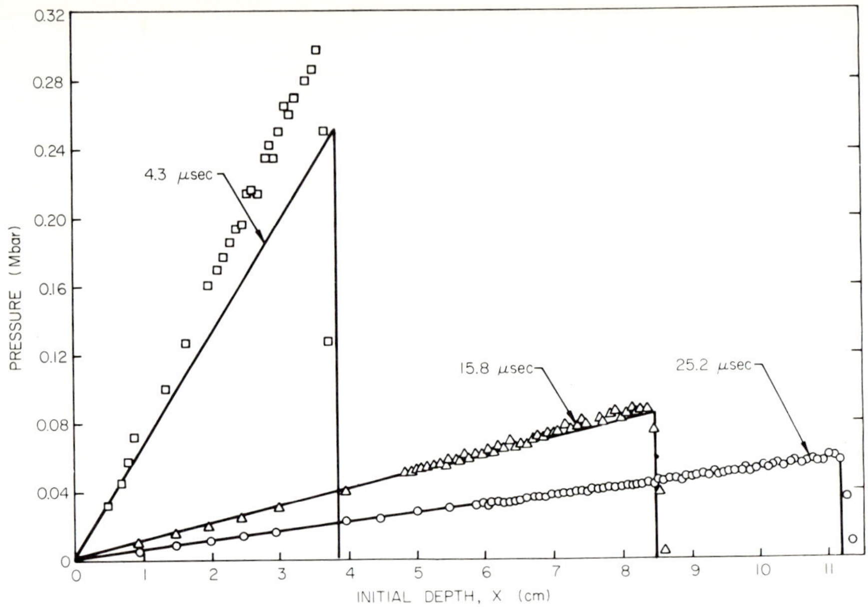

FIG. 12. Comparison of pressure in the finite-slab impact problem with the similarity solution for $\gamma = \frac{7}{5}$ at three times; x is measured from the free surface. (□, △, ○) SPUTTER code, (—) analytical result.

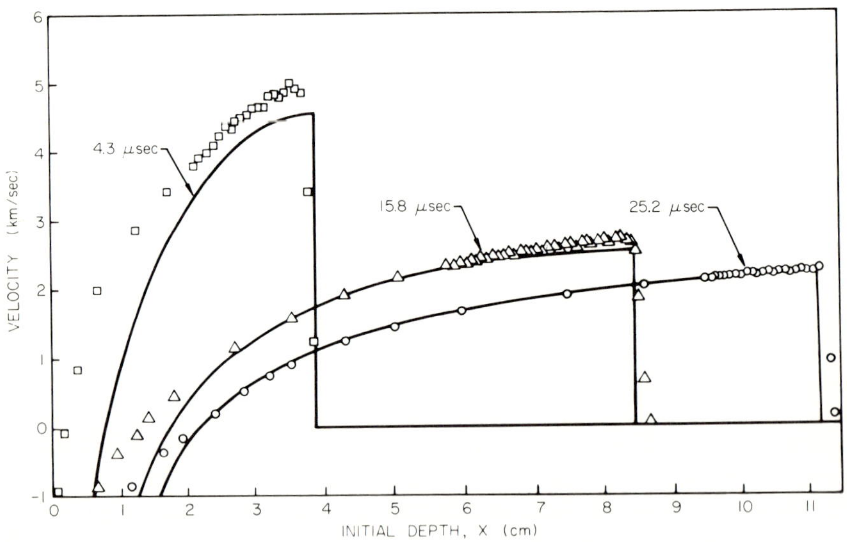

FIG. 13. Comparison of velocity in the finite-slab impact problem with the similarity solution for $\gamma = \frac{7}{5}$ at three times; x is measured from the free surface. (□, △, ○) SPUTTER code, (—) analytical results.

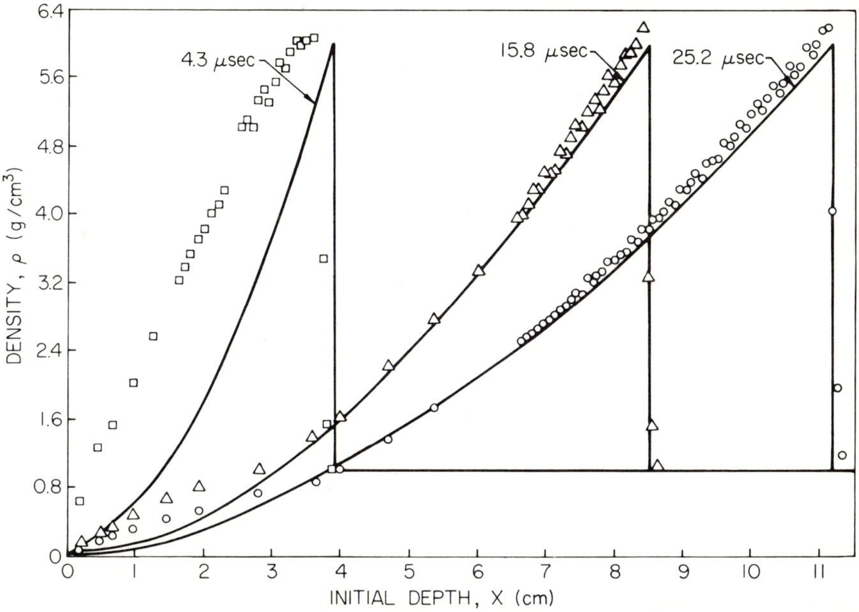

FIG. 14. Comparison of density in the finite-slab impact problem with the similarity solution for $\gamma = \frac{7}{5}$ at three times; x is measured from the free surface. (□, △, ○) SPUTTER code, (—) analytical results.

solution at corresponding times. In the figures, time for the finite-slab solution is given from the instant of contact. The time t used to obtain the analytical solution can be obtained by subtracting 0.08 μsec from the numbers that are given. This choice of the time origin for the analytic solution was chosen so that the shock fronts for the finite impact problem and for the similarity solution would both be 2.4 g/cm² from the free surface at 2 μsec. All data points obtained from the SPUTTER code solution in the neighborhood of the shock are shown. However, only representative data points are given where the oscillations found near the shock front have died out. This example shows that the finite impact solution does approach the similarity solution at late times.

3. *Reduction to a First-Order Equation and Comparison with Blast Waves*

Eqs. (13)–(15) can be reduced to a single first-order differential equation. To do this, the equations are solved for the logarithmic derivatives of $\bar{p}$, $\bar{V}$, and $\bar{u}$:

$$\frac{d\bar{p}}{\bar{p}} = -\frac{2}{\alpha}\frac{d\eta}{\eta}\frac{1 + [\gamma/(\gamma - 1)]\, z}{1 - [2\gamma/(\gamma - 1)](z/y)} \tag{22}$$

$$\frac{d\bar{V}}{\bar{V}} = \frac{1}{\alpha}\frac{2}{\gamma - 1}\frac{d\eta}{\eta} z \frac{1 + 2/y}{1 - [2\gamma/(\gamma - 1)](z/y)} \tag{23}$$

$$\frac{d\bar{u}}{\bar{u}} = -\frac{1}{\alpha}\frac{d\eta}{\eta}\frac{1 + 2/y}{1 - [2\gamma/(\gamma - 1)](z/y)} \tag{24}$$

where

$$y = \bar{u}\eta/\bar{p}, \qquad z = \bar{u}/\eta\bar{V} \tag{25}$$

Taking the logarithmic derivative of the expressions in Eq. (25) leads to a first-order equation in the new variables y and z:

$$\frac{dz}{dy} = \frac{z}{y}\frac{2 + (1 + \alpha)y - [2/(\gamma - 1)](\gamma\alpha - 2 - y)z}{2 - (1 + \alpha)y - [2\gamma/(\gamma - 1)](y - \alpha)z} \tag{26}$$

Comparable equations have been derived by Häfele (*19*) and by Adamskii (*20*). The choice of variables in this paper leads to a simpler equation which is amenable to an approximate solution near the free surface and also in the neighborhood of the origin, $y = z = 0$, where the material velocity changes sign. (Only the free surface approximation is considered here.) The appropriate boundary condition for Eq. (26) is that at the shock front

$$y = 1, \qquad z = 1 \tag{27}$$

One solution to this equation with the boundary condition (27) is known; namely, the blast wave problem for a sheet explosion. The complete solution was given by Sedov (*21*) in parametric form and was obtained by a different method. A derivation is given in this paper to show the relation between the impact and blast problems. For an explosion with a given energy, a problem in which total momentum plays no part, α must evidently be two to make the solution energy dependent. It will be shown that the solution in the variables y and z is, for any γ,

$$\alpha = 2, \qquad z = 1/(2 - y) \tag{28}$$

It will follow from the solution that at the origin the velocity vanishes and the pressure is nonzero, as required by symmetry considerations.

The solution given by Eq. (28) can be found by noting that a self-similar flow with finite total energy will always have the same fraction of the total energy between the origin and a plane at a constant fraction λ of the distance to the right-going shock wave. The distance to such a plane is

$$X = \lambda x_s \tag{29}$$

where X is the distance from the origin and λ is an Eulerian similarity variable. The velocity of the plane at constant λ, which contains the constant fraction of the energy, is $\lambda \dot{x}_s$. The material inside this plane does work $pu\,\Delta t$ on the material outside, which is just compensated by the inward flux of material $\rho(\lambda \dot{x}_s - u)$ with total energy per unit mass $E + u^2/2$. Then

$$pu\,\Delta t = \rho(\lambda \dot{x}_s - u)(E + u^2/2)\,\Delta t$$

For a perfect gas, $E = pV/(\gamma - 1)$; using Eq. (9) we find

$$\bar{p}\bar{V} = \bar{u}^2 \frac{[(\gamma + 1)/2]\,\lambda - \bar{u}}{\gamma \bar{u} - [(\gamma + 1)/2]\,\lambda} \tag{30}$$

To convert this equation into Lagrangian variables, it is necessary to obtain a relation between λ and η. Differentiating Eq. (29), noting that λ is a function of η only, $\lambda = \lambda(\eta)$, and using Eq. (6),

$$\lambda - \eta\lambda' = [2/(\gamma + 1)]\,\bar{u}$$

Combining with the continuity equation (13) and integrating,

$$\lambda = \frac{2}{\gamma + 1}\,\bar{u} + \frac{\gamma - 1}{\gamma + 1}\,\bar{V}\eta \tag{31}$$

This result can be combined with the energy integral (30) to obtain the solution given by Eq. (28) to the sheet-explosion problem. The complete solution is given below in parametric form and is equivalent to that given by Sedov, although his parameter is not y as in the equations below:

$$\eta = (2 - y)^{\gamma(/\gamma-2)} y^{\gamma(2\gamma-1)}\, Y^{-\gamma-1} \tag{32}$$

$$\bar{p} = (2 - y)^{\gamma/(\gamma-2)}\, Y^{1-2\gamma} \tag{33}$$

$$\bar{u} = y^{(\gamma-1)/(2\gamma-1)}\, Y^{2-\gamma} \tag{34}$$

$$\bar{V} = y^{-1/(2\gamma-1)} (2 - y)^{-2/(2-\gamma)}\, Y^3 \tag{35}$$

$$\lambda = y^{(\gamma-1)/(2\gamma-1)} y^{2-\gamma} (2\gamma + 1 - \gamma y)/(\gamma + 1) \tag{36}$$

where

$$Y = \left[\frac{2(2\gamma - 1) - 3(\gamma - 1)\,y}{\gamma + 1}\right]^{(5\gamma-4)/3(\gamma-2)(2\gamma-1)} \tag{37}$$

A comparison of the pressure profiles in the impact and in the blast problems for $\gamma = \frac{7}{5}$ is given in Fig. 15.

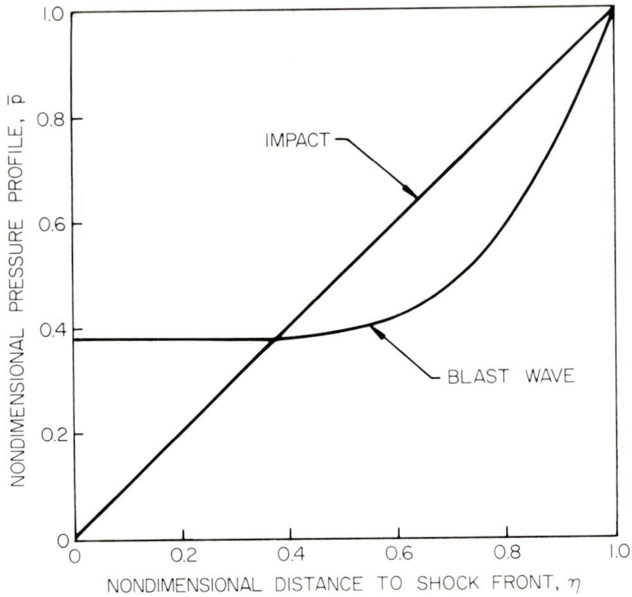

FIG. 15. Comparison of the pressure distribution for a blast wave and for an impact with $\gamma = \frac{7}{5}$, $\alpha = \frac{3}{2}$.

4. *Proof that EI^λ is Finite for Impacts with General γ*

The behavior of the flow near the free surface in the impact problem can be obtained by examining the differential equation (26) for large y, which becomes

$$\frac{dz}{dy} = -\frac{z}{y}\frac{1 + \alpha + [2/(\gamma - 1)]\, z}{1 + \alpha + [2\gamma/(\gamma - 1)]\, z}$$

This equation can be integrated by separation of variables:

$$yz(1 + \alpha + [2/(\gamma - 1)]\, z)^{\gamma-1} = k$$

where k is an arbitrary constant.

For large y, the only nonvanishing solution is

$$z = -[(1 + \alpha)/2](\gamma - 1) \tag{38}$$

which gives the asymptotic value for z. This free surface solution can be used to prove that EI^λ is finite for arbitrary γ, even though the total kinetic energy in the shocked material is infinite and the momentum I is zero. To do this, the asymptotic values of the pressure, specific volume, and velocity near the free surface are found by substituting the above result into Eqs. (22)–(24):

$$\bar{p} \sim A\eta^{((1+\alpha)\gamma-2)/\alpha} \tag{39}$$

$$\bar{V} \sim B\eta^{-(1+\alpha)/\alpha} \tag{40}$$

$$\bar{u} \sim C\eta^{-1/\alpha} \tag{41}$$

where A, B, and C are undetermined constants. The energy and momentum to the right of x, assumed small, is approximately

$$E = \int_x^{x_s} \left(\frac{u^2}{2} + \frac{pV}{\gamma - 1}\right)\frac{dx}{V} \sim -\frac{C^2}{2}\frac{u_s^2}{V_1} x_s \frac{\eta^{1-2/\alpha}}{1 - 2/\alpha} \tag{42}$$

$$I = \int_x^{x_s} u_s \frac{dx}{V} \sim \frac{C\alpha}{\alpha - 1}\frac{u_s}{V_1} x_s \eta^{1-1/\alpha} \tag{43}$$

Then the product EI^λ is finite, where

$$\lambda = (2 - \alpha)/(\alpha - 1) \tag{44}$$

The term EI^λ is independent of η and t and, in fact, depends only on a constant C, the initial specific volume V_1, and α:

$$EI^\lambda = \frac{C^{2+\lambda}}{2(2/\alpha - 1)\, V_1^{1+\lambda}}\left(\frac{\alpha}{\alpha - 1}\right)^\lambda \tag{45}$$

5. *Discussion of the Solution for General* γ

As y goes to infinity, z approaches the asymptotic value given in Eq. (38). In the z–y plane, illustrated in Fig. 16 for $\gamma = \frac{7}{5}$, the solution curve for large y lies above the locus of zero slopes of Eq. (26), as given by

$$z_1 = \frac{\gamma - 1}{2}\frac{2 + (1 + \alpha)\, y}{\gamma\alpha - 2 - y}$$

but approaches it asymptotically. The solution curve lies below the locus of infinite slopes:

$$z_2 = \frac{\gamma - 1}{2\gamma}\frac{2 - (1 + \alpha)\, y}{y - \alpha}$$

for large y. Since these two loci, z_1 and z_2, cross at the saddle point $[y = -2, z = -(\gamma - 1)/\gamma]$ the solution curve must also pass through the saddle point or else it would have an infinite slope at some point, and

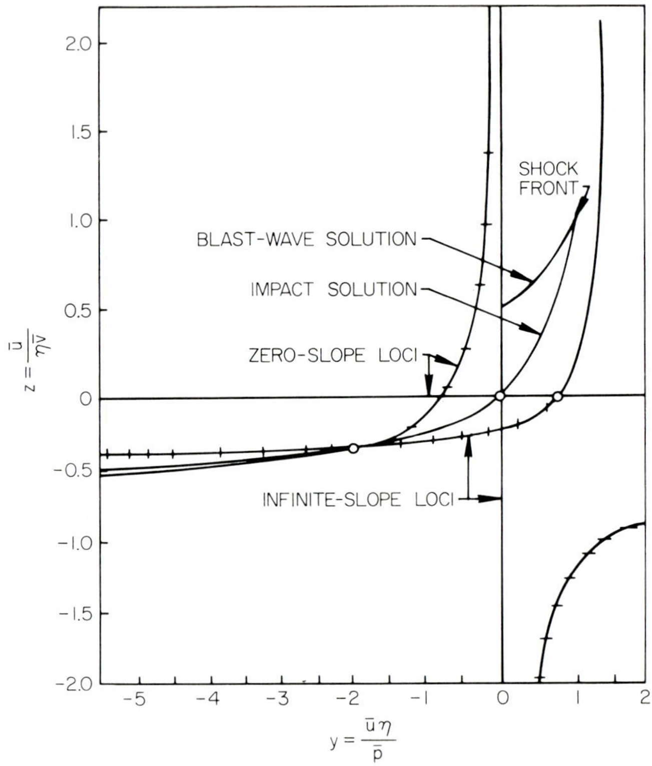

FIG. 16. The solution curves in the z–y plane for blast and impact in the case $\gamma = \frac{7}{5}$. (○) Saddle points.

near that point z would be double valued. [The flow is also unstable at points of infinite slope, as discussed by Häfele (*19*)]. These topological considerations are best visualized by referring to a diagram such as Fig. 16. By a similar argument, the solution curve must pass through the origin, which is also a saddle point. Hence the free surface condition can be replaced by the condition that the solution curve pass through the saddle points, which leads, in principle, to a determination of α. This represents a convenient method of finding solutions numerically without having to deal wth the free surface. The analyses of previous authors were directed to finding the exponent n in the expression $x_s(t) = At^n$, satisfying this condition. Their results can be used to establish the relation between α and γ, since a straightforward comparison with Eq. (8) shows that

$$\alpha = n/(1 - n)$$

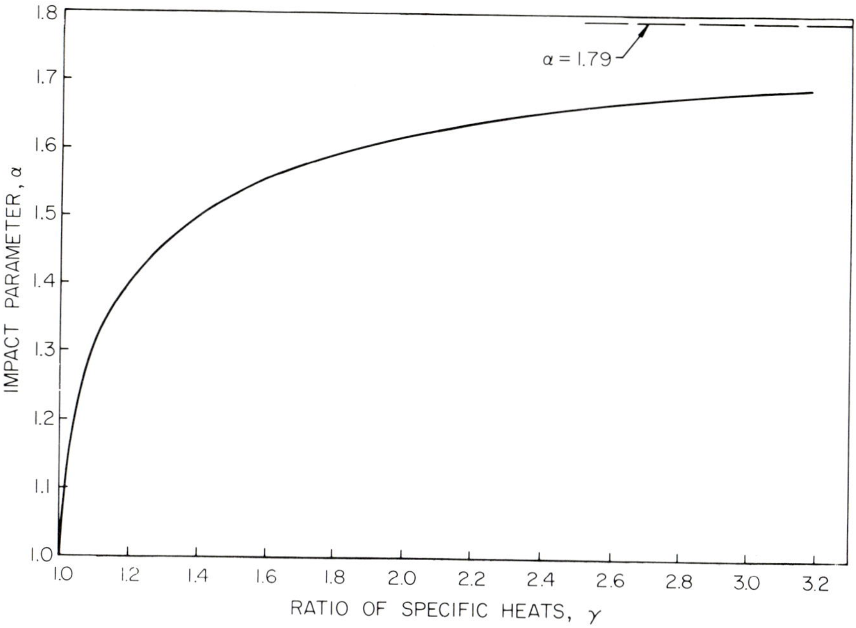

FIG. 17. The impact parameter α as a function of γ for one-dimensional impact.

The most detailed tabulation of n in terms of γ is given by Rae and Kirchner (*18*). Their summary is used to obtain the curve in Fig. 17, which shows how α increases with γ.

6. *Summary of the Plane Wave Analysis*

Analysis of the hydrodynamic equations for one-dimensional impact shows that the flow is characterized by a quantity Lv_0^{α} intermediate between energy and momentum. The value of α depends on the equation of state of the target material and varies between 1 and 1.79 in the case of an ideal-gas target. The solutions resemble in some respects the blast wave solution, for which $\alpha = 2$, but the zero-pressure boundary condition at the free surface in the impact problem leads to a more rapid decay of the shock wave than is found in the blast wave case. In the blast wave problem, the shock pressure goes down with the inverse power of distance into the target, whereas in the impact problem the shock pressure goes down as the inverse $2/\alpha$ power of distance from the explosion, where $2/\alpha$ varies from 1.12 for $\gamma = \infty$ to 2 for $\gamma = 1$.

An example shows that the flow from a finite-slab impact with finite energy and momentum approaches at late times a similarity flow whose energy is infinite and whose momentum is zero. The difference between the finite-impact solution and the similarity solution is confined to a small region near the free surface at late times and is not noticeable over most of the flow profile. Only that portion of the flow near the free surface contains infinite energy in the similarity solution, and whenever that region can be ignored the similarity solution is physically realizable.

Although the energy E is infinite and the momentum I is zero for the similarity flow, the product $EI^{(2-\alpha)/(\alpha-1)}$ is finite for any γ. When $\gamma = \frac{7}{5}$, an exact solution is known, and then $\alpha = \frac{3}{2}$, $EI^{(2-\alpha)/(\alpha-1)} = EI$, and when set equal to the energy–momentum product of the impacting slab, an accurate numerical solution is obtained. It has not been shown that such a procedure is valid for general γ, and this must remain a plausible hypothesis. Assuming this is true, it implies that flows with finite energy and momentum can approach a similarity flow with infinite energy and zero momentum, while conserving the value of EI^{λ}.

IV. Oil: An Eulerian Hydrodynamic Code

Unsteady, axisymmetric flow problems can be calculated with OIL, a computer program which integrates numerically the Eulerian equations of motion. This computer program was developed by Johnson (*4*) as a modification of the particle-in-cell (PIC) code originally developed by Evans and Harlow (*3*). In the original PIC code the region of integration was divided into cells of rectangular cross section fixed in space. In each time step the particles were moved, and some of them crossed cell boundaries. As the particles moved from cell to cell they transported with them appropriate mass, momentum, and energy. This scheme successfully describes flows involving a number of materials so long as the density is high. However, the discrete changes in cell density when a particle crosses a cell boundary result in pressure fluctuations which are typically several hundred thousand atmospheres. Thus, the PIC code is not suitable for calculations of impacts or explosions which must be followed to a time when the pressures are below such values.

In the continuous Eulerian code, arbitrarily small masses can cross a cell boundary. The fluctuations observed in PIC are substantially reduced. However, the simplicity with which multimaterial problems are handled by PIC is lost, and OIL is basically a one-material code. (Multimaterial versions are under study, but the conceptual simplicity of PIC will probably not be retrieved.)

The OIL code consists of two principal subroutines, or phases. Phase 1 accounts for the effect of the pressure gradient and phase 2 for the transport of material from cell to cell. These phases will be discussed separately below. Other special subroutines determine the time step (CDT), compute the equation of state (EOS), set up the initial configuration (SETUP), redefine the cells when the disturbance reaches the grid boundary (REZONE), display the results in periodic prints describing the state of the flow (EDIT), and perform other miscellaneous functions (INPUT, CARDS, ERROR, MAIN).

The equations of motion for axisymmetric flow can be written in the (conservative) form

$$\frac{\partial \rho}{\partial t} = -\frac{1}{r}\frac{\partial}{\partial r}(\rho u r) - \frac{\partial}{\partial z}(\rho v)$$

$$\frac{\partial(\rho u)}{\partial t} = \frac{1}{r}\frac{\partial}{\partial r}[r(-p - \rho u^2)] - \frac{\partial}{\partial z}(\rho u v)$$

$$\frac{\partial(\rho v)}{\partial t} = \frac{1}{r}\frac{\partial}{\partial r}(-\rho u v) - \frac{\partial}{\partial z}(p + \rho v^2)$$

$$\frac{\partial(\rho E)}{\partial t} = \frac{1}{r}\frac{\partial}{\partial r}\{(-p - \rho E)\, u r\} - \frac{\partial}{\partial z}\{(p + \rho E)\, v\}$$

$$p = f(\rho, I), \qquad E = I + \tfrac{1}{2}\rho(u^2 + v^2\}$$

In these equations ρ is the density, u and v are the radial and axial velocities, I is the internal energy per unit mass, E is the total energy per unit mass, and p is the pressure. The terms on the right can be separated into pressure (phase I) terms and transport (phase II) terms. This will be done below. To put the equations into a form useful for numerical work it is convenient to integrate over a finite volume (a cell) and make explicit the mass in the cell and the flux of mass, momentum, and energy across the cell boundaries. When this is done, one obtains

$$\frac{\partial}{\partial t}(M) = \frac{\partial}{\partial t}(M)_{\mathrm{I}} + \frac{\partial}{\partial t}(M)_{\mathrm{II}}$$

$$\frac{\partial}{\partial t}(Mu) = \frac{\partial}{\partial t}(Mu)_{\mathrm{I}} + \frac{\partial}{\partial t}(Mu)_{\mathrm{II}}$$

$$\frac{\partial}{\partial t}(Mv) = \frac{\partial}{\partial t}(Mv)_{\mathrm{I}} + \frac{\partial}{\partial t}(Mv)_{\mathrm{II}}$$

$$\frac{\partial}{\partial t}(ME) = \frac{\partial}{\partial t}(ME)_{\mathrm{I}} + \frac{\partial}{\partial t}(ME)_{\mathrm{II}}$$

The mass, momentum, and energy changes to be computed in phase I are

$$\Delta_1(M) = 0$$

$$\Delta_1(Mu) = -pA \big|_1^2 \Delta t$$

$$\Delta_1(Mv) = -pA \big|_5^6 \Delta t$$

$$\Delta_1(ME) = -puA \big|_1^2 \Delta t - pvA \big|_5^6 \Delta t$$

The surfaces on which the area A is to be evaluated are illustrated in Fig. 18. The internal energy, for use in the equation-of-state calculations and several special tests, is gotten by

$$I = E - \tfrac{1}{2}(u^2 + v^2)$$

The mass and density are related by

$$M = \rho\pi(r_2^2 - r_1^2)(z_6 - z_5)$$

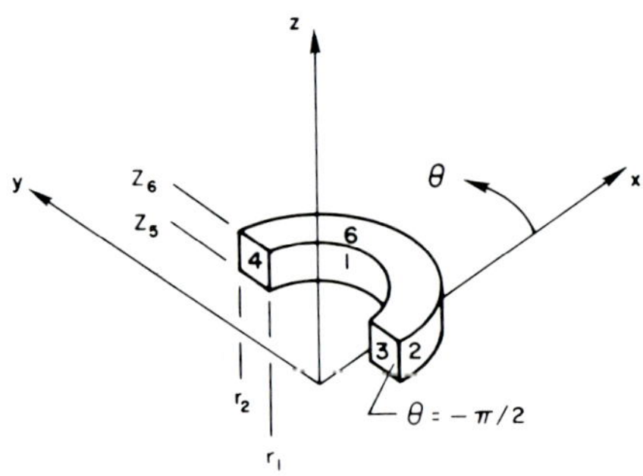

FIG. 18. The shape of an element of volume $\tilde{V}$, appropriate for the Eulerian finite-difference equations in cylindrical coordinates.

Surface number	Components of $\bar{n} = (n_1, n_2, n_3)$ parallel to		
	r	θ	z
1	−1	0	0
2	1	0	0
3	0	−1	0
4	0	1	0
5	0	0	−1
6	0	0	1

and the areas are

$$A_1 = 2\pi r_1(z_6 - z_5)$$
$$A_2 = 2\pi r_2(z_6 - z_5)$$
$$A_6 = A_5 = \pi(r_2^2 - r_1^2)$$

as indicated in Fig. 18. (The term A_3 and A_4 are reserved for the cross-sectional area of the torus.)

In phase II the changes due to transport are calculated from the equations

$$\Delta_2(M) = -\rho u A \,|_1^2\, \Delta t - \rho v A \,|_5^6\, \Delta t$$
$$\Delta_2(Mu) = -\rho u^2 A \,|_1^2\, \Delta t - \rho u v A \,|_5^6\, \Delta t$$
$$\Delta_2(Mv) = -\rho u v A \,|_1^2\, \Delta t - \rho v^2 A \,|_5^6\, \Delta t$$
$$\Delta_2(ME) = -\rho E u A \,|_1^2\, \Delta t - \rho E v A \,|_5^6\, \Delta t$$

A. Equation of State

The pressure is computed from an equation of state nearly of the Mie–Grüneisen form which is described in detail by Tillotson (*17*). The report includes the Hugoniots and isentropes for nine metals. Subsequent reports have given additional data for geologic materials and plastics (*22*, *23*). The equation for pressure of a condensed material is given by

$$p = f(I, \rho)\, I\rho + A\mu + B\mu^2$$

where

$$\mu = \rho/\rho_0 - 1$$

and

$$f(I, \rho) = a + \frac{b}{(I/I_0)(\rho_0/\rho)^2 + 1}$$

In the computer program this equation is modified at low density and high internal energy to approach the ideal-gas equation of state $p = aI\rho$ in a continuous manner. Under certain conditions the pressure may become negative. This is allowed, in general, but a minimum density can be set below which the pressure is set to zero. This density corresponds to the material expansion at the cavitation or spall pressure of the material.

B. Comparison with Taylor Blast Theory

The accuracy of the integration scheme can be determined by comparison with analytical solutions. One of the most suitable for this purpose is the explosion theory due to Taylor (*10*) and discussed extensively by Sedov (*21*). This solution concerns an explosion from a point source in which the leading edge of the disturbance is a strong shock. The symmetry of the solution serves as a good check on the accuracy of the integration scheme. In Fig. 19 the pressure profile at several times is compared with the theoretical solution tabulated by Sedov. At early times, when the shock wave radius is only a few times the radius R_0 of the original sphere of hot gas, the solution has not become self-similar, and agreement between the calculation and Taylor's

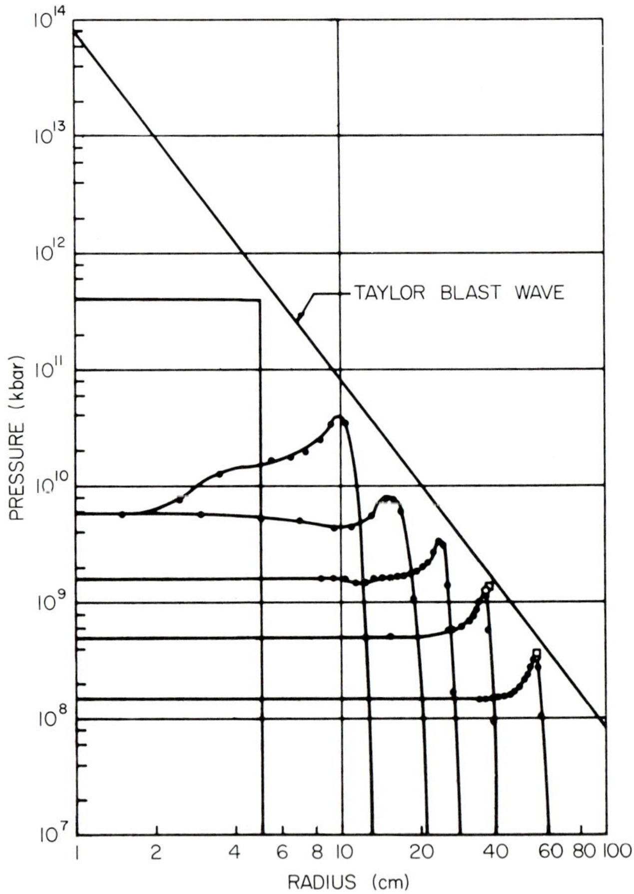

FIG. 19. The pressure from a 5.2×10^{14}-erg explosion (10^{12} ergs/g, sphere radius = 5 cm, $\rho = 1$ g/cm^3) as computed by OIL and compared with the Taylor solution. (●) Axis of symmetry; (□) diagonal; (○) plane of symmetry.

asymptotic theory is not to be expected. When the radius of the shock is ten times R_0 (5 cm), the shock pressure is 60% of the value predicted by the Taylor solution, and the solutions are converging. Calculated shock position agrees with the Taylor solution from very early times. The pressure at the center is 43% of the shock pressure, as compared with 46% in the Taylor solution, and the ratio of kinetic to internal energy is 0.26, compared with 0.28 cited by Sedov.

The calculation represents a stringent test on the code since the density should theoretically drop from 6 g/cm³ at the shock down to 0.9 g/cm³ at a radius 90% of the distance to the shock, a distance typically resolved with four cells. Although high precision is not available in regions of rapid change, quantities integrated over the entire flow seem to be reliably calculated.

V. RPM: The OIL Code with Strength

When a medium supports significant shear stresses it is necessary to replace the pressure appearing in the momentum and energy equations by a more general quantity, the stress tensor σ_{ij}. The equations of motion in this more general case are, in cylindrical coordinates,

$$\frac{\partial \rho}{\partial t} = -\frac{1}{r}\frac{\partial(\rho u r)}{\partial r} - \frac{\partial(\rho v)}{\partial z} \tag{46}$$

$$\frac{\partial(\rho u)}{\partial t} = \frac{1}{r}\frac{\partial}{\partial r}(rT_{11}) + \frac{\partial}{\partial z}(T_{13}) - \frac{1}{r}T_{22} \tag{47}$$

$$\frac{\partial(\rho v)}{\partial t} = \frac{1}{r}\frac{\partial}{\partial r}(rT_{13}) + \frac{\partial}{\partial z}(T_{33}) \tag{48}$$

$$\frac{\partial(\rho E)}{dt} = \frac{1}{r}\frac{\partial}{\partial r}\{[(\sigma_{11} - \rho E)u + \sigma_{31}v]r\} + \frac{\partial}{\partial z}\{\sigma_{13}u + (\sigma_{33} - \rho E)v\} \tag{49}$$

where

$$T_{ij} = \sigma_{ij} - \rho u_i u_j \tag{50}$$

These equations can be rearranged to resemble those for the OIL code, except for the presence of additional terms. These involve the deviator stresses S_{ij}, which are defined by subtracting out the isotropic component from the total stress tensor, i.e.,

$$S_{ij} = \sigma_{ij} - p\delta_{ij} \tag{51}$$

The additional terms on the right due to the existence of deviator stresses (denoted below by the subscript 3) are simply

$$\frac{\partial(\rho u)_3}{\partial t} = \frac{1}{r}\frac{\partial}{\partial r}(rS_{11}) + \frac{\partial}{\partial z}(S_{13}) - \frac{1}{r}S_{22} \tag{52}$$

$$\frac{\partial(\rho v)_3}{\partial t} = \frac{1}{r}\frac{\partial}{\partial r}(rS_{13}) + \frac{\partial}{\partial z}(S_{33}) \tag{53}$$

$$\frac{\partial(\rho E)_3}{\partial t} = \frac{1}{r}\frac{\partial}{\partial r}\{[S_{11}u + S_{31}v]\, r\} + \frac{\partial}{\partial z}\{S_{13}u + S_{33}v\} \tag{54}$$

A. The Finite-Difference Equations

Equations (52)–(54) can be integrated over the finite volume indicated in Fig. 18, to obtain the changes in velocity and internal energy in a single time step. This results in the additional terms

$$\Delta_3(Mu) = S_{11}A\,|_1^2\,\Delta t - S_{22}A_4\,\Delta t + S_{13}A\,|_5^6\,\Delta t \tag{55}$$

$$\Delta_3(Mv) = S_{31}A\,|_1^2\,\Delta t + S_{33}A\,|_5^6\,\Delta t \tag{56}$$

$$\Delta_3(ME) = (S_{11}u + S_{13}v)\,A\,|_1^2\,\Delta t + (S_{13}u + S_{33}v)\,A\,|_5^6\,\Delta t \tag{57}$$

The stresses on each cell surface are computed in such a way that they account in a symmetrical manner for the velocities on the two sides of the interface. The momentum and energy subtracted from one cell are added to the others, so that energy and momentum are rigorously conserved, except for roundoff.

The stresses required in the above equations must be obtained in general from the strains and strain rates (or, more precisely, the deviator stress tensor is calculated from the deviator strain tensor) by means of a constitutive equation which describes the material properties under shear. A rather general form of constitutive equation is

$$\dot{e}_{ij} = a\frac{D}{Dt}S_{ij} + bS_{ij} + A_{ij} \tag{58}$$

where $\dot{e}_{ij}$ is the deviator strain rate, a and b are, in general, scalar functions of the stress, strain, and strain-rate invariants, and A_{ij} denotes those additional terms which may be required to satisfy invariance criteria. The strain-rate tensor $\dot{\varepsilon}_{ij}$ can be found from the velocity gradients

$$\dot{\varepsilon}_{ij} = \tfrac{1}{2}(u_{i,j} + u_{j,i}) \tag{59}$$

and is used to construct the deviator strain rate by means of the relation

$$\dot{e}_{ij} = \dot{\varepsilon}_{ij} - \tfrac{1}{3}\dot{\varepsilon}_{kk}\,\delta_{ij} \tag{60}$$

By definition, then

$$\dot{\varepsilon}_{kk} = u_{i,i} = -(1/\rho)\, D\rho/Dt \tag{61}$$

The deviator strain rate is, loosely speaking, independent of the compression, and describes essentially the shearing action. Extensive discussions of constitutive equations are given by Hill (*8*), Truesdell and Noll (*24*), Eringen (*25*), and Prager and Hodge (*26*).

A few general comments on the constitutive equation described above will be made in the following paragraph to place the method of the current calculations in context. The two classical types of continua, the elastic and viscous media, are special cases. Linear elasticity is described by putting

$$a = 1/2\mu, \qquad b = 0 \tag{62}$$

where μ is the shear modulus of the material. (The other elastic coefficient, the bulk modulus, is the coefficient A appearing in the equation of state.) Linear viscosity is represented by setting

$$a = 0, \qquad b = 1/2\rho\nu \tag{63}$$

where ν denotes the kinematic viscosity of the material. A third important material description is obtained in the case where neither a nor b vanishes, but are given the finite values cited above. In this case the material is said to be viscoelastic.

Elastic-plastic media are described by letting a have the value $1/2\mu$, as in elasticity, and in addition letting the stresses satisfy the von Mises yield condition

$$S_{ij}S_{ij} = 2Y^2 \tag{64}$$

This condition can be combined with the constitutive equation to obtain the value for b:

$$b = S_{ij}\dot{e}_{ij}/(2Y^2)^{1/2} \tag{65}$$

The result is known as the Prandtl–Reuss equations. A good discussion is given by Hill (*8*).

In the cases where a does not vanish, but represents the shear modulus, the terms A_{ij} are required to account for the effect of material rotation. This value is determined by considerations concerning the invariance of the equations, and is discussed in texts on continuum mechanics such as those cited above. Their presence, and in fact their exact value, can be accounted for as follows. The stress tensor may be written in the form

$$S_{ij} = S^l a_{il} a_{jl} \tag{66}$$

where S^l denotes the principal stresses and the a_{il} are direction cosines of the principal axes with respect to a fixed set of axes. [A general discussion is given by Sokolnikoff (*27*).] This expression can be differentiated to obtain a derivative

$$\hat{S}_{ij} = \dot{S}^l a_{il} a_{jl} + S^l \dot{a}_{il} a_{jl} + S^l a_{il} \dot{a}_{jl}$$

which accounts for the rotating axes. This can be written

$$\hat{S}_{ij} = \dot{S}_{ij} - \omega_{ik} S_{jk} - \omega_{jk} S_{ik} \tag{67}$$

where

$$\omega_{ik} = -\dot{a}_{il} a_{kl} \tag{68}$$

and the orthogonality relation

$$a_{ki} a_{kj} = \delta_{ij} \tag{69}$$

has been used. The appropriate value for A_{ij} in Eq. (58) is therefore $-a(\omega_{ik} S_{jk} + \omega_{jk} S_{ik})$. Thus the rate of change of stress $\dot{S}_{ij}$ obtained by differentiating the stress components in fixed axes must be corrected to account for rotation of the material in which the principal axes are embedded. This correction is made in many Lagrangian hydrodynamic codes, such as the one described by Wilkins (*2*).

A physical justification for the linear combination of stress and stress-rate tensors in Eq. (58) is that it makes the stress and strain tensors parallel at low strains (Hooke's law), whereas for large strains the material memory for the original distortion is forgotten and the strain-rate tensor is parallel to the stress-rate tensor. The form of the equation is such that when material is near the yield point, there is a continuous variation in the direction of the stress tensor between the strain and strain-rate directions.

An important simplification to the constitutive equation is obtained by adopting the plastic-rigid model, in which μ goes to infinity. In that case the constitutive equation becomes simply

$$\dot{e}_{ij} = b S_{ij}$$

When combined with the von Mises criterion, Eq. (64), the expression above can be put in the form

$$S_{ij} = \frac{2^{1/2}Y}{(\dot{e}_{kl}\dot{e}_{kl})^{1/2}}\,\dot{e}_{ij} \tag{70}$$

which allows the stresses to be computed directly from the strain rates. As a result the stresses themselves need not be stored in the computer memory, as would be necessary when calculating with the full Prandtl–Reuss equations, but they can be calculated anew at each time step. This approximation is a good one for flows involving large deformations, that is, those in which the shear strains are substantial compared to the shear strain at yield. To illustrate, the strain at yield lies typically in the range from 0.001 to 0.01, whereas the strain in the neighborhood of a crater can be arbitrarily large, and, in the example of Fig. 23, diminishes to 0.1 at two crater radii and 0.01 at six crater radii. Thus, except in the far field, the plastic-rigid approximation is a good one. Further discussion of the theory is given by Hill (*8*).

Since in the cratering problem the elastic regime is not of great importance, it is not accounted for in the calculation method. Although it is feasible in principle to do so, the increase in complexity over the rigid-plastic model in computing time and memory would be substantial since the stress tensor has to be stored, stress rates computed, and the rates of rotation ω_{ij} determined. Furthermore, it has not been shown that by use of the full Prandtl–Reuss equations the plastic flow will be accurately described. In view of the computational difficulties and the uncertainties in the physical process it seems reasonable to adopt the simplest constitutive relations.

B. Calculation of Strain Rates

In the general discussion above the motivation for calculating the shear stresses by means of Eq. (70) was discussed. The necessary strains are determined from the velocities by Eq. (59) and the matrix of strain rates for axisymmetric flow is, consequently,

$$(\dot{\varepsilon}_{ij}) = \begin{pmatrix} u_r & 0 & \frac{1}{2}(u_z + v_r) \\ 0 & u/r & 0 \\ \frac{1}{2}(u_z + v_r) & 0 & v_z \end{pmatrix} \tag{71}$$

The deviator strain rates are found by subtracting out the compression rate, that is, $\dot{e}_{ij} = \dot{\varepsilon}_{ij} - \delta_{ij}\dot{\varepsilon}_{kk}/3$. Tensor manipulations then lead to the convenient expression

$$\dot{e}_{ij}\dot{e}_{ij} = \dot{\varepsilon}_{ij}\dot{\varepsilon}_{ij} - \tfrac{1}{3}\dot{\varepsilon}_{ii}\dot{\varepsilon}_{jj} \tag{72}$$

for the second invariant. The calculation of strain rates at the interfaces is straightforward, and accounts symmetrically for the (cell-centered) velocities in the adjacent cells. The change in internal energy is computed by means of the formula

$$\Delta_3(MI) = 2\pi\,\Delta t\,\{r\,\Delta r\,u_i S_{i3}\,|_5^6 + \Delta z\,r u_i S_{i1}\,|_1^2\} - (M/2)\{(u + \Delta_3 u)^2 - u^2 + (v + \Delta_3 v)^2 - v^2\} \tag{73}$$

which expresses the change in internal energy as the change in total cell energy less the change in cell kinetic energy, as determined by the momentum equations.

The total work done by the plastic stresses is determined by accumulating the work done in each cell at each time step, and this quantity is printed out with the other parameters at the edit intervals.

C. Variations in Yield Strength

The yield stress is defined as the stress at which the material behavior changes rapidly from elastic to plastic. For the purpose of this article the expression "flow stress" describes the resistive effect of plasticity to continuing deformation better than "yield stress," although the terms are somewhat interchangeable. The flow stress is not a fixed quantity, but depends on the thermodynamic condition of the material (say pressure and temperature), on the rate of strain, and on the history of the loading (work hardening). Since there is available virtually no information on the combined effects of pressure, temperature, and strain rate, an approach suggested by its simplicity is to consider the effects as multiplicative. That is,

$$Y = Y_0 f_1(p) f_2(\dot{e}, T) f_3(e) \tag{74}$$

where Y_0 is the nominal yield stress under reference conditions, e is the strain, $f_1(p)$ describes the pressure dependence, $f_2(\dot{e}, T)$ represents the dependence on strain rate and temperature, and $f_3(e)$ accounts for the dependence on strain (work hardening).

The dependence of yield strength on pressure has been measured by Bridgeman (*28*) and more recent results for metals have been summarized by Thomas (*29*) for iron, Vereshchagin (*34*) for a variety of metals, and Robertson (*30*) for geologic materials. The simplest assumption is that yield strength varies linearly with pressure

$$Y = Y_0 + ap \tag{75}$$

with a typically equal to 0.06 for metals and 1.0 for geologic materials.

An expression for $f_2(\dot{e}, T)$ can be obtained from activation energy ideas such as those cited by Zhurkov (*35*). Such an expression is obtained by solving the equation

$$\dot{e} = \nu \exp[(\gamma Y - U_0)/kT]$$

for the flow stress. Here k is Boltzmann's constant, ν is a frequency near that of atomic vibrations, U_0 is the sublimation energy of the material, and γ is an empirical constant. The importance of this expression is that it allows us to estimate strain-rate effects, which may be responsible for a small error if linear scaling is assumed to be valid. This effect has not yet been introduced into impact calculations, but it seems to be a more reasonable way to account for rate effects than to assign a viscosity to a solid material, as is done in some models. When the equation is solved for Y, and we put $C_v = 3kN_0/m$ for the heat capacity of a solid, we find

$$Y = Y_0(1 - (C_v T/3E_0) \ln \nu/\dot{e})$$

At zero temperature there is no strain-rate effect, and similarly it disappears at high strain rates. However, at near normal conditions appropriate values for aluminum are $\dot{e} \sim 1\ \text{sec}^{-1}$, $\nu \sim 10^{13}\ \text{sec}^{-1}$, $T \sim 300°\text{K}$, $E_0 \sim 1.5 \times 10^{11}$ ergs/g $C_v \sim 0.93 \times 10^7$ ergs/g, and the quantity in parentheses is about 0.58. Thus the zero-temperature yield stress Y_0 would be about 72% greater than the value observed under normal test conditions, and this is also the value that should be associated with high strain rates. The data cited by Zhurkov is for pure aluminum, which is known to exhibit a strain-rate effect, but alloys may not follow such a simple model.

A simpler approach was taken in the calculations, in which the yield strength goes to zero at the melting point linearly with internal energy.

D. Setup of Cratering Problems

The impact of an aluminum sphere on an aluminum target is a natural example for testing the success of the computational method, since a great deal of experimental data has been accumulated for aluminum on aluminum impacts in the velocity range from 2 to 10 km/sec. A systematic presentation of one series of results is given by Halperson (*31*).

The properties of aluminum required for the calculation are the equation-of-state constants, for use in the pressure calculation, and the shear strength, for use in the calculation of shear stresses. The constants for the equation of state described by Tillotson (*17*) are reproduced in Table II in cgs units.

TABLE II

a	b	A	B	ρ_0	I_0
0.5	1.63	0.75×10^{12}	0.65×10^{12}	2.7	5×10^{10}

The shear strength of the aluminum is represented by the equation

$$Y = (Y_0 + Y_1\theta + Y_2\theta^2)(1 - I/I_1) \tag{76}$$

where $\theta = \rho/\rho_0 - 1$, Y_0 is the yield stress in shear at zero pressure, and the constants Y_1, Y_2 were derived from plane wave experiments by Erkman and Christensen (*32*) through considerations of the elastic relief wave. Their values lead to

$$\bar{Y}_1 = 4.2 \times 10^{10}, \qquad \bar{Y}_2 = 7.2 \times 10^{10}$$

in cgs units for the yield stress in plane strain. The deviator stress in simple shear is obtained by multiplying the plane strain value by $\sqrt{3}/2$. The term I_1 denotes the internal energy of aluminum at the melting point.

The impacting sphere is represented in setting up the problem by filling each cell intersected by the sphere to the mass within the intersected volume. Consequently the free surface of the sphere is not resolved to better than a cell dimension. The diameter of the spheres used in the experiments was 0.4763 cm ($\frac{3}{16}$ in.). In the caclulations the cell size was chosen to be 0.0397 cm so that the diameter would fit exactly 12 cells.

E. Comparison with Experiment

Two calculations were run with the same initial conditions and parameters which were identical except for Y_0, the normal yield strength of aluminum. The calculations were selected to match two experiments (*31*) which were similar except for the target material. The soft aluminum target material was 1100-F and the impact velocity was 7.39 km/sec. The hard aluminum was 2014-T6 and was impacted at 7.35 km/sec. These alloys are represented in the calculations by putting $Y_0 = 0.75 \times 10^9$ and 2.39×10^9 dyn/cm^2, respectively, corresponding to handbook values of yield strength of 18,800 and 60,000 psi in simple tension. The impact velocity in both calculations was 7.35 km/sec.

The time steps were on the order of 0.02 μsec. At 5 μsec, the crater is 0.9 cm deep in both the hard and soft aluminum, and is approaching its final size and shape. At this time the strength is beginning to affect the calculation significantly. In the hard aluminum the work done by the

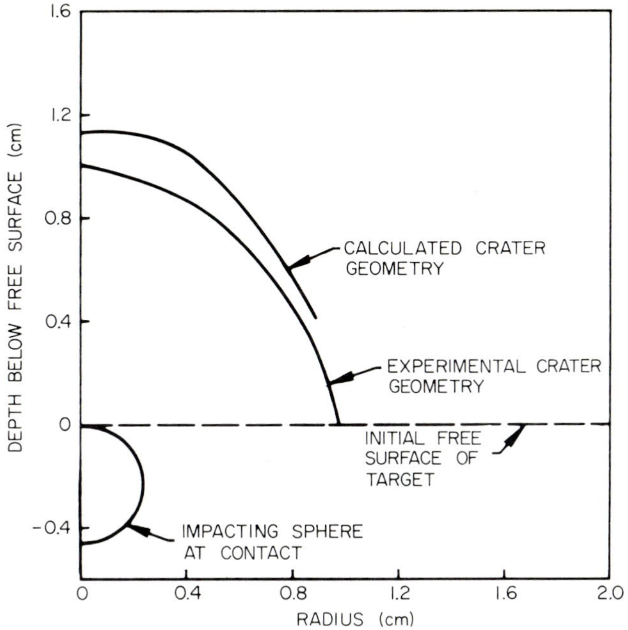

FIG. 20. Comparison of theoretical and experimental craters formed by the impact of an aluminum sphere on a hard aluminum target at 7.35 km/sec. The strength of the 2014-T6 aluminum was taken to be 2.39 kbar in shear.

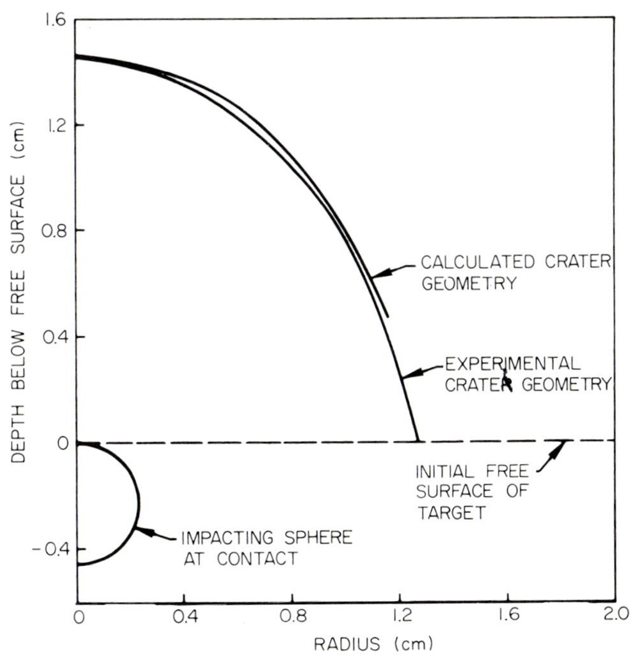

FIG. 21. Comparison of theoretical and experimental craters formed by the impact of an aluminum sphere on a soft aluminum target at 7.35 km/sec. The strength of the 1100-F aluminum target was taken to be 0.75 kbar in shear.

plastic stresses has reached 19% of the total energy and it is 12% of the total in soft aluminum.

The final calculated crater depth in the hard aluminum (2014-T6) is 1.12 cm as compared with the 1.00-cm experimental value. For soft aluminum (1100-F) the calculated depth is 1.48 cm, which compares very favorably with the 1.46-cm experimental value. In spite of the good agreement, however, in neither case is the yield strength for the particular sample tested known well enough to claim a calculational accuracy better than 5%. (Assuming for the moment that crater volume varies inversely with strength, a yield strength uncertainty of 15% would correspond to a crater depth uncertainty of 5%.)

Crater shapes are shown in Figs. 20 and 21 for the hard and soft aluminum targets. The best estimate of the radius is found by arguing that the crater is nearly hemispherical, and the radius and depth are about equal. This is found to be the case for observed craters resulting from like-material impacts by compact bodies. (A calculational difficulty is that material near the crater lip may still have high velocities when the depth has stopped increasing. Thus, we can have greater confidence in calculated depth than in calculated radius.)

In Fig. 22 the growth of craters as a function of time is shown for the two types of aluminum. The time for the crater depth to reach 90% of its terminal value is twice as great in soft aluminum, 16 μsec, as in the hard alloy, where it is 8 μsec. Although it appears on the graph that the

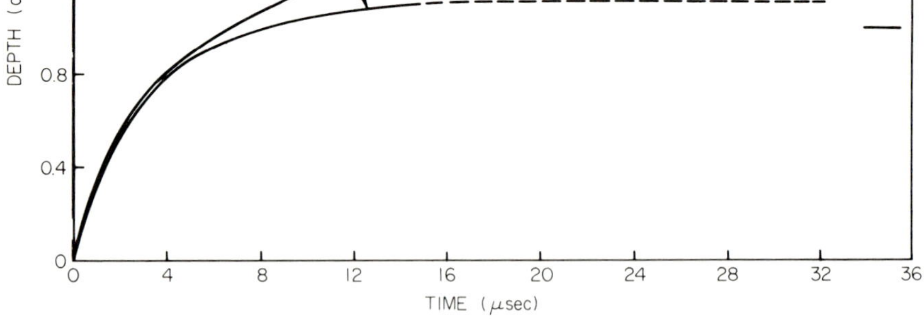

FIG. 22. Crater depth as a function of time for impacts onto hard and soft aluminum, and comparison of terminal values with experiment.

crater could still be growing, the material velocity at the bottom of the crater is down to about 10^3 cm/sec. Thus, even if there were no further deceleration, in an additional 10 μsec the crater would grow only 0.01 cm. The deformation of the target at four equally spaced times is shown in Fig. 23.

Charest (*9*) has measured the decay of shock particle velocity as a function of distance into the target. In his experiments spheres of the same size as those described above were fired onto 1100-O targets at

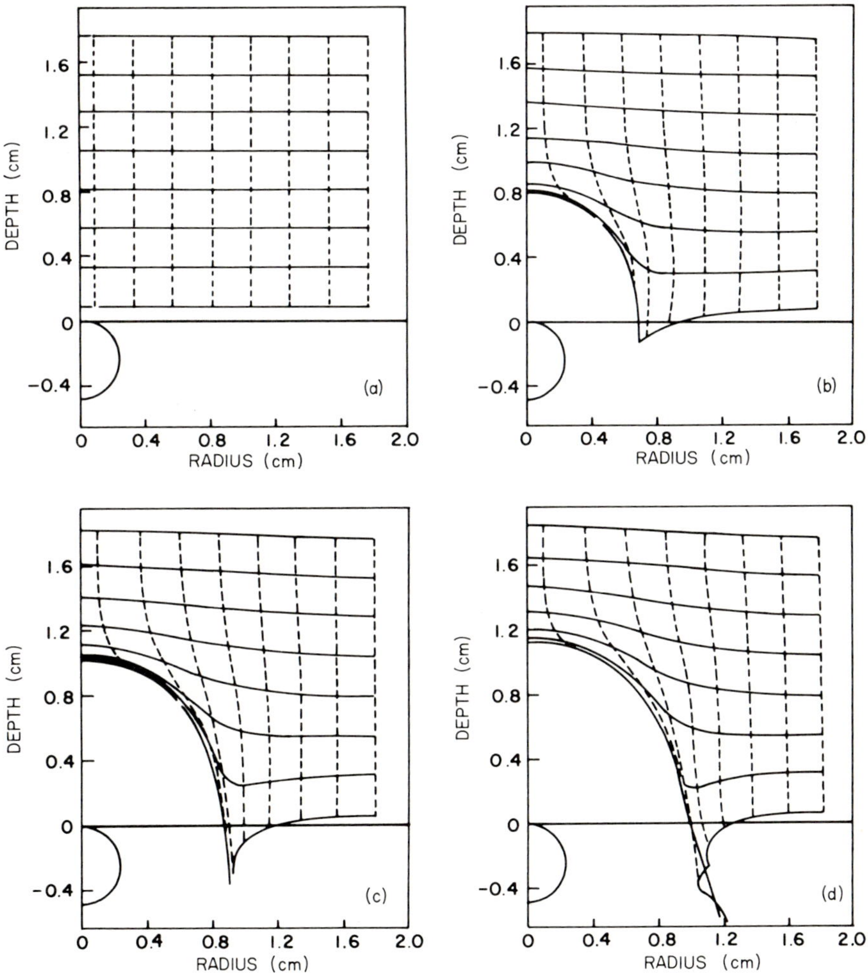

FIG. 23. Deformation of the target at 4-μsec intervals after impact. (a) $t = 0$; (b) $t = 4$ μsec; (c) $t = 8$ μsec; (d) $t = 12$ μsec.

velocities around 7.3 km/sec. The results are compared with the calculated values for 1100-O aluminum in Fig. 24. A small correction factor of $v_0/7.3$ was divided into the measured velocities to correct for the variations in impact velocity v_0 between shots (7.03–7.62 km/sec). The agreement between experiment and calculation is within 20% for velocities above 10^4 cm/sec but at lower velocities the agreement is not so good. Since some instability in the calculation is observed at late times, and the lowest velocity experimental point shown in the figure

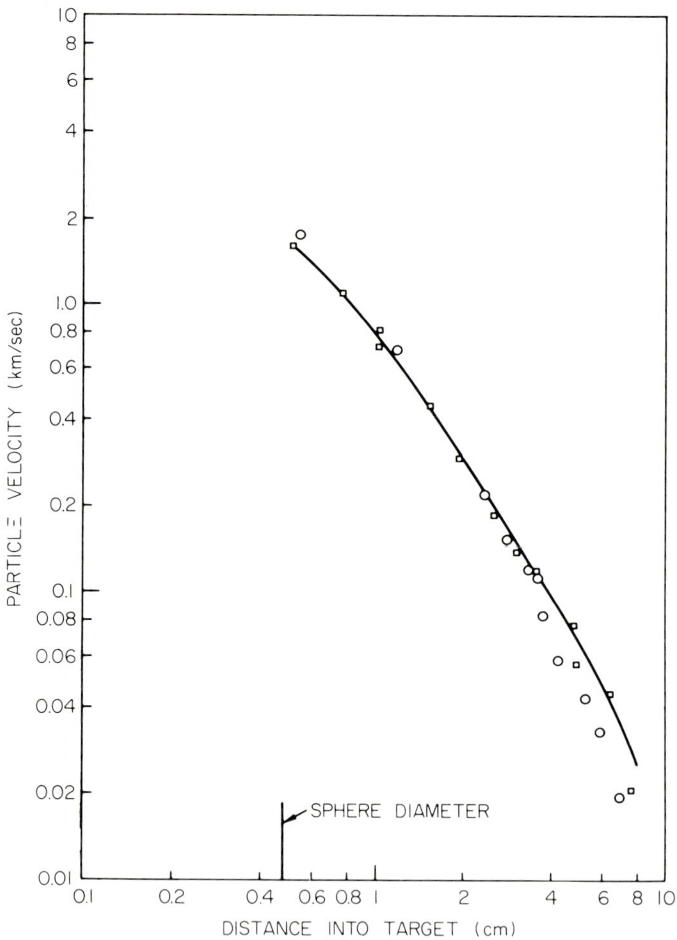

FIG. 24. Comparison of measured and calculated shock particle velocity due to impacts into soft aluminum by a 0.4763-cm sphere. (○) Computer solution; (□) experimental data.

(7×10^3 cm/sec) appears anomalous, the good agreement between the calculated and experimental velocity at a distance 8 cm (16 sphere diameters) into the target is probably not significant.

F. Velocity Scaling

According to the penetration formula

$$p/d = K(v_0/c_0)^\alpha \tag{77}$$

cited by a number of authors and discussed in Section II, crater depth p should increase as the α power of impact velocity v_0. Here d denotes sphere diameter, c_0 is the speed of sound, and K is a nondimensional constant. A fit by Eichelberger and Gehring (*33*) has $\alpha = \frac{2}{3}$ and is one of the few that includes the effect of material hardness in the coefficient. (If their formula is written in the form given above, one finds $K = 0.922 \times 10^{-3} \times \rho_0 c_0{}^2/B$, where B is the Brinell hardness and ρ_0 the projectile density.)

The term α is 0.58 at velocities higher than 10 km/sec, according to the results of Section II. Although the late-stage equivalence rule is based on purely hydrodynamic considerations and applies accurately to impacts at velocities above 10 km/sec, one might expect the 0.58 exponent to be approximately valid at lower velocities, even though strength affects the flow before exact equivalence is attained. Even in the absence of strength, however, pairs of impacts at 5 and 10 km/sec with the same $Lv_0^{0.58}$ do not show agreement of better than 20% between their pressure profiles. Thus the precise equivalence observed at high velocities is not to be expected at velocities around 10 km/sec.

To test the validity of the $\alpha = 0.58$ exponent at the lower velocities, a pair of calculations was carried through for impacts with the same $Lv_0^{0.58}$ but with the diameters of the impacting sphere differing by a factor of two. The smaller-diameter sphere must have a velocity of 24.28 km/sec to make $Lv_0^{0.58}$ the same as in the calculations described in the previous section. The zoning was chosen so that after one rezone of the 24.28-km/sec problem the cell size, grid boundaries, and free surfaces would be identical with the slower one. The velocity and pressure profiles for the two impact problems at 5 μsec after impact are shown in Figs. 25 and 26. The discrepancy between the two profiles amounts to about 7% in shock position (the shock generated by the faster projectile is more advanced) and 7% in peak velocity and pressure. This agreement might be somewhat improved by finding pairs of times at which the two profiles are as similar as possible.

An indication of the effect of strength on the pressure and velocity

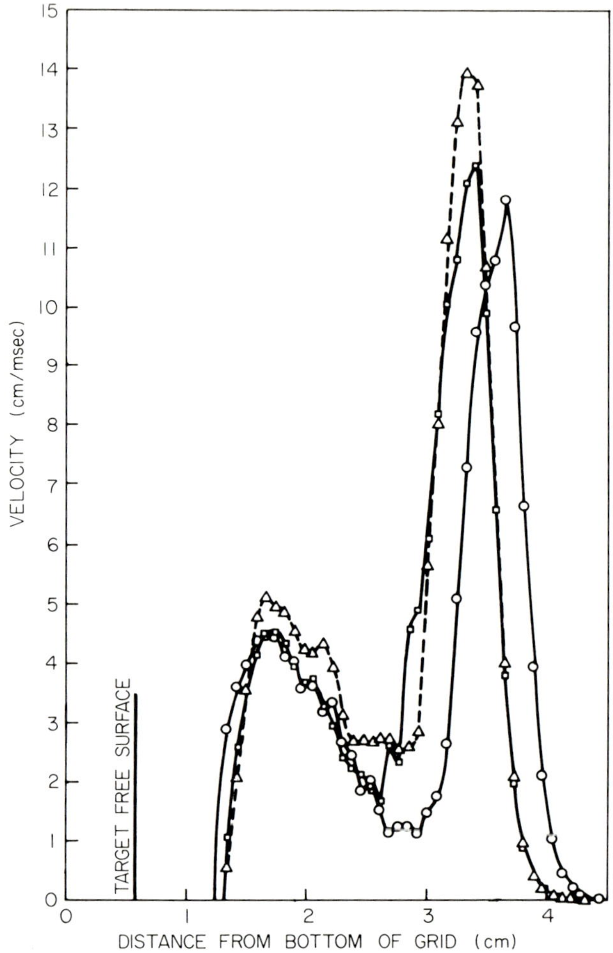

FIG. 25. Comparison of velocity profiles resulting from spheres differing in diameter by a factor of 2 but with the same value of $Lv_0^{0.58}$. The low-velocity impact is the same as described in earlier figures (7.35 km/sec). The high velocity is 24.28 km/sec and the time after impact in all cases is 5 μsec. An impact into soft aluminum at 7.35 km/sec results in the dashed profile. (□) Low-velocity impact into hard aluminum; (○) equivalent high-velocity impact into hard aluminum; (△) low-velocity impact into soft aluminum.

profiles at these velocities is provided by the comparison of the pressure profiles in hard aluminum and soft aluminum on the same graphs. The difference in shock pressure in the two materials is 1.5 kbar (which is very nearly the same as the difference in shear strength) and there is essentially no difference in shock positions.

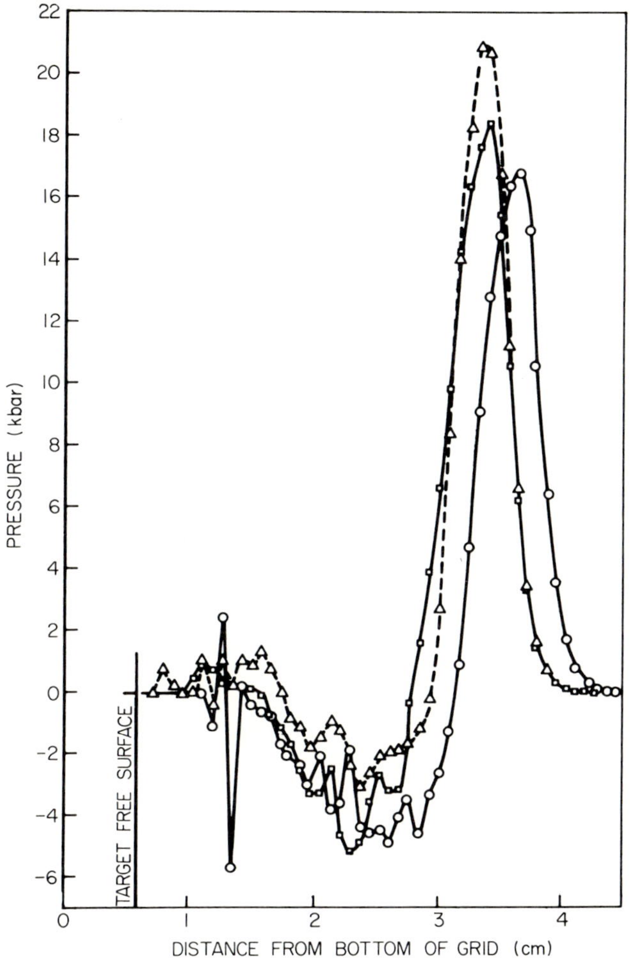

FIG. 26. Comparison of pressure profiles resulting from spheres differing in diameter by a factor of 2 but with the same value of $Lv_0^{0.58}$. The low-velocity impact is the same as described in earlier figures (7.35 km/sec). The high velocity is 24.28 km/sec and the time after impact in all cases is 5 μsec. An impact into soft aluminum at 7.35 km/sec results in the dashed profile. (○) Low-velocity impact into hard aluminum; (□) equivalent high-velocity impact into hard aluminum; (△) low-velocity impact into soft aluminum.

Crater depth as a function of time for the fast and slow impacts is shown in Fig. 27. The terminal crater size is 1.14 cm for the fast impact and 1.10 cm for the slower one. Thus the prediction of equivalent craters from the two projectiles is validated within the estimated accuracy of the

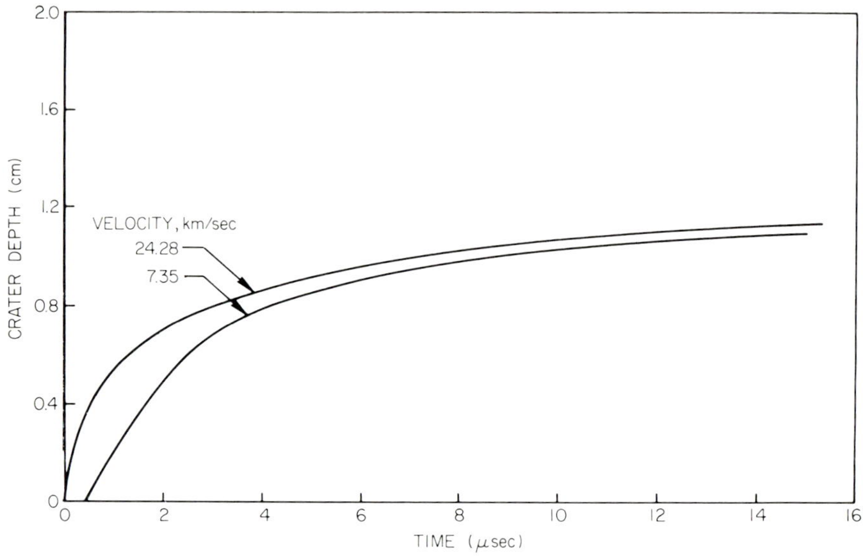

FIG. 27. Growth of crater depth as a function of time for two impacts into hard aluminum with the same $Lv_0^{0.58}$ (late-stage equivalent). The low-velocity impact is the same as described in the previous comparison. The high-velocity sphere has half the diameter and a velocity of 24.28 km/sec.

calculation. If energy scaling were valid, i.e., $\alpha = \frac{2}{3}$, the predicted crater depth would be 14% greater for the high-speed projectile, or 1.25 cm.

ACKNOWLEDGMENTS

Programming of the codes and the various calculations were carried out primarily by W. E. Johnson and later by M. W. Evans, and it is a pleasure to acknowledge their important contributions. The work was monitored by the Ballistic Research Laboratories for the Advanced Research Project Agency under ARPA Order 71-62. F. E. Allison and later S. M. Taylor acted as technical monitors and contributed through many valuable discussions.

REFERENCES

1. B. Alder, S. Fernbach, and M. Rotenberg (eds.), *Methods Comput. Phys.* **3**, (1964).
2. M. L. Wilkins, Calculation of elastic-plastic flow, *Methods Comput. Phys.* **3**, 211 (1964).
3. M. W. Evans and F. H. Harlow, The particle-in-cell method for hydrodynamic calculations, Los Alamos Scientific Lab. Rept. LA-2139 (1957).

4. W. E. Johnson, OIL: A continuous two-dimensional Eulerian hydrodynamic code, General At. Rept. GAMD-5580 (1965).
5. R. Courant and K. O. Friedricks, "Supersonic Flow and Shock Waves." Wiley (Interscience), New York, 1948.
6. R. L. Bjork, Effects of meteoroid impact on steel and aluminium in space, Rand Corp. Paper, 1662 (1958).
7. T. D. Riney, Visco-plastic solution of hypervelocity impact cratering phenomona, *Proc. Symp. Hypervelocity Impact 6th, Cleveland, Ohio* **2**, Part 2 (1963).
8. R. Hill, "The Mathematical Theory of Plasticity." Oxford Univ. Press, London and New York, 1950.
9. J. A. Charest, Measurement of shock wave pressures generated by hypervelocity impacts in aluminum, General Motors Defense Res. Lab. Rept. TR-64-58 (1964).
10. G. I. Taylor, The formation of a blast wave by a very intense explosion, *Proc. Roy. Soc. (London)* **A201**, 159 (1950).
11. K. P. Stanyukovich and V. V. Fedynskiy, The destructive action of meteorite impacts, *Dokl. Akad. Nauk SSSR* **57**, No. 2, 129 (1947); Transl. ASTIA AD-124-240.
12. F. L. Whipple, The meteoric risk to space vehicles, "Vistas in Astronautics 1958," pp. 115–125. Pergamon Press, New York, 1958.
13. J. M. Walsh and J. H. Tillotson, Hydrodynamics of hypervelocity impact, *Proc. Symp. Hypervelocity Impact, 6th, Cleveland, Ohio* **2**, Part 1, 59 (1963).
14. W. Herrmann and O. F. Jones, Survey of hypervelocity impact information, Massachusetts Inst. Technol. Rept. ASRL 99-1 (1961).
15. A. E. Olshaker and R. L. Bjork, Scaling laws for dissimilar materials, *Proc. Symp. Hypervelocity Impact, 6th, Cleveland, Ohio* **1**, 185 (1962).
16. J. H. Kineke, Jr., and L. G. Richards, Influence of target strength on hypervelocity crater formation in aluminum, *Proc. Symp. Hypervelocity Impact, 6th, Cleveland, Ohio* **2**, Part 2, 513 (1963).
17. J. H. Tillotson, Metallic equations of state for hypervelocity impact, General Atomic Rept. GA-3216 (1962).
18. W. J. Rae and H. P. Kirchner, Final report on a study of meteoroid impact phenomona, Cornell Aeronaut. Lab. Rept. No. RM-1655-M-4 (1963).
19. W. Häfele, Zur analytischem behandlung ebener, starker, instationer stosswellen, *Z. Naturforsch.* **10a**, 1006 (1955).
20. V. B. Adamskii, Integration of a system of autosimulating equations for the problem of a short duration shock in a cold gas, *Soviet Phys. Acoust. (English Transl.)* **2**, No. 1, 1 (1957).
21. L. I. Sedov, "Similarity and Dimensional Methods in Mechanics." Academic Press, New York, 1959.
22. R. T. Allen, Equation of state of rocks and minerals, General Atomic Rept. GAMD-7834 (1967).
23. J. M. Walsh, J. H. Tillotson, and W. E. Johnson, Theory of Hypervelocity Impact, Quart. Progr. Rept., General Atomic Rept. GACD-4518 (1963).
24. C. Truesdell and W. Noll, The non-linear field theories of mechanics, "Handbuch der Physik," Vol. III/3. Springer, Berlin, 1965.
25. A. C. Eringen, "Non-Linear Theory of Continuous Media." McGraw-Hill, New York, 1962.
26. W. Prager and P. G. Hodge, "Theory of Perfectly Plastic Solids." Wiley, New York, 1951.
27. I. S. Sokolnikoff and R. D. Specht, "Mathematical Theory of Elasticity." McGraw-Hill, New York, 1946.

28. P. W. Bridgman, Studies in large plastic flow and fracture, "Metallurgy and Metallurgical Engineering Series." McGraw-Hill, New York, 1952.
29. T. Y. Thomas, Theoretical effect of large hydrostatic pressures on the tensile strength of materials, *Proc. Natl. Acad. Sci.* **57**, 1195 (1967).
30. E. C. Robertson, Experimental study of the strength of rocks, *Bull Geol. Soc. Am.* **66**, 525 (1955).
31. S. M. Halperson, Some phenomona associated with impacts into aluminum, *Proc. Symp. Hypervelocity Impact, 6th, Cleveland, Ohio*, **2**, Part 2, **525** (1963).
32. J. O. Erkman and A. B. Christensen, Attenuation of shock waves in aluminum, *J. Appl. Phys.* **38**, No. 13, 5395 (1967).
33. R. J. Eichelberger and J. W. Gehring, Effects of meteoroid impacts on space vehicles, *ARS J.* **32**, No. 10, 1583 (1962).
34. L. F. Vereshchagin and V. A. Shapochkin, The effect of hydrostatic pressure on resistance to shear in solids, *Fiz. Metall. i Metalloved.* **9**, No. 2, 258 (1960).
35. S. N. Zhurkov and T. P. Sanfirova, A study of the time and temperature dependences of mechanical strength, *Soviet Phys. Solid State (English Transl.)* **2**, No. 6, **933** (1960).

CHAPTER IV

THEORY OF IMPACT ON THIN TARGETS AND SHIELDS AND CORRELATION WITH EXPERIMENT

JOHN W. GEHRING, JR.
IMPACT PHYSICS LABORATORY
GENERAL MOTORS CORPORATION
AND
AC ELECTRONICS DEFENSE RESEARCH LABORATORIES
SANTA BARBARA, CALIFORNIA

I. Introduction

Results are presented from a theoretical and experimental program to investigate the damage that would be inflicted upon a missile, satellite, or space vehicle by the impact of a hypervelocity projectile.

In this chapter all of the discussion will be directed toward the impact of a projectile on a simple two-sheet structure, i.e., a rear or backup sheet which is protected by a thin shield. The problem is treated theoretically by examining the wave motion in an impacting projectile and shield, and consideration is given to heating effects.

Two approaches have been taken to theoretically treat the problem. The first is a one-dimensional analysis examining the wave motion, and the second is a two-dimensional analysis using a hydrodynamic treatment. Each of these analyses, when combined with experiments, shows that the damage mechanisms to be considered are the breakup and dispersion of the projectile and shield debris at high velocities and the gross deformation, tensile failure, and spallation of the rear sheet. Gross deformation and tensile failure were treated with blast-loaded, thin-shell analysis, and spallation with a two-dimensional elastic–plastic treatment.

Upon striking a thin sheet a particle or projectile may undergo a variety of processes depending upon impact conditions such as the particle velocity, the particle material and composition, the angle of impact, the material strength, and the thickness of the thin sheet. (A thin sheet as used herein will be defined as a sheet whose thickness is equal to or less than the diameter of the projectile.) The particle may be stopped by the sheet, may pass through the sheet essentially undamaged, or may pass through the sheet fractured, molten, or vaporized. The last two cases are the cases of interest for meteoroid impacts as the velocities are sufficiently high to cause melting or vaporization.

If the thin sheet is penetrated the debris from the projectile and the shield then travel across the space between the sheets and strike a second sheet. Upon striking the second sheet a shock wave is generated within, and traverses, the second sheet. Depending upon the intensity and the structure of this shock an internal fracture or spall may form, resulting in some cases in complete detachment of some material from the rear surface of the sheet.

In addition the second sheet will be given an impulsive load by the impact of the particle–shield debris. This load is applied over a very short period of time (a few microseconds) and results in a second sheet moving with some velocity. The sheet can then fail from this load by tensile failure or shear failure.

The above phenomena will be treated in detail in this chapter.

II. Theoretical Model—One-Dimensional Analysis

Consider a right circular cylinder, whose length is equal to its diameter, impacting a thin shield at hypervelocity. The estimated wave pattern

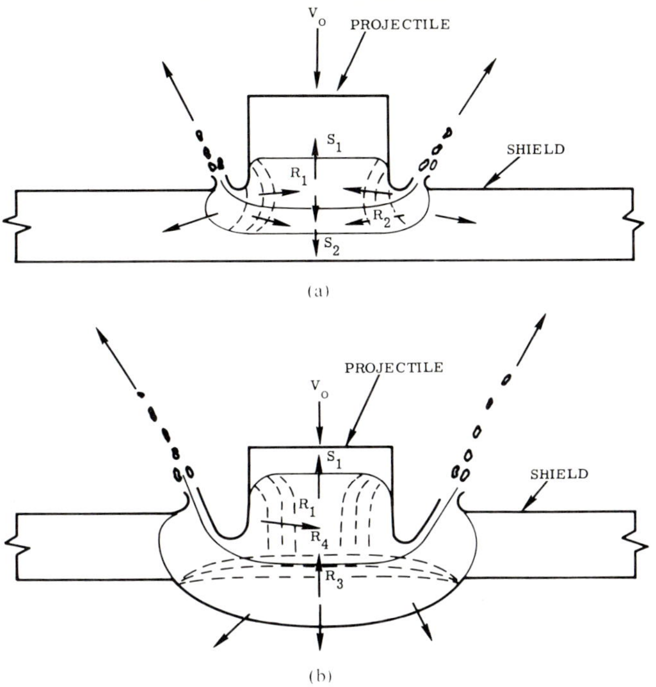

FIG. 1. Estimated wave pattern. (a) In a projectile and shield soon after impact; (b) after reflection of shock from bottom face of shield.

shortly after impact is shown schematically in Fig. 1. It is seen (Fig. 1a) that two shock waves S_1 and S_2 have propagated away from the interface I, and because the projectile is finite in diameter, rarefaction waves R_1 and R_2 have been transmitted toward the axis of symmetry. Also, the formation of these rarefactions has resulted in the ejection of both projectile and shield material in a rearward direction. Now consider the situation shortly after the shock S_2 has reflected from the back face of the shield (Fig. 1b). In order to satisfy the boundary condition of zero pressure, the shock is reflected as a rarefaction wave R_3. The resultant particle velocities behind R_3 are such that the profile of the back face of the shield is as shown in the figure. As the process continues, the bubble grows through the addition of material from the shield and projectile.

From a physical viewpoint, the rarefactions R_1, R_2, R_3, and R_4, generated to satisfy boundary conditions, can be looked upon as being tension waves. Hence, fracture will occur if the net tensile stress at any point in the projectile or shield exceeds the fracture stress. In addition, rarefactions will be produced to satisfy boundary conditions at any new fracture surfaces, and these rarefactions can lead to further fractures. Thus, the whole process of fracture of a projectile and a thin shield can be interpreted as a multiple spalling phenomenon that starts at the free surfaces. Hence, the significance of a shield is that it can fragment the projectile, spread the fragments radially, and significantly reduce the velocity of many of the fragments below the velocity of the original projectile.

A. Heating Effects Due to Impact

An examination of Fig. 1 shows that during the impact process each element of the projectile and the shield is first shocked to some pressure and is then brought back to ambient pressure by rarefaction waves. This process is represented on a pressure–volume plot in Fig. 2. Each element

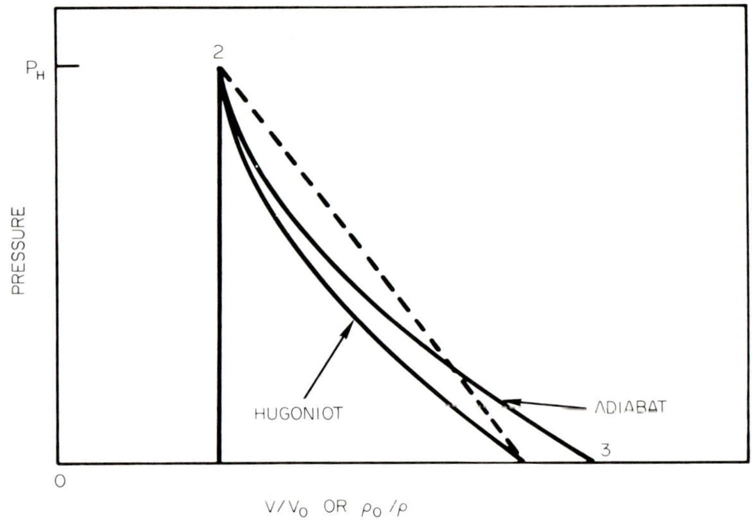

Fig. 2. Pressure–volume curve.

of material starts off at some initial density ρ_0 and is then compressed on the Hugoniot (points 1–2) to some pressure P_H by a shock wave. The material is then brought back to ambient conditions adiabatically (points 2–3) by rarefaction waves. The initial shocking process is a nonisentropic one, whereas the release process is isentropic. Thus, the entropy of the material has been increased by the impact process. In other words, the element of material in its final state will be heated. The specific internal energy e_H of the shock-compressed material is equal to the area of the dashed triangle $P_H(V_0 - V_H)/2$ in Fig. 2, whereas the energy that is returned by the material in pressure–volume work on expansion is the area under the adiabat. This latter area is less than the area under the triangle, and the difference between the two areas is proportional to the residual heat left in the element. Shocks of low strength will leave the material heated but in the solid state. As the shock strength is increased, the entropy excess increases rapidly and can lead to melting, heated liquid, vaporization, or superheated vapor in the final state.

A more complete discussion of this phenomenon has been made by Olshaker and Bjork (*1*) and McQueen and Marsh (*2*). These workers have calculated, for a number of materials, the values of shock pressure sufficient to cause incipient melting, complete melting, and vaporization when the materials are returned to atmospheric pressure. Some estimated values of these pressures are tabulated in Table I for materials of interest in the present paper. Note that, in many cases, appropriate shock pressures for the preceding changes are not available. Also note that the

TABLE I

SHOCK HEATING EFFECTS

Material	Melting temperature (°C)	Vaporization temperature (°C)	Pressure to cause incipient melting (Mbar)	Pressure to cause complete melting (Mbar)	Pressure to cause vaporization (Mbar)
Aluminum	660	2057	0.6	0.9	—
Cadmium	321	767	0.4	0.46	0.8
Copper	1083	2336	1.4	>1.8	—
Gold	1063	2600	1.5	1.6	—
Iron	1535	3000	—	2.0	—
Lead	327	1620	0.3	0.35	1.0
Magnesium	651	1107	—	—	—
Nickel	1455	2900	$\gg 1.5$	—	—
Titanium	1800	>3000	>1.0	—	—

preceding heating effects will be enhanced at low velocities by material viscosity.

It is considered that the forementioned heating effects are important in determining the size of particles in the bubble. If the final debris is in the solid state, the size of spall fragments will decrease with increasing temperature because of a decrease in fracture strength. If the material is molten, only surface tension forces need to be overcome to create droplets; these forces, and hence the droplet size, will become smaller as the liquid becomes hotter. Finally, if the heating effects are great enough, the debris becomes vapor.

B. Maximum Impact Pressure

The flow between the shocks S_1 and S_2 is one dimensional until reached by rarefactions R_1 and R_2. The situation in this one-dimensional region immediately after impact is shown in Fig. 3. In this figure, all

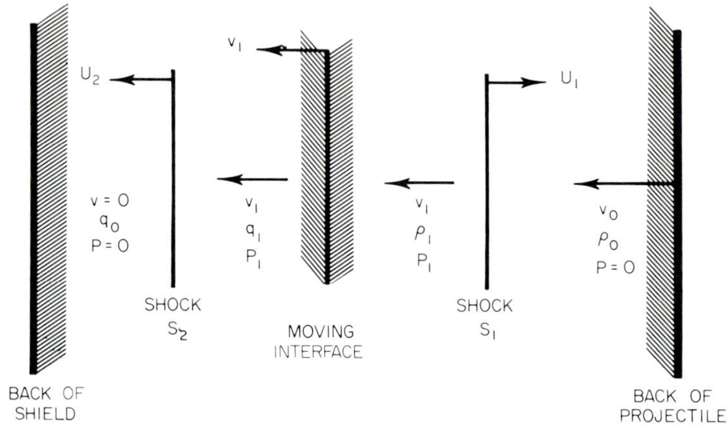

Fig. 3. One-dimensional situation immediately after impact upon shield.

velocities are measured with respect to a set of coordinates fixed in the shield. The following symbols are used:

v_0 = impact velocity,
v_1 = velocity of material between shocks,
U_1 = velocity of shock S_1 in projectile,
U_2 = velocity of shock S_2 in the shield,
P_1 = pressure in region between the shocks,
ρ_0 = density of uncompressed projectile material,
ρ_1 = density of projectile material at pressure P_1,
q_0 = density of uncompressed shield material,
q_1 = density of shield material at pressure P_1.

It can be shown that use of the normal-shock relationships across each shock, plus continuity considerations at the interface, produce the following equations:

$$P_1 = \frac{\rho_0 v_0^2}{1 - (\rho_0/\rho_1)} \left[1 + \left(\frac{\rho_0[1 - (q_0/q_1)]}{q_0[1 - (\rho_0/\rho_1)]}\right)^{1/2}\right]^{-2} \tag{1}$$

$$v_1 = v_0 \left[1 + \left(\frac{q_0[1 - (\rho_0/\rho_1)]}{\rho_0[1 - (q_0/q_1)]}\right)^{1/2}\right]^{-1} \tag{2}$$

$$U_1 = \frac{v_0 - (\rho_1/\rho_0)\, v_1}{(\rho_1/\rho_0) - 1} \tag{3}$$

$$U_2 = \frac{v_1}{1 - (q_0/q_1)} \tag{4}$$

In order to solve a specific example, the Hugoniot equations for the projectile and shield materials are necessary. Such equations have been determined experimentally by United States (*2*, *3*) and Soviet scientists (*4*, *5*) for several materials up to 2–5 Mbar. For higher pressures, the theoretical relationships of Walsh and Tillotson (*6*) may be used.

C. Effect of Shield Thickness

Figure 1b shows conditions existing after the shock in the shield has reflected as rarefaction wave R_3. The flow in the volume that has not been reached by the rarefactions R_1 and R_2 is one-dimensional and is represented on a time–distance diagram in Fig. 4. The shock S_2 is

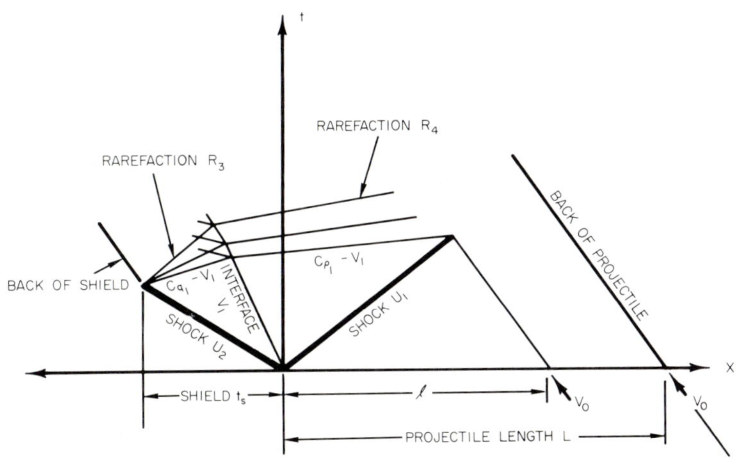

Fig. 4. Time–distance diagram illustrating how the rarefaction from the back of shield overtakes the shock S_1 in a projectile.

assumed to reflect from the back of the shield as the centered rarefaction wave R_3. The leading edge of this rarefaction travels at velocity c_{q_1} relative to the shield material (moving at velocity v_1), where c_{q_1} is the adiabatic speed of sound in the shield material at pressure P_1. Also, reflected and transmitted waves are produced when the rarefaction reaches the interface. The transmitted wave is rarefaction R_4 with the leading edge traveling at velocity c_{p_1} relative to the projectile material moving at velocity v_1. Here again, c_{p_1} represents the adiabatic speed of sound in the projectile material at pressure P_1. Note that, if the projectile and shield are made of the same material, no reflected wave is produced at the interface.

If the shield is thin enough, rarefaction R_4 can overtake and weaken the shock in the projectile. It can be shown that the leading edge of the rarefaction overtakes the shock at a point in the projectile given by

$$\frac{l}{L} = W\frac{\rho_0}{q_0}\left(\frac{c_{q_1} + U_2 - v_1}{c_{p_1} - U_1 - v_1}\right)\left(\frac{c_{p_1}}{c_{q_1}}\right)\left(\frac{U_1 + v_0}{U_2}\right) \tag{5}$$

where l is measured from the front face of the uncompressed projectile (length L), t_s is the shield thickness, and W = (weight of shield/unit area)/(weight of projectile/unit area).

In order to calculate l/L from Eq. (5), it is necessary to know the relevant adiabatic speed of sound c_p and c_q for each material. If the equations of state of the materials are known, then the speeds of sound can be found from the relationship $c^2 = (\partial P/\partial \rho)S$, where S denotes differentiation at constant entropy. Unfortunately, even though the Hugoniots of many materials are known, their equations of state are not available. However, the Russians (*7*) have measured values of c for several materials and indicate that, for metals up to about 5 Mbar, a reasonable approximation for the adiabatic speed of sound at pressure P behind a shock is given by

$$c_P = D\{0.49 + [(D - v)/D]^2\}^{1/2} \tag{6}$$

where D is the shock velocity relevant to a pressure P, and v is the particle velocity behind the shock. In this equation, both velocities must be measured relative to the undisturbed material considered at rest. Thus, in the coordinates of Fig. 3, $D = U_2$ and $v = v_1$ for the shield material, and $D = U_1 + v_0$ and $v = v_0 - v_1$ for the projectile material.

If, for a specific example, use of Eq. (5) shows that l is less than the length of the projectile L, then the rarefaction R_4 weakens the shock in the projectile. For the impact of like materials and one-dimensional flow, Fowles (*8*) has presented a method that can be used to calculate how the shock wave S_1 is attenuated by the centered rarefaction wave R_3. In this method, the materials are treated as compressible inviscid fluids, and it is assumed that the entropy is everywhere constant, including across the shock transition. This analysis, on the axis of symmetry, is only valid up until the time the cylindrical rarefaction R_1 (Fig. 1) overtakes the shock S_1 on the axis of symmetry. It can be shown that this occurs at a point in the projectile given by $l_1 = 0.72d$, where l_1 is measured from the front face of the initially uncompressed projectile and d is the projectile diameter. For a projectile with $L/d = 1$, it is felt that the error obtained in using the preceding one-dimensional analysis is small.

D. Experimental Results

In support of the theoretical analysis previously described a number of experimental test firings were conducted on the General Motors Defense Research Laboratories light gas gun ranges to test the theory. The results of these experiments and the extent to which they prove or disprove the theoretical analysis will now be described.

1. *Role of Heating Effects Due to Impact*

Shield materials of cadmium, copper, and nickel were chosen for tests, as they have almost identical densities but very different thermal properties. In all cases the protected targets were 6.4-mm aluminum (2024-T3) plates spaced 5.08 cm behind the shields. Also, in all the tests, the shield thickness was kept constant at 1.016 mm.

The first series of tests was conducted using 6.35-mm aluminum spheres at 4.16 km/sec. The targets from these tests are shown in the top rows of Figs. 5a and 5b. It is seen immediately that the frontal damage to the targets decreases in the shield order of nickel, copper, and cadmium. Tables I and II help explain this result. They show that, at 4.16 km/sec, the maximum impact pressure is only sufficiently high to melt the cadmium debris from the shield, and no melting of copper or

TABLE II

Characteristics of Impacts in Cd, Cu, and Ni

Projectile	Velocity (km/sec)	Shield	Pressure P (Mbar)
Aluminum	4.2	Cd	0.67
		Cu	0.72
		Ni	0.75
Aluminum	7.3	Cd	1.50
		Cu	1.57
		Ni	1.67
Steel	7.1	Cd	2.60
		Cu	2.90
		Ni	3.05

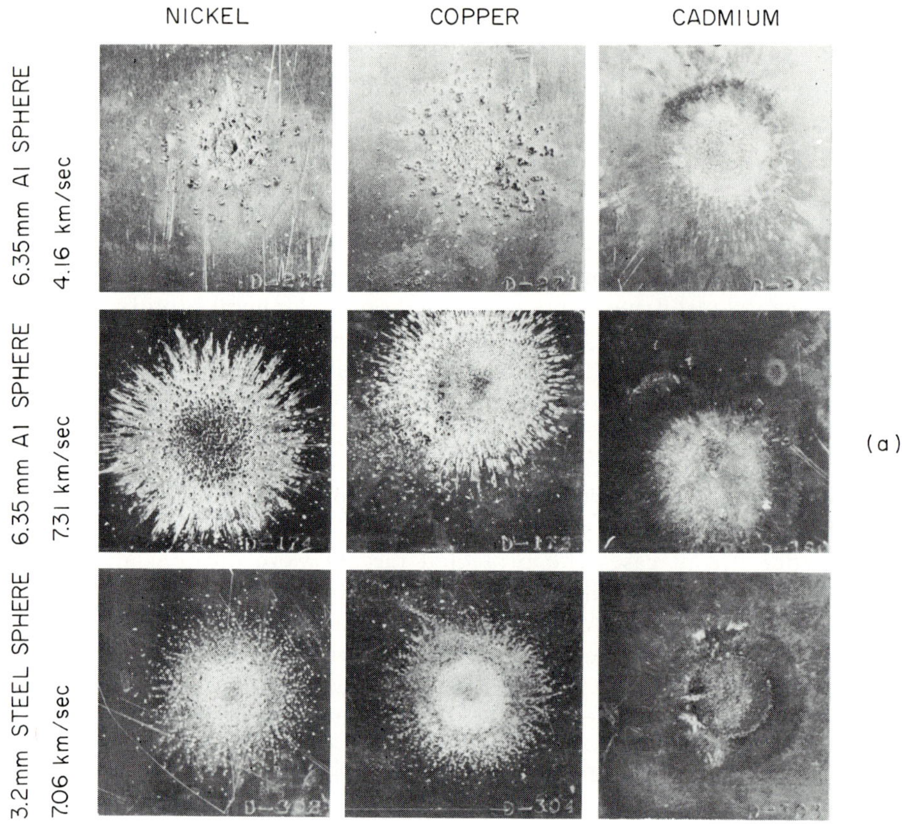

FIG. 5. Effect of shield melting temperature. (a) Front of target; (b) rear of targets. 1.016-mm shields, 5.08-cm spacing, 6.35-mm 2024-T3 Al backup.

nickel fragments will occur. As a result of this, the frontal damage to the backup target with a cadmium shield is reduced significantly. This is because the molten cadmium particles are extremely small, whereas, with the copper and nickel shields, larger solid fragments cause frontal damage to the targets. The x-rays from the preceding tests, Fig. 6, show the fragmentary nature of the debris from the nickel and copper shields. Of course, for each test, the debris from the aluminum projectile also causes damage to the backup target; however, the impact pressures for the three shield materials are such that this damage is expected to be least with the nickel, next with the copper, and greatest with cadmium. Thus, the differences in the damage in Fig. 5 are attributable directly to the state of the shield debris.

In the second series of tests, 6.35-mm aluminum spheres were

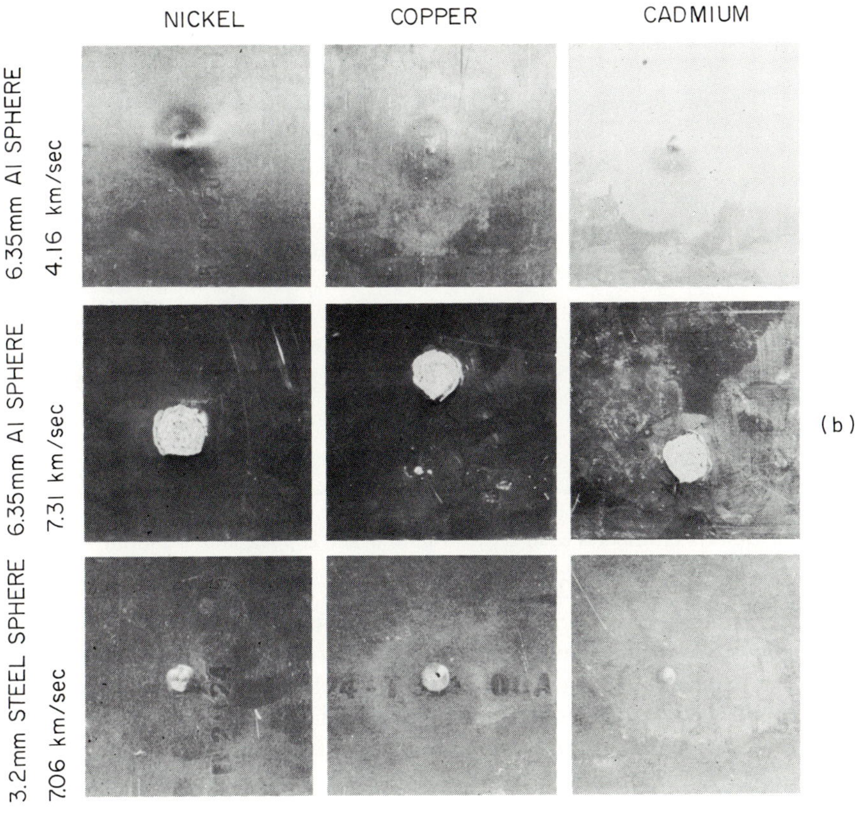

FIG. 5b.

impacted at 7.31 km/sec. Table II shows that the maximum impact pressures generated in these tests are sufficient to cause complete melting of parts of the aluminum projectiles. Also, the pressure is sufficient to vaporize the cadmium and just cause incipient melting of the copper. The targets (second rows of Fig. 5) tend to confirm these theoretical predictions. For instance, a "splash" of aluminum can be observed on all three targets. Also, the frontal damage to the target with a cadmium shield is almost negligible, whereas the presence of pitting on the target with a nickel shield suggests that the nickel fragments are not molten upon impact. Only the x-ray for the cadmium shield is shown in Fig. 6 for this series of tests. The x-ray definitely confirms that the debris in the bubble exists in other states than solid fragments.

The tests in the third series were such that the maximum impact pressure (Table II) was increased even further than in the first two test

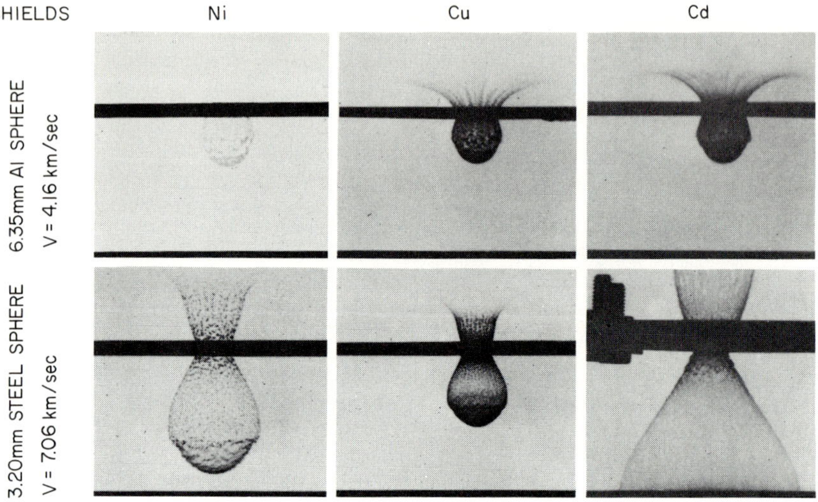

FIG. 6. X-rays of impacts against 1.016-mm nickel, copper, and cadmium shields.

series. In these tests, 3.2-mm steel spheres were fired at 7.06 km/sec The targets are shown in the bottom rows of Fig. 5, and the corresponding x-rays are shown in the bottom row of Fig. 6. Once again the cadmium shield is the best; however, now the copper is shocked to a sufficient pressure to become molten, so that the cadmium and copper shields are alsmost equally effective. Also, both the x-ray and target with the nickel shield indicate that the nickel debris is still solid. The x-ray for the cadmium shield is such that it is difficult to distinguish any particulate debris in the bubble.

Two points of interest from the foregoing tests should be emphasized. The first point is that both the debris from the projectile and shield cause damage to the backup target. The debris from the shield can be particularly damaging because a large portion of it will travel at near the initial impact velocity. Hence, the state of the debris from the shield is very important in determining the depth of penetration in the backup target. The second point arises from examination of the rear of the backup targets (Fig. 5b). It is seen that the spall-producing load is essentially independent of the state of the material in the bubble.

2. *Hole Size in Shield*

The hole diameter in the shield was measured in all the experiments performed in the study. The most revealing results were obtained from the test series on the effects of velocity. The results from this series are shown in Fig. 7. It is seen that, for a specific shield thickness, the hole

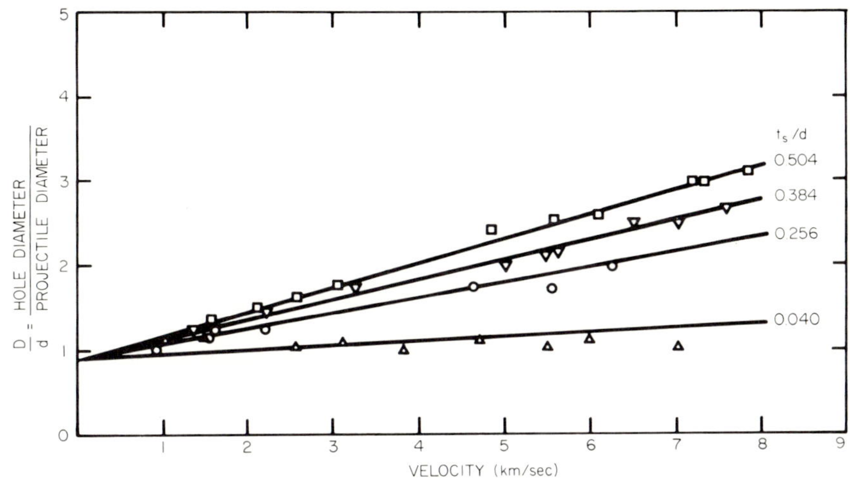

FIG. 7. Hole size in shields for aluminum projectiles and shields. Experimental points; 3.2-mm Al spheres, 2024-T3 Al shields; (—) predictions; $D/d = 0.45v_0(t_s/d)^{2/3} + 0.90$.

diameter D is approximately a linear function of velocity. Further investigation reveals that the equation

$$D/d = 0.45v(t_s/d)^{2/3} + 0.90 \tag{7}$$

where v is impact velocity in kilometers per second, gives a surprisingly good fit to the experimental data, as can be seen in Fig. 7.

It was also found that Eq. (7) is in quite good agreement with the hole size data obtained from the tests using dissimilar shield and projectile materials. These tests indicated [as does Eq. (7)] that hole diameter, at a specific velocity, is fairly independent of shield material and is approximately dependent upon shield thickness alone.

III. Theoretical Model—Two-Dimensional Analysis

A. Projectile–Shield Interaction

In the preceding section of this chapter the simple one-dimensional analysis of the projectile–shield interaction was examined. This treatment similar to that of Sandorff (9), considered a right circular cylinder striking a thin sheet and the one-dimensional shock structure which results along the projectile–shield centerline. It was found that the rarefaction waves from the outside of the cylinder overtake the shock front 0.72 diameter from the interface in the projectile. The shock will reflect from the rear surface of the shield as a rarefaction wave and, depending upon the thin-sheet thickness, may overtake the shock front in the projectile. Both treatments concluded that relief from high pressures resulted in the destruction of the impacting particle.

The one-dimensional treatment also established an optimum shield thickness based upon the centerline shock strength being sufficient, when relieved, to melt or vaporize the impacting projectile. This criterion of optimum shield thickness assures minimal penetration of the second sheet. Also, this criterion calls for decreasing shield thickness with increasing impact velocity.

Sandorff (*9*) established a criterion for an optimum shield thickness on the basis that the rarefaction from the rear surface of the shield would not overtake the shock in the projectile within the projectile length. This criterion requires increasing shield thickness with increasing velocity.

Both of these criteria are based on a somewhat artificial basis in that they require only that the debris that passes through the shield be in some particular form, and say nothing of the debris distribution. Sandorff requires that the projectile be shocked to the maximum pressure possible; whereas the aforementioned requires only that the projectile be shocked sufficiently to melt or vaporize it. Other than to minimize penetration, neither of these criteria explicitly accounts for the reaction of the second sheet.

An accurate evaluation of the effectiveness of a shield should account for spallation and gross-deformation failure of the second sheet. It is clear that, to be effective, the shield must at least either break the impacting particle into small pieces or cause melting of the projectile to insure that no significant penetration will take place.

An accurate description of the state of the debris that passes through the shield requires a complete assessment of the pressure–time history of the projectile and shield. This is conveniently possible only through finite difference computations utilizing a high-speed computer. High-speed impacts upon thin sheets have been considered in this matter by Bjork (*10*), Riney and Heyda (*11*), and Walsh and Johnson (*12*). Bjork and Riney and Heyda utilized particle-in-cell (or PIC) computational schemes while Walsh and Johnson have used an Eulerian scheme. These authors have presented results in terms of centerline densities, velocities, and momentum versus spray-angle profiles. The results show that, as the sheet thickness is increased, the debris is spread more and the momentum per unit solid angle is decreased.

These treatments are hydrodynamic in nature and make no provision for the strength of the shield material. For the thin-sheet case this treatment is sufficiently accurate for the material of interest (i.e., that debris that passes through the shield at high velocity) which has been subjected to pressures many times greater than the strength of the materials. For impacts in the range of meteoroid velocities the debris is expected to be molten or vaporized. The strength of the shield does influence the size of the hole formed. The shock in the shield outside the projectile–shield interface is attenuated very rapidly because of the close proximity of the shield's free surfaces, and very little material is ejected with significant velocity from this area. Thus the hydrodynamic treatments can provide a good description of the high-velocity, hot debris that strikes the second sheet, and the small amount of discrete, low-velocity shield debris can be, in general, ignored.

It was necessary then to develop a discription of the debris that passed through the shield in terms of momentum distribution and density–velocity profiles. To accomplish this a two-dimensional hydrodynamic code, CAMEO, developed by J. Tillotson, was utilized. This is an Eulerian code similar to that used by Walsh and Johnson (*12*). In this system a set of grids or cells is fixed in space and the material being treated flows through the cells continuously. The problems treated are normal impacts of axisymmetric particles; thus cylindrical coordinates are appropriate. The cells are of rectangular cross section and of a toroidal shape, with the cell volume becoming greater with increasing distance from the axis of symmetry. This code is described by Walsh and Johnson (*12*).

There are some limitations in this treatment. Because the material moves in a continuous manner it is not possible to follow the position of the particle–shield interface. Because of this only like-upon-like material impacts can be treated.

Second, there is a tendency to "smear" or diffuse material and average densities, velocities, and internal energies. The termination of the calculations was arbitrarily set as the time at which the momentum that had passed through the shield reached the maximum and the kinetic energy of the debris was a maximum. At such time the debris was essentially "set" with very low pressures and with the momentum distribution changing no further with time.

The CAMEO results scale directly, i.e., equal scale changes of linear dimensions and time produce the same result (*12*). It was found, however. that if the problem were initially stated with small masses, computer roundoff error would result in relatively large mass, momentum, and energy errors. To prevent this type of error the problems were computed with scaled-up projectiles and shields. Thus the aluminum cylinders and spheres used for computation had masses of about 45 g; the aluminum rods 205 g; and the cadmium spheres 146 g. With masses of these sizes the energy error was of the order of 1%.

Upon checking the CAMEO runs to minimize the error described above and to determine the effect of cell size upon the computations, it was found that the smallest cell sizes or the finest zoning reduced the magnitude of error. Coarse zoning tends to lose detail but the gross features tend to the same values. Figure 8 presents the total axial and radial momenta of the debris as a function of time for problems CAMEO 5 and 6. CAMEO 6 was a factor of three finer in zoning than was CAMEO 5; i.e., the projectile in problem 6 consisted of nine times as many cells as CAMEO 5, and the shield was nine cells thick in CAMEO 6 and three cells thick in CAMEO 5. As can be seen the momenta initially are not the same but approach the same values very rapidly, with perhaps a time lag.

The total kinetic energy, as presented in Fig. 9, also rapidly approaches the same values, with again a time lag. From these tests the conclusion was drawn that, although the coarse zoning lacked fine-time precision, the overall gross effects were essentially not affected.

On the basis of these test cases, it was decided to perform the early-time calculations with as fine zoning as possible. This decision conflicted with the restricted number of zones due to limitations imposed by the size of the available computer memory core and an auxiliary memory system could not be used with existing facilities. There were two solutions to this problem: to rezone at some stage in the computation (e.g., convert n^2 zones into n zones), or to increase the size of the zones with increasing distance from the impact area.

It was determined that, since the debris would be in a highly expanded form in the cases of interest and would be undergoing little interaction, and also that rezoning would present difficulties in maintaining both conservation of momentum and energy, it would be best to increase zone sizes with increasing distance from the impact point. A typical zoning procedure of this type is presented in Fig. 10. As can be observed, the zones are all of the same size in the area surrounding and including the impact point and, in addition, are of square cross-sectional area. At greater distances from the impact area, the zone sizes increase and become rectangular in cross section. The zoning procedure utilized in Fig. 10 was used for CAMEO 7–10. Similar schemes were utilized for all other CAMEO data computations.

Several of the CAMEO computations were compared with experiment to provide checks upon the accuracy of the computational results. The outputs from CAMEO 4 and 19 are presented in Figs. 11 and 12, along with experimental results of total axial momentum as a function of the tangent

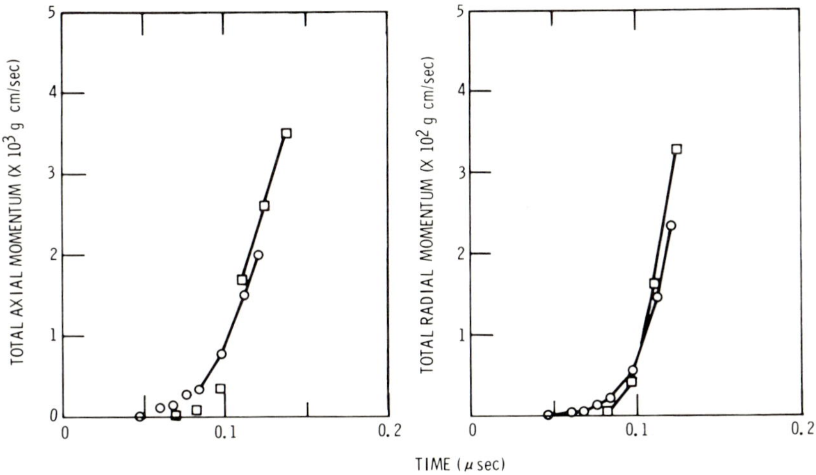

FIG. 8. Computed debris momenta versus time. (○) CAMEO 5; (□) CAMEO 6; 0.045-g Al cylinder, 7.2 km/sec, $t_s/d = 0.30$.

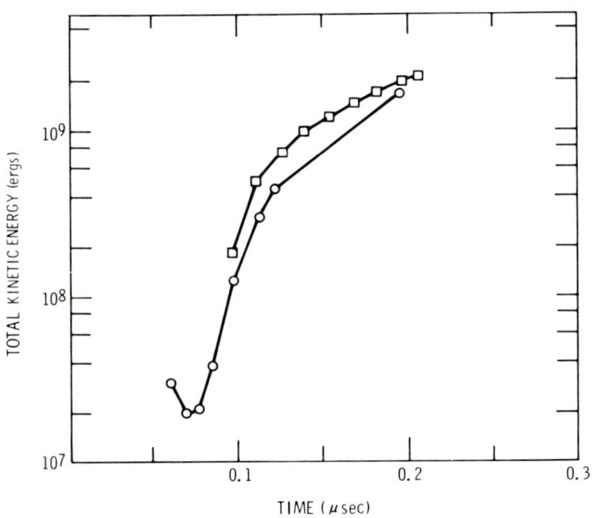

FIG. 9. Computed debris energies versus time. (○) CAMEO 5, (□) CAMEO 6; 0.045-g Al cylinder, 7.2 km/sec, $t_s/d = 0.30$.

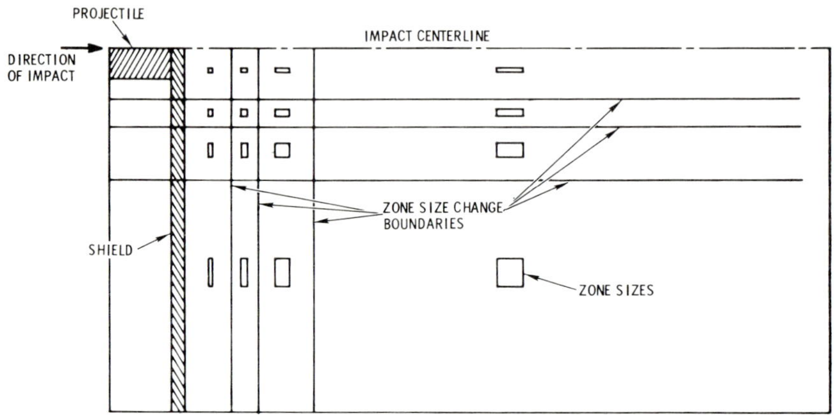

FIG. 10. Zoning procedure for CAMEO 7–10.

of the spray half-angle. The experimental points were obtained by means of a double ballistic pendulum shown in Fig. 13. The shield is mounted upon the pendulum support structure and the second sheet is mounted in two sections on the two pendulums. The debris from the projectile and shield strikes the second sheet and imparts its momentum to the two pendulums. The motions of the pendulums are monitored with small neon lights mounted upon the pendulums; the lights flash at a known constant frequency and expose the film in a stationary open-shutter camera. The record obtained thus gives a direct measure of the distance the pendulum travels versus time. A typical camera trace for a dual-pendulum test is shown in Fig. 14 along with a segmented second sheet.

It is important to note that the momentum measured utilizing the ballistic pendulum is the momentum felt by the sheet that is struck by the debris, and it is not the momentum of the debris. The CAMEO output is the momentum of the debris, and the total axial momentum is about the same as the momentum of the original impacting particle. The debris may rebound upon striking the second sheet, resulting in an increase in the momentum felt by the second sheet. If the debris were completely vaporized and underwent perfectly elastic collisions with the second sheet, it would have twice the axial momentum of the debris. Experiments performed up to 8 km/sec give momentum increases of up to about 1.5 times the impacting particle momentum.

Thus, the comparisons made in Figs. 11 and 12 required correction for the momentum multiplication effect. In Fig. 11 the CAMEO output is presented (as triangles) together with the output multiplied by the measured multiplication factor from experiment (as circles). The agree-

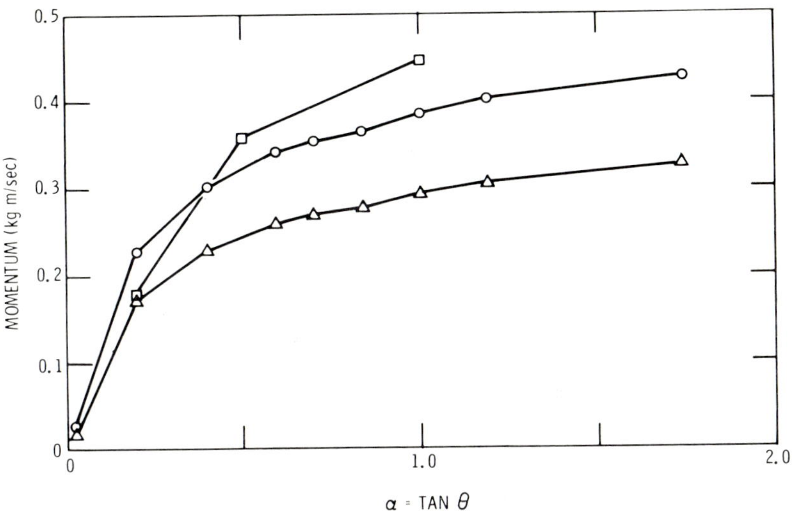

FIG. 11. Momentum versus spray half-angle (aluminum impact). (○) CAMEO 4 × 1.32, (□) experiment, (△) CAMEO 4; 3.18-mm Al sphere, 1100-O Al shield, $v = 7.2$ km/sec, $t_s/d = 0.2$.

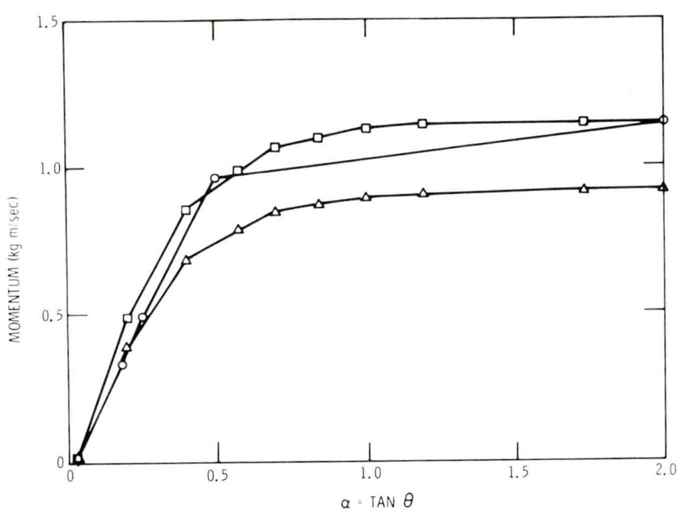

FIG. 12. Momentum versus spray half-angle (cadmium impact). (○) Experiment, $v = 5.61$ km/sec, (□) CAMEO 19 × 1.25, (△) CAMEO 19 × 1.31 μsec, 3.18-mm Cd sphere, Cd shield, $t_s/d = 0.2$.

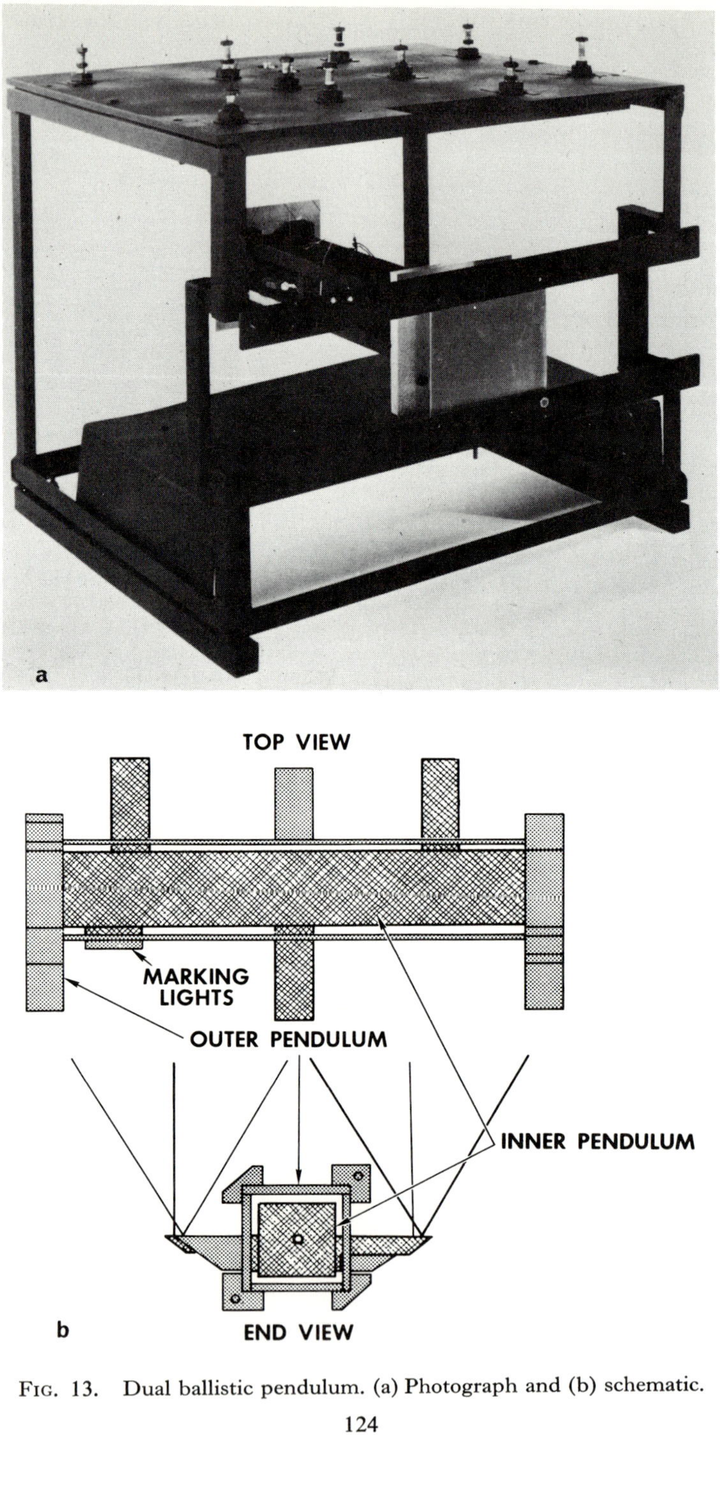

FIG. 13. Dual ballistic pendulum. (a) Photograph and (b) schematic.

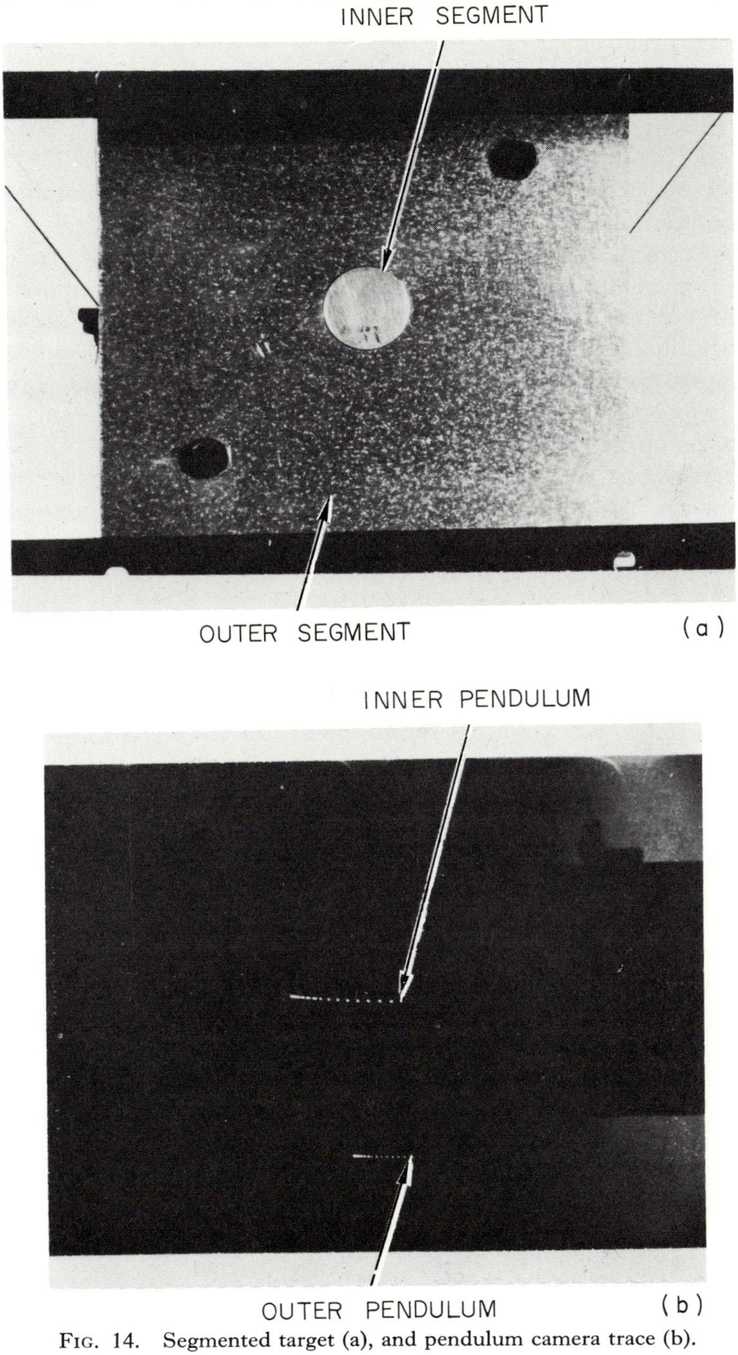

FIG. 14. Segmented target (a), and pendulum camera trace (b).

ment is fairly good, with two experimental points within 15% of the computed points and the point at the smallest angle within 25%. The smallest angle experimental point is the least well defined in terms of angle, thus it would be expected to have the largest error associated with it and would probably be low.

Figure 12 presents a comparison of momentum distribution for a cadmium–cadmium impact. Cadmium with its low melting and vapor ization points (320 and 767°C), low sonic velocity, and low strength is better represented by the CAMEO hydrodynamic calculations than aluminum at experimental velocities. For the conditions of impact, all the debris should be vaporized. In this figure, the computed momentum distribution is fitted to the experimental total momentum, thus the points at $\alpha = 2$ must agree. The points at smaller values of α agree with the computed values quite well.

A further test of the CAMEO computations was made by examining rod impacts. The method of comparison was to predict the amount of rod fractured or "used" in passing through a thin sheet. Because the hydrodynamic calculations do not include strength effects and do not provide for fracture of material, a criterion had to be established for the determination of flow and fracture. The criterion used was a "cutoff" pressure; i.e., if the material was subjected to a pressure above some value, it would be fractured. This approximation ignores the shape of the pressure profile, which would have a significant effect upon fracture. However, in the case of a rod impacting a thin sheet, it would be expected that the shock or pressure pulse would be quite sharp (short in time duration) in both the rod and the thin sheet because of the close proximity of free surfaces. The shock reflects from these free surfaces as rarefactions which rapidly overtake the shock.

Two cases were considered:

(1) The impact at 5 km/sec of a rod three times as long as its diameter upon a thin sheet 0.5 times the rod diameter in thickness.

(2) The impact at 5 km/sec of a rod ten times as long as its diameter upon a sheet 0.76 times the rod diameter in thickness.

Both the rods and the thin sheets were of 2024-T3 aluminum. The first step was to utilize the diameter of the hole formed in the $t_s/d = 0.5$ sheet to establish the "cutoff" pressure. By examining the CAMEO cells at the edge of the shield hole a "cutoff" pressure of between 40 and 55 kbar* was established. Using this pressure range to assign the amount of rod broken, the length of the rod remaining after impact was

* 1 kbar = 10^9 dyn/cm^2; 1 Mbar = 10^3 kbar = 10^{12} dyn/cm^2.

predicted from CAMEO. The ratio of the rod length remaining to the original rod length (L_r/L) was predicted to be between 0.536 and 0.584; the experimental ratio determined from flash x rays was 0.54.

Proceeding to the second impact, a ratio of hole diameter to rod diameter (D/d) of between 2.66 and 2.84 was predicted from the CAMEO output as compared to an actual value of 2.75. The predicted values of L_r/L in this case were 0.857–0.865 and the experimental ratio was 0.84.

The excellent agreement of the CAMEO-predicted values and the experimentally determined values for the rod impacts taken with the momentum comparisons establishes a high level of confidence in the use of CAMEO computations for thin-sheet impacts.

Prior to performing data computations it was necessary to establish that termination of the computations upon reaching the maximum debris axial momentum and kinetic energy would not result in a lack of definition at later times. Examination of the axial momentum distribution of the debris at various times showed little change after the termination described above.

Finally the question of the equivalence of impacts of cylinders and spheres was examined. Computations were made with aluminum spherical projectiles of the same mass as cylinders. The spherical projectiles were not true spheres but approximation to spheres made with the toroidal zones described previously. The spheres were essentially larger than the cylinders of the same mass having larger characteristic diameters.

Figure 15 shows equivalent-mass aluminum cylinder and sphere impacts in terms of the total internal energy of debris that has passed through the shield as a function of time. As can be seen, the internal energy through the shield for the cylinder is greater at any given time than for the sphere. The time behavior of the internal energies is very similar, however, with the sphere curve lagging about 0.055 μsec behind the cylinder. Figure 16 presents the debris kinetic energy for the same two impacts; the behavior is similar to that of the internal energy, again with a time lag of about 0.55 μsec. After about 0.15 μsec the ratio of internal energy to kinetic energy is the same, indicating that after this time the debris passing through the shield is in the same state.

The momentum distributions for equivalent-mass cylinder and sphere impacts at 30 km/sec (Fig. 17) show very little differences at a comparatively late time. Examination of the debris kinetic and internal energies at this time showed the kinetic energies to be within 1% and the sphere internal energy to be about 9% higher than that of the cylinder. The debris mass was also about 9% higher for the sphere, showing that

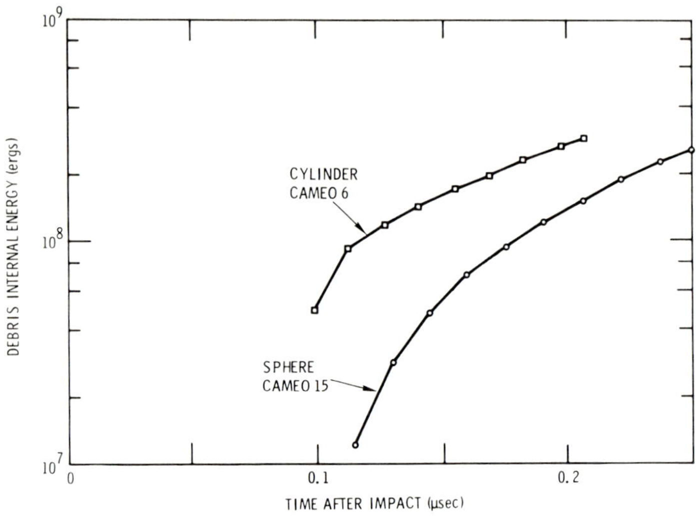

FIG. 15. Debris internal energy versus time for cylinder and sphere. Projectiles 0.045-g Al spheres impacting at 7.2 km/sec, 0.83-mm Al shield.

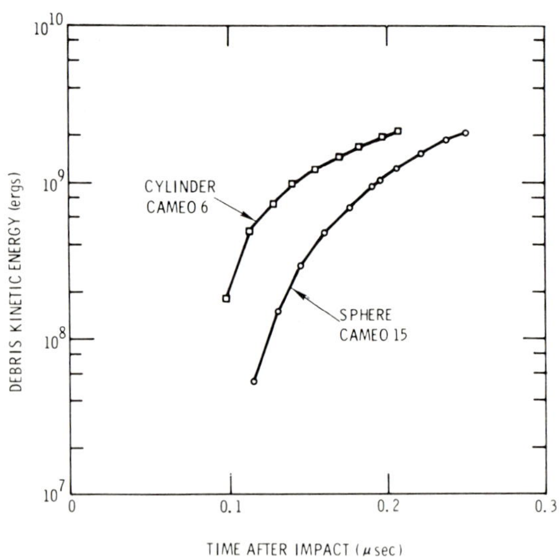

FIG. 16. Debris kinetic energy versus time for cylinder and sphere. Projectiles 0.045-g Al impacting at 7.2 km/sec, 0.83-mm Al shield.

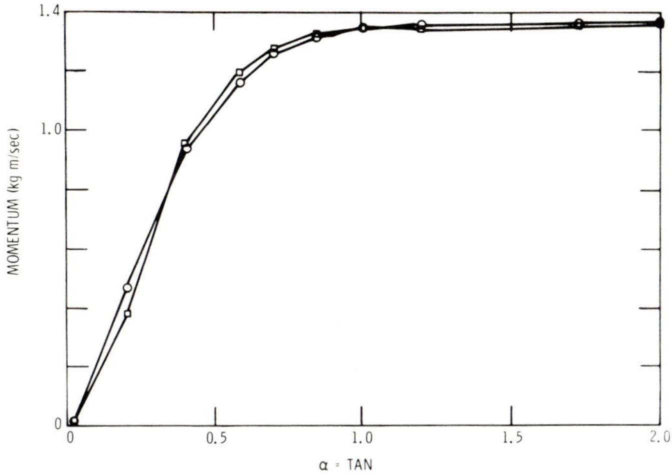

FIG. 17. Momentum versus α for 30 km/sec cylinder and sphere. (○) Sphere, $+0.414\ \mu$sec, $t_s/d = 0.20$; (□) cylinder, $+0.413\ \mu$sec, $t_s/d = 0.228$; Al shield and projectile; $v_0 = 30$ km/sec, $m = 0.045$ g.

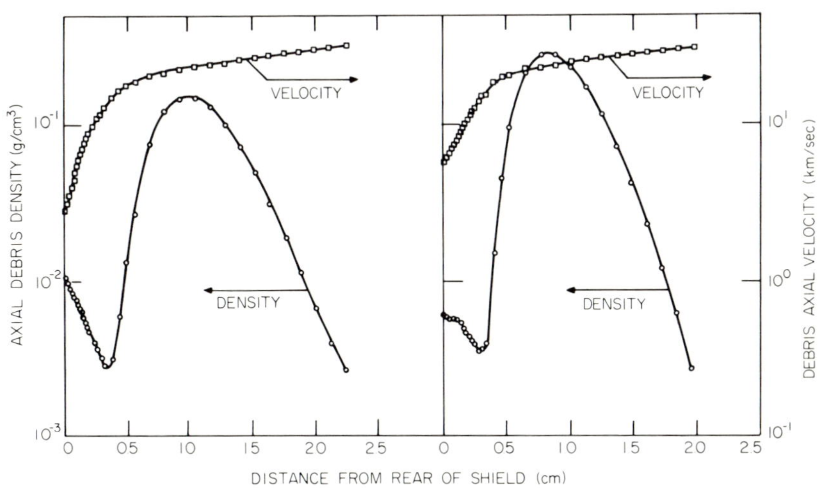

FIG. 18. Cylinder and sphere centerline debris profiles. (a) CAMEO 8, 0.045-g Al cylinder impacting at 30 km/sec, 0.635-mm Al shield, 0.485 μsec; (b) CAMEO 17, 0.045-g Al sphere impacting at 30 km/sec, 0.635-mm Al shield, 0.414 μsec.

more of the shield debris has been ejected as debris for the sphere impact but that its energy is primarily internal. The centerline debris densities and velocities for these impacts are shown in Fig. 18. Because a computer printout is not made for every cycle of computation, the times and distance of travel relative to the rear surface of the shield are not the same. Note that the density and velocity structure is essentially the same for the main body of the debris. When the debris from the sphere impact reaches the same point as the cylinder impact debris, the peak density is about the same due to further expansion of the debris. (This conclusion is reached by examination of later printouts and interpolation.) Since the momentum will control the damage to the second sheet, there seemed to be no preference between the use of the sphere or the cylinder.

Three categories of data runs were performed: (1) aluminum cylinder impacting the same shield at varying velocities; (2) aluminum cylinders impacting shields of various thicknesses at a constant velocity; and (3) cadmium spheres impacting cadmium shields of various thicknesses at a constant velocity.

The results of the first category show that the momentum tends to be concentrated within larger half-angles as the impact velocity is increased through 22.5 km/sec. At velocities higher than this for the given shield thickness, the momentum again becomes more concentrated near the centerline. With increasing impact velocities the axial velocity of the debris is closely related to the impact velocity, while the radial velocity is not. Thus the debris becomes more concentrated at small half-angles. In none of the cases considered does any appreciable amount of momentum fall outside $\alpha = 1$, or 45°.

The second category of impacts considered was aluminum–aluminum impacts at 30 km/sec upon shields varying in thickness from 0.111 times the projectile diameter to the full diameter of the projectile. It was found that the total amount of momentum increases as the shield thickness is increased with an enhancement of about 44% for the thickest shield. This increase in debris momentum is due to ejection of material from the front face of the shield. The amount of this ejecta is related to the shield thickness or hole size in the shield. As the shield is increased in thickness the total axial momentum increases and the amount of momentum at small angles decreases. The overall effect is to reduce the momentum load per unit area. The momentum load at $t_s/d = 0.5$ is almost uniform to $\alpha = 0.7$, and for $t_s/d = 1.0$ the load is more concentrated at $\alpha = 0.7$ than it is at the center of the loaded area. Once again almost all of the momentum is within $\alpha = 1.0$ ar a 45° half-angle.

Figure 19 shows the centerline densities and velocities of the

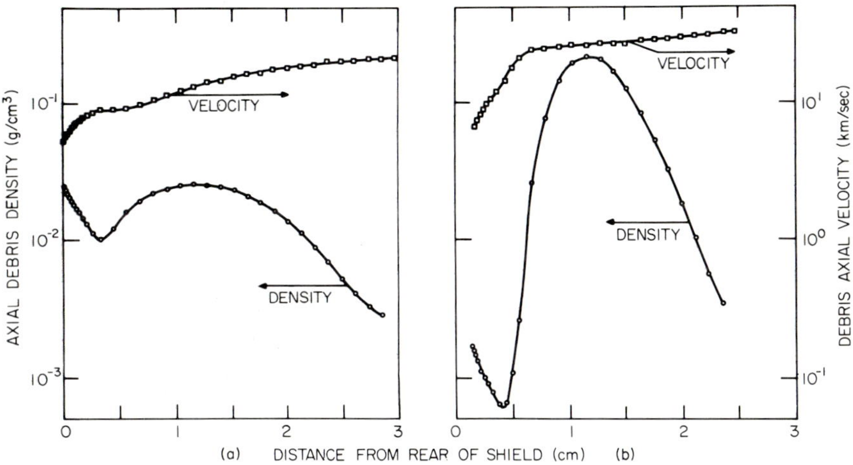

FIG. 19. Centerline debris profiles for two-shield thicknesses. (a) CAMEO 14, 0.045-g Al cylinder impacting at 30 km/sec, $t_s/d = 1.0$, 1.11 μsec; (b) CAMEO 11, 0.045-g Al cylinder impacting at 30 km/sec, $t_s/d = 0.111$, 0.498 μsec.

$t_s/d = 0.111$ and $t_s/d = 1.0$ impacts. Even allowing for further expansion of the debris from the thinner-shield impact, the peak density will be almost a factor of ten greater than that from the thick-shield debris. The total result will be a more intense, shorter-duration pressure pulse generated in the second sheet by the debris from the thin shield.

Finally a group of cadmium–cadmium impacts at 6 km/sec were considered. Again the shield thicknesses were varied. The axial momentum distributions from these computations are qualitatively the same as for the high-velocity aluminum impacts. As the shield is increased in thickness the momentum is more effectively dispersed to larger half-angles.

With these computations, consideration will be given to the reaction of the second sheet of a thin-sheet structure. The results of the aluminum–aluminum impacts were employed to evaluate the resistance of structures to meteoroid impact. As was said earlier, the expected meteoroid is of the order of 0.5 g/cm³. Impacts of low-density plastic and aluminum projectiles have been considered by Walsh and Johnson (*12*) and Wagner *et al.* (*13*). These results show much greater heating of the projectile for underdense projectiles, and lower pressures throughout. The aluminum–aluminum impacts that have been considered herein show all the debris to be vaporized at impact velocities of 30 km/sec and greater. One computation for the impact of an underdense projectile was performed. In this computation the projectile was taken to be expanded aluminum vapor with a density of 0.483 g/cm³, or 0.179 times the normal aluminum density. The impact was at 30 km/sec upon a shield 0.635 mm thick. The projectile had the same mass as the normal-density aluminum cylinders discussed earlier; only the projectile density was altered. The first point of difference in the results is the maximum pressure of the impact, with the underdense impact showing a maximum pressure of about 2 Mbar as opposed to a pressure of about 10 Mbar for the normal-density impact. After about 0.4 μsec the kinetic energy of the debris is comparable with the low-density impact, having about 90% the kinetic energy of the normal-density impact. The internal energy, however, is more than three times as great in the debris of the low-density impact. A comparison of the momentum distributions at this time is shown in Fig. 20. Even though

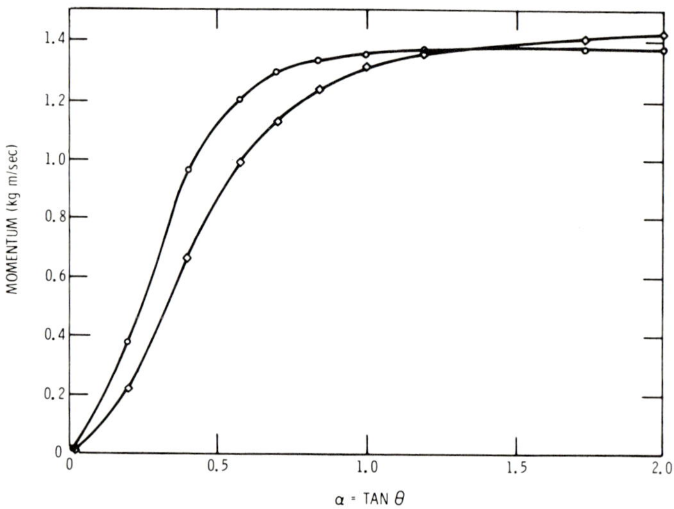

FIG. 20. Momentum versus α for 30-km/sec aluminum impacts. (○) $\rho_0 = 2.70$ g/cm³, 0.413 μsec; (◇) $\rho_0 = 0.483$ g/cm³, 0.423 μsec; 0.045-g Al cylinders impacting 0.635-mm Al shield at 30 km/sec.

the debris from the low-density impact contains about 1.4 times as much mass as the normal-density debris, the total axial momentum is about the same. This means that the axial velocity of the debris from the low-density impact is less than for the normal-density debris. In both cases the debris would be, for the most part, vaporized.

B. Gross-Deformation Failure of the Second Sheet

A high-speed framing camera sequence of the failure of a thin-sheet target is shown in Fig. 21. The second sheet is given a velocity increment and the sheet starts to deform very soon after the load is applied. For the case shown, a tensile failure has occurred after a matter of a few tens of microseconds.

Figure 22 shows a flash x-ray of a second-sheet failure. It can be seen that the central portion has failed through tension and several cracks have started to propagate in the sheet. Note that the sheet has failed through tension and no spall has formed.

To treat this problem theoretically, such factors as the magnitude, distribution, and duration of the load applied to the target must first be determined. Then the large-deformation dynamic response of the target has to be calculated, taking into account the elastic and plastic behavior of the target.

A description of the distribution of the load is available from the hydrodynamic calculations. These distributions have been utilized to determine the target response; the treatment used will be described chronologically in order to describe it as clearly as possible.

The following assumptions were initially made:

(1) The load is uniformly distributed over a circular area with a diameter equal to one-half the spacing.

(2) The load is applied so quickly that the loaded area is effectively given an initial velocity increment.

(3) The momentum transferred to the loaded area is equal to twice the momentum of the original particle.

The main justifications for assumptions (1) and (2) come from Fig. 21, whereas a momentum multiplication factor of 2, assumption (3), has been chosen for the following reasons: Since the impulsive load applied to a target increases with increasing impact velocity, the inner hull for a spacecraft must be designed to resist a meteoroid at the maximum expected velocity of 30 km/sec. At this velocity, most of the debris coming through the shield will be in gaseous form. Hence, if it is assumed that the momentum of the debris is equal to the momentum of the original particle and that perfectly elastic collisions occur between each gas atom and the target, then the momentum multiplication factor should be two.

Given the loading, the next part of the problem is to determine the response of the target (assumed to be a thin shell). However, the problem of determining dynamic deformations and stresses in thin shells involves,

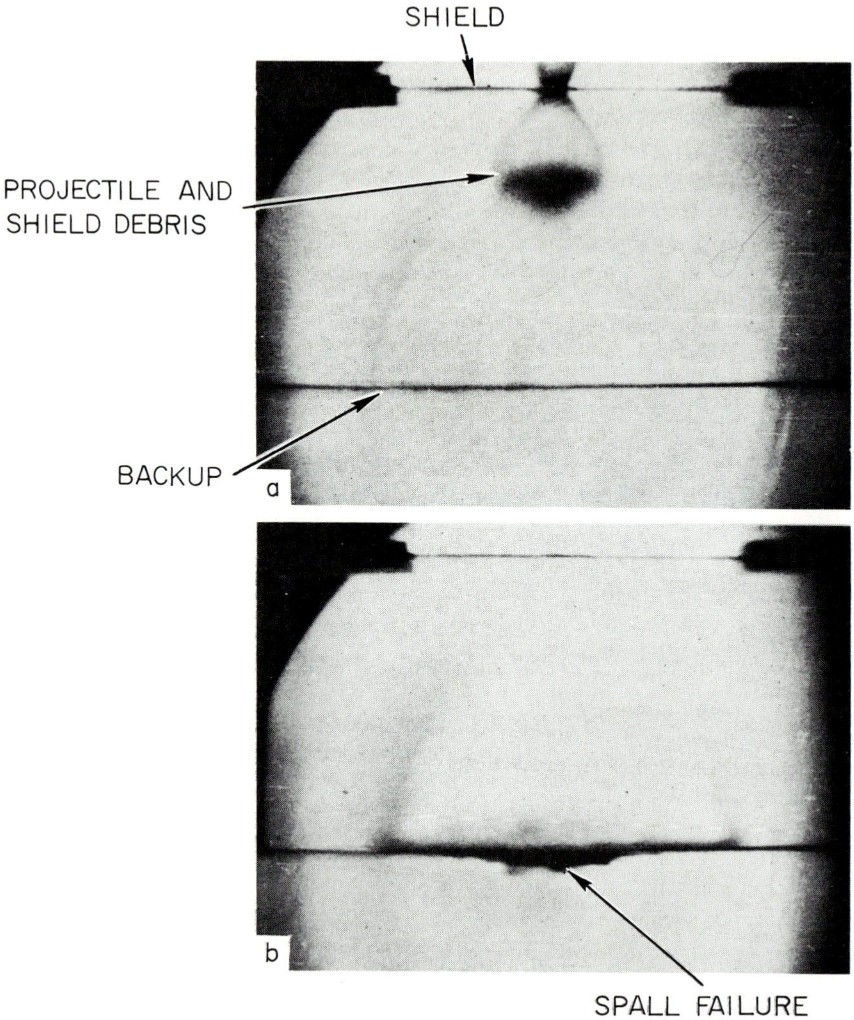

FIG. 21. Framing camera sequence of thin sheet failure: (a) +3.0 sec; (b) +11 sec; (c) +27 sec; (d) +66 sec. Projectile 3.2-mm Al sphere impacting a 0.53-mm 1100-O Al shield at 8.07 km/sec, 0.81-mm 7075-T6 Al backup, 5.08-cm spacing.

in general, a complex system of nonlinear differential equations. For these problems that involve both large deflections and plasticity effects, a numerical technique has been developed by Witmer *et al.* (*14*). This technique is based on a finite difference approximation for the original differential equations. These finite difference equations are then used

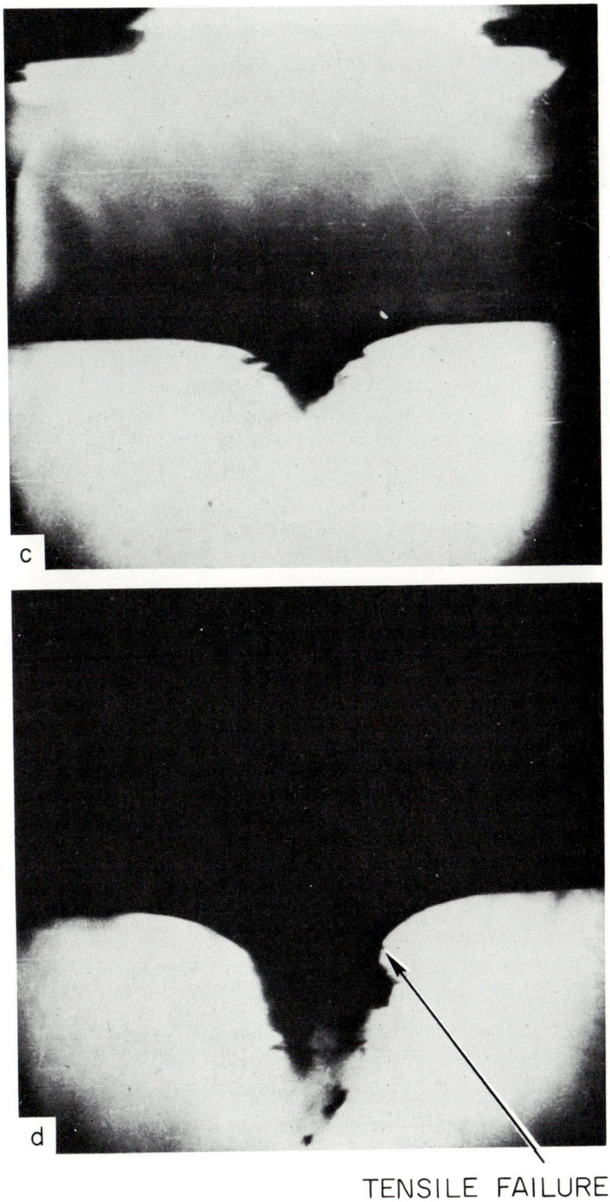

FIG. 21c and d.

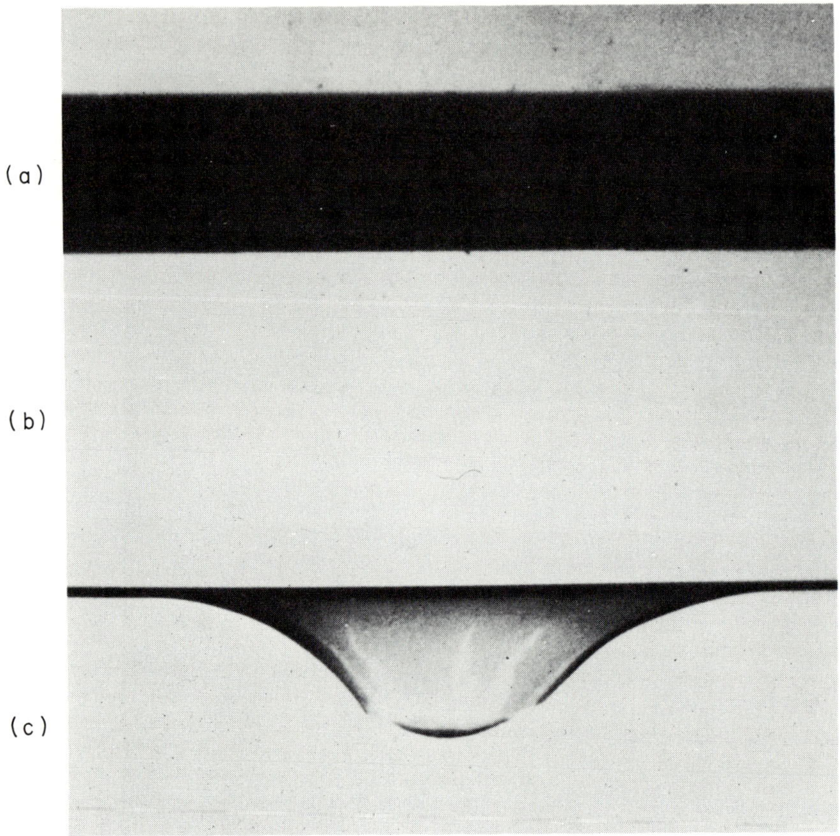

FIG. 22. X-ray picture of thin sheet failure. Projectile 2.56-mm Cd sphere impacting a (a) 0.48-mm Cd shield at 6.49 km/sec, 5.08-cm spacing; (b) 0.46-mm 7075–T6 Al backup 55 μsec after shield impact.

to describe an equivalent lumped-parameter model. For the timewise step-by-step numerical analysis, the increments in stress resultants and stress couples are determined by idealizing the shell thickness as consisting of n concentrated layers (six layers were used in all the present calculations). Also, the material behavior used to determine the above increments can include elastic–perfectly plastic, elastic–strain-hardening, or elastic–strain-hardening strain-rate-sensitive behavior.

The success of the above technique has been shown in Witmer's paper (*14*) in a comparison of theory and experiment for explosively loaded beams.

Figure 23 shows the situation that was first investigated to determine

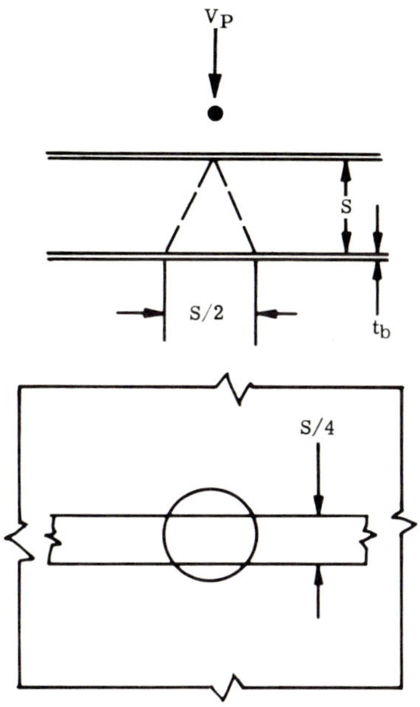

FIG. 23. Strip approximation.

the motion and stress in the backup sheet (*14*). In this approximation a strip (or beam) of material through the center of the loaded area was considered. The rear sheet material was taken as 7075-T6 aluminum. This material was assumed to behave in an elastic–perfectly plastic manner with a yield strength σ_0 of 4.67×10^9 dyn/cm² (70,000 lb/in.²) and a percentage elongation to fracture of 11%. The first step is to calculate the initial velocity v_i imparted to the center portion of the strip. From the previous assumptions this is given by

$$v_i = 32M_pV_p/\pi S^2\rho_bt_b \tag{8}$$

where M_p and V_p are the mass and velocity of the impacting particle, S is the spacing, and ρ_b and t_b are the density and thickness, respectively, of the backup target. For the initial calculation, the following parameters were used: $M_p = 0.045$ g and $t_b = 0.8$ mm.

The results of the calculation are shown in Figs. 24 and 25. The centerline deflection, the tensile strain at the edge of the loaded area, and the tensile strain at the center of the load are shown in these figures. Figure 24 includes the experimental curve of centerline deflection determined from the framing-camera photos in Fig. 21, indicating good

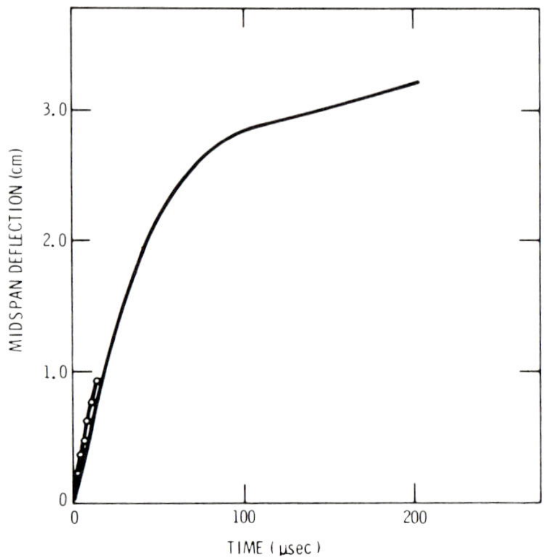

FIG. 24. Centerline displacement versus time. (—) Elastic–perfectly plastic, (○) experiment; 3.2-mm Al sphere at 7.62 km/sec, 5.08-cm spacing, 0.8-mm 7075-T6 Al backup, momentum multiplication = 2.

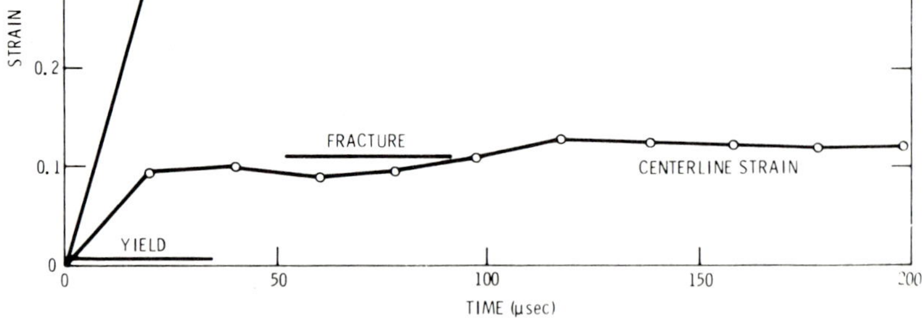

FIG. 25. Centerline and edge strain versus time. Projectile 3.2-mm Al sphere at 7.62 km/sec, 5.08-cm spacing, 0.8-mm 7075-T6 Al backup, momentum multiplication = 2.

agreement between theory and experiment. Note that in Fig. 25 the fracture strain is first reached at the edge of the loaded area after about 7 μsec; after this time the solution is academic. There is also some evidence of spalling in Fig. 21, and such initial wave effects have been ignored for the moment.

The above calculations were repeated for a backup sheet 1.6 mm thick. The results showed that the peak strain occurs at the center of the sheet, and that the sheet should not fracture. Experiment showed that only a small perforation occurs, and this is due to spallation.

With the above reasonable agreement between theory and experiment, it was decided to use the strip approximation further. Because solutions are required for an effectively infinite rear sheet, it was considered necessary to find out how long a strip is effectively infinite for the time of interest in the present problem. To do this, the first analysis (25.4-cm span) was repeated for a 50.8-cm span, with the result that up to about 100 μsec no difference was found in centerline deflection, edge, or center strain for the two cases. Since in most situations of interest either the maximum or the fracture strain would occur in less than 100 μsec, it was decided that a 50.8-cm span would correspond to an effectively infinite sheet and would be used in all subsequent calculations.

The strip approximation was used to determine the response of backup sheets of various thicknesses for impacts of 3.2-mm particles at velocities of 7.62, 15.2, 22.8, and 30.4 km/sec. The results at 7.62 km/sec (Figs. 26–28) show that there is a large difference between backup-sheet thicknesses required for no-yield failure (maximum strain less than 0.7%) and no-fracture failure (maximum strain less than 11%). Also, it can be observed that for sheet thicknesses above about 1.6 mm the maximum strain occurs at the edge of the loaded area. Similar solutions to Figs. 26–28 have been obtained for the other three impact velocities.

The resulting curves of sheet thickness against impact velocity for failure criteria of yield, maximum strain of 2, 4, and 6%, and fracture are presented in Fig. 29. The surprising aspect of this figure is that for each failure criterion the curve of t_b against velocity is nearly linear. This is surprising in view of the fact that the basic mechnisms are nonlinear. It is also seen from this figure that, above about 4% maximum strain at a given velocity, a small decrease in thickness produces a large strain increment. Thus, it would appear wise to design for no more than a few percent maximum strain.

Figure 30 shows results similar to those of Fig. 29, but for 1.02-mm aluminum spheres. The same comments can be made concerning the results for this particle that were made for the 3.2-mm particle.

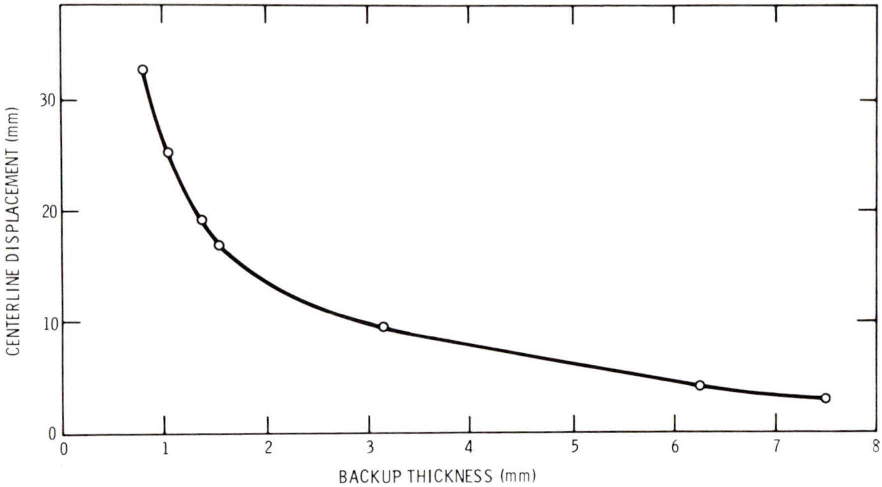

FIG. 26. Centerline displacement versus thickness at 200 μsec. Projectile 3.2-mm Al sphere at 7.63 km/sec, 5.08-cm spacing, 7075-T6 Al backup, momentum multiplication = 2.

Note that the crowding of the constant maximum-strain lines is seen to increase with decreasing particle size.

From Figs. 29 and 30, the backup-sheet thicknesses for the yield and fracture criteria can be plotted against particle diameter at a number of velocities. When this is done it is found that, for either criterion,

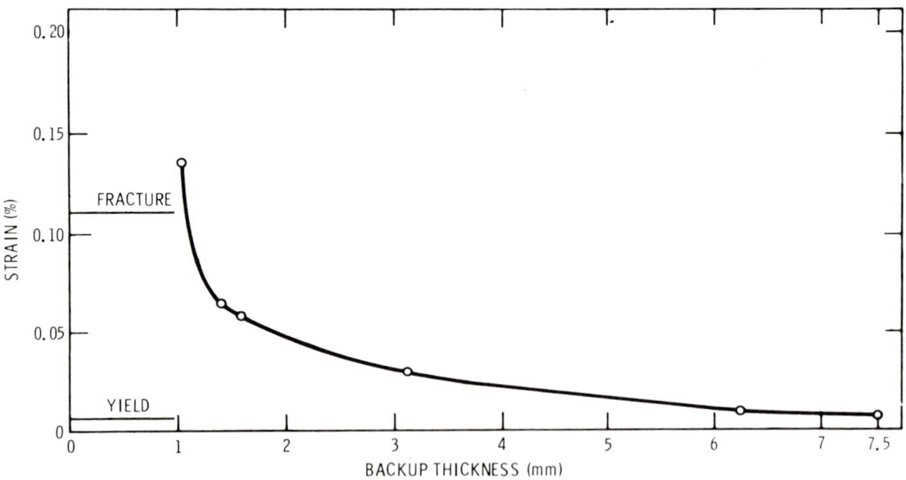

FIG. 27. Maximum centerline strain thickness. Projectile 3.2-mm Al sphere at 7.62 km/sec, 5.08-cm spacing, 7075-T6 Al backup, momentum multiplication = 2.

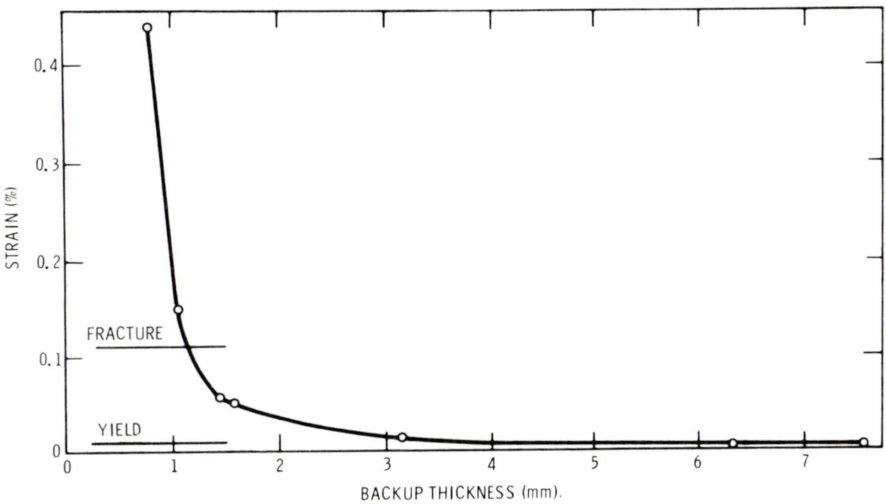

FIG. 28. Maximum edge strain versus thickness. Projectile 3.2-mm Al sphere at 7.62 km/sec, 5.08-cm spacing, 7075-T6 Al backup, momentum multiplication = 2.

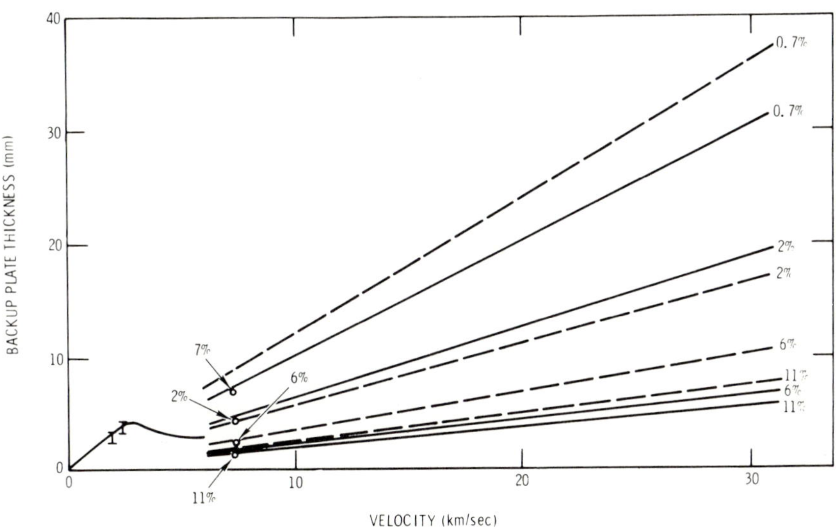

FIG. 29. Backup thickness versus velocity. (—) Strip approx, (– –) plate approx, (○) plate approx measured distribution; 3.2-mm Al sphere, 5.08-cm spacing, 7075-T6 Al backup, momentum multiplication = 2.

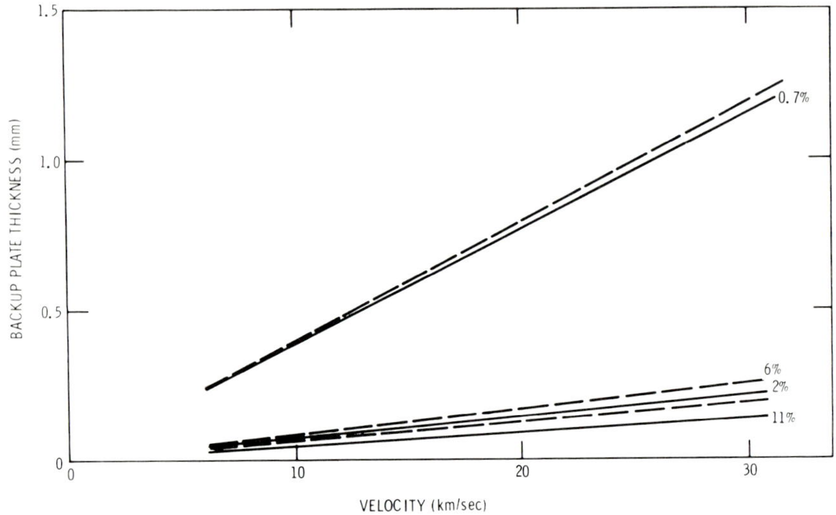

FIG. 30. Backup thickness versus velocity. (—) Strip approx, (– –) plate approx, 1.02-mm Al sphere, 5.08-cm spacing, 7075-T6 Al backup, momentum multiplication = 2.

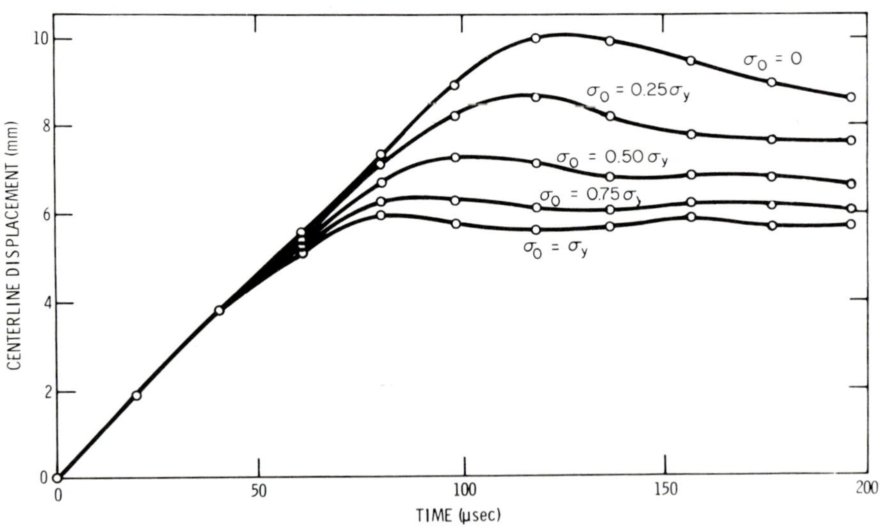

FIG. 31. Centerline displacement versus time–pretensioned beam: projectile 1.02-mm Al sphere at 30.4 km/sec, 7075-T6 Al backup.

t_b is approximately proportional to the cube of the particle diameter. Thus, under these conditons, Eq. (8) indicates that v_i, the initial velocity of the loaded area, is roughly constant.

Spacing of 2.54, 5.08, and 10.16 cm were investigated for an Apollo-scale particle (1.02-mm diam) at 30.4 km/sec. The results for both the yield and fracture criteria show that the sheet thickness necessary for either decreases approximately with the inverse square of spacing. This result, plus those of the preceding sections, gives rise to an approximate equation for rear-sheet thickness. For 7075-T6 aluminum this equation is given by

$$t_b = C(M_p V_p / S^2) \tag{9}$$

where $C = 415 \pm 140$ and 82 ± 14 for the yield and fracture criteria, respectively, t_b is in millimeters, M_p is in grams, V_p is in kilometers per second, and S is in centimeters. [Note: because the exact solutions do not require v_i in Eq. (8) to be exactly constant, there is not a constant value for C at a specific maximum strain.]

In many space applications, the structure to be protected will be a pressurized fuel tank; hence, it was decided to investigate the effect of pre-tensioning the rear sheet (in one direction only). The computer code for the strip approximation can accommodate such pre-tensioning, and solutions were obtained for a 1.04-mm particle at 30.4 km/sec with pre-tensioning of 25, 50, 75, and 100% of the static yield stress. The centerline displacements against time are shown in Fig. 31.

These results reveal that the thickness required for the yield criterion is not very sensitive to pre-tension, the value of the yield stress being less than 10% greater than that at zero (see Fig. 32). The thickness based on the fracture criterion is sensitive to the amount of pre-tension increasing by more than 60% of the yield stress.

The strip approximation has been shown to be in good agreement with experimental results; however, the complete solution should consider a centrally loaded circular plate. The numerical technique of Witmer *et al.* (*14*) has been applied to the plate analysis and the results agree well with those from the strip approximation. Such comparisons are shown in Figs. 29 and 30. Both the strip and plate treatments are more fully described by Maiden and McMillan (*15*) and McMillan (*16*).

Using the plate analysis, the strain versus rear-sheet thickness was studied for a number of shield thicknesses. The hydrodynamic momentum distributions were used to determine the thickness of the rear sheet required to prevent fracture.

When combined, the shield thickness plus the backup-sheet thickness define the total structure thickness. The required total thickness

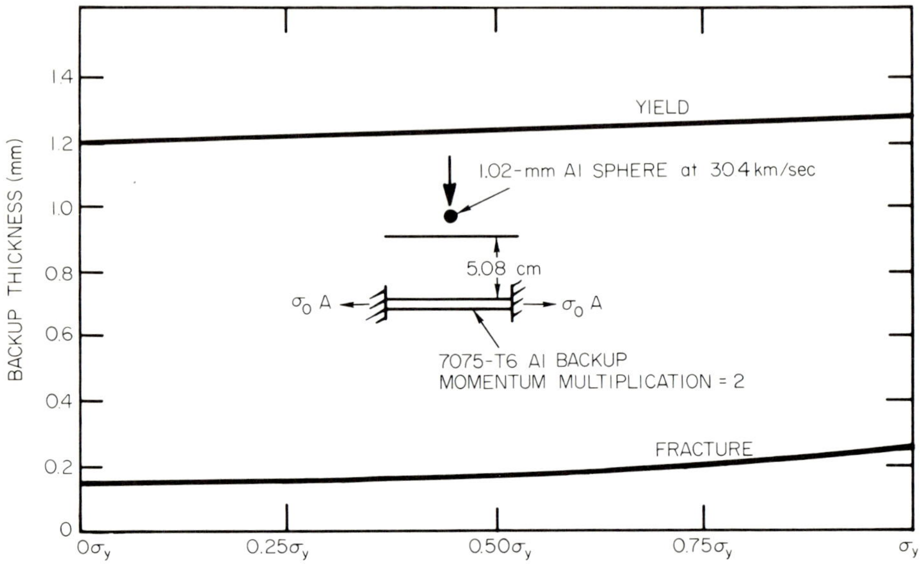

FIG. 32. Backup thickness versus pre-tension stress.

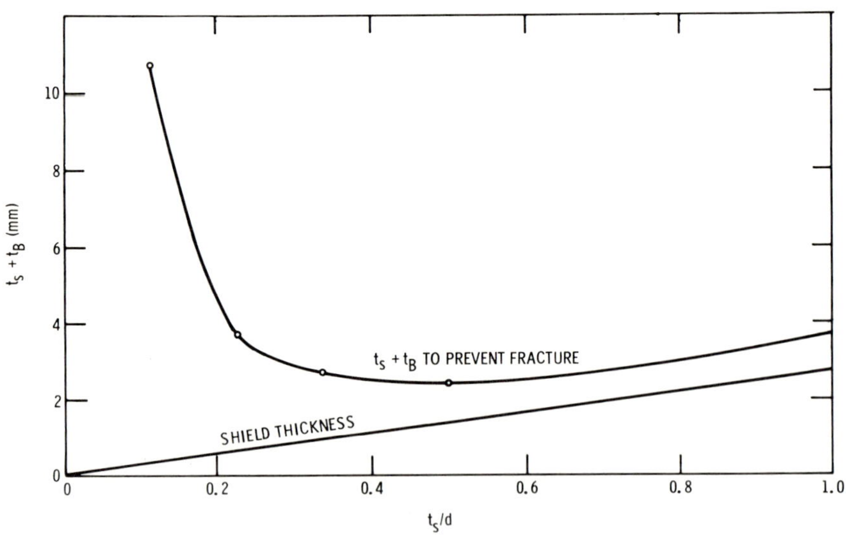

FIG. 33. Total thickness versus t_s/d. Momentum distributions from CAMEO, 0.045-g Al projectiles impacting Al shields at 30 km/sec, 7075-T6 Al backup, 5.08-cm spacing.

$(t_s + t_b)$ can be plotted as a function of the ratio of shield thickness to projectile diameter (t_s/d) as in Fig. 33.

The next problem considered was whether the plate analysis results could be scaled with projectile mass and intersheet spacing. If not, it would be necessary to compute each case of interest. The strain versus thickness was computed for a number of impact conditions. For these comparisons the "scaled input" points were obtained by scaling the input momentum distribution to the plate analysis proportional to the impacting particle mass. The original computational results were scaled directly with the particle mass to obtain the "scaled output" points. It was found that there is practically no difference between these results, leading to the conclusion that the backup thickness for particular t_s/d and spacing is directly proportional to the impacting projectile mass.

The effect of spacing with the CAMEO momentum distributions was considered in the same manner as the effect of projectile mass. The original results were scaled as one over the square of the spacing. Again there was essentially no difference between the "scaled input" and "scaled output" results, with the conclusion that the results can be scaled as backup thickness proportional to one over the square of the spacing. The combination of these results leads to the backup thickness for a sparticular t_s/d being directly proportional to the impacting particle mass divided by the square of the spacing.

The above results are based on the assumption that the debris that strikes the second sheet is in a liquid or vapor form with a density much less than that of the impacting particle or shield. This will not be the situation if the spacing between the sheets is small compared to the particle size. This corresponds to the CAMEO calculations for the early-time flow before the debris is "set." For the debris in this form, cratering of the second sheet will take place. Second, it has been assumed that the velocity vectors of the debris all pass through the impact centerline exactly at the rear surface of the shield. This is not the case in the CAMEO results, where the debris appears to be coming from along the centerline within the shield or particle. Thus the debris is not spreading quite as rapidly as has been assumed. The results as presented should not be applied to large particles impacting thin shields with short spacings to the second sheet (less than eight particle diameters) or to very large spacings.

Using the scaling results, it is possible to express the thickness to prevent fracture for impacts at 30 km/sec as a function of the shield thickness, particle size, and spacing as

$$t_t = \frac{M_p}{0.045}\left(\frac{5.08}{S}\right)^2\left[\frac{0.0102}{(t_s/d)^2} + 0.079\right] \tag{10}$$

where M_p is the particle mass in grams, S the spacing in centimeters, t_s the shield thickness in centimeters, and t_f the second-sheet fracture thickness in centimeters.

Equation (10) shows that the dominant factors in determining the thickness to prevent fracture are the projectile mass and the spacing, with the shield thickness to projectile diameter ratio becoming important only if it is small. Use of the equation is restricted to 30-km/sec impacts in which the spacing is fairly large (greater than eight projectile diameters, the debris is underdense and molten or vaporized, and the shield thickness ratio is in the range of about 0.1 to 1.0.

This result applies to aluminum-on-aluminum impacts with the second-sheet material of 7075-T6 aluminum. The formula can be used to evaluate the resistance to fracture of the second sheet of a two-sheet structure with the t_s/d ratio computed on the basis of the equivalent-mass aluminum particle. It would be unwise to design a two-sheet structure on the basis that it just resist fracture, because of the sensitivity of the maximum strain near the fracture point.

The combination of the CAMEO hydrodynamic computations and the plate analysis were applied to cadmium–cadmium impacts to provide the basis for an experimental check. In these computations a momentum multiplication factor of 1.4 was used.

This value was determined from experiment. Four shield thicknesses were computed for an impact velocity of 6 km/sec. The fracture thicknesses versus shield-thickness ratios are shown in Fig. 34, together with

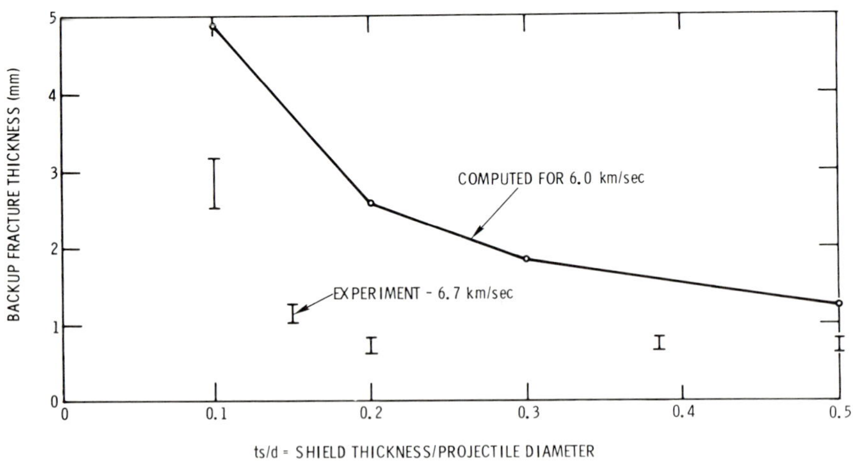

FIG. 34. Backup thickness versus t_s/d. Projectiles 3.18-mm Cd spheres impacting Cd shields, 7075-T6 Al backups, 5.08-cm spacing.

experimentally determined fracture thicknesses. In all cases the fracture thicknesses found experimentally are much less than those predicted.

It is felt that this discrepancy is probably due to the way in which the debris transfers its momentum to the second sheet. A momentum transfer ratio of 1.4 has been observed experimentally and it was assumed that the debris delivered 1.4 times its momentum to the second sheet independent of spray half-angle. The debris is more concentrated at small spray half-angles and there is an indication, from framing camera records (Fig. 35), that the leading edge of the debris reflects after striking the second sheet. In this figure the reflected front can be seen after the debris has struck the second plate. This material travels back to the shield and some of it passes through the hole and overtakes the debris ejected from the shield, causing light to be given off (the last two frames). The passage of the front of the debris back through the later debris would be expected to reduce the momentum transfer. Since the interaction would be greatest where the debris is most dense, the small spray half-angle debris would be affected the most. The reaction of the loaded plate is very dependent upon the small spray half-angle load, and this interaction is probably the cause of the low experimental values. Stated another way, the momentum multiplication factor is probably a function of spray half-angle.

A number of scaling experiments were also performed with cadmium–cadmium impacts. When the projectile mass was reduced by one-half with a shield thickness ratio of 0.2, the fracture thickness was reduced from 0.64–0.81 mm to 0.46–0.64 mm. The results of varying the fracture thickness with spacing are given in Table III. These results are in good agreement with the mass and spacing scaling predictions.

TABLE III

Cadmium–Cadmium Impacts at 6.7 km/sec

t_s/d	t_f (mm) (5.08-cm spacing)	t_f (mm) (10.2-cm spacing)	t_f (mm) (2.54-cm spacing)
0.1	2.28–3.18	0.41–0.64	—
0.2	0.64–0.81	0.25–0.64	—
0.385	0.64–0.81	0.25–0.38	2.54–3.18

C. Spall Failure

The impact of the projectile–shield debris upon the second sheet generates a shock wave in the second sheet. This wave upon reflection can cause fracture or spall of the second sheet. Analysis of this problem

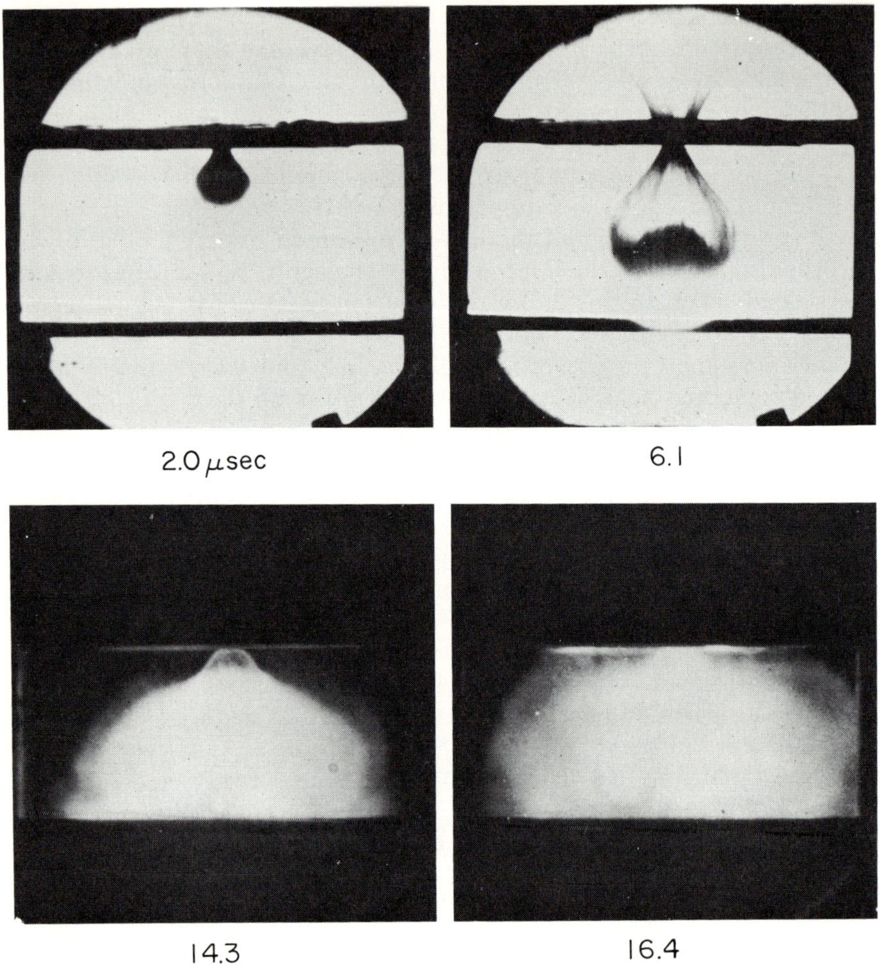

FIG. 35. Framing camera sequence of cadmium impact. Projectile 3.18-mm Cd sphere im-

requires consideration of the strength of the second sheet. This problem has been treated by use of the two-dimensional elastic–plastic analysis described by McMillan (*16*) and Jones and Kriech (*17*). The debris distributions from the CAMEO computations were utilized as the input to the spall calculations. The second-sheet treatment included the effects of strength, while the impacting cloud of debris was treated as purely hydrodynamic.

The results of CAMEO were used to determine if the gross-deformation failure thickness determined above would spall. The CAMEO run

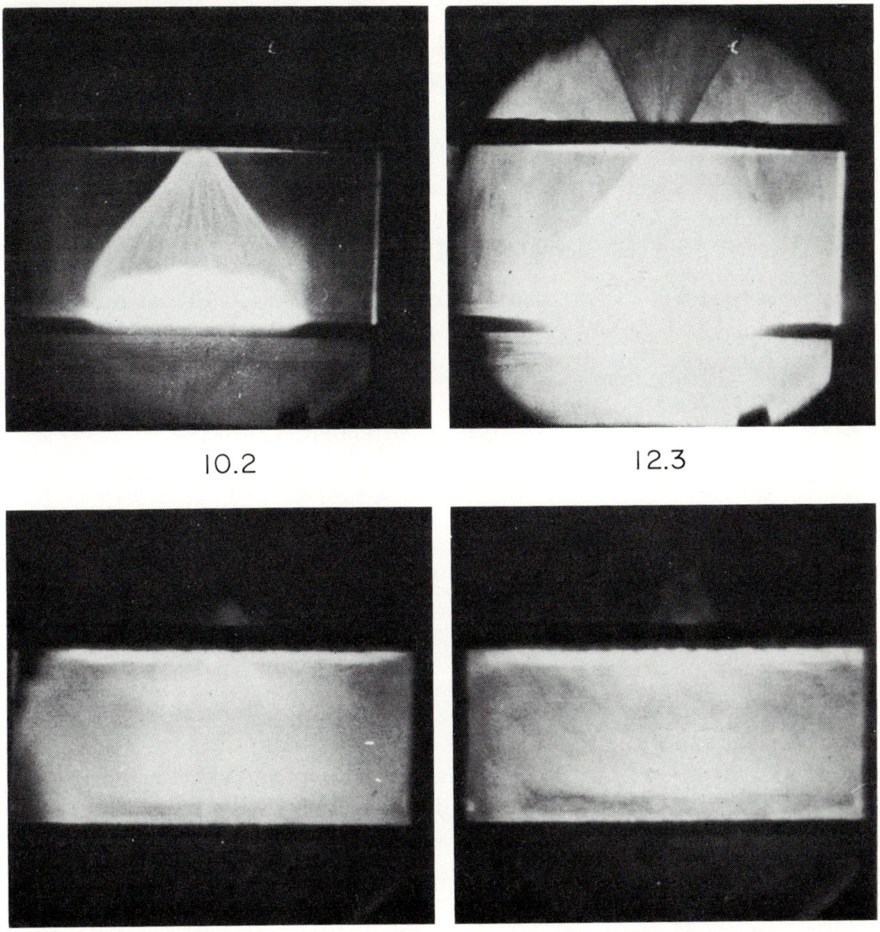

pacting a 0.33-mm Cd shield at 6.49 km/sec, 4.75-mm 7075-T6 Al backup, 5.08-cm spacing.

considered the impact of a 0.045-g aluminum projectile upon an aluminum sheet 0.635 mm thick. A second sheet of 7075-T6 aluminum is spaced 5.08 cm from the first sheet or shield. The second-sheet thickness is 3.18 mm. This computation showed a rise to a compressive stress of about 70 kbar in the second sheet. The pulse duration was very long and multiple reflections occurred within the sheet. The maximum tensile stress was found to be about 7 kbar. The 7075-T6 aluminum would spall with a tensile stress of about 13 kbar.

These results show that the shape and duration of the pressure pulse applied to the sheet have overriding importance in determining whether the sheet will spall. This pulse is determined by the debris cloud that strikes the plate. From the computation it is expected that for this impact.

(1) Less spacing will cause spall; greater spacing will not.
(2) Larger projectile masses will cause spall; less mass will not.
(3) Thinner (t_s/d) shields will cause spall; thicker shields will not.
(4) Higher projectile densities will cause spall; lower densities will not.

Spall here refers to fracture of the material. This does not mean that material will be detached from the surface, but only that there will be an internal fracture. This provides a conservative estimate of spall damage.

D. Interaction of Debris with a Shielded Target—Experimental Results

Experimental studies in support of the theoretical studies described in the previous section have been performed with a light-gas gun at impact velocities up to 8 km/sec. This section describes some of the results of these experiments. Raw data from all the experiments are given by A. R. McMillan (*16*).

1. *Momentum Multiplication*

Extensive experiments utilizing a ballistic pendulum were carried out to measure the momentum transfer to the second sheet as a function of impact parameters. Initial experiments involved the measurement of momentum applied to the backup while varying the bumper thickness and the impact velocity in aluminum–aluminum impacts. The results of these experiments are summarized in Fig. 35. The thin bumpers (0.305 mm, 0.635 mm, and 1.02 mm) gave very similar results. The ratio of the measured momentum applied to the backup sheet divided by the incident projectile momentum increased with velocity up to about 5 km/sec, then remained constant throughout the rest of the experimental range. The thicker shield, $t_s/d = 0.5$, exhibited a continued increase in the momentum multiplication ratio throughout the experimental range. Since it has been argued that the momentum multplication ratio for the thin-sheet impact case has an upper bound of 2.0, and semi-infinite impact has no upper bound (*12*), it is to be expected that the momentum multiplication ratio should show an increase for the thicker targets.

The momentum multplication values presented in Fig. 36 cover the range of solid-debris impact through completely melted debris. There appears to be no sharp change in behavior in this transition, with perhaps

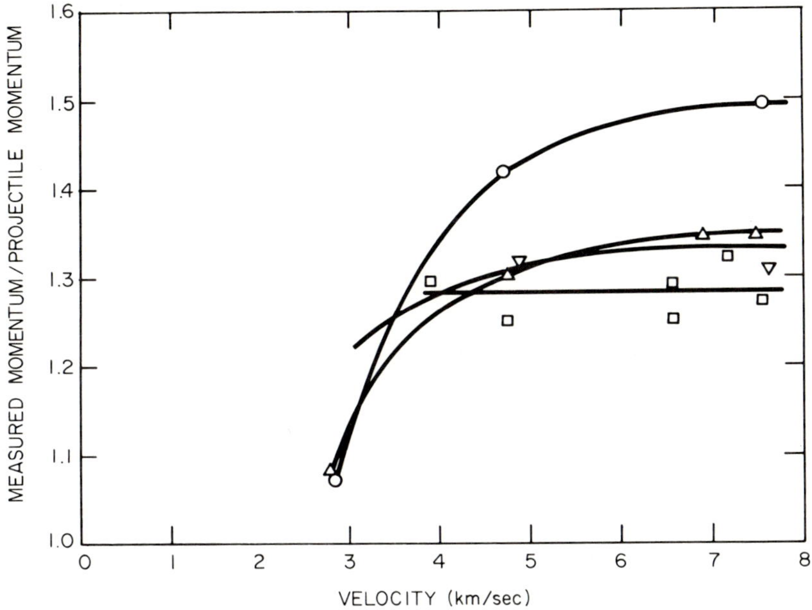

FIG. 36. Momentum multiplication versus velocity. (○) $t_s/d = 0.5$, (△) = 0.3, (□) = 0.2, (▽) = 0.1; 3.2-mm Al spheres, 1100-O shields, 7075-T6 backup, 5.08-cm spacing.

the exception of a change of slope at about 5 km/sec corresponding to the onset of melting in aluminum–aluminum impact.

Experiments were performed in which the second sheet was of 1100-O aluminum in place of the 7075-T6 aluminum used in all other texts. No difference in momentum multiplication was observed.

As a test of the effect of spacing upon the momentum multiplication ratio, experiments were conducted with spacings less than the 5.08-cm spacing of the previously described experiments. The values obtained in these experiments were identical, within the $\pm 4\%$ error of the measurements, to the values obtained with the 5.08-cm spacing. On the basis of the experiments it was concluded that spacing has no effect on momentum transfer for spacings greater than eight projectile diameters.

The value of the momentum multiplication factor attained for the aluminum-aluminum thin-sheet impact case was much less in the case of the three thinnest shields than the value of 2.0 postulated as the upper bound. The aluminum–aluminum impact represented a liquid-debris impact and it was felt that perhaps the impact of vapor debris would more closely approach the value of 2.0. Experiments were performed with cadmium and lead projectiles and shields in which the debris striking the backup sheet was in vapor form. The results of these experiments are shown in Table IV. As can be seen, the values attained are again much less than 2.0.

The experiments performed with 0.64-mm cadmium shields represented a transition from liquid debris striking the backup (the test at 3.18 km/sec) to vapor debris. The value of the momentum multiplication ratio reached a value of 1.4 at 5.38 km/sec and remained unchanged throughout the experimental range. The experiments performed with the 0.33-mm cadmium shields covered the transition from solid debris to liquid debris. Again the highest value attained is

TABLE IV

CADMIUM–CADMIUM AND LEAD–LEAD IMPACTS

Shot no.	Shield thickness (mm)	Impact velocity (km/sec)	Momentum multiplication
3.18-mm cadmium spheres, 5.08-cm spacing, cadmium shield:			
D-1230	0.33	3.60	1.26
1453	0.33	5.18	1.28
1019	0.33	5.61	1.34
1889	0.33	6.83	1.42
1884	0.48	7.01	1.39
1046	0.64	3.18	1.21
1324	0.64	5.38	1.41
1326	0.64	5.70	1.40
1327	0.64	5.76	1.38
1045	0.64	6.40	1.41
1454	0.64	6.40	1.39
1580	1.22	3.69	1.39
1581	1.22	5.18	1.48
1888	1.22	6.89	1.52
1883	1.22	7.01	1.48
2.57-mm cadmium spheres, 5.08-cm spacing, cadmium shield:			
1891	0.48	4.91	1.31
1890	0.48	6.89	1.39
3.18-mm lead spheres, 5.08-cm spacing, lead shield:			
1820	1.78	4.51	1.43
1819	1.78	4.85	1.48
1822	0.89	4.18	1.24
1821	0.89	4.36	1.24

far less than 2.0. The 1.22-mm cadmium shields and the lead tests represent vapor-debris impacts, and again the momentum ratio is much less than 2.0.

2. *Low-Density Projectiles*

Four sets of experiments were performed using low-density and underdense materials for projectiles. The first series of tests was performed using Inlyte as the projectile material. This material is composed of very small hollow glass spheres held together in a plastic matrix. The projectiles used had a bulk density of 0.73 g/cm³. The values of momentum multiplication obtained for impacts against aluminum shields (Table V) are about the same as observed for aluminum–aluminum impacts. Examination of the backup sheets showed fragment damage (probably from shield fragments) and no evidence of melting of the aluminum.

TABLE V

LOW-DENSITY IMPACTS

Shot no.	Projectile	Shield thickness (mm)	Impact velocity (km/sec)	Momentum multiplication
D-1391	4.90-mm Inlyte (0.73 g/cm³)	0.64	4.69	1.20
1394		0.64	7.91	1.24
1393		0.305	4.68	1.25
1575	4.19-mm nylon (1.18 g/cm³)	0.64	4.11	1.45
1561		0.64	7.50	1.52
1349		1.02	6.31	1.55
1371		1.02	7.13	1.51
1653	2.62-mm foam Cu (4.88 g/cm³)	0.64	5.99	1.36
1344		1.02	3.29	1.46
1345		1.02	5.15	1.24
1652		1.02	6.61	1.39
1650	2.62-mm foam Ni (4.88 g/cm³)	0.64	6.50	1.43
1346		1.02	5.03	1.35

The second series utilized foamed copper and nickel as the projectile material. These foamed metals have approximately one-half the density of the fully dense materials. With the projectiles, the momentum ratios were about the same as for aluminum–aluminum impacts. The backup targets showed some evidence of melted aluminum from the shield and some cratering. The foam material with its many internal surfaces would not be expected to transmit the shock very well, and incomplete projectile breakup results with a fairly thin sheet. With the 1.02-mm shield this was less evident than with the 0.64-mm shield.

Nylon projectiles were used in the last series of tests. The nylon is fully dense with a density of 1.18 g/cm^3. In contrast to the underdense Inlyte, copper, and nickel, the momentum ratio was quite high (Table V). The second sheets showed damage similar to that observed with the Inlyte projectiles with many small craters apparently caused by shield fragments. Here there was no evidence of melted aluminum. Because of the similarity in the damage to the second sheet for the nylon and Inlyte impacts, it must be concluded that the differing values of momentum transfer must be due to some property of the projectile.

Throughout all the momentum measurements no value of momentum multiplication over 1.55 was observed for impacts with intersheet spacings of eight projectile diameters or greater. For intersheet spacings of four projectile diameters two values of momentum multiplication of about 1.7 were observed. In no case was a momentum multiplication ratio near 2.0 observed. The lack of attainment of momentum multiplication ratios near the limit of two can be accounted for by a probable lack of perfectly elastic vapor impact on the second sheet and internal interaction of the debris cloud as it strikes the second sheet.

IV. Discussion

The findings of the analysis described in this chapter suggest the following procedure to design a simple spacecraft structure consisting of a meteor bumper and a main hull. Two damage mechanisms must be considered. The first mechanism is complete penetration of the structure by solid particulate debris and is more severe at low impact velocities. This type of damage can be avoided by calculating the optimum bumper for the lowest expected meteoroid velocity. This calculation can be made most easily if the meteoroid and bumper material are assumed the same; hence the properties of the meteoroid have to be estimated. This calculated bumper thickness can then be converted, on an equiweight basis, to the thickness required for any other more suitable bumper material. Such a design procedure will insure that the debris impacting the main hull will be extremely small and will have little penetrating power. At higher impact velocities, this bumper will be even more effective in melting or vaporizing the meteoroid and shield debris.

The second damage mechanism is impulsive failure of the main hull due to the collective impact of the debris that passes through the bumper. This latter type of failure includes both spalling and petaling of the rear sheet or main hull.

An attempt has been made to critically review the theoretical approaches to the thin-sheet impact problems and to verify these approaches by means of laboratory experiments. The analysis and the results presented treated only the case of a simple two-sheet structure. No attempt was made to review the use of fillers betweeen the sheets multiple shields, honycombs, oblique impact, and many other unique designer problem areas. These items are treated in Chapter IX of this treatise. Also in Chapter IX an engineering approach to the design of a two-sheet structure is described in brief detail such that a quick approach to the problem areas mentioned above can be included in the design analysis of a space vehicle hull.

Aknowledgments

The author wishes to express his gratitude to his many co-workers at the General Motors Defense Research Laboratories who have, over the recent years, contributed to the research programs summarized in this chapter. Particular appreciation is extended to A. R. McMillan, C. J. Maiden, B. L. Lathrop, and R. E. Sennett. Much of the work leading to this manuscript was supported by NASA, The Manned Spacecraft Center in Houston, Texas, under the direction and with the expert advice of Mr. Burton Cour-Palais.

REFERENCES

1. A. E. Olshaker and R. L. Bjork, Hydrodynamics applied to hypervelocity impact—II. The role of melting and vaporization in hypervelocity impact, *Proc. Symp. Hypervelocity Impact, 5th, Colorado School of Mines, Colorado, 1961* **1**, Part 1, 225–239 (1961).
2. R. G. McQueen and S. P. Marsh, Equations of state for nineteen metallic elements from shock wave measurements to two megabars, *J. Appl. Phys.* **31**, 1253 (1960).
3. J. M. Walsh, M. H. Rice, R. G. McQueen, and F. L. Yarger, Shock wave compressions of twenty-seven metals. Equations of state of metals, *Phys. Rev.* **108**, 196 (1957).
4. L. V. Al'tshuler, K. K. Krupnikov, and M. I. Brazhnik, Dynamic compressibility of metals under pressures from 400,000 to 4,000,000 atmospheres, *Soviet Phys. JETP (English Transl.)* **34**, 614 (1958).
5. L. V. Al'tshuler, S. B. Kormer, A. A. Bakanova, and R. F. Trunin, Equation of state for aluminum, copper, and lead in the high pressure region, *Soviet Phys. JETP (English Transl.)* **11**, 573 (1960).
6. J. M. Walsh and J. H. Tillotson, Hydrodynamics of hypervelocity impact, *Proc. Symp. Hypervelocity Impact, 6th, Firestone Tire and Rubber Co., Akron, Ohio*, **2**, 59 (1963).

7. L. V. Al'tshuler, S. B. Kormer, M. I. Brazhnik, L. A. Vladimirov, M. P. Speranskaya, and A. I. Funtikov, The isentropic compressibility of aluminum, copper, lead, and iron at high pressures, *Soviet Phys. JETP (English Transl.)* **11**, 766 (1960).
8. G. R. Fowles, Attenuation of the shock wave produced in a solid by a flying plate, *J. Appl. Phys.* **31**, 655 (1960).
9. P. Sandorff, A meteoroid bumper design criterion, *Symp. Hypervelocity Impact, 6th*, **3**, AD-423-063 (1963).
10. R. L. Bjork, Review of physical processes in hypervelocity impact and penetration, *Symp. Hypervelocity Impact, 6th*, **2**, 1, AD-423-063 (Aug. 1963).
11. T. D. Riney and J. F. Heyda, Hypervelocity impact calculations, *Symp. Hypervelocity Impact, 7th*, **2**, 77, AD-463-228 (Feb. 1965).
12. J. M. Walsh and W. E. Johnson, On the theory of hypervelocity impact, *Symp. Hypervelocity Impact, 7th*, **2**, 1, AD-463-228 (Feb. 1965).
13. M. H. Wagner, N. B. Brooks, and R. L. Bjork, Impact of a porous aluminum projectile on aluminum at 20 and 72 km/sec, *Symp. Hypervelocity Impact, 7th*, **3**, 1, AD-463-229 (Feb. 1965).
14. E. A. Witmer, H. A. Balmer, J. W. Leech, and T. H. H. Pian, Large dynamic deformation of beams, rings, plates and shells, *AIAA J.* **1**, No. 8, 1848 (1963).
15. C. J. Maiden and A. R. McMillan, An investigation of the protection afforted a spacecraft by a thin shield, *AIAA J.* **2**, No. 11, 1992 (1964).
16. A. R. McMillan, Experimental investigations of simultated meteoroid damage to various spacecraft structures, GM TR66-67 (Nov. 1966).
17. A. H. Jones and C. E. Kriech, TEDDY—A two-dimensional Lagrangian code for determining the wave propagation in elastic–plastic media, GM TR66-85 (Dec. 1966).

CHAPTER V

*NUMERICAL EVALUATION OF HYPERVELOCITY IMPACT PHENOMENA**

T. D. RINEY

SPACE SCIENCES LABORATORY, GENERAL ELECTRIC COMPANY
KING OF PRUSSIA, PENNSYLVANIA†

* This work, performed in the period from June 1963 to August 1967, was directed in part by the Ballistics Division of AFATL under Contract No. AF 08(635)-3781, and in part by the Airbreathing Engines Division of the NASA Lewis Research Center under Contract No. NAS 3-8512.

† Present address: Systems, Science and Software, La Jolla, California.

I. Introduction

Extensive use has been made in recent years of numerical techniques and high-speed computers for solving complex equations governing models of material response to intense impulsive loading. One application is the study of hypervelocity impact phenomena. In this paper the choice of an appropriate numerical scheme for the various phases of the impact deformation process is discussed, but no detailed description of the numerical techniques is given. Instead, the bulk of the paper is devoted to summarizing the numerical results that have been obtained at this laboratory using computer programs based on these schemes.

Impact calculations are presented for three classes of axisymmetric projectile–target configurations:

1. Five problems consider thin-plate targets composed of two layers of distinct materials to determine the effect of the material discontinuity on the breakup of the projectile and the subsequent spreading of the debris in the spallation cone.
2. Eight problems treat the impact of hollow cylindrical and hollow spherical projectiles onto thin plates and examine the mechanisms of shock wave focusing, jet formation, and dispersal of the resulting debris cone.
3. Seven problems consider thick targets impacted by a low-density projectile simulating a meteoroid projectile to determine the effect of porosity on the final crater dimensions.

Calculations for each of these classes of problems have previously been described in more detail in separate publications (*1–3*).

II. Regimes of Material Response

A. Separation of Response Regimes

In the case of impact into thick targets the distinction is often made between an early stage in which the materials respond as compressible fluids and a late stage when strength effects cannot be ignored. The decomposition into separate regimes is more apparent, however, for impact into finite plate targets, e.g., the meteor bumper system depicted in Fig. 1. This concept of using a thin outer shell to break up the projectile and spread the debris before it reaches the main structure was first suggested by Whipple (*4*).

Upon impact, shocks are produced in the projectile and bumper materials which propagate away from the moving contact surface,

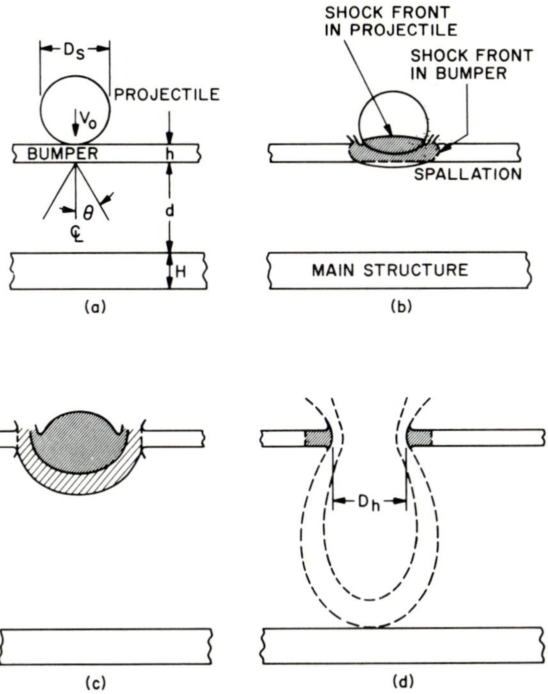

FIG. 1. Schematics depicting (a) the impact onto a thin bumper plate, (b) its penetration, (c) the subsequent formation of a spallation cone, and (d) the loading transmitted by the cone to the main structure downstream.

Fig. 1b. The pressure pulse contained between the shock fronts is relieved from the outset by rarefaction waves originating from the periphery of the contact surface, inducing a backsplash of ejected material. When the shock front reaches the opposite free surface of the bumper, the pressure pulse is relieved, causing a rapid expansion of the compressed material accompanied by a decrease in the internal energy and a corresponding increase in the kinetic energy. The central region of the projectile–bumper configuration is torn loose from the bumper, Fig. 1c, and no longer interacts with the perforated plate.

This disengagement of the spallation debris from the bumper allows the subsequent deformation of the bumper and the dispersion of the debris downrange to be treated separately. The response of the main target to the loading produced by the debris represents yet another regime that may be treated separately.

The interaction of the projectile with the bumper, during which the projectile and central region of the plate are processed by shock and rarefaction waves, is adequately described by a hydrodynamic model provided the shock processing to which the material has been subjected leaves it in a molten or vapor state.

At a given instant and at a specified distance behind the bumper, Fig. 1d, a unit volume of the spallation cone would, under conditions for which the spallation debris is fluid, contain radial momentum and axial momentum, given by

$$M_r = \rho u, \qquad M_z = \rho v \tag{1}$$

where ρ is the local mass density and u and v are the corresponding radial and axial velocity components of the debris. The radial momentum serves to spread the debris as it moves downrange; only the axial component of momentum contributes to the loading imposed on the main structure by the impinging debris, Fig. 1d.

The axial momentum density M_z represents the impulse that would be delivered by a unit volume of debris to a surface located at the specified position in space and time. The axial momentum flux,

$$M_f = vM_z$$

or

$$\begin{aligned} M_f &\equiv M_f(r, z, t) \\ &= \rho v^2 \end{aligned} \tag{2}$$

represents the momentum delivered to a unit area of the surface in unit time.

The loading imposed on the main structure by the spallation cone, though very large, acts for a very short time. A shock wave is produced at the surface of the main structure which propagates across its thickness. Provided the shock transmitted through the structure is large enough, there will be a spall failure at its opposite free surface. This type of failure, observed by Maiden *et al.* (*5*), can also be accompanied by cracking or rupturing of the main structure. When the standoff distance is large the intensity of the loading imposed on the main structure may be less than that required for a pressure-pulse-induced spall failure. In this case the second plate may still fail due to the gross deformation which the pressure loading of the debris produces in the structure. Petaling failure has also been observed (*5*), and it may be the predominant mode of failure in bumper-structure systems with reasonable spacing subjected to meteoroid velocities.

The radially propagating shock wave in the perforated bumper is rapidly attenuated by rarefaction waves issuing from the lateral free surfaces of the bumper. There is a transition from the hydrodynamic behavior in the central region of the projectile–bumper configuration to a regime in which the pressure has attenuated to the point that the strength of the bumper becomes effective and the outward plastic flow of the bumper material stops, thus forming the final hole size. The final hole diameter is found to increase approximately linearly with impact velocity and an empirical relation has been presented to Maiden and McMillan (*6*) relating the final hole diameter D_H to the impact parameters,

$$D_H/D_S = 0.45v_0(h_1/D_S)^{2/3} + 0.90 \tag{3}$$

where D_S is the projectile diameter and h_1 is the bumper thickness.

No numerical treatment of the late stage deformation of the penetrated bumper has been obtained, but Heyda (*7*) has presented an effective analytic description of the stress wave propagation problem through use of the method of characteristics. The stresses and particle velocities in both the elastic and elastic–plastic zones of the penetrated plate are readily calculated by this method using Tresca's yield criterion and the assumption of plane stress.

B. Numerical Methods

When a particular choice for the constitutive relation governing the continuum response of a material to intense impulsive loading is incorporated into the conservation relations for mass, momentum, and energy, a system of partial differential equations is obtained. A solution to these equations for realistic geometries and boundary conditions cannot be obtained without resorting to large-scale computer codes. Of many possible choices of numerical schemes, three basic procedures have been applied to problems of this nature. In schemes based on a Lagrangian description, a coordinate system or grid is embedded into the configuration to be studied. The deformation and flow of the projectile–target configuration is then monotored with reference to this deformed grid-work. In Eulerian numerical schemes the coordinate system is not embedded into the configuration but, instead, is fixed in space, and the calculation follows the material that happens to be in a given computational cell at that particular time. A third type of numerical method is the particle-in-cell scheme which is a combination insofar as both the Eulerian and Lagrangian coordinate systems are utilized in each cycle of the calculation. The Lagrangian representation, however, is approximate in the particle-in-cell scheme since it consists of discrete mass particles whose positions are monitored rather than a grid whose deformation is monitored. Each of these marker particles carries its share of cellwise mass, energy, and momentum for that time cycle as it moves through the fixed mesh.

Each of these numerical schemes has advantages and disadvantages. The Lagrangian scheme treats material interfaces and free surfaces in a straightforward fashion and also permits the use of constitutive relations for the material in which the stress history of each piece of material is taken into account. However, it is sensitive to the distortions in the flow field being studied and instabilities are introduced if the embedded cells are deformed too severely. A Lagrangian description would be inappropriate for the study of the spallation cone formation and dispersal since the materials are subjected to severe distortions, but would be appropriate for a study of the response of the main structure to the loading imposed by the spallation debris.

In the case of Eulerian schemes large distortions cause no problem, but it is difficult to accurately treat more than a single material and free surfaces cannot be treated without resorting to artificial constraints on the flow.* This is a consequence of the fact that material entering a cell of the fixed Eulerian mesh is considered to be immediately diffused uniformly throughout the cell volume. The capability of treating unlike materials and special projectile geometries with varied free surfaces is required for many of the problems considered in the sequel. Moreover, the expansion of a spallation cone envelope would be affected by the artificial constraints imposed to reduce the false diffusion inherent in Eulerian schemes. An Eulerian treatment is a more appropriate choice for cratering studies for thick targets.

The combined Eulerian and Lagrangian coordinate systems used in the particle-in-cell scheme readily permits the treatment of the large distortions and multiple materials. The representation of the materials by the mass particles, whose positions are monitored at each state of the calculation, allows free surfaces to be located and avoids the false diffusion problem. Although the particle-in-cell scheme does not permit the stress history of individual pieces of the material to be taken into account, this is of no importance in the treatment of the spallation cone kinetics which are governed by the compressible fluid model. For these reasons a particle-in-cell code is a logical choice for the study of the spallation cone kinetics in meteor bumper problems.

* Significant improvements in the early Eulerian schemes have been made by introduction of tracer particles to monitor free surfaces and material interfaces as they move through the fixed grid. This Lagrangian flavor has been used in the five-material, hydrodynamic and elastic–plastic HELP code, developed since this paper was written, by L. J. Hageman and J. M. Walsh at Systems, Science and Software.

At the Space Sciences Laboratory, computer codes based on Lagrangian, Eulerian, and particle-in-cell schemes are all utilized in order to best treat different phases of the impact process. The nomenclature for these codes is given in Table I. These codes were developed

TABLE 1

NOMENCLATURE FOR TWO-DIMENSIONAL, TIME-DEPENDENT COMPUTER CODES DISCUSSED IN THE TEXT

VISTA	Third generation particle-in-cell code (successor to PICWICK I and PICWICK II)
PICWICK III	Eulerian code
CRAM	Lagrangian continuum code
DEPROSS	Lagrangian structural code
VISTA–CRAM	Linked codes for studies of meteor bumper systems in which spallation failure of second plate is anticipated
VISTA–DEPROSS	Linked codes for meteor bumper studies in which petaling of main structure is expected failure mode
VISTA–PICWICK III	Linked codes used for thick-target studies

at the Space Sciences Laboratory except for DEPROSS which was developed by Balmer (*8*), at the Massachusetts Institute of Technology, for the study of dynamic behavior of structures. The VISTA and CRAM codes are based on numerical schemes originated by Harlow (*9*) and Wilkins (*10*), respectively. The only essential difference between the Eulerian scheme used in PICWICK III and the particle-in-cell scheme is in the treatment of the mass transport phase of each time cycle. PICWICK III employs an approximation of the continuity equation to determine the diffusion of mass, momentum, and energy across the fixed cell boundaries. It is similar to the OIL code developed independently by Johnson (*11*).

The VISTA code, used for the applications described in the following sections, provides a history of the density and velocity of the particles in the spallation cone behind the bumper, Fig. 1d. This information can be saved on magnetic tape and then read in as initial data defining the loading conditions that the debris subsequently imposes on the main structure. The linking of the particle-in-cell and Lagrangian numerical codes in this manner has been successfully accomplished (*12*). If the loading intensity is large enough that wave propagation effects are important the VISTA–CRAM linking would be required for the treatment of the main structure. For less intense loading conditions, where petaling failure rather than spallation is the anticipated failure mode, the main structure may be considered incompressible and the VISTA–DEPROSS linking may be employed.

Another use of linked computer codes for hypervelocity impact investigations is made in Section VI. A VISTA–PICWICK III combination is used to predict final crater size in a thick aluminum target impacted by reduced-density projectiles simulating stony meteoroids. The particle-in-cell scheme provides an appropriate choice for the early stages of the process where it is important to adequately account for material interfaces and free surfaces. The Eulerian scheme is more appropriate for later calculations where the discrete mass representation of the particle-in-cell scheme does not provide sufficient resolution and the projectile has delivered its momentum and energy to the target and, consequently, its material characteristics no longer play an important role.

III. Governing Equations

The simplifying assumption that the impacted materials behave like inviscid compressible fluids, valid during the early stages of an impact process, leads to the following system of equations for the axisymmetric case:

$$p = f(\rho, I) \tag{4}$$

$$\partial\rho/\partial t + u\,\partial\rho/\partial r + v\,\partial\rho/\partial z + \rho \operatorname{div} \bar{u} = 0 \tag{5}$$

$$\rho(\partial u/\partial t + u\,\partial u/\partial r + v\,\partial u/\partial z) = -\partial p/\partial r \tag{6}$$

$$\rho(\partial v/\partial t + u\,\partial v/\partial r + v\,\partial v/\partial z) = -\partial p/\partial z \tag{7}$$

$$\rho(\partial I/\partial t + u\,\partial I/\partial r + v\,\partial I/\partial z) = -p \operatorname{div} \bar{u} \tag{8}$$

Here ρ, p, $\bar{u} = (u, 0, v)$, and I denote the density, pressure, velocity, and specific internal energy, respectively.

The dependence of the pressure on the specific mass and internal energy, of the form of Eq. (4), can be constructed from theoretical considerations and from experimental studies of one-dimensional shocks induced by plate–plate impact. The particular formulation used here has been developed by R. K. Osborne at the Los Alamos Scientific Laboratory:

$$\begin{gathered} p = [1/(e + \varphi_0)]\{\zeta(a_1 + a_2|\,\zeta\,|) + e[b_0 + \zeta(b_1 + b_2\zeta) + I(c_0 + c_1\zeta)]\} \\ e = \rho_0 I, \qquad \zeta = \rho/\rho_0 - 1 \end{gathered} \tag{9}$$

The material constants ρ_0, a_1, a_2, b_0, b_1, b_2, c_0, c_1, φ_0 occurring in Eq. (9) are listed in Table II for a number of materials for which the data are available.

TABLE II

MATERIAL PARAMETERS OCCURRING IN THE LOS ALAMOS EQUATION OF STATE[a]

Material	ρ_0	a_1	a_2	b_0	b_1	b_2	c_0	c_1	φ_0
Beryllium	1.845	0.9512	0.3453	0.9269	2.9484	0.5080	0.5644	0.6204	0.800
Boron	2.34	1.8212	4.3509	0.3764	0.3287	1.0801	0.5531	0.6346	0.250
Graphite	2.25	0.1608	0.1619	0.8866	0.5140	1.4377	0.5398	0.5960	0.500
Magnesium	1.735	0.5665	0.3343	2.2178	0.8710	0.4814	0.4163	0.5390	1.500
Titanium	4.51	1.9428	0.6591	1.8090	2.6115	1.7984	0.4003	0.5182	1.800
Water	1.00	0.000384	0.001756	0.01312	0.06265	0.21330	0.5132	0.6761	0.020
Plexiglas	1.18	0.006199	0.015491	0.14756	0.05619	0.50504	0.5575	0.6151	0.100
Polystyrene	1.04	0.038807	0.043646	0.77420	0.03610	0.46048	0.5443	0.6071	0.500
Polyethylene	0.913	0.007841	0.009766	0.19257	0.10257	0.31592	0.5748	0.6230	0.100
Micarta	1.39	0.016164	0.023579	0.34261	0.15107	0.43434	0.0540	0.0612	0.150
Silastic	1.43	0.004794	0.04684	0.33969	0.02377	0.50767	0.4925	0.5721	0.300
Aluminum	2.702	1.1867	0.7630	3.4448	1.5451	0.96430	0.43382	0.54873	1.5000
Copper	8.90	4.9578	3.6884	7.4727	11.519	5.5251	0.39493	0.52883	3.6000
Iron	7.86	7.78	31.18	9.591	15.676	4.634	0.3984	0.5306	9.00

[a] The parameters are fitted for the gram-centimeter-microsecond system of units and are due to R. K. Osborne.

Equation (9) provides an adequate description of the response of normal-density materials to intense compression. It has been found, however, to be quite unreliable for predicting states far off the Hugoniot of normal-density materials as is evidenced by its failure to check experimental data obtained by impacting porous solids. Equation (9) cannot therefore satisfactorily describe the response of the reduced-density aluminum δ projectile in the VISTA impact calculations presented in Section VI.

A theoretically correct equation of state for aluminum, based on more complete considerations from solid-state physics, thermodynamics, and some quantum mechanical calculations, has been developed by Kormer *et al.* (*13*), McCloskey (*14*), and Bjork *et al.* (*15*). The equation is correct in the sense that it is the best that can be written based on current physical theory and has been formulated to fit the available Hugoniot data generated from both normal-density and porous aluminum. The formulation, however, is presented in tabular form generated from complex computer computations and is most unsuitable for use in the VISTA calculations.

An empirical equation of state, developed by Heyda and described by Riney and Heyda (*3*), based on a modification of the plate-gap model of a porous solid, is employed to represent a reduced-density projectile in the VISTA calculations. The simple model together with normal density ρ_0 data for a material leads to remarkably accurate predictions for the shock behavior of the material if initially it is of reduced density $\bar{\rho}_0$. The form of the equation of state, corresponding to Eq. (4), is as follows:

$$p = \rho_0(2u - (2I)^{1/2})[A + B(2u - (2I)^{1/2})] \tag{10}$$

where

$$u = \frac{\sqrt{\Delta} - A + (B/2)[3(2I)^{1/2} + 3A - B(2I)^{1/2}((\eta - 1)/\eta)]}{2B[2 - B((\eta - 1)/\eta)]} \tag{11}$$

$$\Delta = [A + (B/2)\{(A + B(2I)^{1/2})((\eta - 1)/\eta) - 3(2I)^{1/2}\}]^2 + 4AB(2I)^{1/2} \tag{12}$$

For aluminum the dependence of A, B on $\eta = \bar{\rho}_0/\rho$ is given by

$$\begin{aligned} A(\eta) &= 0.3452 + 0.1276\eta \quad \text{cm}/\mu\,\text{sec} \\ B(\eta) &= 1.8237 - 0.2933\eta \end{aligned} \tag{13}$$

IV. Laminated Meteor Bumpers

A. Background

One of the first systematic experimental investigations of the effectiveness of a meteor bumper in reducing the total damage imparted by a high-velocity projectile was reported by Humes *et al.* (*16*). Although these tests were restricted to a maximum velocity of $v_0 = 0.41$ cm/μsec, the investigation definitely established that the spallation debris was effectively spread and that the damage from the high-velocity projectiles can be greatly reduced by using a properly selected bumpter shield.

Maiden and McMillan (*6*) studied the penetration of thin plates by a projectile at velocities up to $v_0 = 0.8$ cm/μsec. These experiments demonstrated the effect that the thermodynamic state of the debris has on the damage inflicted on the main structure. The residual heat in the shock processed projectile and target debris materials should be sufficient for the material to be completely liquified, otherwise the larger solid fragments cause frontal damage to the structure. At a given impact velocity, etc., shields of equal mass are equally effective in breaking up the projectile and spreading the debris. More recent experiments by Humes (*17*) at velocities up to $v_0 = 0.6$ cm/μsec also show that the total penetration damage is primarily a function of the mass per unit area of the bumper target, M_t. The material of the bumper apparently plays a role only relative to the homogeneity of the loading imparted to the main structure.

For thin plates and high velocities, the inviscid compressible fluid model provides a valid description of the breakup of the projectile and central region of the bumper and the subsequent dispersal of the debris downrange. In this section particle-in-cell calculations for five problems involving impact into thin-plate targets composed of two layers of distinct materials will be summarized. Their purpose is to determine the effect of the material discontinuity on the effectiveness of the plate as a meteor bumper. The response of the main structure to the spallation debris is not treated.

B. Impact Conditions

The impact conditions for the VISTA calculations were selected to ensure that the breakup of the projectile and central region of the thin plate and the subsequent dispersal of the debris downrange are governed by a compressible fluid model. In all five problems studied the impacting projectile is a solid aluminum cylinder of unit aspect ratio, the impact velocity is held at $v_0 = 1.5$ cm/μsec, the total plate thickness equals the radius of the projectile, and the two layers of the plate are of equal thickness, Fig. 2. In four of the problems the material of the second layer is aluminum and the first layer was varied as follows: Problem A, lead; B, beryllium; C, Plexiglas; and D, aluminum. In the fifth problem, C′, the impact situation is identical to Problem C except the order of the Plexiglas and aluminum layers is reversed. The five problems afford a wide variation of those factors that might be expected to influence the effectiveness of a meteoroid shield.

By combining the equations expressing the conservation of mass and momentum across a shock, the well-known relation

$$p = \rho_0 U u \tag{14}$$

between the pressure p, undisturbed density ρ_0, shock velocity U, and particle velocity u is obtained. In the special case of a limiting weak shock the propagation velocity will be a constant of the material, $U = c$. The coefficient $\rho_0 U = \rho_0 c$ in Eq. (14) is then well defined and is known as the "shock impedance" of the material.

The pressure and particle velocity must be continuous even at interfaces of distinct materials. Accordingly, the product $\rho_0 U$ would have to be preserved at each pressure if the simple wave model on which Eq. (14) is

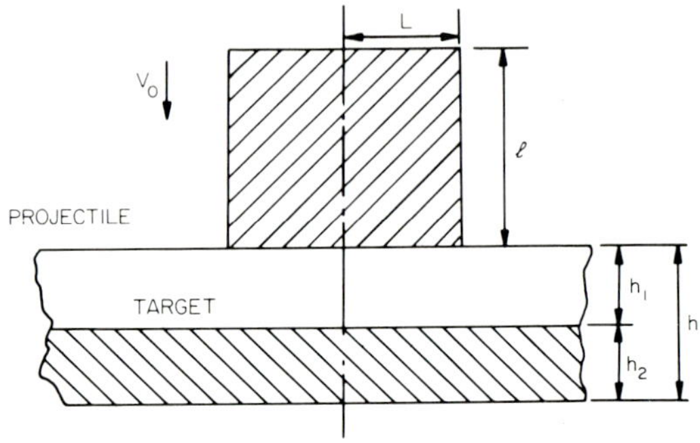

FIG. 2. Layered plate target configuration studied.

	Materials			Dimensions				
Problem	Projectile	Layer 1	Layer 2	l	$2L/l$	h/l	h_1/h_2	v_0
A	Al	Pb	Al	0.3057	1	0.5	1	1.5
B	Al	Be	Al	0.3057	1	0.5	1	1.5
C	Al	Plexiglas	Al	0.3057	1	0.5	1	1.5
D	Al	Al	Al	0.3057	1	0.5	1	1.5
C′	Al	Al	Plexiglas	0.3057	1	0.5	1	1.5

based is to be valid. Such is not the case for two distinct materials and an impinging shock wave is always decomposed into refracted and reflected components. The mismatch of the shock impedances of the two materials provides a rough measure of the shock decomposition to be expected. The values of the shock impedance ratio across the interface of the plate for each of the five problems treated are listed in Table III.

TABLE III

SHOCK IMPEDANCE MISMATCH, IMPACT PRESSURE, TRANSMITTED PRESSURE, AND MASS PER UNIT AREA FOR LAYERED PLATE TARGETS

Problem	$\frac{(\rho_0 c)_1}{(\rho_0 c)_2}$	$S_p = S_1$ (Mbar)	S_2 (Mbar)	M_t (g/cm²)	c_1 (cm/μsec)
A	1.49	5.0	3.06	1.080	0.207
B	0.93	2.7	3.1	0.348	0.798
C	0.19	1.85	2.9	0.297	0.259
D	1	3.1	3.1	0.413	0.585
C′	5.16	3.1	1.85	0.297	0.585

The relations expressing conservation of mass, momentum, and energy across a shock front have been combined with the equation of state, Eq. (9), for each of the four materials to compute the shock polars presented in Fig. 3. The solid curves, depicting the variation of pressure with particle velocity, are graphical representations of Eq. (14) for each of the indecated materials. The nonlinearity of the curves is illustrative of the variations in the coefficient $\rho_0 U$ for the corresponding material. The points of intersection of the right-running and left-running shock polars in Fig. 3 determine the initial pressure produced in the corresponding materials upon impact at $v_0 = 1.5$ cm/μsec. The amplitude of the shock propagating from the contact surface into the projectile S_p and from the contact surface into the first layer S_1 for each of the five problems of interest is listed in Table III.

An estimate can also be made from Fig. 3 of the decomposition of the shock S_1 into reflected wave R_W and a wave S_2 transmitted into the second layer of the target. The schematic in Fig. 4 depicts the one-dimensional idealization as well as the procedure followed for approximating the initial interaction at the interface of the Plexiglas and aluminum layers in Problem C. The initial shock S_1 produced in the Plexiglas layer by the impacting projectile is denoted as Point 1 in Fig. 4. Upon reaching the layer interface a right-running shock S_2 is transmitted into the second layer and a left-running wave R_W is reflected into the first layer. Since the pressure and particle velocity in the two layers must be conserved across the interface the intersection of the left-running Plexiglas shock polar through 1 and the right-running aluminum shock polar through the origin determines Point 2 which corresponds to S_2 and R_W. In this case $S_2 = S_1 + R_W \approx 2.9$ Mbar so that R_W is a shock wave of amplitude approximately equal to 1.05 Mbar.

A similar construction may be made for Problem B. The approximation is less accurate for Problems A and C′ since R_W is a rarefaction wave and a shock polar was used in determining Point 2 from Point 1. Since in Problem D there is no material interface, $R_W = 0$ and $S_1 = S_2$. These estimates for S_2 are listed in Table III.

Also listed in Table III is the mass per unit area of each of the layered plates. To assess the importance of the lateral transfer of energy during the penetration of the first layer of the plate, the corresponding peak shock velocity, c_1, for each of the problems is also listed.

The central region of the projectile–bumper configuration is left with residual internal energy as a consequence of the processing by the nonisentropic shock compression followed by an isentropic release of the pressure by rarefaction waves. For aluminum and lead the shock pressure required to leave the material in a completely molten state has

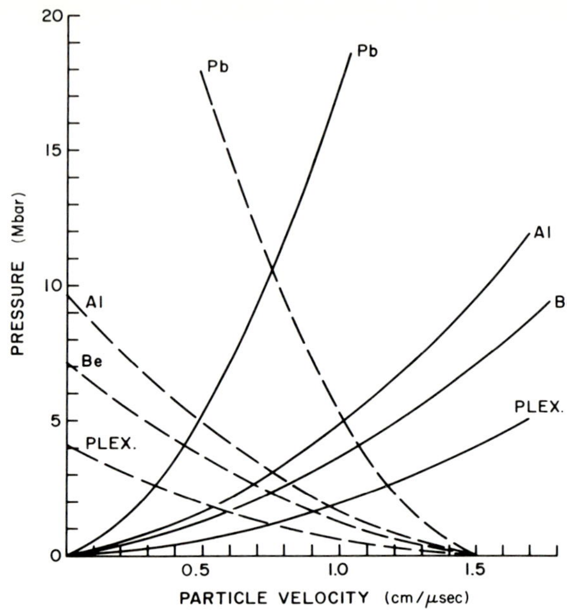

FIG. 3. Right-running and left-running shock polars for indicated materials.

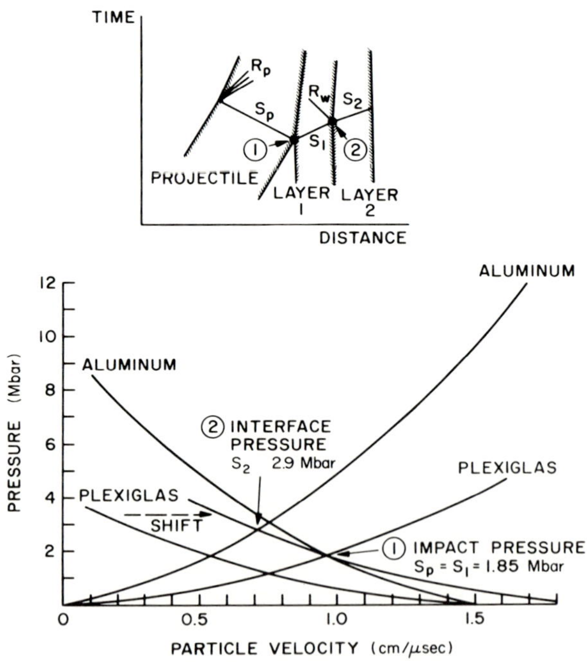

FIG. 4. Construction determining approximate pressure transmitted across Plexiglas–aluminum interface in Problem C.

been estimated to be approximately 1.0 and 4.0 Mbar, respectively, by Urlin (*18*). A comparison with the values for S_p listed in Table III shows that the aluminum projectile will be completely liquified in all five cases. From the values of S_1 and S_2 it is also clear that the central region of the aluminum and lead components of the targets will be liquified. Although estimates are not available, the pressure required to liquify Plexiglas should be smaller and these components in Problems C and C′ may be assumed to be left in a fluid state. Even if this were not the case for the beryllium in Problem B, its brittle behavior should cause the debris to be spalled into fine particles.

C. Results of Calculations

The capability of VISTA to utilize the SC-4020 to provide graphical displays of the status of the impact process at any stage of the calculations was employed extensively. Both mass particle plots depicting the projectile–target configuration and velocity plots describing the current velocity field were obtained. In Figs. 5a–h representative computer displays are presented at four stages of the process for Problem C′. Mass particles representing the aluminum first layer of the bumper target are denoted by plus signs. Different symbols, asterisks and dots, denote the mass particles representing the Plexiglas second layer of the bumper and the aluminum projectile. The propagation of the shock waves across the thickness of the bumper and up into the projectile are clearly delineated.

These flow fields together with the particle plots illustrate the overall impact and penetration process quite clearly. The combination of the small plate thickness and the high impact velocity in this case results in a rapid punching out of a region of the plate equal to about twice the original projectile diameter. The highly compressed region emerges from the rear of the target and will subsequently expand and disperse as it propagates downrange. Once the central region of the projectile–target configuration has torn loose from the target it no longer interacts with the perforated plate.

To compare the effectiveness of the various bumpers in causing the projectile to break up and disperse as it propagates downrange, it is of interest to monitor the total forward axial momentum and total outward radial momentum in the system:

$$Z_+ = \sum_{i,j} M_i^j (v_+)_i^j, \qquad R_+ = \sum_{i,j} M_i^j (u_+)_i^j \tag{15}$$

Here the plus indicates that the sums are extended only over positive values of u and v. The term Z_+ provides a rough indication of the total

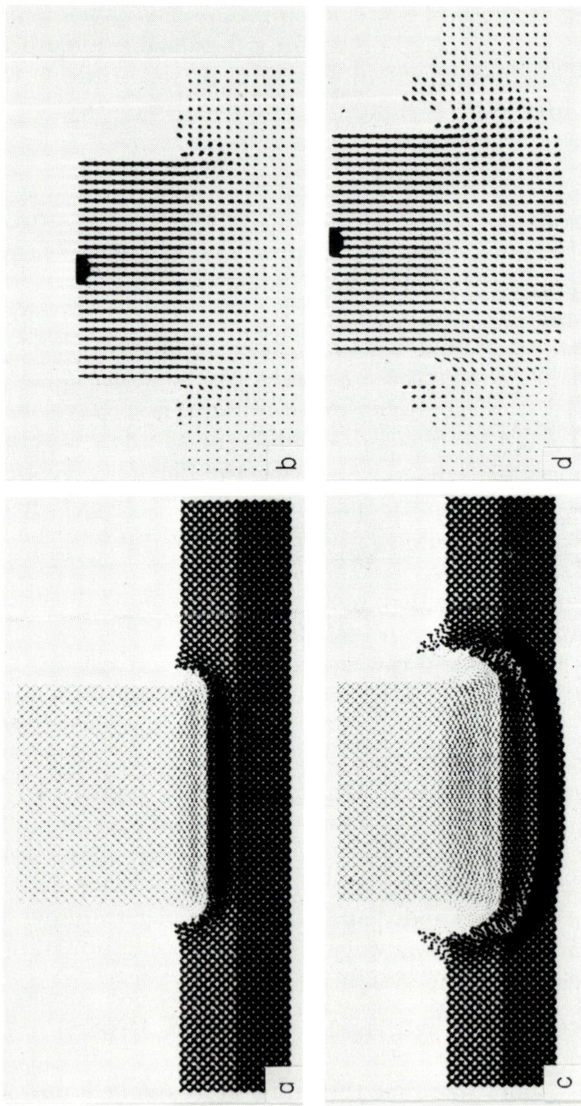

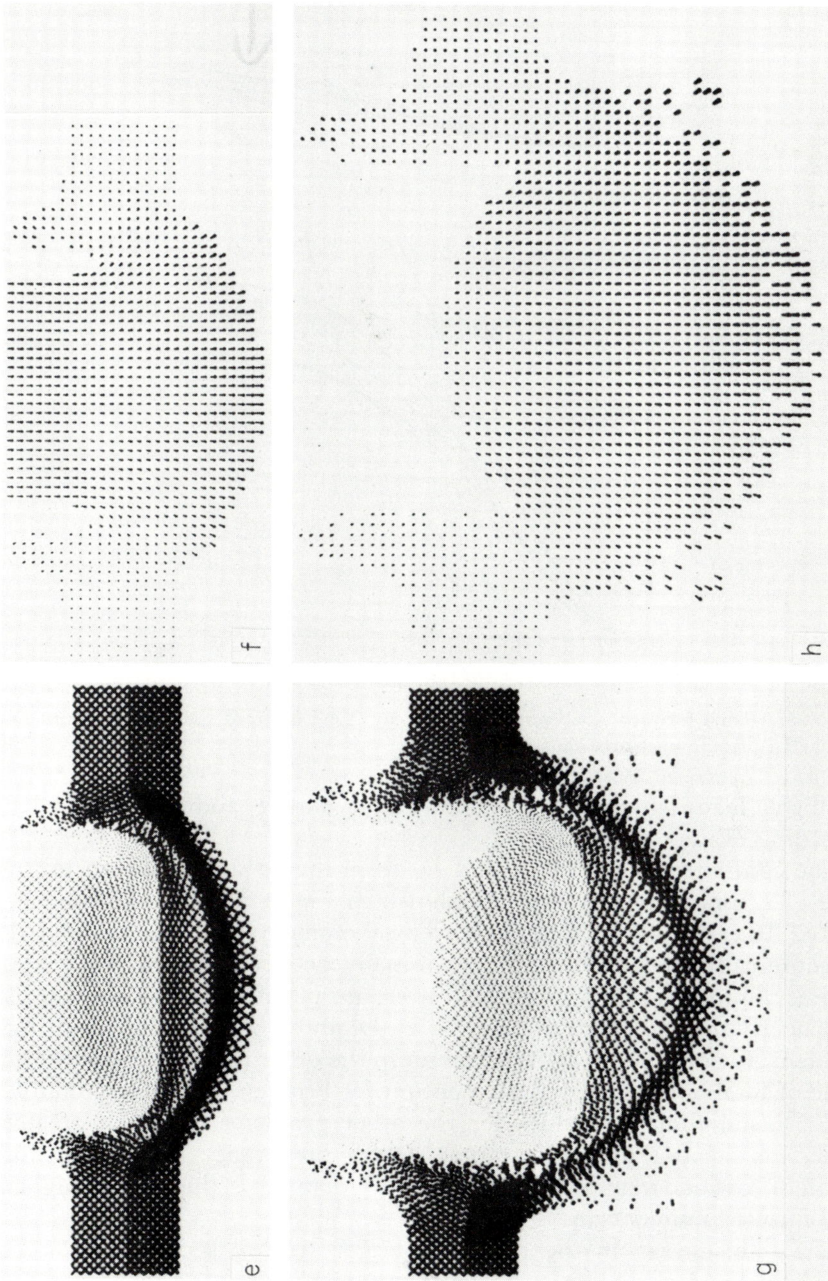

FIG. 5. Projectile–target configuration and velocity field in Problem C′. (a) and (b) $t = 0.05$ μsec; (c) and (d) $t = 0.10$ μsec; (e) and (f) $t = 0.15$ μsec; (g) and (h) $t = 0.27$ μsec.

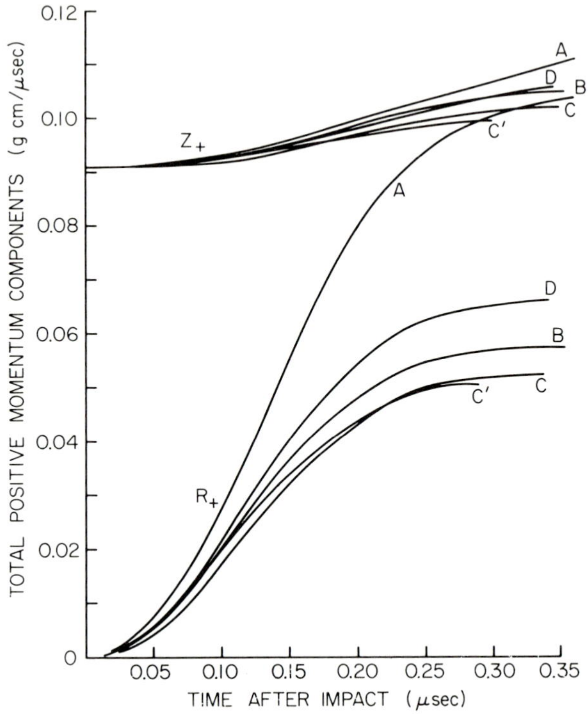

FIG. 6. Total forward axial momentum Z_+ and total outward radial momentum R_+ for layered plate impact problems.

impulsive load delivered to a second target and R_+ indicates the radial spreading of the total load.

The computed values of Z_+ and R_+ from the instant of impact to the time of complete penetration are shown in Fig. 6 for all five problems treated. In each case the total forward axial momentum is seen to increase monotonically from the magnitude initially delivered by the projectile. The asymptotic value approached after the penetration is complete is very nearly the same for Problems B, C, D, and C′. The value of Z_+ for Problem A is only about 10% greater. On the other hand, the asymptotic value of total outward radial momentum for Problem A is nearly double the corresponding R_+ values for the other problems. A much greater dispersion of the projectile–target debris downrange will occur for Problem A which may be expected to more than offset the slightly higher total forward momentum.

The total energy in the projectile–target system is constant for all five problems. In Problem A, however, there is significantly more internal energy, and consequently less kinetic energy, than for any of the other

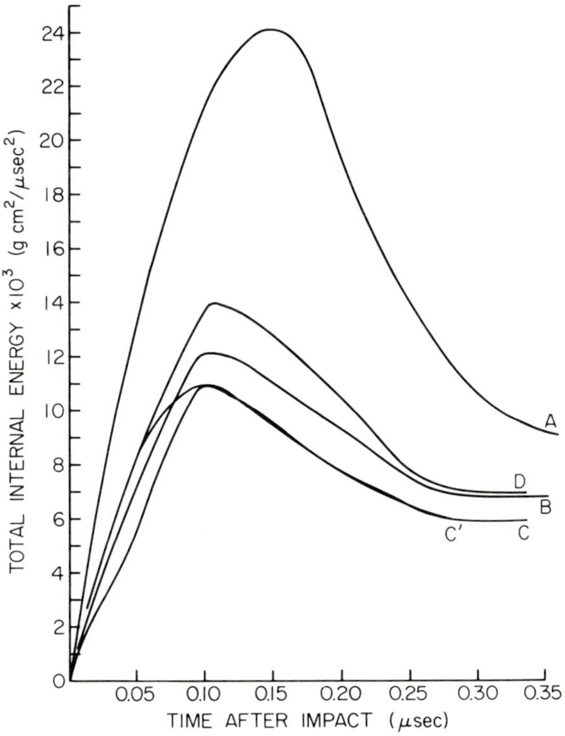

FIG. 7. Total internal energy for the indicated layered plate impact problems.

four. This fact is illustrated in Fig. 7. The dramatic decrease in the total internal energy from the peak value corresponds to the point in time that the shock in the target reaches the opposite free surface. The change from internal to kinetic energy upon the release of the contained pressure pulse is illustrated by the increase of the particle velocity in the target near the free surface that is apparent by a comparison of Figs. 5b and f.

The momentum and energy content of the projectile material alone was also monitored as a function of time. Here again asymptotic values are approached in each impact as the penetration process is completed and it is found that the bumper configuration of Problem A is much more effective in defeating the momentum and energy of the projectile than are the other four configurations.

The second layer of the target plays no part in the process until the shock in the target, originating at the impact surface, propagates to the material interface at time

$$t = h_1/\dot{R}_H \tag{16}$$

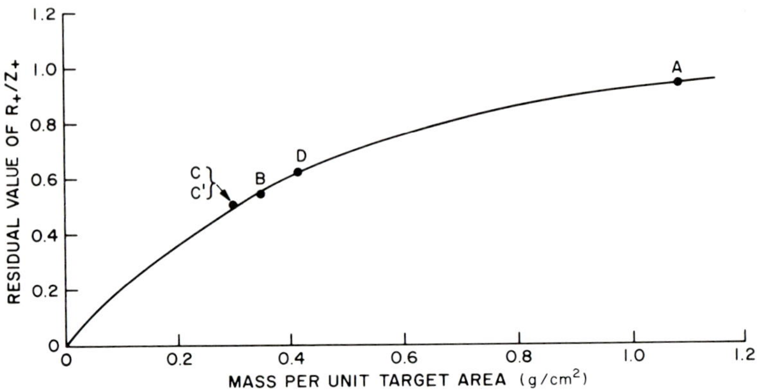

FIG. 8. Ratio of residual outward radial momentum in debris cone to residual axial momentum in the debris plotted against mass per unit area of bumper.

Here $\dot{R}_H$ is the shock wave velocity in the target. For aluminum–aluminum impact at 1.5 cm/μsec, $\dot{R}_H = 1.51$ cm/μsec and, consequently, the calculations for Problems D and C′ should be identical until $t = 0.051$ μsec. It is interesting to note that results depicted in Figs. 6 and 7 for Problem C′ are indeed the same as for Problem D until

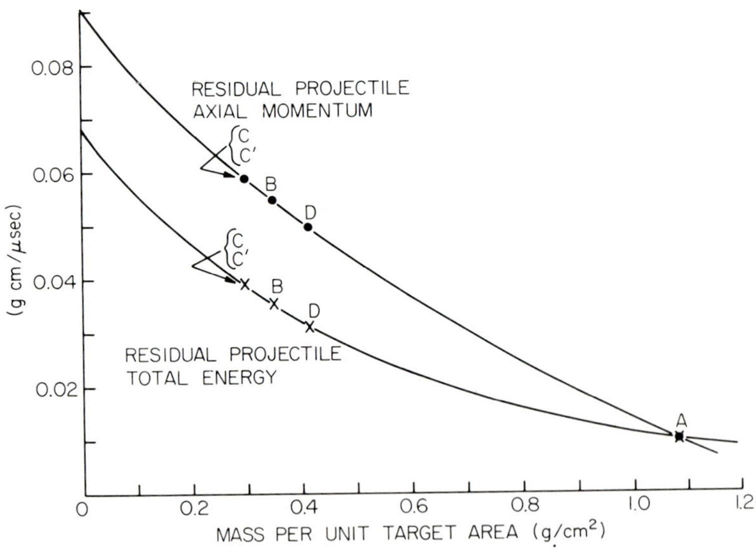

FIG. 9. Residual axial momentum and energy in the projectile plotted against mass per unit area of bumpers.

a time greater than $t = 0.051$ μsec, but gradually they approach asymptotic values very close to those attained in Problem C.

The ratio of the residual value of total outward radial momentum to the residual value of the total axial momentum contained in the spallation cone

$$\lim_{t \to \infty} R_+/Z_+ \tag{17}$$

may be used as an index of the effectiveness of a bumper in dispersing the debris. This ratio has been approximated from Fig. 6, and of the factors listed in Table III only the mass per unit area of the target is found to correlate with this index, Fig. 8.

The residual axial momentum and the residual total energy carried by the projectile material alone can also be obtained. These values, which indicate the relative effectiveness of the five bumper configurations in defeating the projectile, have also been found to correlate only with the mass per unit area of the target, Fig. 9.

It is concluded that the predominant factor in the defeat of the projectile by a meteoroid bumper and the dispersal of the debris is the mass per unit area of the bumper. The peak impact pressures experienced by the projectile and bumper target, the impedance mismatch at the interface of the target layers, and the wave velocity of the shock in the target material have only secondary effects for the impact conditions studied. This conclusion is based on the hypothesis that the peak pressures realized in the central region of the projectile–bumper configuration are sufficiently large for the debris cone materials to be in a molten state due to the residual heat after the shock processing.

These conclusions are in general agreement with the experimental results for single-material targets reported by Maiden *et al.* (*5*). For thicker composite layered targets and lower impact velocities, however, experimental studies have shown that the "secondary" effects cannot be neglected.

V. Hollowed Projectiles

A. Configurations Studied

In this section theoretical calculations are presented describing the the hypervelocity impact and penetration of thin-plate targets by projectiles simulating hollow spheres, hollow cylinders, hollow cylinders with a central disk, and a cylinder with a conical cavity bored out of the impacting end. The purpose of the study is to provide a quantitative understanding of the effect that the projectile geometry has on the penetration mechanisms. The importance of these effects has been experimentally demonstrated by Warnica and Christman (*19*) who have presented results comparing the penetration of a thin plate by various projectile shapes: disk, solid sphere, rod, and hollow sphere.

Both projectile and target are aluminum in all eight of the problems treated in this section. Two impact velocities, $v_0 = 0.76$ and $v_0 = 1.5$ cm/μsec were considered. As discussed in Section IV.B, Urlin (*18*) has estimated that aluminum shock processed at a pressure greater than 1.0 Mbar will be left in a molten state and at an impact velocity of $v_0 = 1.5$ cm/μsec the central region of the projectile–target configuration should be left in fluid state. For $v_0 = 0.76$ cm/μsec the peak pressure 1.08 Mbars and the debris from the central punched-out region should be mixed, liquid and solid. The assumption that the material responds as a compressible fluid appears to be a reasonable one for the conditions studied here. For the investigation of the complicated flows resulting from impact by these hollowed projectiles, only the particle-in-cell method for solving the compressible fluid equations is applicable and the VISTA code was applied.

The complete specifications of the eight impact problems treated are given in Figs. 10–12. For three of the problems treating the impact of an aluminum hollow cylinder, designated as Problems E, F, and G in Fig. 10, the same outside projectile dimensions were maintained:

$$l_0 = 2L_0 = 4L_i = 0.3057 \quad \text{cm}$$

The hollowed core reduces the mass to 0.0453 g which is the same as that of a $\frac{1}{8}$-in-diam solid aluminum sphere. In Problem E, which serves as our standard for comparison purposes, the aluminum plate thickness equals the inner diameter of the projectile, $h = 2L_i$, and the impact velocity is $v_0 = 1.5$ cm/μsec. In Problem F the impact velocity is lowered to $v_0 = 0.76$ cm/μsec; in G the plate thickness is increased to $h = 3L_i$. In Problem J the conditions are identical to E except a central disk has been added. The length of the hollow cylinder in Problem K is double that of E.

Calculations were also made for an aluminum projectile with a hollow segment, bored out of the impacted end, in the shape of an inverted cone, Fig. 11. The impact velocity and the length of the projectile in this problem, L, were the same as that of K. The dimensions of the hollow spherical projectiles, Problems H and I in Fig. 12, were selected to have the same mass as the hollow cylinders in Problems E, F, and G. The thickness of the targets are also the same as in Problems E and F and the same two impact velocities are considered, $v_0 = 0.76$ and $v_0 = 1.5$ cm/μsec.

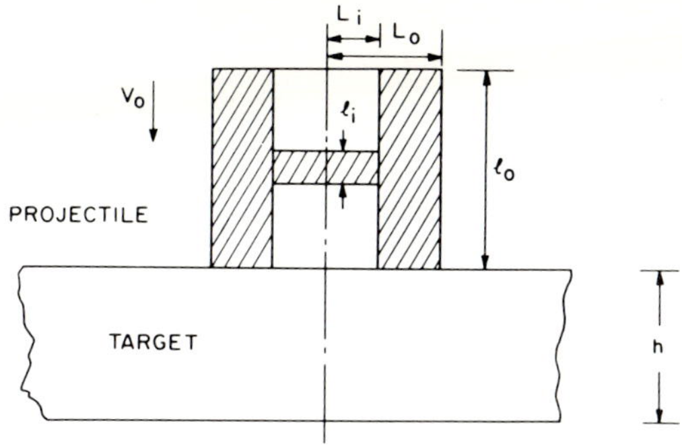

FIG. 10. Hollowed cylindrical projectiles and targets studied.

Problem	Materials		Dimensions					
	Projectile	Target	l_0	l_i/l_0	$2L_0/l_0$	L_i/L_0	h/L_0	v_0
E	Al	Al	0.3057	0	1	0.5	1	1.5
F	Al	Al	0.3057	0	1	0.5	1	0.76
G	Al	Al	0.3057	0	1	0.5	1.5	1.5
J	Al	Al	0.3057	0.25	1	0.5	1	1.5
K	Al	Al	0.6114	0	0.5	0.5	1	1.5

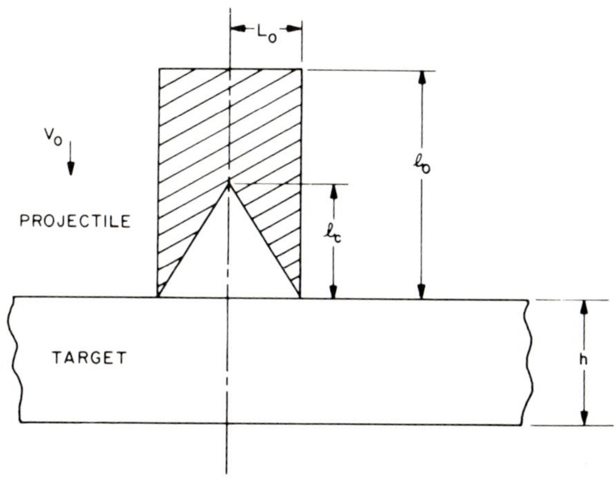

FIG. 11. Cylindrical projectile with conical cavity bored out.

Problem	Materials		Dimensions				
	Projectile	Target	l_0	l_c/l_0	$2L_0/l_0$	h/L_0	v_0
L	Al	Al	0.6114	0.5	0.5	1	1.5

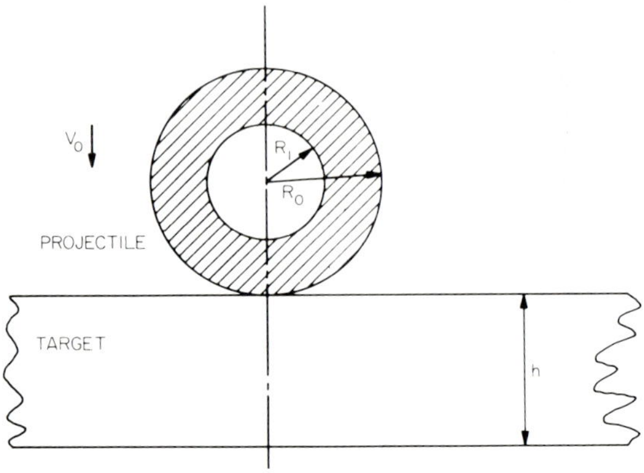

FIG. 12. Hollow spherical projectiles and targets studied.

Problem	Materials: Projectile	Materials: Target	Dimensions: h	Dimensions: R_i/R_0	v_0
H	Al	Al	0.153	0.5	0.76
I	Al	Al	0.153	0.5	1.5

B. IMPACT MECHANICS

The SC-4020 was utilized to provide graphical displays of the VISTA calculations for all eight impact problems. The calculated configurations and velocity fields at four stages of the process for Problem E are depicted in Fig. 13. From the instant of impact, rarefaction waves initiate from the inner and outer peripheries of the annular contact surface setting up radial pressure gradients resulting in inward and outward flow, Figs. 13a and b. The pressure in the outward flowing decreases due to geometrical divergence effect. The inward flowing material, however, has less available volume as it approaches the axis of symmetry. This essentially cylindrical focusing of the shock wave causes the pressure to be increased to a peak value of $p_{\max} \approx 10.1$ Mbar compared with an impact value of $p_H = 3.1$ Mbar. The location and time when the maximum pressure is attained is indicated in Fig. 13d. The subsequent expansion of the material from this highly compressed region augments the backward jet and produces a forward central jet as is demonstrated in Figs. 13e–h.

The mass particles representing material that has been splashed upstream are followed only to the point that they reach the top of the finite difference mesh. Thereafter they do not enter the calculations but their point of exit from the mesh is denoted in the particle plots to illustrate the amount of material that has left the mesh. This results in a gathering of points along the top mesh boundary as observed in Fig. 13h.

The largest calculated velocity for particles in the forward-moving jet was 2.0 cm/μsec. These few particles are seen in Fig. 13g to be breaking away from the rest of the jet which moves at a velocity of from 1.5 to 1.9 cm/μsec. The bulk of the backward jet travels upstream at about 0.7–0.9 cm/μsec.

It is interesting to note that the mechanisms involved in this impact and penetration process are very similar to those operative in the formation of a shaped-charge jet. In that case the radial focusing that causes large pressures in the metal liner is produced by the detonating wave in the encompassing explosive rather than by pressures generated by impact of a hollow cylinder.

The lower impact velocity in Problem F gives results remarkably similar to those in E. The shock focusing produces a peak pressure $p_{\max} \approx 3.5$ Mbar at $t = 0.15$ μsec after impact compared with an impact value of $p_H = 1.08$ Mbar. The location of the focusing is the same and the form of the distorted configuration is similar.

In Problem G the target is half again as thick as in E. The increased distance of the opposite surface from the location of $p_{\max}$ allows the toroidal shock front in the target to become very nearly ellipsoidal by the time it arrives there. As a consequence of the flatness of the shock acting on the rear surface and the increased distance of the free surface from the peak pressure region, the shock focusing is less effective in producing a forward-moving jet. The maximum velocity of the particles in the forward-moving central region computed for Problem G is 1.40 cm/μsec.

Except for the addition of the central disk in the projectile, the impact situation in Problem J is identical to that of Problem E, Fig. 10. The disk is intended to inhibit the flow of the backward jet from the impacted surface. At time $t = 0.11$ μsec after impact the peak pressure, $p_{\max} \approx 11.4$ Mbar, is realized. This value is higher than the peak value of 10.1 Mbar computed for Problem E and is a result of the constraint that the central disk imposed on the backward jet. Comparison of the flow fields for Problems J and E at corresponding times shows that the formation and the characteristics of the forward jet are not significantly affected by the presence of the central disk. The disk does effectively retard the backward jet, however, reducing the velocity of

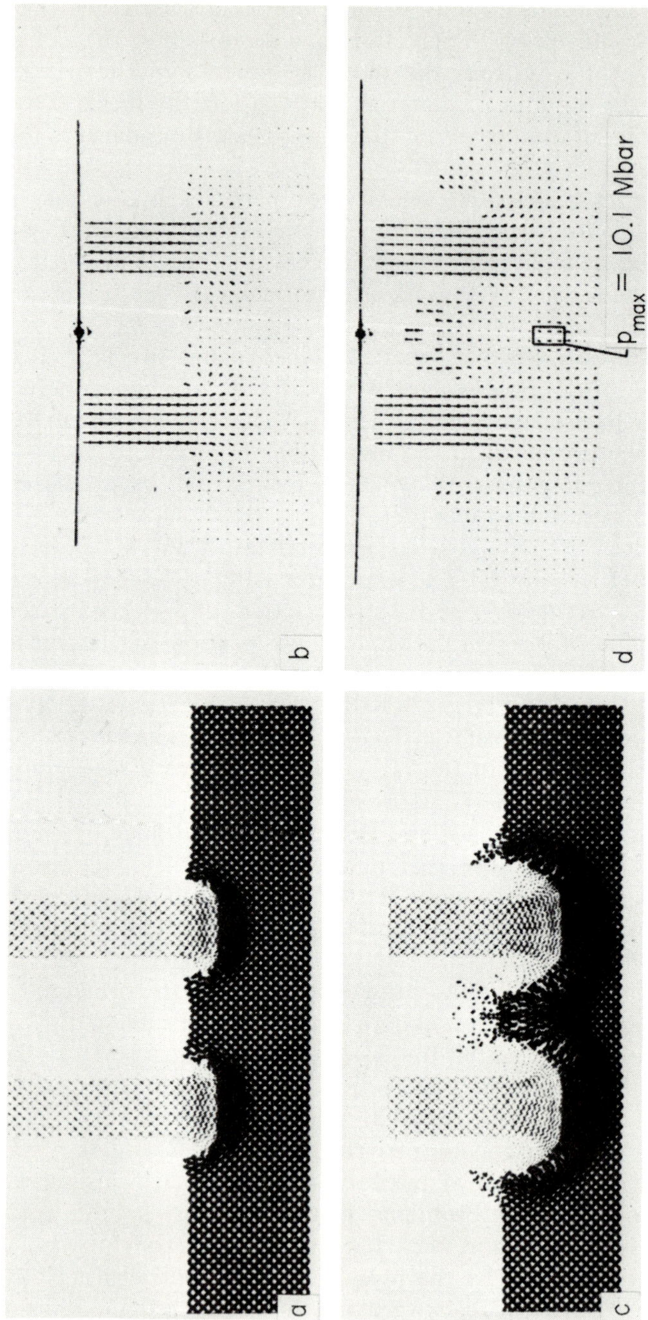
a
b
c
d
p_{max} = 10.1 Mbar

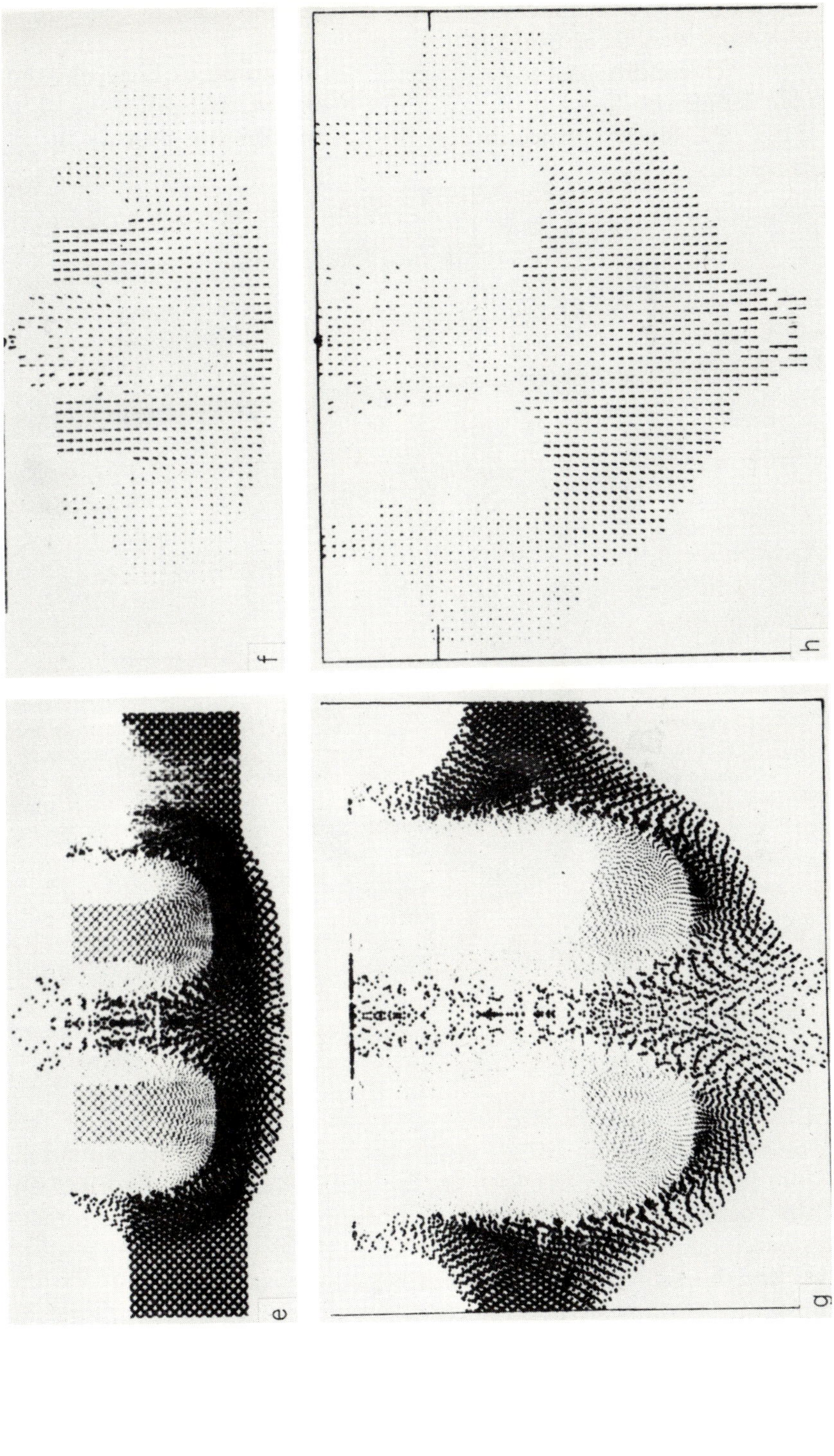

FIG. 13. Projectile–target configuration and velocity field at indicated time for Problem E. (a) and (b) $t = 0.05$ μsec; (c) and (d) $t = 0.10$ μsec; (e) and (f) $t = 0.15$ μsec; (g) and (h) $t = 0.32$ μsec.

the bulk of the material from approximately $v_0/2$ in the case of Problem E to approximately $v_0/5$ for Problem J.

The impact conditions for Problem K are identical to E except the projectile length is doubled. A comparison of Fig. 14a with Fig. 13g shows that the longer projectile has little effect on the process up to

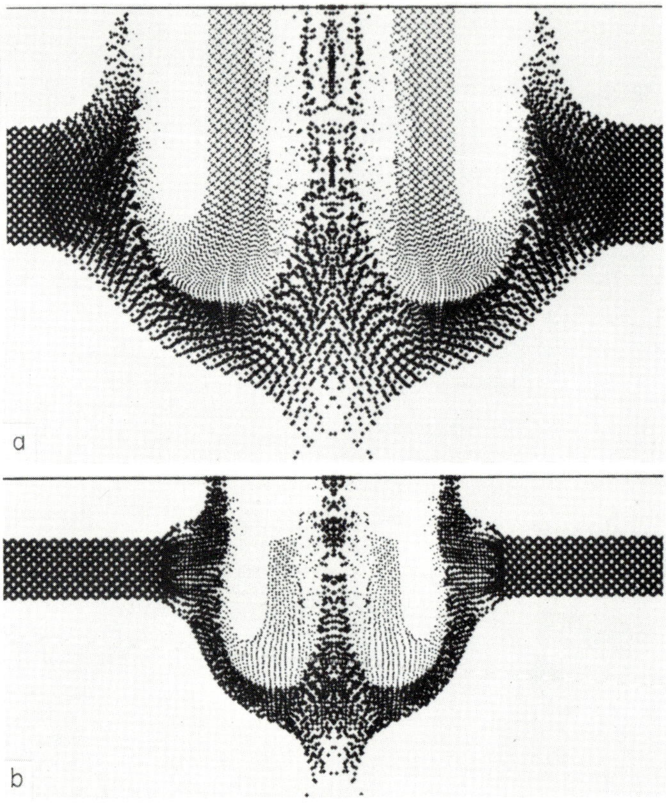

FIG. 14. Projectile–target configuration for Problem K. (a) $t = 0.28\ \mu$sec; (b) $t = 0.41\ \mu$sec.

time $t = 0.28\ \mu$sec. Soon after this the flow reached the side of the finite difference mesh and the mesh was rezoned. The representation of the situation at $t = 0.41\ \mu$sec in the rezoned mesh is depicted in Fig. 14b. Approximately half the length of the projectile remains intact.

The outer dimensions of the projectile in Problem L and the impact velocity are the same as for K. The inverted conical cavity bored from the impacting surface of the projectile causes the shock in the target to be focused along the axis of symmetry and the solid portion of the upper half of the projectile completely stops the backward jet, Fig. 15. At a time slightly greater than $t = 0.32\ \mu$sec the flow reached the side of the finite difference mesh. In order to follow the process further the mesh was rezoned automatically by coalescing four cells into one thereby doubling the linear dimensions of the mesh, Fig. 15c. A peak pressure $p_{\max} \approx 11.0$ Mbar is realized at $t = 0.17\ \mu$sec after impact and the forward-moving material behind the plate reaches a maximum value of 2.0 cm/μsec.

The effectiveness of a hollowed cylinder in producing a jet depends severely on the alignment of its axis at the instant of impact. No such problem would exist, however, for the hollow spherical projectiles.

The flow field induced by the hollow spheres in Problems H and I are quite different from those for the hollow cylinders. In neither case is focusing of the shock obtained. The graphical displays of the situation at various stages of the deformation process for Problem I are presented in Fig. 16. Inward radial flow required for shock focusing at the axis does not occur. The material ejected from the inner surface of the projectile opposite the center of impact, Fig. 16a, has impinged upon the diametrically opposite inner surface, by $t = 0.25\ \mu$sec, Fig. 16b. This produces a pressure at the apex of the sphere which in turn causes outward radial flow from the region. The projectile is rapidly torn apart and the spallation bubble expands rapidly as it moves downrange. The same mechanisms are observed in the plots of the calculations for Problem H.

C. Numerical Results

The formation of the central jet due to the shock focusing, when it occurs, inhibits the dispersion of the debris. Since the momentum carried by the emerging projectile–target debris may impinge on a second target downrange, it is of interest here, as in Section IV.C, to compute the total forward axial momentum Z_+ and total outward radial momentum R_+ of the configuration. The computed values of Z_+ and R_+ as a function of time have been monitored for all eight problems treated in this section.

In Fig. 17 the variations of Z_+ and R_+ with time since impact are given for Problems E, G, and I. After the penetration process is completed the curves level out and asymptotic values with time are approached. That this is the case is ever more clearly demonstrated in Fig. 18 when Z_+ and R_+ for Problems J, K, and L are plotted. The asymptotic values are attained in these problems for which the

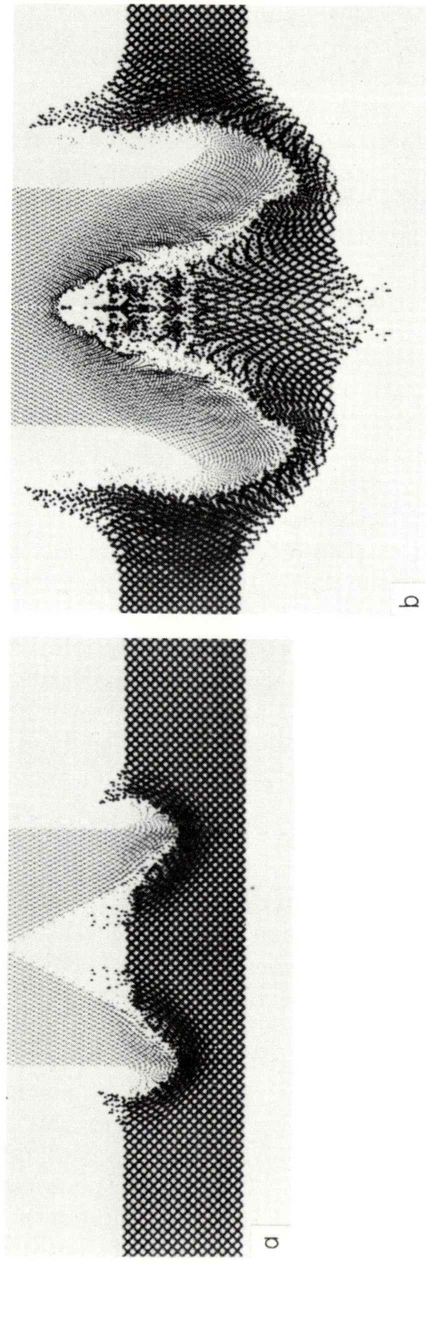

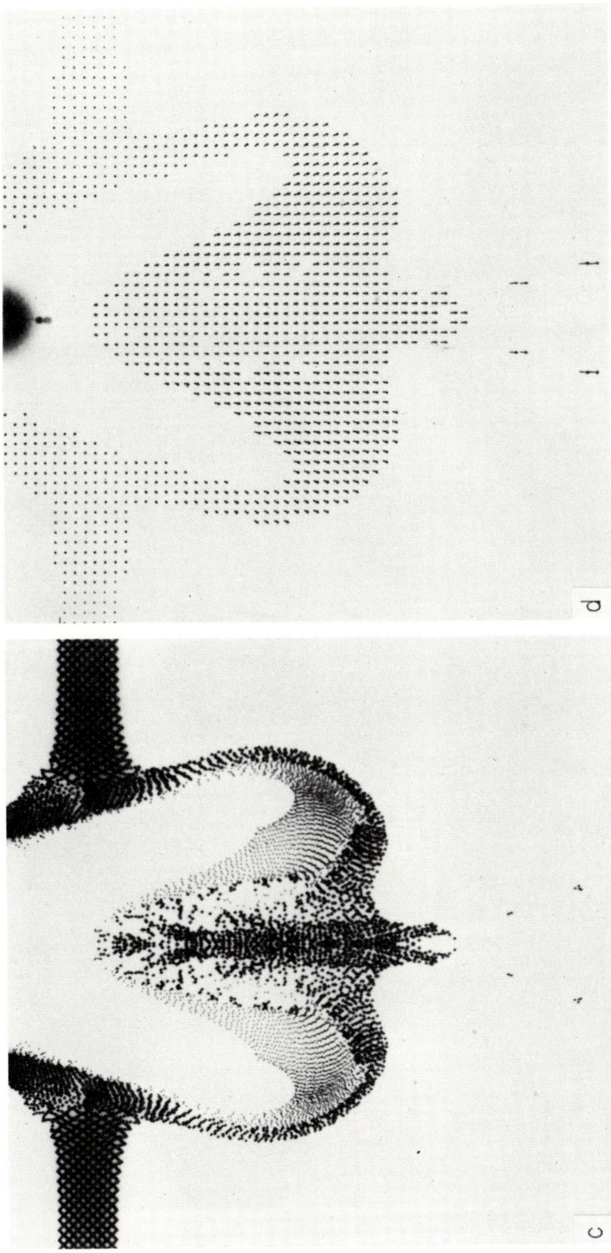

FIG. 15. Projectile–target configuration and velocity field at indicated time for Problem L. (a) $t = 0.10$ μsec; (b) $t = 0.28$ μsec; (c) and (d) $t = 0.80$ μsec. The top of the projectile is still entering the finite difference mesh in (a) and (b).

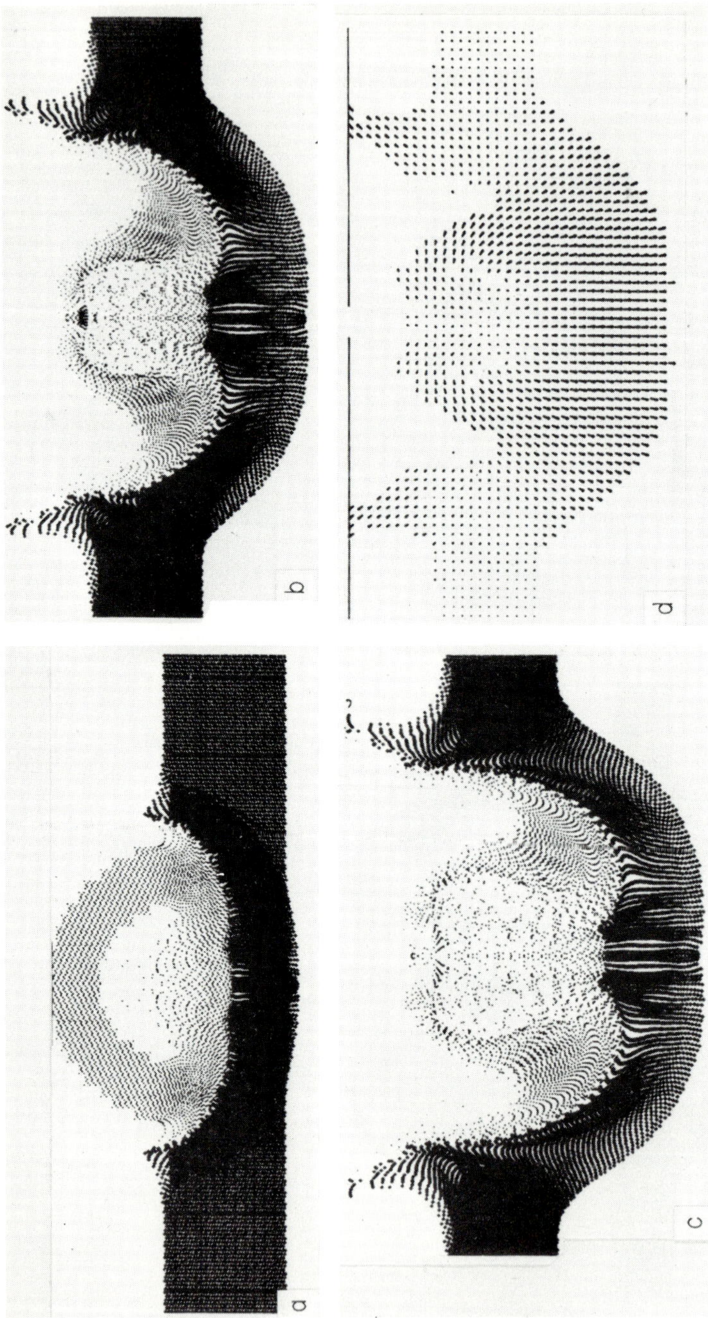

FIG. 16. Projectile–target configuration and velocity field at indicated time for Problem I. (a) $t = 0.13$ μsec; (b) $t = 0.25$ μsec; (c) and (d) $t = 0.31$ μsec.

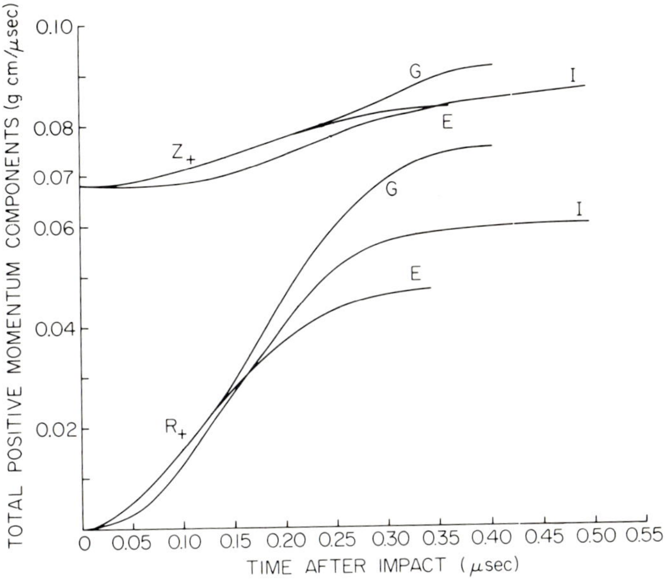

Fig. 17. Total forward axial momentum Z_+ and total outward radial momentum R_+ for indicated problems.

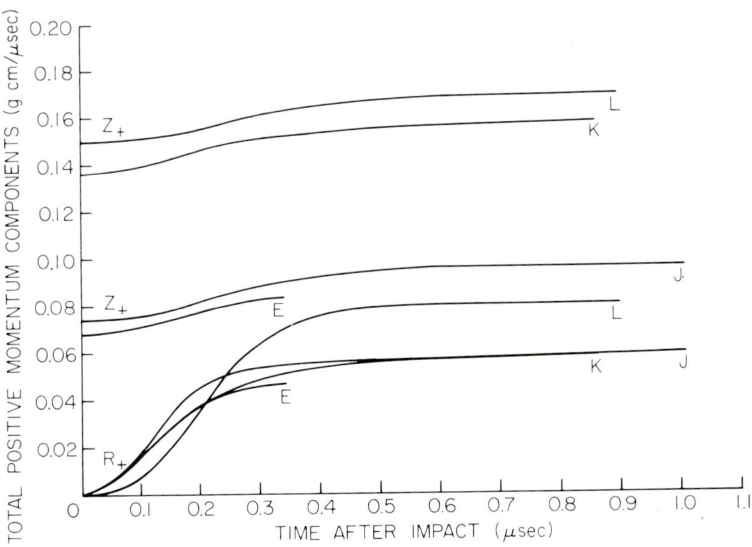

Fig. 18. Total forward axial momentum Z_+ and total outward radial momentum R_+ for indicated impact problem.

TABLE IV

THE INITIAL AND FINAL AXIAL MOMENTUM OF THE PROJECTILE AND THE LIMITING VALUES OF Z_+, R_+, AND Z_+/R_+[a]

	Projectile		Projectile–target		
Problem	Z_0	Z_f	Z_+^∞	R_+^∞	Z_+^∞/R_+^∞
E	0.068	0.035	0.083	0.048	1.73
F	0.034	0.018	0.047	0.027	1.74
G	0.068	0.021	0.093	0.077	1.21
H	0.034	0.016	0.046	0.033	1.39
I	0.068	0.030	0.092	0.060	1.53
J	0.074	0.037	0.096	0.058	1.66
K	0.136	0.102	0.157	0.058	2.71
L	0.150	0.106	0.169	0.080	2.11

[a] Units are in the gram-centimeter-microsecond system.

calculations were extended much later in time. In Table IV approximations to these limiting values at the time the calculations were concluded, Z_+^∞ and R_+^∞, are listed for all eight problems. The ratio Z_+^∞/R_+^∞ provides a rough index of the effectiveness of the projectile in confining the spallation debris to a central stream that may act on a second target downrange. The larger the index the more concentrated will be the deposition of the forward momentum carried by the debris to the second target.

It is also of interest to monitor the time variation of the axial momentum of the projectile material alone. The value in each case decreases from impact value Z_0 to a limiting value Z_f. Values of Z_f for each problem treated are given in Table IV.

Specifically, the calculations summarized in Table IV demonstrate the following:

(1) The effectiveness of the shock focusing on the formation of the forward jet is strongly dependent on the thickness of the plate. The velocity of jet decreases and the leading surface of the spallation cone becomes flatter as the thickness is increased.

(2) The shape of the forward jet and the ratio of its velocity to the impact value is relatively insensitive to the impact velocity. Similar results were obtained at $v_0 = 0.76$ and $v_0 = 1.5$ cm/μsec.

(3) The fact that no shock focusing is obtained using the spherical projectiles for the impact conditions studied has the consequence that

the spallation debris disperses much more rapidly than for corresponding impact conditions using hollow cylindrical projectiles.

(4) Calculations for hollow cylindrical projectiles with a solid central portion show that the additional material effectively inhibits the formation of the backward jet of target material. The forward jet is till formed, however, and the spallation debris is effectively confined along the impact axis.

(5) Doubling the length-to-diameter ratio of the hollowed cylindrical projectile does not strongly affect its interaction with the plate. About half the length of the projectile remains very nearly intact to impact a possible second plate downrange and the overall dispersal of the debris is, therefore, drastically reduced.

VI. Crater Predictions for Thick Targets

A. Crater Termination Criteria

In order to study the effect that the reduced density of meteoroids may have on their penetrating ability, numerical calculations have been made for a thick aluminum alloy target, 2024-T6, impacted by projectiles simulating stony meteoroids. The projectiles were chosen to correspond to a solid one of normal density, 2.7 g/cm³, two heterogeneous configurations with distributed mass kernels of normal density and average bulk density of 0.5 g/cm³, and a homogeneous projectile of reduced density equal to 0.5 g/cm³. These four configurations, designated, respectively, as α, β, γ, and δ projectiles, as depicted in Fig. 19, all have a mass equal to the mass of a $\frac{1}{8}$-in.-diam sphere. A total of seven impact situations were treated, Table V. The basic material of the projectile

TABLE V

Seven Impact Problems Treated and a Summary of the Predicted Crater Dimensions[a]

Problem	Projectile	v_0	P_c/l	R_c/l	P_c/R_c	P_c	R_c	V_c	$(V_c/E)^{1/3}$	$P_c/M^{1/3}$
N1	α	0.735	2.4	2.2	1.09	0.66	0.61	0.52	3.49	1.87
N2	α	2.0	4.4	4.0	1.10	1.22	1.11	3.15	3.30	3.43
N3	β	2.0	2.1	2.4	0.87	1.03	1.17	2.95	3.20	2.89
N4	γ	2.0	2.3	2.5	0.92	1.12	1.22	3.50	3.38	3.15
N5	δ	2.0	2.4	2.7	0.89	1.17	1.32	4.27	3.61	3.28
N6	β	0.735	1.0	1.2	0.83	0.49	0.59	0.35	3.07	1.38
N7	β	5.0	3.9	4.3	0.91	1.91	2.10	17.6	3.15	5.36

[a] Units are in the gram-centimeter-microsecond system.

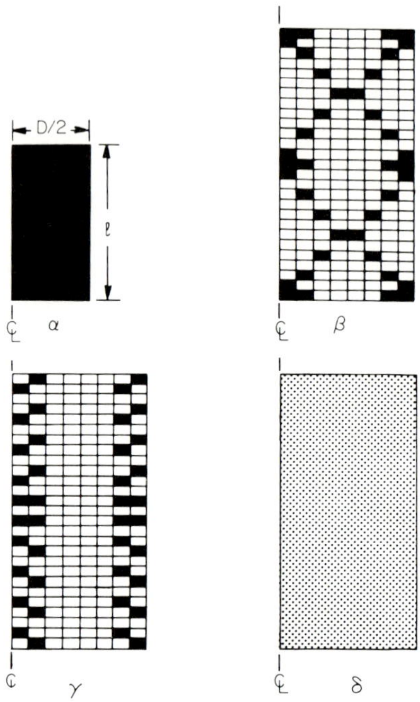

FIG. 19. Axial sections depicting mass distribution in the four projectile configurations treated. The symbols ρ_0 and $\bar{\rho}_0$ denote the local and average densities, respectively.

Projectile	$l = D$ (cm)	ρ_0 (g/cm^3)	$\bar{\rho}_0$ (g/cm^3)	Mass (g)
α	0.2774	2.7	2.7	0.04525
β	0.4886	2.7 or 0	0.5	0.04525
γ	0.4886	2.7 or 0	0.5	0.04525
δ	0.4886	0.5	0.5	0.04525

is assumed to be aluminum which has Hugoniot characteristics similar to stone.

The numerical treatment selected for solving the axisymmetric compressible fluid equations employs a particle-in-cell computer code, VISTA, for the early stages of the process in which it is desired to follow the distortion of the various projectile configurations in detail. Once the projectile has delivered momentum and energy to the target the computational data is written out on magnetic tape in a form suitable for continuing the calculations with an Eulerian computer code, PICWICK III. Since the liked VISTA–PICWICK III calculations employ a compressible fluid model it is necessary to select a strength-dependent criterion for predicting the final crater surface. Indirect procedures are still required since current numerical treatments cannot describe the full range of response regimes operative, and the cratering process is less amenable to decomposition into separate regimes than is the meteor bumper deformation sequence.

Two criteria have been considered in predicting final crater depth in thick targets subjected to hypervelocity impact. One is based on the concept of late-stage equivalence of two flow fields and the other is based on a comparison of the dynamic pressure in the flow field with the yield stress of the target.

The principle of late-stage equivalence does not predict the final crater dimension directly, but rather it provides a basis for comparing flow fields [Walsh and Tillotson (*20*)]. Since the pressure pulse in the target is the basic cause of the cratering mechanism, the flow fields produced by two projectiles in a specified target material are compared at times when the locations of the shock wave in the target are an equal distance below the target surface. For equivalence it is required that at these corresponding times the total forward axial momentum Z_+, the total outward radial momentum R_+, and the amplitude of the pressure pulse become equal prior to the onset of strength and strain-rate effects (*21*).

If two impact conditions produce equivalent flow fields in a target then the final craters are assumed identical. If late-stage equivalence is not realized between the impact condition of interest and one for which the final crater is known, however, the criteria cannot serve as a basis for crater predictions. It cannot be applied for the cases of heterogeneous projectiles of primary interest in this study.

The dynamic pressure criterion for predicting the final crater dimensions is based on the detailed nature of the flow field established in the impacted target, [Bjork *et al.* (*15*)]. By the time the projectile has expended itself and the cratering process in the target is in progress, the flow field is separated into two principal regions since the shock wave in the target propagates at a higher velocity than the rate of crater growth. This disengagement of the two regions means that after a certain point in time the shock wave has no further direct effect on the cratering process. The final crater depends on the momentum content of the target material in the neighborhood of the forming crater and its interaction with the residual strength of the material which has been processed by the earlier passage of the shock wave.

At a given location in the flow field surrounding the forming crater the velocity vector has the magnitude and direction defined by

$$v_n = (u^2 + v^2)^{1/2}, \qquad \tan \varphi = u/v$$

The local mass flux is given by

$$M_f = \rho v_n = \rho(u^2 + v^2)^{1/2}$$

The assumption that the momentum contained in this flux is perfectly absorbed by a surface normal to the velocity vector leads to the following expression for the dynamic pressure exerted on the surface at that location:

$$p_{\text{dyn}} = \tfrac{1}{2}\rho v_n^{\,2} = \tfrac{1}{2}\rho(u^2 + v^2) \tag{18}$$

In predicting the location of the final crater surface it is assumed that the dynamic pressure exerted there must equal the yield stress of the target material:

$$p_{\text{dyn}} = \sigma_y(T) \tag{19}$$

The target material will be left with a residual temperature T in the wake of the shock processing and, consequently, its yield strength will be decreased.

The criterion in Eq. (19) represents an extension to hypervelocity impact calculations of a criterion earlier employed to predict the depth of a crater produced in a target by a metallic jet formed by a shaped charge [Eichelberger (*22*)]. The jet penetration theory, however, is steady state and one dimensional and Bernoulli's law may be applied with more confidence.

In the present application the dynamic pressure distribution in the region of the target surrounding the forming crater changes with time and, consequently, Eq. (19) does not define a unique surface. Nevertheless, in the calculations performed in the present studies it is found that the surface thus defined remains very nearly stationary for an extended time and Eq. (19) does indeed locate a surface which will be interpreted as the predicted final crater surface.

Before proceeding to the specific results for the normal-density α projectiles it is convenient to indicate the manner in which the temperature dependence of the yield stress was determined. For a given shock amplitude to which the aluminum is subjected, there will correspond a residual temperature after the material is released to zero pressure which may be computed from the equation of state and thermodynamic data. The temperature variation of the yield strength $\sigma_y(T)$, of aluminum alloy Al 2024-T3 is available from the producer's handbook. Elimination of the temperature between these two relations yields a single curve depicting the yield strength of Al 2024-T3 after processing by a shock of amplitude p_H, Fig. 20. The curve is also assumed valid for Al 2024-T6.

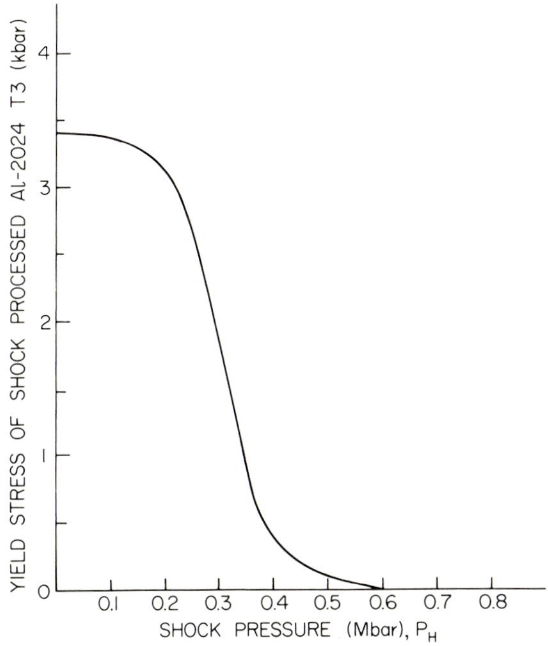

FIG. 20. Yield stress of aluminum alloy Al 2024 subsequent to the passage of a shock wave.

B. NORMAL-DENSITY PROJECTILES

The α projectile, normal-density aluminum, was treated for the case of impact into a thick aluminum alloy Al 2024 target at $v_0 = 0.735$ cm/μsec, Problem N1, and at $v_0 = 2.0$ cm/μsec, Problem N2, Table V. The VISTA–PICWICK III results for Problem N1 permit the calculated prediction of crater depth, based on the criterion in Eq. (19), to be compared with experimental data. The results for Problem N2 permit a test of the prediction procedure in the meteoroid velocity range. The latter prediction can be tested against the prediction based on the late-stage equivalence of the flow fields computed for N1 and N2.

Representative SC-4020 graphical displays of the projectile–target configuration and velocity field for the VISTA calculations are shown in Fig. 21 for Problem N1. At $t = 0.434\ \mu$sec, the VISTA data were read out on magnetic tape and the calculation then resumed with PICWICK III. The representative displays of the mass flux field, $(\rho u, \rho v)_{ij}$, for Problem N1, shown in Fig. 22, demonstrate the separation of the flow field in the target into two principal regions in which the mass flux,

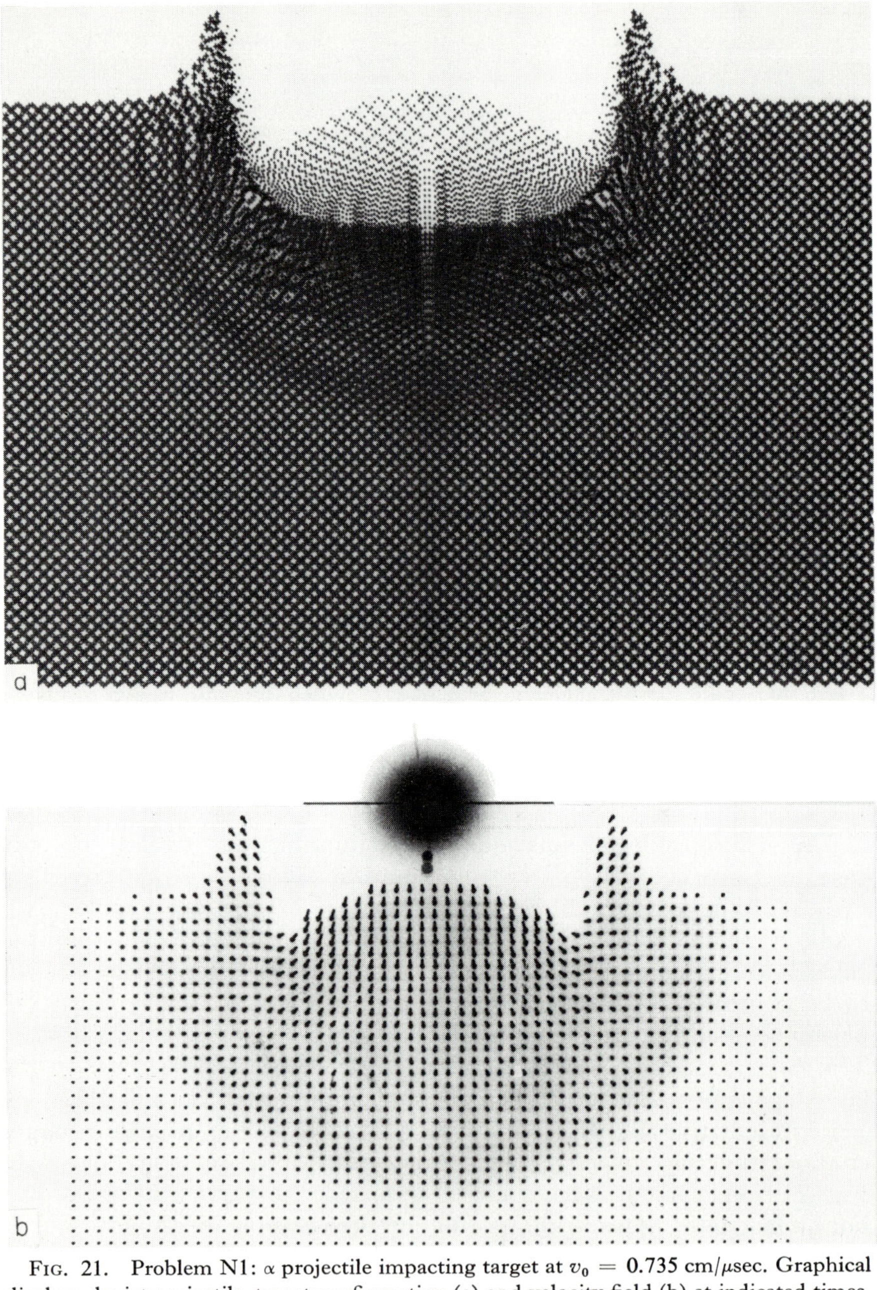

FIG. 21. Problem N1: α projectile impacting target at $v_0 = 0.735$ cm/μsec. Graphical displays depict projectile–target configuration (a) and velocity field (b) at indicated times. (a) $t = 0.452$ μsec, (b) $t = 0.415$ μsec.

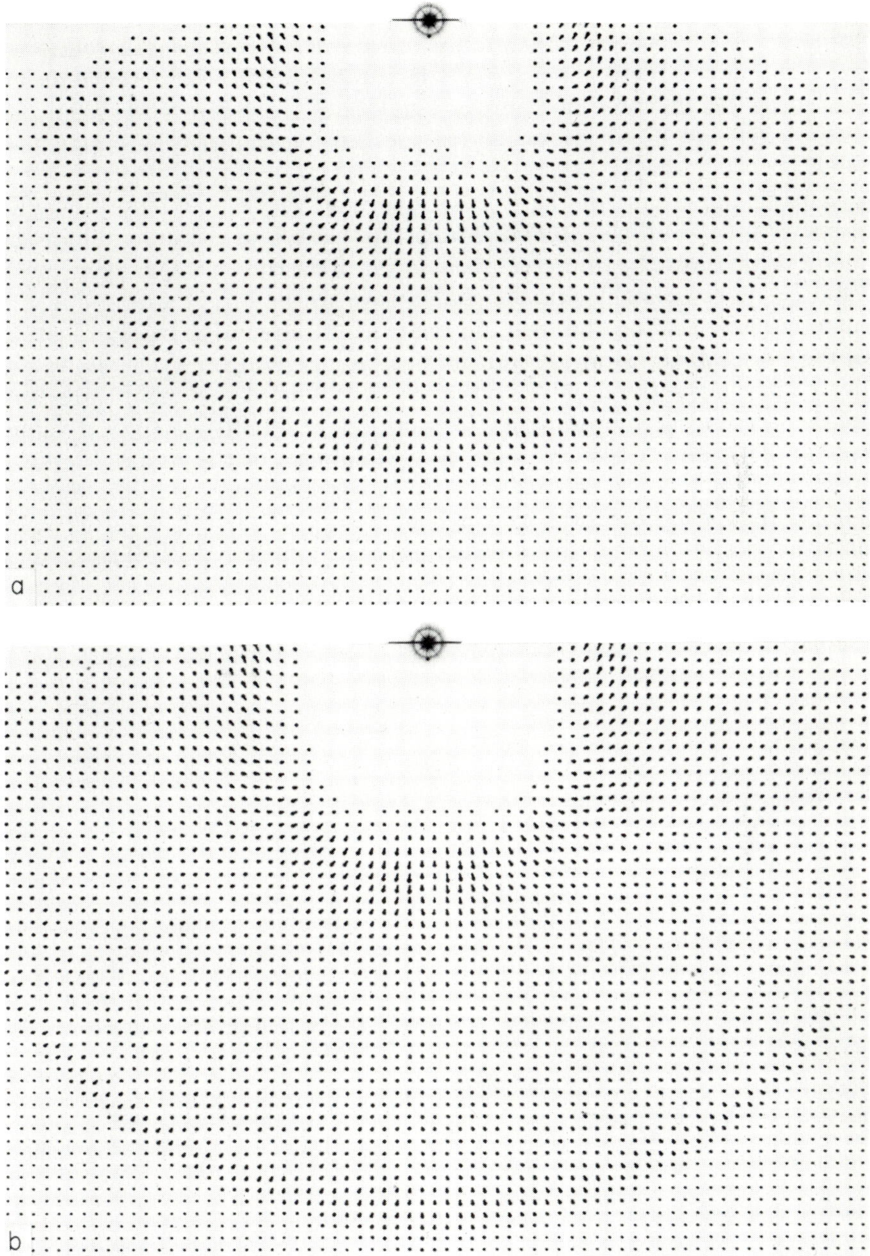

FIG. 22. Problem N1: α projectile impacting target at $v_0 = 0.735$ cm/μsec. Graphical displays depict mass flux field at indicated times. (a) $t = 1.623$ μsec, (b) $t = 2.297$ μsec.

and hence the dynamic pressure, has peak values. One region is associated with the receding shock front and the other with the material surrounding the forming crater. It was necessary to rezone the finite difference mesh, four cells being combined into a single cell, between times depicted in Figs. 21 and 22, in order for the mesh to encompass the entire flow field.

During the course of the calculations with PICWICK III the dynamic pressure was monitored in each computational cell of the finite difference mesh. Much lower values are attained between the two regions of peak values, one associated with the shock wave in the target and one in the material surrounding the forming crater. Consequently, an examination of the PICWICK III computer output data at a given time cycle allows the determination of the depth into the target at which the dynamic pressure in the region surrounding the forming crater attains a specified value. In Fig. 23 the lowest such point in each column of the finite difference mesh is depicted for two distinct times for Problem N1. The dots in Fig. 23 indicate the depths at which $p_{\mathrm{dyn}} = 3.4$ kbar for $t = 1.62\ \mu$sec, and the dashes locate this delineation criterion at $t = 2.30\ \mu$sec after

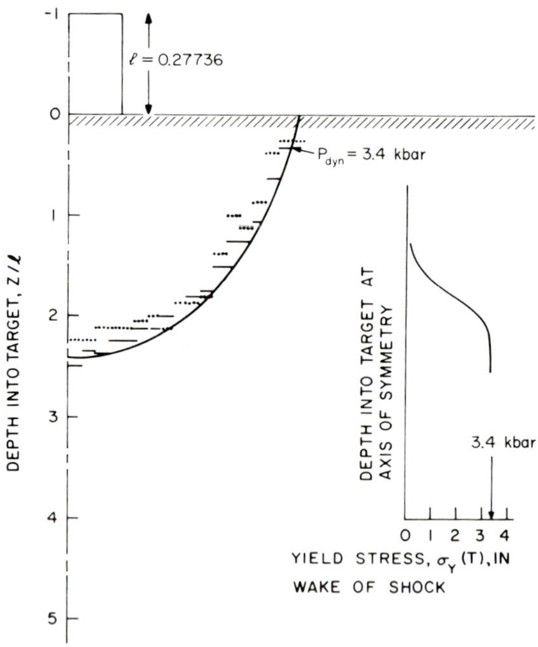

FIG. 23. Problem N1: α projectile impacting target at $v_0 = 0.735$ cm/μsec. Crater prediction based on dynamic pressure criterion.

impact. There is very little variation for this extended time interval and a smooth surface has been sketched representing the approximate envelope of these depths.

The computed axial peak pressure $p_{\max}$ attained at varying distances into the target was also determined during the course of the calculations. By combining this data with the curve in Fig. 20, it is possible to estimate the residual yield stress $\sigma_y(T)$ of the target material, at various depths, in the wake of the shock to which the material at that depth has been subjected. These values of $\sigma_y(T)$ are also plotted in Fig. 23. Since the $p_{\text{dyn}} = 3.4$-kbar surface at the axis corresponds to a value $\sigma_y(T) = 3.4$ kbar, the surface will be assumed to represent the final crater surface according to the criterion of Eq. (19). Consequently, the final depth P_c and radius R_c of the crater are predicted on this basis to be

$$\begin{gathered} P_c/l = 2.4, \qquad R_c/l = 2.2, \qquad P_c/R_c = 1.09 \\ P_c = 0.66 \quad \text{cm}, \qquad R_c = 0.61 \quad \text{cm} \end{gathered} \tag{20}$$

Similar results obtained for Problem N2 are presented in Fig. 24. The upper dots and dashes in Fig. 24 indicate the depths surrounding the forming crater at which $p_{\text{dyn}} = 3.4$ kbar for $t = 1.69$ μsec and $t = 2.70$ μsec after impact, respectively. The dashed curve is the approximate envelope of these locations and would represent the predicted final crater if there were no decrease in the yield stress $\sigma_y(T)$ for the shock processed material on this surface. Along the axis, the $p_{\text{dyn}} = 3.4$-kbar surface corresponds to a yield stress of $\sigma_y(T) = 1.0$ kbar and, consequently, does not satisfy the criterion of Eq. (19). The solid curve in Fig. 24 represents the approximate envelope of the depths corresponding to $p_{\text{dyn}} = 2.0$ kbar. Since it corresponds to the yield stress, $\sigma_y(T) = 2.0$ kbar, of the shock processed material at that depth along the axis, it represents the predicted crater surface on the basis of Eq. (19):

$$\begin{gathered} P_c/l = 4.4, \qquad R_c/l = 4, \qquad P_c/R_c = 1.1 \\ P_c = 1.22 \quad \text{cm} \qquad R_c = 1.11 \quad \text{cm} \end{gathered} \tag{21}$$

It is of interest to see what exponent α, for the velocity dependence of the penetration depth, is implied by Eqs. (20) and (21), $P_c \sim v_0{}^{\alpha}$:

$$(2.0/0.735)^{\alpha} = 4.4/2.4$$

where

$$\alpha = \log 1.83/\log 2.72 = 0.60 \tag{22}$$

A value of $\alpha = 0.67$ would correspond to pure energy scaling, whereas a value of $\alpha = 0.33$ would correspond to pure momentum scaling of the

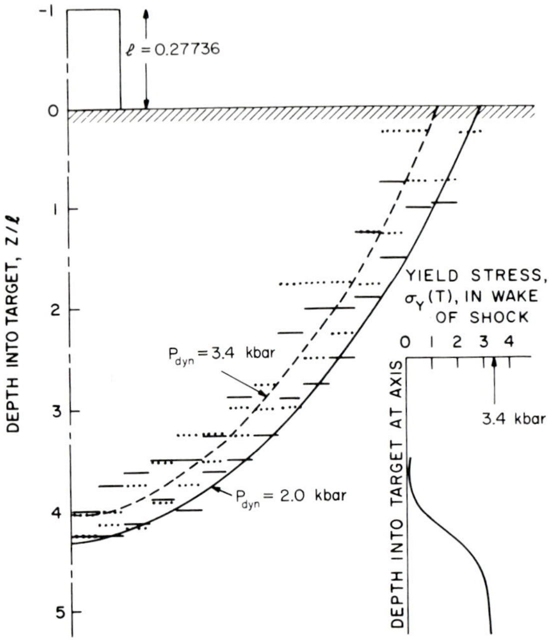

FIG. 24. Problem N2: α projectile impacting target at $v_0 = 2.0$ cm/μsec. Crater prediction based on dynamic pressure criterion.

velocity effect on penetration depth. The sensitivity of the estimate for α becomes apparent when it is observed that a change of P_c/l in Eq. (21) to 4.3 would yield an estimate of $\alpha = 0.58$; if the change were simultaneous with a decrease in the estimate of P_c/l in Eq. (20) to 2.3 then one would obtain $\alpha = 0.54$.

C. Verification of Predictions

The prediction for Problem N1, Eq. (20), can be tested against available experimental data. A survey of the literature reveals six data points with impact conditions very close to that simulated in this calculation (*23–25*). Two of the thick targets are aluminum alloy Al 2024-T6 and four are Al 2014-T6. These data are listed in Table VI. To provide a direct comparison with the prediction of Eq. (20) it has been necessary to adjust the data to allow for small variations in the projectile size and velocity and to assume that the crater resulting from a sphere is the same as a unit aspect cylinder of the same mass. The slight adjustments, made under the assumption of energy equivalence, are

TABLE VI

EXPERIMENTAL DATA GIVING CRATER DEPTH PRODUCED IN THICK ALUMINUM ALLOY TARGETS BY ALUMINUM SPHERICAL PROJECTILES[a]

Ref.	Shot no.	Target	v_0	D_s	P_c/D_s	v_0'	$(P_c/D_s)'$	$(P_c/l)'$	P_c'
23	—	Al 2014-T6	0.695	0.635	2.02	0.735	2.09	2.39	0.663
24	5-268	Al 2014-T6	0.775	0.4763	2.06	0.735	2.14	2.45	0.680
24	5-281	Al 2014-T6	0.735	0.4763	2.00	0.735	2.00	2.29	0.636
24	5-283	Al 2014-T6	0.697	0.4763	2.01	0.735	2.08	2.38	0.660
25	D-906	Al 2024-T6	0.753	0.3175	2.17	0.735	2.21	2.53	0.702
25	D-898	Al 2024-T6	0.789	0.3175	2.09	0.735	2.14	2.45	0.680

[a] The primes denote the data extrapolated over the small changes of size and velocity required for comparison with the predicted value for Problem N1. Units are in the gram-centimeter-microsecond system.

also listed in Table VI. The prediction $P_c = 0.66$ cm, Eq. (20), is seen to lie near the midpoint of the adjusted experimental observations.

It is also of interest to reexamine the flow field calculations for problems N1 and N2 on the basis of the principle of late-stage equivalence. Since the hydrodynamic equations may be scaled, the calculations actually apply for any geometrically similar aluminum–aluminum impact configuration at corresponding velocity. The results of Problem N1 will be compared with those of Problem N2 on an equal energy basis:

$$\begin{aligned} v_0 &= 2.0 \quad \text{cm}/\mu\text{sec}, \qquad l = D = 0.27736 \quad \text{cm} \\ t'/t &= l'/l = D'/D = (v_0/v_0')^{2/3} = (2.721)^{2/3} = 1.949 \\ \text{mass}'/\text{mass} &= (l'/l)^3 = (v_0/v_0')^2 = 7.404 \\ v_0' &= 0.735 \quad \text{cm}/\mu\text{sec}, \qquad l' = D' = 0.5406 \quad \text{cm} \end{aligned} \tag{23}$$

Distances and times computed for Problem N1 are thus multiplied by 1.9494; the computed total forward momentum Z_+ and the total outward radial momentum R_+ present within the system at any instant of time, Eq. (15), are multiplied by 7.404. The scaled N1 problem will be denoted by N1′.

In Fig. 25a the time-dependent values of Z_+ and R_+ are compared for Problems N1′ and N2. The time-dependent positions of the shock front in the target, for each of the two equal energy cases, have also been plotted. These latter curves show that the shock produced by the larger but slower projectile, Problem N1′, lags behind the one produced by

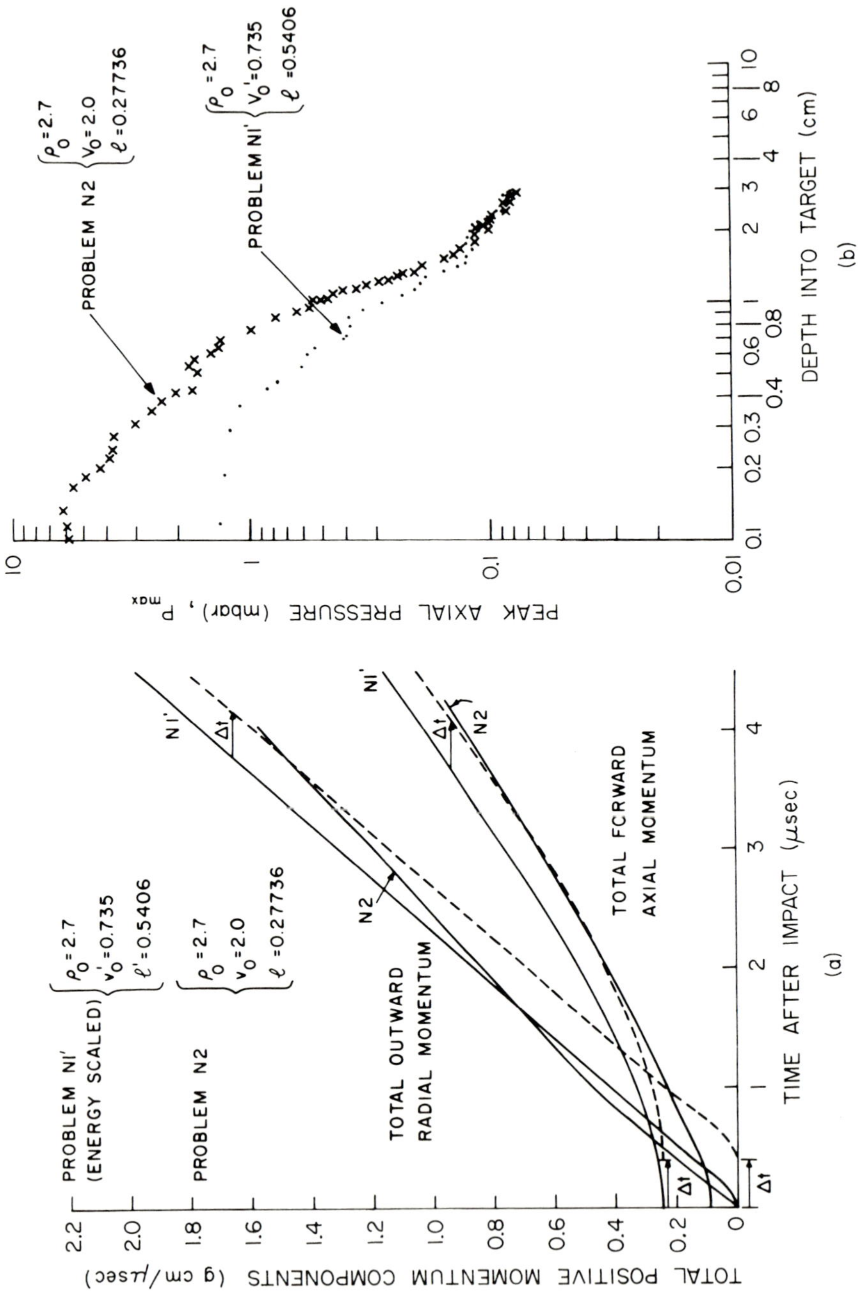

FIG. 25. Test of late-stage equivalence of Problem N1 when scaled for comparison with Problem N2 on equal energy basis. (a) Total positive momentum, (b) axial pressure amplitude.

the smaller but higher-velocity projectile, Problem N2, by $\Delta t \doteq 0.4$ μsec. The results with this correction, corresponding to a translation along the time axis of the R_+ and Z_+ curves for Problem N1′, are also shown in Fig. 25a.

The calculated peak pressures realized at various target depths are depicted in Fig. 25b for Problems N1′ and N2. The values attained are very close for depths greater than 1.5 cm, corresponding to pressures less than 110 kbar.

The late-stage equivalence criterion is seen from Figs. 25a–b to be very nearly satisfied by energy scaling of the velocity. This implies that the exponent α, for the velocity dependence of the penetration depth, $P_c \sim v_0^\alpha$, is very near to $\alpha = 0.67$. Actually, the corrected values of R_+ and Z_+ for Problem N1′ are somewhat greater than the corresponding curves for Problem N2 and a slightly smaller value of α would be within the accuracy of the computations. The application of the late-stage equivalence criterion thereby leads to results compatible with the prediction $\alpha = 0.60$ obtained by directly applying the dynamic pressure criterion to Problems N1 and N2 [Eq. (22)].

D. Predictions for Low-Density Projectiles

The β, γ, and δ projectiles, each of average bulk density of 0.5 g/cm³, were all treated for the case of impact at $v_0 = 2.0$ cm/μsec, Table V. These problems, N3–N5, respectively, are set up to study the effect of the mass distribution within the projectile. The β projectile was also considered at impact velocities of $v_0 = 0.735$ and 5.0 cm/μsec. The corresponding problems, N6 and N7, respectively, together with N3, are examined to determine the effect of impact velocity for heterogeneous projectiles.

Typical graphical displays of the flow processes are presented in Figs. 26 and 27 for Problem N3. The initial representation of the β projectile and deformed projectile–target configuration predicted by the VISTA calculations are shown in the former. The shape of the shock front in the target becomes more regular as it propagates and at the time of the link to PICWICK III, $t = 0.703$ μsec, the distortions are apparent only near the surface of the forming crater. At this time the partition of the momentum and energy between projectile and target materials is varying smoothly with time. The radial momentum content of the projectile then is less than 10% and the axial momentum content is actually negative. Much of the total energy carried by the projectile material, approximately 25%, is contained in the blowoff or backsplash of the projectile, Fig. 26b. Mass flux plots of the subsequent PICWICK

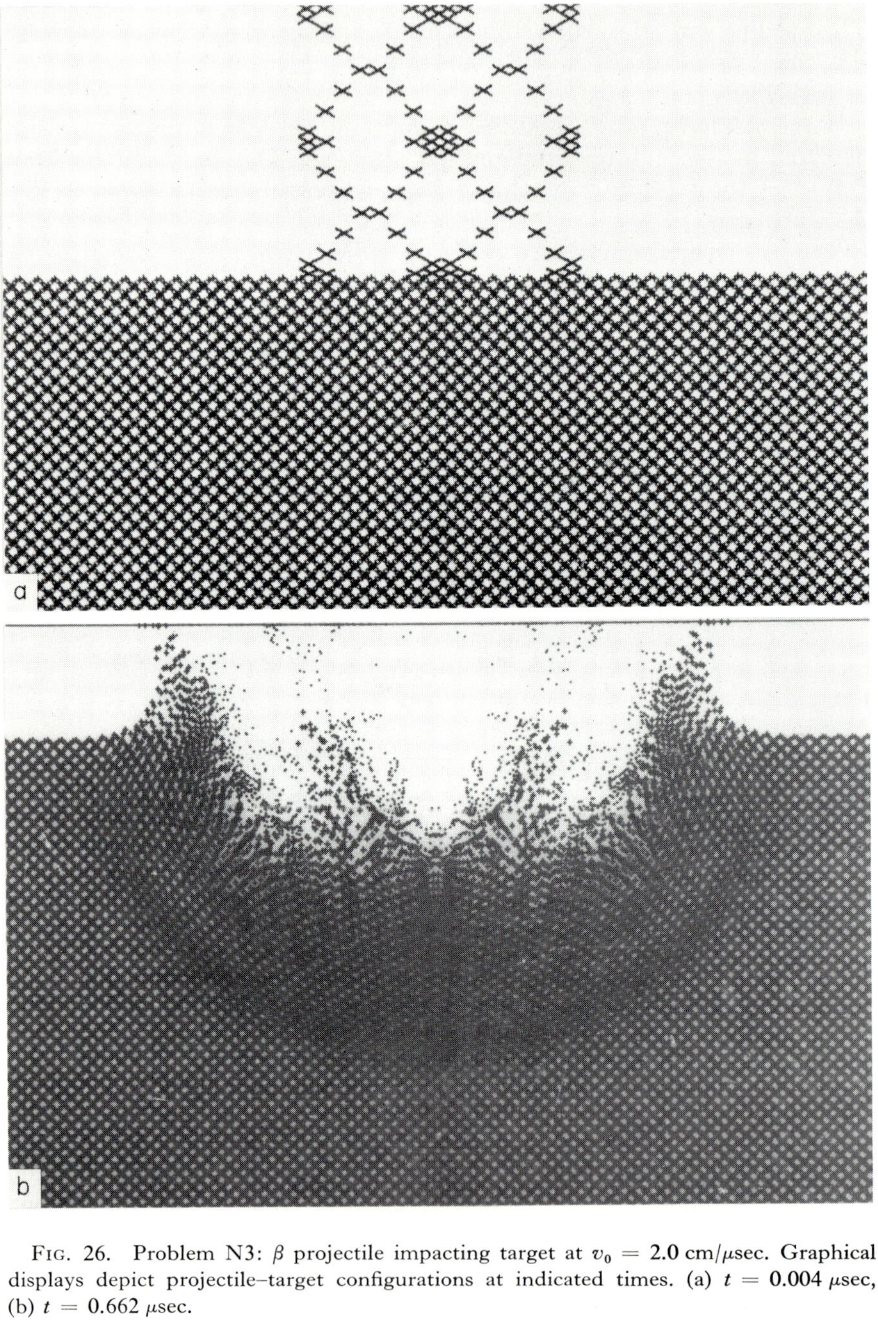

FIG. 26. Problem N3: β projectile impacting target at $v_0 = 2.0$ cm/μsec. Graphical displays depict projectile–target configurations at indicated times. (a) $t = 0.004$ μsec, (b) $t = 0.662$ μsec.

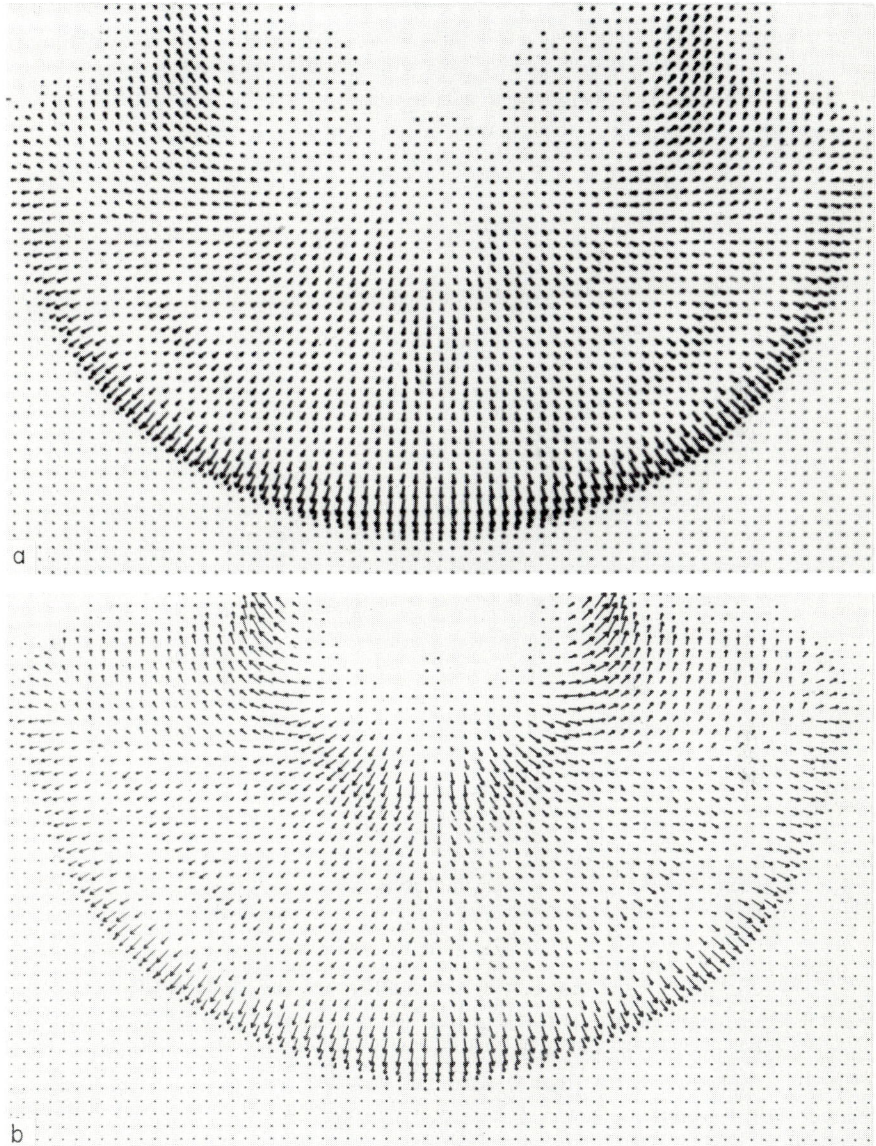

FIG. 27. Problem N3: β projectile impacting target at $v_0 = 2.0$ cm/μsec. Graphical displays depict mass flux at indicated times. (a) $t = 1.121$ μsec, (b) $t = 2.719$ μsec.

calculations, Fig. 27, show a relatively smooth flow field which separates into two regions of peak dynamic pressure.

As discussed in Section VI.B for normal-density projectiles, an examination of the PICWICK III data at a given time cycle allows the determination of the depth into the target at which the dynamic pressure in the region surrounding the forming crater attains a specified value. In Fig. 28 the upper dots indicate the depths at which $p_{\text{dyn}} = 3.4$ kbar

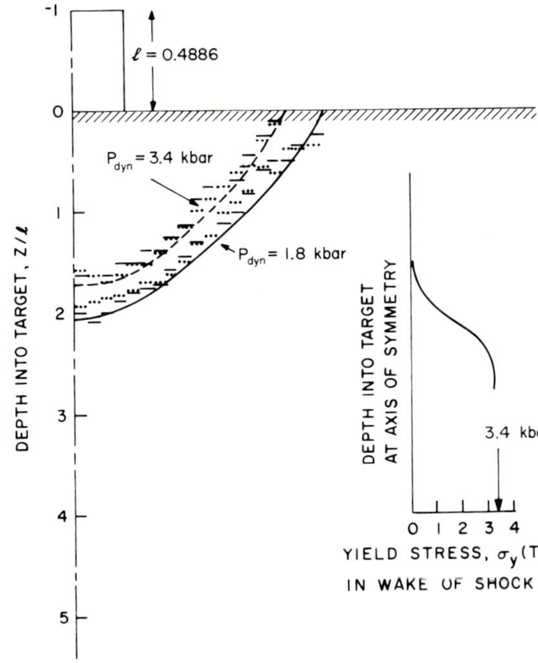

FIG. 28. Problem N3: β projectile impacting target at $v_0 = 2.0$ cm/μsec. Crater prediction based on dynamic pressure criterion.

for $t = 1.85\ \mu$sec, and the upper dashes locate the delineation criterion for $t = 3.17\ \mu$sec after impact. The dashed curve forming the approximate envelope of these locations would represent the predicted final crater if there were no decrease in the yield stress $\sigma_y(T)$ for the shock processed material. Along the axis, however, the $p_{\text{dyn}} = 3.4$-kbar surface corresponds to a yield stress of $\sigma_y(T) = 0.3$ kbar and, consequently, does not satisfy Eq. (19). The solid curve in Fig. 28 represents the approximate envelope of the depths corresponding to $p_{\text{dyn}} = 1.8$ kbar and, since it corresponds to a residual yield stress of approximately the same value, it represents the predicted crater surface according to Eq. (19):

$$P_c/l = 2.1, \qquad R_c/l = 2.4, \qquad P_c/R_c = 0.87$$
$$P_c = 1.03 \quad \text{cm}, \qquad R_c = 1.17 \quad \text{cm} \tag{24}$$

The only difference in Problems N2 and N3 is the distribution of the mass in the α and β projectiles, respectively. Comparison of Eqs. (21) and (24) shows that the low bulk density of the projectile in Problem N3 results in a decrease in crater depth of about 20%, but an increase in the radius of the crater of approximately 6%. A comparison of the time-dependent values of Z_+ and R_+ was made for these two problems. It was found that even with the correction for the time lag Δt the corresponding R_+ and Z_+ are far apart. Thus late-stage equivalence is not attained for these two problems even though the momentum and energy delivered by the projectiles are identical. This negative result is consistent with the crater predictions arrived at through the application of the dynamic pressure criterion. For, if late-stage equivalence were actually obtained, the final crater shapes and dimensions for Problems N2 and N3 would be predicted to be the same.

TABLE VII

PRESSURE, SHOCK VELOCITY, AND COMPRESSION IN PROJECTILE AND TARGET AT IMPACT[a]

	$v_0 = 0.735$	$v_0 = 2.0$	$v_0 = 5.0$
A. Projectile and target of normal-density aluminum:			
p_H	1.032	4.904	24.0
$\dot{R}_H$	1.039	1.810	3.550
ρ_H/ρ_0	1.547	2.242	3.390
B. Target of normal density, projectile of reduced density 0.5 g/cm^3:			
p_H	0.238	1.498	8.408
$\dot{R}_H$ (target)	0.712	1.174	2.248
$\dot{R}_H$ (projectile)	0.777	1.961	4.649
ρ_H/ρ_0 (target)	1.210	1.672	2.601
ρ_H/ρ_0 (projectile)	0.869	0.838	0.833

[a] Values computed using equations of state (9) and (10), respectively, for normal-density and reduced-density aluminum. Units are in the gram-centimeter-microsecond system.

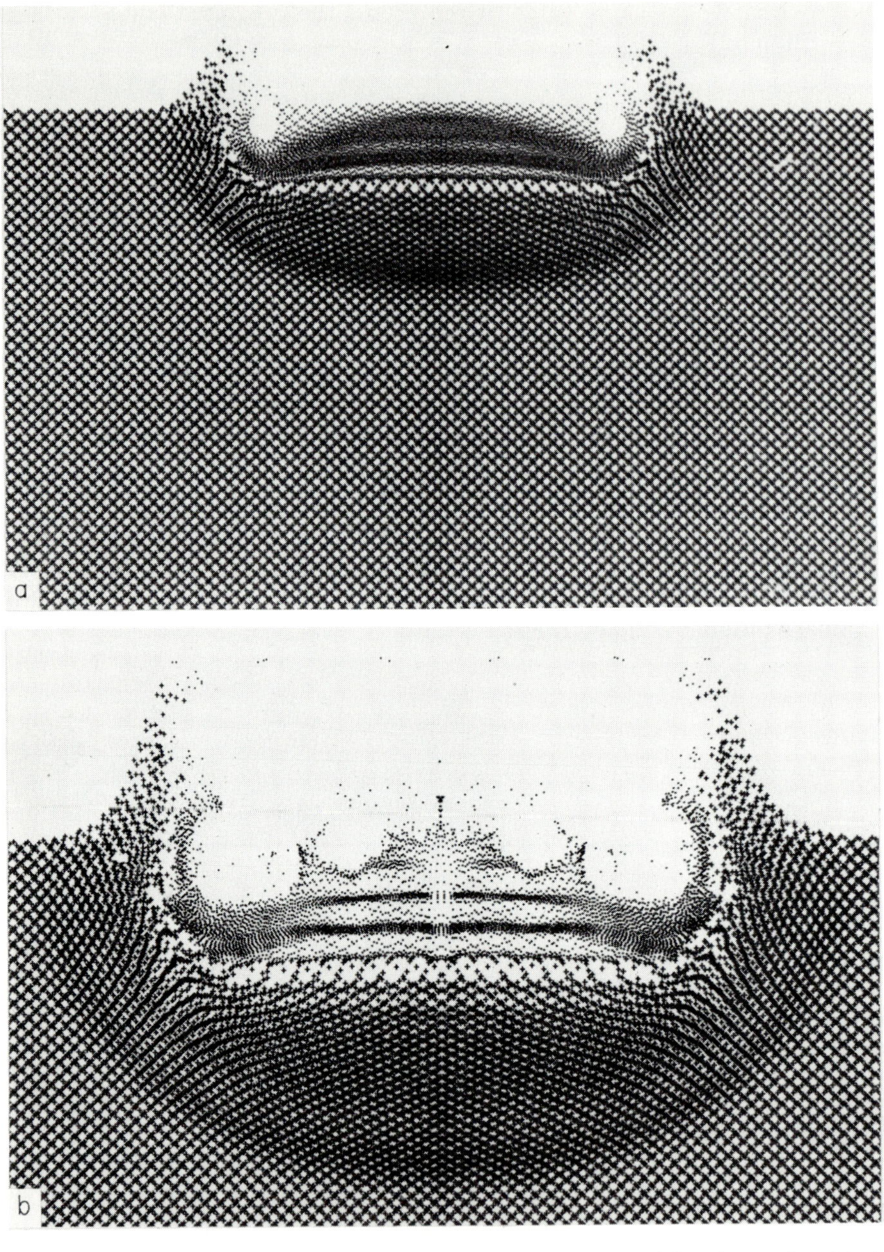

FIG. 29. Problem N5: δ projectile impacting target at $v_0 = 2.0$ cm/μsec. Graphical displays depict projectile–target configuration at indicated times. (a) $t = 0.238$ μsec, (b) $t = 0.418$ μsec.

The results for Problem N4, γ projectile impacting at $v_0 = 2.0$ cm/μsec, are similar to those for N3. The heterogeneity of the projectile produces an irregular flow field in the early stages that becomes smoother with time.

Quite different early-stage mechanisms are produced by the homogeneous low-density δ projectile impacting at $v_0 = 2.0$ cm/μsec. The lower-density projectile, $\bar{\rho}_0 = 0.5$ g/cm^3, results in an impact pressure less than one-third that produced by a projectile of normal impact pressure density, Table VII. In this case, Problem N5, the projectile is not rapidly torn apart by inner and outer radial flow, but is almost uniaxially compressed until the crush-up wave travels upward to the rear surface of the projectile, Fig. 29. Equal pressures are obtained in the projectile and target at the interface, $p_H = 1.498$ Mbar, but the mass density of the target material remains greater than the compressed projectile material. The specific internal energy in the projectile material is correspondingly higher than it is in the target material. The rarefaction wave originating from the rear surface of the crushed-up projectile leaves the low-density material with a very high residual internal energy causing a rapid blowoff of the projectile material, Fig. 29b. By the time the link to PICWICK III, $t = 0.7117$ μsec, only 5% of the total radial and axial momentum remains in the projectile material and much of the 25% of the energy remaining is contained in the projectile material that is blown off from the forming crater.

In Table V the predicted dimensions of the final crater based on the dynamic pressure criterion for these two problems are listed together with the three discussed earlier. Also listed are the predictions deduced from similar treatments of the β projectile impacting at $v_0 = 0.735$ cm/μsec, Problem N6, and at $v_0 = 5.0$ cm/μsec, Problem N7.

E. Effect of Projectile Mass Distribution

Table V also lists the approximate volume of each of the craters, computed from

$$V_c \doteq \tfrac{2}{3}\pi P_c R_c^2 \tag{25}$$

These values divided by the projectile energy, V_c/E, represent the cratering efficiency. Since an uncertainity of 5% in the linear dimensions P_c and R_c results in an uncertainity of 15% in V_c/E, however, the values of $(V_c/E)^{1/3}$ are given in Table V. The listed values for the predicted crater depths, normalized through division by the cube root of the projectile mass, are also displayed graphically in Fig. 30a. The results

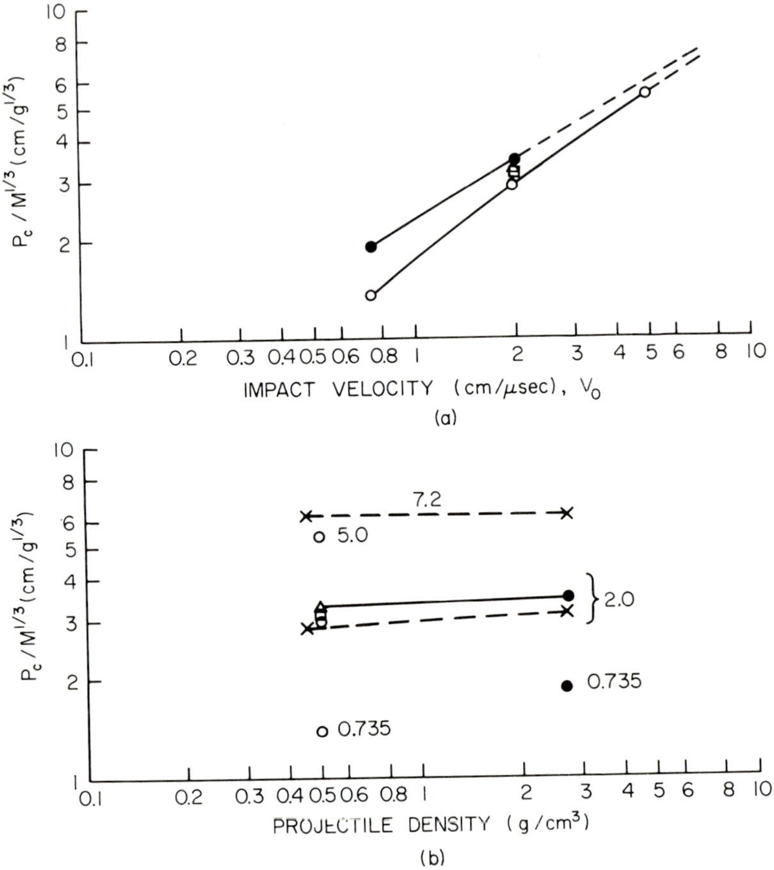

FIG. 30. Predicted crater depth divided by cube root of projectile mass depicted as function of (a) impact velocity and (b) projectile density. Numbers in (b) refer to impact velocity in centimeters per microsecond. (●) α projectile, (○) β projectile, (□) γ projectile, (△) δ projectile, (×) Bjork *et al.* (*15*).

demonstrate that the mass distribution of a projectile impacting an aluminum alloy Al 2024-T6 target has the following effects:

(1) At $v_0 = 0.735$ cm/μsec the heterogeneous β projectile of average bulk density 0.5 g/cm³ produces a crater approximately 75% as deep as that produced by the normal-density, $\rho_0 = 2.7$ g/cm³, α projectile of the same mass.

(2) At $v_0 = 2.0$ cm/μsec the heterogeneous β and γ projectiles and the homogeneous δ projectile, all of average bulk density 0.5 g/cm³, produce craters of depths from 85 to 95% as deep as that produced by

the normal-density α projectile. The 85% corresponds to the β projectile and the 95% corresponds to the homogeneous δ projectile.

(3) At $v_0 = 5.0$ cm/μsec the heterogeneous β projectile produces a crater approximately 90% as great as the extrapolated value of crater depth for the normal-density α projectile.

(4) The depth of the crater for the normal-density α projectile exceeds its radius by about 10%, Table V. For the low-density β, γ, and δ projectiles, however, the opposite is true, the depth being from 83 to 92% of the radius.

(5) The parameter $(V_c/E)^{1/3}$ in Table V indicates the crater volume produced per unit of impact energy. For a given projectile geometry the value tends to decrease slightly with increasing impact velocity. At $v_0 = 2.0$ cm/μsec, the homogeneous δ projectile removes a greater amount of target material than do the other three configurations.

The results here may be compared, where there is overlap, with those reported by Bjork *et al.* (*15*). There the dynamic pressure criterion was employed in the particle-in-cell calculations for homogeneous reduced-density projectiles impacting thick steel and aluminum targets. Homogeneous aluminum projectiles of densities 0.44 and 2.7 g/cm³ were treated for impact onto aluminum targets (Al 2024-T3) at $v_0 = 2.0$ and 7.2 cm/μsec. These four crater estimates, together with the present predictions, are displayed in Fig. 30b where dashed lines join points corresponding to low-density and normal-density homogeneous projectiles of the same impact velocity. The slopes of the dashed lines are quite small as is the slope of the solid line joining the present predictions for the δ and α projectiles at impact velocity of $v_0 = 2.0$ cm/μsec Problems N2 and N5. Although there is agreement relative to the insensitivity of the penetration of homogeneous projectiles in the impact range, $v_0 \geqslant 2.0$ cm/μsec, the present study predicts craters approximately 15% deeper for corresponding impact conditions.

REFERENCES

1. T. D. Riney and E. J. Halda, *AIAA J.* **6**, 338 (1968).
2. T. D. Riney and S. M. Briggs, *in* "Developments in Mechanics" (J. E. Cermak and J. R. Goodman, eds.), Vol. IV, p. 703. Pergamon Press, Oxford, 1968.
3. T. D. Riney and J. F. Heyda, Topical Rept., Contract No. NAS 3-8512, NASA Lewis (1968).
4. F. L. Whipple, "Physics and Medicine of the Upper Atmosphere" (C. S. White and O. O. Benson, Jr., eds.), p. 137. Univ. of New Mexico Press, Albuquerque, 1952.
5. C. J. Maiden, A. R. McMillan, and R. E. Sennett, General Motors DRL Rept. TR-64-61 (1964).

6. C. J. Maiden and A. R. McMillan, Preprint No. 64-95, AIAA Aerospace Sciences Meeting (1964).
7. J. F. Heyda, *in* "Developments in Mechanics" (J. E. Cermak and J. R. Goodman, eds.), Vol. IV, p. 741. Pergamon Press, Oxford, 1968.
8. H. A. Balmer, Mass. Inst. Tech. Rept. ASRL TR-128-3 (1965).
9. F. H. Harlow, Los Alamos Scientific Lab. Rept. LA-2301 (1959).
10. M. L. Wilkins and R. Giroux, Lawrence Radiation Lab. Rept. UCRL-7322 (1963).
11. W. E. Johson, General At. Document No. GAMD-5580 (1965).
12. R. T. Sedgwick and S. R. Woodall, Interim Rept., Contract No. F-08635-67-C-0079, Eglin AFB (1967).
13. S. B. Kormer, A. I. Funtikov, V. D. Urlin, and A. N. Kolesnikova, *Soviet Phys. JETP (English Transl.)* **15**, 477 (1962).
14. D. J. McCloskey, Rand Corp. Rept. RM-3095-PR (1964).
15. R. L. Bjork, K. N. Kreyenhagen, and M. H. Wagner, NASA CR-757 (1967).
16. D. H. Humes, R. D. Hopko, and W. H. Kinard, *Proc. Hypervelocity Impact Symp. 5th* **1**, Part 2, 567 (1962).
17. D. H. Humes, NASA TN-D-3104 (1965).
18. V. D. Urlin, *Soviet Phys. JETP (English Transl.)* **2**, 341 (1966).
19. R. L. Warnica and D. R. Christman, *Proc. Hypervelocity Impact Symp. 7th* **7**, 143 (1965).
20. J. M. Walsh and J. H. Tillotson, *Proc. Hypervelocity Impact Symp., 6th* **2**, Part 1, 59 (1963).
21. T. D. Riney, *AIAA J.* **3**, 52 (1965).
22. R. J. Eichelberger, *J. Appl. Phys.* **27**, 63 (1956).
23. S. M. Halperson, *Proc. Hypervelocity Impact Symp., 6th* **2**, Part 2, 525 (1963).
24. S. M. Halperson, *Proc. Hypervelocity Impact Symp., 7th* **5**, 235 (1965).
25. A. B. Wenzel and N. Clough, NASA TM (to be published).

CHAPTER VI

*ANALYTICAL STUDIES OF IMPACT-GENERATED SHOCK PROPAGATION: Survey and New Results**

WILLIAM J. RAE

CORNELL AERONAUTICAL LABORATORY, INC.

BUFFALO, NEW YORK

* The preparation of this review was supported by the Cornell Aeronautical Laboratory Internal Research Program. The writer is very grateful to Dr. Ronald G. Rehm, Dr. T. R. Sundaram, and Dr. George Rudinger for many helpful comments.

List of Symbols

a speed of sound
a, b constants related to plane wave loading history, Eq. (67)
A^2 normalized sound speed (squared), Eq. (185)
A constant in power-law shock trajectory
B amplitude coefficient, Eq. (65)
A_1, A_2 constants in the asymptotic solution of the plane wave, perfect-gas case, at large negative ζ
B_1, B_2 constants in the solution near the shock–free–surface intersection
c weak-wave speed in the linear shock-speed–particle-speed relation
c_p specific heat at constant pressure
$D(\tilde{M}; s, \delta)$ function defined in Eq. (160)
e internal energy per unit mass
$\hat{e}_3$ unit vector normal to the z, y plane
E projectile kinetic energy, axisymmetric case
$\mathscr{E}$ kinetic energy per unit area, plane wave case
f normalized pressure, $p/\rho_0\dot{z}_{s0}^2$
$F(\gamma)$ amplitude of power-law shock trajectory, axisymmetric case
$F(p, \rho)$ functional form of equation of state
$F(z, y, t)$ function whose zero locus is the shock location
F_1, F_2, F_3, F_4 similitude functions, defined in Eqs. (164)–(166)
fcn$(\cdot\cdot, \cdot\cdot, \cdot\cdot)$ denotes functional dependence on the arguments displayed
g normalized internal energy, $e/\dot{z}_{s0}^2$
$G(\gamma)$ function defined in Eq. (62)
$\mathscr{G}(\cdot\cdot)$ loading function, Eq. (66)
$H(\gamma)$ function defined in Eq. (A21)
$I(\gamma)$ integral defined in Eq. (141)
k_1, k_2, k_3 constants in shock-shape expression
K constant in approximate off-axis pressure distribution, Eq. (135)
K_1 constant used in specifying boundary of neglected mass, Appendix
l coordinate giving distance from shock–free-surface intersection
$\mathscr{L}_0$ axisymmetric scaling length, Eq. (29)
$\mathscr{L}_1$ plane wave scaling length, Eq. (64)
L_0 projectile scale $(M/\rho_0)^{1/3}$: also used as the thickness of the impacting slab in the plane wave case.
L_1 initial depth of a plane of particles whose trajectories are found in Section IV.C.3
m_0 neglected mass
M projectile mass
M_s shock Mach number, q_s/c
M_{s0} shock Mach number along symmetry axis, $\dot{z}_{s0}/c$
$\tilde{M}$ value which the shock Mach number had when the shock crossed a particle of given entropy
N characteristic exponent in power-law shock trajectory; also used (Section IV.C.4) to denote the component, normal to a constant-β line, of the pseudovelocity vector
p pressure
$p_L(t)$ loading function, Eq. (66)

P	projectile momentum, axisymmetric case
P_{max}	amplitude of loading function, Eq. (66)
$\mathscr{P}$	momentum per unit area, plane wave case
q	particle velocity
q_s	shock velocity
r	spherical radius
R	y/L_0
R_0	scaling length for spherically symmetric explosions, Eq. (187)
R_s	spherical radius of shock
s	slope parameter in linear shock-speed–particle-speed relation
t	time
t_s	time when a given particle is processed by the shock
t_1	time when z_{s0} is equal to L_1
T	component of pseudovelocity vector tangent to constant-β lines
T_{max}	duration of loading function, Eq. (66)
u	z component of particle velocity
U	projectile speed before striking the target
v	y component of particle velocity
V	$(\partial\omega/\partial\sigma)_{\sigma=0}$
x	parameter in Enig's equation of state, Eq. (155)
y	cylindrical radius
z	axial coordinate
z_{s0}	distance from origin to the shock, along the symmetry axis
$\dot{z}_{s0}$	shock speed along the symmetry axis
Z	z/L_0
α	similitude parameter
$\bar{\alpha}$	coefficient of volume expansion
β	angle measured from free surface
γ	specific heat ratio
Γ	Grüneisen factor
δ	$\rho C_P/\bar{\alpha}\rho_0 c^2$, Eq. (161)
$\Delta(\rho)$	function defined in Eq. (148)
ζ	normalized axial coordinate, z/z_{s0}
η	normalized spherical radius, r/z_{s0}
θ	spherical polar angle
λ	similitude parameter; see Eqs. (68) or (87) for connection between λ and N in the plane and axisymmetric cases, respectively
ρ	mass density
σ	normalized cylindrical radius, y/z_{s0}
σ_1	value of σ at the shock–free-surface intersection
τ	Ut/L_0
ϕ	normalized axial component of particle velocity, $u/\dot{z}_{s0}$
$\Phi(\Psi)$	function defined in Eqs. (153)–(155)
$\varphi(\rho)$	used in Eq. (24) to denote general density dependence in state equation that allows self-similar solutions
ψ	normalized density, ρ/ρ_0
Ψ	function defined in Eqs. (153)–(155)
ω	normalized radial component of particle velocity, $v/\dot{z}_{s0}$
$(\ )_c$	denotes cohesive contribution
$(\ )_H$	denotes evaluation on the Hugoniot
$(\ )_{UL}$	denotes upper limit of integration
$(\ )_{cs}$	denotes conditions at a control surface
$(\ )_R$	denotes head of rarefaction fan
$(\ )_{s0}$	denotes conditions at the shock, along the axis of symmetry
$(\ )_s$	denotes conditions at a general point on the shock
$(\ )_{0,1}$	denotes conditions ahead of, behind, the shock
$(\ )_{(0),(1),\dots}$	denotes order of the coefficient in an expansion in powers of l
$(\dot{\ })$	denotes d/dt
$(\hat{\ })$	denotes a unit vector

I. Introduction

Many authors have treated the motion of a half-space of a condensed medium caused by the impact of a high-speed projectile or by the detonation of an explosive at the surface. Most of the available data are the results of experiments or of computer calculations. The purpose of the present work is to present a synthesis of what is known, based on the analytical properties of the governing differential equations.

In some theoretical treatments of shock propagation, the phenomena occurring are represented by an extremely simplified model, for which a solution can be written down without difficulty. Papers of this sort, which Eichelberger classified as "quasi-theoretical" in his 1963 review (*1*), are mentioned only briefly in the present work. The primary emphasis is given to theoretical studies that start from a full formulation of the problem and then proceed through a sequence of approximations to a simpler set of equations whose relation to the original problem can be clearly understood.

When solids collide at speeds on the order of several kilometers per second, the pressures generated reach megabar proportions. Because these pressures exceed the strengths of materials by many factors of ten, there is a regime where it is possible to neglect the effect of strength, and to treat the solid as an inviscid, compressible fluid. The steps leading to this approximation, and the resulting set of differential equations, are reviewed in Section II. The entire content of this work is devoted to the solution of these equations under a variety of further approximations. Thus the solutions described here apply only to that portion of the impact process where strength can be neglected. In any impact, of course, the pressures eventually decay, and the material strength then becomes the dominant factor, determining the final configuration in which the material comes to rest. The inviscid solutions of the present work offer no direct information on this final configuration. In some cases, they have been used to infer information about the final deformation, but a clear distinction must be made between what is found from the inviscid solution in its regime of validity, and what is inferred about the subsequent, strength-dominated response of the solid.

Much of our present understanding of the propagation of strong shock waves in gases was provided by the methods of self-similarity. Perhaps the most famous example of this approach is the treatment of the point-source explosion problem, by von Neumann (*2*), Sedov (*3*), Taylor (*4*), and others. In seeking solutions of the impact problem, it is natural to turn again to the similarity method. Unfortunately, there are two features of the impact problem that prevent this method from enjoying the same degree of success that it achieved in the earlier blast wave problems. The first is that the flow contains two spatial dimensions, and thus is described by a set of *partial* differential equations, even when similarity is assumed. The second is that the intensity of impact-generated shock waves is usually too small to permit a similarity assumption; flows of this type are characteristically non-self-similar. The full set of requirements for self-similarity is discussed in Section III.

One case for which self-similarity can be assumed occurs when the equation of state of the target is taken to be that of a perfect gas. A solution for this case was calculated by Walsh *et al.* (*5*). These numerical results were very generously made available to the present writer by Dr. Walsh, and have been used to determine the full two-dimensional, self-similar solution. This solution, described in Section IV, provides a great deal of information about the two-dimensional properties of the flow field, for the self-similar case.

The effects of nonsimilarity are considered in Section V. These effects can be attributed to the form of the state equation appropriate for a solid; thus the section begins with a review of various expressions for the equation of state. Following this, exact solutions are considered. It is shown that the "late-stage equivalence" similitude discovered by Walsh (*5–7*) holds for very general conditions, and has a rigorous analytical basis. The available details of the solution are reviewed, and several approximate treatments of the nonsimilar problem are discussed.

The concluding remarks, in Section VI, contain a summary of the present state of our knowledge of impact-generated shock propagation, some comments on the application of this knowledge to crater size predictions, and a list of some of the more important problems that remain to be solved.

Many of the results appearing below are new, and are presented here for the first time. Specifically, the new material includes:

(1) The self-similar flow field functions and associated particle paths (Section IV).

(2) The analytic solution for the region near the shock–free-surface intersection point (Section IV).

(3) The derivation of the general similitude properties of the basic differential equations (Section V).

II. Formulation of the Problem

A. Restriction to Axisymmetric Flow

The only case considered in this work is that in which the target is a half-space and the path of the projectile is normal to the original target surface. The projectile is assumed to have an axis of symmetry coincident with its path. Thus, the flow taking place after the initial contact is axisymmetric. This restriction is made for the purpose of simplicity in the mathematical treatment of the problem. In addition, it has been observed experimentally that the crater resulting from an oblique impact has the same shape as that due to a normal impact, for angles of obliquity up to a maximum value that depends on a number of factors (*8*). Thus, the results described here may also be applicable to cases of oblique incidence, although the similarity of the final crater shape does not constitute sufficient proof that the shock propagation histories are the same in the two cases.

Furthermore, the projectile is represented as a point source of energy and momentum throughout most of this work. Thus, the solutions presented can be applied to an actual impact situation only when the main shock has propagated to a depth many times the significant dimension of the projectile.

B. Fluid-Mechanical Model

The net force acting on a small element of the target material is contributed to by gradients of the normal and shear stresses. The relative magnitudes of these two contributions can be estimated by comparing a typical normal stress with a typical shear stress, since the gradients of these quantities in any two perpendicular directions are of the same order of magnitude.

The proper order for the normal stress is the pressure generated at the shock, which is initially on the order of $\rho_0 U^2$, and which decays as the shock spreads out into the target. The ultimate strength of the target material appears to be a suitable order of magnitude for the shear stress. Comparison of these two quantities for impact of typical solids at speeds on the order of several kilometers per second suggests that the material strength effect is less than the normal-stress contributions by several factors of ten. Thus there is a regime in which the equations used to describe the target response can be taken as the familiar equations of compressible, inviscid-fluid flow,* expressing the conservation of mass, momentum, and energy.

While this approximation has been used for more than a decade, there has always been an awareness that it must fail at sufficiently low pressure (*9*, *10*). Recent years have seen significant advances toward the determination of the pressure level at which the strength effect becomes important, as well as useful methods for predicting the effect—see, for example, Deal (*11*). Throughout the present work, however,

* This approximation is often referred to as the "hydrodynamic" model. Use of this term is avoided in the present work because of its association in many texts with incompressible flow.

attention is restricted to results obtained in the zero-strength approximation.

In prescribing the energy equation, energy changes arising from viscous dissipation and heat conduction are omitted; this is consistent with their neclect in the momentum equation. The transport of energy by radiation is also neglected.

C. Equations of Motion

A cylindrical coordinate system is chosen with its axis along the approach path of the projectile (Fig. 1).

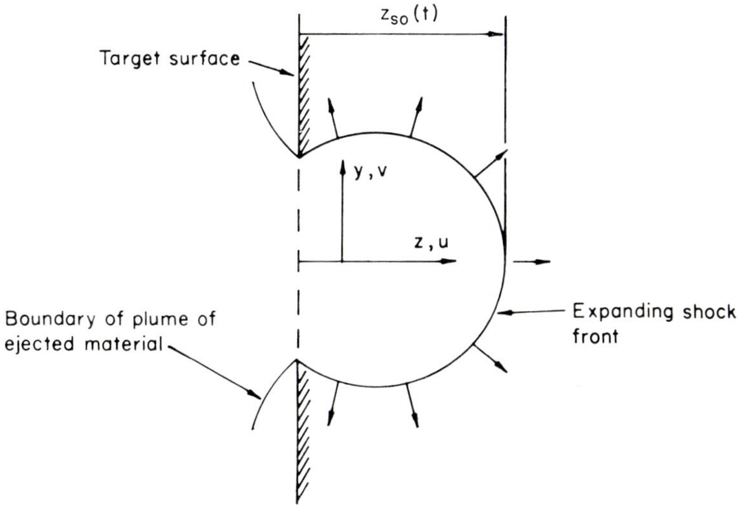

FIG. 1. Coordinate system.

In these coordinates, the equations of motion are

$$\partial\rho/\partial t + u\,\partial\rho/\partial z + v\,\partial\rho/\partial y + \rho(\partial u/\partial z + v/y + \partial v/\partial y) = 0 \quad (1)$$

$$\partial u/\partial t + u\,\partial u/\partial z + v\,\partial u/\partial y + (1/\rho)\,\partial p/\partial z = 0 \quad (2)$$

$$\partial v/\partial t + u\,\partial v/\partial z + v\,\partial v/\partial y + (1/\rho)\,\partial p/\partial y = 0 \quad (3)$$

$$\partial e/\partial t + u\,\partial e/\partial z + v\,\partial e/\partial y - (p/\rho^2)(\partial\rho/\partial t + u\,\partial\rho/\partial z + v\,\partial\rho/\partial y) = 0 \quad (4)$$

$$e = F(p, \rho) \quad (5)$$

The specific functional dependence appearing in the equation of state is left arbitrary for the moment. The use of an equation of state implies an assumption of thermodynamic equilibrium; thus kinetic processes such as condensation or phase change are excluded from consideration.

D. Boundary Conditions

Along the shock, the Rankine–Hugoniot conditions apply. For a small planar element of the shock, moving at speed q_s into a medium at rest, these conditions are (Fig. 2)

$$\rho_0 q_s = \rho_1(q_s - q_1) \tag{6}$$

$$p_1 - p_0 = \rho_0 q_s q_1 \tag{7}$$

$$e_1 - e_0 = \tfrac{1}{2}(p_0 + p_1)(1/\rho_0 - 1/\rho_1) \tag{8}$$

A second boundary condition is that, at the edge of the plume of fluid being ejected from the target, the pressure in the plume should equal the ambient static pressure surrounding the target (usually this pressure is taken to be zero). These boundary conditions are very difficult to apply, since the time-dependent location and shape of the shock and of the plume boundary cannot be specified in advance.

$$p_1, \rho_1, e_1, q_1 \;\Big|\xrightarrow{q_s} p_0, \rho_0, e_0; \quad q_0 = 0$$

FIG. 2. Shock wave notation.

III. Similarity Requirements

A. General Considerations

In studying a problem as complex as the present one, it is useful to see whether there are any similitudes that might be used to correlate solutions for different impact situations. The purpose of this section is to discuss the conditions under which a similitude can be found.

In its most general form, a similitude identifies scaling factors for both the dependent variables (pressure, density, particle velocity) and the independent ones (two spatial coordinates and the time), such that for different impacts the scaled pressure distribution, for example, would be the same function of the scaled spatial coordinates, provided the same scaled time were chosen for both impacts. At a later value of the scaled time, the two scaled pressure distributions would again be identical to each other, although different from the distribution that existed earlier.

A simpler similitude is sometimes found in which the scaled distributions are the same for all values of the scaled time; solutions of this type are said to be self-similar. It is shown below that a self-similar solution can be found when the state equation of the target is approximated by that of a perfect gas. If an equation of state more realistic for a solid is used, the solution is not self-similar, but a more complex similitude can be shown to exist.

B. Normalization

As a first step in deriving the similitude properties of the solution, it is useful to normalize the spatial coordinates by the distance $z_{s0}(t)$ to which the shock has propagated along the axis of symmetry:

$$\zeta = z/z_{s0}(t), \qquad \sigma = y/z_{s0}(t) \tag{9}$$

The time variable is replaced by the Mach number of the shock along the axis of symmetry:

$$M_{s0}(t) = \dot{z}_{s0}/c \tag{10}$$

Here c is the speed at which small disturbances travel into the undisturbed medium. The dependent variables are redefined as follows:

$$u(z, y, t) = \dot{z}_{s0}(t)\,\phi(\zeta, \sigma, M_{s0}) \tag{11}$$

$$v(z, y, t) = \dot{z}_{s0}(t)\,\omega(\zeta, \sigma, M_{s0}) \tag{12}$$

$$p(z, y, t) = \rho_0[\dot{z}_{s0}(t)]^2 f(\zeta, \sigma, M_{s0}) \tag{13}$$

$$\rho(z, y, t) = \rho_0\,\psi(\zeta, \sigma, M_{s0}) \tag{14}$$

$$e(z, y, t) = [\dot{z}_{s0}(t)]^2 g(\zeta, \sigma, M_{s0}) \tag{15}$$

In writing the functional dependences on the right-hand sides of these equations, the point-source assumption has been made. If this were not done, it would be necessary to add a separate dependence on the scale of the projectile $L_0/z_{s0}(t)$.

In terms of these variables, the equations of motion become

$$(\phi - \zeta)\frac{\partial\psi}{\partial\zeta} + (\omega - \sigma)\frac{\partial\psi}{\partial\sigma} + \psi\left(\frac{\partial\phi}{\partial\zeta} + \frac{\omega}{\sigma} + \frac{\partial\omega}{\partial\sigma}\right)$$
$$= \frac{d\ln\dot{z}_{s0}}{d\ln z_{s0}}\,\frac{\partial\psi}{\partial\ln M_{s0}} \tag{16}$$

$$\frac{d \ln \dot{z}_{s0}}{d \ln z_{s0}} \phi + (\phi - \zeta) \frac{\partial \phi}{\partial \zeta} + (\omega - \sigma) \frac{\partial \phi}{\partial \sigma} + \frac{1}{\psi} \frac{\partial f}{\partial \zeta}$$

$$= - \frac{d \ln \dot{z}_{s0}}{d \ln z_{s0}} \frac{\partial \phi}{\partial \ln M_{s0}} \tag{17}$$

$$\frac{d \ln \dot{z}_{s0}}{d \ln z_{s0}} \omega + (\phi - \zeta) \frac{\partial \omega}{\partial \zeta} + (\omega - \sigma) \frac{\partial \omega}{\partial \sigma} + \frac{1}{\psi} \frac{\partial f}{\partial \sigma}$$

$$= - \frac{d \ln \dot{z}_{s0}}{d \ln z_{s0}} \frac{\partial \omega}{\partial \ln M_{s0}} \tag{18}$$

$$2 \frac{d \ln \dot{z}_{s0}}{d \ln z_{s0}} g + (\phi - \zeta) \frac{\partial g}{\partial \zeta} + (\omega - \sigma) \frac{\partial g}{\partial \sigma} - \frac{f}{\psi^2} \left\{ (\phi - \zeta) \frac{\partial \psi}{\partial \zeta} + (\omega - \sigma) \frac{\partial \psi}{\partial \sigma} \right\}$$

$$= - \frac{d \ln \dot{z}_{s0}}{d \ln z_{s0}} \left\{ \frac{\partial g}{\partial \ln M_{s0}} - \frac{f}{\psi^2} \frac{\partial \psi}{\partial \ln M_{s0}} \right\} \tag{19}$$

The Rankine–Hugoniot conditions become

$$\psi_1 \left(1 - \frac{\dot{z}_{s0}}{q_s} \frac{q_1}{\dot{z}_{s0}} \right) = 1 \tag{20}$$

$$f_1 - \frac{p_0}{\rho_0 \dot{z}_{s0}^2} = \frac{q_s}{\dot{z}_{s0}} \frac{q_1}{\dot{z}_{s0}} \tag{21}$$

$$g_1 - e_0/\dot{z}_{s0}^2 = \tfrac{1}{2}(f_1 + p_0/\rho_0 \dot{z}_{s0}^2)(1 - 1/\psi_1)$$

$$= F(\rho_0 \dot{z}_{s0}^2 f_1 , \rho_0 \psi_1)/\dot{z}_{s0}^2 \tag{22}$$

C. Requirements for Self-Similarity

If the flow is to be self-similar, it must have the same normalized spatial distributions at all times. The conditions necessary for this to occur can be seen by inspection of the above equations.

The first requirement is the point-source assumption, i.e., that $L_0/z_{s0}(t)$ be negligible. This assumption has already been made. The second requirement is that the quantity $d \ln \dot{z}_{s0}/d \ln z_{s0}$ be a constant; this is satisfied if the shock scale grows as a power of the time:

$$z_{s0} = At^N \tag{23}$$

The third, fourth, and fifth sources of time dependence enter through the boundary conditions at the shock. The third is introduced by the function $q_s/\dot{z}_{s0}$; the flow can be self-similar only if this quantity is independent of the time. Thus the shock must preserve the same shape as it grows. The fourth condition for self-similarity is that the pressure ahead of the shock be negligible in comparison to $\rho_0 \dot{z}_{s0}^2$. The latter quantity is of the order of the pressure behind the shock, and is large compared with p_0 whenever the fluid-mechanical approximation is valid. The fifth possibility of time dependence enters by way of the state equation. As pointed out by Kynch (*12*) and Sedov (*13*), self-similar solutions are possible when the equation of state has the form

$$e = p\,\varphi(\rho) \tag{24}$$

where $\varphi(\rho)$ is any function of the density. When the state equation is of this form, when the term $p_0/\rho_0 \dot{z}_{s0}^2$ is negligible, and when the shock shape is constant, it is possible to solve the third of the Rankine–Hugoniot conditions for the density ratio across the shock, ψ_1, which remains constant:

$$2p_0\,\varphi(\rho_0\psi_1) - 1 + 1/\psi_1 = 0 \tag{25}$$

The equation of state of a perfect gas

$$e = p/(\gamma - 1)\,\rho \tag{26}$$

is of the form of Eq. (24). Thus, departures from self-similarity for a perfect gas arise from the counterpressure term $p_0/\rho_0 \dot{z}_{s0}^2$. The situation is reversed for shock propagation in a solid. There, the counterpressure term is negligible, but the state equation is not of the proper form, except in the limit of extremely high pressures. Only the very early portion of an impact at extremely high speed will satisfy the self-similarity conditions in a solid. Thus, impact-generated shock propagation in solids is characteristically non-self-similar.

If all of the conditions for self-similarity are met, then the functions f, ψ, g, ϕ, and ω depend only on the two similarity coordinates ζ and σ*:

$$p/\rho_0 \dot{z}_{s0}^2\,,\ \rho/\rho_0\,,\ e/\dot{z}_{s0}^2\,,\ u/\dot{z}_{s0}\,,\ v/\dot{z}_{s0} = \text{fcns}(\zeta, \sigma) \tag{27}$$

The theoretical existence of this similarity is of little use, however, until the actual dependence of f and the other functions on ζ and σ has been found, and until the constant A and the exponent N appearing in the shock trajectory, Eq. (23), have been related to the energy and momentum of the projectile. Dimensional analysis can be used to construct a characteristic length and time from the quantities E and P. It is convenient to include the quantities ρ_0 and c in this formulation as well. The result is the following form of the power-law shock trajectory:

$$z_{s0}/\mathscr{L}_0 = F(\gamma)(ct/\mathscr{L}_0)^N \tag{28}$$

* They also depend on any constants appearing in the state equation, such as γ in the perfect-gas case.

where the scaling length $\mathscr{L}_0$ is defined as

$$\mathscr{L}_0 = \left\{ \frac{E}{\rho_0 c^2} \left(\frac{P}{\rho_0 c} \right)^{(2-5N)/(4N-1)} \right\}^{(4N-1)/3(1-N)} \tag{29}$$

Walsh and his co-workers (*5*, *7*) present a discussion of the self-similar solution from a slightly different point of view. They first observe that a dimensional analysis of the problem, characterized by the initial density ρ_0, the projectile size L_0, the impact speed U, and the specific heat ratio γ, shows that the solution must be of the form

$$p/\rho_0 U^2, \rho/\rho_0, u/U, v/U = \text{fcns}(z/L_0, y/L_0, Ut/L_0; \gamma) \tag{30}$$

They then show that a simpler scaling can be found if one introduces the assumption of "late-stage equivalence." This assumption states that the solutions for two different impacts will be identical, after early transient differences have died out, provided the quantity $L_0 U^\alpha$ is the same for the two impacts, α being a constant. The "late" aspect of this statement is identical with the point-source assumption, i.e., it simply requires that the shock scale be large compared with the projectile scale. The requirement that $L_0 U^\alpha$ be a constant is the same as the statement that the scaling length $\mathscr{L}_0$ be the same in the two cases, where α and N are related by

$$\alpha = N/(1 - N) \tag{31}$$

Thus, for the perfect-gas case, late-stage equivalence is the same as self-similarity.

The functional dependences employed by Walsh *et al.* are

$$\frac{pZ^{2/\alpha}}{\rho_0 U^2}, \frac{\rho}{\rho_0}, \frac{uZ^{1/\alpha}}{U}, \frac{vZ^{1/\alpha}}{U} = \text{fcns}\left(\frac{Z^{1+\alpha}}{\tau^\alpha}, \frac{R^{1+\alpha}}{\tau^\alpha}; \gamma\right) \tag{32}$$

where

$$\tau = Ut/L_0, \qquad Z = z/L_0, \qquad R = y/L_0 \tag{33}$$

It will be seen in Section IV that this form of the solution is identical with the self-similar form described above.

D. Similitude for Finite Shock Speed

The equation of state of a solid approaches that of a perfect gas only at very high pressures. Thus, the self-similarity described above can be expected to apply to solids only under the limiting condition of very large shock strength ($M_{s0} \to \infty$). As M_{s0} becomes finite, the equation of state no longer approximates that of a perfect gas, and the flow becomes non-self-similar. (Specifically, the second, third, and fifth of the requirements listed above are not met.) However, the only new parameter that must be considered is M_{s0} *; the quantity $d \ln \dot{z}_{s0}/d \ln z_{s0}$, the shock shape, and the boundary values at the shock depend only on M_{s0}. Thus, the solution can be written as

$$p/\rho_0\dot{z}_{s0}^2,\ \rho/\rho_0,\ e/\dot{z}_{s0}^2,\ u/\dot{z}_{s0},\ v/\dot{z}_{s0} = \text{fcns}(\zeta, \sigma, M_{s0}, \ldots) \tag{34}$$

where the shock trajectory is of the form

$$z_{s0}/\mathscr{L}_0,\ M_{s0} = \text{fcns}(ct/\mathscr{L}_0, \ldots) \tag{35}$$

Here the three dots indicate the presence of various constants from the equation of state.

The important feature of these results is that no new scaling lengths enter the problem when M_{s0} becomes finite. The scaling length $\mathscr{L}_0$, which originates in the self-similar limit, continues to be the appropriate similitude parameter throughout the non-self-similar part of the flow. In particular, the parameter N (or, equivalently, α) in the formula for $\mathscr{L}_0$ depends only on the high-pressure limit of the state equation.

The similitude expressed in Eqs. (34) and (35) was first observed by Walsh *et al.* (*6*), who called it "late-stage equivalence." These authors pointed out that dimensional analysis would yield the following form of the solution:

$$p/\rho_0 U^2,\ \rho/\rho_0,\ u/U,\ v/U = \text{fcns}(z/L_0,\ y/L_0,\ Ut/L_0,\ U/c) \tag{36}$$

They demonstrated numerically that in two impacts for which $L_0 U^\alpha$ is the same (where α is the same constant that produces self-similarity in the perfect-gas limit) the solution is the same function of time in both cases. The solution common to both is different at different times, of course, since this flow is not self-similar. They went on to show that the existence of this similitude implies that the solution can be rewritten in the form

$$\frac{pZ^{2/\alpha}}{\rho_0 U^2},\ \frac{\rho}{\rho_0},\ \frac{uZ^{1/\alpha}}{U},\ \frac{vZ^{1/\alpha}}{U} = \text{fcns}\left(\frac{Z^{\alpha+1}}{\tau^\alpha},\ \frac{R^{\alpha+1}}{\tau^\alpha},\ \frac{(U/c)^{\alpha+1}}{\tau}\right) \tag{37}$$

The connection between this correlation and that of Eqs. (34) and (35) is discussed in Section V.

* In addition, parameters other than γ appear in the equation of state. The similitude described here can in principle be used only to correlate solutions for a given material. Actually, the evidence described in Section V indicates that the differences between various materials are relatively minor, at least in axisymmetric flows.

Rather than use the same term (late-stage equivalence) for the self-similar and the non-self-similar portions of the flow, it seems preferable to speak of both of these equivalences as "point-source similitude," noting that the similitude has self-similar behavior in the limit $M_{s0} \to \infty$. This terminology is adopted throughout the present work.

It should be noted that the point-source assumption is not related to any assumptions made regarding the equation of state. All the solutions discussed in the present work employ the point-source model, but different forms of the equation of state are used in various places.

IV. The Perfect-Gas Case

The purpose of this section is to present a complete picture of the two-dimensional character of the flow, in a case where self-similarity applies. Self-similarity is possible with any equation of state of the form indicated in Eq. (24). The reason for devoting special attention to the perfect-gas case is that it approximates the high-pressure form of the state equation for many solids.

A. Basic Equations

Under the conditions of self-similarity, the shock scale grows as a power of the time; thus

$$d \ln \dot{z}_{s0}/d \ln z_{s0} = -(1 - N)/N \tag{38}$$

If this is used in Eqs. (16)–(19), and if the right-hand sides are set equal to zero, the result is

$$(\phi - \zeta)\frac{\partial \psi}{\partial \zeta} + (\omega - \sigma)\frac{\partial \psi}{\partial \sigma} + \psi\left[\frac{\partial \phi}{\partial \zeta} + \frac{\omega}{\sigma} + \frac{\partial \omega}{\partial \sigma}\right] = 0 \tag{39}$$

$$-\frac{1-N}{N}\phi + (\phi - \zeta)\frac{\partial \phi}{\partial \zeta} + (\omega - \sigma)\frac{\partial \phi}{\partial \sigma} + \frac{1}{\psi}\frac{\partial f}{\partial \zeta} = 0 \tag{40}$$

$$-\frac{1-N}{N}\omega + (\phi - \zeta)\frac{\partial \omega}{\partial \zeta} + (\omega - \sigma)\frac{\partial \omega}{\partial \sigma} + \frac{1}{\psi}\frac{\partial f}{\partial \sigma} = 0 \tag{41}$$

$$-2\frac{1-N}{N}f + (\phi - \zeta)\frac{\partial f}{\partial \zeta} + (\omega - \sigma)\frac{\partial f}{\partial \sigma} - \frac{\gamma f}{\psi}\left[(\phi - \zeta)\frac{\partial \psi}{\partial \zeta} + (\omega - \sigma)\frac{\partial \psi}{\partial \sigma}\right] = 0 \tag{42}$$

It can be shown that these equations are either elliptic or hyperbolic in character, depending on whether

$$(\phi - \zeta)^2 + (\omega - \sigma)^2 - \gamma f/\psi \lesseqgtr 0 \tag{43}$$

The reason why a mixed mathematical character arises can be seen by following the path of a small-disturbance front, originating at Point A in Fig. 3a. The front of this wavelet travels at a velocity equal to the vector

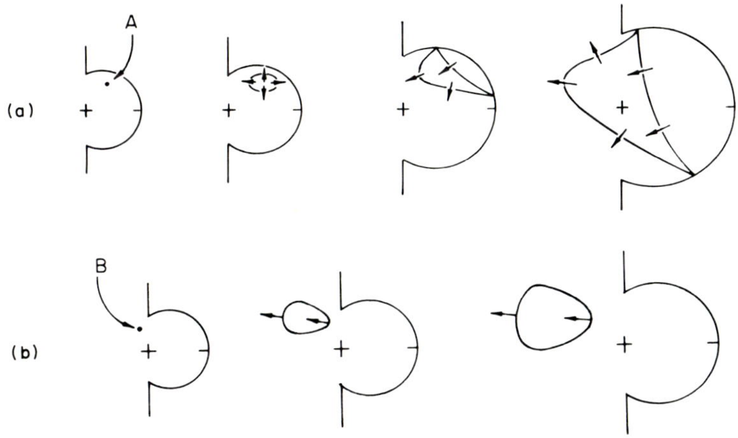

FIG. 3. Paths of disturbance fronts.

sum of the local sound speed and particle velocity, and is located, at a succession of later instants, as shown in the figure. This figure is a schematic illustration of what might happen if the disturbance originates at a point around which the vector velocity sum is distributed in such a manner that the disturbance can overtake the shock wave, where it is reflected; it is eventually felt at all points in the region enclosed by the shock. If each of the instants in the figure above is now reduced to the same scale, and if the solution in these scaled coordinates is to be the same for each instant, then clearly the governing differential equations must have the property that the solution at Point A affects the solution at all the points traversed by the disturbance. The interdependence of the points in part of the flow field is manifested in the partially elliptic character of Eqs. (39)–(42).

If the disturbance originates at Point B (see Fig. 3b), where the wavelet does not reflect from the shock wave, then the disturbance has only a finite zone of influence. When the transformation of these sketches to the self-similar plane is made, the finite zone-of-dependence property is assured by the partially hyperbolic nature of the solution.

Finally, because disturbances made in the elliptic region are ultimately transmitted to the hyperbolic region, it follows that the solutions in the two regions must be matched along their common boundary. It will be seen below that the matching can be achieved by the proper choice of the parameter N.

The boundary conditions at the shock are

$$\psi_1 = \frac{\gamma + 1}{\gamma - 1}, \qquad f_1 = \frac{2}{\gamma + 1}\left(\frac{q_s}{\dot{z}_{s0}}\right)^2, \qquad \frac{q_1}{\dot{z}_{s0}} = \frac{2}{\gamma + 1}\frac{q_s}{\dot{z}_{s0}} \tag{44}$$

The two components ϕ_1 and ω_1 are found from $q_1/\dot{z}_{s0}$ by taking its components in the ζ and σ directions:

$$\mathbf{q}_1/\dot{z}_{s0} = \phi_1\hat{\zeta} + \omega_1\hat{\sigma} \tag{45}$$

The determination of these components cannot be carried out until the shock shape is specified.

At the time of the present writing, no solutions have been calculated directly from Eqs. (39)–(42), because of their mixed hyperbolic–elliptic nature, and also because of the unknown value of N, and the unknown location of the shock surface and the plume boundary. A further problem arises when an attempt is made to satisfy the conservation of the total energy and momentum. Because of its crucial importance, this subject is discussed fully in the paragraphs below.

B. Conservation of Energy and Momentum

1. *The Axisymmetric Case*

The integrals expressing the total energy and momentum of the system are

$$E = 2\pi \int_{z=-\infty}^{z_{s0}(t)} \int_{y=0}^{y_{UL}(z,t)} [e - e_0 + \tfrac{1}{2}(u^2 + v^2)]\,\rho y\,dy\,dz \tag{46}$$

$$P = 2\pi \int_{z=-\infty}^{z_{s0}(t)} \int_{y=0}^{y_{UL}(z,t)} \rho u y\,dy\,dz \tag{47}$$

The upper limit for the integration on y (see Fig. 4) is the shock surface, for $z > 0$, and the plume boundary, for $z < 0$.

The choice of the lower limit $z = -\infty$ requires some explanation. In an actual impact with a projectile of nonzero size, the particles lying originally on the free surface would be located at a finite value of z, for any finite time after the impact. An upper limit to their distance from the free surface can be calculated by assuming that, at $t = 0$, they

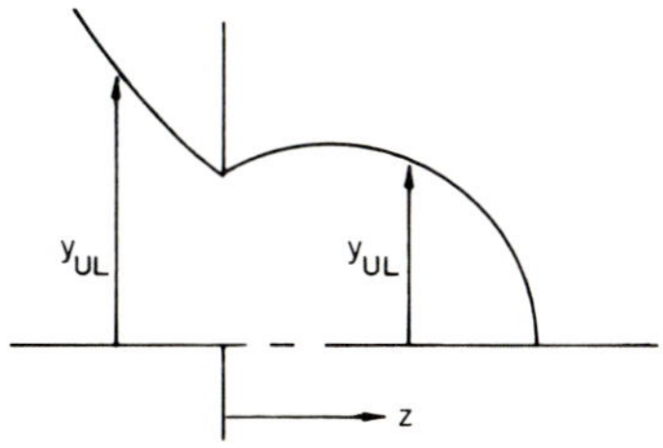

FIG. 4. Integration limits.

immediately begin moving to the left at the full escape speed, which for a perfect gas has the value

$$u_{\text{escape}} = [2/(\gamma - 1)](\gamma p_1/\rho_1)^{1/2} \tag{48}$$

Here p_1 and ρ_1 are the pressure and density generated at the impact point. This estimate represents an upper limit to their distance from the free surface since these particles are actually carried below the free surface at first; only later do they reverse their velocity, and asymptotically approach the escape speed in the $-z$ direction. If the point-source idealization is now made, then the projectile, of vanishing size, must carry a nonzero amount of energy and momentum. To do this, its velocity must be infinite; thus the pressure generated at impact, as well as the speed of the escape front, will also be infinite. For this reason, the lower limit in the integrals above is set at $z = -\infty$.

If the self-similar functions are now introduced into Eqs. (46) and (47), the result is

$$E = 2\pi\rho_0 \dot{z}_{s0}^2 z_{s0}^3 \int_{-\infty}^{1} \int_{0}^{\sigma_{UL}(\zeta)} [f/(\gamma - 1)\,\psi + \tfrac{1}{2}(\phi^2 + \omega^2)]\,\psi\sigma\,d\sigma\,d\zeta \tag{49}$$

$$P = 2\pi\rho_0 \dot{z}_{s0} z_{s0}^3 \int_{-\infty}^{1} \int_{0}^{\sigma_{UL}(\zeta)} \phi\psi\sigma\,d\sigma\,d\zeta \tag{50}$$

It is clear that a single value of N cannot render both of these expressions independent of time. Constancy of energy would require $N = \frac{2}{5}$, while constant momentum requires $N = \frac{1}{4}$. The solution of this dilemma was provided, for the one-dimensional case, by Zeldovich (*14*). It was subsequently extended to the axisymmetric problem by Rayzer (*15*) and by the present writer (*16*). Zeldovich's solution is given below.

2. *The One-Dimensional Case*

If the projectile is assumed to be a slab of infinite length in the y direction, or if the disturbance is caused by an explosive slab of infinite length in the y direction, the response of the target can be assumed to be a function only of z and t. The solution of this problem has been discussed in several papers (*14*, *16–22*). An excellent review is given by Mirels (*23*). The features important for the present case are repeated below.

The equations of motion are Eqs. (1), (2), and (4), with the y derivatives omitted, and the term v/y absent from the continuity equation. The self-similar forms of these equations are found by dropping their right-hand sides, to yield

Continuity:

$$(\phi - \zeta)\psi' + \psi\phi' = 0 \tag{51}$$

Momentum:

$$-[(1 - N)/N]\phi + (\phi - \zeta)\phi' + f'/\psi = 0 \tag{52}$$

Energy:

$$-2[(1 - N)/N]f + (\phi - \zeta)f' - (\gamma f/\psi)(\phi - \zeta)\psi' = 0 \tag{53}$$

Here the prime denotes $d/d\zeta$, and the boundary conditions are

$$\psi(1) = (\gamma + 1)/(\gamma - 1), \qquad \phi(1) = f(1) = 2/(\gamma + 1) \tag{54}$$

In both the impact and explosive-slab cases, the target acquires a nonzero energy and momentum. For the impact case, the magnitudes of these quantities can be related to the energy and momentum of the projectile. In the explosive case, the energy and momentum acquired by the target depend on particulars of the loading history at the surface.

The integrals expressing the conservation of the total energy and momentum (per unit area in the plane of the target surface) are

$$\mathscr{E} = \rho_0 \dot{z}_{s0}^2 z_{s0} \int_{-\infty}^{1} (f/(\gamma - 1) + \psi\phi^2/2)\, d\zeta \tag{55}$$

$$\mathscr{P} = \rho_0 \dot{z}_{s0} z_{s0} \int_{-\infty}^{1} \psi\phi\, d\zeta \tag{56}$$

The lower limits in these integrals have been taken at minus infinity, since the projectile or explosive slab is considered to be of vanishing thickness. Note that the one-dimensional problem presents the same dilemma regarding the choice of N: the energy integral requires $N = \frac{2}{3}$, while the momentum integral requires $N = \frac{1}{2}$.

However, there is a much more important consideration affecting the choice of N: namely, Eqs. (51)–(53) have an acceptable solution only for one value of N. The reason for this can be seen by solving them for the derivatives:

$$\phi' = -\frac{1-N}{N}\frac{2f/\psi - \phi(\phi - \zeta)}{(\phi - \zeta)^2 - \gamma f/\psi} \tag{57}$$

$$\psi' = -\frac{1-N}{N}\frac{(\phi - \zeta)\,\psi\phi - 2f}{(\phi - \zeta)[(\phi - \zeta)^2 - \gamma f/\psi]} \tag{58}$$

$$f' = -\frac{1-N}{N} f \frac{\gamma\phi - 2(\phi - \zeta)}{(\phi - \zeta)^2 - \gamma f/\psi} \tag{59}$$

These equations become singular at the point where $(\phi - \zeta)^2 - \gamma f/\psi$ changes sign. This point divides the zones in the similarity plane in which small disturbances can or cannot be felt all the way to the shock wave. For a given value of γ, Eqs. (57)–(59) have nonsingular solutions only for a single value of N. For that value, the numerators of these equations are also zero at the singular point; resolution of the indeterminate form yields the slope of the solution that passes smoothly through the singularity. Table I lists the values of $N(\gamma)$ that lead to regular

TABLE I

$N(\gamma)$ FOR ONE-DIMENSIONAL BLAST WAVES

γ	N	Ref.
1	0.5	*23*
1.1	0.56888	*23*
1.4	0.6	*23*
$\frac{5}{3}$	0.61073	*23*
2.8	0.626704	*23*
∞	0.6416	*23*
1.001	0.5106	*16*
1.1	0.5683	*16*
1.2	0.5843	*16*
1.3	0.5935	*16*
1.4	0.6000	*16*
1.8	0.6143	*16*
2.0	0.6182	*16*
2.5	0.6244	*16*
3.0	0.6279	*16*
4.0	0.6321	*16*
5.0	0.6341	*16*
7.0	0.6365	*16*
10.0	0.6385	*16*
100.0	0.6412	*16*
1000.0	0.6416	*16*

solutions. The distributions of density, pressure, and velocity that are obtained are shown in Fig. 5 for the case $\gamma = 1.4$, which has the closed-form solution (*17*, *18*)

$$\psi = 6(5 - 4\zeta)^{-5/2}, \qquad \phi = \tfrac{5}{6}(2\zeta - 1), \qquad f = \tfrac{5}{6}(5 - 4\zeta)^{-3/2} \tag{60}$$

There can be no question that the nonsingular solutions are the only acceptable ones. But if N is chosen so as to achieve a smooth crossing of the singularity, it is then impossible to satisfy either of Eqs. (55) and (56). The problem involved in the choice of N is not created simply by the fact that there is a second quantity (the momentum) to be conserved. Even if the total momentum were to be zero, the constant-energy value $N = \frac{2}{3}$ still would not produce a physically acceptable solution.

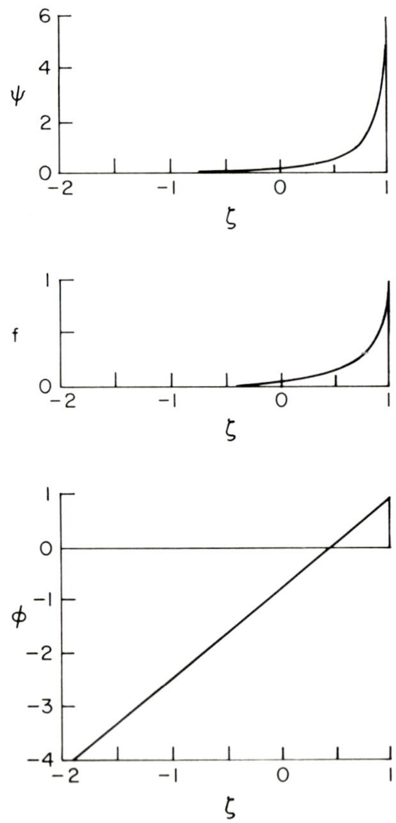

FIG. 5. Self-similar, plane wave solution (perfect gas, $\gamma = 1.4$).

A resolution of this difficulty was proposed by Zeldovich (*14*); it is also discussed in the recent text by Zeldovich and Rayzer (*24*). Their suggestion is that a small portion of the mass be omitted in calculating the total energy and momentum. This small mass initially lies at the free surface and is strongly compressed during the initial stages of the impact or explosion. It acquires a certain entropy during this compression, which it retains thereafter, since the flow is particle isentropic. However, the entropy acquired during the initial stages is in general different from the entropy that would be assigned by the self-similar solution. That solution depends only on conditions at the shock and on the properties of the self-similar differential equations; it contains no information about the details of how the shock was initiated. The small amount of mass processed during the early phase always bears the imprint of this phase, and is never correctly described by the self-similar solution. Zeldovich's argument is that, in seeking a self-similar solution, only that mass should be included whose motion is expected to be properly described by such a solution.

An analogous interpretation has been developed in recent years for the problem of hypersonic flow over a blunted slender body [see, for example, Mirels (*23*), Yakura (*25*), and Guiraud (*26*)].* Far downstream of the nose, the flow near the shock can be self-similar if the coordinates, transverse to the free-stream direction, of the body and the shock grow as a power of the streamwise coordinate. The layers of fluid near the body, however, are those which have passed through the blunt part of the bow shock; there they acquired an entropy distribution that reflects the details of the shock shape near the nose. This distribution is in general different from that given by the self-similar solution.

To correctly account for the effect of the entropy layer, separate analyses must be carried out for the high-entropy region and the self-similar region, and the two analyses must then be joined; for example, by the method of matched asymptotic expansions (*27*).

If an analysis of this type were done for the present problem, it would be possible to derive a formula for the shock trajectory which at large time would be of the form $z_{s0} = At^N$. The exponent N would be that given by the nonsingular solution of the self-similar equations, while the amplitude A would reflect the details of the shock-initiation process—i.e., A would depend on the amount of mass neglected. Zeldovich outlined this approach in his original paper, recommending that the amplitude A be determined by matching the self-similar solution to a

* The writer is very grateful to his colleague, Dr. T. R. Sundaram, for calling his attention to this analogy, and for many interesting discussions of its application to the impact problem.

numerical solution of the complete, nonsimilar, partial differential equations at some intermediate time. Since the appearance of Zeldovich's paper, the method of matched asymptotic expansions has undergone considerable development, and could now be used to derive an explicit formula for the amplitude coefficient A in terms of the loading history at the target surface. To the writer's knowledge, the full details of this approach have not yet been carried out.

Zeldovich's result can be expressed in the form (see Appendix)

$$z_{s0}(t) = H(\gamma)\frac{m_0}{\rho_0}\left[\frac{\rho_0{}^2\mathscr{E}t^2}{m_0{}^3}\right]^{N/2} \tag{61}$$

where the indeterminate length scale has been expressed in terms of the neglected mass m_0. The present writer observed (*16*) that a determinate result could be found by making a slight modification to Zeldovich's analysis. This result was presented by Rae and Kirchner (*16*) with the erroneous indication that it was an exact solution of the problem, equivalent to what would be derived by the method of matched asymptotic expansions. Further study has made it clear that such is not the case; however, the results derived in this manner have been found to be a very accurate and simple approximation, and are worthy of mention.

If the length scale is made determinate in the manner indicated above, the solution for the slab-impact case has the form (see Appendix for details)

$$z_{s0}(t) = C(\gamma)\frac{\mathscr{P}^2}{\rho_0\mathscr{E}}\left[\frac{t}{\mathscr{P}^3/\rho_0\mathscr{E}^2}\right]^N \tag{62}$$

This can be rewritten in terms of a scaling length $\mathscr{L}_1$ as

$$z_{s0}/\mathscr{L}_1 = G(\gamma)\,(ct/\mathscr{L}_1)^N \tag{63}$$

where

$$\mathscr{L}_1 = \left\{\frac{\mathscr{E}}{\rho_0 c^2}\left(\frac{\mathscr{P}}{\rho_0 c}\right)^{(2-3N)/(2N-1)}\right\}^{(2N-1)/(1-N)} \tag{64}$$

For $\gamma = 1.4$, the constant $G(\gamma)$ has the value $(128/9)^{1/5}$.

The accuracy of this formula can be estimated by comparing its predictions with the numerical results of Zhukov and Kazhdan (*21*). These authors calculated the response of a half-space of gas, with $\gamma = 1.4$, to a square-wave pressure loading on the surface. At late time, their solution displayed the same behavior as that of the self-similar solution, and they determined the amplitude coefficient from their numerical results. In terms of the peak pressure at the shock, Zhukov and Kazhdan designate this amplitude by the symbol B:

$$p_1 = BP_{\max}[(T_{\max}/z_{s0})(P_{\max}/\rho_0)^{1/2}]^{2(1-N)/N} \tag{65}$$

Here the symbols $P_{\max}$ and $T_{\max}$ are parameters that characterize the loading. In the general case, the free surface is considered to be subjected at $t = 0$ to a sharply rising pressure pulse $p_L(t)$ of peak amplitude $P_{\max}$ and duration $T_{\max}$ (see Fig. 6):

$$p_L(t) = P_{\max}\, \mathscr{G}(t/T_{\max}) \tag{66}$$

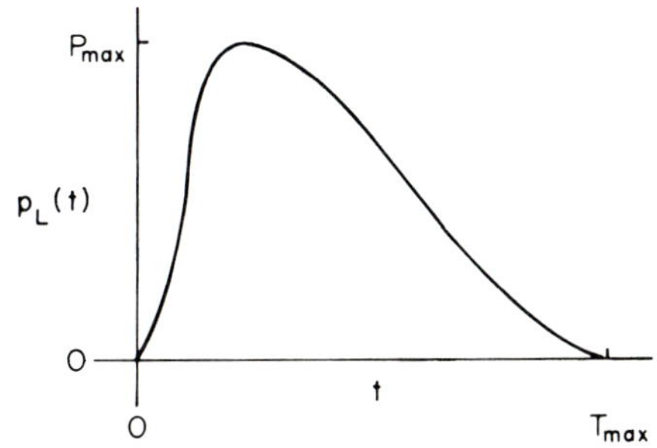

FIG. 6. Surface loading function.

The momentum and energy imparted to the target during this loading are

$$\begin{aligned} \mathscr{P} &= aP_{\max}T_{\max}\,, \qquad a = \int_0^1 \mathscr{G}(t/T_{\max})\, d(t/T_{\max}) \\ \mathscr{E} &= bT_{\max}\frac{P_{\max}^{3/2}}{\rho_0^{1/2}}\,, \qquad b = \int_0^1 \mathscr{G}(t/T_{\max})\, u_L(t/T_{\max})/(P_{\max}/\rho_0)^{1/2}\, d(t/T_{\max}) \end{aligned} \tag{67}$$

where u_L denotes the velocity of the loaded surface. In particular, for a square-wave loading, $a = 1$, $b = [2/(\gamma + 1)]^{1/2}$. Use of these values in Eq. (62) leads to the prediction that $B = 1.657$. The number actually found by Zhukov and Kazhdan is 1.8224.* Thus the approximation presented above is in error by about 10% in pressure, or about 5% in shock location.

* The amplitude coefficient B depends only on the shape of the loading function, as expressed in the constants a and b. Thus it would have the same constant value for a slab of any thickness and velocity, and a different constant value for a square-wave loading of any magnitude.

Other evidence for the accuracy of this solution can be found in the numerical calculations of Dienes (*22*) and of Chou and Burns (*28*). The scale of the shock trajectory in both of these calculations is predicted with great accuracy by Eq. (62).*

Dienes' calculations were done for the case of zero initial pressure in the target, an initial density of 1.0 g/cm³, and for a slab whose thickness–density product was 1.2 g/cm², moving initially at 10^6 cm/sec. Not only is the shock location accurately given by Eq. (62) but also the distributions of pressure, density, and particle velocity are found to be matched by Eq. (60).

The calculations presented by Chou and Burns were done by the method of characteristics, for a series of six impact conditions. The case singled out for special attention here used a slab thickness of 3.319 cm, traveling initially at 9.144 km/sec. The initial pressure in the target was taken to be 1 atm, and so the solution at very late time becomes non-self-similar because of the effect of counterpressure. There is a time interval, however, before the onset of the nonsimilar effect, and sufficiently long after the termination of the early impact phase, during which Chou and Burns' calculations are closely matched by Eq. (62). The agreement is shown in Fig. 7, where it can be seen that the error in shock location is less than 5%.

Dienes (*22*) observed that the approximate solution discussed above could also be derived by assuming that the self-similar solution matches neither $\mathscr{E}$ nor $\mathscr{P}$ individually, but that it produces a flow having the same value of $\mathscr{E}\mathscr{P}^\lambda$ as did the impacting slab, where λ is defined as

$$\lambda = (2 - 3N)/(2N - 1) \tag{68}$$

He too compared this solution with his own numerical results, but, like the present writer, was unaware of the fact that this solution would be shifted slightly in amplitude if used in the correct sense outlined by Zeldovich.

To summarize this discussion of energy and momentum conservation, it should be repeated that at a large time after the loading of the front surface of a half-space of gas, and before the onset of the effect of counterpressure, the flow field is correctly described by the nonsingular, self-similar solution, except for those portions that were set in motion during the loading process and shortly afterwards. The power-law dependence of the solution, and its scaling with respect to the parameters ρ_0, $\mathscr{E}$,

* The agreement, in the case of Dienes' calculations, is partially obscured by the fact that he made a slight shift in the origin of the time scale of the theoretical solution, so as to force the theoretical and numerical shock locations to coincide at 2.0 μsec. The agreement is still very good, even if no such shift is introduced.

and $\mathscr{P}$, is that given by Eq. (63). The amplitude coefficient appearing in the solution depends on γ and also on the details of the time history of the loading process.

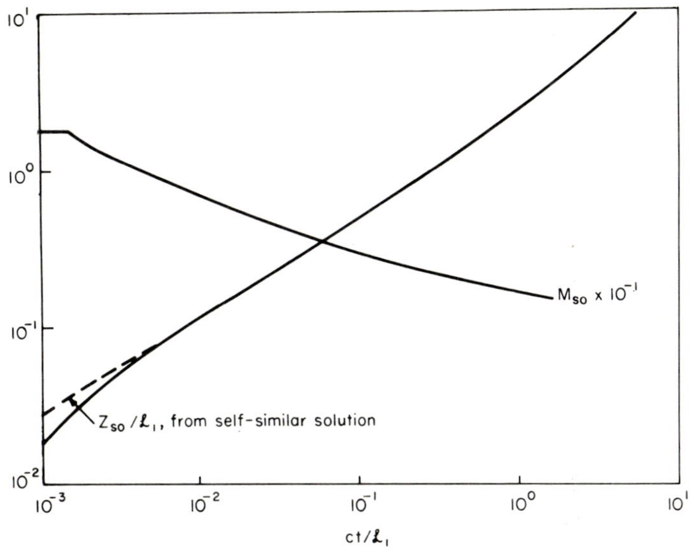

FIG. 7. Plane wave shock-trajectory similitude functions. Solid curves taken from data of Chou and Burns (*28*).

The considerations given above for one dimension can be extended to the case of two spatial dimensions. The axisymmetric impact case has been considered by the present writer (*16*), while Rayzer (*15*) has treated both impacts and explosions of either a point source or a line source. In these cases also, the solution at late time has a self-similar, power-law dependence throughout most of the flow field. The criterion for selecting the exponent is that it be the value which allows a smooth transition between the hyperbolic and elliptic regions. The scaling with respect to the total energy and momentum and the initial density are given below. Simultaneous conservation of energy and momentum is achieved by neglecting the small portion of the mass affected during the initiation of the flow (or, more precisely, by finding the correct solution for this part of the flow and by then joining it to the self-similar part of the flow by the matched asymptotic expansion procedure). The amplitude coefficient in the expression for the shock trajectory depends on the specific heat ratio γ and also on the details of the loading history.

C. The Axisymmetric Solution for $\gamma = 1.5$

The direct solution of the self-similarity equations for the axisymmetric case is extremely difficult, because of the mixed hyperbolic-elliptic nature of the problem, the unknown value of N, and the unknown shape of the shock and of the plume boundary. At the time of this writing, no direct solution of these equations has been reported.

However, the solution of the original time-dependent equations of motion has been reported by Walsh *et al.* (*5*, *7*) for the case of a right-circular cylinder striking a half-space of perfect gas, with $\gamma = 1.5$, zero initial pressure, and a density of 1.0 g/cm³. As noted above, these authors observed the development of the self-similar solution, and indicated a functional form which the solution could take. The actual distributions of the self-similar functions were not presented, however. Dr. Walsh made available to the present writer the complete numerical output for this problem; the self-similar distributions have been found, and are described below.

1. *Shock Shape*

Walsh's calculations were done with a computer code (*29*) that was an improvement over the particle-in-cell method in that the mass distribution was represented in a continuous manner rather than by a number of discrete particles. The projectile was taken to be 36 cm in both length and diameter, traveling initially at 10^6 cm/sec. After the flow becomes self-similar, any instant can be chosen as the one to be used for evaluating the similarity functions. The one chosen here was the largest time for which results were printed (5.40×10^{-3} sec). At this point in the calculation, the cell size was 8 cm in the axial and radial directions. To locate the shock, the pressure was graphed as a function of distance, assuming that the pressure in each cell could be assigned at the midpoint of that cell. Figure 8 shows the result for the row of cells lying nearest to the symmetry axis. It is characteristic of numerical calculations of this type that the shock is spread out over several cells. The location of the shock was guessed, usually between the second and third points on the front portion of the pressure pulse. The shock location points, estimated from curves of this type (p vs. z for $y \leqslant 180$ cm, and p vs. y for $z \leqslant 236$ cm), were then graphed, and a curve was faired through these points. This faired curve could be closely approximated by the ellipse (see Fig. 9):

$$\left(\frac{z - z_1}{z_{s0} - z_1}\right)^2 + \left(\frac{y}{y_0}\right)^2 - 1 = 0 \tag{69}$$

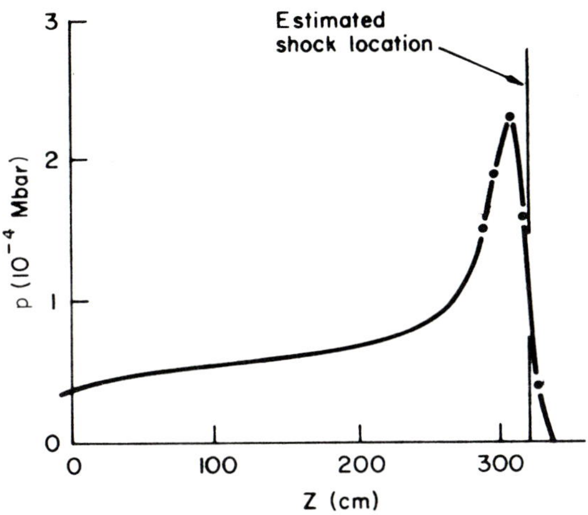

FIG. 8. Pressure distribution in the row of cells nearest to the axis of symmetry.

with $z_{s0} = 321$ cm, $z_1 = 96$ cm, and $y_0 = 241$ cm. In terms of the similarity coordinates, this becomes

$$\left(\frac{\zeta - k_1}{k_2}\right)^2 + \left(\frac{\sigma}{k_3}\right)^2 = 1 \tag{70}$$

where

$$k_1 = \frac{z_1}{z_{s0}} = 0.2991, \qquad k_2 = 1 - k_1 = 0.7009, \qquad k_3 = \frac{y_0}{z_{s0}} = 0.7508 \tag{71}$$

It is possible to find the angle between the shock and the free surface at the point where they meet, independently of the numerical results.

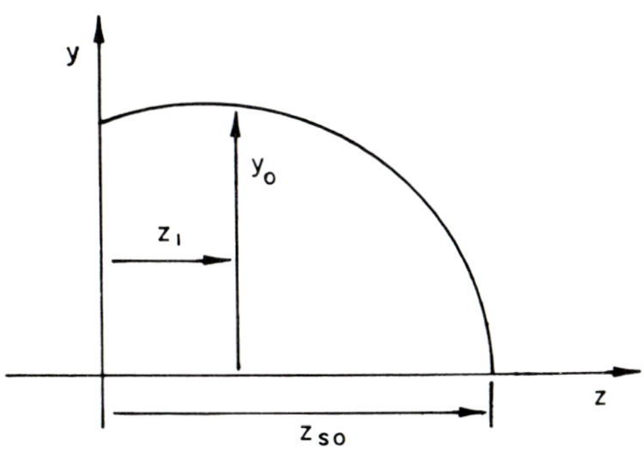

FIG. 9. Notation used in shock-shape formula.

The basis for this solution was given in a number of Russian papers (*30–35*), which are reviewed by Collins (*36*) and by Collins and Holt (*37*). The reasoning is that if the solution is to be self-similar then the shock must meet the free surface at an angle such that the wave system generated behind the shock does not cause any change in the shock shape. This condition is satisfied when the speed of the rarefaction front behind the shock just matches the velocity of the point of intersection of the shock and the free surface, a condition which is referred to as "critical reflection."

The velocity of an arbitrary point on a rarefaction front is equal to the vector sum of the sound speed (taken in the direction normal to the front) and the particle velocity, where both the sound speed and particle velocity are evaluated at the conditions just ahead of the advancing front. The expansion wavelet that originates at time t at a given point on the free surface is located, Δt later, as the locus of points which have been displaced, normal to the front, by a distance proportional to the local value of this vector-velocity sum. In particular, the small arc of the wave front near the free surface is traveling with a velocity equal to the vector sum of the postshock particle velocity, and the postshock sound speed taken in the direction normal to the wave front Thus, this small arc can be located as part of a circle, which is found from the construction (*38*) shown in Fig. 10.

The use of this construction is not meant to imply that the surface particle which was struck by the shock at time t has actually followed the path designated $q_1 \Delta t$ in Fig. 10, nor that the entire expansion wavelet

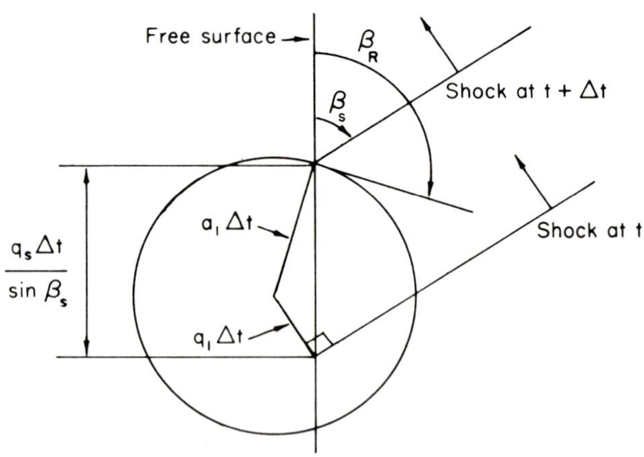

FIG. 10. Construction for determining intersection angle between shock and free surface.

has remained circular in shape. In fact, the trajectory of any surface particle is affected by the expansion itself, and the shape of the wavelet is dictated by the distribution of sound speed and particle velocity around the periphery of the wave. The only use to which the construction is put is to facilitate the vector addition of velocities and to establish the wave front orientation near the free surface.

Use of Fig. 10 leads to the formula

$$\sin \beta_s = \{2(q_1/q_s) - (q_1/q_s)^2 + (a_1/q_s)^2\}^{-1/2} \tag{72}$$

For an infinite-strength shock in a perfect gas, this becomes

$$\sin \beta_s = [(\gamma + 1)/2\gamma]^{1/2} \tag{73}$$

When $\gamma = 1.5$, $\beta_s = 65.9°$. The value that is found from Eq. (70) is 63.2°. The difference between these numbers is well within the accuracy inherent in the process of fairing the numerical data.*

The papers by Collins (*36*) and Collins and Holt (*37*) present a different formula for β_s. Instead of adding the velocity components q_1 and a_1 vectorially, as in the figure above, these authors equate the sum of the absolute values of these components to the speed of the shock–free-surface intersection point [see Eq. (3.20) of Refs. (*36*) and (*37*)]. Their formula for the shock angle, expressed in the present notation, is

$$\sin \beta_s]_{\text{C-H}} = \frac{\gamma + 1}{2 + [2\gamma(\gamma - 1)]^{1/2}} \tag{74}$$

For $\gamma = 1.5$, this formula gives $\beta_s = 50.8°$, which differs considerably from Walsh's data. Collins and Holt have pointed out in a private communication that they regard both their formula and that of the present work as approximations; they feel that the presence of the expansion fan makes it impossible to determine precisely the particle velocity to be used in finding β_s. The present writer would agree that the particle velocity will vary across the expansion fan, but is of the opinion that the particle velocity appropriate to the leading edge of the fan is uniquely the postshock value.

The shock velocity distribution corresponding to Eq. (69) was found by the usual method (39); Eq. (69) was written in the form

$$F(z, y, t) = 0 \tag{75}$$

* The writer did not become aware of the formula for β_s until after Eq. (70) had been established. It would be possible to adjust the constants so as to match Eq. (73) exactly, but this has not been done here.

The expression for dF/dt can be written as

$$q_{sz}\,\partial F/\partial z + q_{sy}\,\partial F/\partial y + \partial F/\partial t = 0 \tag{76}$$

where the components of the shock velocity are

$$\mathbf{q}_s = q_{sz}\hat{z} + q_{sy}\hat{y} \tag{77}$$

Along lines perpendicular to the shock, the shock velocity components have the ratio

$$q_{sy}/q_{sz} = (\partial F/\partial y)/(\partial F/\partial z) \tag{78}$$

Thus Eqs. (76) and (78) can be rewritten as

$$q_{sz} = -\frac{(\partial F/\partial t)/(\partial F/\partial z)}{1 + [(\partial F/\partial y)/(\partial F/\partial z)]^2} \tag{79}$$

and

$$q_s = q_{sz}\left[1 + \left(\frac{\partial F/\partial y}{\partial F/\partial z}\right)^2\right]^{1/2} = \frac{\partial F/\partial t}{[(\partial F/\partial y)^2 + (\partial F/\partial z)^2]^{1/2}} \tag{80}$$

In terms of the similarity coordinates, then, the shock velocity distribution is

$$\frac{q_s}{\dot{z}_{s0}} = \frac{(\zeta/k_2)((\zeta - k_1)/k_2) + (\sigma/k_3)^2}{[(1/k_2^2)((\zeta - k_1)/k_2)^2 + (1/k_3^2)(\sigma/k_3)^2]^{1/2}} \tag{81}$$

The components are

$$\left\{\frac{q_{sz}}{q_s}, \frac{q_{sy}}{q_s}\right\} = \frac{\{(\zeta - k_1)/k_2^2, \sigma/k_3^2\}}{[(1/k_2^2)((\zeta - k_1)/k_2)^2 + (1/k_3^2)(\sigma/k_3)^2]^{1/2}} \tag{82}$$

Using these velocities, the similarity functions can be evaluated at the shock from

$$\psi_1 = \frac{\gamma + 1}{\gamma - 1}, \quad f_1 = \frac{2}{\gamma + 1}\left(\frac{q_s}{\dot{z}_{s0}}\right)^2, \quad \phi_1 = \frac{2}{\gamma + 1}\frac{q_{sz}}{\dot{z}_{s0}}, \quad \omega_1 = \frac{2}{\gamma + 1}\frac{q_{sy}}{\dot{z}_{s0}} \tag{83}$$

Elsewhere in the flow field, the similarity functions were found from Eqs. (11)–(14), using as normalizing quantities

$$\dot{z}_{s0} = 2.20 \times 10^4 \quad \text{cm/sec}, \qquad \rho_0\dot{z}_{s0}^2 = 4.84 \times 10^{-4} \quad \text{Mbar}$$

The value of $\dot{z}_{s0}$ was found from the relation

$$\dot{z}_{s0} = (N/t)\, z_{s0} \tag{84}$$

with $N = 0.37$, $t = 5.40 \times 10^{-3}$ sec, $z_{s0} = 321$ cm.

These numbers can also be used to evaluate the constant appearing in the shock trajectory. Analogously to Eq. (62), the formula for the axisymmetric case is

$$z_{s0}(t) = F(\gamma)\left(\frac{P^2}{\rho_0 E}\right)^{1/3}\left[\frac{t}{(P^5/\rho_0 E^4)^{1/3}}\right]^{0.37} \tag{85}$$

From the above data, $F(1.5)$, is found to be 1.6415. As in the plane wave case, this amplitude coefficient also depends on the details of the initiation process. The figure given above applies for a right-circular cylinder of length-to-diameter ratio unity. The caluclations reported by Walsh *et al.* (*5*, *7*) indicate that the coefficient would be essentially the same for a sphere, or for a right-circular cylinder of length-to-diameter ratio between the limits $\frac{1}{3}$ and 3.

Equation (85) can be rewritten in terms of a scaling length $\mathscr{L}_0$ as

$$z_{s0}/\mathscr{L}_0 = F(\gamma)(ct/\mathscr{L}_0)^N \tag{86}$$

where

$$\mathscr{L}_0 = \left\{\frac{E}{\rho_0 c^2}\left(\frac{P}{\rho_0 c}\right)^{\lambda}\right\}^{1/3(1+\lambda)}, \qquad \lambda = \frac{2-5N}{4N-1} \tag{87}$$

2. *The Similarity Functions*

Contour plots, showing the distributions of ψ, f, ϕ, and ω in the ζ, σ plane were prepared from the numerical data. These results are shown in Figs. 11–14.

The density contours show a resemblance to what would be found for a spherically symmetric blast wave (*2*–*4*). The density falls very rapidly with distance normal to the shock, but it changes slowly with angular position.

The constant-pressure contours show considerably greater variation with angular position. They also display the feature, in common with the spherically symmetric explosion problem, that the pressure drops rapidly near the shock but quickly levels off to a relatively constant value.

The two velocity components display a linear variation for ζ greater than about 0.6; the distributions become more complicated at lower values of ζ, with the appearance of a region where the radial component is directed toward the axis.

The numerical method produces a nonzero value for all of the dependent variables along the free surface, outside the target. It will be seen below that the solution near the shock–free-surface intersection should be of the centered-wave type, and that the plume boundary, along which the pressure and density fall to zero, should not coincide

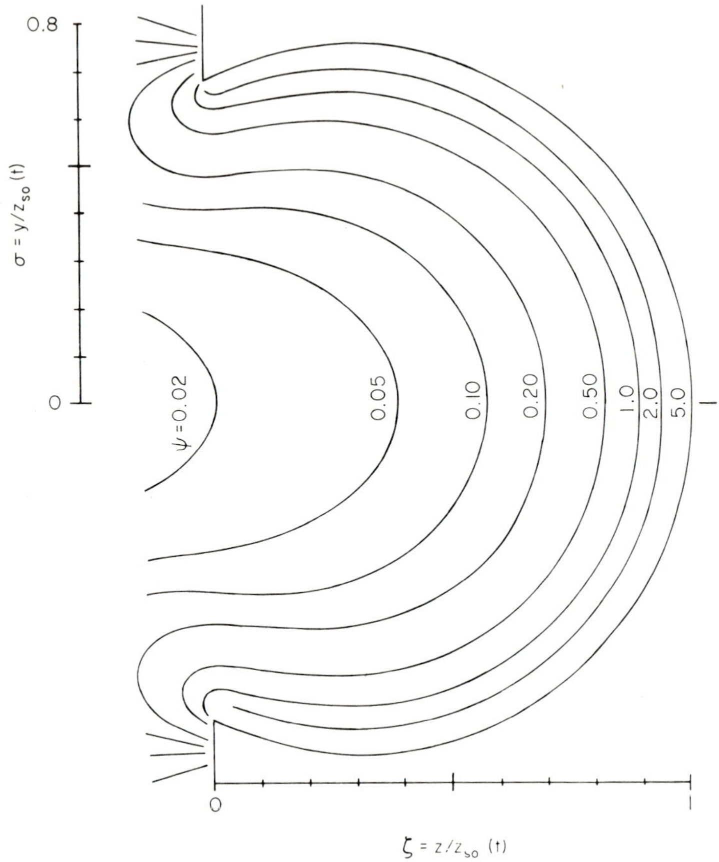

FIG. 11. Contours of constant-density ratio ($\psi = \rho/\rho_0$, $\gamma = 1.5$).

with the free surface for $\gamma = 1.5$. The centered-wave nature of the solution is partially masked by the effect of the nonzero grid size.

The numerical result for the boundary between the elliptic and hyperbolic regions is shown in Fig. 12. This boundary should intersect the shock–free-surface juncture. Its failure to do so is another demonstration that the numerical method loses some of the details of the centered-wave flow structure in this region.

It must be emphasized that a considerable amount of fairing and smoothing has gone into this reduction of Walsh's data, and that future research will undoubtedly lead to some adjustments in the locations of these contours. It is not possible to estimate their accuracy from the data on hand.

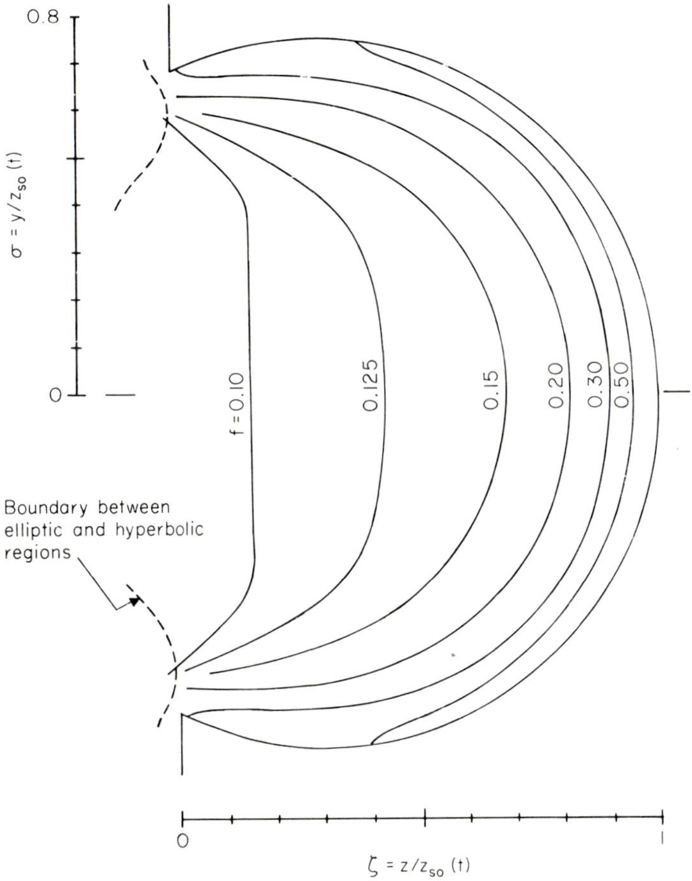

FIG. 12. Contours of constant pressure ($f = p/\rho_0 \dot{z}_{s0}^2$, $\gamma = 1.5$).

3. *Particle Paths*

For the self-similar inviscid flow considered here, it is possible to display the path taken by any particle after it is processed by the shock. This is done by expressing the fact the entropy of the particle remains constant, and equal to the value it acquired at the shock.* Since the entropy depends only on p/ρ^γ, this becomes

$$p/\rho^\gamma = p_1/\rho_1{}^\gamma \tag{88}$$

* It is assumed that there are no shock waves other than the main one.

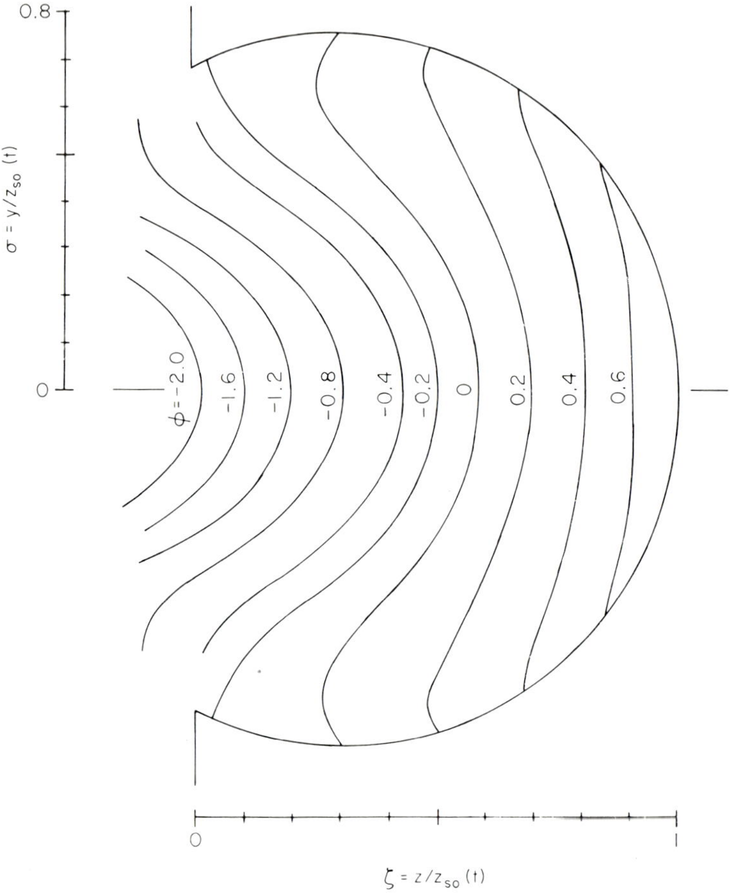

FIG. 13. Contours of constant axial velocity component ($\phi = u/\dot{z}_{s0}$, $\gamma = 1.5$).

where the subscript denotes conditions immediately after passage through the shock. In terms of the similarity functions, the relation is

$$\left(\frac{\dot{z}_{s01}}{\dot{z}_{s02}}\right)^2 = (f/f_1)(\psi_1/\psi)^\gamma \tag{89}$$

Here $\dot{z}_{s02}$ denotes the shock speed at the present instant, when the particle being considered has the dimensionless pressure and density f and ψ (without subscripts), and $\dot{z}_{s01}$ refers to the earlier time t_s when the particle in question was processed by the shock. This equation can be rewritten as

$$\frac{t}{t_s} = \left[\frac{f/\psi^\gamma}{f_1/[(\gamma+1)/(\gamma-1)]^\gamma}\right]^{1/2(1-N)} \tag{90}$$

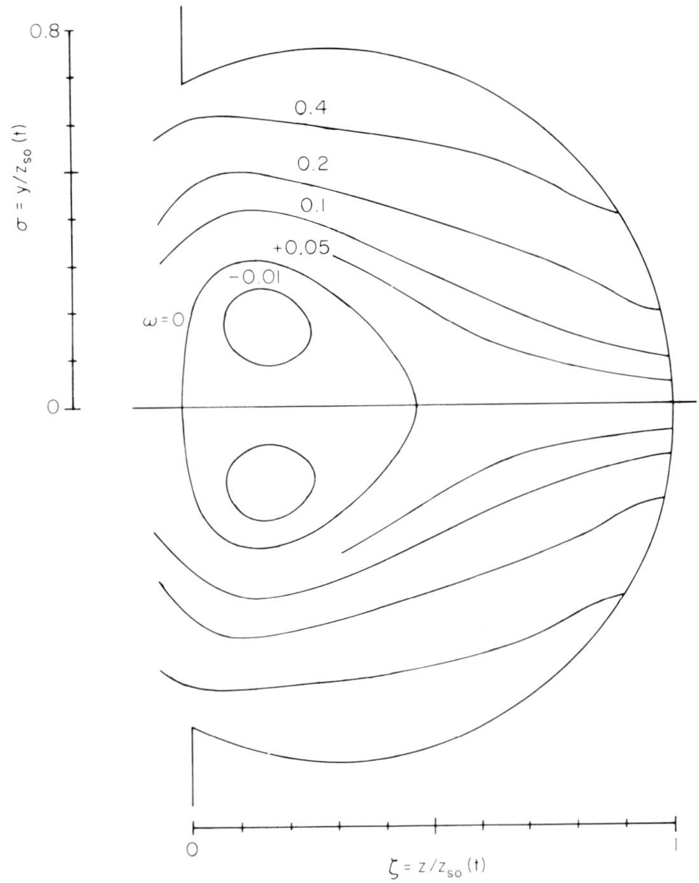

FIG. 14. Contours of constant radial velocity component ($\omega = v/\dot{z}_{s0}$, $\gamma = 1.5$).

Along the axis of symmetry, this relation alone can be used to find t/t_s at each value of ζ, since the axis is also a particle path. At points off the axis of symmetry, the particle path must be found explicitly. In a time interval dt, the position of the particle changes by $dy = v\,dt$ and $dz = u\,dt$; in terms of the similarity functions, these become

$$\begin{aligned} dy &= \sigma\,dz_{s0} + z_{s0}\,d\sigma = \omega\dot{z}_{s0}\,dt = \omega\,dz_{s0} \\ dz &= \zeta\,dz_{s0} + z_{s0}\,d\zeta = \phi\dot{z}_{s0}\,dt = \phi\,dz_{s0} \end{aligned} \tag{91}$$

Thus displacements in the similarity plane take place at a rate proportional to the pseudovelocity components $\omega - \sigma$ and $\phi - \zeta$:

$$d\sigma = (\omega - \sigma)\,d\ln z_{s0}, \qquad d\zeta = (\phi - \zeta)\,d\ln z_{s0} \tag{92}$$

and the particle paths are the family of lines satisfying

$$d\sigma/d\zeta = (\omega - \sigma)/(\phi - \zeta) \tag{93}$$

Solutions of this equation were found by the method of isoclines, and are shown as the dashed lines in Fig. 15. Contours of constant t/t_s are

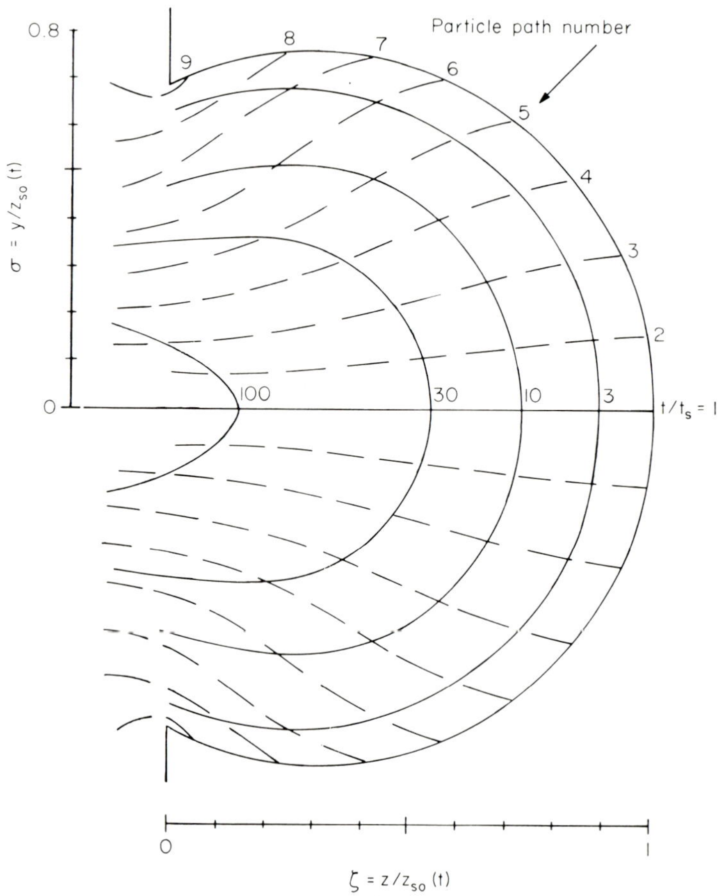

FIG. 15. Particle paths ($\gamma = 1.5$).

shown as solid lines. Table II lists the conditions at the points where the particles enter the shock.

Several interesting features can be observed in Fig. 15. The first is the time history of the path of particle which lies on the axis of symmetry, at a distance L_1 below the free surface, before being shocked

TABLE II

INITIAL CONDITIONS FOR PARTICLE PATHS

Path no.	1	2	3	4	5	6	7	8	9
ζ_s	1.0	0.984	0.930	0.832	0.709	0.567	0.418	0.245	0.040
σ_s	0	0.160	0.330	0.488	0.612	0.697	0.742	0.751	0.700

at $t = t_s$. This particle follows the shock with decreasing speed, until $t/t_s = 24.5$, at which time its velocity changes sign. This occurs (see Fig. 13) at $\zeta = 0.585$; thus the maximum depth to which this particle travels is

$$\frac{z_{\max}}{L_1} = \frac{z_{\max}}{z_{s0}(t)} \cdot \frac{z_{s0}(t)}{z_{s0}(t_s)} = 0.585(24.5)^{0.37} = 1.91 \tag{94}$$

Thereafter the particle moves toward the free surface, crossing it at $t/t_s = 125$. For comparison, the exact solution of the one-dimensional problem for $\gamma = 1.4$ shows that the maximum penetration of $2.60L_1$ occurs at $t/t_s = 15.6$, and that the particle crosses the original free surface when $t/t_s = 55.9$.

Next, it is of interest to follow the motion of a layer of particles, all of which lie initially in a plane a distance L_1 below the free surface. Let t_1 denote the instant when the center of the shock arrives at $z = L_1$. [For a chosen value of L_1 the quantity t_1 can be related to the mass and velocity of the projectile for any specific case by using Eq. (85)]. Somewhat later, the point on the shock at which particle path number 2 begins (i.e., $\zeta = 0.984$, $\sigma = 0.160$) will reach the depth L_1; i.e., the shock scale at that instant will be

$$z_{s0}(t_s)/L_1 = 1/0.984 = 1.018$$

Thus, the y coordinate of this point will be

$$\frac{y_s}{L_1} = \frac{y_s}{z_{s0}(t_s)} \cdot \frac{z_{s0}(t_s)}{L_1} = \frac{0.160}{0.984} = 0.163$$

The time at which this particle is shocked is

$$\frac{t_s}{t_1} = \left[\frac{z_{s0}(t_s)}{z_{s0}(t_1)}\right]^{1/N} = \left[\frac{z_{s0}(t_s)}{L_1}\right]^{1/N} = (1.018)^{1/0.37} = 1.049$$

The location of this particle at any subsequent time can now be found. For instance, if a later time t is chosen such that $z_{s0}(t)/L_1 = 5$ (i.e.,

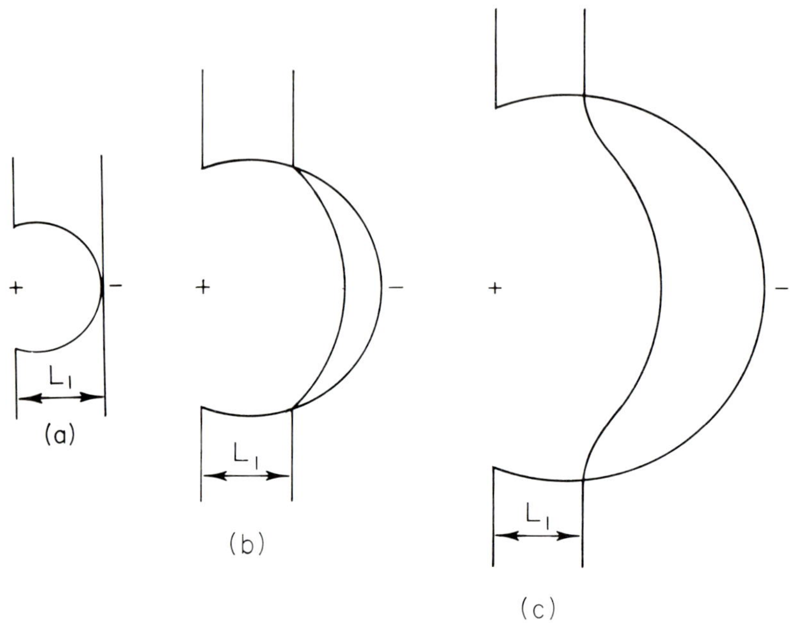

FIG. 16. Postions of particles located initially in the plane $z = L_1$. (a) At $t = t_1$ ($z_{s0} = L_1$); (b) at $t = 6.5t_1$ ($z_{s0} = 2L_1$); (c) at $t = 19.5t_1$ ($z_{s0} = 3L_1$); (d) at $t = 42.2t_1$ ($z_{s0} = 4L_1$); (e) at $t = 77t_1$ ($z_{s0} = 5L_1$).

$t/t_1 = (5)^{1/0.37} = 77$] then the time after shock crossing for this particle would be

$$t/t_s = (t/t_1)/(t_s/t_1) = 77/1.049 = 73.5$$

The location of this particle can then be found, from Fig. 15, as $\zeta = 0.25$, $\sigma = 0.08$, which gives $z/L_1 = 1.25$, $y/L_1 = 0.40$.

Calculations of this type were carried out for all nine particle paths, and for times such that $z_{s0}(t)/L_1$ was equal to 2, 3, 4, and 5 (the corresponding times are $t/t_1 = 6.5$, 19.5, 42.2, and 77). The results are shown in Fig. 16. The layer of particles is carried down into the target initially. Later on, particles which are sufficiently far from the axis begin to acquire a significant radial component, with those at very large values of y_s/L_1 being carried toward the free surface by the shock itself. At this later time, the particles which originally lay near the axis have reversed their downward motion and are traveling toward the free surface.

This motion bears a resemblance to the type of deformation observed by Frasier and Karpov (*40*) in a target composed of alternate light and dark layers of wax. At some time before the motion was brought to rest by the strength effects, the particles near the axis were carried downward,

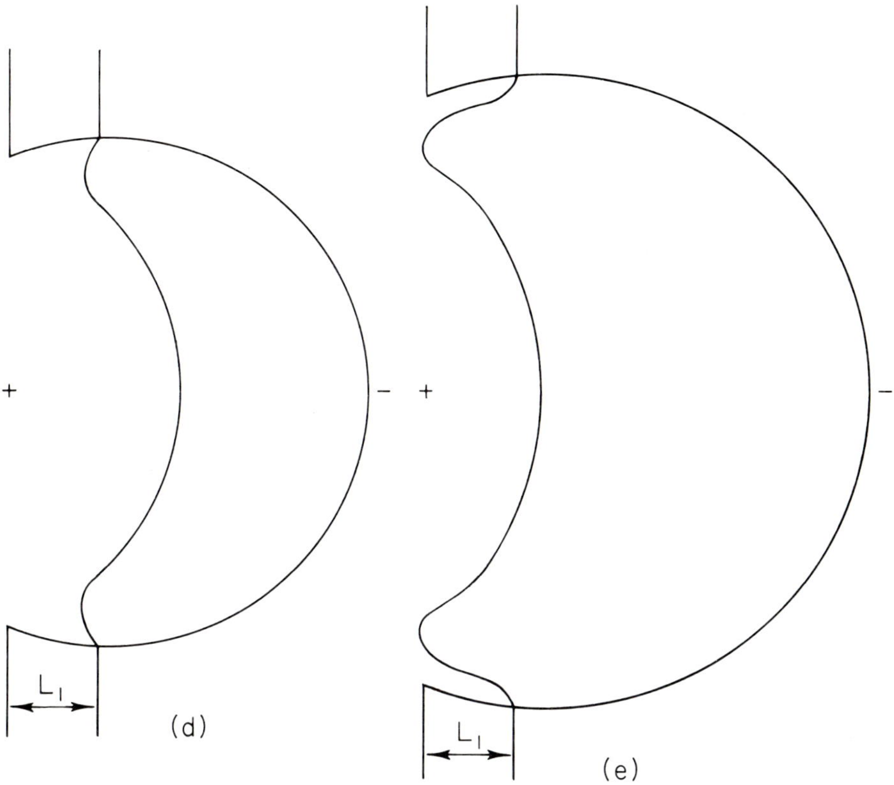

FIG. 16d and e.

while those away from the axis were sheared into a wavelike pattern. The present results display, at least qualitatively, the same sort of behavior, although they obviously have no quantitative relation to the wax experiments (which involve both nonsimilarity and strength effects). However, it is interesting to note that an inviscid treatment contains a mechanism for generating the type of deformation observed. The question of how this pattern is affected, and ultimately arrested, by the effects of nonsimilarity and of material strength remains a subject for further study.

4. *Solution near the Shock–Free-Surface Intersection*

It is possible to derive an exact solution for the properties of the flow in the vicinity of the point where the shock meets the free surface. The derivation below follows that given in Section 108 of Courant and Friedrichs (*41*).

a. Differential Equations. The similarity equations are expressed in a coordinate system whose origin is located at the shock–free-surface intersection (see Fig. 17):

$$\begin{aligned} \zeta &= l \sin \beta \\ \sigma &= \sigma_1 + l \cos \beta \end{aligned} \tag{95}$$

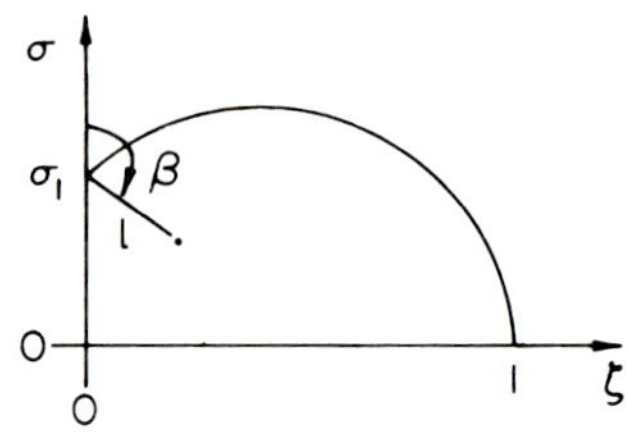

FIG. 17. Coordinate system near the shock–free-surface intersection.

All of the dependent variables are expanded as a series in l; for example,

$$\phi(l, \beta) = \phi_{(0)}(\beta) + l\phi_{(1)}(\beta) + l^2\phi_{(2)}(\beta) + \cdots \tag{96}$$

If these expansions are now substituted into the full similarity equations [Eqs. (*39*)–(*42*)] the leading terms (coefficients of l^{-1}) are found to be (a prime denotes $d/d\beta$)

Continuity:

$$[\phi_{(0)} \cos\beta - (\omega_{(0)} - \sigma_1) \sin\beta]\, \psi'_{(0)} + \psi_{(0)}[\phi'_{(0)} \cos\beta - \omega'_{(0)} \sin\beta] = 0 \tag{97}$$

Axial momentum:

$$[\phi_{(0)} \cos\beta - (\omega_{(0)} - \sigma_1) \sin\beta]\, \phi'_{(0)} + \frac{f'_{(0)}}{\psi_{(0)}} \cos\beta = 0 \tag{98}$$

Radial momentum:

$$[\phi_{(0)} \cos\beta - (\omega_{(0)} - \sigma_1) \sin\beta]\, \omega'_{(0)} - \frac{f'_{(0)}}{\psi_{(0)}} \sin\beta = 0 \tag{99}$$

Energy:

$$[\phi_{(0)} \cos\beta - (\omega_{(0)} - \sigma_1) \sin\beta] \left[f'_{(0)} - \frac{\gamma f_{(0)}}{\psi_{(0)}} \psi'_{(0)} \right] = 0 \tag{100}$$

Since these equations are homogeneous in the derivatives, they will have a nontrivial solution only if the determinant of the coefficients vanishes. This condition is

$$N_{(0)}^2[N_{(0)}^2 - \gamma f_{(0)}/\psi_{(0)}] = 0 \tag{101}$$

where

$$N_{(0)} = \phi_{(0)} \cos\beta - (\omega_{(0)} - \sigma_1) \sin\beta \tag{102}$$

This notation is prompted by the fact that the quantity on the right-hand side of Eq. (102) is just the zeroth-order component, normal to a constant-β line, of the vector whose components are $\phi - \zeta$ and $\omega - \sigma$ (T is the tangential component; see Fig. 18):

$$\begin{aligned} N &= (\phi - \zeta)\cos\beta - (\omega - \sigma)\sin\beta \\ T &= (\phi - \zeta)\sin\beta + (\omega - \sigma)\cos\beta \end{aligned} \tag{103}$$

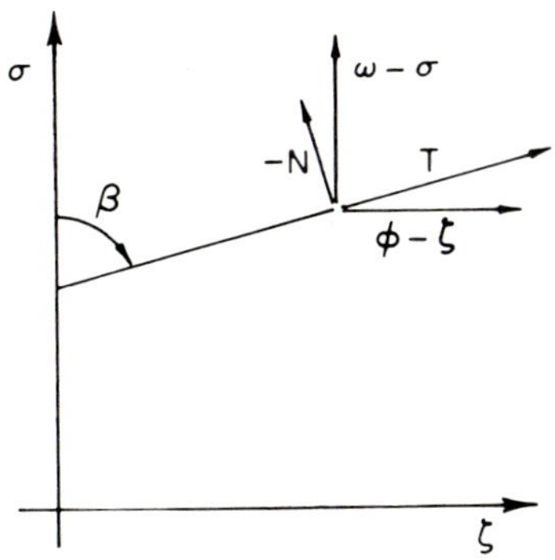

FIG. 18. Significance of the quantities N and T.

To make the determinant of the coefficients zero, the choice is made

$$N_{(0)}^2 = \gamma f_{(0)}/\psi_{(0)} \tag{104}$$

This expression is the counterpart, for the present problem, of the usual Prandtl–Meyer relation, which states that the velocity component normal to the rays is sonic (*42*).

Since $N_{(0)}$ is in general not zero, the energy equation can be written as

$$f'_{(0)} - (\gamma f_{(0)}/\psi_{(0)})\,\psi'_{(0)} = 0 \tag{105}$$

whose solution is the isentropic relation

$$f_{(0)} = \text{const}\,\psi_{(0)}^{\gamma} \tag{106}$$

The continuity equation can be written

$$N_{(0)}\psi'_{(0)}/\psi_{(0)} + (N'_{(0)} + T_{(0)}) = 0 \tag{107}$$

The quantity $\psi'_{(0)}/\psi_{(0)}$ appearing here is evaluated from Eqs. (104) and (106), giving

$$\psi'_{(0)}/\psi_{(0)} = [2/(\gamma - 1)](N'_{(0)}/N_{(0)}) \tag{108}$$

Thus the continuity equation becomes

$$N'_{(0)} = -[(\gamma - 1)/(\gamma + 1)]\, T_{(0)} \tag{109}$$

A second relation between $N_{(0)}$ and $T_{(0)}$ can be derived by adding the axial momentum equation, multiplied by $\sin\beta$, to the radial momentum equation, multiplied by $\cos\beta$; the result is

$$T_0' = N_{(0)} \tag{110}$$

This is just a statement of the irrotationality condition, i.e.,

$$\operatorname{curl} \mathbf{q} = \hat{e}_3(\dot{z}_{s0}/lz_{s0})\{\omega'_{(0)} \cos\beta + \phi'_{(0)} \sin\beta\} + O(l^0) \tag{111}$$

Equations (109) and (110) can now be combined to give

$$T''_{(0)} + [(\gamma - 1)/(\gamma + 1)]\, T_{(0)} = 0 \tag{112}$$

Thus

$$T_{(0)} = B_1 \sin\left(\frac{\gamma - 1}{\gamma + 1}\right)^{1/2} \beta + B_2 \cos\left(\frac{\gamma - 1}{\gamma + 1}\right)^{1/2} \beta \tag{113}$$

$$N_{(0)} = T'_{(0)} = \left(\frac{\gamma - 1}{\gamma + 1}\right)^{1/2} \left\{B_1 \cos\left(\frac{\gamma - 1}{\gamma + 1}\right)^{1/2} \beta - B_2 \sin\left(\frac{\gamma - 1}{\gamma + 1}\right)^{1/2} \beta\right\} \tag{114}$$

b. Boundary Conditions. In order to determine B_1 and B_2, boundary conditions on $N_{(0)}$ and $T_{(0)}$ must be assigned at some value of β. Immediately behind the shock, where $\beta = \beta_s$ and where the component of the shock velocity along the surface is $\sigma_1 \dot{z}_{s0}$, the dependent variables have the values [see Eq. (83)]

$$\psi_{(0)} = (\gamma + 1)/(\gamma - 1) \tag{115}$$

$$f_{(0)} = [2/(\gamma + 1)](\sigma_1 \sin\beta_s)^2 = \sigma_1^2/\gamma \tag{116}$$

$$\phi_{(0)} = -[2\sigma_1/(\gamma + 1)] \sin\beta_s \cos\beta_s = -(\sigma_1/\gamma)[(\gamma - 1)/(\gamma + 1)]^{1/2} \tag{117}$$

$$\omega_{(0)} = [2\sigma_1/(\gamma + 1)] \sin^2\beta_s = \sigma_1/\gamma \tag{118}$$

The quantities ψ, f, ϕ, and ω remain constant at these values, for the region between the shock and the head of the expansion wave, i.e., for $\beta_s < \beta < \beta_R$. Thus, in this region, $N_{(0)}$ and $T_{(0)}$ vary with β according to

$$N_{(0)} = \sigma_1 \left\{-\frac{1}{\gamma}\left(\frac{\gamma - 1}{\gamma + 1}\right)^{1/2} \cos\beta - \left(\frac{1}{\gamma} - 1\right) \sin\beta\right\} \tag{119}$$

$$T_{(0)} = \sigma_1 \left\{-\frac{1}{\gamma}\left(\frac{\gamma - 1}{\gamma + 1}\right)^{1/2} \sin\beta + \left(\frac{1}{\gamma} - 1\right) \cos\beta\right\} \tag{120}$$

The value of β_R can be found, as the tangent to the rarefaction wave circle in Fig. 10:

$$\sin \beta_R = \left\{1 - \left(\frac{q_1}{a_1} \cos \beta_s\right)^2\right\}^{1/2}, \qquad \cos \beta_R = -\frac{q_1}{a_1} \cos \beta_s \tag{121}$$

For an infinite-strength shock in a perfect gas, this becomes

$$\sin \beta_R = (\gamma^2 - 1)^{1/2}/\gamma, \qquad \cos \beta_R = -1/\gamma \tag{122}$$

Thus, the initial conditions are

$$T_{(0)}(\beta_R) = 0, \qquad N_{(0)}(\beta_R) = \sigma_1[(\gamma - 1)/(\gamma + 1)]^{1/2} \tag{123}$$

After determining the constants B_1 and B_2 to match these boundary conditions, the solution can be written in the form

$$T_{(0)} = \sigma_1 \sin\{[(\gamma - 1)/(\gamma + 1)]^{1/2}(\beta - \beta_R)\} \tag{124}$$

$$N_{(0)} = \sigma_1[(\gamma - 1)/(\gamma + 1)]^{1/2} \cos\{[(\gamma - 1)/(\gamma + 1)]^{1/2}(\beta - \beta_R)\} \tag{125}$$

c. Properties of the Solution. The two components of the particle velocity can now be found from these relations. To find the density and pressure, the constant in Eq. (106) is evaluated, from conditions at β_R:

$$f_{(0)} = \frac{\sigma_1^2}{\gamma} \left(\frac{\gamma - 1}{\gamma + 1}\right)^\gamma \psi_{(0)}^\gamma \tag{126}$$

Then, using Eq. (104), the density is found as

$$\psi_{(0)} = \left\{\left(\frac{N_{(0)}}{\sigma_1}\right)^2 \left(\frac{\gamma + 1}{\gamma - 1}\right)^\gamma\right\}^{1/(\gamma-1)} \tag{127}$$

and the pressure is calculated from Eq. (126).

The value of β at which the density and pressure fall to zero is found by setting $N_{(0)} = 0$, i.e.,

$$[(\gamma - 1)/(\gamma + 1)]^{1/2}(\beta - \beta_R) = \pi/2 \tag{128}$$

Thus the "vacuum" boundary is

$$\beta_{\text{vacuum}} = \beta_R + [(\gamma + 1)/(\gamma - 1)]^{1/2}(\pi/2) \tag{129}$$

The total change in β from the front of the rarefaction fan to the vacuum boundary is $[(\gamma + 1)/(\gamma - 1)]^{1/2} (\pi/2)$, in complete correspondence to the total Mach-angle change for a steady Prandtl–Meyer flow (*43*).

The variations of β_s, β_R, and β_{vacuum} with γ are shown in Fig. 19; it is interesting to note that the vacuum boundary interacts with the free surface for γ less than about 1.37, and lies above it for values of γ greater than 1.37.

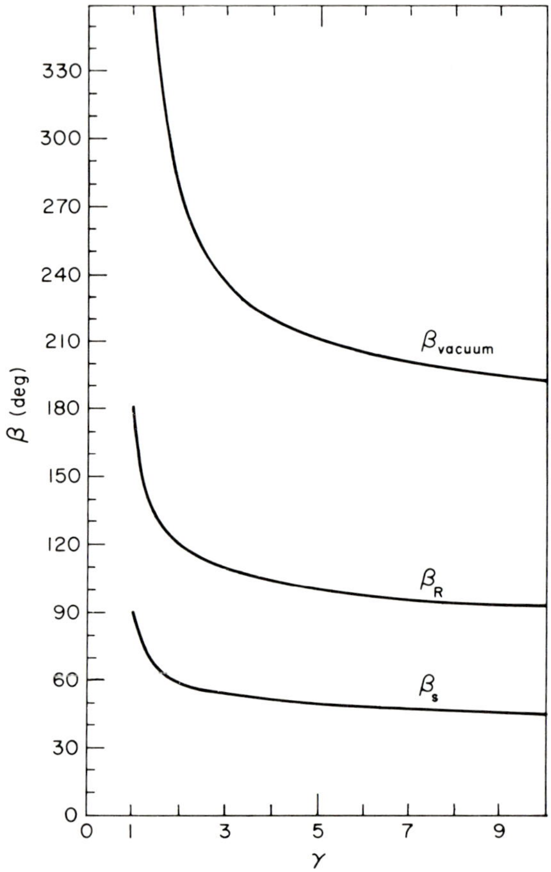

FIG. 19. Angular position of shock, rarefaction front, and zero-pressure surface.

The above solution reveals the angle along which the various contours in Figs. 11–14 should approach the shock–free-surface intersection. These angular locations are in reasonable agreement with the results inferred from Walsh's solution, except in the region outside the target, as noted earlier. The singular characteristic, which divides the elliptic and hyperbolic regions, can be shown from the solution above to lie along $\beta = \beta_R$. As mentioned in the discussion of Fig. 12, the numerical data do not recover this location very well.

It would be of interest to extend this solution to the next order in l, as done for steady isentropic flow by Johannesen and Meyer (*44*), and for flows having entropy gradients* and discontinuities of streamline curvature by Hakkinen (*45*, *46*). The solution is required to this order, if the details of the free surface trajectory are to be found.

D. Approximate Solutions

Many authors have attempted to uncover some of the features of the perfect-gas solution by resorting to various approximations. The most significant of these is the work of Rayzer (*15*), who identified the analogy with the one-dimensional case, and extended Zeldovich's arguments (*14*) concerning the simultaneous conservation of energy and momentum. He worked out an axisymmetric solution based on the approximation that (in the present notation) ϕ and ω should be linear functions of ζ and σ, respectively. This assumption produces a constant value (in the similarity coordinates) for the divergence of the velocity field, a condition which Rayzer suggests as a suitable generalization of the nearly constant value of $d\phi/d\zeta$ that is found in the plane wave case (it is exactly constant for $\gamma = 1.4$). To work out the details of the solution, he assumes a crude shape for the shock, and is able to find a pair of values of N and γ. The point-impact case he presents has $N = 0.386$, for $\gamma = 1.205$.

The distributions which he obtains for the various functions are admittedly crude, but they resemble in many respects the numerical solution described earlier. For example, the density ratio ψ is predicted as 0.0187 along the $\zeta = 0$ plane; reference to Fig. 11 shows that this is a reasonable value.

Rayzer presents a sketch of the shock shape that resembles the one found by Walsh, but he does not cite any numerical evidence for this shape.

Finally, it should be mentioned that Rayzer has also studied the line-impact case. His results again reveal a shock trajectory law that is very close to the constant-energy solution. In connection with the line–explosion problem, mention should be made of the experiments reported by Deribas and Pokhozhaev (*48*), who used an exploding wire at the surface of a tank of water. They present a brief discussion of some theoretical scaling concepts and indicate that their results favor a similitude based on the momentum acquired by the fluid. Later experi-

* Dr. Sundaram has pointed out the interesting possibility that the solution to this order might reveal the existence of a lip shock, which is known to occur in steady supersonic flow when a highly vortical region undergoes a sudden expansion (*47*).

ments by Minin (*49*), however, indicate that the effects of gravity were important in the Deribas–Pokhozhaev data. Minin's measurements showed that when this effect is eliminated the growth rate of the funnel or plume [essentially $\sigma_1 z_{s0}(t)$] has a power-law dependence, with the exponent 0.47, which is just slightly less than the value 0.5 that would apply for a constant-energy, self-similar solution. In addition, Minin observed a growth rate of $t^{0.38}$ for the case when the exploding wire was arranged as a small flat coil, so as to generate essentially a point explosion.

Another approximate treatment of the point-source, perfect-gas problem was given by the present writer (*16*). In that work, the similarity equations were written along the axis of symmetry*:

$$(\phi - \zeta)\,\psi' + \psi(\phi' + 2V) = 0 \tag{130}$$

$$-[(1 - N)/N]\,\phi + (\phi - \zeta)\,\phi' + f'/\psi = 0 \tag{131}$$

$$-2[(1 - N)/N]\,f + (\phi - \zeta)\,f' - (\gamma f/\psi)(\phi - \zeta)\,\psi' = 0 \tag{132}$$

Here the prime denotes $d/d\zeta$, and the function V is defined as

$$V = (\partial\omega/\partial\sigma)_{\sigma=0} \tag{133}$$

If information could now be provided about the function V, the solution near the axis of symmetry could be carried out in complete analogy to the one-dimensional case. In particular, Eqs. (130)–(132) become singular at the point where $(\phi - \zeta)^2 = \gamma f/\psi$, which is obviously the point where the boundary between the elliptic and hyperbolic zones intersects the axis of symmetry.

In order to provide an estimate of the function V, the radial momentum equation was differentiated, giving

$$-\frac{1 - N}{N}\,V + (\phi - \zeta)\,V' + (V - 1)\,V + \frac{1}{\psi}\left(\frac{\partial^2 f}{\partial\sigma^2}\right)_{\sigma=0} = 0 \tag{134}$$

The off-axis pressure distribution was approximated by

$$(\partial^2 f/\partial\sigma^2)_{\sigma=0} = f'/\zeta - (K/\zeta^2)(1 - \zeta)\,f \tag{135}$$

where the parameter K is a measure of the rate of decrease of the pressure with distance from the axis. In the blunt-body case, this parameter is directly related to the shock shape; in the present problem, it is much

* This approach was motivated by the success of analogous methods in treating the hypersonic flow over a blunt body (*50*).

more difficult to estimate. The particular form of this approximation corresponds, in a spherical coordinate system (see Fig. 20), to

$$[(\partial^2 f/\partial \theta^2)_\eta]_{\theta=0} = -K(1-\eta)f \tag{136}$$

where the coordinates are defined as

$$\eta = r/z_{s0}(t) \tag{137}$$

$$\zeta = \eta \cos \theta \tag{138}$$

$$\sigma = \eta \sin \theta \tag{139}$$

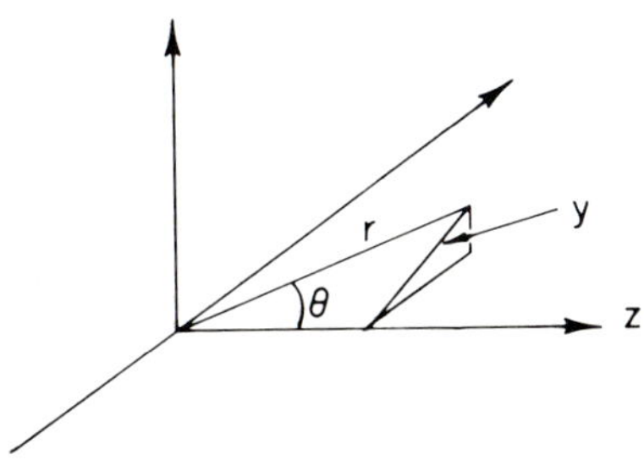

FIG. 20. Spherical polar coordinates.

The η factor in Eq. (136) was chosen so as to make $\partial^2 f/\partial\theta^2$ zero at $\eta = 1$, since it has already been assumed by Rae and Kirchner (*16*) that the shock was a hemisphere with origin located at the impact point on the free surface.

It was found that for values of γ from 2.0 to 6.0, and K set equal to 1.0 and 10.0, the value of N required for a smooth transition through the singular point of Eqs. (130)–(132) always lay between 0.367 and 0.4. These results are in good agreement with Rayzer's approximate treatment, and with Walsh's calculations. In all three of these papers, the value of N is found to be near, but slightly less than $\frac{2}{5}$, the value that would apply for a symmetric, constant-energy solution.

A direct comparison between the function V predicted by the approximation above and that revealed by Walsh's solution is difficult to make, partly because of the hemispherical shock-shape assumption, and partly because of the loss of accuracy encountered in differencing the numerical data. The general shape and magnitude of the V distritutions are in general agreement, however.

Because of the high density at the shock, most of the mass processed is concentrated in a thin shell at the shock. This fact, together with the closeness of the approximate values of N to the value 0.4, suggested that the flow could be approximated as one-half of a spherically symmetric, constant-energy solution,* and this further approximation is advocated by Rae and Kirchner (16).

To calculate the shock trajectory in this approximation, it is assumed that all of the kinetic energy of the projectile goes into the flow behind a hemispherical shock. Then an energy balance gives

$$E = 2\pi\rho_0 R_s{}^3 \dot{R}_s{}^2 I(\gamma) \tag{140}$$

where the radius of the shock (equal to z_{s0} along the symmetry axis) is denoted by $R_s(t)$, and where

$$I(\gamma) = \int_0^1 (f/(\gamma-1)\,\psi + \tfrac{1}{2}\phi^2)\,\psi\eta^2\,d\eta, \qquad \eta \equiv r/R_s(t) \tag{141}$$

To evaluate this integral, the distributions of f, ψ, ϕ must be found. The solution for these quantities was found numerically by G. I. Taylor (*4*), and a closed-form solution was found by von Neumann (*2*), and by Sedov (*3*). [The closed-form solution was also derived, somewhat later, by J. L. Taylor (*51*), Latter (*52*), and Sakurai (*53*).] Values of $I(\gamma)$ based on these papers can be found in Rae (*54*) for $1.2 \leqslant \gamma \leqslant 100$.

The solution of Eq. (140) for the shock trajectory is

$$R_s(t) = \left(\frac{25}{8\pi\, I(\gamma)}\,\frac{Et^2}{\rho_0}\right)^{1/5} \tag{142}$$

Using the value $I(1.5) = 0.335$, and using a projectile of mass M and velocity U would give

$$R_s(t)]_{\gamma=1.5} = 2.25\left(\frac{2M}{\rho_0}\right)^{1/3}\left[\frac{Ut}{(2M/\rho_0)^{1/3}}\right]^{0.4} \tag{143}$$

For comparison, Walsh's data [see Eq. (85)] give

$$z_{s0}(t)]_{\gamma=1.5} = 1.270\left(\frac{2M}{\rho_0}\right)^{1/3}\left[\frac{Ut}{(2M/\rho_0)^{1/3}}\right]^{0.37} \tag{144}$$

Thus, the symmetric solution approximation overestimates the shock penetration depth, except for extremely small values of t.

The first application of the symmetric solution to the impact problem was made by Davids *et al.* (*55*), and by Davids and Huang (*56*). At the time of those papers, the relation of the symmetric model to the true solution was not clear, and consequently the proper basis for choosing the

* In addition, the symmetry axis distributions of f, ψ, and ϕ agree fairly well with those found from the symmetric solution.

parameter N was not appreciated. As a result, some of the effort in Refs. (*55*) and (*56*) was directed toward finding solutions for $N \neq 0.4$. However, it appears that the only nonsingular solution of the symmetric problem is that for $N = 0.4$.

V. The Real-Fluid Case

Impact-generated shock propagation in solids rarely takes place at speeds large enough to permit the idealization that the material behaves thermodynamically like a perfect gas. There is a large range of conditions for which the inviscid-fluid model is still valid, but where a more realistic state equation is needed. In this section, some of the equations of state that more accurately describe a compressed solid are described. Following this, exact numerical solutions are discussed, with special reference to similitude considerations. Finally, some approximate solution methods are given.

A. The Equation of State

For the range of thermodynamic conditions encountered in hypervelocity impact, the Mie–Grüneisen equation of state [see, for instance, Rice *et al.* (*57*)] provides a suitable description. Workers at various laboratories in the United States have evolved fairly complicated expressions of this equation, especially for metals. [See, for example, the equations used by Bjork (*58*), Tillotson (*59*), and Riney (*60*), and the comparison of these given by Rae (*54*).] For the purposes of this article, a somewhat simpler expression will suffice.

1. *General Form*

The Mie–Grüneisen equation is usually written as (*57*)

$$e(p, \rho) - e_c(\rho) = [p - p_c(\rho)]/\rho\, \Gamma(\rho) \tag{145}$$

where the subscript c denotes the cohesive contributions, and where $\Gamma(\rho)$ is the Grüneisen factor, which depends weakly on ρ.

Along the Hugoniot, Eq. (145) takes the form

$$e_{\mathrm{H}}(\rho) - e_c(\rho) = [p_{\mathrm{H}}(\rho) - p_c(\rho)]/\rho\, \Gamma(\rho) \tag{146}$$

Subtracting the left and right sides of this from the corresponding sides of Eq. (145) gives

$$e = p/\rho\, \Gamma(\rho) - \Delta(\rho) \tag{147}$$

where

$$\Delta(\rho) = p_H(\rho)/\rho\, \Gamma(\rho) - e_H(\rho) = p_c(\rho)/\rho\, \Gamma(\rho) - e_c(\rho) \tag{148}$$

The presence of the second term in Eq. (147) precludes a self-similar solution [compare with Eq. (24)] except in the limit of extremely high pressure where the quantity $\Delta(\rho)$ is negligible compared to the leading term.

If this state equation is now substituted into the Rankine–Hugoniot conditions, the result is

$$\left[1 + \frac{2\,\Delta(\rho_1)}{q_s^2}\right]\left(\frac{\rho_1}{\rho_0}\right)^2 - 2\left[\frac{1}{\Gamma(\rho_1)} + 1\right]\frac{\rho_1}{\rho_0} + 1 + \frac{2}{\Gamma(\rho_1)} = 0 \tag{149}$$

This relation could be used to find the dependence of ρ_1/ρ_0 on q_s, if information were available about the functions $\Gamma(\rho_1)$ and $\Delta(\rho_1)$.

2. *Specific Form for a c, s Material*

This process can be reversed and the measured Hugoniot data used to infer information about $\Gamma(\rho_1)$ and $\Delta(\rho_1)$. These relations take on a simple form for a substance whose Hugoniot displays a linear relation between the shock speed and particle speed:

$$q_s = c + sq_1 \tag{150}$$

The weak-wave speed c is approximately equal to the bulk dilatational wave speed in many cases (*57*). Such a substance is referred to here as a c, s material. Many solids are well approximated by this relation; Rice *et al.* (*57*), for example, list values of c and s for various substances.

The shock wave relations for a c, s material can be written in terms of the shock Mach number $M_s = q_s/c$ as

$$\begin{aligned} \frac{\rho_1}{\rho_0} &= \frac{sM_s}{1 + (s-1)\,M_s}, & \frac{e_1 - e_0}{c^2} &= \frac{(M_s - 1)^2}{2s^2} \\ \frac{p_1}{\rho_0 c^2} &= \frac{M_s(M_s - 1)}{s}, & \frac{q_1}{c} &= \frac{M_s - 1}{s} \end{aligned} \tag{151}$$

These relations can now be used to evaluate $\Delta(\rho)$; the resulting form of the equation of state is

$$\frac{e_1 - e_0}{c^2} = \frac{p/\rho_0 c^2}{(\rho/\rho_0)\,\Gamma(\rho/\rho_0)} - \frac{(\rho/\rho_0 - 1)[2/\Gamma(\rho/\rho_0) + 1 - \rho/\rho_0]}{2[s - (s-1)\,\rho/\rho_0]^2} \tag{152}$$

At very high shock strengths, the Grüneisen factor must be taken as $2(s - 1)$, in order to match the limiting density ratios of Eqs. (149)

and (151). In the work reported by Rae (*54*), this same constant value was used for all densities, although it would be more appropriate to allow Γ to vary with ρ in such a way as to reach the value $2s - 1$ at normal density (*57*).

Whatever dependence is chosen for $\Gamma(\rho/\rho_0\,;s)$, Eq. (152) reveals that equation-of-state data for many materials can be correlated in terms of the quantities $p/\rho_0 c^2$, ρ/ρ_0, and $(e - e_0)/c^2$, with only s remaining as a parameter. Correlations of this type have been presented by Walsh (*6*), Gylden (*61*), and Gogolev *et al.* (*62*), among others.

It was observed earlier that the Mie–Grüneisen equation approaches the perfect-gas equation of state whenever the pressure becomes large enough that the leading term dominates. Eq. (152) can be used to make a quantitative assessment of the pressure level required. In particular, at the shock, use of Eqs. (151) shows that, as $M_s \to \infty$, the leading term is of order M_s^2 while the second term is of order M_s. At points removed from the shock, the relative order of the leading term is even greater (M_s^2 times as large) than that of the term $\Delta(\rho)$ [see Eq. (183) below]. Thus in the infinite-Mach-number limit, Eq. (152) can be replaced by the perfect-gas equation of state, with γ equal to $2s - 1$.

Strictly speaking, Eq. (152) is not valid at pressures so high that the linear q_s, q_1 relation ceases to apply. A more serious limitation is that it sometimes leads to negative values of the pressure* and the square of the sound speed, for points along the isentropes at less than normal density (*54*). However, it does provide an adequate and simple representation for the high-pressure regions of the flow.

3. *Enig's Formulations*

Two other expressions for the state equation of a c, s material have been presented by Enig (*63*), and by Enig and Petrone (*64*). These expressions are more satisfactoy, at densities below normal, than the one described above. However, they appear to have escaped the attention of workers in the field of hypervelocity impact.

The basis of the first expression (*63*) was to pursue the logical consequences of the "mirror-image" approximation, which states that the pressures and velocities experienced by a particle in an isentropic expansion process lie along a mirror image of the Hugoniot curve (see Fig. 21). This construction is consistent with the well-established approximation that the escape speed acquired by isentropic expansion from a point on the Hugoniot is close to twice the particle speed on the

* Negative pressures are also predicted for some conditions by the state equations of Bjork (*58*), Tillotson (*59*), and Riney (*60*).

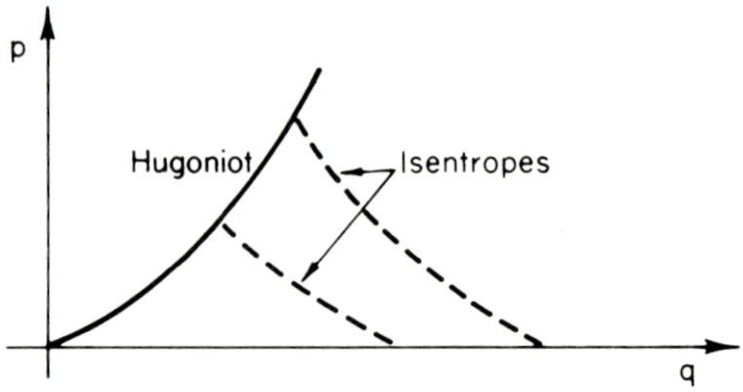

FIG. 21. The mirror-image approximation.

Hugoniot (*65*). Enig noted that the mirror-image assumption implies an equation of state, and he derived a closed-form expression of this equation for the case of a c, s material. In the present notation, his result (*63*) is

$$4s^2 \frac{e - e_0}{c^2} = \frac{sp}{\rho_0 c^2} + s\left(\frac{\rho_0}{\rho} - 1\right) + \Phi(\Psi) \tag{153}$$

where

$$\Psi = (1 + 4sp/\rho_0 c^2) \exp[4s(\rho_0/\rho - 1)] \tag{154}$$

and where the functional dependence of Φ on Ψ is given parametrically by

$$\Psi = (2x + 1)^2 \exp[-4x/(1 + x)], \qquad \Phi = x^3/(1 + x) \tag{155}$$

The parameter x is a monotonic function of the entropy; specifically, it denotes the velocity that a particle of given entropy had immediately after being processed by the shock:

$$x = sq_1/c = \tilde{M} - 1 \tag{156}$$

Here the symbol $\tilde{M}$ stands for the value of the shock Mach number at the instant when the particle in question was processed by the shock.

Enig observed that in calculating the temperature, the simultaneous assumptions of the mirror-image approximation, the linear q_s, q_1 relation, and the constancy of the thermal coefficient of volume expansion at low pressure lead to a rather singular behavior for the specific heat at constant pressure. The latter problem does not, however, affect the usefulness of the expression involving only e, p, and ρ.

The state equation developed by Enig and Petrone (*64*) employs the isentropes that had previously been shown by Walsh and Christian (*65*) to yield the maximum possible free surface velocity for expansion from a given point on the Hugoniot. These isentropes are based on the approximation that the quantity $-\rho^2(\partial e/\partial \rho)_p$ is a constant. This quantity is related to the density ρ, the specific heat at constant pressure C_P, and the coefficient of volume expansion $\bar{\alpha}$ by

$$-\rho^2(\partial e/\partial \rho)_p = \rho C_P/\bar{\alpha} \qquad \text{along} \quad p = 0 \tag{157}$$

In the present notation, this second equation of state is

$$\frac{\rho_0}{\rho} - 1 = -\frac{\tilde{M} - 1}{s\tilde{M}} - D(\tilde{M}; s, \delta)\frac{p/\rho_0 c^2 - \tilde{M}(\tilde{M} - 1)/s}{p/\rho_0 c^2 + \delta} \tag{158}$$

$$\begin{aligned}\frac{e - e_0}{c^2} = {} & \frac{\tilde{M} - 1}{s\tilde{M}}\left[\frac{\tilde{M}(\tilde{M} - 1)}{2s} - \frac{p}{\rho_0 c^2}\right] - \frac{p}{\rho_0 c^2}\left(\frac{\rho_0}{\rho} - 1\right) \\ & - D(\tilde{M}; s, \delta)\left\{\frac{p}{\rho_0 c^2} - \frac{\tilde{M}(\tilde{M} - 1)}{s} - \left[\frac{\tilde{M}(\tilde{M} - 1)}{s} + \delta\right] \right. \\ & \left. \times \ln\frac{p/\rho_0 c^2 + \delta}{\tilde{M}(\tilde{M} - 1)/s + \delta}\right\}\end{aligned} \tag{159}$$

The function D is defined as

$$D(\tilde{M}; s, \delta) = -\frac{\tilde{M} - 1}{s\tilde{M}} \frac{\delta + \tilde{M}(\tilde{M} - 1)/2s}{\delta + [\tilde{M}(\tilde{M} - 1)/s]\ln[1 + \tilde{M}(\tilde{M} - 1)/s\delta]} \tag{160}$$

and

$$\delta = \rho C_P/\bar{\alpha}\rho_0 c^2 \tag{161}$$

Both of these formulations lead again to the conclusion that equation-of-state data are correlated by the quantities $(e - c_0)/c^2$, $p/\rho_0 c^2$, and ρ/ρ_0, with s and δ as parameters. They have the advantages of being relatively simple, and of describing the less than normal density states in a manner consistent with observed free surface behavior. The second equation has been applied successfully to the problem of shock initiation of liquid explosives. However, to the writer's knowledge, neither of these equations has yet been applied to the hypervelocity impact problem.

B. Exact Solutions

The particle-in-cell method (*66*) and various modifications of it (*67*) have been used successfully to calculate the response of a half-space to either an impact or an explosion at the surface. Such calculations have been reported by Bjork (*68*), Brode and Bjork (*69*), Walsh *et al.* (*5–7*) and by Riney and Heyda (*70–76*). These calculations have included various projectile shapes and have covered a wide range of impact speeds and types of material. In general, the response begins with a phase during which the shock propagates at a constant speed, and with a shape that is dictated by the shape of the projectile. Later, the shock takes on a roughly hemispherical shape, after the projectile itself has been severely deformed, and the shock has engulfed an amount of mass several times that of the projectile. Throughout the process, the particle-in-cell method spreads the shock over several cell widths. Except for this region and the vincinity of the shock–free-surface intersection, the solutions are satisfactory. In some cases, severe oscillations of the solution are encountered, particularly at low density. These erratic distributions can be partially eliminated by reducing the cell size, and an improvement is achieved by the continuous transport-of-mass modification (*29*).

1. *Similtude*: *Late-Stage Equivalence*

The solutions described above can be applied to conditions other than those for which they were specifically done, by using the similitude discovered by Walsh (*6*), which he named "late-stage equivalence." The general features of this similitude were described earlier in Section III.D; they are formally derived in this section, in order to show their general applicability.

a. Functional Forms. The "late-stage" aspect of the similitude is simply the point-source assumption; i.e., the scaling law will accurately represent the flow caused by a projectile of nonzero size only when z_{s0} is large compared to the scale of the projectile. The fact that the similitude is that appropriate to a point source means that a nonzero amount of energy and momentum are deposited at $t = 0$ over an area of vanishing size. The early stages of the target's response to this excitation are characterized by self-similar behavior in which M_{s0} is infinite, the equation of state is approximated by that of a perfect gas with $\gamma = 2s - 1$, and the quantity $d \ln \dot{z}_{s0}/d \ln z_{s0} = (N - 1)/N$ is the constant that will permit a smooth joining of the elliptic and hyperbolic regions.

When the solution enters the non-self-similar phase, the only new parameter that appears is M_{s0}. The shock shape and the boundary values of f, ψ, ϕ, etc., change, but they depend only on M_{s0}. The quantity $d \ln \dot{z}_{s0}/d \ln z_{s0}$ also changes as a function of M_{s0}, and is found as a part of the numerical solution of Eqs. (16)–(19).

One specific way of finding $d \ln \dot{z}_{s0}/d \ln z_{s0}$ is the following: an increment in $1/M_{s0}$ is chosen (thus giving the value that M_{s0} will have at the end of the step) and an average value for $d \ln \dot{z}_{s0}/d \ln z_{s0}$ over the step is assumed. The method of characteristics can be used to calculate the entire flow field at the end of the step. In particular, one part of the solution is a new value for M_{s0} ; if this new value does not agree with the value corresponding to the end of the step, a new value of $d \ln \dot{z}_{s0}/d \ln z_{s0}$ (which affects the slopes of the characteristics) is used, and the process is repeated until satisfactory agreement is achieved.

After the solution is determined down to a desired minimum value of M_{s0}, the shock trajectory to that point must then be found, as indicated in Eq. (35). This is done by integrating the distribution found for $d \ln \dot{z}_{s0}/d \ln z_{s0}$:

$$z_{s0} = (z_{s0})_{\text{init}} \exp\left\{\int_{(M_{s0})_{\text{init}}}^{M_{s0}} \frac{d \ln z_{s0}}{d \ln \dot{z}_{s0}} \, d \ln M_{s0}\right\} \tag{162}$$

The initial conditions can be evaluated by taking $(M_{s0})_{\text{init}}$ to be very large, so that the self-similar solution applies. Use of Eq. (86) leads to

$$\frac{z_{s0}}{\mathscr{L}_0} = \frac{N^{N/(1-N)} F(\gamma)^{1/(1-N)}}{(M_{s0})_{\text{init}}^{N/(1-N)}} \exp\left\{\int_{(M_{s0})_{\text{init}}}^{M_{s0}} \frac{d \ln z_{s0}}{d \ln \dot{z}_{s0}} \, d \ln M_{s0}\right\} \tag{163}$$

The important point to be noted is that, even though the scaling length $\mathscr{L}_0$ originates in the self-similar flow, it nevertheless continues to be the appropriate length throughout the non-similar portions of the flow. Thus the value of N, or, equivalently, the value of the parameter α used by Walsh, depends only on the limiting perfect-gas solution.

Equation (163) can be written as

$$z_{s0}/\mathscr{L}_0 = F_1(M_{s0}; s, \ldots) \tag{164}$$

where the three dots in the argument allow for the presence of any equation-of-state parameters other than s. In principle, this can be solved for $\dot{z}_{s0}$:

$$\frac{\mathscr{L}_0}{c} \frac{d(z_{s0}/\mathscr{L}_0)}{dt} = F_2\left(\frac{z_{s0}}{\mathscr{L}_0}; s, \ldots\right) \tag{165}$$

Integration would then give

$$\frac{z_{s0}}{\mathscr{L}_0} = F_3\left(\frac{ct}{\mathscr{L}_0}; s, \ldots\right), \qquad M_{s0} = F_4\left(\frac{ct}{\mathscr{L}_0}; s, \ldots\right) \tag{166}$$

For $N = \frac{2}{5}$, the quantity $\mathscr{L}_0$ is proportional to the cube root of the energy release divided by the characteristic pressure, which is the scaling length appropriate in symmetric explosion problems (*77*).

Walsh *et al.* employ the quantities L_0 and U, instead of E and P, in presenting the correlation. The correspondence is

$$\mathscr{L}_0 = (1/2)^{\alpha - 1/3} L_0 (U/c)^{\alpha} \tag{167}$$

where

$$L_0 = (M/\rho_0)^{1/3}, \qquad \alpha = N/(1 - N) \tag{168}$$

Thus the similitude variables used by Walsh are related to those of the present treatment by

$$\frac{\tau}{(U/c)^{\alpha+1}} = \left(\frac{1}{2}\right)^{1/3-\alpha} \frac{ct}{\mathscr{L}_0}; \frac{\{Z, R\}^{\alpha}}{\tau^{\alpha}} = \{\zeta, \sigma\}^{\alpha+1} \left(\frac{1}{2}\right)^{\alpha-1/3} \left(\frac{\mathscr{L}_0}{ct}\right)^{\alpha} \left(\frac{z_{s0}}{\mathscr{L}_0}\right)^{\alpha+1} \tag{169}$$

The last two factors in this expression depend on M_{s0} and s. Finally, the correspondence of the dependent variables is

$$pZ^{2/\alpha}/\rho_0 U^2 = f(\zeta, \sigma; M_{s0})\, M_{s0}^2 \zeta^{2/\alpha}(\tfrac{1}{2})^{2-2/3\alpha}(z_{s0}/\mathscr{L}_0)^{2/\alpha} \tag{170}$$

$$\{u, v\}Z^{1/\alpha}/U = \{\phi, \omega(\zeta, \sigma; M_{s0})\}\, M_{s0}\zeta^{1/\alpha}(\tfrac{1}{2})^{1-1/3\alpha}(z_{s0}/\mathscr{L}_0)^{1/\alpha} \tag{171}$$

In both of these, the final factor on the right is a function of M_{s0} , s, and the other parameters in the state equation.

To conclude this discussion of the point-source similitude, it should be repeated that other equation-of-state parameters, in addition to s, may be important for some materials. Even when they are, different impacts into a single material will all be correlated by $\mathscr{L}_0$, since the other parameters remain the same for each of the different impacts. If different impacts into a series of *different* materials are considered, and if each of these materials has different values for its other equation-of-state parameters, then there is no guarantee that $\mathscr{L}_0$ alone will succeed in correlating the results. It appears that the effect of these parameters is stronger in the plane wave case than in axisymmetric flow, but a quantitative estimate of their importance requires further study.

b. Evaluation of the Similitude Functions. The published shock trajectory data can be used to estimate the function F_3 . Figure 22 shows a selection of the available results.* All of these trajectories begin with a constant-speed phase, during which the projectile is being deformed. The trajectory then makes a transition to the point-source type of behavior. If the impact speed is extremely large (such that M_{s0} is initially on the order of 30–100), then the point-source solution will lie in the self-similar range, where the slope $d \ln z_{s0}/d \ln t$ is around 0.37. For impacts at more moderate speeds (such that M_{s0} is initially on the

* The numerical data were taken from the collection given by Rae (*54*). The results of Bjork *et al.* for porous aluminum striking solid aluminum at 72 km/sec were taken directly from Bjork *et al.* (*78*). The original calculations are presented only at certain discrete points. For purposes of clarity in this presentation, solid lines faired through the data points are shown.

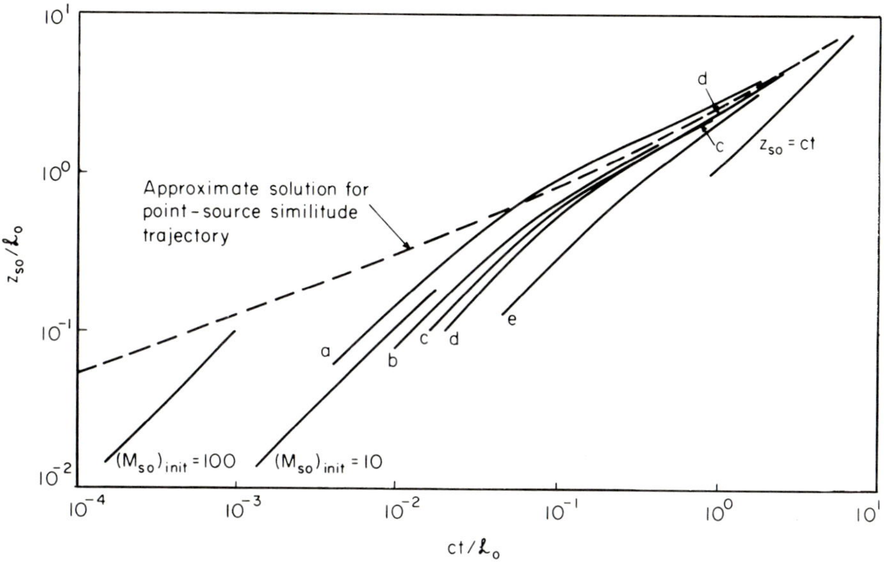

FIG. 22. Approximate shock trajectory for point-source similitude, compared with numerical calculations.

Curve	Source	Projectile–target	M (g)	U (km/sec)	$\mathscr{L}_0$ (cm)	c (km/sec)
a	Bjork (*58*)	Iron–tuff	1.065×10^{10}	30	7.12×10^{3}	2.21
b	Walsh and Tillotson (*6*)	Iron–iron	38.6	40	5.45	4.0
c	Heyda and Riney (*74*)	Al–Al	0.1525	60	1.247	5.85
d	Bjork *et al.* (*78*)	Porous Al–solid Al	345	72	18.2	5.85
e	Heyda and Riney (*74*)	Al–Al	0.1525	20	0.659	5.85

order of 10–30), the early-time trajectory joins the point-source solution in a region where the logarithmic slope is more nearly 0.5 or 0.6. Finally, if the impact speed is such that M_{s0} is initially on the order of 3, the late-stage or point-source phase of the solution will lie very close to the weak-wave limit, in which $z_{s0} \sim t$. The effectiveness of the correlation derived above can be seen by comparing Walsh's data for iron striking iron at 40 km/sec with those of Heyda and Riney for aluminum–aluminum impact at 60 km/sec.

It appears that the self-similar phase is limited to the regime $ct/\mathscr{L}_0 < 10^{-2}$, while the weak-wave limit is reached when $ct/\mathscr{L}_0 > 10$. Unfortunately, there are not enough data available in published form to evaluate the complete point-source trajectory. The dashed curve in Fig. 22 is based on an approximation described in Section V.C.1.

There are also not enough published data to evaluate the contour plots of the dimensionless functions, such as $f(\zeta, \sigma; M_{s0})$. This lack is particularly unfortunate, since it leaves unresolved the question of what portions of the flow are correctly represented by the point-source solution, as a function of the time after impact. Most of the evidence on which the point-source similitude is based consists of quantities like the shock trajectory or the total momentum. The integrals which enter these quantities are dominated by the high-density layers near the shock, however. Thus, as pointed out by Rae (*54*) and Bjork *et al.* (*78*), the possibility remains that two flows whose shock trajectories have con-converged to the point-source behavior might still contain large regions in which the flow patterns differ, and still reflect the details of the initial impact phase. Walsh *et al.* (*5*, *7*) have presented some data* which show that similitude is achieved if one waits long enough, but there is still a need for a more precise definition of the manner in which the transition from the early phase to the similitude phase occurs.

A second consequence of the lack of these contour plots for finite M_{s0} is that the similitude cannot be used to yield results for cases that have not yet been calculated. If, for example, one wanted to know the pressure distribution in the y direction, at a depth of 10 cm in a beryllium target, at 100 μsec after being struck by an aluminum sphere, 1 cm in diameter moving at 60 km/sec, there would be no recourse other than to do the calculation *ab initio*. However, if the contour plots mentioned above were available, the desired information could be taken directly from them.

Thus, the present state of the exact solution for the axisymmetric, nonsimilar case can be summarized by stating that a powerful similitude exists, which can be derived directly from the differential equations for nonsimilar motion due to a point source, and is an extension of the self-similar, perfect-gas solution. However, the detailed structure of the nonsimilar solutions has not been published in sufficient detail to allow its application to a general set of conditions.

c. The Plane Wave Case. The propagation of plane waves in real materials has been studied extensively [see, for example, Rice *et al.* (*57*)], chiefly for the purpose of interpreting measurements of shock waves driven by explosives or by impacting slabs. Most of the attention in these

* See Figs. 14, 15, 17, and 18 of Walsh *et al.* (*5*).

studies has been devoted to the early-time interval during which effects of the rarefactions coming from the loading side are not yet felt. Later studies [Fowles (*79*), for example] have examined the attenuation of the shock due to these rarefactions, and the possible influence of material strength on this attenuation at lower shock intensities (*80*).

If one waits long enough after the impact of a slab of finite thickness, the solution approaches a planar, "thin-sheet" similitude analogous to the point-source similitude that applies in the axisymmetric case. The existence of this similitude follows directly from the equations for nonsimilar plane wave motion (i.e., Eqs. (16)–(19), with ω and derivatives with respect to σ set equal to zero) once the scale of the impacting slab is neglected. Following an analysis similar to that of Eqs. (162)–(166), it can be shown that

$$z_{s0}/\mathscr{L}_1\,,\ M_{s0} = \text{fcns}(ct/\mathscr{L}_1\,, \gamma, \ldots) \tag{172}$$

where γ is the specific heat ratio appropriate to the high-pressure limit of the equation of state, and where the three dots denote the presence of equation-of-state parameters other than γ.

The scaling length $\mathscr{L}_1$ is defined as

$$\mathscr{L}_1 = \left\{\frac{\mathscr{E}}{\rho_0 c^2}\left(\frac{\mathscr{P}}{\rho_0 c}\right)^{(2-3N)/(2N-1)}\right\}^{(2N-1)/(1-N)} = \left(\frac{1}{2}\right)^{\alpha-1} L_0 \left(\frac{U}{c}\right)^{\alpha}, \qquad \alpha = \frac{N}{1-N} \tag{173}$$

Here N is the characteristic exponent that produces a nonsingular solution in the high-pressure, or self-similar limit, where the solution [Eq. (62)] can be written in the power-law form shown in Eq. (63).

Chou and Burns (*28*) presented method-of-characteristics calculations for a perfect gas, with $\gamma = 1.4$. The pressure ahead of the shock was taken as 1 atm, and consequently the solution ultimately becomes non-self-similar, because of counterpressure. At sufficiently early time, and after termination of the impact phase, the solution is self-similar, and the values of N and $G(\gamma)$ are corroborated by the numerical results, as noted earlier. Chou and Burns reported the results of six calculations, which showed that, except for the early impact phase, the entire solution, all the way to the acoustic limit, is correlated by the thin-sheet similitude of Eqs. (172) and (173), where the parameter α is the same as that given by the limiting self-similar solution. The results of the particular case in which U was equal to 9.144 km/sec were used to determine the similitude functions for the shock trajectory. These are shown in Fig. 7.

These authors also presented calculations for the case of aluminum and copper, using part of the equation of state developed by Tillotson (*59*): namely,

$$p = \left(a + \frac{b}{(e/E_0)(\rho_0/\rho)^2 + 1}\right)e\rho + A\left(\frac{\rho}{\rho_0} - 1\right) + B\left(\frac{\rho}{\rho_0} - 1\right)^2 \qquad (174)$$

The values used for the parameters a, b, E_0, A, and B are listed in Chou and Burns (*28*). At very high pressure, this equation approaches that of a pefect gas, with $a = \gamma - 1$, as pointed out by Tillotson (*59*). Chou and Burns use the value $a = 0.5$ for both aluminum and copper. The analysis above predicts that the similitude parameter α would be that corresponding to $\gamma = 1.5$, which is found by interpolating in Table I to be 1.53. The results for copper are correlated by the value $\alpha = 1.50$, in satisfactory agreement with this prediction.

Their results for aluminum are anomalous, however, for they show that similitude is most nearly approached for the range of their calculations if α is 1.28. This result agrees with the earlier calculations, and experiments as well, of Chou and Allison (*81*).

In summary, it can be said that the calculations of Chou and Allison (*81*) and of Chou and Burns (*28*) comply with the thin-sheet similitude outlined above, both for a perfect gas with counterpressure and for copper. The similitude which they found for aluminum, however, is not the same as the one that originates in the high-pressure limit. Further study of aluminum, including a direct comparison with the thin-sheet similitude solution itself, is required for an understanding of these results.

2. *Method of Characteristics*

Recently, the method of characteristics has been successfully applied to axisymmetric hypervelocity impact problems by Madden (*82*), and by Madden and Chang (*83*). These calculations have thus far been limited to relatively early times, where the shock traveling back into the projectile has not yet encountered its rear face. The results compare favorably with those of Walsh, and they provide considerable detail about the structure of the flow field. Future applications of this technique can be expected to add significant information in regions where the particle-in-cell and associated methods tend to smear out the details of the flow. The method should prove to be of particular value in sorting out the differences in the amplitude coefficient associated with various loading histories.

C. Approximate Solutions

There have been many attempts to develop an approximate theory for the nonsimilar phase of impact-generated shock propagation.

Several of these, which have enjoyed some success, are briefly described below.

1. *Oshima's Method*

Oshima (*84–86*) developed a method for approximating the nonsimilar effect due to counterpressure in a perfect gas.* The present writer (*54*) adapted this method to the conditions of the impact problem, where the nonsimilarity arises from the form of the state equation.

The only cases to which Oshima's method has been applied are those in which the problem is specialized to the axis of symmetry. For this case, Eqs. (16)–(19) can be written as

$$(\phi - \zeta)\frac{\partial\psi}{\partial\zeta} + \psi\left(\frac{\partial\phi}{\partial\zeta} + 2V\right) = -\frac{d \ln \dot{z}_{s0}}{d \ln z_{s0}}\frac{\partial\psi}{\partial \ln M_{s0}} \tag{175}$$

$$\frac{d \ln \dot{z}_{s0}}{d \ln z_{s0}}\phi + (\phi - \zeta)\frac{\partial\phi}{\partial\zeta} + \frac{1}{\psi}\frac{\partial f}{\partial\zeta} = -\frac{d \ln \dot{z}_{s0}}{d \ln z_{s0}}\frac{\partial\phi}{\partial \ln M_{s0}} \tag{176}$$

$$2\frac{d \ln \dot{z}_{s0}}{d \ln z_{s0}}g + (\phi - \zeta)\frac{\partial g}{\partial\zeta} - \frac{f}{\psi^2}(\phi - \zeta)\frac{\partial\psi}{\partial\zeta} = -\frac{d \ln \dot{z}_{s0}}{d \ln z_{s0}}\left\{\frac{\partial g}{\partial \ln M_{s0}} - \frac{f}{\psi^2}\frac{\partial\psi}{\partial \ln M_{s0}}\right\} \tag{177}$$

The essential part of Oshima's method is to make what amounts to a "local-similarity" approximation of the derivatives with respect to M_{s0}:

$$\partial \ln \mathscr{F}/\partial M_{s0})_{\text{any } \zeta} = \partial \ln \mathscr{F}/\partial M_{s0})_{\zeta=1} \tag{178}$$

where $\mathscr{F}$ is any of the functions f, ψ, g, ϕ, or ω.

In Rae (*54*), the terms are evaluated for a material having a linear shock-speed–particle-speed relation:

$$f(1, 0; M_{s0}) = \phi(1, 0, M_{s0}) = (1/s)(1 - 1/M_{s0})$$

$$\psi(1, 0; M_{s0}) = sM_{s0}/[1 + (s - 1)\, M_{s0}]$$

$$g(1, 0; M_{s0}) = (1/2s^2)(1 - 1/M_{s0})^2 \tag{179}$$

* Applications of the method are given by Lewis (*87*), and some comments on the approximation are made by Rae (*54*). The method has recently been extended to imploding shocks and detonations by Lee (*88*).

For this case, Eqs. (175)–(177) become

$$(\phi - \zeta)\frac{d\psi}{d\zeta} + \psi\left(\frac{d\phi}{d\zeta} + 2V\right) = -\frac{d\ln \dot{z}_{s0}}{d\ln z_{s0}}\frac{\psi}{1 + (s-1)M_{s0}} \tag{180}$$

$$\frac{d\ln \dot{z}_{s0}}{d\ln z_{s0}}\phi + (\phi - \zeta)\frac{d\phi}{d\zeta} + \frac{1}{\psi}\frac{\partial f}{d\zeta} = -\frac{d\ln \dot{z}_{s0}}{d\ln z_{s0}}\frac{\phi}{M_{s0} - 1} \tag{181}$$

$$2\frac{d\ln \dot{z}_{s0}}{d\ln z_{s0}}g + (\phi - \zeta)\frac{dg}{d\zeta} - \frac{f}{\psi^2}(\phi - \zeta)\frac{d\psi}{d\zeta}$$
$$= -\frac{d\ln \dot{z}_{s0}}{d\ln z_{s0}}\left\{\frac{2g}{M_{s0} - 1} - \frac{f/\psi}{1 + (s-1)M_{s0}}\right\} \tag{182}$$

The function g can now be expressed in terms of f and ψ by using the equation of state.* In Rae (*54*), this was taken as Eq. (152) with Γ replaced by $2(s-1)$:

$$g = \frac{1}{2(s-1)}\left\{\frac{f}{\psi} - \frac{\psi - 1}{M_{s0}^2[s - (s-1)\psi]}\right\} \tag{183}$$

With this choice, Eq. (182) becomes

$$df/d\zeta - A^2\, d\psi/d\zeta$$
$$= -\frac{d\ln \dot{z}_{s0}}{d\ln z_{s0}}\frac{2[M_{s0}/(M_{s0} - 1)]\, g(f, \psi) - f/\psi[1 + (s-1)M_{s0}]}{(\phi - \zeta)(\partial g/\partial f)_\psi} \tag{184}$$

where A^2 is a dimensionless sound speed,

$$A^2 = \frac{a^2}{\dot{z}_{s0}^2} = \frac{f/\psi^2 - (\partial g/\partial \psi)_f}{(\partial g/\partial f)_\psi}$$
$$= (2s-1)\frac{f}{\psi} + \frac{\psi}{M_{s0}^2[s - (s-1)\psi]^2} \tag{185}$$

If an approximation for V is now provided, Eqs. (180), (181), and (184) can be solved by the same method that is used for the plane wave, perfect-gas equations; i.e., the parameter $d\ln \dot{z}_{s0}/d\ln z_{s0}$ is adjusted until a solution free of singularities is found. These singularities occur at the points where $(\phi - \zeta)^2 = A^2$ and where $\phi = \zeta$.

* It should be noted that the right-hand side of Eq. (182) was obtained by applying Eq. (178) to the boundary value of g. The final result would be different if the state equation had first been used to eliminate g, and if Eq. (178) had then been applied to the boundary values of f and ψ. Thus, the form of the energy equation in Oshima's method is not unique.

The fact that Oshima's form of the nonsimilar equations must be solved by the same method used in the self-similar limit may appear somewhat surprising, since the original equations that apply to the non-self-similar regime can be solved directly by the method of characteristics and the quantity $d \ln \dot{z}_{s0}/d \ln z_{s0}$ evaluated as part of the solution. However, because Oshima's method makes a local-similarity approximation, the solution acquires the same property manifested by the self-similar solution; i.e., the flow fields at a series of adjacent instants must be related to each other in a manner consistent with the assumed similarity. Consequently, this approximation leads to elliptic and hyperbolic zones, which can be smoothly joined only by proper choice of the eigenvalue, $d \ln \dot{z}_{s0}/d \ln z_{s0}$.

Solutions of this type were reported by Rae (*54*) for the case $s = 2$, with $M_{s0} = 8$, 4, 2, and 1.5, and with V found from Eqs. (134) and (135), with $K = 10$. The results* are summarized in Table III, in terms of the quantity $d \ln \dot{z}_{s0}/d \ln z_{s0}$. The solution for $M_{s0} = \infty$ is the perfect-gas solution for $\gamma = 3$.

TABLE III

RESULTS OF OSHIMA'S METHOD

M_{s0}	∞	8	4	2	1.5
$-d \ln z_{s0}/d \ln \dot{z}_{s0}$	0.600	0.718	0.843	1.22	1.751

These results were used, in Eq. (163), to find the approximate shock trajectory shown in Fig. 22. The initial conditions were taken from Walsh's perfect-gas solution for $\gamma = 1.5$ (although, strictly speaking, this is not the correct perfect-gas limit for $s = 2$):

$$(M_{s0})_{\text{init}} = 50, \qquad \left(\frac{z_{s0}}{\mathscr{L}_0}\right)_{\text{init}} = 0.1230, \qquad \left(\frac{ct}{\mathscr{L}_0}\right)_{\text{init}} = 9.09 \times 10^{-4}$$

The distribution of M_{s0} with $z_{s0}/\mathscr{L}_0$ is shown in Fig. 23. This figure can be used, in conjuction with the Rankine–Hugoniot relations, to predict

* Rae (*54*) also contains some results for the nonsimilar, spherically symmetric, constant-energy case. The values of $d \ln \dot{z}_{s0}/d \ln z_{s0}$ for this case have approximately a constant ratio to those for the asymmetric case. The constancy of this ratio was erroneously interpreted by Rae (*54*) as evidence for the validity of Walsh's similitude. As noted above, the validity of the similitude for a point-source disturbance follows without approximation directly from the governing differential equations.

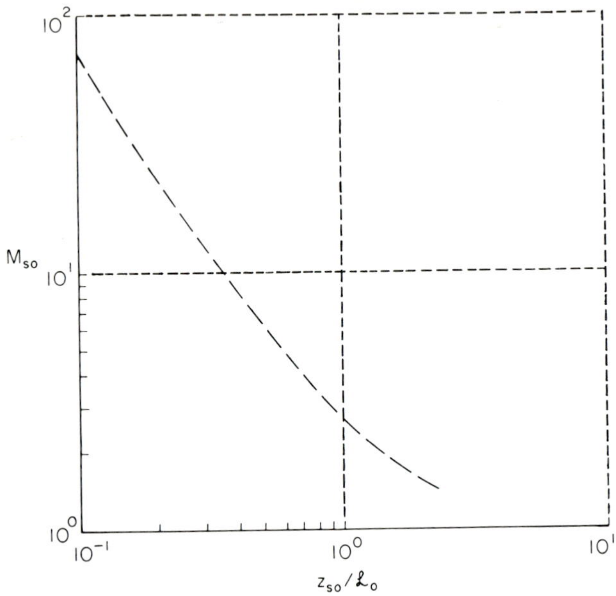

FIG. 23. Approximate prediction of the shock Mach number along the symmetry axis for point-source similitude.

the variation of the peak pressure (and other peak quantities on the axis of symmetry) with distance into the target. This prediction is shown in Fig. 24, and is compared with the calculations of Walsh *et al.* (*6*), Heyda and Riney (*74*), and with the experimental data* of Charest (*89*).

2. *Variable-γ Method*

One approach to the nonsimilar problem is to treat it as a succession of perfect-gas solutions, in which γ is allowed to vary in such a way as to always produce the correct conditions at the shock. This approximation is usually coupled with an assumption that the shock shape is hemispherical, with center at the impact point; thus variations around the periphery of the shock are neglected. Finally, the approximation is usually carried one step further, by replacing the flow field by one-half of a spherically symmetric, constant-energy solution.

When all of these approximations are made, the only remaining problem is that of choosing the rule for allowing γ to vary. Davids *et al.* (*55–56*) chose a constant value of γ, in the vicinity of 7.0, to represent aluminum. The present writer (*54, 90*) worked out the solution for the

* All of the data in Fig. 24 are shown as discrete symbols. Only Charest's results are experimental, however, while the other symbols are the results of numerical calculations.

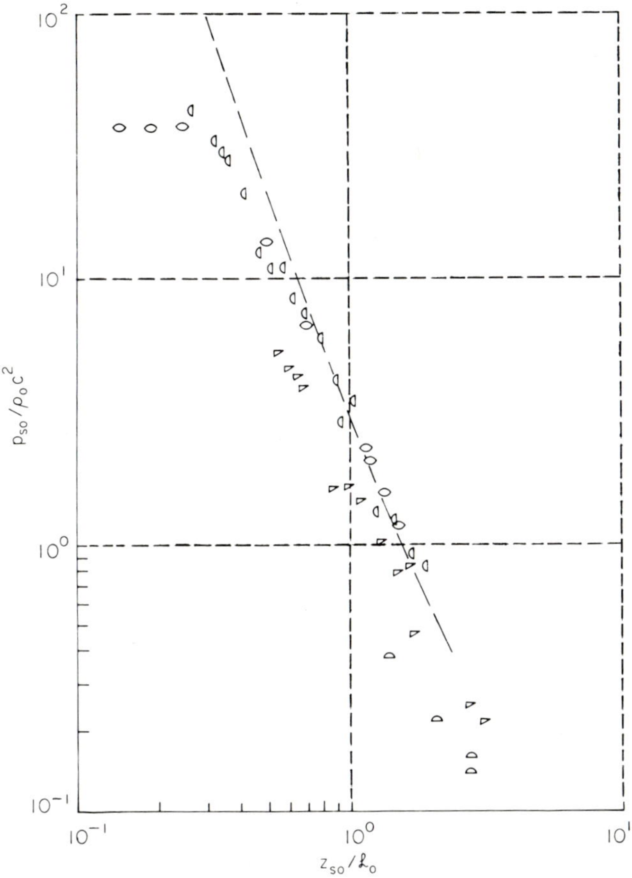

FIG. 24. Peak pressure on the symmetry axis, compared with approximate point-source–similitude prediction.

Symbol	Source	Projectile–target	M (g)	U (km/sec)	$\mathscr{L}_0$ (cm)	$\rho_0 c^2$ (Mbar)
○	Walsh and Tillotson (*6*)	Iron–iron	38.6	40	5.45	1.259
▱	Heyda and Riney (*74*)	Al–Al	0.1525	20	0.659	0.924
⌓	Charest (*89*)	Al–Al	0.1525	7.32	0.368	0.924
◖	Heyda and Riney (*74*)	Lead–lead	0.641	20	1.208	0.484

case where γ was allowed to vary with M_{s0}, so as to match the quantities at the shock:

$$\gamma(M_{s0}) = 2\frac{q_1}{q_s} - 1 = 2\frac{\rho_0 q_s^2}{p_1} - 1 = \frac{\rho_1/\rho_0 + 1}{\rho_1/\rho_0 - 1} = \left\{\frac{2q_s^2}{e_1}\right\}^{1/2} - 1 \quad (186)$$

This functional dependence is then inserted in Eq. (140), which gives a relation between M_{s0} and z_{s0}/R_0, and a quadrature yields ct/R_0 as a function of z_{s0}/R_0, where the scaling length is defined as

$$R_0 = (E/2\pi\rho_0 c^2)^{1/3} \quad (187)$$

Except for a constant factor, this is what $\mathscr{L}_0$ would reduce to for $N = \frac{2}{5}$.

Calculated shock trajectories are correlated by these parameters nearly as well as by $\mathscr{L}_0$ and c, as might be anticipated from the relatively small value of $2N - 5$. In addition, the theoretical curves deduced from the quadrature mentioned above showed good agreement with the calculated data and led to the same conclusions (*54*, *91*), regarding the various phases of the solution, as were presented in Section V.B.1 above. However, these symmetric-solution results are chiefly of historical interest today, in view of the general theoretical basis for the similitude based on $\mathscr{L}_0$.

3. *Porzel–Zaker Method*

Another approximate solution which attempts to account for non-similarity was developed by Porzel (*92*) for strong shock motion in gases, and was later adapted by Zaker (*93*) to the problem of a point-source explosion in a solid. The essence of the method is to assume the density distribution; the continuity equation can then be used to find the particle velocity distribution. These two are then used in the momentum equation to find the pressure distribution.

The present writer treated this method briefly (*54*), using the equation of state of *c*, *s* material. This treatment was carried out in full detail by Bach and Lee (*94*), who showed that the method gives a more realistic particle velocity distribution than those of the other approximate methods mentioned above. However, its predictions of the shock trajectory are essentially the same as those of the latter methods.

VI. Concluding Remarks

A. Present Status of Theory

Many of the features of the response of a half-space of an inviscid, compressible, Mie–Grüneisen fluid to a point release of energy and momentum on its free surface can now be said to be well understood. The solution begins with a self-similar regime, in which the shock scale grows as a power of the time, where the power is slightly less than the value appropriate to a symmetric point-source explosion. This phase continues up to the point where $ct/\mathscr{L}_0$ is approximately 10^{-2}. Later, the logarithmic slope of the shock-scale–time curve begins to increase as the solution enters the nonsimilar regime, and tends toward one as the weak-wave limit is reached. This occurs at approximately $ct/\mathscr{L}_0 = 10$. Throughout the solution, results for all cases are correlated by the parameters $\mathscr{L}_0$ and c, with a relatively minor additional dependence on equation-of-state parameters. (This latter dependence is much greater in the plane wave case.)

In an actual impact, the projectile has nonzero size, and the shock trajectory begins with a constant-speed phase, which persists for a period on the order of the time required for a shock to travel through the projectile. After this period, the projectile becomes severely deformed and the solution makes a transition to the point-source trajectory. The point at which this transition is completed depends on the initial speed of the shock at the impact point. Unless this initial value of M_{s0} is on the order of 50 or greater, the self-similar phase is not observed. This extremely large value is not likely to occur, except for impact speeds on the order of 50 km/sec or greater. For most of the speed range of interest for meteoroid impact, the transition to the point-source solution occurs in the nonsimilar range. A quantitative estimate of how long it takes to complete this transition is not available at present. After the transition is completed, the appropriate scaling factors are $\mathscr{L}_0$ and c, which originate in the self-similar phase.

Details of the two-dimensional flow pattern are available only in the self-similar regime, and even there, the structure of the flow near the shock and near the shock–free-surface intersection is not known with great precision. In the nonsimilar regime, there are essentially no details of the flow pattern available. Thus, despite the existence of a powerful similitude, the fact remains that calculations of the flow pattern must be repeated for any specific case of interest, if anything more than the shock trajectory is desired.

B. Implications for Crater Size Predictions

It must be emphasized that all of the solutions described in this review deal exclusively with an inviscid fluid. There is no mechanism by which such a fluid can be permanently brought to rest at a finite time after impact. Consequently, these solutions offer no direct information regarding the ultimate configuration in which the target comes to rest.

However, the inviscid solution has often been used as a basis for inferring the crater size by various auxiliary criteria. In applying such criteria, a clear distinction must be made between what is explicitly given by the inviscid solution, and what is inferred from it on the basis of the auxiliary criterion. A failure to recognize this distinction led to a lengthy disagreement (*1*) between Bjork (*68*) and Walsh (*6*) over the dependence of the crater size on impact velocity. In fact, the source of their difference lay in the crater formation criteria, since the evidence then available indicated that the shock trajectories found by both of these writers could be correlated by the same scaling law (*90*), a law that was very close to the constant-energy, spherically symmetric result. The further evidence that has subsequently become available (some of which is shown in Fig. 22) serves to confirm this conclusion.

Walsh (*6*) adopts as his criterion the hypothesis that, if two different impacts generate the same flow pattern during the late stages of the inviscid motion, then the subsequent deformation during the strength-affected phases would also be the same. Thus, crater size would be scaled by the same parameters as those which govern the inviscid similitude; i.e., crater size would be proportional to impact speed raised to the α power.

Bjork in his earlier papers (*68*) identified the crater boundary as a locus of points where the pressure was low, and the particle velocities randomly oriented, a condition which he interpreted [Bjork (*68*, p. 512)] as a finite-difference representation of a stationary region of zero pressure. From this criterion, he arrived at the conclusion that the crater size would grow with the one-third power of the impact speed.

As noted above, a permanent zero-velocity region is not possible in an inviscid-fluid model; on the other hand, the locus corresponding to any nonzero pressure would scale with the α power of U, since a given value of $p_1/\rho_0 c^2$ implies a fixed value of M_{s0}, which in turn yields a single value of $z_{s0}/\mathscr{L}_0$.

In later work, Bjork and his associates (*78*) have adopted a different criterion, defining the crater boundary as the locus along which the kinetic energy density $\frac{1}{2}\rho(u^2 + v^2)$ is equal to the (temperature-dependent) yield stress of the material. In presenting their results, these authors specifically caution against the inference of any velocity exponents as a basis for extrapolation; and they stress that the complexity of the physical processes which play a role in the problem makes it unlikely that a simple functional dependence would apply over a broad range of parameters. In addition, it is important to note the suggestion made by Bjork *et al.* [Appendix D of Bjork (*78*)] that the crater takes shape in a region to which the point-source similitude does not apply. If this assertion is true in general, it clearly invalidates the criterion used by Walsh.* To decide whether the assertion is true, it is necessary to compare the flow field predicted by the point-source similitude with that calculated for a specific case. Unfortunately, the point-source solution is not available at present in sufficient detail to allow such a comparison. Thus it is not possible to make a definitive assessment of the boundaries in space and time within which the point-source solution is valid, nor how this region of validity varies with the size, shape, speed, and composition of the projectile.

To put this discussion in proper perspective, it must be stressed that all of these attempts to predict a crater size on the basis of the inviscid solution represent a substantial further approximation. In the real case, quantities such as the temperature distribution in the flow are themselves affected by the material strength. To estimate the strength on the basis of an inviscid solution for the temperature is a complete decoupling of these effects; the error introduced by this decoupling cannot be judged on the basis of results available at the present time.

The skill and ingenuity displayed in successfully achieving numerical solutions of the Euler equations in three independent variables have properly been the object of considerable acclaim. It is sometimes difficult to appreciate the limits of validity of these solutions and, specifically, to realize that inferences drawn about crater formation do not rest on the same firm theoretical basis as do the inviscid solutions themselves.

In the present writer's opinion, there is no reliable basis in the inviscid solutions available at present for selecting a dependence of crater size on impact velocity. A satisfactory resoltution of this question will require solutions in which the effects of material strength are accounted for.

C. Remaining Problems

Many problems remain to be solved before the inviscid flow solution itself can be called complete. The two-dimensional pattern of flow contours in the nonsimilar range must be found. In addition, an improved definition of the flow near the shock and near the shock–free-surface intersection is needed.

* In addition, it would imply that the crater boundary should be sensitive to the shape of the projectile.

The effects of projectile shape play a dominant role in the period during which the transition to the point-source flow is occurring. The method of characteristics (*82*, *83*) may prove particularly valuable in better defining this portion of the flow.

There are also certain target-shape effects which merit attention. This survey has ignored entirely the thin-plate perforation problem [see Friend *et al.* (*95*) and Maiden *et al.* (*96*), for example] and the intermediate-thickness regime in which spall fracture is the major damage mechanism (*97*), but in both these areas there are many unsolved problems. Even for a semiinfinite target, one unknown effect is that due to surface curvature, where some evidence (*98*) suggests that a convex surface leads to a more rapid decay of the shock than that encountered when the free surface is a plane.

The discussion above deals only with state equations of the Mie–Grüneisen type, and thus does not include many materials of technical importance, such as porous solids or materials which undergo a change of phase (*99*). The theory of impact-generated shock propagation in materials such as these is still very incomplete.

Finally, a continuing theoretical attack on the response of materials to high-speed impact must always look toward the development of solutions in which the material strength is accounted for. Many efforts have already been made in this direction (*100–102*), but a satisfactory understanding of the problem still lies some years in the future. In the meantime, it is hoped that the present survey will provide a convenient point of reference for evaluating future progress.

Appendix: Approximate Solution of the Plane Wave, Perfect-Gas Case

A determinate solution can be found if the small amount of mass neglected and the constant in Eq. (23) are used as two parameters to match the total energy and momentum. To acutally evaluate these parameters, it is necessary to begin by finding the asymptotic form of the solution at large negative values of ζ. The fact that the pressure approaches zero in this region means that f' can be neglected in Eq. (52), leading to the approximate solution

$$\phi \approx \zeta/N \tag{A1}$$

This relation is now inserted in Eq. (51), to find the density:

$$\psi \approx A_1 |\zeta|^{-1/(1-N)} \tag{A2}$$

Finally, use of ϕ and ψ in Eq. (53) gives the asymptotic pressure formula

$$f \approx A_2 |\zeta|^{2-\gamma/(1-N)} \tag{A3}$$

The constants A_1 and A_2 are not independent. Lees and Kubota (*103*) have shown that the similarity equations have the first integral

$$\psi^{\gamma-2(1-N)/N} |\phi - \zeta|^{-2(1-N)/N} = \frac{\gamma+1}{2} \left(\frac{\gamma+1}{\gamma-1}\right)^{\gamma} f \tag{A4}$$

Use of the asymptotic formulas in this relation shows that

$$A_2 = \frac{A_1^{\gamma}}{[(\gamma+1)/2][(\gamma+1)/(\gamma-1)]^{\gamma}[(1-N) A_1/N]^{2(1-N)/N}} \tag{A5}$$

The value of A_1 must be found by integrating the similarity equations out to a large negative value of ζ. For example, in the case $\gamma = 1.4$, the exact solution shows that $A_1 = \frac{3}{16}$. In general, A_1 depends on γ.

The small amount of mass to be neglected, m_0, will be to the left of a point $\zeta_0(t)$:

$$m_0 = \rho_0 A t^N \int_{-\infty}^{\zeta_0} \psi \, d\zeta \tag{A6}$$

The integral appearing here can be evaluated directly. The expression stating the conservation of the mass lying between the shock and a control-surface position $z_{cs}(t)$ is

$$\frac{\partial}{\partial t} \int_{z_{cs}}^{z_{s0}} \rho \, dz = \rho_1(\dot{z}_{s0} - u_1) - \rho(z_{cs})[\dot{z}_{cs} - u(z_{cs})] \tag{A7}$$

Sedov (*3*) has pointed out that this relation, expressed in terms of the self-similar functions, leads to the integral

$$\int_{\zeta_a}^{\zeta_b} \psi \, d\zeta = -[\psi(\phi - \zeta)]_{\zeta_a}^{\zeta_b} \tag{A8}$$

The same integral can be derived by integrating Eq. (51) by parts. If this relation is now used in Eq. (A6), the result is

$$m_0 = \rho_0 A t^N [(1-N)/N] A_1 |\zeta_0|^{1-1/(1-N)} \tag{A9}$$

The quantity m_0 is made independent of time by choosing

$$\zeta_0 = -K_1 t^{1-N}, \qquad K_1 > 0 \tag{A10}$$

This gives

$$m_0 = AA_1K_1^{-N/(1-N)}(1 - N)\,\rho_0/N \tag{A11}$$

The parameters K_1 and A are the two that are used to match the quantities $\mathscr{E}$ and $\mathscr{P}$.

The energy conservation integral, with m_0 neglected, is

$$\mathscr{E} = \rho_0 A t^N N^2 A^2 t^{2N-2} \int_{\zeta_0}^{1} \left(\frac{f}{\gamma - 1} + \tfrac{1}{2}\psi\phi^2\right) d\zeta \tag{A12}$$

The integral appearing here can be evaluated from the expression for the conservation of the energy between the shock and the control-surface position:

$$\begin{aligned}\frac{\partial}{\partial t}\int_{z_{cs}}^{z_{s0}} \rho(e + \tfrac{1}{2}u^2)\,dz &= -[\rho(e + \tfrac{1}{2}u^2)]_{z_{s0}}(u_1 - \dot{z}_{s0}) - [\,pu]_{z_{s0}} \\ &\quad + [\rho(e + \tfrac{1}{2}u^2)]_{z_{cs}}[u(z_{cs}) - \dot{z}_{cs}] + [\,pu]_{z_{cs}}\end{aligned} \tag{A13}$$

Rewriting this in terms of the self-similar functions yields (*3*)

$$\int_{\zeta_a}^{\zeta_b} \left(\frac{f}{\gamma - 1} + \tfrac{1}{2}\psi\phi^2\right) d\zeta = \frac{-N}{3N - 2}\left[(\phi - \zeta)\left(\frac{f}{\gamma - 1} + \tfrac{1}{2}\psi\phi^2\right) + f\phi\right]_{\zeta_a}^{\zeta_b} \tag{A14}$$

As before, this integral can also be found by manipulation of the similarity equations.

If this integral and the asymptotic solutions are now used in Eq. (A12), the result is

$$\mathscr{E} = \rho_0 A^3 \frac{A_1}{2}\frac{1 - N}{2 - 3N} K^{(2-3N)/(1-N)} \tag{A15}$$

Finally, the momentum conservation integral

$$\mathscr{P} = \rho_0 A t^N N A t^{N-1} \int_{\zeta_0}^{1} \psi\phi\,d\zeta \tag{A16}$$

is evaluated by expressing the conservation of momentum for the fluid lying between the shock and a control surface:

$$\begin{aligned}\frac{\partial}{\partial t}\int_{z_{cs}}^{z_{s0}} \rho u\,dz &= \rho_1 u_1(\dot{z}_{s0} - u_1) - [\rho u]_{z_{cs}}[\dot{z}_{cs} - u(z_{cs})] \\ &\quad - [\,p]_{z_{s0}} + [\,p]_{z_{cs}}\end{aligned} \tag{A17}$$

When this is rewritten in terms of the self-similar functions, the result is [again shown by Sedov (*3*) and derivable by manipulation of the similarity equations]

$$\int_{\zeta_a}^{\zeta_b} \psi\phi \, d\zeta = [-N/(2N-1)][f + \phi\psi(\phi - \zeta)]_{\zeta_a}^{\zeta_b} \tag{A18}$$

With the use of this result, the momentum conservation integral becomes

$$\mathscr{P} = \rho_0 N A^2 t^{2N-1} \frac{A_1}{N} \frac{1-N}{2N-1} \mid \zeta_0 \mid^{2-1/(1-N)} \tag{A19}$$

As $\zeta \to -\infty$, $\mathscr{P}$ vanishes. Thus, if all the mass were to be included in this integral, and described by the self-similar solution, its total momentum would be zero, as pointed out by Zeldovich (*14*). He takes the viewpoint that the solution should be interpreted in this light, namely, as one in which the input energy is matched to that of the flow by neglecting some of the mass, and in which the momentum acquired by the flow is considered to be zero, a condition guaranteed by including all of the mass in the momentum conservation integral. If this viewpoint is taken, Eqs. (A11) and (A15) are solved for A and K_1 in terms of $\mathscr{E}$ and m_0; the resulting expression for the shock trajectory is

$$z_{s0}(t) = H(\gamma) \frac{m_0}{\rho_0} \left(\frac{\rho_0{}^2 \mathscr{E} t^2}{m_0{}^3} \right)^{N/2} \tag{A20}$$

where

$$H(\gamma) = \left[\frac{2(2-3N)}{1-N} \right]^{N/2} \frac{1}{A_1^{1-N}} \left(\frac{N}{1-N} \right)^{(2-3N)/2} \tag{A21}$$

An approximate result in which the length scale is determinate can be found (*16*) by neglecting m_0 in the momentum conservation integral, which gives

$$\mathscr{P} = \rho_0 A^2 A_1 \frac{1-N}{2N-1} K_1^{(1-2N)/(1-N)} \tag{A22}$$

Solving for A and K_1 in terms of $\mathscr{E}$ and $\mathscr{P}$ gives

$$A = G(\gamma) \, (\mathscr{P}^2/\rho_0 \mathscr{E})(\rho_0 \mathscr{E}^2/\mathscr{P}^3)^N \tag{A23}$$

$$K_1 = \left[\frac{4A_1(1-N)(2-3N)^2}{(2N-1)^3} \frac{\mathscr{E}^2 \rho_0}{\mathscr{P}^3} \right]^{1-N} \tag{A24}$$

where

$$G(\gamma) = \frac{2(2-3N)}{2N-1} \left[\frac{(2N-1)^3}{4A_1(1-N)(2-3N)^2} \right]^{1-N} \tag{A25}$$

This solution then takes the form

$$z_{s0}(t) = G(\gamma) \frac{\mathscr{P}^2}{\rho_0 \mathscr{E}} \left[\frac{t}{\mathscr{P}^3/\rho_0 \mathscr{E}^2} \right]^N \tag{A26}$$

For $\gamma = 1.4$, the values $A_1 = \frac{3}{16}$ and $N = 0.6$ lead to $G(1.4) = (128/9)^{1/5}$.

REFERENCES*

1. R. J. Eichelberger, Summary: theoretical and experimental studies of crater formation. *Proc. Hypervelocity Impact Symp., 6th* **2**, Part 2, 683–704, AD 423064 (August, 1963).
2. H. A. Bethe, K. Fuchs, J. O. Hirschfelder, J. L. Magee, R. E. Peierls, and J. von Neumann, Blast wave, Los Alamos Scientific Lab. Rept. No. LA-2000 (Aug. 1947). Available from CFSTI.
3. L. I. Sedov, "Similarity and Dimensional Methods in Mechanics." Academic Press, New York, 1959.
4. G. I. Taylor, The formation of a blast wave by a very intense explosion. I. Theoretical discussion, *Proc. Roy. Soc. (London)* **A201**, 159–174 (1950).
5. J. M. Walsh, W. E. Johnson, J. K. Dienes, J. H. Tillotson, and D. R. Yates, Summary report on the theory of hypervelocity impact, General Atomic Div. General Dynamics Corp., Rept. No. GA-5119, AD 436251 (March 31, 1964). Also published "The Fluid Dynamics of Space Flight," AGARDograph 87, Vol. 1, pp. 343–387. Gordon and Breach, New York, 1966.
6. J. M. Walsh and J. H. Tillotson, Hydrodynamics of hypervelocity impact, General Atomic Div. General Dynamics Corp., Rept. No. GA-3827, AD 401023 (Jan. 22, 1963). Also *Proc. Symp. Hypervelocity Impact, 6th* **2**, Part I, 59–104, AD 423063 (Aug. 1963).
7. J. M. Walsh and W. E. Johnson, On the theory of Hypervelocity impact, *Proc. Hypervelocity Impact Symp., 7th* **2**, 1–75 AD 463228 (Feb. 1965).
8. See, for example: G. M. Bryan, Oblique impact of high velocity steel pellets on lead targets, *Proc. Symp. Hypervelocity Impact, 5th* **1**, Part 2, 511–534 AD 284280 (April 1962).
9. R. J. Eichelberger and J. W. Gehring, Effects of meteoroid impacts on space vehicles, *Am. Rocket Soc. J.* **32**, 1583–1591 (1962).
10. R. J. Eichelberger, Hypervelocity Impact, *in* "Behavior of Materials under Dynamic Loading" (N. J. Huffington, ed.), pp. 155–187. ASME, New York, 1965.
11. W. E. Deal, Shock wave research on inert solids, *Proc. Symp. Intern. Detonation, 4th* 321–345. Available as Office Naval Res. Rept. No. ACR-126, Superintendent of Documents, U.S. Govt. Printing Office, Washington, D.C.
12. G. J. Kynch, Blast Waves, "Modern Developments in Fluid Dynamics" (L. Howard, ed.), Vol. I, Part IV, Section II. Oxford Univ. Press (Clarendon), London and New York, 1953.

* Much of the hypervelocity-impact literature that has originated in the United States has appeared in the form of company reports. In most cases, these are available from the Clearinghouse for Federal Scientific and Technical Information (CFSTI) of the United States Department of Commerce, according to the listed N- and AD-serial numbers.

13. L. I. Sedov, "Similarity and Dimensional Methods in Mechanics," p. 154. Academic Press, New York, 1959.
14. Ya. B. Zeldovich, Motion of a gas due to a pressure of short duration (shock), *Akust. Zh.* **2**, 28–38 (1956); *Soviet Phys. Acoust. (English Transl.)* **2**, 25–35 (1956).
15. Yu. P. Rayzer, Motion of a gas under the influence of a point-impact shock on its surface (an explosion on a surface), *Zh. Prikl. Mekh. i Tekhn. Fiz.* No. 1, 57–66 (Jan.–Feb. 1963). Translation, by M. J. Nowak, available as General Atomic Div., General Dynamics Corp., Rept. No. GA-tr-5081, AD 440513 (May 28, 1964), and by Air Force Systems Command Foreign Technol. Div., as Rept. No. FTD MT 63-80, AD 434954 (Sept. 24, 1963).
16. W. J. Rae and H. P. Kirchner, Final report on a study of meteoroid impact phenomena, Cornell Aeron. Lab. Rept. No. RM-1655-M-4, N 63-16887 (Feb., 1963).
17. W. Häfele, Zur analytischen Behandlung ebener, starker, instationärer Stosswellen, *Z. Naturforsch.* **10a**, 1006–1016 (1955).
18. S. v. Hoerner, Losung der hydrodynamischen Gleichungen mit linearem Verlauf der Geschwindigkeit, *Z. Naturforsch.* **10a**, 687-692 (1955).
19. V. B. Adamskii, Integration of a system of autosimulating equations for the problem of a short duration shock in a cold gas, *Akust. Zh.* **2**, 3–9 (1956); *Soviet Phys. Acoust. (English Transl.)* **2**, 1–7 (1956).
20. V. B. Adamskii and N. A. Popov, The motion of a gas under the action of a pressure on a piston, varying according to a power law, *Prikl. Mat. i Mekh.* **23**, 564–573 (1959); *Appl. Math. Mech.* **23**, 793–806 (1959).
21. A. I. Zhukov and Ya. M. Kazhdan, Motion of a gas due to the effect of a brief impulse, *Akust. Zh.* **2**, 352–357 (1956); *Soviet Phys. Acoust. (English Transl.)* **2**, 375–381 (1957).
22. J. K. Dienes, Late-stage equivalence and similarity theory for one-dimensional impacts, General Atomic Div., General Dynamics Corp. Rept. No. GA-5755, AD 608870 (Nov. 11, 1964). Reprinted as *Hypervelocity Impact Symp., 7th* **2**, 187–219, AD 463228 (Feb. 1965).
23. H. Mirels, Hypersonic flow over slender bodies associated with power-law shocks, "Advances in Applied Mechanics" (H. L. Dryden and T. von Karman, eds.), Vol. VII, pp. 1–54. Academic Press, New York, 1962.
24. Ya. B. Zeldovich and Yu. P. Rayzer, "Physics of Shock Waves and High-Temperature Hydrodynamic Phenomena" (W. D. Hayes and R. F. Probstein, eds.), Vol. II. Academic Press, New York, 1967.
25. J. K. Yakura, Theory of entropy layers and nose bluntness in hypersonic flow, Hypersonic flow research, *Progr. Astronautics Rocketry*, **7**, 421–470 (1962).
26. J. P. Guiraud, Asymptotic theory in hypersonic flow, *in* "Fundamental Phenomena in Hypersonic Flow" (J. G. Hall, ed.), pp. 70–84. Cornell Univ. Press, Ithaca, New York, 1966.
27. M. D. Van Dyke, "Perturbation Methods in Fluid Mechanics." Academic Press, New York, 1964.
28. P. C. Chou and B. P. Burns, Late-stage equivalence in one-dimensional impacts, *J. Appl. Phys.* **38**, 553–560 (1967).
29. W. E. Johnson, OIL: A continuous two-dimensional Eulerian hydrodynamics code, General Atomic Div., General Dynamics Corp. Rept. No. GAMD-5580, AD 477240 (Jan. 7, 1965).
30. A. A. Grib, A. G. Ryabinin, and S. A. Khristianovich, On the reflection of a plane shock wave in water from a free surface, *Prikl. Mat. i Mekh.* **20**, 532–544 (1956).

31. O. S. Ryzhov and S. A. Khristianovich, On nonlinear reflection of weak shock waves, *Prikl. Mat. i Mekh.* **22**, 586–599 (1958); *Appl. Math. Mech.* **22**, 826–843 (1958).
32. A. A. Grib, O. S. Ryzhov, and Khristianovich, Theory of short waves, *Zh. Prikl. Mekh. i Tekhn. Fiz.* No. 1, 63–74 (1960).
33. B. I. Zaslavskii, Some particular solutions of the equations of "short waves," *Zh. Prikl. Mekh. i Tekhn. Fiz.* No. 1, 34–38 (1962).
34. B. I. Zaslavskii, On the reflection of a spherical shock wave in water from a free surface, *Zh. Prikl. Mekh. i Tekhn. Fiz.* No. 6, 50–58 (1963).
35. B. I. Zaslavskii, On the nonlinear interaction of a spherical shock wave, arising as the result of the explosion of a submerged charge, with a free water surface, *Zh. Prikl. Mekh. i Tekhn. Fiz.* No. 4, 57–65 (1964).
36. R. Collins, Intense explosions at the ocean surface, College of Engineering, Univ. of California, Berkeley, Rept. No. AS-66-9, AD 800347 (Aug. 1966).
37. R. Collins and M. Holt, Intense explosions at the ocean surface, *Phys. Fluids* **11**, 701–713 (1968).
38. A. Hertzberg and A. Kantrowitz, Studies with an aerodynamically instrumented shock tube, *J. Appl. Phys.* **21**, 874–878 (1950).
39. See, for example: V. L. Streeter, "Fluid Dynamics," 1st ed., Chapter 2, Section 8. McGraw-Hill, New York, 1948.
40. J. T. Frasier, Hypervelocity impact studies in wax, Ballistic Res. Lab. Rept. No. 1124, AD 255772 (Feb. 1961). See also: B. G. Karpov, Transient response of wax targets to pellet impact at 4 km/sec, Ballistic Res. Lab. Rept. No. 1226, AD 428221 (Oct. 1963); J. T. Frasier and B. G. Karpov, The transient response of wax targets subjected to hypervelocity impacts, *Exp. Mech.* **5**, 305–312 (1965); J. T. Frasier, B. G. Karpov, and L. S. Holloway, The behavior of wax targets subjected to hypervelocity impacts, *Proc. Hypervelocity Impact Symp., 7th* **5**, 123–160, AD 463231 (Feb. 1965); J. T. Frasier, The transient response of targets subjected to hypervelocity impacts, Ballistic Res. Lab. Rept. No. 1290, AD 474604 (July 1965).
41. R, Courant and K. O. Friedrichs, "Supersonic Flow and Shock Waves." Wiley (Interscience), New York, 1948.
42. K. P. Stanyukovich, "Unsteady Motion of Continuous Media," p. 362. Pergamon Press, Oxford, 1960.
43. H. W. Liepmann and A. Roshko, "Elements of Gasdynamics," p. 99. Wiley, New York, 1957.
44. N. H. Johannesen and R. E. Meyer, Axially symmetrical supersonic flow near the centre of an expansion, *Aeron. Quart.* **2**, 127–142 (1950).
45. R. J. Hakkinen, Supersonic flow near two-dimensional and axially symmetric convex corners and curvature discontinuities, Douglas Aircraft Co. Rept. No. SM-27747 (July 1958).
46. R. J. Hakkinen, Further results on supersonic flow near convex corners, Douglas Aircraft Co. Rept. No. SM-35992 (March 1960).
47. R. F. Weiss and S. Weinbaum, Hypersonic boundary-layer separation and the base flow problem, *AIAA J.* **4**, 1321–1330 (1966).
48. A. A. Deribas and S. I. Pokhozhaev, The problem of an explosion on the surface of a liquid, *Dokl. Akad. Nauk SSSR* **144**, 524–526 (1962); *Soviet Phys. Doklady (English Transl.)* **7**, 383–384 (1962).
49. V. F. Minin, Explosion on the surface of a liquid, *Zh. Prikl. Mekh. i Tekhn. Fiz.* No. 3, 159–161 (1964); translation available as Air Force Systems Command Foreign Technol. Div. Rept. No. MT-64-319, 275–281, AD 636992 (Feb. 28, 1966).

50. W. D. Hayes and R. F. Probstein, "Hypersonic Flow Theory." Academic Press, New York, 1959. See especially Chapter 4.
51. J. L. Taylor, An exact solution of the spherical blast-wave problem, *Phil. Mag.* **46**, 317–320 (1955).
52. R. Latter, Similarity solution for a spherical shock wave, *J. Appl. Phys.* **26**, 954–960 (1955).
53. A. Sakurai, An exact solution of the blast-wave problem, *J. Phys. Soc. Japan* **10**, 827–828 (1955).
54. W. J. Rae, Nonsimilar solutions for impact-generated shock propagation in solids, Cornell Aeron. Lab. Rept. No. AI-1821-A-2, N 65-21625 (March 1965).
55. N. Davids, Y. K. Huang, and W. Juanzemis, Some theoretical models of hypervelocity impact, *Proc. Symp. Hypervelocity Impact, 5th* **1**, Part I, 111–132, AD 284282 (April 1962).
56. N. Davids and Y. K. Huang, Shock waves in solid craters, *J. Aerospace Sci.* **29**, 550–557 (1962).
57. M. H. Rice, R. G. McQueen, and J. M. Walsh, Compression of solids by strong shock waves, *Solid State Phys.*, **6**, 1–63 (1958).
58. R. L. Bjork, Analysis of the formation of Meteor Crater, Arizona: A preliminary report, *J. Geophys. Res.* **66**, 3379–3387 (1961).
59. J. H. Tillotson, Metallic equations of state for hypervelocity impact, General Atomic Div., General Dynamics Corp. Rept. No. GA-3216, AD 486711 (July 18, 1962).
60. T. D. Riney, Solution of visco-plastic equations for axisymmetric hypervelocity impact. Second Summary Rept., Nov. 3, 1961–Nov. 2, 1962, APGC TDR-62-74, AD 294959 (Dec. 1962).
61. N. Gylden, On the similarity between compressible, nonviscous flows in metals, *J. Appl. Phys.* **36**, 2082 (1965).
62. V. M. Gogolev, V. G. Myrkin, and G. I. Yablokov, Approximate equation of state of solid bodies, *Zh. Prikl. Mekh. i Tekhn. Fiz.*, No. 5, 93–98 (Sept.–Oct. 1963). Translation in Air Force Systems Command Foreign Technol. Div. Rept. No. FTD-MT-64-61, 139–148, N 65-30413 (Feb. 8, 1965).
63. J. W. Enig, A complete E, P, V, T, S, theromodynamic description of metals based on the P, u mirror-image approximation, *J. Appl. Phys.* **34**, 746–754 (1963).
64. J. W. Enig and F. J. Petrone, Equation of state and derived shock initiation criticality conditions for liquid explosives, *Phys. Fluids* **9**, 398–408 (1966).
65. J. M. Walsh and R. H. Christian, Equation of state of metals from shock wave measurements, *Phys. Rev.* **97**, 1544–1556 (1955).
66. F. H. Harlow, The particle-in-cell method for numerical solution of problems in fluid dynamics, *Proc. Symp. Appl. Math.*, **15**, 269–288 (1963).
67. For a sampling of articles describing the particle-in-cell method, various modifications of it, and other methods as well, see Fundamental methods in hydrodynamics, "Methods in Computational Physics" (B. Alder, S. Fernbach, and M. Rotenberg, eds.), Vol. 3. Academic Press, New York, 1964.
68. R. L. Bjork, Effects of a meteoroid impact on steel and aluminum in space, *Proc. Intern. Congr. Astronaut., 10th* **2**, 505–514 (1960).
69. H. L. Brode and R. L. Bjork, Cratering from a megaton surface burst, RAND Corp. Rept. RM-2600, AD 250380 (June 30, 1960).
70. T. D. Riney, Theoretical hypervelocity impact calculations using the PICWICK code, Interim Rept., June 24, 1963–December 23, 1963, ATL-TDR-64-8, AD 434220 (March 1964). Also available as General Electric Space Sciences Lab. Rept. R64SD13, AD 430606 (February 1964).

71. T. D. Riney, Behavior of metals during hypervelocity impact cratering, "Solid Mechanics," Vol. 2, Part 2, pp. 419–445. Pergamon Press, New York, 1965.
72. T. D. Riney and J. F. Heyda, Hypervelocity impact calculations and their correlation with experiment, General Electric Space Sci. Lab. Rept. No. R64SD64, AD 606123 (September 1964).
73. T. D. Riney and J. F. Heyda, Hypervelocity impact calculations, *Proc. Symp. Hypervelocity Impact, 7th* **2**, 77–186, AD 463228 (February 1965).
74. J. F. Heyda and T. D. Riney, Peak axial pressures in semi-infinite media under hypervelocity impact, *Proc. Symp. Hypervelocity Impact, 7th* **3**, 75–122, AD 463229 (Februari 1965). Also available as General Electric Space Sci. Lab. Rept. No. R64SD87, AD 452991 (February 1964).
75. J. F. Heyda and T. D. Riney, Peak pressures in thick targets generated by reduced density projectiles, NASA Contractor Rept. No. CR-609, N66-36369 (September 1966).
76. T. D. Riney, Depth of penetration of hypervelocity projectiles, *AIAA J.* **3**, 52–60 (1965).
77. R. H. Cole, "Underwater Explosions." Princeton Univ. Press, Princeton, New Jersey, 1948.
78. R. L. Bjork, K. N. Kreyenhagen, and M. H. Wagner, Analytical study of impact effects as applied to the meteoroid hazard, NASA Contractor Rept. CR-757, N67-29028 (May 1967).
79. G. R. Fowles, Attenuation of the shock wave produced in a solid by a flying plate, *J. Appl. Phys.* **31**, 655–661 (1960).
80. D. R. Curran, Nonhydrodynamic attenuation of shock waves in aluminum, *J. Appl. Phys.* **34**, 2677–2685 (1963).
81. P. C. Chou and F. E. Allison, Strong plane shock produced by hypervelocity impact and late-stage equivalence, *J. Appl. Phys.* **37**, 853–860 (1966).
82. R. Madden, The application of the method of characteristics in three independent variables to the hypervelocity impact problem, Ph. D. thesis, Virginia Polytech. Inst. (June 1967).
83. R. Madden and T. S. Chang, Axially symmetric hypervelocity impact calculations using the method of characteristics, North Carolina State Univ. Rept. No. TSC-68-1 (Feb. 1968).
84. K. Oshima, Blast waves produced by exploding wire, Aeron. Res. Inst., Univ. of Tokyo, Rept. No. 358 (July 1960).
85. K. Oshima, Blast waves produced by exploding wires, *in* "Exloding Wires," (W. G. Chace and H. K. Moore, eds.), Vol. 2, pp. 159–180. Plenum Press, New York, 1962.
86. K. Oshima, Quasi-similar solutions of blast waves, Aeron. Res. Inst., Univ. of Tokyo, Rept. No. 386 (March 1964).
87. C. H. Lewis, Plane, cylindrical, and spherical blast waves based upon Oshima's quasi-similarity model, Arnold Engineering Develop. Center TN 61-157, AD 268654 (Dec. 1961).
88. B. H. K. Lee, Nonuniform propagation of imploding shocks and detonations, *AIAA J.* **5**, 1997–2003 (1967).
89. J. A. Charest, Measurements of shock-wave pressures generated by hypervelocity impacts in aluminum, General Motors Corp., Defense Res. Lab. Tech. Rept. No. TR 64-58, NASA Contractor Rept. CR-78399, N 66-37523 (Nov. 1964).
90. W. J. Rae and H. P. Kirchner, A blast-wave theory of crater formation in semi-infinite targets, *Proc. Symp. Hypervelocity Impact, 6th* **2**, Part 1, 163–227, AD 423063 (August 1963).

91. W. J. Rae, A critical review of impact theories, Paper presented at the Highwater Conf. on Meteoroid Impact, Highwater, Quebec, (July 14–15, 1966). Available from the Library, Cornell Aeron. Lab., and from CFSTI: AD 815234.
92. F. B. Porzel, Height of burst for atomic bombs, 1954. Part I, The free-air curve, Los Alamos Sci. Lab. Rept. LA-1664 (May 1954). Available from CFSTI.
93. T. A. Zaker, Point source explosion in a solid, Armour Res. Foundation, Illinois Inst. of Techn. Rept. No. ARF 4132-6 (Nov. 1959). Available from CFSTI.
94. G. G. Bach and J. H. Lee, Shock propagation in solid media, AIAA Paper No. 67-141 (Jan. 1967).
95. W. H. Friend, C. L. Murphy, and I. Shanfield, Review of meteoroid-bumper interaction studies at McGill University, McGill Univ., Space Res. Inst. Rept. No. SRI-R-15, NASA Contractor Rept. No. CR-54857, N 67-35001 (Aug. 1966).
96. C. J. Maiden, J. W. Gehring, and A. R. McMillan, Investigation of fundamental mechanism of damage to thin targets by hypervelocity projectiles, General Motors Defense Res. Lab. Rept. No. TR 63-208, AD 404274 (March 1963).
97. W. J. Rae, Comments on the solution of the spall-fracture problem in the approximation of linear elasticity, Cornell Aeron. Lab. Rept. No. AI-1821-A-3, NASA Contractor Rept. CR-54250, N 65-19848 (January 1965).
98. I. J. Loeffler, S. Lieblein, and N. Clough, Meteroid protection for space radiators, of "Power Systems for Space Flight" (M. A. Zipkin and R. N. Edwards, eds.), pp. 551–579, "Progress in Astronautics," Vol. II. Academic Press, New York, 1963.
99. For an excellent summary, see D. G. Doran and R. K. Linde, Shock effects in solids, *Solid State Phys.* **19**, 229–290 (1966).
100. H. G. Hopkins, Dynamic expansion of spherical cavities in metals, *Progr. Solid Mech.* **1**, 83–1640 (1960).
101. W. Herrmann, A. H. Jones, and J. H. Percey, The inclusion of material strength in hydrodynamic calculations, U. S. Air Force, Special Weapons Center, Rept. No. TDR-63-12, AD 410386 (April 1963).
102. E. L. McDowell, Deviatoric effects in high intensity stress waves, U. S. Air Force Weapons Lab. Rept. No. TR-65-15, AD 620334 (Aug. 1965).
103. L. Lees and T. Kubota, Inviscid hypersonic flow over blunt-nosed slender bodies, *J. Aeron. Sci.* **24**, 195–202 (1957).

CHAPTER VII

THE EQUATION OF STATE OF SOLIDS FROM SHOCK WAVE STUDIES

R. G. MCQUEEN, S. P. MARSH, J. W. TAYLOR, J. N. FRITZ, AND W. J. CARTER

LOS ALAMOS SCIENTIFIC LABORATORY
UNIVERSITY OF CALIFORNIA
LOS ALAMOS, NEW MEXICO

I. Introduction

During the past several years Los Alamos Scientific Laboratory has had a continuing effort to measure the equation of state of solids by shock wave techniques. A large part of this effort was devoted to establishing reliable standards so that work on other materials could be given the highest degree of credibility possible. Varieties of materials have been measured for various reasons and are reported here so that these data can be made available to any who might need them.

In general these data have not appeared in the open literature, at least not in tabular form. In some instances we have continued work on materials previously reported. For these materials we have also presented the old data corrected on the basis of our recent work in recalibrating our equipment and the new determination of our standards. This represents only a very small fraction of the data.

The results of some thermodynamic calculations are also summarized. It is obvious that all such results cannot be published, but it is felt that a need does exist for a compilation of a few quantities. It is hoped that these tables are of sufficient detail to be of general use. It was also felt that for completeness a brief description of the current calculational techniques should be included, even though these have appeared to some extent in other reports.

In addition to the equation of state measurements, a rather extensive program designed to study details of the shock wave structure has also been pursued. Such studies, usually done at relatively low pressure, not only give valuable information on the effect of rigidity in the shock wave but also give an almost direct measurement of the dynamic yield point in tension of the material under various loading conditions. The results of these experiments are best described by displaying the reduced data records. However, in those cases where a single number can be obtained to describe some important feature the results are only presented in tabular form.

A detailed discussion of the general theory of dislocation motion as applied to the problem of elastic-plastic flow in plane wave geometry has also been given.

II. Theoretical Considerations and Calculational Techniques

A. Basic Equations for Hugoniots and Isentropes

The theory of shock waves and experimental techniques have been developed and discussed previously by many workers and we will review only briefly those concepts necessary for an understanding of this work (*1–3*).

With the assumption that the shock wave profile is steady in time, the Rankine–Hugoniot equations have been developed. These equations relate the pressure (P), internal energy (E), specific volume (V), or density (ρ) behind the shock wave to these same quantities in front of the shock wave in terms of the shock velocity (u_s) and the particle velocity (u_p) due to the shock wave. These equations express the conservation of mass, momentum, and energy, and, for a coordinate system in which the material in front of the shock wave is at rest, they are, respectively,

$$\rho_0 u_s = \rho(u_s - u_p) \tag{1}$$

$$P - P_0 = \rho_0 u_s u_p \tag{2}$$

$$\left[(E - E_0) - \frac{u_p^2}{2}\right] \rho_0 u_s = P_0 u_p \tag{3}$$

The zero subscript refers to the initial state and is often referred to as the centering point of that Hugoniot. Equations (1) and (2) can be combined to give

$$u_s = V_0[(P - P_0)/(V_0 - V)]^{1/2} \tag{4}$$

and

$$u_p = [(P - P_0)(V_0 - V)]^{1/2} \tag{5}$$

By substituting Eqs. (4) and (5) into Eq. (3) the well-known Hugoniot energy equation

$$E - E_0 = (P + P_0)(V_0 - V)/2 \tag{6}$$

is obtained.

The measurement of any pair of the variables listed in Eqs. (1)-(6), when coupled with the known initial conditions, suffices to determine a point on the Hugoniot. In fact, all variables except the internal energy have been used if some current indirect pressure-measuring techniques are considered. The shock wave loci, or Hugoniots, can also be presented as a relation between any pair of these variables. However, one pair of variables, the shock velocity and particle velocity, is particularly attractive in describing the experimental results. It has been found that the Hugoniots of many materials can be adequately represented by a linear u_s–u_p relationship:

$$u_s = c_0 + su_p \tag{7}$$

Departures from linearity can usually be traced to porosity, large elastic waves, or phase transitions. In a few cases a better representation can be obtained by including a quadratic term, qu_p^2. However, if such a term is very large the data should probably be reexamined with the possibility of one or more of the previously mentioned effects being present. The shock velocity is basically a very sensitive parameter, being in essence a differential measurement (Eq. 4). Thus a small change in slope in the u_s–u_p relation in some region could indicate the existence of a second-order transition, an effect quite difficult to observe in the P–V plane.

An additional argument for using Eq. (7) as a description of the shock wave equation of state is that the leading coefficient can be determined by direct measurements of the elastic wave velocities in an isotropic medium. Hence the zero particle velocity intercepts of the shock wave data can be compared with this quantity, or this important quantity can be used in fitting the data.

By using the linear u_s–u_p Hugoniot, the Hugoniot equation of state of many materials, centered at zero pressure, may be described by the three parameters ρ_0, c_0, and s. The shock wave locus can then be found for any other pair of variables by the use of Eq. (7) with the Rankine–Hugoniot equations. For this purpose it is useful to introduce the compression η, defined by

$$\eta = (V_0 - V)/V_0 = 1 - \rho_0 V \tag{8}$$

Combining this with mass conservation, Eq. (1), yields

$$\eta = u_p/u_s \tag{9}$$

The linear u_s–u_p Hugoniot, Eq. (7), then gives

$$u_p = c_0\eta/(1 - s\eta) \tag{10}$$

$$u_s = c_0/(1 - s\eta) \tag{11}$$

Conservation of momentum, Eq. (2), then yields a pressure–volume shock locus, or, equivalently, the following P–η relations:

$$P_{\mathrm{H}} = \rho_0 c_0^2\eta/(1 - s\eta)^2 \tag{12}$$

$$dP_{\mathrm{H}}/d\eta = \rho_0 c_0^2(1 + s\eta)/(1 - s\eta)^3 \tag{13}$$

The derivative, Eq. (13), is necessary for calculating sound speeds and isentropes. The energy equation, with E_0 set equal to zero, becomes

$$E_{\rm H} = \eta P_{\rm H}/2\rho_0 \tag{14}$$

For many applications it is necessary to extend the shock wave data to regions off the Hugoniot. One example of this is the need to make shock wave impedance match calculations where reflected shocks and rarefaction release waves must be known. For such calculations it is most convenient to work through the general energy–pressure equation

$$E = E_{\rm H} + (P - P_{\rm H})/(\rho\gamma) \tag{15}$$

This equation states that the pressure and energy at any particular volume can be related to these quantities on the Hugoniot, designated here and later by the subscript H, through some average value of the Grüneisen parameter

$$\gamma = V(\partial P/\partial E)_V \tag{16}$$

For present applications γ is always considered to be independent of pressure and is usually determined from the relationship

$$\rho\gamma = \rho_0\gamma_0 \tag{17}$$

where the zero subscripts represent room-temperature zero-pressure values.

Reflected shocks have been calculated from the following equation:

$$P_2 = \frac{P_{\rm H} - (\gamma/V)_2[P_{\rm H} - P_1)(V_0 - V_2)/2]}{1 - (\gamma/V)_2(V_1 - V_2)/2} \tag{18}$$

which has been derived from Eqs. (6) and (15). In this equation the subscript 2 refers to the second shock locus. Here $P_{\rm H}$ is the pressure on the original Hugoniot at the volume V_2. The point P_1, V_1 is the centering point for the second shock and lies on the original Hugoniot, which is centered at zero pressure and the volume V_0.

In the absence of dissipative mechanisms, such as material rigidity, viscosity, etc., the thermodynamic equation of state loci of the pressure release waves are isentropes.

The thermodynamic law

$$dE = T\,dS - P\,dV \qquad \text{with} \quad dS = 0 \tag{19}$$

can be coupled with the known values of the pressure and energy along the Hugoniot through the Grüneisen parameter to calculate the isentropes. This yields the difference equation

$$P_i = \frac{P_H - (\gamma/V)_i[P_{i-1}\,\Delta V/2 + E_H - E_{i-1}]}{1 + (\gamma/V)_i\,\Delta V/2} \tag{20}$$

where

$$E_i = E_{i-1} - (P_i + P_{i-1})\,\Delta V/2 \tag{21}$$

which, by successive application, gives the P, V, E loci of these isentropes. The subscript H used in the above equation refers to values on the Hugoniot at V_i (see Fig. 1). Since a weak shock is equivalent to

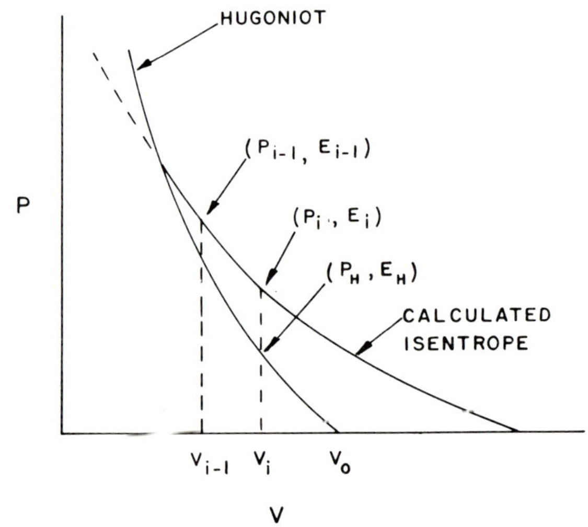

FIG. 1. Pressure–volume schematic for calculating isentropes. The calculations are usually started on the Hugoniot which establishes the first set of P, V, E points.

an isentropic pressure wave (*1*), the differential form of Eq. (5)

$$du_p = (-dP\,dV)_S^{1/2} \tag{22}$$

can be used to compute particle velocities along these isentropes.

The net increase in material velocity due to the rarefaction wave from some point on the Hugoniot to zero pressure is given by the Riemann integral along an isentrope:

$$u_r = \int_{V(P_H)}^{V(P=0)} [-(\partial P/\partial V)_S]^{1/2}\,dV \tag{23}$$

which when added to the particle velocity imparted by the shock wave gives the free surface velocity

$$u_{fs} = u_p + u_r \tag{24}$$

a quantity amenable to experimental observation.

B. Inclusion of Temperature and Entropy in the Equation of State

In general it is impossible to measure the temperature of the material behind the shock wave, and, except for a few transparent materials (*4*, *5*), this quantity must be calculated. This is usually done through the thermodynamic identity

$$dT/T = -\gamma\, dV/V + dS/C_V \tag{25}$$

Here T is the temperature and C_V is the specific heat at constant volume. If the isentropes have been located, Eq. (25) is readily integrated using values along the zero-pressure axis as initial conditions. This method utilizes the available experimental data to the utmost. It is also possible to measure the residual temperature of the shock-loaded nontransparent materials (*6*, *7*) to verify that the correct initial values are indeed being used in the computations.

For many materials sufficient thermodynamic data does not exist to warrant this treatment and temperatures are calculated directly along the Hugoniot. To do this, the Rankine–Hugoniot equation for the energy is combined with the thermodynamic law, Eq. (19), to yield an equation for the entropy along the Hugoniot:

$$2T\, dS = (V_0 - V)\, dP + (P - P_0)\, dV \tag{26}$$

Given an experimental Hugoniot and values of the specific heat and the Grüneisen parameter along the Hugoniot, the coupled differential equations (25) and (26) may be integrated to give the temperature and entropy along the Hugoniot.

For many solids an adequate representation of the specific heat can be obtained from one of the simplest forms of the Debye theory (*8*). The form we shall use here is characterized by a single Debye theta, $\Theta(V)$, which is a function only of volume. The thermal energy, $E_T = E - E_K$ (where E_K is the energy of the solid at 0°K and the same volume), is given by

$$E_T = 3nkTD_3(x) \tag{27}$$

$$D_3(x) = \frac{3}{x^3}\int_0^x \frac{z^3\, dz}{e^z - 1}, \qquad x = \Theta/T \tag{28}$$

Here, n is the number of atoms per gram and k is Boltzmann's constant. The specific heat and the entropy are given by

$$C_V = 3nk[4\, D_3(x) - 3x/(e^x - 1)] \tag{29}$$

$$S = 3nk[\tfrac{4}{3}\, D_3(x) - \ln(1 - e^{-x})] \tag{30}$$

This form for the specific heat is consistent with the assumption that the Grüneisen parameter depends only on the volume, since it can be seen from Eqs. (29) and (30) that the specific heat depends only on the entropy. Indeed, from these two equations and Eq. (25) the usual function for γ is obtained:

$$d \ln \Theta / d \ln V = -\gamma \tag{31}$$

Since we wish to use these thermal equations to approximate the specific heat as well as possible, we have chosen Θ so that Eq. (29) will give the correct room-temperature zero-pressure value for the specific heat.

For some materials, where both C_V and n were not readily available, the approximation

$$k\Theta = \hbar c_0 (6\pi^2 n \rho)^{1/3} \tag{32}$$

was used. In the cases where C_V was known but n was uncertain, we used $n \approx C_V/3k$, provided that the sound speed and density were low enough to expect a low Debye theta.

This simplified choice for the specific heat does not represent the detailed behavior of the specific heat of most materials, particularly at high temperatures. The electronic contribution to the specific heat (important for metals at high temperatures) and the anharmonic contribution of the lattice vibrations to the specific heat (linearly increasing with temperature) are not taken fully into account. The optical modes for some of these solids could be better represented by an Einstein theta, Θ_E. The choice of specific heat used here does give a reasonable way to extrapolate to the zero-Kelvin isentrope at zero pressure. The error in the $P_K(V)$ curve produced by the inaccuracy in C_V is probably less than or comparable to the uncertainty in the Hugoniot curve. An exception to this is the zero-Kelvin isentrope calculated from metal Hugoniots in the high-pressure very-high-temperature region, where the neglect of the electronic specific heat produces an appreciable error. Although isentropes do not explicitly depend on the specific heat, the increasing contribution of the electrons to the specific heat at high temperatures causes the assumption that the Grüneisen gamma is a function only of volume to become invalid.

A fourth-order Runge–Kutta numerical method was used to integrate Eqs. (25) and (26). The entropy obtained from the integration agreed with the entropy given by Eq. (30) to better than one part in 10^6. Given the temperature along the Hugoniot, the thermal energy can be calculated from Eq. (27). This energy and the corresponding thermal pressure, $P_T = E_T\gamma/V$, can then be removed from the Hugoniot curve to yield the zero-Kelvin isentrope. In a similar fashion, increments of the thermal energy and pressure can be subtracted from the Hugoniot to yield a room-temperature isotherm. Any isentrope is also readily calculated from tabulated values of pressure, energy, and the Debye theta along the zero-Kelvin isentrope by evaluating the thermal energy, Eq. (27), and pressure for a fixed value of x, i.e., constant entropy.

Although it might appear that the isentropes calculated in this fashion depend on the forms we have assumed for the specific heat, they do not. The family of isentropes obtained by the above prescription are identical to those produced by Eq. (20). The isentropes depend only on the Hugoniot curve and the particular form chosen for the volume dependence of the Grüneisen parameter. The shape of the zero-Kelvin isentrope does not depend on specific heat values. The specific heat serves only to specify which of the family of isentropes is to be the one at zero Kelvin. When coupled with γ and the measured Hugoniot, the specific heat then determines the zero-Kelvin density.

A general expression for the sound speed, except for the assumption of a volume-dependent Grüneisen parameter, in the pressure–volume plane can be obtained by equating a general change in energy from Eq. (15) with the isentropic energy change, Eq. (19):

$$c^2 = -V^2 \frac{dP_{\mathrm{H}}}{dV}\left[1 - \frac{\gamma}{V}\frac{V_0 - V}{2}\right] + V^2 \frac{\gamma}{V}\left[\frac{P_{\mathrm{H}} - P_0}{2}\right] + V^2(P - P_{\mathrm{H}})\left[\frac{\gamma}{V} + \frac{d\ln(V/\gamma)}{dV}\right] \tag{33}$$

The first two terms are the square of sound speed along the Hugoniot. The last term contains the dependence of the sound speed on the pressure offset from the Hugoniot. Equation (33) can be used to compute the sound speeds for the isentropes and isotherms once the pressure along these curves is known. The isentropes and the sound speeds along them may be used to evaluate the Riemann integral, Eq. (23), and the free surface velocity, Eq. (24), produced when a shock encounters a zero-pressure boundary. For most of the calculations a small extrapolation of the u_s–u_p Hugoniot into the negative particle velocity region has been used. The second-order contact existing between the Hugoniot and the isentrope at the centering point in the P–V plane justifies this extrapolation when the volume increase produced by the irreversible shock heating is small. Actually for many materials, particularly for metals, an extrapolation of the linear u_s–u_p Hugoniot all the way down to zero shock velocity (infinite volume) yields a reasonable estimate of the binding energy of the solid (*9*). The state of the medium at the foot of the release wave, as well as the free-surface velocity, has been calculated for various initial pressures on the Hugoniot and listed for the many materials considered here. This sort of a calculation does not take into account phase changes, such as melting and vaporization, which can occur upon release from very large shock pressures.

C. Phase Changes and Hugoniots

Although a linear relation in the u_s–u_p plane is adequate for simple materials, phase changes produce discontinuities both in value and slope in the u_s–u_p relation. These are usually caused by the two-wave structure that accompanies most phase changes. It was pointed out by Bethe (*1*) that, for a stable shock wave, the sound velocity must increase with pressure. A similar requirement is that the shock velocity must increase with pressure. This means that if the shock velocity should decrease with pressure, then the shock front would break up into two or more waves, or possibly one wave with a continuously smeared front. A P–V Hugoniot with a break and the type of shock profile that would result is sketched in Fig. 2. For a shock wave profile (P versus distance) to be constant in time, that is, for a stable shock to exist, it is required that

$$u_{s2} \geqslant u_{s1} \tag{34}$$

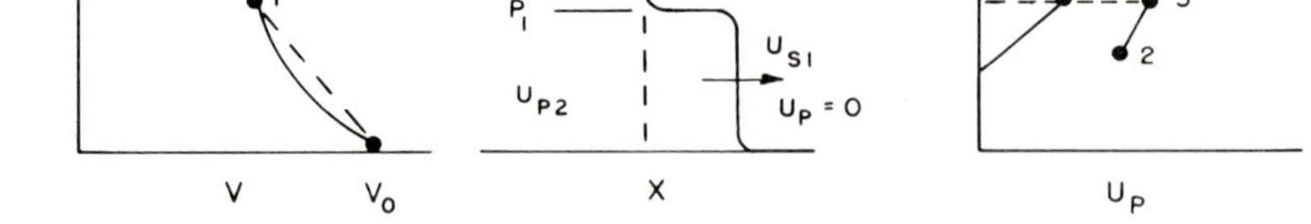

Fig. 2. The type of Hugoniot necessary to produce a two-wave shock structure and the resulting wave profile. This type of Hugoniot will in general give a u_s–u_p locus as shown with a flat region of constant shock velocity. Point 2 will not be observed with techniques that measure only the first arrival of the shock wave.

If Eq. (4) for the shock velocities is substituted into Eq. (34), the following equation is obtained:

$$u_{p1} + V_1[(P_2 - P_1)/(V_1 - V_2)]^{1/2} \geqslant V_0[(P_1 - P_0)/(V_0 - V_1)]^{1/2} \quad (35)$$

Subtracting u_{p1}, as written in Eq. (5), from each side gives the result

$$V_1[(P_2 - P_1)/(V_1 - V_2)]^{1/2} \geqslant V_1[(P_1 - P_0)/(V_0 - V_1)]^{1/2} \quad (36)$$

which means that to have a single stable shock wave, all points on the P–V Hugoniot must lie below the ray connecting the initial and final states. In Fig. 2 the Hugoniot is drawn with a cusp at Point 1. If the slope of the Hugoniot above this decreased continuously, $(\partial^2 P/\partial V^2)_H < 0$, the second wave would not condense as shown in the figure but would spread out, a situation that apparently does occur in some materials.

In the flash-gap type of experimentation reported here, details of the u_{s2} wave are not observed because the first wave with velocity u_{s1} and pressure P_1 will usually close the flash gaps and cause the gas to light before the second wave reaches the free surface. This situation continues until Point 2 is on the extension of the ray through the origin and Point 1, at which time the second wave is moving at the same velocity as the first wave. This gives rise to a u_s–u_p Hugoniot curve (Fig. 2) with three segments. The flat central region does not necessarily represent an equation of state locus. The impedance match solution used to obtain the particle velocity is based on the shock velocity of the first wave (u_{s1}), whereas it should be made on the velocities associated with the unknown pressure (P_2). However, at the intersection labeled 1 in Fig. 2, the particle velocities and shock velocities are meaningful and can be used to determine the pressure of the first wave which, if equilibrium conditions exist, is the pressure where a phase change occurs.

If the pressure of the first wave (P_1) is small compared to the pressure of the second wave (P_2), it is possible that the second wave can reach the free surface and be effective in closing the flash gap before the first weak wave does. Such a record could indicate a shock transit time closely associated with a single wave propagating at the slower velocity with a pressure like that of P_2. Thus, weak high-velocity waves will not necessarily be recorded with the flash-gap technique.

A first-order phase change is characterized by the sign of the slope, dP/dT, of the phase line. For $dP/dT > 0$, it is possible to shock either from the high-density to the low-density phase (Fig. 3a) or from the low-density to the high-density phase (Fig. 3b), depending upon whether the temperature along the Hugoniot increases more or less rapidly than along the phase line. The first case is typical for melting

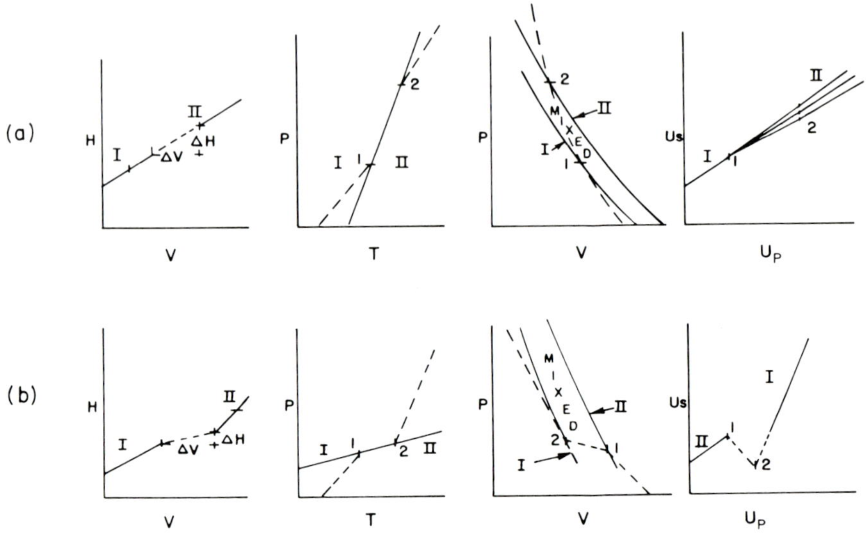

FIG. 3. Effect of a phase change on the u_s–u_p Hugoniot. In this figure the case where $dP/dT > 0$ is considered. Here and in the next figure H vs. V is plotted at constant pressure. The pressure–temperature locus of the phase line and of the Hugoniot (dashed lines) are shown. P vs. V is plotted for the phase line boundaries and the Hugoniot loci (dashed line); finally the type of u_s–u_p Hugoniot (solid line) that might be observed with the experimental technique described here is plotted.

or vaporization transformations. It is assumed that the original state of the Hugoniot lies in Region I (the low-temperature phase) and that it crosses the phase line as indicated. At the present time we do not know of any clear examples of the detection of such a transition solely by u_s–u_p data. It is possible that the Hugoniot could enter the mixed-phase region and finally come out again in Region I. Figure 3b is representative of a solid–solid transformation (graphite–diamond, for example). The slope of the phase line is still positive, but the Hugoniot is centered in the high-temperature phase (Region II). When this Hugoniot crosses the mixed-phase region, there is ample experimental and thermodynamic data to show that the Hugoniots as drawn in the P–V and u_s–u_p planes can exist and will give rise to a two-wave structure as previously discussed.

It is possible in either of these two cases that the Hugoniot P–T locus will not intersect the phase line. It would be useful if the onset of melting could be detected on the Hugoniot; however, when the equation of state parameters on either side of the phase line are so similar (i.e., dH/dV is nearly continuous across the phase change) the

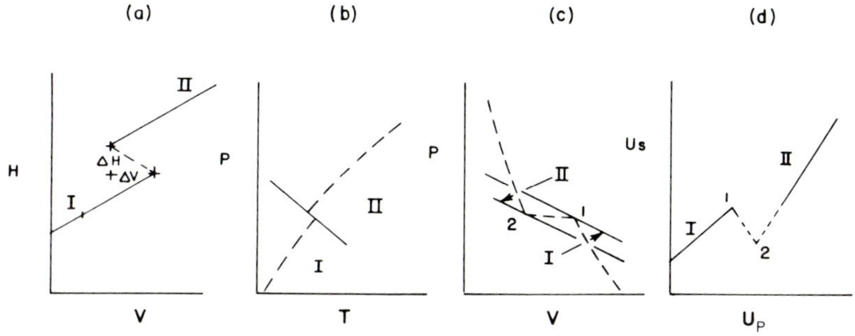

FIG. 4. Phase change with $dP/dT < 0$.

u_s–u_p Hugoniot will be only slightly affected. A study of this has been undertaken on *Pb*, where the onset of melting on the Hugoniot is thought to be reasonably well established by calculations.

The other type of transformation ($dP/dT < 0$) could correspond to anomalous melting (e.g., Ge and Si) or crystallographic phase changes (e.g., Fe and Zr). This case is described in Fig. 4. If the phase change $dP/dT < 0$ exists, the Hugoniot must eventually cross the phase line, provided that the Hugoniot is centered in the low-temperature phase. Although the P–V boundaries of the phase lines are different in this figure and the previous one, the resulting P–V and u_s–u_p Hugoniots have identical characteristics.

From Fig. 3 or 4 it can be seen that when a material is shocked from a low-density phase to a high-density phase there can be a noticeable effect on the u_s–u_p Hugoniot. In general, the larger the volume change, the longer the flat central region in this plot. Although an estimate can be made of the size of the volume change, it is unfortunate that no

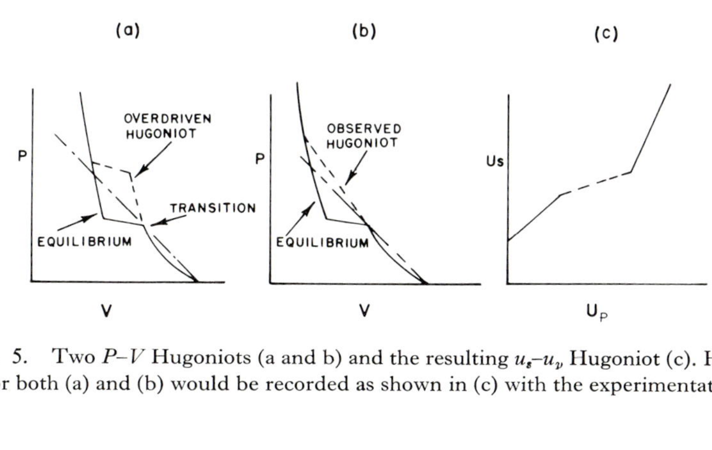

FIG. 5. Two P–V Hugoniots (a and b) and the resulting u_s–u_p Hugoniot (c). Hugoniot data for both (a) and (b) would be recorded as shown in (c) with the experimentation used here.

information can be obtained on the size of the entropy charge. In fact it is impossible to ascertain the sign of the slope of the phase line.

Additional discussion is required for the materials that do not have a flat central u_s–u_p Hugoniot. With the flash-gap technique used here, it is not possible to state whether or not a two-wave shock structure exists. In Fig. 5 are drawn two P–V Hugoniots and the u_s–u_p locus that would result from each. In Model (a) it is assumed that the original state can persist in the second phase region, and it does not begin to transform at the equilibrium conditions; moreover, it assumes that the larger the original overdrive pressure, the more the material will be overdriven by an amount depending on the initial shock strength; however, with time, the material transforms and the pressure decays. S. Minshall has observed this behavior in Sb (unpublished work). In other words, there is a relaxation effect that could be time dependent (not checked here) and/or pressure dependent. In case (b) it has been assumed that the transformation begins immediately but does not go to completion. This means that although the pressure lies slightly above the Rayleigh ray through the mixed-phase region, the slope does not increase as rapidly as it would if the material had stayed in the initial phase. Although (a) can give rise to a two-wave structure in the shock front, (b) of course will not. Hence, in principle the situation is resolvable. Several calculations based on equilibrium thermodynamics have not yet revealed a case where the calculated Hugoniot fell above the extension of the Rayleigh ray through the transition pressure for this type of transformation. In either case a phase change is indicated. Since the mixed-phase region is characterized by a slightly increasing shock velocity, a two-wave structure will not necessarily exist. The data in the mixed-phase region can be in a very nonequilibrium state. This indeed seems to be the case for carbon when graphite is shocked into the stable diamond field. With this system (graphite–diamond) enough is known about the thermodynamic and equation of state parameters to calculate the equilibrium Hugoniot curve. These calculations show that the shock wave data are far from being in an equilibrium state. The important thing to note is that transformations are indicated by the kink in the u_s–u_p Hugoniot that can be caused by only a very small percentage of the material undergoing a phase change. If it were necessary to choose between Model (a) or (b), the former seems preferable and has been verified for crystalline quartz by Wackerle (*10*).

It should be noted that the one-dimensional nature of our type of experimentation also gives rise to a two-wave shock structure. For the relatively low-pressure shocks, the first wave is due to deformation in one dimension only and hence it must propagate at longitudinal sound velocity. Its amplitude depends on several things (the rigidity modulus, dislocation density, and impurities to mention a few) and has been observed to be as high as 0.1 Mbar in quartz (*10*). When the yield point is exceeded, the material relaxes toward its hydrodynamic equilibrium state. Although it is not obvious how close to this state the material can reach in the short times available in the shock front, comparison of some shock wave data (*11*) with Bridgman's work (*12*) indicates that the hydrodynamic assumption for metals is a good first approximation.

In many cases, whether the cause of a two-wave structure is elastic-plastic structure or a phase transition can be readily resolved by examining the high-pressure data. If these data extrapolate to a value below the bulk sound velocity, then it must be concluded that a transformation exists. If the high-pressure data extrapolate to a velocity higher than the zero-pressure bulk sound velocity, a transition is still indicated but possibly of a different kind. For example, this would be compatible with a second-order phase change where the material has become more compressible. Other phase changes are not excluded however. Even though the high-pressure data extrapolate to the zero-pressure bulk sound velocity, low-pressure phase changes can exist. For these cases and where the data do not go to high enough pressure to resolve the problem through the previous considerations, it is still possible to decide whether or not a transition exists. To do this, first examine the flat segment. If this does not occur near the measured longitudinal wave velocity, one is not observing the effect of a two-wave structure due to elastic-plastic flow. If the shock velocity does happen to be in agreement with the longitudinal wave velocity, one must then check to see if there is any lower-pressure data that extrapolate toward the bulk sound velocity. If so, it must be concluded that agreement with the longitudinal velocity is purely coincidental because if it were indeed due to an elastic wave, the low-pressure data would have to extrapolate to the zero-pressure longitudinal value. It should be remembered that if any of the above conditions are satisfied the existence of a phase change is established. Two necessary conditions for establishing that the wave structure is due to an elastic wave are that the low-pressure data extrapolate to the longitudinal sound speed or that the flat region be at or slightly higher than the longitudinal sound speed and also that the high-pressure region extrapolate reasonably well to the bulk sound velocity.

It is desirable in those materials having a phase change to transform the high-pressure shock wave data to a Hugoniot centered at $P = 0$ but in the high-density phase, a metastable state. This has been done for stishovite and some rocks (*13*, *14*), and the same principles applied there are used here. The problem is to find the best zero-pressure density (ρ_0), sound velocity (c_0), and slope (s) of the u_s–u_p Hugoniot in the metastable state. To do this it is necessary to know these three quantities for the original low-density phase (these are estimated for the unknown phase to begin the calculation), the specific heat and Grüneisen parameter for both states, the shock velocity where the material begins to transform, and finally the slope of the P–T phase line. For these calculations the specific heat is taken as constant and, in the absence of any experimental information to the contrary, the same value is used for both phases. For the Grüneisen parameter, the same value of $\rho_0\gamma_0$ and the same volume dependence was used for the high-pressure phase as was used for the low-pressure phase. Slopes of phase lines, when they were not known, were taken to be zero. This is consistent with the small entropy changes usually associated with solid–solid phase changes. Reasonable variations in the slope of the phase line affect the calculated metastable equation of state very little (*14*). The volume change, and the corresponding entropy change from the phase line slope, must be found first so that the difference in the zero-pressure internal energies of the two states can be determined. With reference to Fig. 6 this is done in the following manner:

(1) First the thermodynamic quantities P, T, V, S, and E are found for the original phase at the transition (Point 1 on the figure) on Hugoniot A. This determines the value of the Gibbs' free energy (G).

(2) The isentrope, calculated from the assumed metastable Hugoniot, is found that satisfies the condition that temperatures be the same at the transition pressure ($T_1 = T_2$).

(3) This determines ΔV and hence ΔS.

(4) Since the free energies must be the same on the phase line ($G_1 = G_2$), the energy difference, $E_{0A} - E_{0B}$, the only remaining adjustable parameter, can then be found.

With the difference in internal energies determined, the transposed Hugoniot can be calculated as was done for stishovite (*13*). The equation used to find this curve in the P–V plane is

$$P_C = \frac{P_B[1 - (\gamma/V)(V_{0B} - V)/2] + (\gamma/V)[E_{0A} - E_{0B}]}{1 - (\gamma/V)(V_{0A} - V)/2} \tag{37}$$

This is the curve labeled C in Fig. 6 and is the Hugoniot based on the assumed equation of state of the high-density phase but centered at the original state (V_{0A}, E_{0A}). The P–V curve is then transformed to the u_s–u_p plane through Eqs. (4) and (5) and compared with the experi-

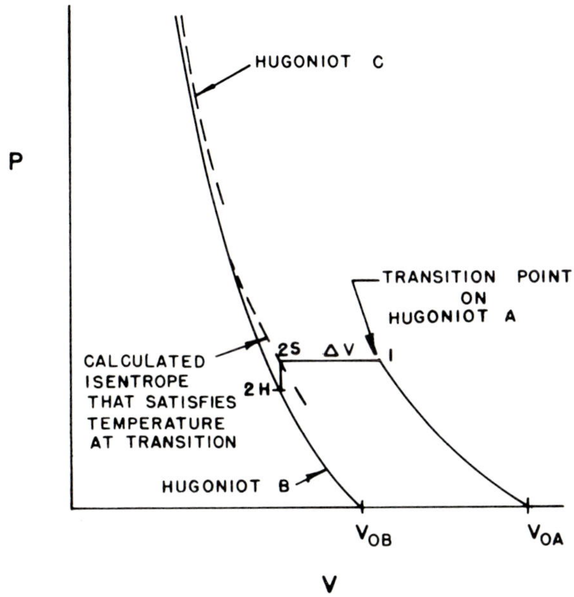

FIG. 6. *P–V* plot to illustrate how the metastable Hugoniot is found: A is the Hugoniot in the original low-density phase; B the derived Hugoniot in the high-density metastable phase; C the Hugoniot calculated from Hugoniot B but centered at $P = 0$ in Phase A. Curve C is compared with the experimental data. Point 1 on Hugoniot A should, in principle at least, be established by experiment. Point 2 corresponds to the calculated isothermal volume change of this transition.

mental data. The difference (Δu_p) in the shock particle velocity between each data point and the calculated Hugoniot at the corresponding shock velocity is determined. In the present work the trial c_0 and s are held fixed and the initial density (ρ_{0B}) varied until the algebraic sum of the velocity differences is zero ($\sum \Delta u_p = 0$). This variation tends to move the calculated curve back and forth more or less parallel to itself. Next, the initial sound velocity (c_0) is varied, again holding the slope fixed, and the previous procedure repeated. A c_{0B} is then found that gives the minimum value for $\sum(\Delta u_p)^2$. This variation tends to rotate the calculated curve through the center of the data points. Finally, $\sum(\Delta u_p)^2$ is further minimized by varying the slope (s), and repeating the previous calculations for new c_0's and ρ_0's. For reasonable slopes, this last minimization changes the curvature of the calculated curve. Even with previous conditions satisfied, this is a relatively insensitive parameter. It is arbitrary whether the iterations are performed on the densities or the sound speeds first; however, it is felt that the slope should be done last because it is the more insensitive parameter. In fact a couple of bad data points could cause one to iterate to a physically meaningless slope.

III. The Equation of State of the Standards

A. Experimental Techniques

The definition of the word standard is a rather nebulous thing. In general any material that can be used as a basis for determining the equation of state of other materials is in principle a standard. For the present discussion we are being slightly more restrictive in that we will consider a material a standard only if its equation of state has been determined in some manner that does not require the use of the equation of state of some other material.

It has long been recognized that Hugoniot data could be obtained by simultaneously measuring the impact velocity of a rapidly moving plate and the subsequent shock velocity induced in a stationary plate. If the driver plate and target plate are made of the same material and are in the same thermodynamic state, then the symmetry of the collision requires not only that the pressure be the same in the driver and in the target plate but also that the particle or material velocity behind the shock wave be exactly one-half the driver velocity. When using gun devices the projectile velocity can be measured with extreme precision by monitoring its free-flight motion over relatively long distances by electronic pins or other suitable devices. This has been done here for a few measurements in the low-pressure region, and over the entire pressure region by others (*15*) using two-stage gun devices. However, the bulk of the experimental data obtained here has been obtained using explosively driven driver plates, which complicates accurate measurement of the driver plate velocity, u_d. The biggest difficulty is due to the fact that the driver plates are not moving at constant velocity at the time of impact, which means that the velocity must be determined at a particular position over a very short interval of motion. Pin techniques have been used successfully in explosive systems (*16*, *17*). However, for these experiments we have used a sweeping image camera to record shock wave arrival by the flash-gap technique developed by Walsh and Christian (*18*). For the work reported here, all cameras have had their writing speeds redetermined by a new technique which uses a photomultiplier tube to indicate the time of arrival of a slit image sweeping past precisely known positions on the film plane. This technique allows the instantaneous writing speed to be measured across the entire film plane. This has resulted in a decrease of about 1% in writing speed for our equipment. Writing speed is defined as the velocity at which an image is swept across the film plane at a particular angular velocity of the rotating mirror. Errors in the earlier measurements have been attributed to consistent errors in the timing pulses delivered to bridge wires which were used to put light pulses on the image plane. Details of the technique are to be described elsewhere.

The shock velocity is measured in the usual way by determining the shock wave arrival times at different levels in the target plate. In order to circumvent the fabrication difficulties encountered in making precise velocity measurements of the driver plate directly, a small groove is machined in the impact side of the target plate which allows a differential measurement of the shock wave velocity and driver-plate velocity when the shock wave arrival is observed over a plane area from the free surface direction.

Details of the assembly can be seen in Fig. 7. The upper part of Fig. 7 shows the arrangement of the plastic flash blocks which in effect give four independent sets of u_s–u_d data points. The left-hand side of the assembly is used to measure the shock velocity and the other side the driver velocity. Each set of blocks is viewed through four or five slits. The cross-sectional views below are designs used in different pressure regions. The first two designs are for the low-pressure regime; the upper one has the u_d groove in the correct position. At higher pressure u_d grooves are not as readily pinched off and can be made as deep as the target plate thickness allows. For the high-pressure shots the target plate thickness is limited by the very thin drivers (0.9 mm) required.

The resulting photographic record is shown in Fig. 8. This is a relatively low-pressure shot and the uppermost cross-sectional design was used. In this record, time increases downward; hence, the early traces on the left-hand side represent the shock wave arrival at the bottom of the narrow groove. The corresponding reference traces establish the wave arrival at the top of the plate. Since the flash blocks on the right-hand side are at a lower level, the reference traces for the u_d measurements arrive earlier. The offsets represent the difference in the driver–shock transit times through the small gap machined in the bottom of the plate.

As simple as such a system appears to be, considerable care must be used in choosing the correct groove depths so that optimum precision can be obtained. In these experiments the width and lateral location of the grooves were always the same. In low-pressure shocks the u_d traces are pinched off by the sidewise rarefactions caused by the shock wave which runs considerably ahead of the projectile. Thus, the driver velocity can only be measured over a run of about 1 mm. This limitation is not

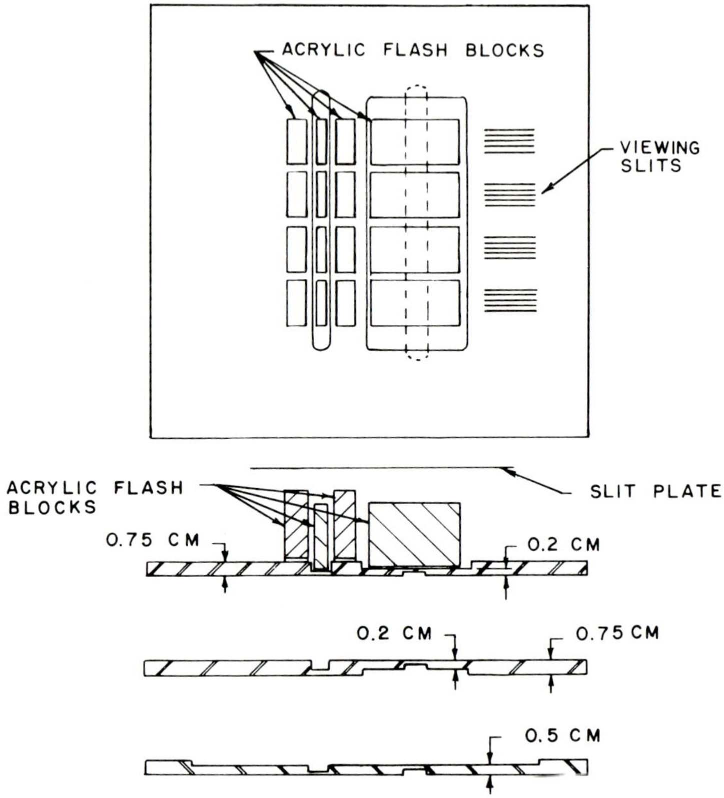

FIG. 7. Design of the target plate used to obtain shock particle velocity data.

too important since the u_s–u_p Hugoniot is relatively insensitive to errors in the particle velocity measurement at low pressures. For example, in aluminum a 5% error in the shock particle velocity at 1.0 mm/μsec corresponds to an error of only 1% in shock velocity. At low pressures there is no actual practical restriction on the depth of the shock velocity groove since runs of 5 or 6 mm are sufficient to give adequate space and time measurements. At high pressures the u_d traces are not as readily pinched off; however, the total thickness of the target plate is limited by the thickness of the driver plate, which determines how much material the unperturbed shock wave can traverse before the rarefaction wave from the back surface overtakes it. At low pressures the catch-up ratio, $r = t/d$, where t is the target thickness and d the driver thickness, must be computed using the longitudinal wave velocity. However, at very high pressure where this condition is most stringent, the hydro-

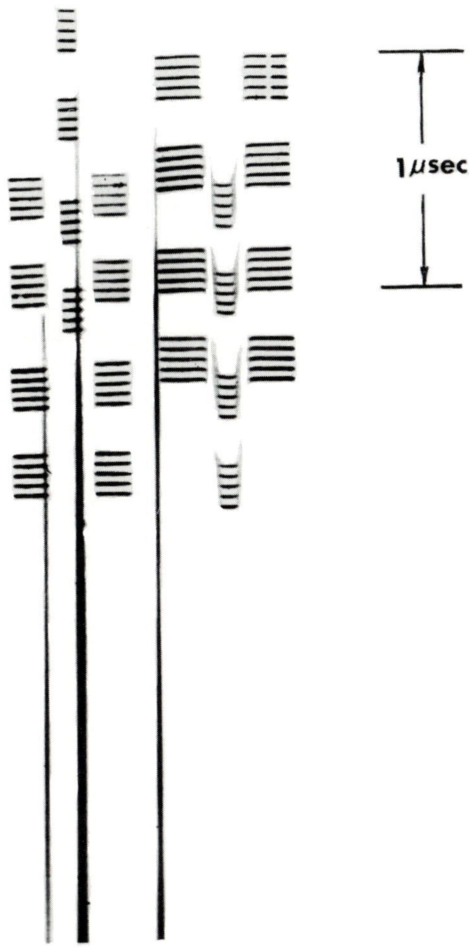

FIG. 8. An enlargement of a photographic record obtained using the assembly described in Fig. 7.

dynamic or bulk sound velocity c_b can be used since the material is most certainly in the liquid phase. Calculations showed that target-to-driver ratios as large as four could be used.

It can be seen in Fig. 7 that for low-pressure systems the u_d groove could be located at different levels relative to the back of the target plate. If the driver plate is moving at a constant velocity, the location of the groove is of no importance. However, accelerations can be expected in plates driven by high explosives (*19*). To minimize the effect of plate acceleration during the measuring interval, the u_d groove is machined directly in the back of the target plate. This is the correct place to locate this groove because very small pressure pulses which might be present in the driver plate at collision time are much less likely to be manifested as an increase in the measured shock wave velocity than as an increase in the driver velocity. Moreover, this guarantees that the initial shock in the u_d side of the plate is the same as on the shock velocity side. Several shots were duplicated, except that the u_d grooves were located at different levels. This resulted in a nontrivial increase in the derived particle velocity. These shots are included in the figures and tables but were not used in fitting the results.

A few special shots were fired to verify experimentally that the shock wave pressure was more or less constant through the target plate. In these experiments the free surface velocity was measured at various levels in simulated target plates. All of these experiments showed that within the experimental accuracy the pressure of the shock wave was uniform through the assembly until the overtaking rarefaction wave caught up with the shock wave, in agreement with earlier calculations.

In addition to the above, the thermodynamic state of the driver plate must also be considered. Since shock waves have been used to accelerate the driver plate, it must be assumed that some shock heating has occurred, and hence that the symmetry condition is no longer strictly satisfied. However, if the shock-loading history of the driver is known, it is possible to calculate how much the u_d/u_p ratio differs from 2.0. This is done by calculating a hot Hugoniot for the driver, centered at a lower density and higher energy because of the shock heating, from an assumed Hugoniot of the normal material through Eq. (15). A straightforward impedance match solution then gives the desired results. It was found that the first strong shock from the high explosive was the major contributor to the shock-heating correction with the second shock adding practically all the remainder. For shocks with amplitudes less than 100 kbar the correction in particle velocity is less than 0.1%. In the majority of these experiments an attempt has been made to minimize the initial shock pressure by leaving a small air space between the high explosive and the driver plate. Framing camera studies indicated that the initial acceleration of the driver was much smaller, but that final velocities were not substantially lowered. Hydrodynamic calculations made by C. Mader of this laboratory indicated a similar behavior and also that the tension waves due to the Taylor wave in the high explosive were considerably reduced. Thus, all shots with long free runs had high explosive–driver-plate separations. Earlier work showed that a thin layer of plastic placed behind the driver plate prevented early driver plate breakup. We are currently using a 10-mil layer of polyethylene for this purpose. For low-pressure shots the driver is placed in direct contact with the explosive and the driver is separated from the target by a very small distance; hence, the driver is subjected to one known shock and the heating correction can be made exactly. Large explosive charges, approximately 30 cm long, were used in these experiments to minimize uncertainties in the shock velocity and driver velocity caused by the tension wave in the driver resulting from the Taylor wave in the high explosives.

The photographic record is read with an optical comparator connected to an IBM card punch. A computer program fits the position of the references (shock arrival at the free surface of the base plate) with a cubic and the offsets with a quadratic by means of the method of least squares. It computes the time offset as measured under each slit and makes the appropriate averages. To insure that the film was properly read and interpreted, the computer replots the input data along with the fitted curves of the data which allows a rapid check for errors.

In addition to the shock wave measurements we have attempted to measure the zero-pressure sound velocities for all the materials investigated in this report. The pulse-echo method was used to determine both longitudinal and transverse wave velocities (c_l and c_t) from which the bulk sound velocity (c_b) was calculated from the relation for isotropic material:

$$c_b^2 = c_l^2 - \tfrac{4}{3}c_t^2 \tag{38}$$

Pairs of X-cut (10 MHz) and Y-cut (5 MHz) quartz crystals were used as transducers for the longitudinal and transverse waves, respectively. Coupling of the X-cut quartz to the sample was accomplished with spindle oil; Dow resin 276-V9, a high-viscosity fluid, was used for the Y-cut quartz. Signals from the receiving transducer were amplified by wide-band amplifiers and displayed on a Tektronix 545 oscilloscope. The time between echos was measured using the delay-time multiplier on the scope. The multiplier was calibrated against a 0.2-μsec time mark generator stabilized by a Tektronix Model 180 oven-controlled 10-MHz sine wave generator so that it could be used as a linear time base. Sample sizes varied from 1.25 to 0.75 cm in thickness.

B. Hugoniot Data for 921-T Aluminum, 2024 Aluminum, Copper, Iron, and Uranium–3-Wt%-Molybdenum Alloy

The Hugoniots of the five materials listed above have been determined by measuring the shock and particle velocities by the technique described in the previous section. It was felt that the Hugoniot of at least three materials should be determined: one of relatively low shock impedance; one with high shock impedance; and one with intermediate shock impedance so that other materials could be measured with a standard of comparable shock impedance. Shock impedance is defined as the product of the initial density times the shock velocity, $\rho_0 u_s$. This should minimize any errors introduced when extending the equation of state of the standards for use in impedance match solutions. With three standards it is possible to check the calculated rarefaction release curves at two points for the high-impedance material and two points on the calculated reflected shock curves for the low-impedance material. The material of intermediate shock impedance can be checked at one point on any reflected shock or rarefaction curve.

For the low-shock impedance material, 2024 Al was chosen not only because of its availability and good machining properties but also because a vast amount of shock wave data has been accumulated using it as a standard. Because of possible adverse effects due to the rigidity in the 2024 Al alloy, another aluminum alloy, 921-T Al tooling plate, was also studied. It was found that this material has only a very small amplitude elastic wave, which implies it should have more ideal hydrodynamic behavior.

A deoxidized annealed copper was chosen for the intermediate standard. Unfortunately, its desirable feature of being able to flow plastically prevents it from being accelerated to very high velocities.

To extend the Cu equation of state to higher pressures the Hugoniot of 1018 steel was measured. These materials have quite comparable shock impedances; hence, the solution of the shock wave impedance problem of Fe shocking Cu is almost completely dependent on the Hugoniot locus of the Fe. Moreover, if the Fe Hugoniot is known, the Cu Hugoniot can be obtained directly by using Fe drivers with Cu target plates. The brass standard used at Los Alamos Scientific Laboratory earlier (*11*) was discontinued because we were unable to obtain consistently material uniform in composition and free from porosity. For the high shock impedance material we have used a uranium–3-wt%-molybdenum alloy. This alloy has reasonably good machining characteristics and can be readily fabricated in this laboratory. Some of the important thermodynamic constants for these materials are listed in Table I. The results of the shock wave measurements are plotted in Figs. 9 and 10. The curves drawn through the data in Fig. 10 were calculated from the u_s–u_p fits shown in Fig. 9. In this plane the Hugoniots for the two Al alloys are indistinguishable; the shock impedance of the higher-density 921-T Al alloy has been compensated by the slightly higher shock velocity in the 2024 Al alloy. A substantial

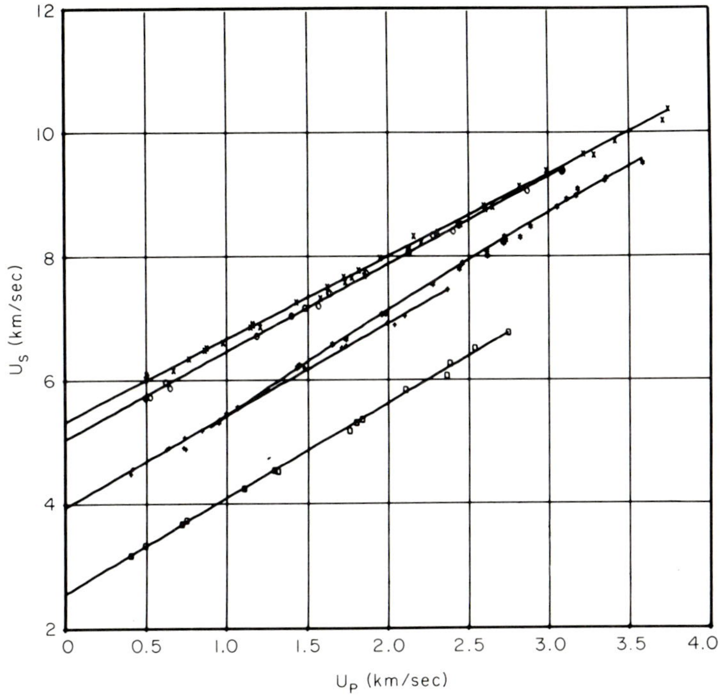

FIG. 9. Hugoniot u_s–u_p data from shock particle velocity experiments. (×) 2024 Al, (○) 921T Al, (*) Fe, (+) Cu, (□) U–3%-Mo alloy. Each set of data points was obtained by independent symmetrical collision experiments. The curves were obtained by the method of least squares.

area of the P–u_p plane is covered by these five materials. The iron Hugoniot is shown starting at about 0.4 Mbar since this is the stable shock threshold. Iron can be used at lower pressures as a standard for

TABLE I

THERMODYNAMIC CONSTANTS AT 293°K

Materials	Density (g/cm³)	Thermal expansion (linear) (/°K)	Specific heat (C_p) (ergs/g°K)	Ultrasonic sound velocities c_l (km/sec)	c_t (km/sec)	c_b (km/sec)	γ
2024 Al	2.785	21.8×10^{-6}	0.879×10^{7}	6.39	3.15	5.25	2.00
921-T Al	2.833	22.9	0.877	6.34	3.14	5.20	2.10
Copper	8.93	16.3	0.386	4.76	2.33	3.93	1.96
Iron	7.85	11.7	0.448	5.94	3.26	4.59	1.69
U–Mo alloy	18.45	12.8	0.121	3.31	1.85	2.53	2.03

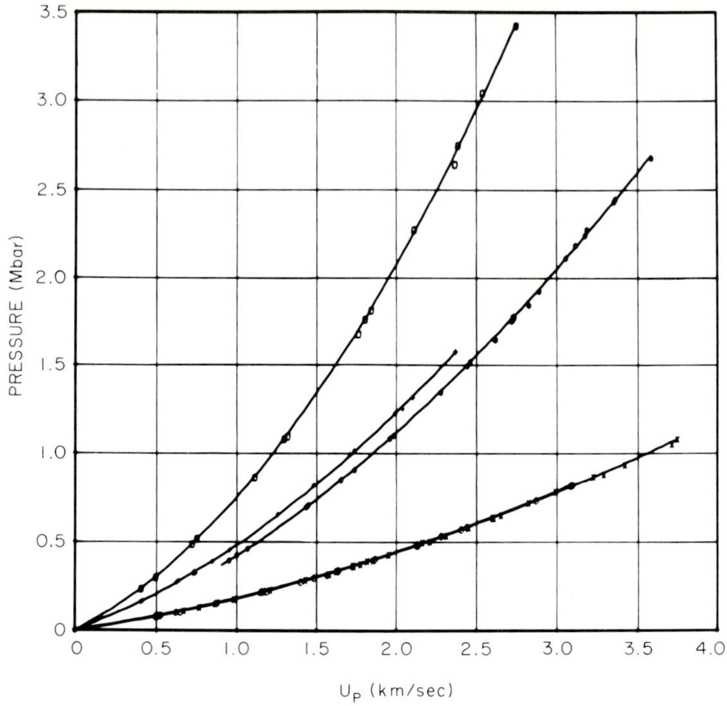

FIG. 10. Hugoniot P–u_p data from shock particle velocity experiments. (×) 2024 Al, (○) 921T Al, (*) Fe, (+) Cu, (□) U–3%-Mo alloy.

materials of lower shock strength if its initial pressure is this large. On compression, the iron transition occurs at about 0.13 Mbar, but little is known about the reverse transition on the release wave. These data and some derived quantities are listed in Appendix A at the end of the volume. The least-squares fits of the u_s–u_p data are given in

TABLE II

COEFFICIENTS OF THE HUGONIOT EQUATION

$$u_s = c_0 + su_p + qu_p^2$$

Material	c_0 (km/sec)	s	q (sec/km)
2024 Al	5.328	1.338	0
921-T Al	5.041	1.420	0
Copper	3.940	1.489	0
Iron	3.574	1.920	—0.068
U–Mo alloy	2.565	1.531	0

Table II. Linear fits of the data were adequate for all materials except iron, where a small amount of curvature or a slight break in the curve can be observed. This lack of linearity has been observed in other materials that have undergone a phase change. In iron the curvature or change in slope could be due to additional phase changes, either the hcp $\rightarrow$ fcc transition, the solid $\rightarrow$ liquid transition, or both. These techniques are not capable of resolving this problem. The zero-pressure sound velocity data measured here, Table I, and reported elsewhere (*20*) were not used in the least-squares fits.

C. The Effect of Rigidity on the Equation of State Measurements

A detailed description of the mechanisms involved in elastic-plastic flow is given in Section VI; however, a brief discussion of the effects of rigidity on plane wave propagation is required in order to explain some of the results described in this section. In plane one-dimensional compressive flow, there is a macroscopic deformation in one coordinate only; hence, the pressure and volume are initially related by a one-dimensional (1-D) equation of state. However, as the pressure increases the deviatoric shear stress increases, which causes the material to relax towards what has been called the three-dimensional (3-D) equilibrium Hugoniot. The hypothetical behavior of a plane shock and rarefaction wave with rigidity present is sketched in Fig. 11. The upper curve in Fig. 11a represents a one-dimensional Hugoniot and could be used to find the instantaneous longitudinal elastic wave velocity through Eq. (33). The lowest curve represents the transverse wave velocity due to the shear modulus along the Hugoniot as a function of particle velocity. It has been established through static experiments that the shear modulus increases with pressure so it has been drawn with a positive slope. However, since the temperature along the Hugoniot increases in an exponential manner, this curve has been drawn with a downward curvature. If the material melts on this Hugoniot this curve will go to zero and the longitudinal and bulk velocity curves will coincide. No effort has been made in this figure to satisfy Eq. (38).

The remaining figures correspond to the lower-pressure region of (a) where rigidity effects will be most pronounced. In Fig. 11b and in the remaining figures, HEL represents the Hugoniot elastic limit on compression and REL the elastic limit on the release wave. If the material behaves as indicated (that is, if it does not relax to the equilibrium curve), then all states on the Hugoniot will be slightly stiffer. This means that measured shock velocities will be higher than

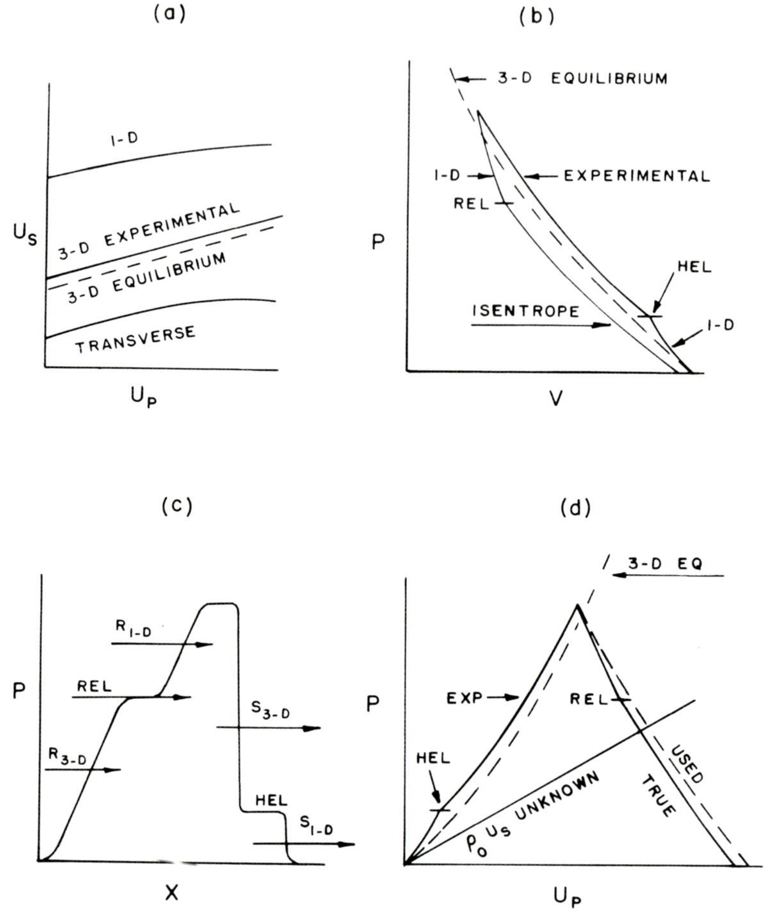

FIG. 11. The effect of rigidity on the Hugoniot curves. (a) Possible curves for the various wave velocity loci as a function of particle velocity. (b) Relationship of the experimental Hugoniot and true release isentrope with the equilibrium Hugoniot. (c) Type of pressure profile curve that would result from the collision of a finite driver plate with a target plate of the same material if its shock wave equation of state were given by (b). (d) The P–u_p transform of (b). In (b)–(d) the relative amplitudes of the various waves are the same.

that obtained if a truly hydrodynamic shock wave were produced. If such a situation exists, then the extrapolated zero-pressure intercept of the u_s–u_p Hugoniot should lie above the measured bulk sound velocity. This is indeed the case for 2024 Al, Cu, and the uranium alloy, where the extrapolated values are, respectively, 2.7, 0.5, and 1.6% too high. These results are not unexpected and are compatible with free surface velocity profiles (Figs. 12 and 13) measured here on 2024 Al and Cu.

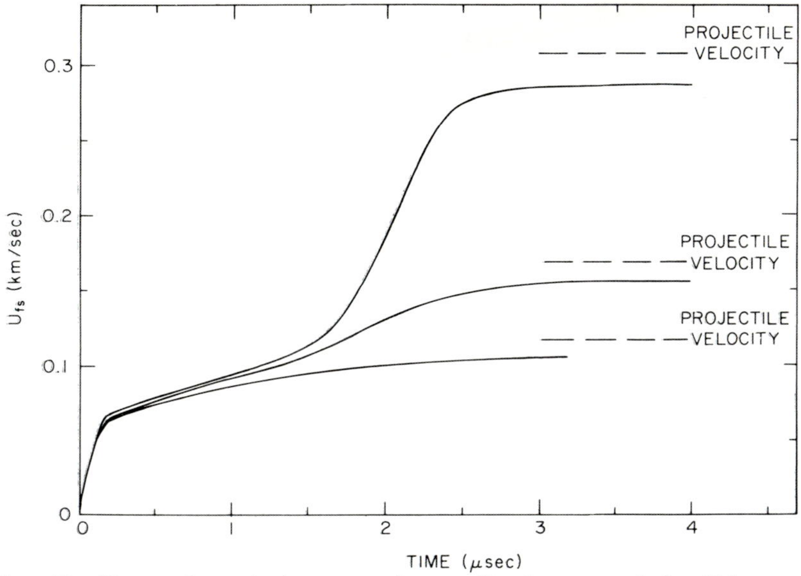

FIG. 12. Free-surface velocity versus time profiles in symmetrical collision cannon experiments on 2024 Al.

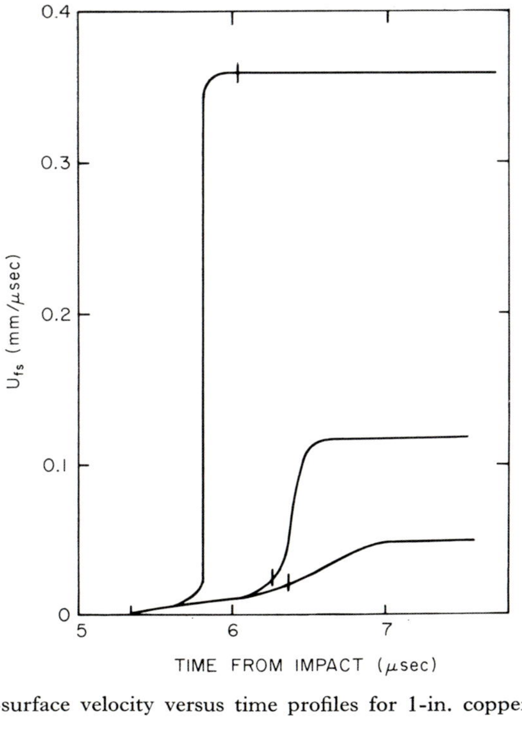

FIG. 13. Free-surface velocity versus time profiles for 1-in. copper target plates.

The experimental curves in Fig. 12 give the velocity–time history of the free surface motion of $\frac{1}{2}$-in. target plates impacted at various velocities. The projectile velocities are indicated on the figure. The actual shock wave profiles will be quite similar to these curves. Here zero time is related to first motion.

In Fig. 13 the time axis represents the total time from impact. The vertical mark on the time axis represents the time for the zero-pressure longitudinal wave to traverse the 1-in. target plates. The vertical marks on the curves represent the ideal shock transit time and were computed from the measured particle velocity and the linear u_s–u_p fit of the shock wave data. There does not appear to be any trend in where these appear on the curves.

The Hugoniot elastic limit can be seen to be about 5.5 kbar for the Al but is practically nonexistent for the Cu. If one assumes that there exists a residual stress of about 2 kbar in the 2024 Al Hugoniot, which should be removed to obtain the equilibrium Hugoniot, the resulting curve goes nicely through the computed bulk sound velocity.

If the material does not reach equilibrium in the shock wave, then it is clear that a similar behavior will occur on the pressure release wave, as is indicated in Fig. 11. Because of the hysteresis in the P–V curve, the elastic precursor release wave can be expected to be larger than the shock precursor (Fig. 11c). In the P–u_p plane (Fig. 11d) it can be seen that this results in a smaller free surface velocity than would be expected if it were assumed that the material was behaving as an ideal fluid. From Fig. 12, it can be seen that the free-surface velocities are less than the projectile velocity for these low-pressure, symmetrical-collision cannon shots. Ideally, u_{fs} and u_d should be equal in the low-pressure regime. A similar effect has been found in 2024 Al using high-explosive driver systems where the driver-plate velocity and resulting free surface velocity were measured on the same shot. It was found that for driver velocities up to 3 km/sec, the free surface velocity was still about 1% less than the driver velocity. On the basis of the sound velocities, the small amplitude of the elastic precursor, and the very steep wave profile observed even at very small amplitude, it has been concluded that Cu can be treated as a hydrodynamic material for present applications.

The measured Cu free surface velocities (Fig. 13) are identical to the projectile velocity except for the intermediate velocity curve (30 kbar), where there is a free surface velocity deficit of about 10%. This has been tentatively associated with the onset of twinning in high-pressure static experiments (*21*) at this pressure, the velocity deficit being attributed to the increase in stacking faults and the resulting increase in stored energy. With stronger shocks this effect becomes less important.

D. Experimental Studies on Porous Samples

By using porous samples it is possible to extend the equation of state to regions of higher internal energy and temperature at a particular volume than can be obtained using samples of crystal density. Such data can be used to find the average value of the Grüneisen parameter at a given density.

To obtain shock wave data on porous materials and on the other materials reported later, the impedance match technique developed by Walsh and Christian (*18*) and the flash-gap technique described earlier has been used. Through this technique, measurement of a single shock wave parameter in a material whose equation of state is known (the standard) suffices to determine a point on the Hugoniot of other materials by measuring only the shock velocity through relatively small samples placed on a plate made of the standard material. The success of the method lies in the fact that in shock wave hydrodynamics the measurement of any two equation of state parameters is sufficient to determine all the others through the Rankine–Hugoniot equations or their modifications, and that the pressure and particle velocity must be the same across an interface. The continuity requirement suggests a P–u_p representation for the equation of state, as is shown in Fig. 14. Here the standard, which has been shocked to a pressure P_1, interacts with the specimen either by a pressure-release wave (rarefaction) or by a reflected shock that increases the pressure. Since for any single shock Eq. (2) defines a straight line in the P–u_p plane with slope $\rho_0 u_s$, the intersection of this line with the calculated cross curve of the standard determines the shock pressure and particle velocity of the specimen of interest. Shock velocity measurements of small samples on the standard establish the pressure of the standard and the $\rho_0 u_s$ lines of the unknowns. The intersection of these lines and the calculated reflected shock curve or rarefaction release curve determines the particle velocity and pressure in the unknowns. The experimental arrangement is described in Fig. 15. The shock transit times are measured through each of the eight samples (crosshatched) placed on the base plate, two of which are used to establish the pressure. Plastic flash-gap blocks are placed on these and on the base plate to record the shock arrival. A multiple-slit system similar to that used to obtain the record shown in Fig. 8 was used.

There is a correction which must be applied to the shock velocity measurements made with the flash-gap technique because of the finite size of the flash gaps. It has been found through other experiments that the light recorded on the photographic film is produced just before the gaps are closed. Hence the difference in gap closing time due to

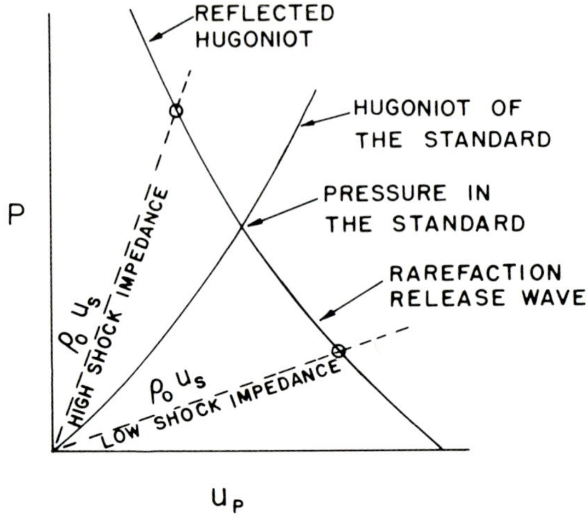

FIG. 14. Schematic for obtaining Hugoniot data by the impedance match technique.

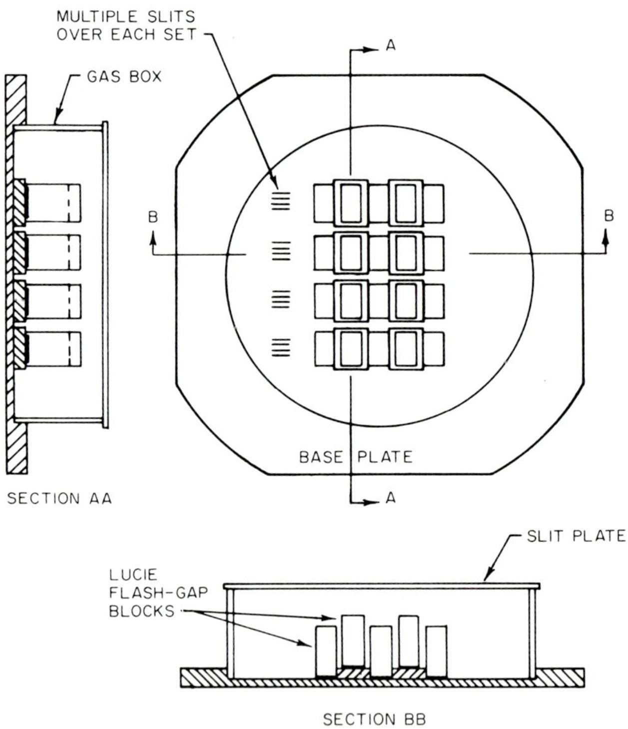

FIG. 15. Details of the assembly used to obtain shock wave data with the impedance match technique.

the difference between the free surface velocity of the base plate material and the individual slugs must be removed to obtain the correct shock transit time. To minimize the size of this correction, shims made of the base plate material are placed under all the flash-gap blocks. This also gives more uniform traces. In addition to this precaution, the actual transit times through the shims and gaps are calculated. This is done by first using the uncorrected shock transit time to determine a trial equation of state point. By using an equation of state based on this point, the shock interaction of the slug with the shim is calculated, thus giving a correction to the observed transit time due to the difference in the shock velocity and free surface velocity of the shims on the base plate and on the various other materials. The procedure is repeated until the corrections do not change. In these experiments gaps about 0.01 cm thick and shims about 0.02 cm thick have been used. Without these corrections systematic errors in shock velocities in excess of 1% would be present.

It is unfortunate that shims of the unknown material cannot be used for the reference traces, since this in principle would automatically take care of these corrections. Iron shims have been used under all flash gaps with copper and uranium-alloy base plates. The required impedance match corrections have been made accordingly. This was done because high-quality steel feeler gage stock is readily available, and also because it was found that the photographic records were considerably improved when steel shims were used.

The Grüneisen parameter can be found directly from these experiments through the generalized equation of state formula, Eq. (15), and the Hugoniot relationships (see Fig. 16).

It would be possible to choose some form for the Grüneisen parameter and by appropriate minimization find the one that would best reproduce the experimental data. However, we have chosen simply to fit the experimental u_s–u_p data with a quadratic which is then used to find the pressure and energy offsets from the fitted crystal-density Hugoniot. The $\partial E/\partial P$'s derived from these fits are plotted in Figs. 17–19. In addition to these curves, the $\partial E/\partial P$'s computed from the crystal-density Hugoniot and the experimental data points are also plotted in these figures. The large scatter in these points reflects the scatter in the experimental data. The shock wave data entertain larger than usual scatter (especially at low density), partially because of larger errors in measuring the initial densities of the samples.

In Figs. 17–19, each data point was computed from the shock wave data for each porous sample and the fit of crystal density data given in Table II. The curves were computed from least-squares fits of the u_s–u_p data for each density group. Samples with lower initial densities

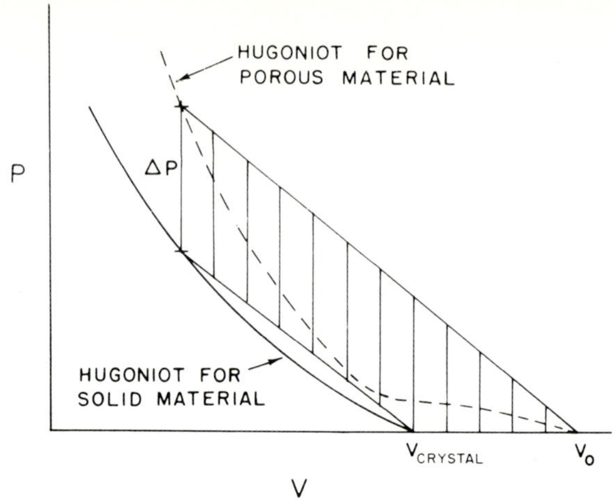

FIG. 16. Schematic to illustrate how porous materials can be used to measure the Grüneisen parameter. The difference in internal energy between the two Hugoniots at V is indicated by the crosshatched area. The corresponding pressure difference is also indicated.

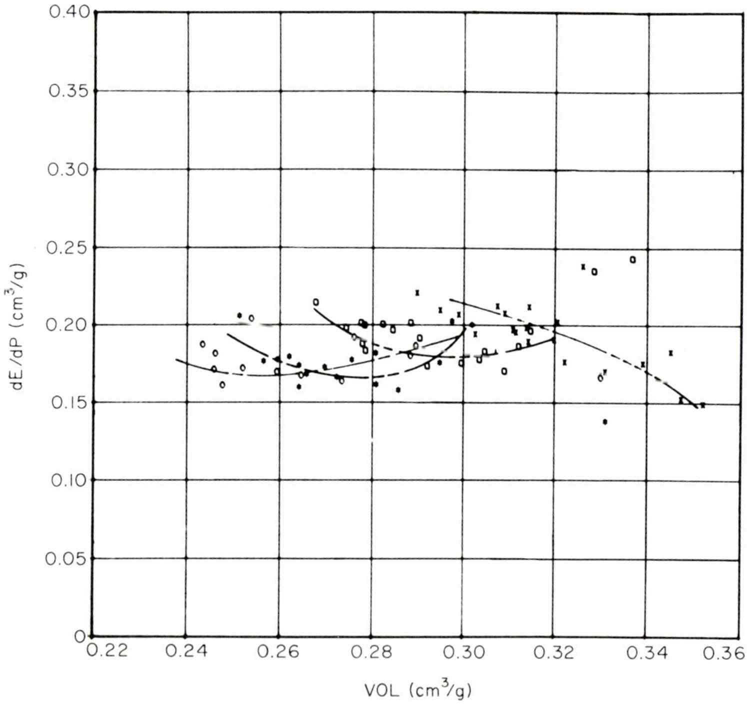

FIG. 17. Experimental values for $dE/dP)_V$ versus volume determined from shock wave studies on porous 2024 Al of fractional crystal densities: (○) 0.91, — - —; (∗) 0.80, — -- —; (□) 0.70, — --- —; (×) 0.59, — ---- —.

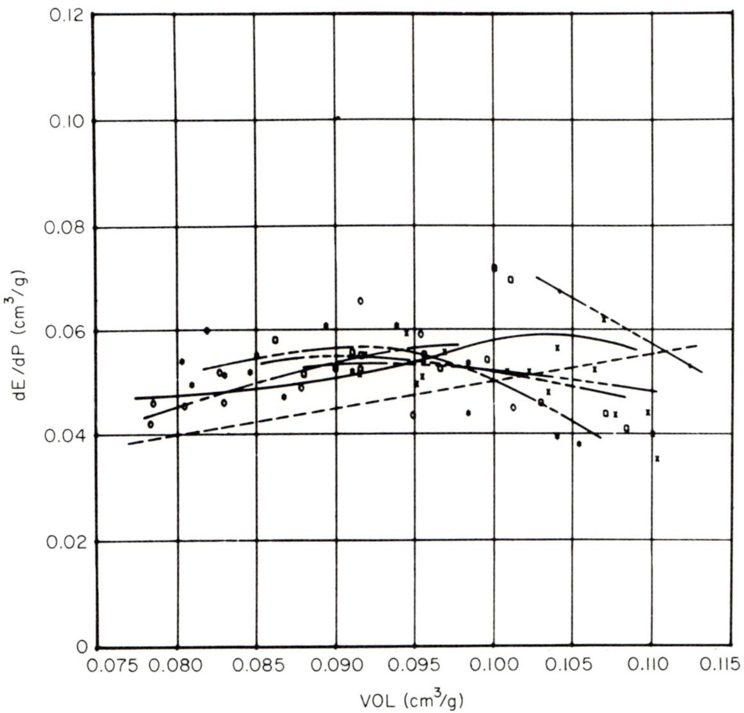

FIG. 18. Experimental values for $dE/dP)_V$ versus volume determined from shock wave studies on porous Cu of fractional crystal densities: (○) 0.88, — - —; (*) 0.82, — -- —; (□) 0.71, — --- —; (×) 0.64, — ---- —; (+) 0.51, — ----- —.

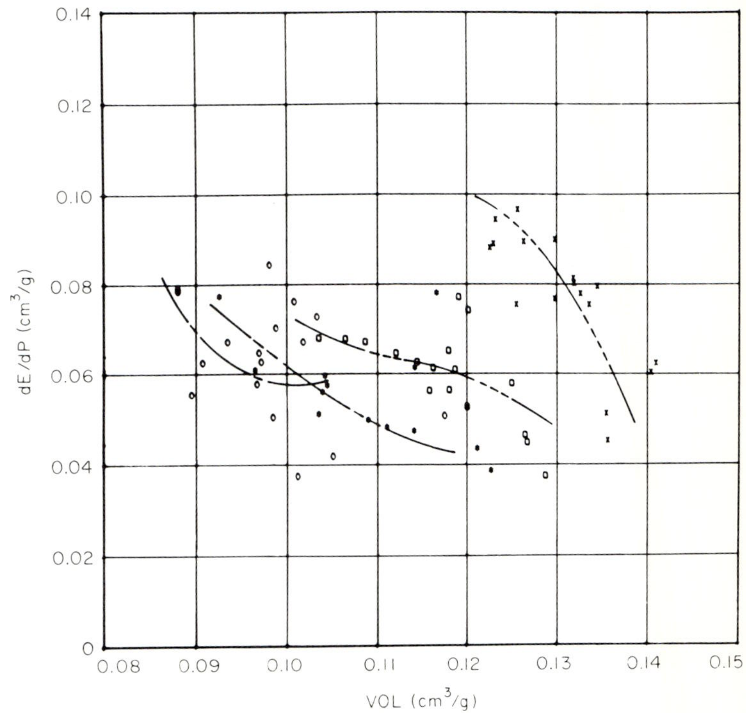

FIG. 19. Experimental values for $dE/dP)_V$ versus volume determined from shock wave studies on porous Fe of fractional crystal densities: (○) 0.89, — - —; (*) 0.76, — -- —; (□) 0.60, — --- —; (×) 0.43, — ---- —.

represent average values of $dE/dP)_V$ measured over larger temperature ranges at a given volume. Within the experimental accuracy there does not appear to be any measurable variation of γ with temperature. The general appearance of the data gives support to using the simple relationship $\rho\gamma$ constant.

In addition to the experimental curves in Fig. 18, the values of $dE/dP)_V$ vs. V obtained from the Dugdale–MacDonald relationship (solid line) and γ constant (dashed line) are also plotted. It can be seen that the data agree quite well with the Dugdale–MacDonald calculation. At small volumes experimental data lie well above the γ constant line, which is known to be a poor representation of the volume dependence of γ because it predicts vanishing sound speeds at large compressions, Eq. (33).

The data in Fig. 19 could indicate an increasing $dE/dP)_V$ (decreasing γ) with increasing temperature. This interpretation, although in accord with theoretical expectation, is somewhat dubious because of the large scatter in the data.

Since differences in pressure and energy are important, the precision requirements to measure γ accurately become quite stringent. An alternative method is to compute the Hugoniot of the porous material from the crystal-density Hugoniot using some smooth volume dependence for γ. The equation

$$P = P_{\mathrm{H}}[1 - (\rho\gamma)(V_0 - V)/2]/[1 - (\rho\gamma)(V_0{}^* - V)/2] \tag{39}$$

gives the pressure as a function of volume for a porous material centered at a volume $V_0{}^*$ in terms of its crystal-density Hugoniot. Equations (4) and (5) can be used to transpose this Hugoniot to the shock particle velocity plane.

The experimental u_s–u_p data and derived P–ρ data are given in Appendix B at the end of the volume and in Figs. 20–25 for porous 2024 aluminum, copper, and iron. The crystal-density Hugoniots are plotted on these graphs with several calculated Hugoniots centered at various initial densities. These Hugoniots were calculated from the crystal-density Hugoniot using $\rho\gamma$ constant and the average initial density of each group of data points as the centering point. It has been assumed that the initial internal energy of porous materials is the same as the internal energy of the corresponding crystal-density material. The good agreement of the calculated Hugoniots attest to the adequacy of using this simple function for the Grüneisen parameter. Scatter in the data is partially due to variations in density.

The curves in Fig. 20 were computed from the least-square fit of the crystal-density Hugoniot and a constant value of $dE/dP)_V$, obtained

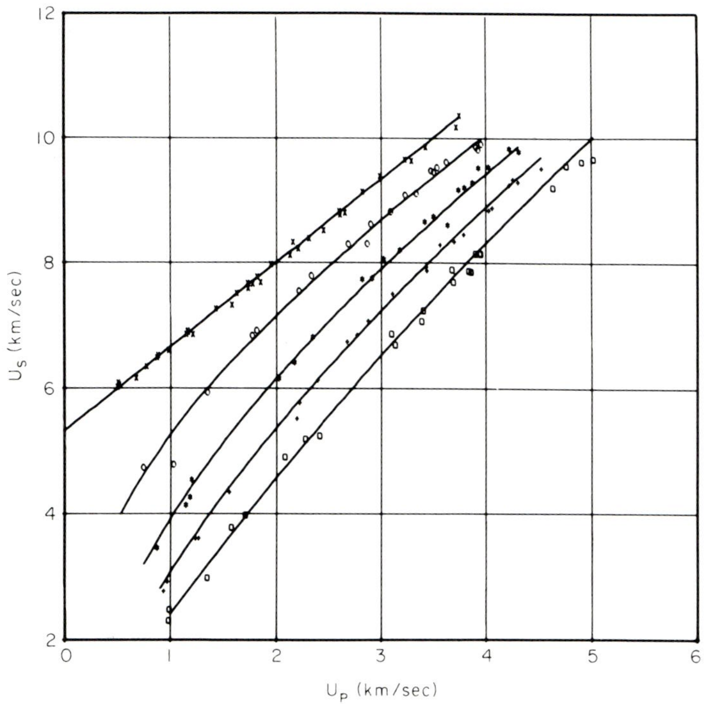

FIG. 20. Shock particle velocity data for porous 2024 Al of fractional crystal densities: (×) 1.00, (○) 0.91, (∗) 0.80, (+) 0.70, (□) 0.59.

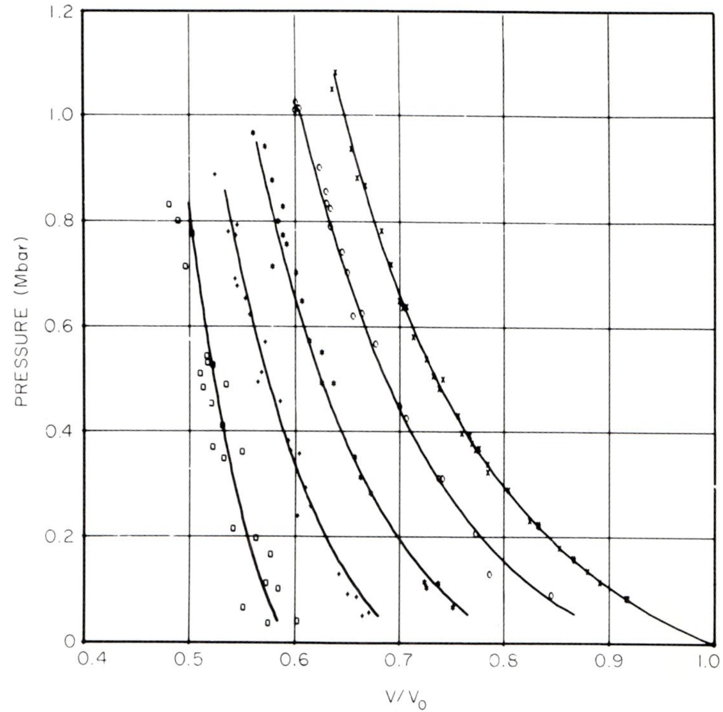

FIG. 21. Pressure versus relative volume for porous 2024 Al of fractional crystal densities: (×) 1.00, (○) 0.91, (∗) 0.80, (+) 0.70, (□) 0.59.

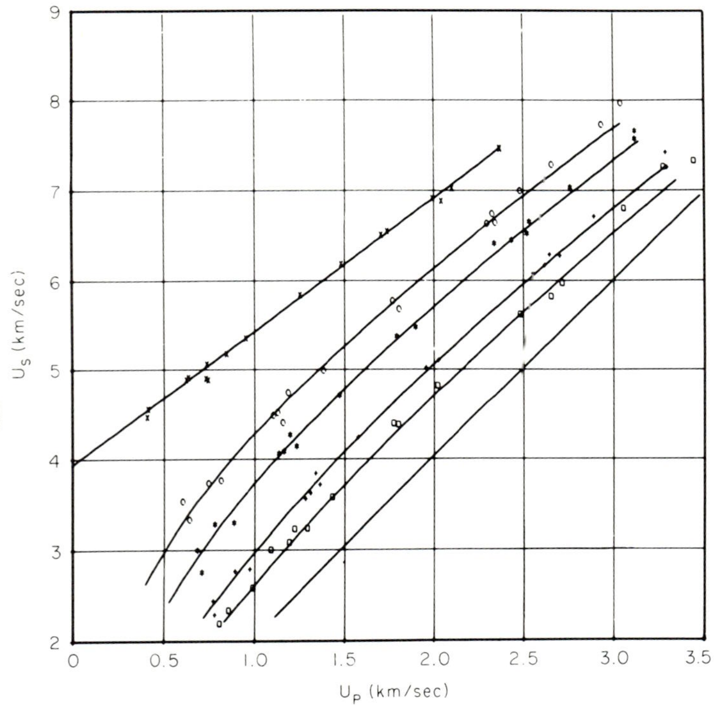

FIG. 22. Shock particle velocity data for porous Cu of fractional crystal densities: (×) 1.00, (○) 0.88, (∗) 0.82, (+) 0.71, (□) 0.64, (·) 0.51.

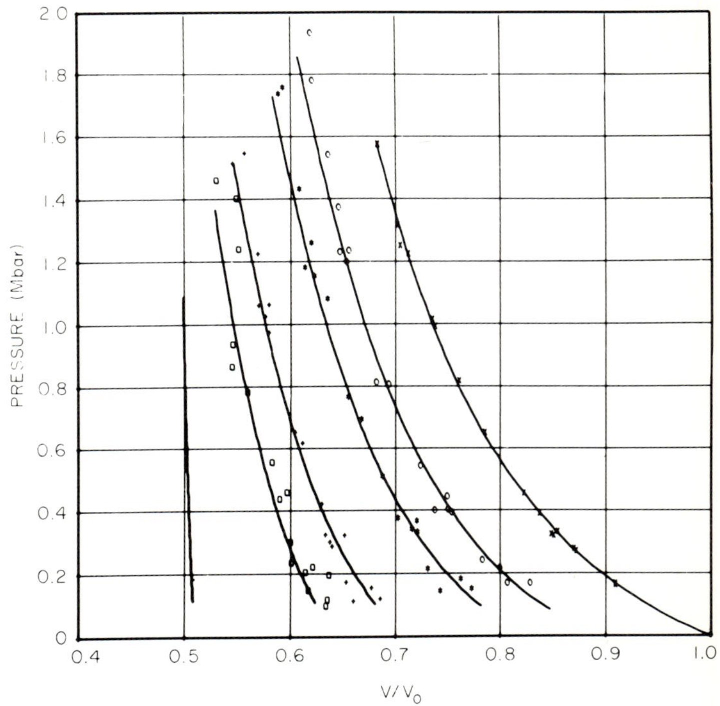

FIG. 23. Pressure versus relative volume data for porous Cu of fractional crystal densities: (×) 1.00, (○) 0.88, (∗) 0.82, (+) 0.71, (□) 0.64, (·) 0.51.

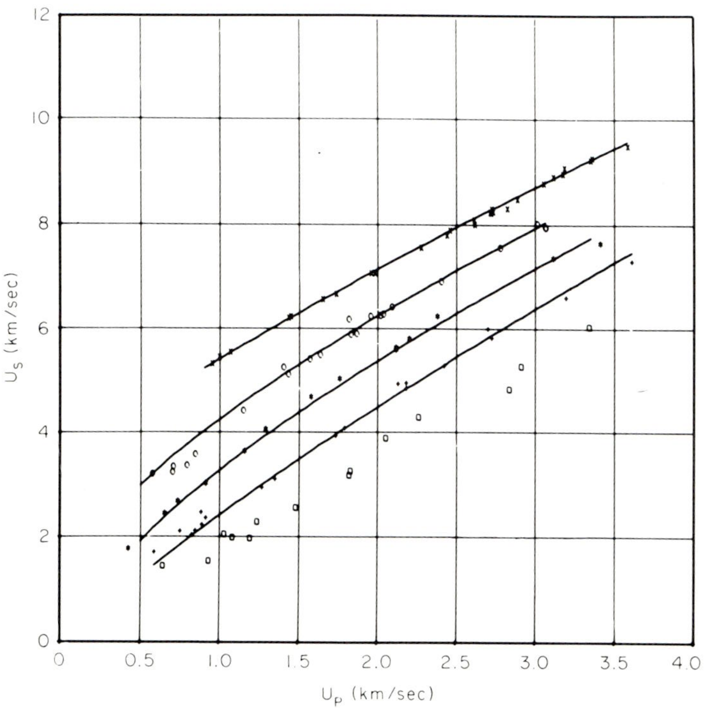

FIG. 24. Shock particle velocity data for porous Fe of fractional crystal densities: (×) 1.00, (○) 0.89, (∗) 0.76, (+) 0.60, (□) 0.43.

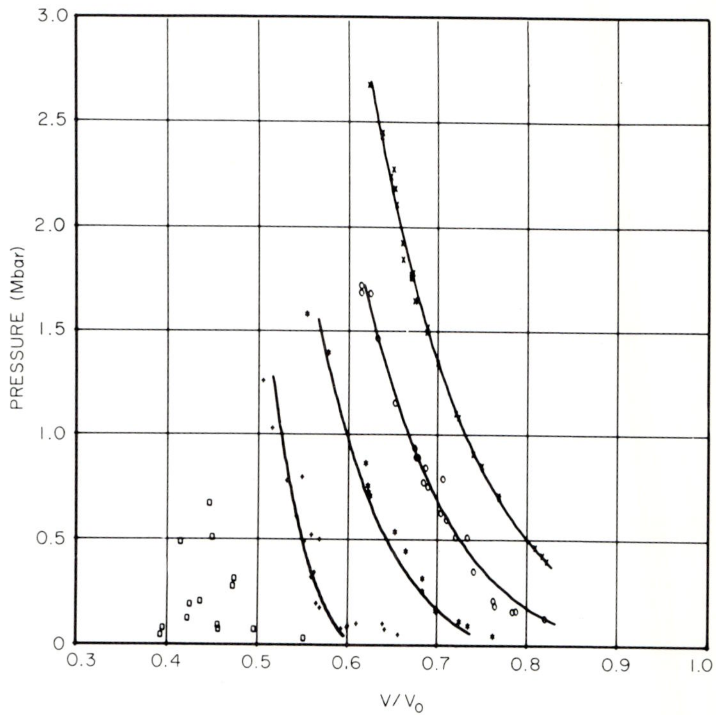

FIG. 25. Pressure versus relative volume data for porous Fe of fractional crystal densities: (×) 1.00, (○) 0.89, (∗) 0.76, (+) 0.60, (□) 0.43. No attempt has been made to fit the very low initial density data. It is possible that some of the low-pressure data is being influenced by the 130-kbar phase change.

from the γ and ρ in Table I, for the average initial density of each group of samples. The data were also corrected for deviations in initial density. In Figs. 20–22 it can be noted that the calculated curves originate at $u_s = u_p = 0$. Thus the higher-density curves have considerably more curvature in the low-pressure region. The low initial density curves become almost linear as they approach the initial density required for the vertical P–V asymptote.

In Figs. 21, 23, and 25 the curves are the P–V transforms of the u_s–u_p fits in the previous figure. The scatter in the data due to variation in density is more pronounced in these plots. V_0 in the figures refer to the average density of each group of samples. Although the compression is large for these low initial density samples, the final densities are not. The approach to the asymptotic density is quite apparent.

On occasion the final density of the shocked porous material does not reach crystal density. Such data points are not applicable in the present determination of γ and are quite probably not true indications of the equilibrium equation of state in the low-pressure regime. High-precision measurements, at low pressure, of the free-surface velocity of porous materials (Fig. 26) show that the wave shape is not constant in time and is generally characterized by a long smeared-out shock front. Such

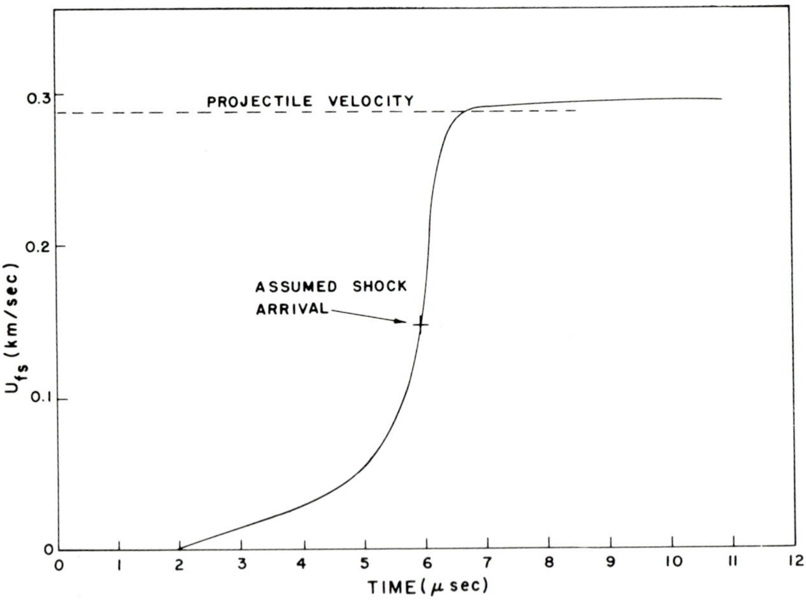

FIG. 26. Free surface velocity versus time profile for shock-loaded PT-0178 graphite. This material has an initial density of 1.55 g/cm³.

behavior is readily explained by the effect rigidity plays in the collapse of the voids. In this regime thermodynamic equilibrium does not exist and one is in essence observing some type of rheological equation of state. The higher the pressure in the shock front, the more the voids will collapse. If the wave front spreads out in time, then a psuedo equation of state, concave downward in the P–V plane as drawn in Fig. 16, is indicated. Flash-gap techniques applied to such systems are almost worthless since the part of the wave profile effective in closing the gap is not known. In addition to the above considerations, the deformation undoubtedly entertains thermal relaxation effects which further complicate the analysis.

The long precursor in Fig. 26 is thought to be due to the collapse of the voids. This experiment indicates that with dynamic loading the voids are effectively collapsed at a pressure of about 5 kbars. Although the pressure is low the long (2 μsec) precursor makes flash-gap experiments of dubious value in the low-pressure regime. It is noteworthy that the toe of the precursor arrives at a time commensurate with the measured longitudinal wave velocity.

There is a remarkable amount of similarity in the appearance of the experimental data obtained on these porous materials. This is partially a reflection of the fact that the Grüneisen parameter can be adequately characterized by the simple relationship given by Eq. (17). In particular all the u_s–u_p curves have a zero intercept at $P = 0$ and $\rho = \rho_{0\text{crystal}}$ if extrapolated by Eq. (39). A particular case of interest is the Hugoniot of a porous material that asymptotically approaches crystal density. It has long been recognized that where dP/dV_{H} becomes infinite on the Hugoniot, the Grüneisen parameter is given by

$$\gamma = 2V/(V_0 - V) \tag{40}$$

For the linear u_s–u_p relationship it can be seen from Eq. (13) that the volume V_a where this occurs is given by

$$V_a/V_0 = (s - 1)/s \qquad (P = \infty) \tag{41}$$

However, any place where the slope s_a of the u_s–u_p Hugoniot satisfies the relationship

$$s_a = u_s/u_p \tag{42}$$

dP/dV_{H} becomes infinite and Eqs. (40) and (41) are applicable. This is just a manifestation of the fact that in the u_s–u_p plane isovolume curves are given by straight lines emanating from the origin. If γ is known, then Eqs. (40) and (41) can be rewritten to give the asymptotic value for the volume

$$V_a/V_0 = \gamma/(\gamma + 2) \tag{43}$$

and slope

$$s_a = (\gamma + 2)/2 \tag{44}$$

In particular if γ is indeed independent of temperature and pressure then in principle there should exist a Hugoniot whose asymptote occurs at crystal density ($V_a = V_{0\text{crystal}}$). Its initial density or porosity should be given by Eq. (43). Since it starts at $P = 0$ it should be linear in the u_s–u_p plane with $C_0 = 0$ and the slope given by Eq. (44). Using the values of γ_0 given in Table I the required values of the porosity for this to occur are 0.50, 0.49, and 0.46 for 2024 aluminum, copper, and iron, respectively, with corresponding values of the slopes of 2.00, 1.98, and 1.85, respectively.

In spite of the large amount of scatter in the high-porosity data, due to a large extent to density variations, it is easy to see in the figures that the data is compatible with this simple model, especially in the u_s–u_p plots where the data can be smoothed in a reasonable manner.

E. Cross Checking the Standards

In the previous sections the Hugoniots of five different materials were obtained by direct determination of their shock particle velocity loci. In addition, the high-pressure values of the Grüneisen parameter were obtained for three of these. However, before using these materials as standards for other materials, it is desirable that they should be tested against each other by the impedance match technique, using each in turn as a standard to remeasure the Hugoniot of all the others. The data obtained, using the same shot assembly as described for the porous materials, are given in Appendix C at the end of the volume and are plotted in Figs. 27–31. The least-squares fits of the shock particle velocity data are also shown in these figures. In these figures one material was used as a standard to obtain Hugoniot data for the other materials with the impedance match technique. The overall agreement is quite satisfactory and shows that all the standards satisfy this necessary condition with the possible exception of the 921-T aluminum alloy.

The new Hugoniot data points for 921-T aluminum obtained by using the other materials as standards lie consistently above the Hugoniot obtained from the u_s–u_p measurements. The disagreement is due to the fact that the original data were obtained using plates in which a small but nontrivial amount of porosity was present, while the small samples

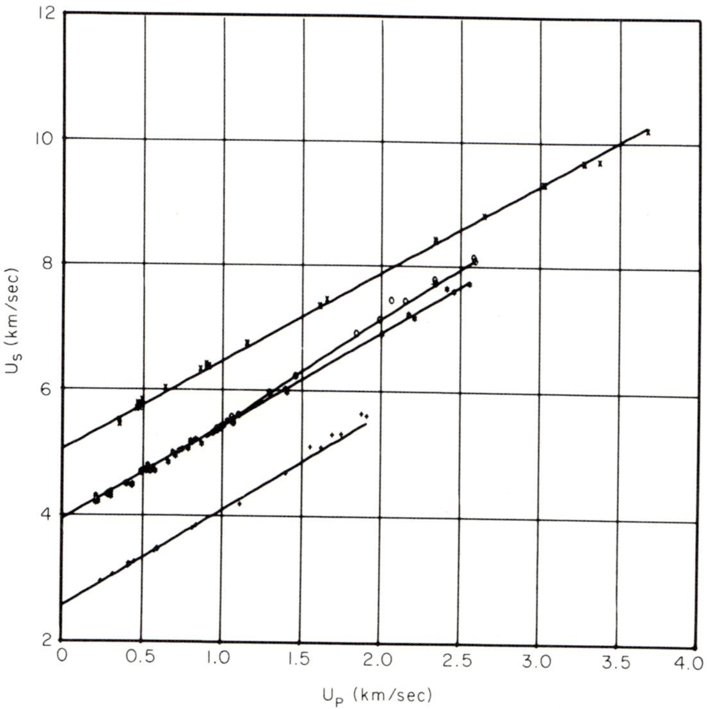

FIG. 27. Shock particle velocity data for (×) 921-T Al, (∗) Cu, (○) Fe, and (+) U–3%-Mo alloy obtained using 2024 Al base plates as a standard.

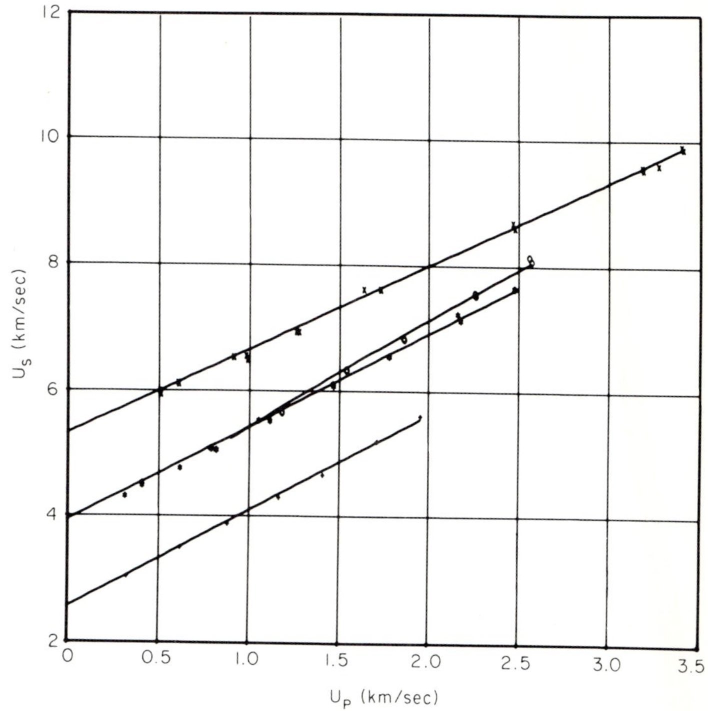

FIG. 28. Shock particle velocity data for the standards obtained by the impedance match technique using 921-T Al base plates as the standard. (×) 2024 Al, (○) Fe, (∗) Cu, (+) U–3%-Mo alloy.

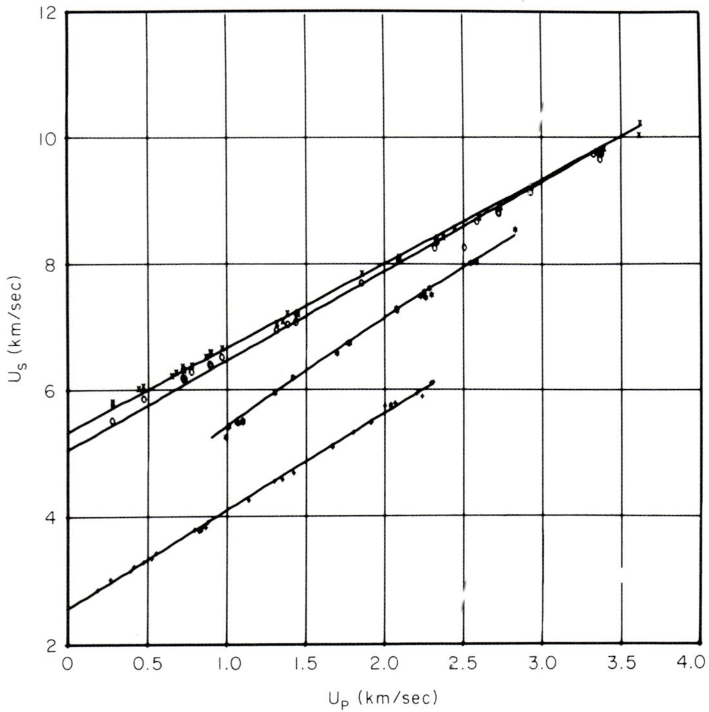

FIG. 29. Shock particle velocity data for the standards obtained by the impedance match technique using Cu base plates as the standard. (×) 2024 Al, (○) 921T Al, (∗) Fe, (+) U–3%-Mo alloy.

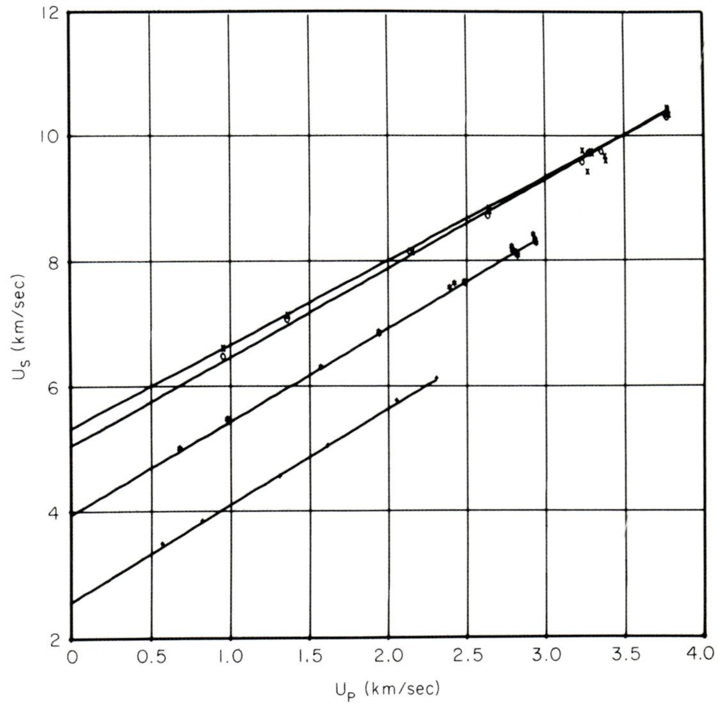

FIG. 30. Shock particle velocity data for the standards obtained by the impedance match technique using Fe base plates as the standard. (×) 2024 Al, (○) 921T Al, (∗) Cu, (+) U–3%-Mo alloy.

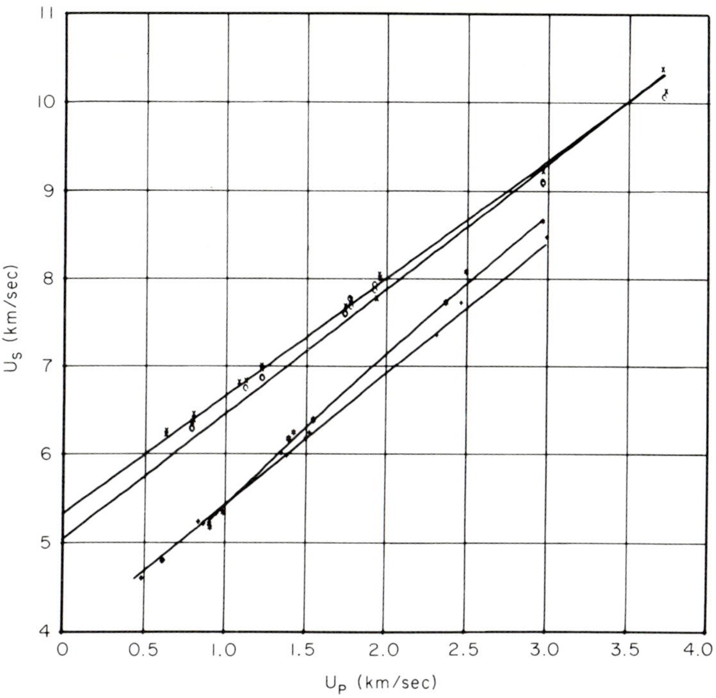

FIG. 31. Shock particle velocity data for the standards obtained by the impedance match technique using U–3%-Mo alloy base plates as the standard. (×) 2024 Al, (○) 921T Al, (∗) Fe, (+) Cu.

used in the cross checks were nearer the ideal density. The disagreement seen in the data obtained with the 921-T base plates (Fig. 29) is due to a similar difficulty. The base plates used in these experiments were slightly low in density but the slugs used to establish the pressure in the base plates were of high density. This results in assigning too high a pressure to the base plate and, subsequently in the impedance match solution, too large a particle velocity for the other materials. This effect can be taken into account but not without a considerable revision of our impedance match codes. This effect is most pronounced in the low-pressure region. The results in the high-pressure region are quite acceptable; in fact, the agreement over the entire pressure range is good for the high shock impedance uranium alloy.

The results of the comparisons do not preclude the possibility that the equation of state of all these materials contain some kind of systematic error. The effect of rigidity on decreasing the free surface velocity has already been mentioned. This effect must necessarily be present when these materials are used as standards for materials of lower shock impedance. It is not obvious exactly how a residual deviatoric stress will affect the standards when used in the reflected shock mode since the actual state of the material is not known. It is hoped that the increase in the shear modulus due to the increase in pressure is compensated by the increase in temperature also encountered in strong shock waves.

F. Free-Surface Velocity Measurements on 2024 Aluminum and Copper

Shock-velocity free-surface velocity measurements have been made on 2024 aluminum and copper. This type of measurement was used to obtain the Hugoniot loci originally used for our standards. At that time the particle velocity was found by an iterative procedure in which calculated free surface velocities were made to agree with experimental free surface velocities by varying the particle velocities. In these calculations the energy at the foot of the rarefaction release wave was

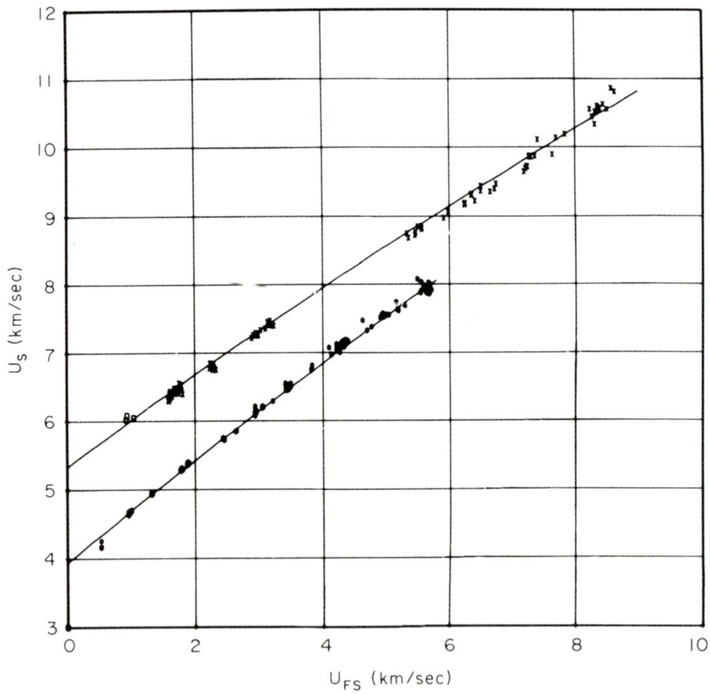

FIG. 32. Shock–free surface velocity data for (×) 2024 Al, (∗) Cu, and (□) 2024 Al (pin data).

made to agree with an internal energy versus density locus along the zero-pressure isobar determined from specific heat and thermal expansion data. The results of such a procedure must naturally reflect errors associated with the zero-pressure data and with improper handling of the isentropes which were known to cross through the melting phase region. Fortunately these errors were small and reasonably good Hugoniots were obtained. With the Hugoniots determined by more direct methods, it is no longer necessary to use the free surface velocity data as primary data. However, this provides an additional check on our calculational procedures and experimental measurements. In Fig. 32 the free-surface velocities calculated from shock-velocity particle-velocity Hugoniots are compared with the experimental data. These data were obtained by flash-gap techniques quite similar to those described earlier. The solid curves were obtained by evaluating the Riemann integral along the isentropes calculated from the shock particle velocity Hugoniots (Table II). The results for copper are quite good, while 2024 aluminum contains regions where the agreement is quite poor. It is felt that these are, in the main, real deviations and reflect phase changes, both melting and vaporization that are inadequately handled in the calculated rarefaction release isentropes. These were observed in earlier work (*3*) and at that time a fourth-order fit of the u_s–u_p Hugoniot was required to fit the free-surface velocity data. This was not thought justified because of known uncertainties in the calculations, and consequently a linear fit was used. Such a linear fit has subsequently, of course, been justified. This illustrates the advantage of using shock velocity measurements to monitor the pressure in the standard.

IV. The Equation of State of Solids

In this section are assembled Hugoniot data, obtained by the impedance match technique with the standards established in Section III, on a number of materials of interest to those working in the field of high-velocity impact. These materials include lightweight metals, refractory elements, precious metals, selected compounds and alloys, and a wide variety of plastics and other synthetics representative of those commonly in use today. The data are presented in the form of u_s–u_p and either P–ρ or P–V/V_0 plots, whichever is more appropriate. For those who may require more detailed information, tables of the experimental data together with a summary of the most important thermodynamic quantities derived from these data are included at the end of the section.

Also included is a summary (Table III) of the high-explosive systems commonly used at the Los Alamos Scientific Laboratory to generate the wide range of base plate pressures necessary for adequate determination of Hugoniots, especially where phase transitions exist. Although many other systems have been used, this list has the virtue of generating

TABLE III

PRESSURE OF HIGH-EXPLOSIVE SYSTEMS

Pressure in base plate (Mbar)		Lens diameter (in.)	Thickness (in.)			Gas
2024 Al	Copper		High explosive	Driver	Base plate (max)	
In-contact system:						
0.08	0.11	12	8 Baratol (0.5 Hg, 0.5 acrylic)		0.5	Xe
0.11	0.15	12	8 TNT (0.5 Hg, 0.5 acrylic)		0.5	Xe
0.17	0.24	12	8 Baratol		0.5	Xe
0.24	0.33	8	8 TNT		0.5	Xe
0.36	0.50	8	8 CB		0.5	A, Xe
0.40	0.56	8	9 Octol		0.5	A, Xe
Free-run system:						
0.45	0.81	12	4 CB	0.250	0.5	A, Xe
0.50	0.93	12	4 CB	0.187	0.3	A, Xe
0.57	1.09	12	6 CB	0.187	0.3	A
0.60	1.13	12	8 CB	0.187	0.3	A
0.65	1.23	12	4 CB	0.125	0.2	A
0.70	1.33	12	6 CB	0.125	0.2	A
0.75	1.43	12	8 CB	0.125	0.2	A
0.80	1.52	12	6 Cyclotol	0.125	0.2	A
0.87	1.65	12	9 Octol, 0.5 CB	0.125	0.2	A
0.94	1.75	12	9 Cyclotol, 0.5 CB	0.094	0.1	A
1.00	1.88	12	8 9404	0.094	0.1	A
1.10	2.07	12	8 9404	0.063	0.1	A

pressure at more or less equally spaced intervals in the target plates over the entire pressure range. Moreover, the free-run systems utilize commercially available stainless-steel sheet stock for driver plates; these plates can be used without additional machining if reasonable care is taken in cutting them. Experience has shown that these systems give the best impact with respect to planarity for the desired pressure.

The systems have also been designed so that the target plates and samples can be of maximum thickness. The plane wave lenses and high explosives are for the most part commercially available.

A. Group II Elements: Be, Mg, and AZ-31B Mg

The beryllium samples used here were sintered, fine grained, isotropic, and very near theoretical crystal density. The BeO content is not known, but was probably small. Earlier work on Be exhibited rather large scatter (as much as 10% in the shock velocity). With better material control, this scatter has disappeared. Shock wave arrival studies on cast beryllium plates with large grains gave very irregular traces, even with shock pressures in excess of 0.3 Mbar, reflecting the anisotropy effects in this material.

The Mg used was cast and quite pure. Quite possibly some of the scatter in the data might be caused by the large (~0.5 cm diam) grains in the samples. This material is of some importance in analyzing the behavior of high explosives, and since porosity-free castings of pure Mg are extremely difficult to obtain, the Hugoniot of the AZ-31B alloy was also measured. This alloy is commercially available as tooling plate; it is mostly magnesium with the largest minor component being 3% aluminum. Aside from a difference in initial densities, the Hugoniots of these two materials are identical.

The u_s–u_p and P–ρ data for these materials are presented in Figs. 33 and 34.

B. Group IV Elements: Ti, Zr, and Hf

The three Group IV elements Ti, Zr, and Hf, in contrast to most of the metallic elements reported here, each show a transition. Although the transitions have very small volume changes, they are reasonably well determined. The transitions are assumed to be polymorphic transitions from the zero-pressure hcp phase to a bcc structure. This assumption is strengthened but not verified by the fact that shock-loaded Ti has been recovered in the bcc phase.

Although there are some zero-pressure thermal expansion and specific heat data available for these materials, they are generally rather uncertain in the high-temperature region where the transition occurs. The thermal expansion data for Ti, for example, are so poor that it is difficult to say whether or not the hcp to bcc transition at zero pressure and 1157°K increases or decreases the density. However, the gross features of these data are undoubtedly correct. The high temperatures,

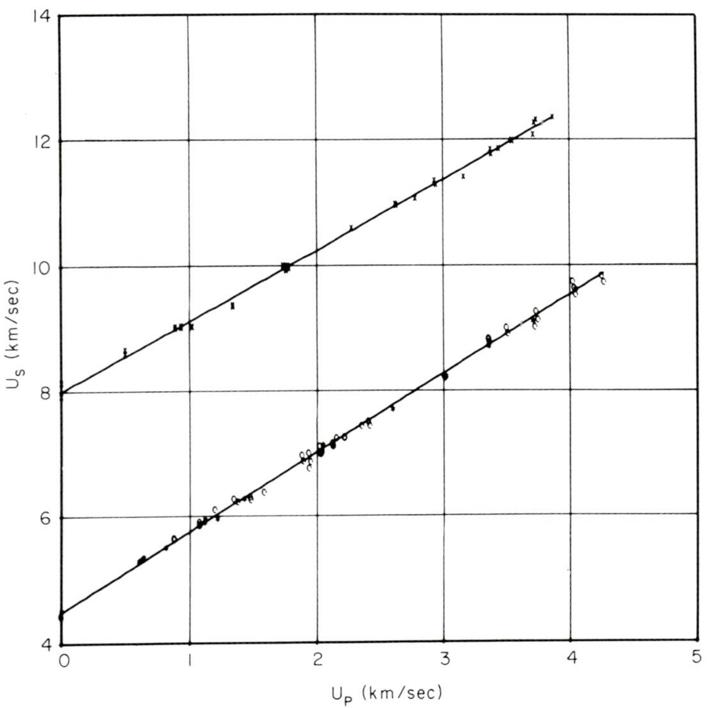

FIG. 33. Hugoniot u_s–u_p data for (×) Be, (○) Mg, and (*) AZ-31B Mg alloy.

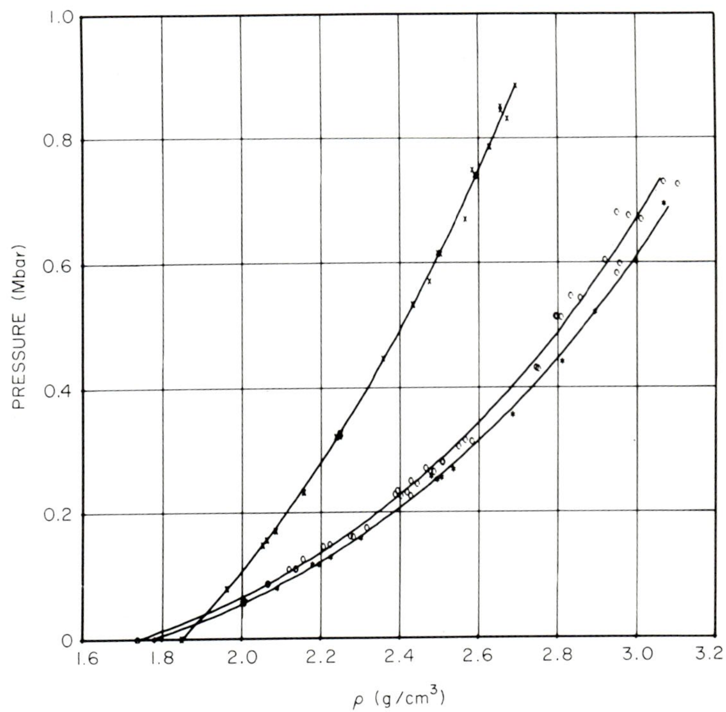

FIG. 34. Hugoniot P–ρ data for (×) Be, (○) Mg, and (*) AZ-31B Mg alloy. The Hugoniots of the pure Mg and the Mg–Al alloy separate in this plane only because of the difference in initial densities.

large latent heats of transition, and very small negative volume changes of the zero-pressure transitions are compatible with the hypothesis that the zero-pressure and high-pressure phase changes are the same.

We have attempted to determine the equation of state of the room-temperature metastable phase by the calculational procedures outlined in Section II. The results of these calculations are summarized in Table IV, together with some thermodynamic quantities at the transition point.

TABLE IV

TRANSITION PROPERTIES OF TITANIUM, ZIRCONIUM, AND HAFNIUM

	Parameters at transition					Metastable Hugoniot		
Material	u_s (km/sec)	P (kbar)	T (°K)	ρ (g/cm³)	$\Delta V/V$	$c_0{}^*$ (km/sec)	s^*	ρ^* (g/cm³)
Titanium	5.74	175	370	5.13	0.011	4.81	1.10	4.54
Zirconium	4.63	260	540	7.98	0.012	2.71	1.35	6.00
Hafnium	3.86	400	725	16.30	0.042	2.43	1.30	12.79

The calculated zero-pressure density for the metastable Hugoniots are all quite close to the density of the original samples. The density of the recovered Ti samples is in good agreement with the calculated value.

The Ti samples available to us were very strongly oriented; although longitudinal and shear wave velocities can be measured with some precision, they cannot be used to calculate a reliable zero-pressure bulk wave velocity.

The u_s–u_p and P–V results are presented in Figs. 35 and 36. The small discontinuity in the data in Fig. 35 just below $u_p = 1.0$ km/sec indicates the onset of the transition. The bulk sound velocities indicated on the graph were computed from Bridgman's compressibility data. The above shows that it is difficult to detect transitions with small volume changes in P—V plots. This is due in part to the fact that true P—V data are not obtained in the two-wave region. Although it appears in this figure that the compressibility decreases for the higher-density materials, the shock wave compressions $\Delta V/V_0$ are actually quite comparable for the three materials. The thermodynamic calculations were only taken up to the vicinity of the transition pressures for these elements.

C. GROUP V ELEMENTS: V, Nb, AND Ta

The Hugoniots of these elements appear to be well determined and are adequately described by the linear u_s–u_p relation (Fig. 37). The

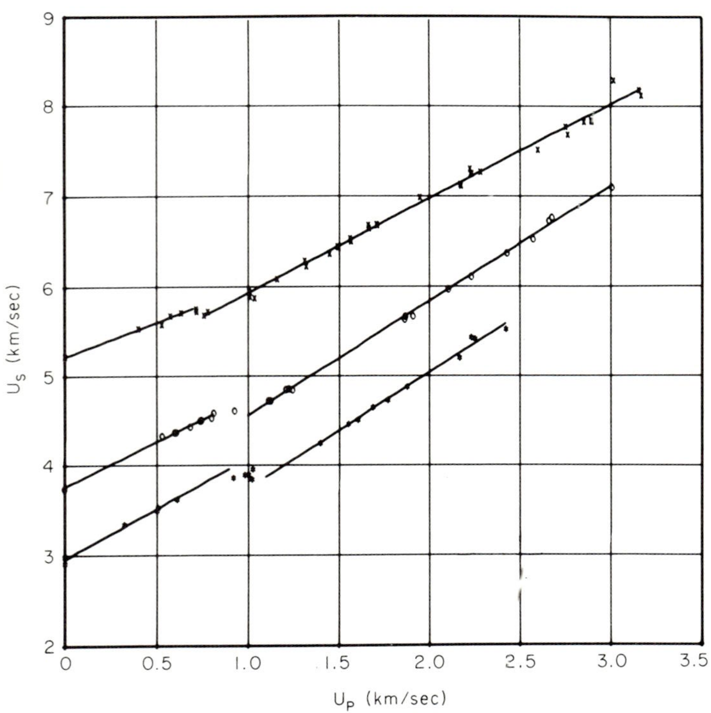

FIG. 35. Hugoniot u_s–u_p data for (×) Ti, (○) Zr, and (*) Hf.

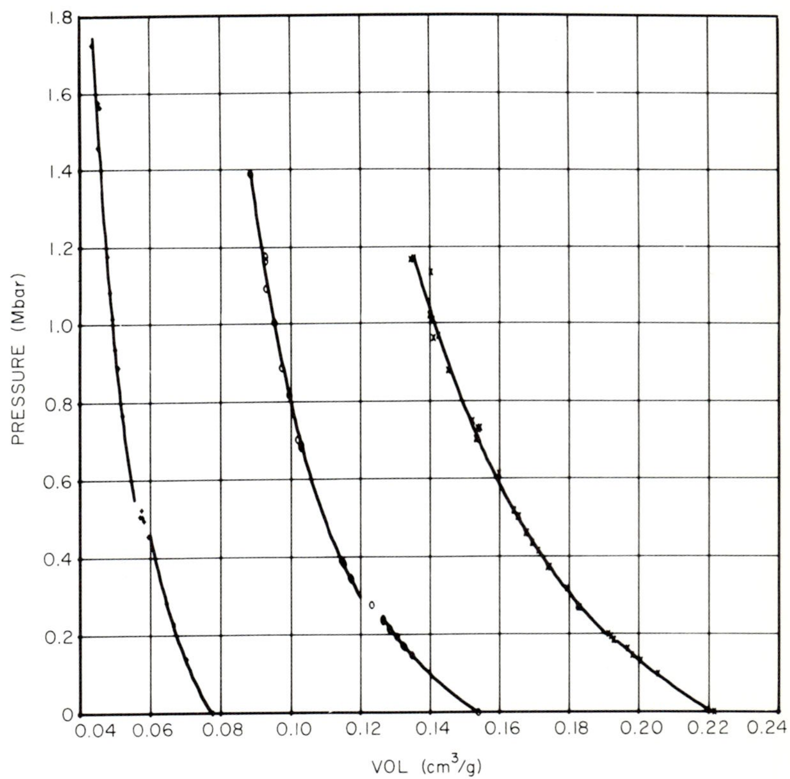

FIG. 36. Hugoniot P–V data for (×) Ti, (○) Zr, and (+) Hf.

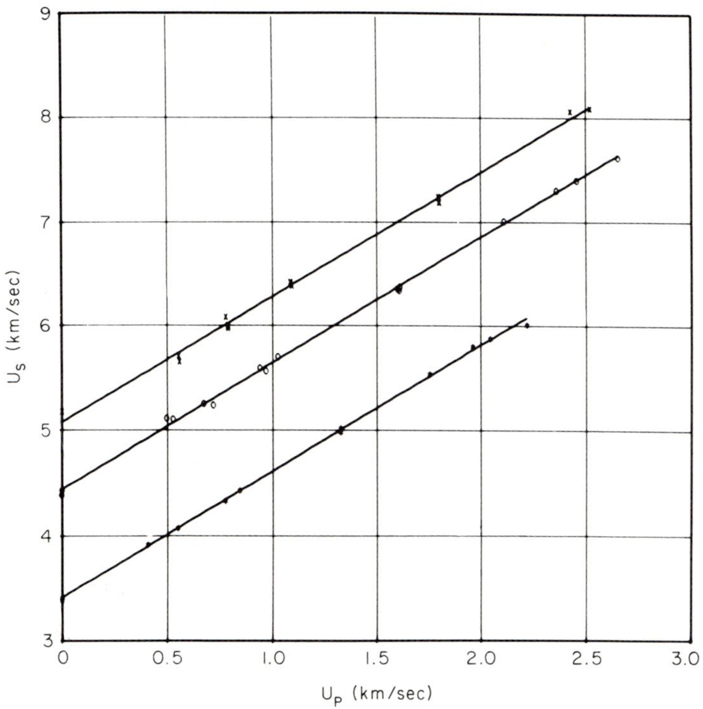

FIG. 37. Hugoniot u_s–u_p data for (×) V, (○) Nb, and (∗) Ta.

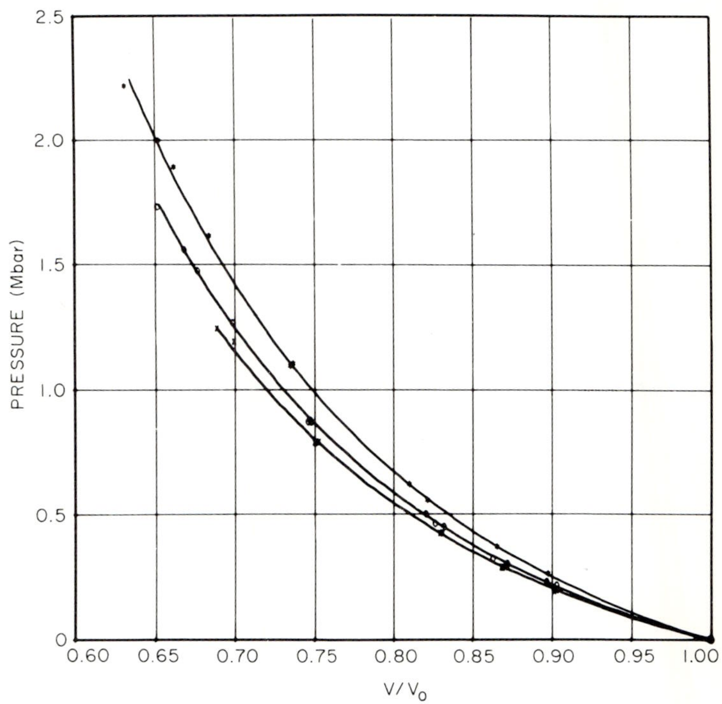

FIG. 38. Hugoniot P–V/V_0 data for (×) V, (○) Nb, and (∗) Ta.

intercepts are quite near the bulk sound speeds, indicating an absence of low-pressure transitions and also that the Hugoniot data are near the hydrodynamic equilibrium value. This is compatible with the low (~8 kbar) Hugoniot elastic limit measured on Ta. As expected, the u_s intercepts decrease with density but the slopes are virtually identical (1.201, V; 1.207, Nb; 1.200, Ta). Even though the relative shock velocities decrease with increasing density for these elements, the P vs. V/V_0 plots (Fig. 38) show that the shock wave compressibilities, although quite comparable, have reversed because of the density variation.

D. Group VI Elements: Cr, Mo, and W

These three elements are quite similar to the three Group V elements V, Nb, and Ta, each having one additional electron. All six elements have the bcc crystal structure. The u_s–u_p data (Fig. 39) are quite similar but do show a little more scatter. It is interesting that increasing Z by one in these Group V–VI elements increases the density and sound velocity by approximately 17%, except for the sound velocity in Cr which is about the same as V. It is also interesting that the slopes of the u_s–u_p Hugoniots for Mo and W are practically identical, 1.233 and 1.237, respectively, and only about 2.5% higher than the slopes of the Group V elements, while the slope of the Cr Hugoniot is 1.49. Since the slope of the u_s–u_p Hugoniot is closely associated with the repulsive forces in the lattice structure, this implies that the potential for Cr is quite different than the other five elements. It can be noted that, in going from Group V to Group VI, Mo and W simply add one *d* electron while Cr adds two *d* electrons at the expense of one *s* electron. While this gives an indication that Cr might behave differently, it should also be noted that this results in Cr and Mo having similar electronic structures. This might be the reason these two elements have approximately the same initial sound velocity.

It is possible that the shock wave data for Cr has been affected by the second-order phase change at the 310-°K Néel temperature. Large changes in the specific heat, thermal expansion coefficient, and elastic constants have been observed in this temperature range (*22*). However, these effects, due to the paramagnetic to antiferromagnetic transition, appear to be limited to a small temperature range about 40°K wide and hence should not cause any large perturbation on the Hugoniot curve since the Hugoniot is the locus of the final states. Another cause of the data scatter could be the existence of fcc and bcc phases quenched in different amounts in various samples. The density of the samples used here was less than the crystal density by slightly more than 1%.

The pressure P vs. V/V_0 data are given in Fig. 40. The curves drawn through the data were derived from the linear fits of the u_s–u_p data.

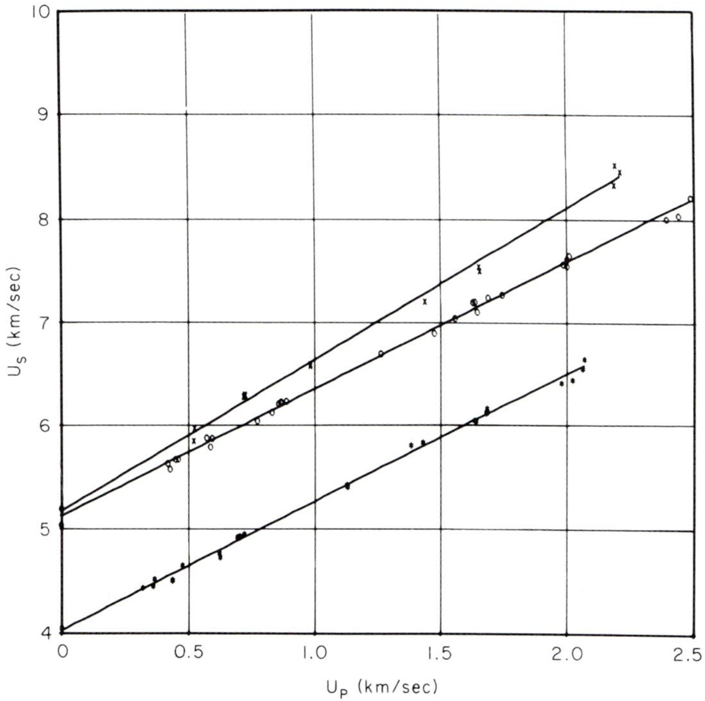

FIG. 39. Hugoniot u_s–u_p data for (×) Cr, (○) Mo, and (∗) W.

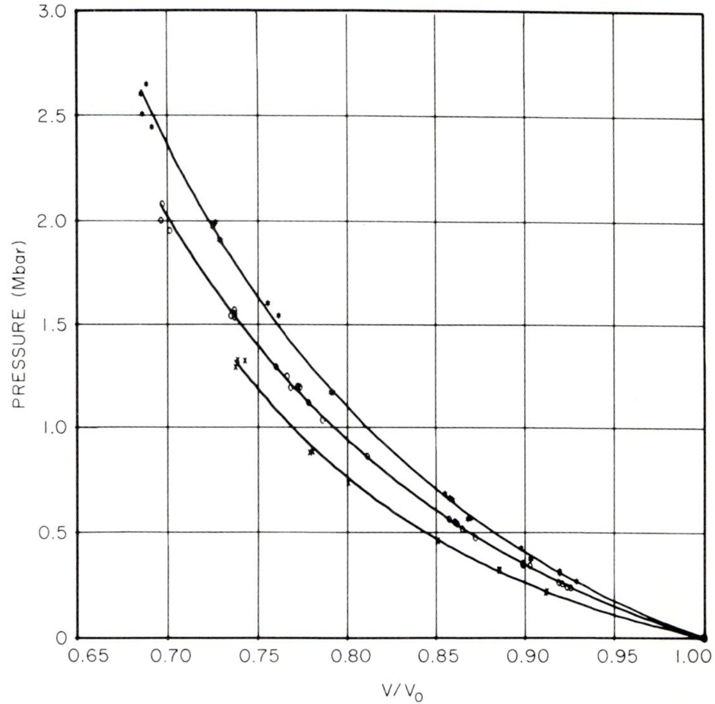

FIG. 40. Hugoniot P–V/V_0 data for (×) Cr, (○) Mo, and (∗) W.

The difference in curvature of these three Hugoniots is barely perceptible in this plot.

The Hugoniot elastic limit has been measured for these three elements. For samples 0.5 cm thick it was found that the longitudinal elastic wave had amplitudes of 15, 16, and 25 kbar for Cr, Mo, and W, respectively. However, there is a thickness effect in Mo and W; the elastic wave decays a few kilobars in going through about a centimeter of material. The Cr and W samples used for these studies were arc melted. The Mo specimens were made from sintered bars.

E. Other Metallic Elements: Re, Rh, Pd, Ir, and Pt

Re is properly classed as a Group VII element, falling into a family with manganese and technetium. However, we have been unable to obtain manganese in a form suitable for high-pressure equation of state experimentation, and there are no stable isotopes of technetium. There were two groups of Re samples, one at crystal density and one at 98% crystal density. Although there is some scatter in the data, there is no reason not to fit the crystal density Hugoniot data by the linear u_s–u_p fit shown in Fig. 41. This linear u_s–u_p fit, together with the thermodynamic value for the Grüneisen parameter, yields the calculated Hugoniot for the porous Re shown in Figs. 41 and 42. The agreement with the experimental data is satisfactory.

The high-density precious metals, rhodium, palladium, iridium, and platinum, exhibit no anomalies in the u_s–u_p plane up to pressures of 3 Mbar. Unfortunately, the rhodium and iridium samples were very anisotropic, and usable ultrasonic sound velocity data could not be obtained. Since Bridgman did not report compressibility values for these two materials, there is no way to check the extrapolated intercept. The extrapolated shock velocity intercepts for Pd and Pt are in agreement with the measured values.

The u_s–u_p and P–V/V_0 data for these materials are presented in Figs. 43–46.

F. Compounds: SiC, B_4C, and WC

These materials are all quite hard and have high longitudinal and transverse sound velocities. One could therefore expect that rigidity effects would influence the Hugoniots, especially in the lower-pressure regions. In the case of crystal-density SiC, the effect is dramatic; the longitudinal sound speed in SiC is 11.73 km/sec, and apparently signals transmitted at this velocity are sufficiently strong to close the flash gap

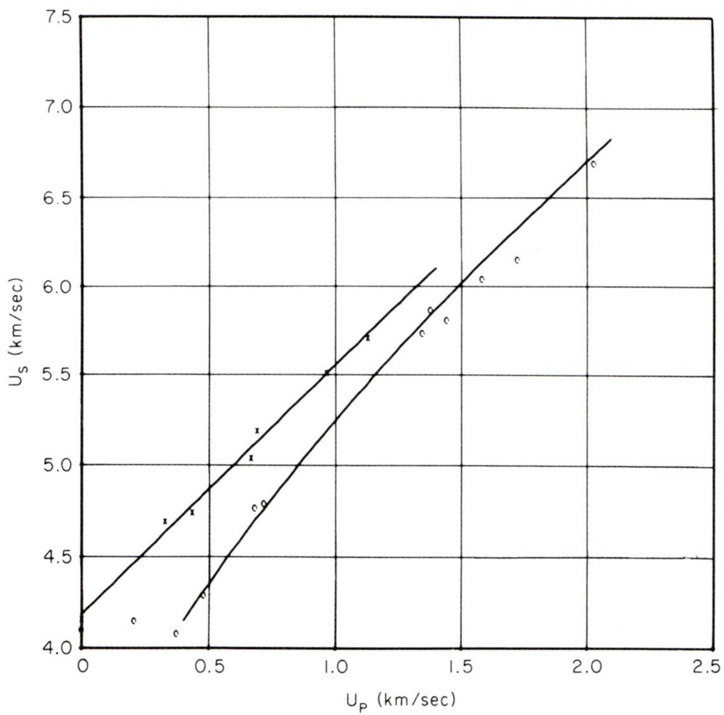

FIG. 41. Hugoniot u_s–u_p data for Re. (×) Crystal density, (○) 0.98 crystal density. The Hugoniot fit for the 0.98 crystal-density material has been calculated by the methods described earlier.

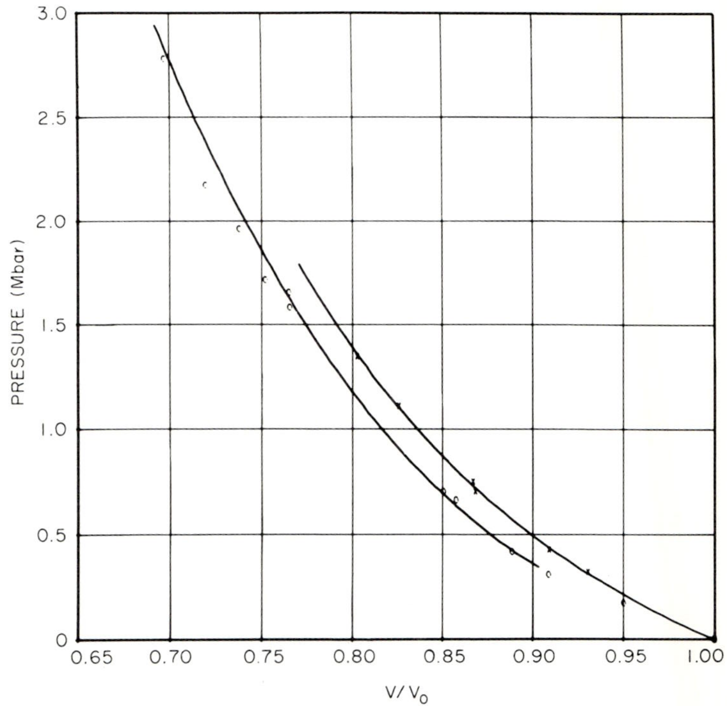

FIG. 42. Hugoniot P–V/V_0 data for Re. (×) Crystal density, (○) 0.98 crystal density. V_0 is the specific volume of each individual sample.

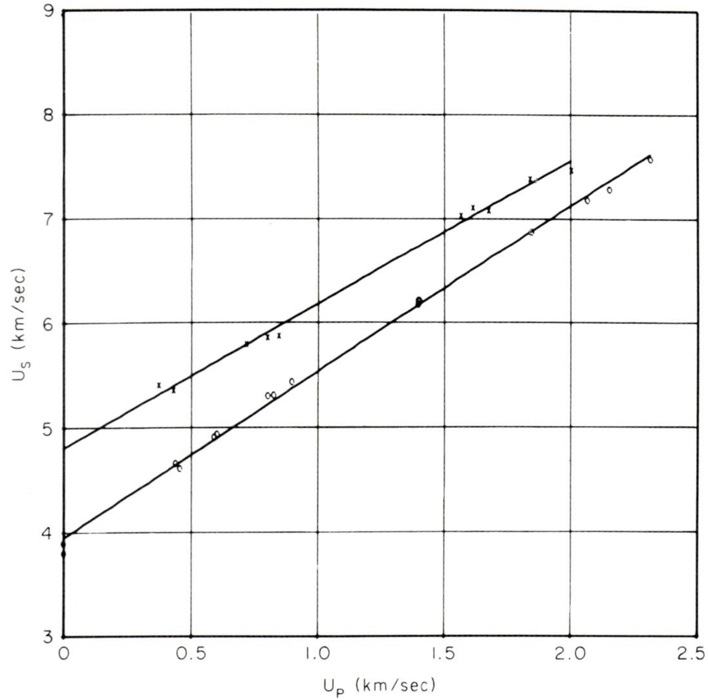

FIG. 43. Hugoniot u_s–u_p data for (×) Rh and (○) Pd.

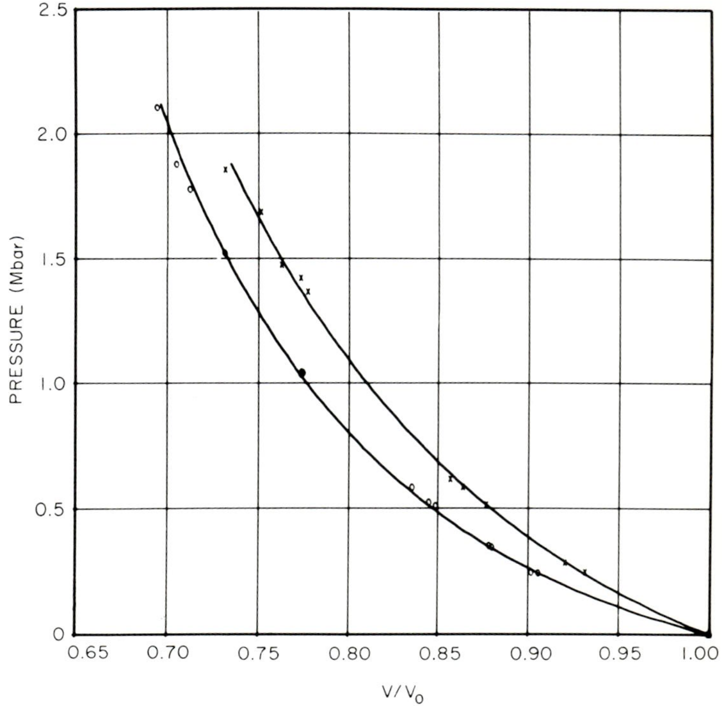

FIG. 44. Hugoniot P–V/V_0 data for (×) Rh and (○) Pd.

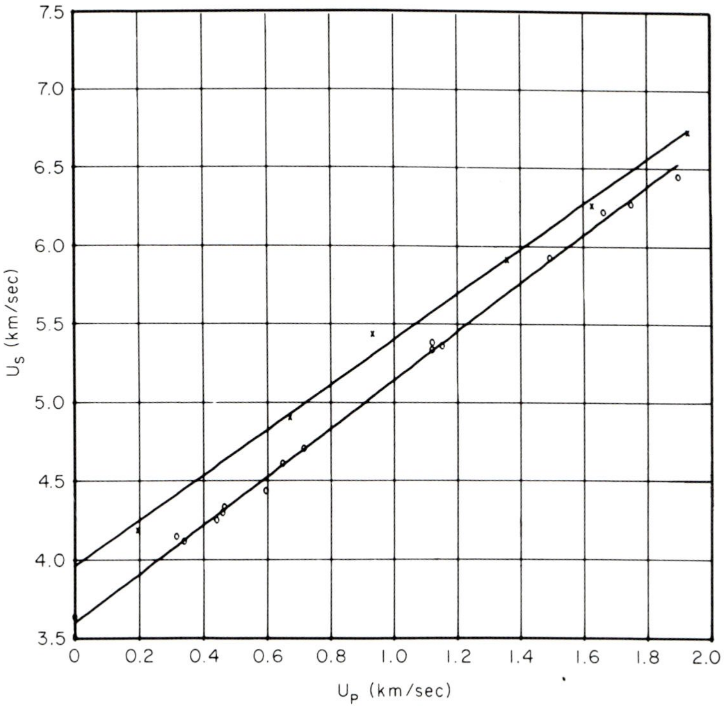

FIG. 45. Hugoniot u_s–u_p data for (×) Ir and (○) Pt.

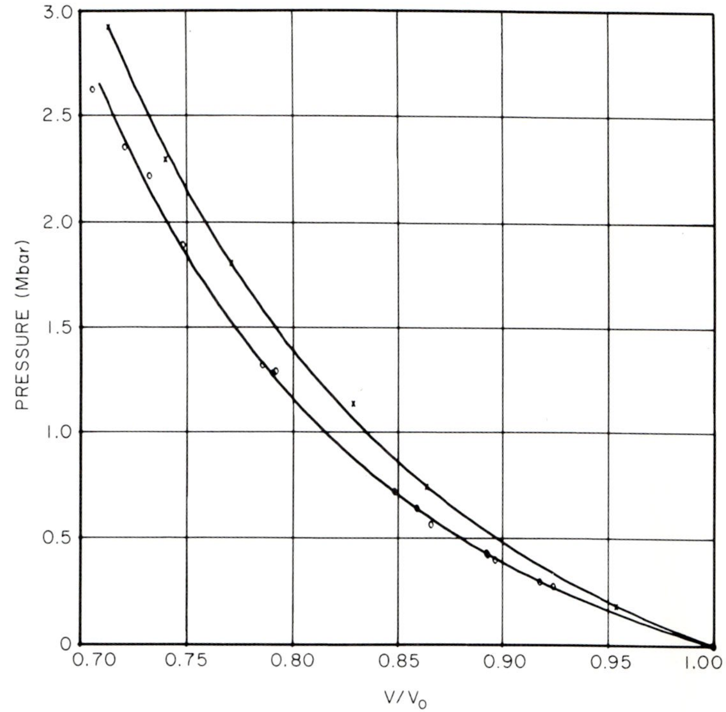

FIG. 46. Hugoniot P–V/V_0 data for (×) Ir and (○) Pt.

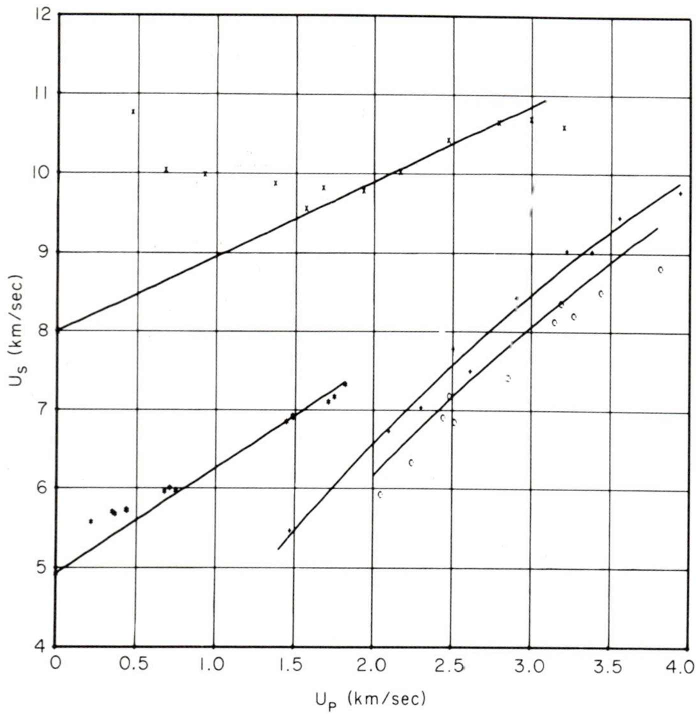

FIG. 47. Hugoniot u_s–u_p data for SiC, B_4C, and WC. (×) SiC, $\rho_0 \sim 3.1$ g/cm^3; (○) SiC, $\rho_0 \sim 2.3$ g/cm^3; (∗) WC; (+) B_4C, $\rho_0 \sim 1.9$ g/cm^3. The low-density silicon carbide curve has been calculated from the assumed crystal-density Hugoniot drawn on the figure.

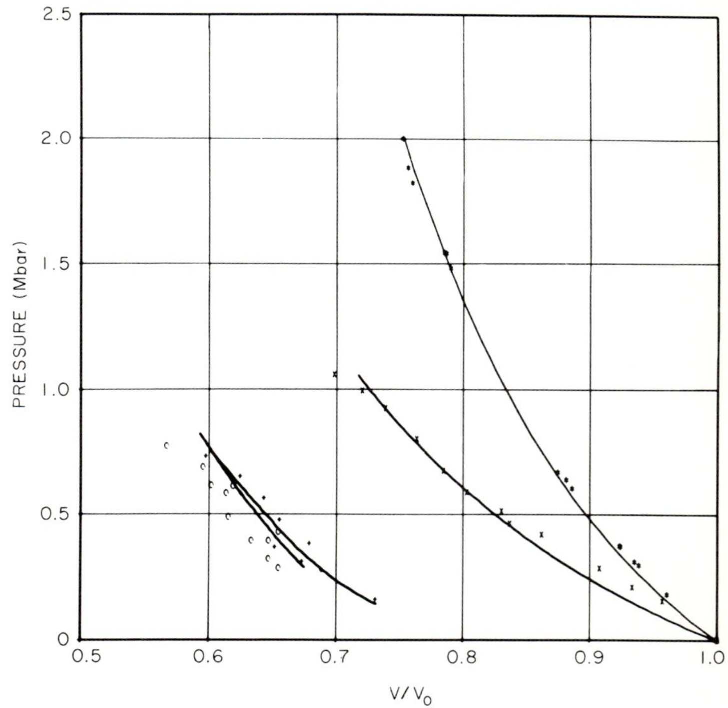

FIG. 48. Hugoniot P–V/V_0 data for SiC, B_4C, and WC. (×) SiC, $\rho_0 \sim 3.1$ g/cm^3; (○) SiC, $\rho_0 \sim 2.3$ g/cm^3; (∗) WC; (+) $\rho_0 \sim 1.9$ g/cm^3.

before the arrival of the 3-D wave (see Fig. 47). This results in a spuriously high measured shock velocity even at pressures as high as 400 kbar. While we usually do not use the zero-pressure bulk sound velocity in fitting the u_s–u_p data, this data is so obviously not representative of the equilibrium equation of state that we feel here it should be used. Thus the intercept of the Hugoniot fit is forced to coincide with the ultrasonic sound speed; the slope of the linear u_s–u_p relation is then determined by the high-pressure data alone. The possibility of a low-pressure phase change in SiC is not discounted, however.

A check on this assumed crystal-density Hugoniot is provided by the low initial density SiC data. Using the assumed crystal-density Hugoniot and the experimental thermodynamic value for the Grüneisen parameter, the Hugoniot was calculated by the methods outlined in Section III for porous materials. The calculated Hugoniot curves are plotted in Figs. 47 and 48. Although the agreement with the data is satisfactory, a better fit would be obtained if the fit for the crystal-density Hugoniot were made slightly softer.

The same effect is seen in the WC data; hydrodynamic equilibrium probably is not obtained until pressures in excess of 700 kbar are reached. Again, only the ultrasonic sound speed and the very-high-pressure data were used for determining the equilibrium Hugoniot.

The B_4C reported here is considerably below the theoretical crystal density of 2.5 g/cm³. As with all such porous materials, the Hugoniot data are best fitted with a quadratic u_s–u_p relation. There has been no attempt to calculate a crystal-density Hugoniot for B_4C, although in principle this could be done from the low-density Hugoniot.

G. Alloys: Stainless Steel

The three stainless steels reported here are typical of the commercial 300 series of steels, which contain nominally 18% chromium and 8% nickel by weight. The Hugoniots of all three alloys appear to be identical and are adequately fitted by the linear u_s–u_p relation shown in Figs. 49 and 50. No transition was expected or observed since these alloys are austenitic.

H. Synthetic Materials

The Hugoniots of a wide variety of materials have been measured here that are probably best categorized as simply miscellaneous synthetic materials. These include several types of plastics such as acrylic resin,

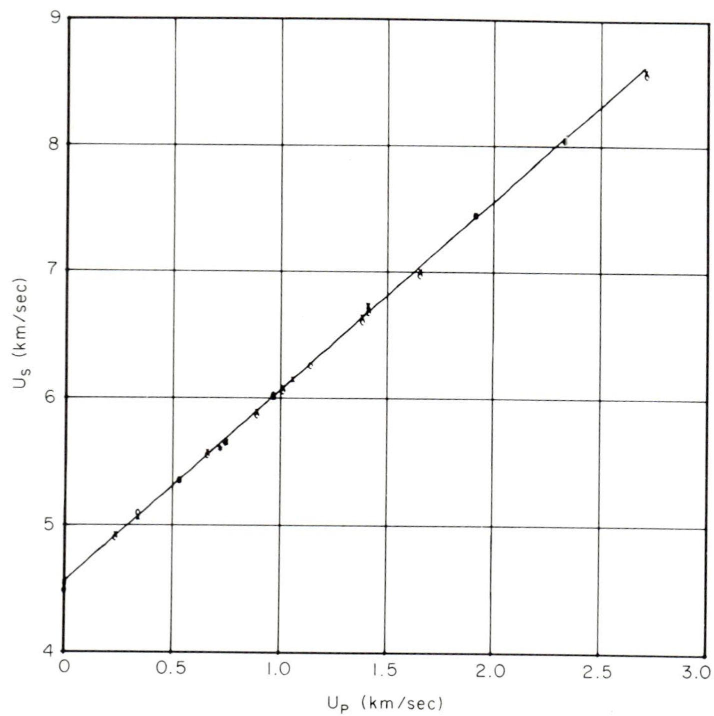

FIG. 49. Hugoniot u_s–u_p data for stainless steel. (×) Type 304, (○) type 304L, (∗) type 347. The Hugoniot fits of all three types are identical.

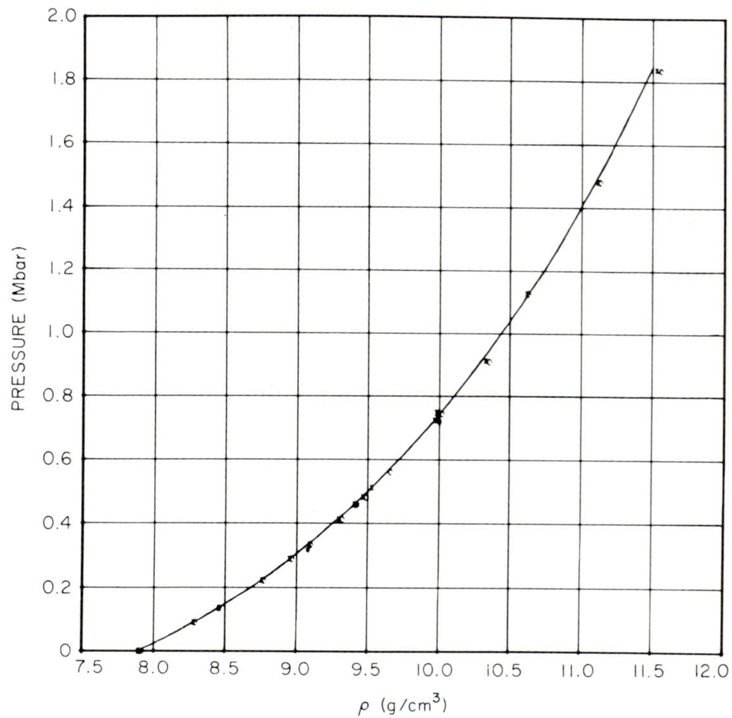

FIG. 50. Hugoniot P–ρ data for stainless steel. (×) Type 304, (○) type 304L, (∗) type 347.

epoxy, phenoxy, polyethylene, and polyurethane; some rubbers; and various plastic-loaded composites and foams. Most of these materials have been of some interest in some aspect of the space program, which implies that their equation of state properties are of some concern in the hypervelocity impact problem. In some instances the materials can be grouped in a logical fashion. However, this is not always the case and the reader can only wonder why some materials have been grouped together.

1. *Adiprene, Neoprene, Polyrubber, and Silastic Rubber*

Except for Silastic, these rubbers appear to be reasonably linear in the u_s–u_p plane, at least over the interval of measurement. It is unfortunate that we were unable to obtain any sound velocity or compressibility data for these materials. Such data would have been very useful in establishing the equation of state in the low-pressure area. The data appear to be reasonably good in spite of the fact that we were unable to machine the samples and had to use samples cut from sheet stock. The behavior of the Silastic (RTV-521) rubber in the u_s–u_p plane (Fig. 51) strongly suggests that this material behaves either as if it had a high degree of porosity (it can be seen that the u_s–u_p curve is heading toward a zero-pressure shock velocity near zero) or as if it had a phase change of some type. Although the Hugoniots of the two different initial density polyrubbers appear to be reasonably well determined in the u_s–u_p plane, it can be seen in Fig. 52 that a small extrapolation of the u_s–u_p fits would actually cause the densities of the two materials to invert.

2. *Polyurethane*

The bulk polyurethane reported here was supplied as a single piece by the CPR International Corporation, and is quite uniform in density. The Hugoniot shows a distinct change in slope in the u_s–u_p plane (Fig. 53) above about 210 kbar; the material is probably well into the liquid phase at this pressure.

The Hugoniot intercept of the lower phase is considerably higher than the measured ultrasonic value. This may indicate another transition below the range of our data, possibly due to melting.

The low-density rigid foams were made at Los Alamos Scientific Laboratory and range in density from 6 to 38% of bulk density. It is very difficult to obtain data on such low-density porous materials by optical techniques because the flash gaps will not close reliably, even with the use of low-impedance shims. The general trend of the data, however, is as expected; the u_s–u_p loci fall lower as density is decreased. It will be noted from the P–ρ plot (Fig. 54) that the low-density foams

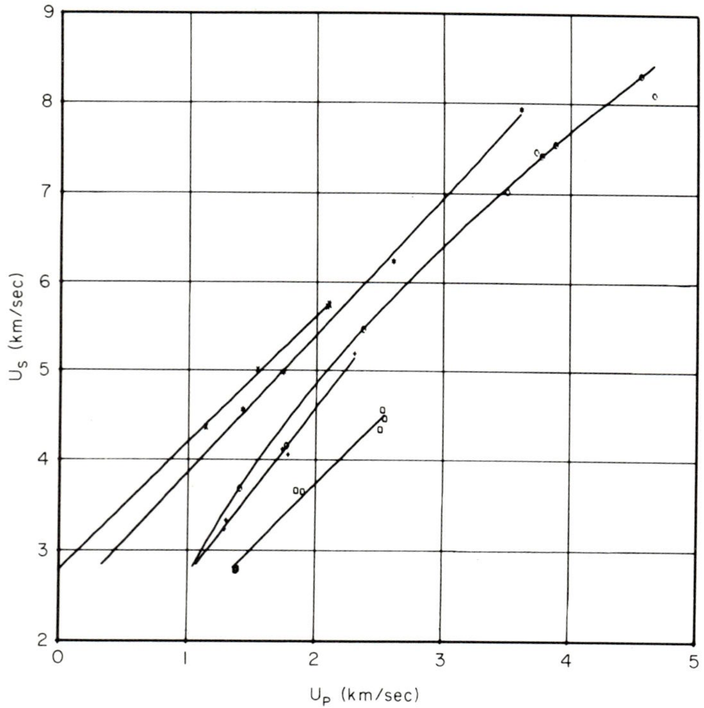

FIG. 51. Hugoniot u_s–u_p data for various rubbers. (×) Neoprene; (○) Silastic rubber (RTV-521); (∗) adiprene, bulk; (+) polyrubber, $\rho_0 = 1.01$ g/cm³; (□) polyrubber, $\rho_0 = 0.77$ g/cm³.

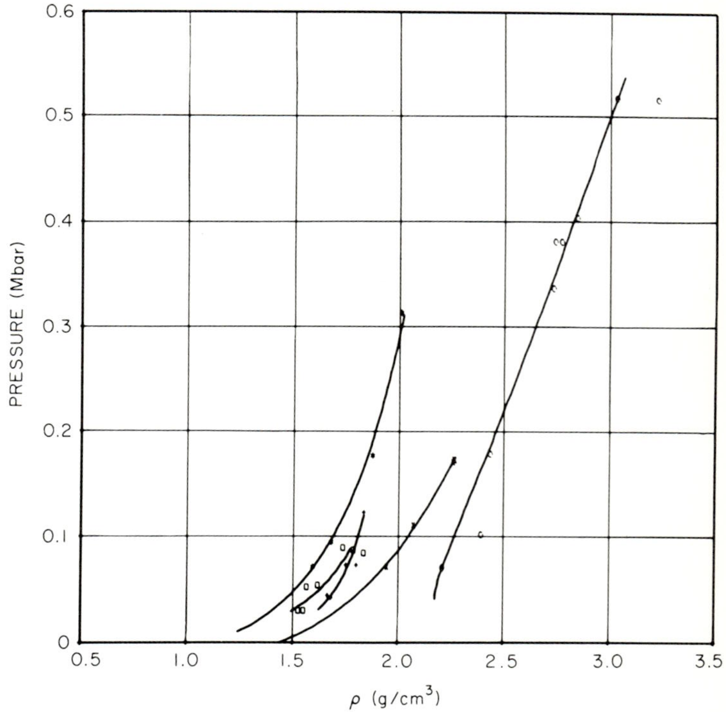

FIG. 52. Hugoniot P–ρ data for various rubbers. (×) Neoprene; (○) Silastic rubber (RTV-521); (∗) adiprene, bulk; (+) polyrubber, $\rho_0 = 1.01$ g/cm³; (□) polyrubber, $\rho_0 = 0.77$ g/cm³.

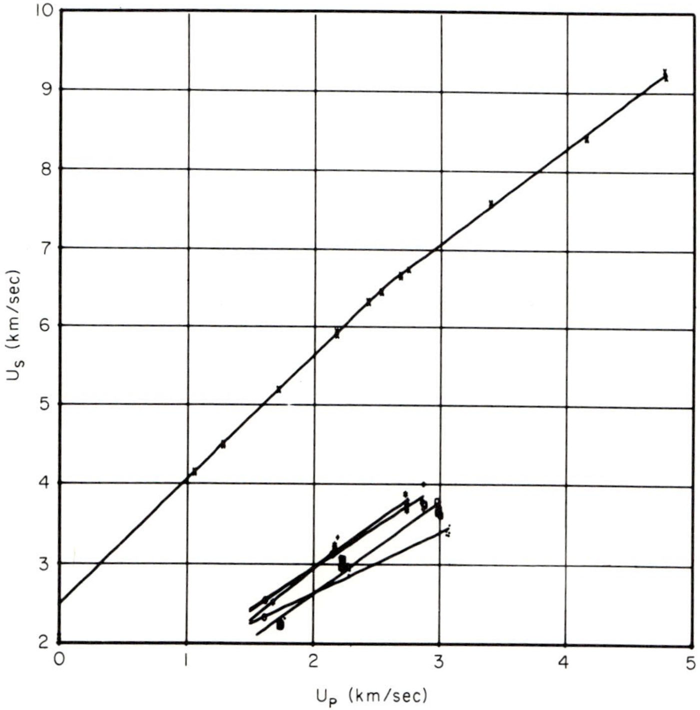

FIG. 53. Hugoniot u_s–u_p data for polyurethane. (×) $\rho_0 =$ 1.265 g/cm³ (bulk), (○) $\rho_0 = 0.35$–0.48 g/cm³, (∗) $\rho_0 = 0.31$–0.34 g/cm³, (+) $\rho_0 = 0.25$–0.30 g/cm³, (□) $\rho_0 = 0.14$–0.21 g/cm³, (·) $\rho_0 =$ 0.07–0.11 g/cm³.

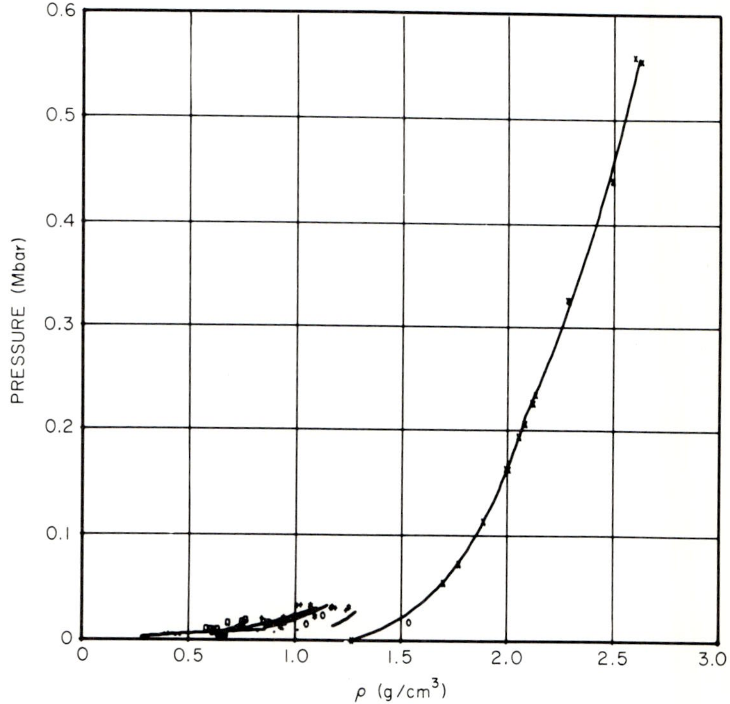

FIG. 54. Hugoniot P–ρ data for polyurethane. (×) $\rho_0 = 1.265$ g/cm³ (bulk), (○) $\rho_0 = 0.35$–0.48 g/cm³, (∗) $\rho_0 = 0.31$–0.34 g/cm³, (+) $\rho_0 = 0.25$–0.30 g/cm³, (□) $\rho_0 = 0.14$–0.21 g/cm³, (·) $\rho_0 =$ 0.07–0.11 g/cm³.

barely achieve bulk density even at the maximum attained pressure of 35 kbar.

3. *Polyethylene and Polystyrene*

The u_s–u_p Hugoniot of polyethylene has been drawn with a single linear fit in Fig. 55, even though the data could be fitted slightly better with a two segment fit having a break near $u_p = 3$ km/sec. The linear least-squares fit of the data gives a zero-pressure intercept of 3.00 km/sec, considerably higher than the bulk sound velocity of 1.9 km/sec, which implies that some type of low-pressure transition has occurred.

Some free surface velocity data have also been obtained for polyethylene. In these experiments one-half of the base plate was used to measure shock velocities, two through the standard material and two through polyethylene samples. The other half of the base plate was used for measuring the free surface velocity of the polyethylene. The results of these experiments are given in Table V. The free surface velocities

TABLE V
RESULTS OF MULTIPLE SLUG–FREE SURFACE VELOCITY EXPERIMENTS ON POLYETHYLENE

			921-T Al (std)			2024 Al (std)		
u_s in std 921-T Al (km/sec)	u_s in polyethylene (km/sec)	u_{fs} (km/sec)	u_p (km/sec)	u_r (km/sec)	u_r/u_p	u_p (km/sec)	u_r (km/sec)	u_r/u_p
6.396	5.012	2.805	1.485	1.320	0.889	1.425	1.380	0.968
6.741	5.598	3.562	1.833	1.729	0.943	1.821	1.741	0.956
7.382	6.581	4.821	2.466	2.355	0.955	2.485	2.336	0.940
7.414	6.654	4.906	2.495	2.411	0.966	2.534	2.372	0.936
7.440	6.632	4.960	2.525	2.435	0.964	2.519	2.441	0.969
8.025	7.538	6.200	3.087	3.113	1.008	3.131	3.069	0.980
9.108	9.058	8.352	4.113	4.239	1.031	4.157	4.195	1.009

listed are the average of ten independent measurements. The particle velocities listed are derived from the impedance match solution with the 921-T Al standard and also from the u_s–u_p Hugoniot measured with 2024 Al as a standard. These velocities are then subtracted from the measured free surface velocities to give u_r. In contrast to the usual behavior, it is found that u_r/u_p is only slightly greater than unity over the entire experimental range.

Several thermodynamic calculations were made using various input parameters to see if this behavior could be reproduced. The results of two problems of interest are shown in Fig. 56. For these calculations

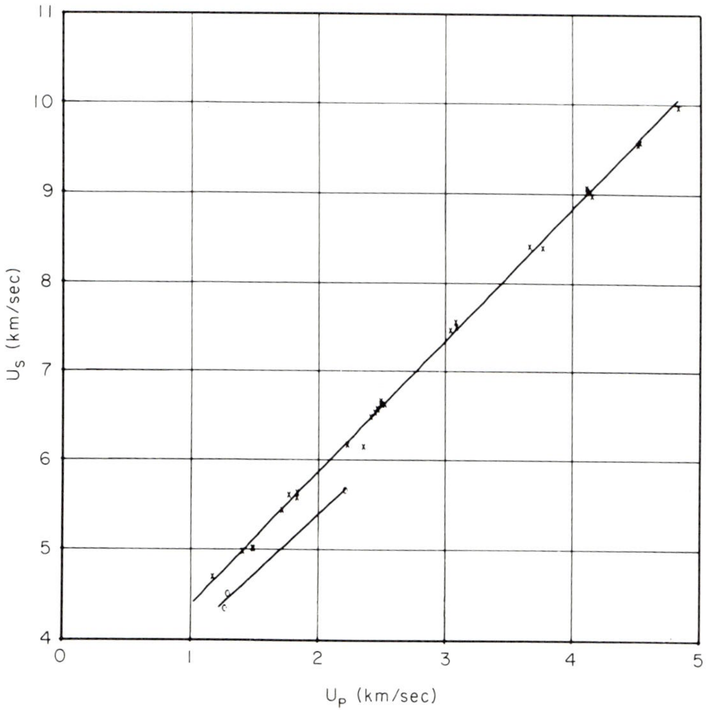

FIG. 55. Hugoniot u_s–u_p data for (×) polyethylene and (○) polystyrene.

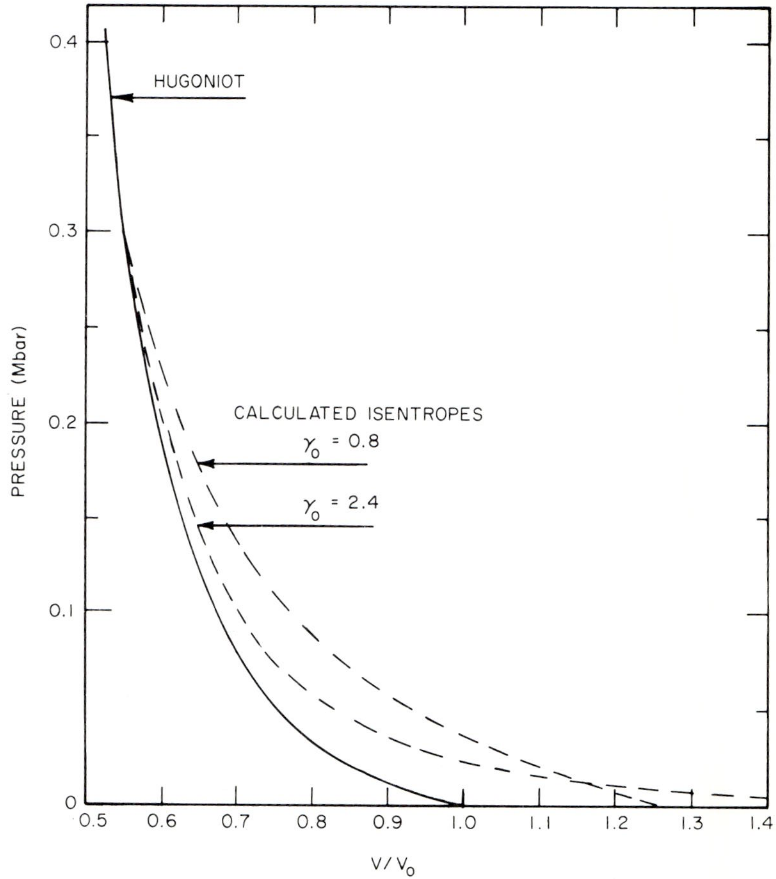

FIG. 56. Pressure versus V/V_0 and calculated isentropes for polyethylene.

the product $\rho\gamma$ was assumed constant. The zero-pressure Grüneisen parameter γ_0 was calculated from the existing thermodynamic data except that the extrapolated bulk Hugoniot intercept $c_0 = 3.0$ km/sec gave $\gamma_0 = 2.4$, while the measured ultrasonic sound velocity $c_0 = 1.9$ km/sec gave $\gamma_0 = 0.8$. As has been noted earlier, the calculated free surface velocities are quite insensitive to variations in γ. Although the u_r/u_p ratios are 1.15 and 1.14, respectively, the release isentropes have much different shapes. The final calculated states are 20% different in density and over 60% different in release temperatures.

It is possible that cross bonding has occurred in the shock process (it is possible to obtain polyethylene at density 0.96 g/cm³) and that the pressure release isentropes tend to be at higher densities. However, at temperatures near 1000°K, which is about what we compute for the release temperature, it is hard to imagine that much if any of the polymerization bonding still remains.

It is encouraging that the shock wave data taken with the 2024 Al standard and the 921-T Al standard are indistinguishable. Although this does not mean that both are correct, it does imply for low-impedance use that their rarefaction release behaviors are similar.

4. *Styrofoam*

The foams investigated here are the common commercially available rigid foamed polystyrene with nominal densities of 2, 4, and 10 lb/ft³. The scatter in the data is extremely large (Fig. 57) and again reflects the difficulty of closing the flash gaps reliably with low-mass samples having nonuniform composition. The data do not admit of detailed interpretation. The reader is cautioned that it really is P–ρ that is plotted in Fig. 58. The behavior of these materials in the P–ρ plane is a reflection of the fact that such low-density materials are very easily shocked up to the asymptotic compression limit and are folding backward. The curves are obviously not meant to represent the exact equation of state of these materials.

5. *Lucite and Plexiglas*

These two transparent materials, although both acrylic resins, are slightly different. Lucite is an acrylic methyl methacrylate resin and Plexiglas is an acrylate and methacrylate resin. Their Hugoniots are only slightly different. The Plexiglas used was Rohm and Haas II; however, samples were obtained from at least two different batches, which might partially account for some of the scatter in the data.

It can be seen from the u_s–u_p data (Fig. 59) that the extrapolated Hugoniot intercept is higher than the bulk sound velocity obtained from

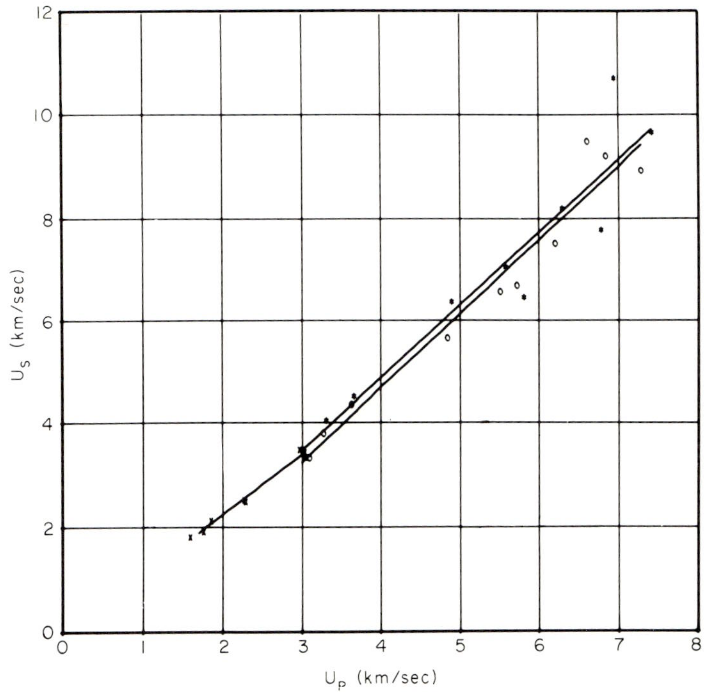

FIG. 57. Hugoniot u_s–u_p data for Styrofoam. (×) $\rho_0=0.174$ g/cm³, (○) $\rho_0=0.063$ g/cm³, (*) $\rho_0 = 0.032$ g/cm³.

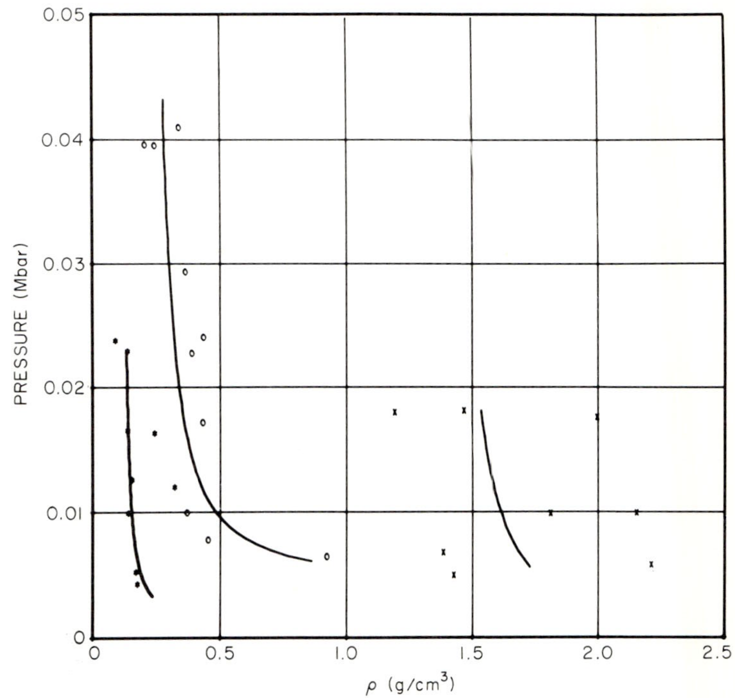

FIG. 58. Hugoniot P–ρ data for Styrofoam. (×) $\rho_0=0.174$ g/cm³, (○) $\rho_0=0.063$ g/cm³, (*) $\rho_0 = 0.032$ g/cm³.

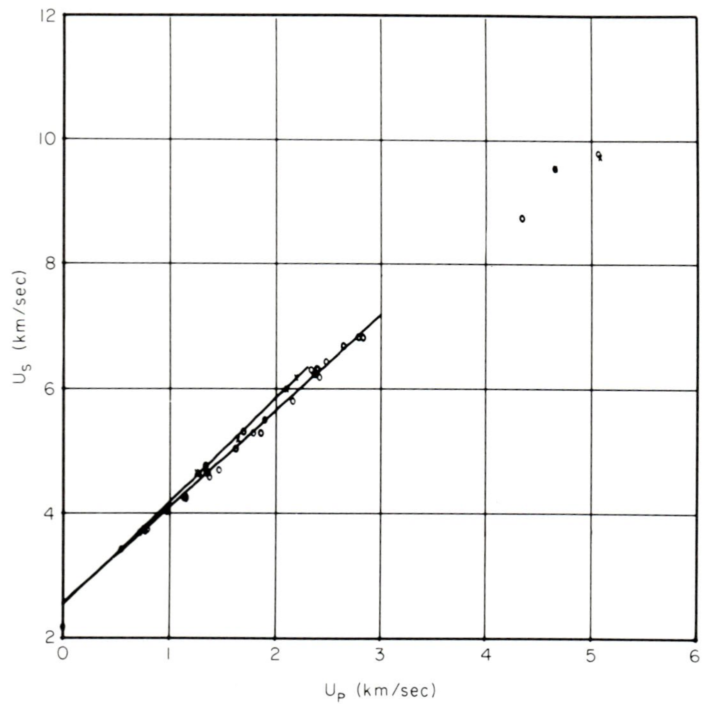

FIG. 59. Hugoniot u_s–u_p data for (×) Lucite and (○) Plexiglas.

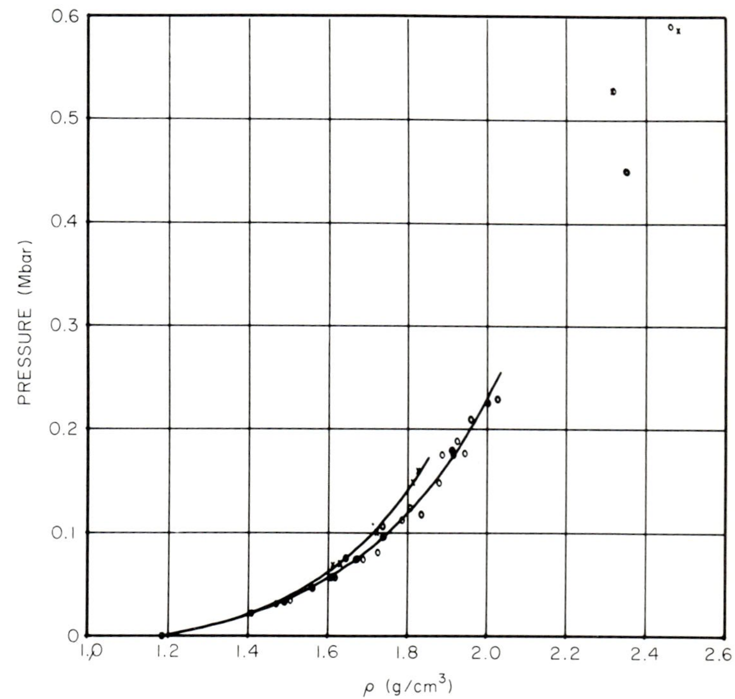

FIG. 60. Hugoniot P–ρ data for (×) Lucite and (○) Plexiglas.

ultrasonic measurements. Precision measurements of the free surface velocity versus time at low pressure show the shock wave structure to be free from any elastic precursor; hence, it must be concluded that some type of low-pressure transition has occurred. This seems to be characteristic of many plastics.

It can also be seen that the linear fits of the lower-pressure data do not extrapolate to the few high-pressure data points. This could be caused by a transition, which would require additional straight-line segments to describe the Hugoniot properly. It is also possible that the u_s–u_p Hugoniot should be drawn with some curvature. Additional high-pressure data is needed to resolve this problem.

The difference in the Hugoniots of these two materials is more apparent in the P–ρ plot given in Fig. 60.

6. *Paraffin, Epoxy, Exon, and Durite*

The u_s–u_p Hugoniots of paraffin and Durite appear to be linear (Fig. 61). There are not enough data to establish the linearity of the Exon Hugoniot. The low-pressure epoxy data do not extrapolate to the two high-pressure data points. Although the ultrasonic data for epoxy is not considered to be very good, it does give a zero-pressure bulk sound velocity considerably lower than the extrapolated shock wave data would indicate. Ultrasonic sound speed data are not available for the other materials. It should be pointed out that the epoxy investigated here was not quite the same as that used in forming the plastic composites discussed later, although the difference is probably trivial. The exact compositions are given in Appendix E at the end of the volume.

The pressure–density Hugoniots of these materials are presented in Fig. 62.

7. *Phenoxy and Phenolic Refrasil*

There are sufficient data for each of these materials to establish the existence, and a fairly good estimate of the location, of a phase change. The u_s–u_p data plot, Fig. 63, shows a break at a shock velocity of about 7 km/sec for phenoxy and at about 5.4 km/sec for the phenolic Refrasil. These correspond to pressures just over 200 kbars for both materials. A small extrapolation of the P–ρ data, Fig. 64, indicates that both transitions have about a 10% density increase. The above considerations tend to suggest that these transitions are related. However, the Durite data did not show any indication of a transition, and this material is also a phenolic-loaded plastic.

Since the silica content of the phenolic Refrasil was 63 wt% (as

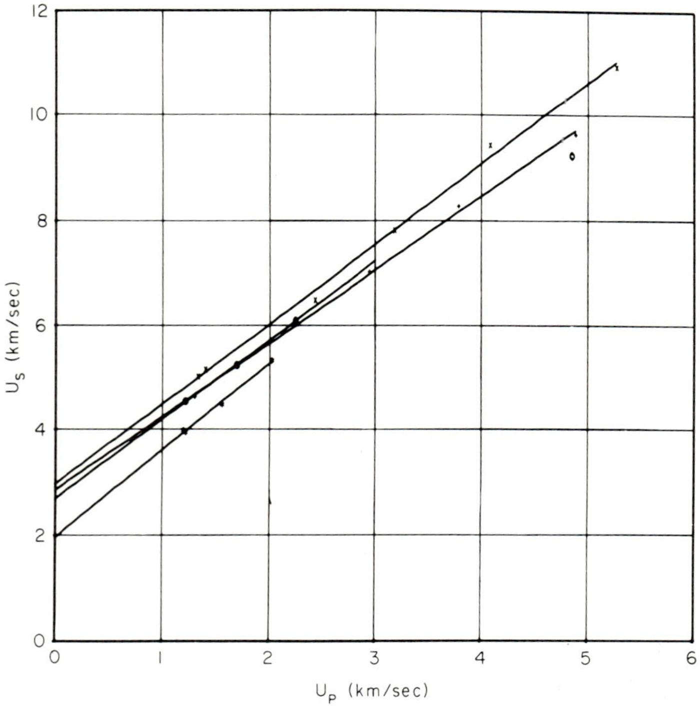

FIG. 61. Hugoniot u_s–u_p data for (×) paraffin, (○) epoxy, (∗) Exon, and (+) Durite.

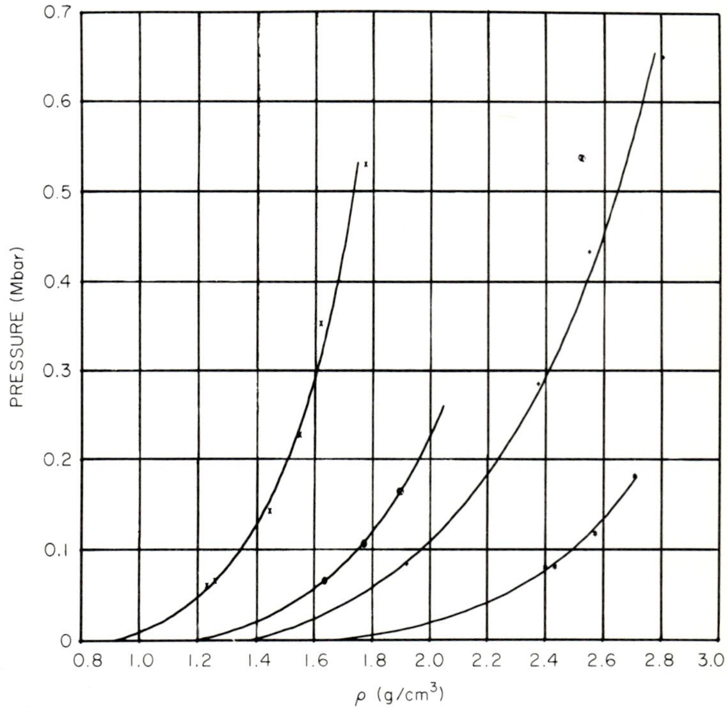

FIG. 62. Hugoniot P–ρ data for (×) paraffin, (○) epoxy, (∗) Exon, and (+) Durite.

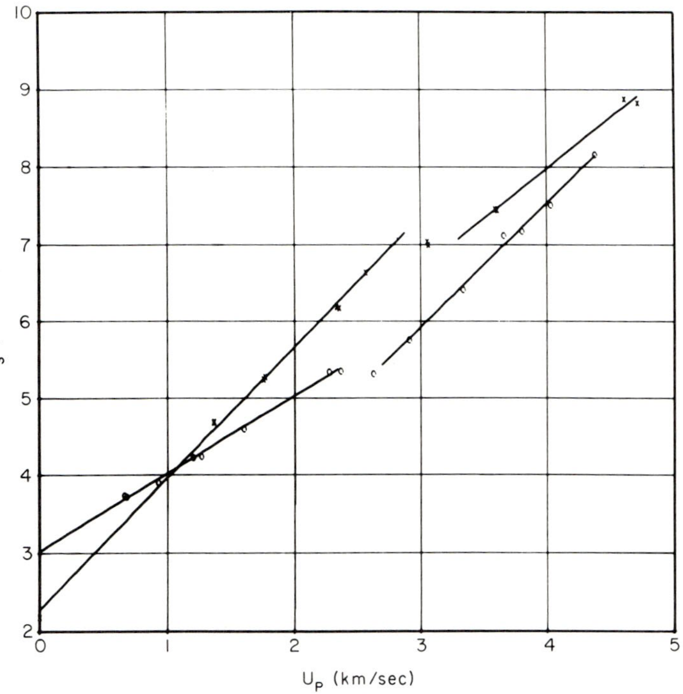

FIG. 63. Hugoniot u_s–u_p data for (×) phenoxy and (○) phenolic Refrasil.

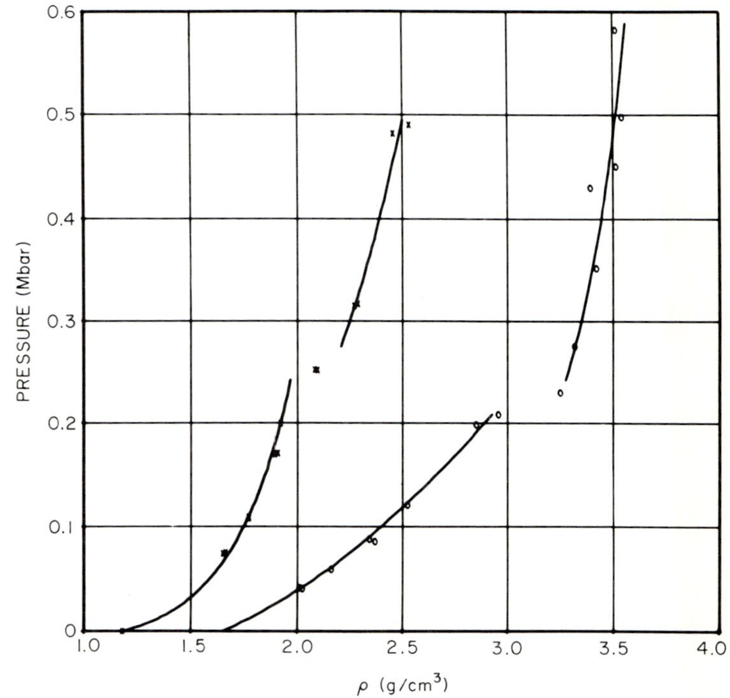

FIG. 64. Hugoniot P–ρ data for (×) phenoxy and (○) phenolic Refrasil.

determined by thermal decomposition), some indication of the 100-kbar quartz-to-stishovite transition was expected to be observed. The quartz transition has a density increase of over 80%. The high-pressure phase appears to be quite incompressible; this is compatible with a mixture of stishovite and plastic. From visual observation of the phenolic Refrasil samples, it was determined that the quartz fibers lie more or less in a plane a few degrees off perpendicular to the direction of wave propagation. Because the samples are obviously not isotropic, no bulk sound velocity is reported. The ultrasonic sound velocity and intercept of the phenoxy Hugoniot are quite close, a situation rather unusual for plastic materials.

8. *Plastic Composites*

These materials were originally studied in an attempt to obtain an insulating material which would match the density and shock impedance of magnesium. Such properties are useful in experiments involving explosively generated electromagnetic fields.

The lithium tetraborate–epoxy mixture is $Li_2B_4O_7 \cdot 5H_2O$ in 10 wt% epoxy plastic. Although this material has been fitted with a quadratic u_s–u_p relationship, it is quite possible that a phase change with small volume change occurs in this material. The u_s–u_p data in Fig. 65 can be fitted very well by two straight lines with a discontinuity at 7.2 km/sec. However, the data for these materials do not appear to be sufficiently reliable to justify the two-segment fit. If a phase change exists, it would probably be related to the transition believed to occur in epoxy somewhere in this pressure range.

Unfortunately, there is not enough data on the copper oxide- and Zerifac-loaded epoxies to check the possibility of a phase change in this pressure region. The data have therefore been fitted with a quadratic u_s–u_p relation. Zerifac (the trade name of a commercially mined lithium aluminum silicate) loaded epoxy is composed of 71 wt% Zerifac and 29 wt% epoxy (Epon 815–diethylenetriamine ratio of 100 to 11 by weight); and CuO-loaded epoxy is composed of 44 wt% CuO and 56 wt% of this same epoxy. The Zerifac-loaded polyurethane foam is composed of 50 wt% Zerifac and 50 wt% polyurethane, the mixture then being foamed to a final density of 40 lb/ft^3. Because this material is below crystal density, the u_s–u_p locus will not remain linear at low pressures. As with other materials much below crystal density, even small shocks result in final densities very near the asymptotic volume. Although the u_s–u_p data are quite adequate, the usual large scatter in the P–ρ plane (Fig. 66) is found for this low-density material.

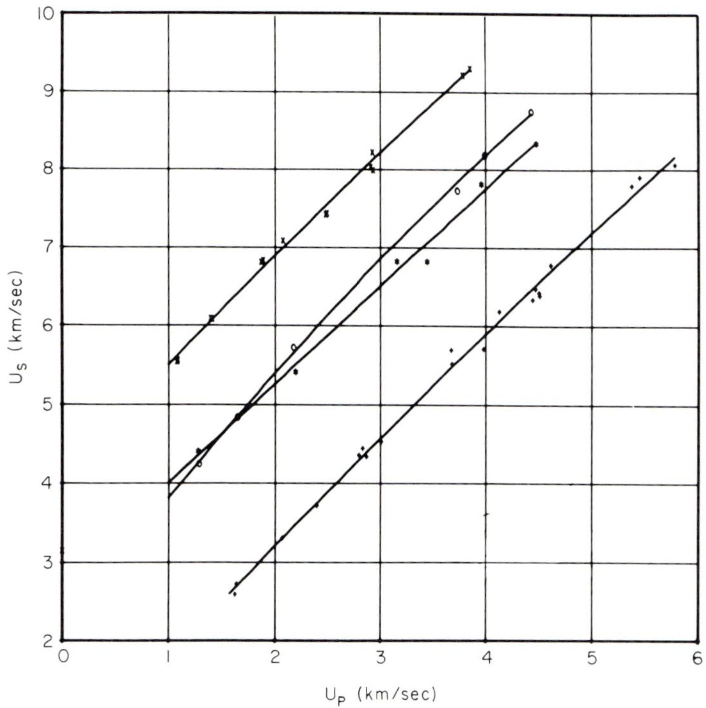

FIG. 65. Hugoniot u_s–u_p data for plastic composites. (×) Lithium tetraborate–epoxy, (○) copper oxide–epoxy, (∗) Zerifac–epoxy, (+) Zerifac–polyurethane foam.

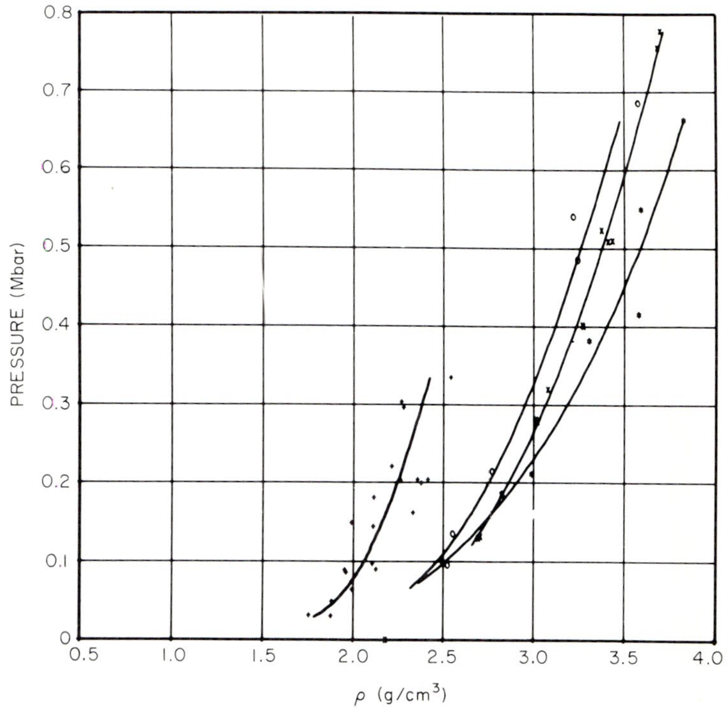

FIG. 66. Hugoniot P–ρ data for plastic composites. (×) Lithium tetraborate–epoxy, (○) copper oxide–epoxy, (∗) Zerifac–epoxy, (+) Zerifac–polyurethane foam.

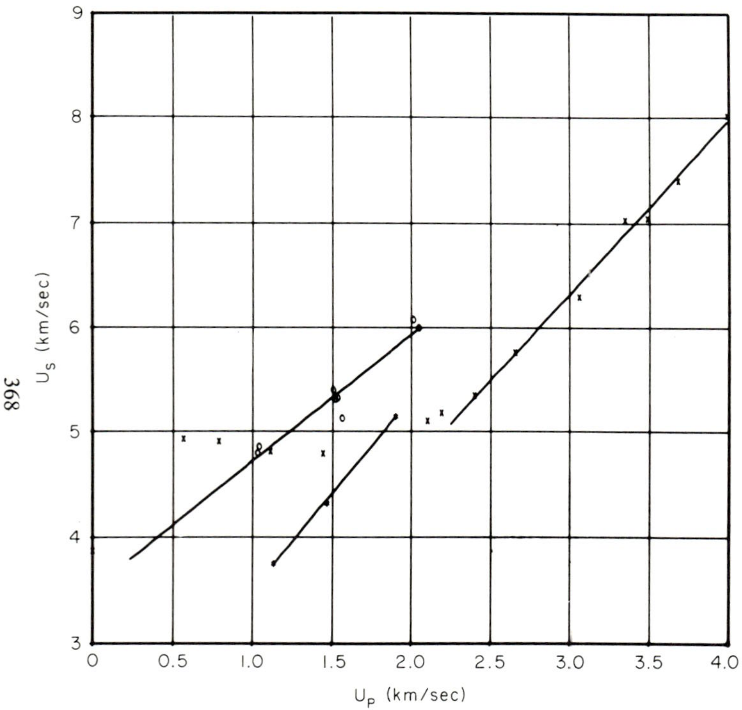

FIG. 67. Hugoniot u_s–u_p data for (×) Pyrex, (○) Melmac, and (∗) Teflon.

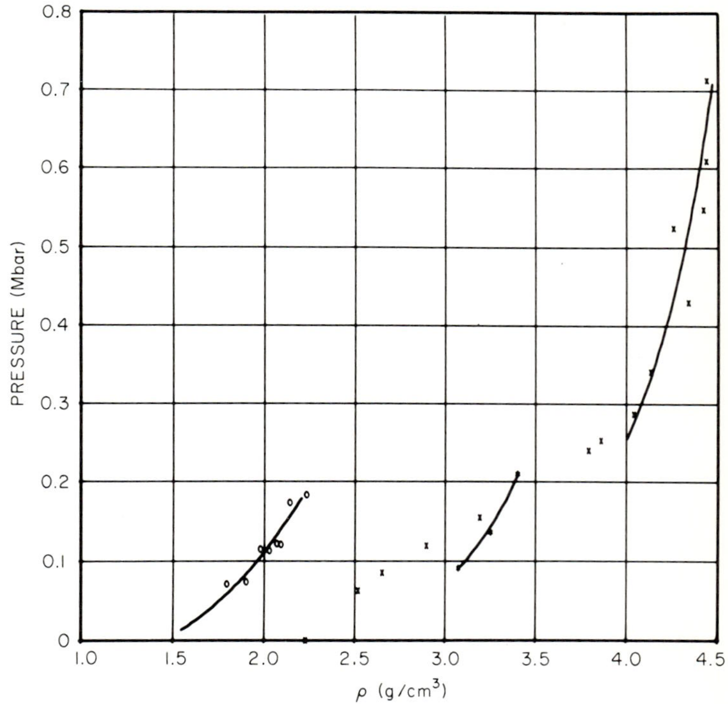

FIG. 68. Hugoniot P–ρ data for (×) Pyrex, (○) Melmac, and (∗) Teflon.

9. *Pyrex, Melmac, and Teflon*

Applications have been found for these three common but dissimilar materials rather different than what was originally anticipated. The Hugoniot data are rather sparse for Melmac and Teflon, as only in-contact shots were fired. The u_s–u_p data (Fig. 67) are reasonably linear over the limited pressures covered. Assuming that no transitions are present, it appears that these two materials have the highest and lowest bulk sound velocities of all the plastics investigated here. The slope of the u_s–u_p Hugoniot for Teflon is higher than usual, apparently compensating for its low initial compressibility.

The Hugoniot data for Pyrex show a long, relatively flat region of constant shock velocity. This is caused by the two-wave structure resulting from the quartz to stishovite transition that has been studied previously. This transition is known to occur when either crystalline quartz or fused quartz is shocked above the transition pressure, which is thought to be about 100 kbar. The quartz transformation is identified with the Pyrex transformation because of the similarity of Pyrex and fused quartz. The Pyrex used here was 80.5% SiO_2, 12.9% B_2O_3, 3.8% NaO, 2.2% Al_2O_3, and 0.4% K_2O by weight; it was a vitreous material with an initial density almost identical to fused quartz. The data above the transition region, actually above the stable single shock threshold (300 kbar), is quite steep in the P–ρ plane (Fig. 68). This is compatible with the high initial sound speed, 10 km/sec, and density, 4.3 g/cm^3, thought to exist for stishovite.

10. *Micarta, Selectron, and Min-K-2000.*

Micarta and Selectron appear to be rather well behaved as plastics go. The u_s–u_p data (Fig. 69) are fitted reasonably well by the single straight-line curves. It is unfortunate that the extrapolated intercepts could not be compared with ultrasonic or compressibility data.

The Min-K-2000 is a fibrous insulating material. The negative slope in the P–ρ plane, Fig. 70, is of course due to the linear least-squares fit of the u_s–u_p data. This fit gives a negative sound velocity intercept, which causes this behavior. The data can actually be fitted quite nicely by a straight line through the origin (u_s–u_p), which then gives a vertical asymptote at a density of approximately 1.2 g/cm^3.

I. Data Tables and Calculated Results

The experimental data for the materials considered here are presented in some detail in this section. In Table VI are listed, when available, the longitudinal, transverse, and bulk sound velocities at zero pressure

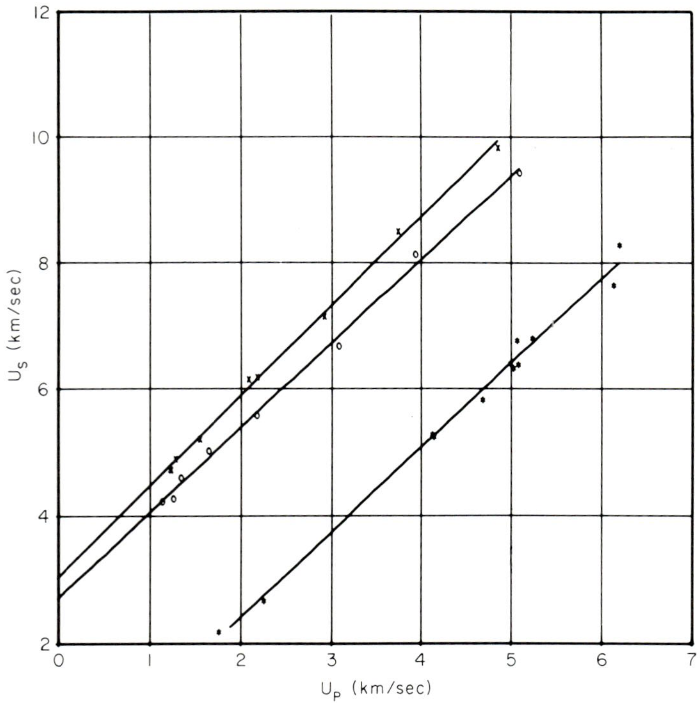

FIG. 69. Hugoniot u_s–u_p data for (×) Micarta, (○) Selectron, and (∗) Min-K-2000.

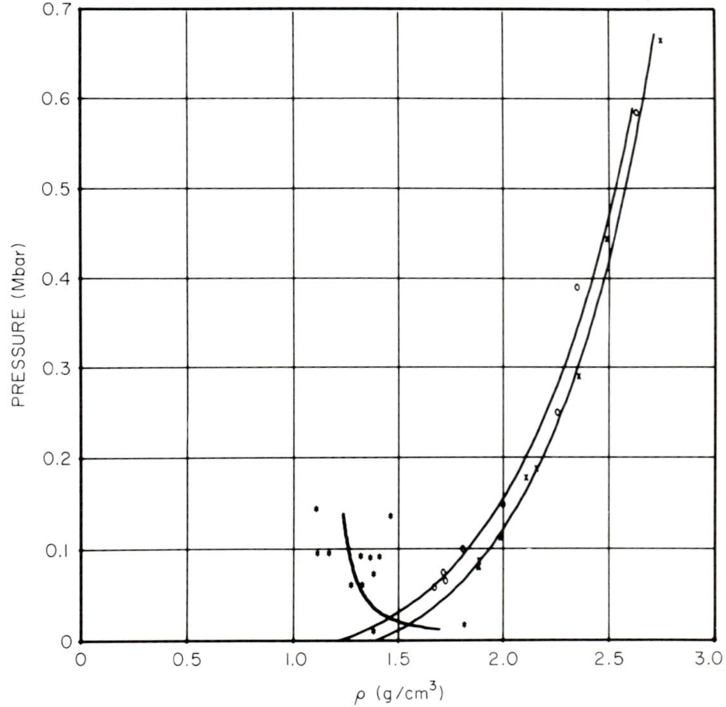

FIG. 70. Hugoniot P–ρ data for (×) Micarta, (○) Selectron, and (∗) Min-K-2000.

TABLE VI

ULTRASONIC AND HUGONIOT PARAMETERS

Material	c_l (km/sec)	c_t (km/sec)	c_b (km/sec)	c_0 (km/sec)	s
Beryllium	13.18	8.98	8.14	7.998	1.124
Magnesium	5.74	3.15	4.44	4.492	1.263
Magnesium (AZ-31 B)	5.73	3.05	4.52	4.522	1.242
Titanium	—	—	—	5.220	0.767
Zirconium	4.77	2.39	3.89	3.757	1.018
Hafnium	3.86	2.12	2.98	2.954	1.121
Vanadium	—	—	—	5.077	1.201
Niobium	5.03	2.11	4.39	4.438	1.207
Tantalum	4.15	2.07	3.39	3.414	1.200
Chromium	6.60	4.32	4.32	5.173	1.473
Molybdenum	6.45	3.47	5.04	5.124	1.233
Tungsten	5.22	2.89	4.02	4.029	1.237
Rhenium	5.33	2.89	4.16	4.184	1.367
Rhodium	—	—	—	4.807	1.376
Palladium	4.57	2.06	3.90	3.948	1.588
Iridium	—	—	—	3.916	1.457
Platinum	4.08	1.79	3.51	3.598	1.544
Silicon carbide	11.73	7.43	8.00	8.000[a]	0.950
Tungsten carbide	6.89	4.18	4.92	4.920[a]	1.339
Stainless Steel					
304	5.77	3.12	4.51	4.569	1.490
304 L	5.79	3.16	4.49	4.569	1.490
Neoprene	—	—	—	2.785	1.419
Adiprene	—	—	—	2.332	1.536
Polyurethane	2.39	1.03	2.07	2.486	1.577
Polyethylene	2.04	0.66	1.90	2.901	1.481
Polystyrene	—	—	—	2.746	1.319
Lucite	2.69	1.38	2.17	2.260	1.816
Plexiglas	2.72	1.40	2.19	2.572	1.536
Paraffin	—	—	—	2.960	1.531
Epoxy	2.87	1.48	2.30	2.678	1.520
Exon	—	—	—	1.952	1.660
Durite	—	—	—	2.847	1.404
Phenoxy	2.51	1.07	2.18	2.266	1.698
Phenolic Refrasil	—	—	—	3.016	1.005
$Li_2B_4O_2 \cdot 5H_2O$ epoxy	4.40	2.67	3.14	4.052	1.486
Pyrex	5.55	3.45	3.88	—	—
Melmac	—	—	—	3.505	1.213
Teflon	—	—	—	1.682	1.819
Micarta	2.67	1.50	2.03	3.048	1.422
Selectron	—	—	—	2.722	1.333

TABLE VI (*continued*)

Material	c_l (km/sec)	c_t (km/sec)	c_b (km/sec)	c_0 (km/sec)	s
Elkonite					
TC-10	5.43	2.95	4.23	—	—
10W3	4.77	2.53	3.77	—	—
3W3	4.75	2.50	3.77	—	—
1W3	4.55	2.66	3.36	—	—
2125 C	4.18	2.15	3.36	—	—

[a] Hugoniot required to go through ultrasonic bulk sound velocity.

as determined by ultrasonic measurements. In some cases, for example in low-density foams or highly oriented crystalline material, it was impossible to transmit a meaningful signal through the sample and the sound velocities could not be measured. The Hugoniot intercept c_0 has been included, when applicable, for comparison with the bulk wave velocities calculated from the ultrasonic data. For easy reference, the slopes of the u_s–u_p Hugoniots are also included for each of these materials. These Hugoniot fits are only valid to the first transition, if a transition exists in the material.

The Hugoniot data are presented in Appendix D at the end of the volume under the following table headings. The directly measured quantities are the initial densities, ρ_0, the shock velocity of the standard material, u_s(STD), and the shock velocity in the material under investigation, u_s. The particle velocity, u_p, and pressure, P, are inferred through the impedance match solution. The conservation relations then yield the density, ρ, and relative volume, V/V_0. The identification number in the column labeled STD refers either to the material used as a standard or the origin of the zero-pressure bulk sound velocity. The tabulated numbers have the indicated significance.

STD	Meaning
1	2024 Al standard
2	921-T Al standard
6	Tu-3 Mo standard
10	Copper standard
11	High-leaded brass standard
12	Iron standard
30	Sound speed derived from Bridgman's isothermal bulk modulus
35	Sound speed derived from elastic constants C_{ij}
40	Isotropic sound speeds from ultrasonic measurements

Included in this table as well are Hugoniot data for some elkonites, two gold–germanium alloys, and pure germanium. These materials are discussed in Section V.

Appendix E presents the thermodynamic quantities on the Hugoniot, the release isentrope, and two isotherms at absolute zero and at room temperature, all calculated by the methods outlined in Section II. We have used handbook values of the zero-pressure thermodynamic parameters required for these calculations; these values have been listed at the heading of the table for each separate material, along with a brief description of the material itself.

V. The Method of Mixtures

Since it is obviously impossible to measure the equation of state of all the materials of interest for various applications, a study has been undertaken to measure the Hugoniots of a few composite materials and to compare these Hugoniots with ones computed from the known equation of state of their constituents. Such materials fall into two large categories: those in which there is no chemical interaction of the constituents, that is, simple mixtures; and those in which there is a chemical reaction or bonding of the constituents on an atomic scale. In the case of simple mixtures probably the most common naturally occurring examples are rocks. Other simple mixtures are synthetic materials such as the various plastic-bonded solids and the commercially available elkonites (mixtures of copper or silver with refractory materials such as tungsten, molybdenum, or tungsten carbide). For these materials there is no problem associated with the bonding and crystal structure. However, these are heterogeneous materials and questions of hydrodynamic and thermodynamic equilibrium behind the shock wave become more complicated. Since the shock wave not only sees particles of different shock impedance but also travels through these particles at different velocities, it is apparent that the specimens under investigation must be made of very small particles if equilibrium conditions, particularly thermal equilibrium, are to be obtained in the short times available.

Alloys are of particular interest for several reasons. One is that a large number of so-called alloys are actually simple mixtures. Also, metallic bonding in general produces a much smaller volume and energy defect than do ionic and covalent bonding. It is then of interest to see how well the theory of ideal mixtures can account for the Hugoniot curves of various alloys.

For a large class of compounds simple mixing models are obviously inadequate, and it would be better to attempt to derive a high-pressure equation of state by other means. For example, if the equation of state of an ionic solid were desired, it would undoubtedly be better to use the known binding energies and some assumed potential rather than to attempt to mix the two components. There are some types of compounds that are somewhat amenable to mixing calculations. For example, to a certain extent the silicates can be considered as mixtures of their basic neutral oxide components. Hence, if the equation of state of the oxides are known, the similarities that exist in the bonding suggest that simple mixing models may be adequate. The possibility of synthesizing the equation of state of minerals from the equation of state of their basic oxide constituents is intriguing and preliminary work indicates that some success will be obtained.

First efforts in this program have been very encouraging. Heterogeneous materials can come sufficiently close to hydrodynamic and thermodynamic equilibrium so that their measured Hugoniots are experimentally indistinguishable from those calculated from the ideal mixing model. It has been found also that mixing techniques offer a new way to study materials that are complicated by rigidity effects.

There should be no reason that mixture calculations could not be used over the entire pressure regions. However, the existence of phase changes and rigidity effects some time prevents an accurate determination of the equation of state in this region. The effect of phase changes can be handled in a completely straightforward manner, but requires additional information. The effect of rigidity appears to be a major area of difficulty in hydrodynamic studies in some materials. It is not immediately apparent how simple mixing ideas will apply to such problems.

A. Calculational Model

1. *Hugoniot Equation of State of Mixtures*

There are several models that could be used in an attempt to predict the Hugoniots of mixtures from the known equation of state of their components. One approach might be to consider the mixture as a laminated structure and to compute an average shock velocity based on the shock transit time through the various layers. Such a method is appealing in that one is attempting to compute the quantity that is subject to direct measurement. Another approach would be to interpolate pressure–volume Hugoniots at constant pressure on the basis of the volume occupied by the various components. This method appears to be a bit more straightforward and is somewhat like the approach we have chosen.

For an ideal mixture, the isothermal equation of state as measured in a high-pressure static device should be precisely that obtained by the addition of the volumes occupied by the various components at that pressure. For a Hugoniot curve, on the other hand, this simple mixing is slightly complicated by the different temperature increases in different constituents resulting from pressure increases produced by a shock. However, this is a small effect compared to the compressional energy stored in each of the components by the shock. The change in pressure resulting from the redistribution of energy because of these temperature differences is less than the uncertainties in pressure changes produced by our limited knowledge of the behavior of the Grüneisen parameter at high pressures. Accordingly, a simplified method of calculating the shock pressure of a mixture is used. A "zero-Kelvin" isotherm is first calculated for each component. For the present application we have assumed both constant specific heat, C_V, and a constant value for the parameter $(\partial E/\partial P)_V$. Once the zero-pressure density is obtained the rest of the zero-Kelvin curve is calculated by simultaneously satisfying the thermodynamic law (Eq. 19), the experimental Hugoniot, and the Grüneisen parameter. The specific heat, of course, vanishes as T goes to zero, but the error in the low-temperature range of this extrapolation should be small relative to the uncertainties in pressure in the Hugoniot data. Since we go down and up in temperature in the same way, the "zero-Kelvin" curve should thus be regarded as an artifice of the calculation, although the resulting P–ρ curves will actually be quite near the real zero-Kelvin curve. A zero-Kelvin curve may now be obtained for the mixture from the following equations:

$$1/\rho_{\mathrm{K}} = V_{\mathrm{K}} = \sum m_i \, V_i(P) \tag{45}$$

$$E_{\mathrm{K}} = \sum m_i \, E_i(P) \tag{46}$$

where ρ, V, and E are the density, specific volume, and specific energy of the mixture on the zero-Kelvin curve; m_i are the mass fractions of the components in the mixture: and V_i and E_i are the volume and energy for the ith component as a function of pressure along the zero-Kelvin curve. The Grüneisen parameter of the mixture, or, equivalently, the quantity $(\partial E/\partial P)_V = V/\gamma$, is needed to extend the energy function up in pressure at constant volume. Also, the specific heat at constant volume of the mixture is needed to center the Hugoniot of the mixture at some temperature other than zero. For these two quantities the equations

$$V/\gamma = \sum m_i (V/\gamma)_i \tag{47}$$

$$C_V = \sum m_i C_{Vi} \tag{48}$$

are used.

The relation $E(P, V)$ for the mixture is now given by Eq. (15). This function is used in the Rankine–Hugoniot relationship to calculate the Hugoniot of the mixture. A slight error is introduced in this procedure because it is necessary to compute pressure offsets at a constant volume of the mixture, while the γ's used in this calculation have been computed at constant pressure from constant volumes of the components. This is not considered to be a limiting defect since the values of γ at high pressure are not accurately known.

2. *Calculation of Elastic Wave Velocities of Mixtures*

In addition to the Hugoniots of mixtures, the propagation of elastic waves in a heterogeneous substance is also subject to experimental investigation. As before, it is desired to sum the specific quantities since these are additive for the mixture.

The purely specific quantity most closely connected with the bulk sound speed c_b is $(-\partial V/\partial P)_T$. It is this quantity which is additive for mixtures provided the intensive variables pressure and temperature have come to equilibrium. However, in a sound wave, achievement of temperature equilibrium between different components of the mixture depends on grain size. If no heat flow occurs, individual grains are compressed and released isentropically. The specific quantity S, the entropy, is constant, and the expression $(-\partial V/\partial P)_S$, which is equal to $1/\rho^2c^2$, is linearly additive; that is, this function should be a linear function of the mass fractions m_i, and therefore a linear function of the specific volume $(1/\rho)$ of the mixture.

The same arguments apply with respect to the longitudinal wave velocity c_l, which lead to the result that the expression $(-\partial L/\partial T)_S$, which is equal to $1/\rho^2c_l^2$, should be a linear function of the mass fractions. Here L is the length/unit mass and T the appropriate tension. In contrast to the bulk compression, we do not expect this function to be exactly linear. It would be if the components were stacked up in series, but they can also occur in parallel. If they are in parallel, the stiffer member of the system can take up more of the tension and $-\partial L/\partial T$ will decrease. This would be an interesting feature to check in some properly prepared specimens where one could vary the orientation of laminations or fibers embedded in materials.

Since shearing motions do not involve specific quantities, these arguments say nothing about the shear waves directly. However, the elastic shear wave velocity c_t is specified by the isotropic relationship given by Eq. (38). Thus in principle it should be possible to estimate the value of the elastic moduli of an isotropic mixture from its constituents.

If these thermodynamic arguments are to be applicable to a material it must be homogeneous for distances larger than a fraction of the wavelength of the elastic wave and heterogeneous on an atomic scale (i.e., chemical reactions should not drastically change the component materials). A material can have voids, have a soft component in a hard matrix, or vice versa. It can be fibered, layered, have preferred orientations of highly anisotropic materials, or a host of other peculiarities that cause it to deviate from results predicted by the simple thermodynamic mixing we have used here. Strong shocks tend to get rid of voids and structure (at least make it less important) and make materials more isotropic. These difficulties should therefore plague a zero-pressure sound speed measurement much more than the Hugoniot data.

As an alternative to the ideal mixing, we consider a layered model. The time for a wave to travel through many layers is simply the sum of the individual times for each layer. Hence the reciprocal average velocity (c) is the sum of the reciprocal component velocities (c_i) weighted by the volume fraction (y_i) of the component:

$$1/c = \sum (y_i/c_i) \qquad \text{with} \quad \sum y_i = 1 \tag{49}$$

Mixing generally tends to lower velocities. The layered structure permits each component to transmit the wave as fast as each component can. The above result usually serves as an upper bound for expected values of measured velocities.

B. Experimental Results on Mixtures and Alloys

1. *The Zero-Pressure Sound Velocity and Hugoniots of Mixtures*

The present investigation on mixtures is limited to a study on some Elkonites. Elkonite is the name of a family of metal composites manufactured by P. R. Mallory and Co., Inc. They are composed of a sintered metal with a high melting point (W, WC, or Mo) into which has been infiltrated a metal having a lower melting point (Ag or Cu). The particular Elkonites (and their composition) which were used in this study are shown in Table VII. The composition was determined by wet analysis. On the basis of the measured composition we have calculated what the density should be and have compared it with the measured density. It can be seen that all samples have porosity. It was also found that for a given type of material the composition varied as well as the amount of porosity. This has caused considerable scatter in the Hugoniot data and in the sound speed measurements. The results of the sound speed measurements are compared with the calculated sound velocities in Figs. 71 and 72.

TABLE VII

ELKONITE COMPOSITIONS

Trade name	ρ_0 (g/cm^3)		Weight percent			
	Meas.	Calc.	Ag	Cu	W	WC
20S	15.21	15.67	27.6	—	72.4	—
35S	14.68	15.16	32.5	—	67.5	—
50S	13.27	13.68	48.9	—	51.1	—
4050	13.14	13.85	46.9	—	53.1	—
2125C	9.69	10.35	—	74.5	25.5	—
1W3	12.40	12.68	—	45.0	55.0	—
3W3	13.80	14.70	—	32.0	68.0	—
10W3	14.85	15.09	—	24.0	76.0	—
TC-10	11.65	11.68	—	42.0	—	58.0

The solid curves in these two figures are the ideal mixing results. The dashed curves are the layer model results. The data favors ideal mixing and, as expected, the layer calculation serves as an upper bound to the measured sound speeds. One trend that can be observed in the data is the increase in sound speed over that predicted by the mixing model as we go toward larger weight percents of tungsten. One could suppose that the rigid material forms continuous paths at these higher densities, and in spite of voids or the lower-density, lower-sound-speed filler material, the sound wave is transmitted by these "bridges" over a path length approximately equal to the thickness of the sample and at a velocity somewhere between that of the longitudinal velocity and the rod velocity (i.e., the one given by Young's modulus) of the tungsten. On the other hand, as we go toward the copper- or silver-rich regions, bridges of tungsten must become more tenuous and less well braced. There may be transmission along these weaker bridges, but it will attenuate more rapidly and let the copper or silver carry the bulk of the wave. Here porosity can have a greater effect, in that it now creates copper or silver bridges loaded with high-density tungsten. When the tungsten no longer forms bridges, it effectively no longer transmits the signal at its higher longitudinal velocity and its high density now becomes a drag on the wave. These topological considerations indicate the complexity of the problem and also indicate that elaborate statistical considerations are necessary to solve the problem for an ensemble of particles. For large grain sizes a considerable scatter is still expected in the results obtained on individual samples about such an average

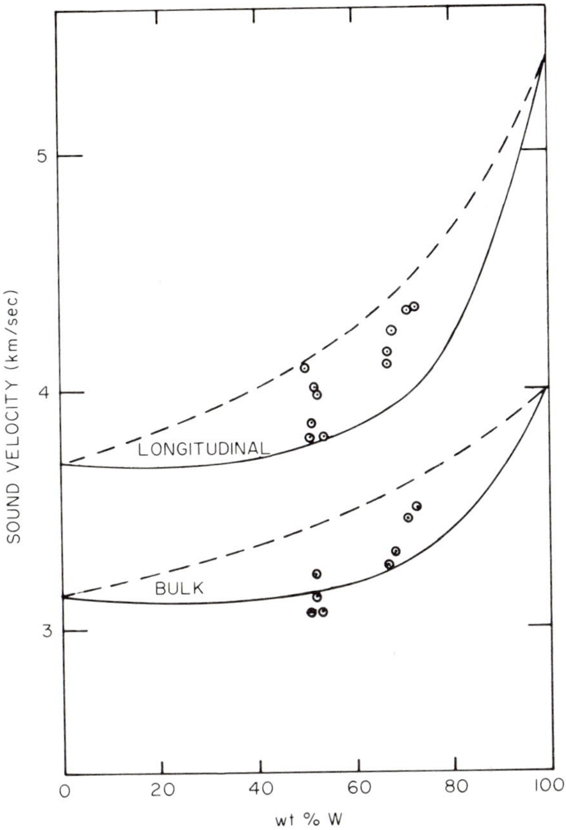

FIG. 71. Longitudinal and bulk sound velocities for the Ag–W Elkonites.

solution even if it were obtained. Although we have not tried to handle the porosity known to be in our samples in our calculations for the sound speed, it has been taken into account in our calculations for the shock wave experiments. Unfortunately we cannot simply extrapolate the shock model to zero compression to get the acoustic limit, for any voids in this model would lead to a zero limit for the sound speed. The hydrodynamic model does not take rigidity into account, and the rigidity, of course, keeps the voids from collapsing, especially in the very-low-pressure regime.

The Hugoniots of four copper–tungsten and one copper–tungsten carbide Elkonites have been measured. The equation of state used for the constituents is given in Tables II and VI.

The copper Hugoniot is one of our shock wave standards. The

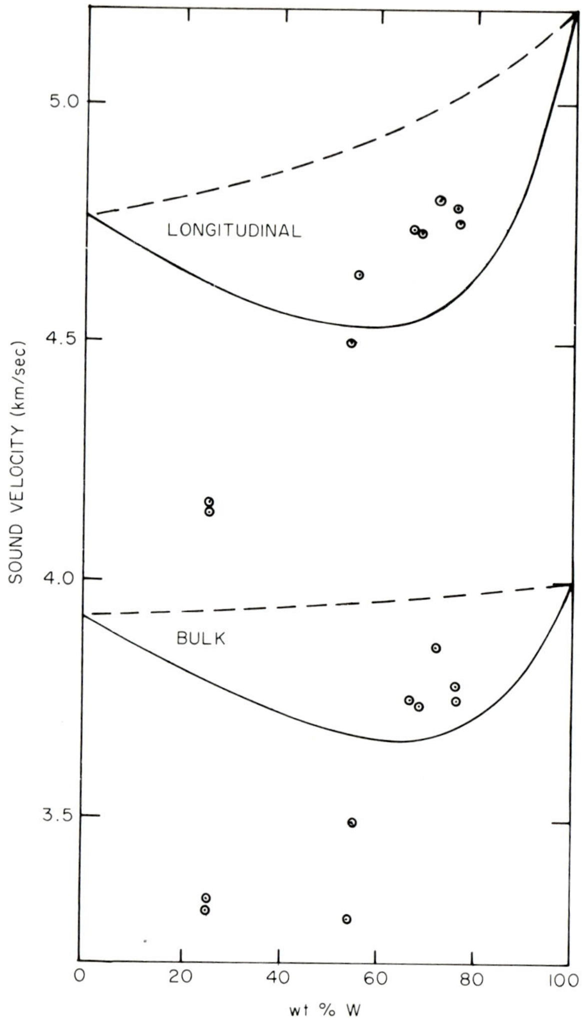

FIG. 72. Longitudinal and bulk sound velocities for the Cu–W Elkonites.

tungsten and tungsten carbide Hugoniots were partially determined using this standard. Hence the three materials should be compatible. The Elkonites were also measured with this standard. The data and calculations are compared in Figs. 73 and 74 in both the u_s–u_p plane and in the P–ρ plane. The agreement of the Cu–W data with the experiments is truly satisfying. In general the scatter in the data is due to variation in density in the samples. Since the composition of each

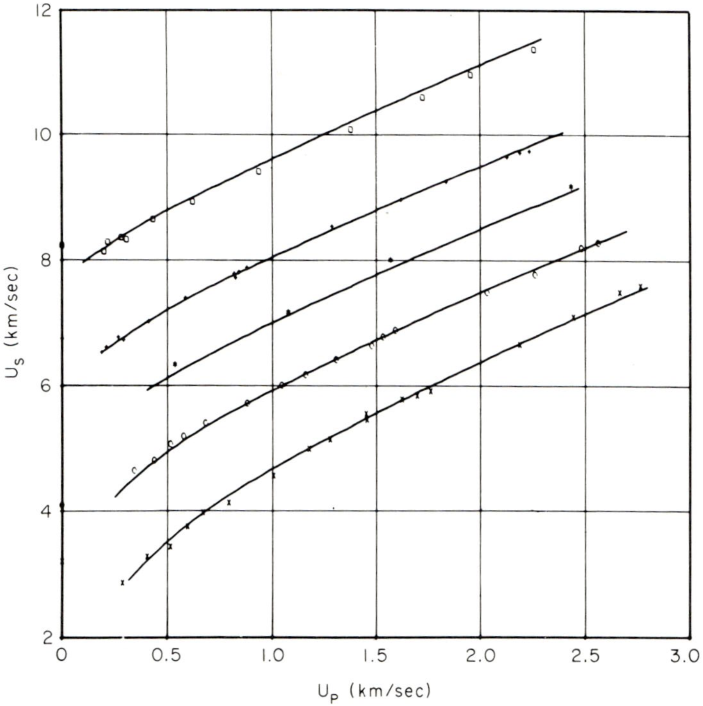

FIG. 73. Comparison of shock particle velocity data for Cu–W and Cu–TC Elkonites with the Hugoniots calculated by the method of mixtures. (×) 2125C (u_s), (○) 1W3 ($u_s + 1$), (∗) 3W3 ($u_s + 2$), (+) 10W3 ($u_s + 3$), (□) TC-10 ($u_s + 4$). Note the various shock velocities have been changed by increments of 1 km/sec.

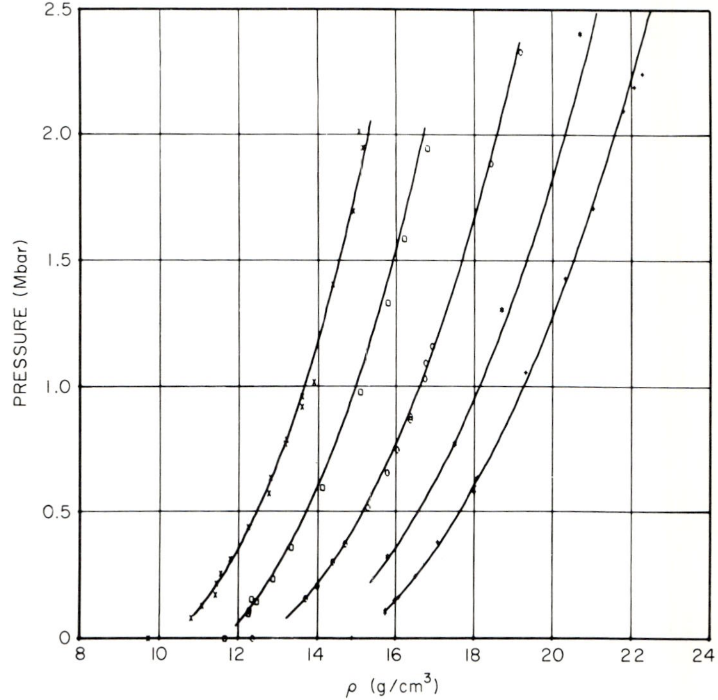

FIG. 74. Comparison of pressure–density Hugoniot data for Cu–W and Cu–TC Elkonites with the Hugoniots calculated by the method of mixtures. (×) 2125C, (○) 1W3, (∗) 3W3, (+) 10W3, (□) TC-10.

sample was not measured, these density variations could be due either to nonuniform composition or to varying degrees of porosity. One feature common to all the materials is the low shock velocity at low pressure. This is due to the porosity and has been observed for many other materials. The calculations account for this in a satisfactory manner.

The data for the TC-10 material does not agree as well with the calculations as the copper–tungsten series. There are two possible explanations for this. One is that the material used for the determination of the tungsten carbide Hugoniot is slightly different than that used in the mixture. It is known that the tungsten carbide for the former contained a cobalt binder. The other reason is that the tungsten carbide Hugoniot reflects the residual shear stress due to the large rigidity in this material. The measured tungsten carbide Hugoniot could very well be in a highly nonhydrostatic state, while in the mixture the copper will flow plastically and allow the rigid material to be compressed more hydrostatically. This is one of the reasons this study was initiated; that is, to develop a technique for measuring the Hugoniots of very rigid materials by imbedding them in materials that deform easily by plastic flow in the shock front. The good results obtained with the copper–tungsten series indicate that this might indeed be a very useful technique. The u_s–u_p Hugoniot of silicon carbide, also very rigid, shows this same effect.

2. *Alloys*

a. The Hugoniot Equation of State of Au–Ge Alloys. The Hugoniots of two gold–germanium alloys have been measured using the standard impedance match method with copper used as the standard. Although these two alloys came from the same casting, their compositions are different because germanium is not soluble in gold and separation occurred in the casting. These alloys, in fact, are in reality almost perfect mixtures.

The weight percents have been calculated from the densities. The measured and calculated sound velocities are given in Table VIII. The data (c_{ij}'s) used to calculate the end members for gold and germanium were taken from Clark's handbook (*23*).

The shock particle velocity data of these two materials are given in Fig. 75. Two features of these Hugoniots should be noted. The most important is that the data do not extrapolate to the zero-pressure sound velocity. The fact that they fall below the sound velocity by over 10% is an indication that a transition has occurred. This transition probably occurs at a particle velocity of about 0.15 km/sec. The other feature

TABLE VIII

MEASURED SOUND VELOCITIES OF Au–Ge ALLOYS

ρ_0 (g/cm^3)	Ge (wt %)	c (km/sec) Meas.	c (km/sec) (Calc.)	c_t (km/sec) Meas.	c_t (km/sec) (Calc.)	c_b (km/sec) Meas.	c_b (km/sec) (Calc.)
15.49	9.3	3.388	(3.49)	1.465	(1.72)	2.94	(2.87)
16.88	5.8	3.326	(3.41)	1.329	(1.56)	2.95	(2.89)

is the nonlinear behavior of the two curves. As with some other materials, these Hugoniots can be drawn with two straight lines with the break occurring near a particle velocity of 1.2 km/sec. To help explain these features we have reproduced the phase diagram ($P = 0$) of the Au–Ge system (*24*) in Fig. 76. It is well known that germanium contracts upon melting, and apparently the Au–Ge alloy at the eutectic also contracts on melting. Shocked germanium undergoes a transition at a pressure near 125 kbar, possibly due to melting. The transition in the gold–germanium alloy at $u_p = 0.15$ km/sec corresponds to about this pressure. This seems to be about correct, considering the effect that gold will have on the slope of the phase line and the pressure–temperature relationship along the Hugoniot of the alloy. Although the Au-Ge alloys start to melt at this pressure, melting will not be complete until a higher temperature is attained; this corresponds to a considerably higher pressure. The break in the u_s–u_p Hugoniot curve at a pressure near 700 kbar, although somewhat high, could indicate the completion of melting.

To check our ability to predict the Hugoniot of mixtures we have taken the linear Hugoniot of gold,

$$u_s = 3.075 + 1.560u_p \quad \text{km/sec}$$

and mixed it with the Hugoniot of transformed germanium. The transformed germanium Hugoniot was obtained from our shock wave data by assuming that the metastable Hugoniot of a transformed material will have a linear u_s–u_p Hugoniot, which allows one to determine the new zero-pressure density and sound speed for selected slopes s. This calculation has been described in detail in the second section of this chapter. These calculations predict the equations of state for high-density germanium listed in Table IX.

The results of these fits are compared with the original germanium

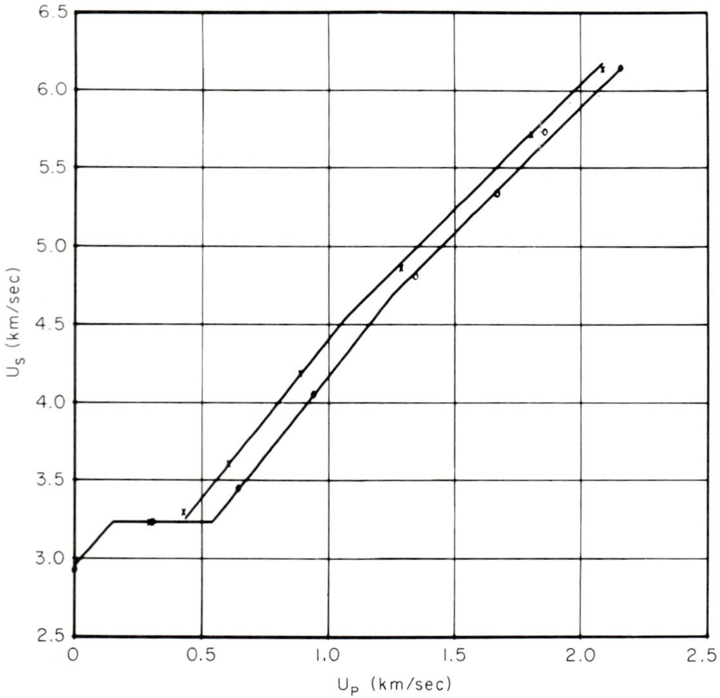

FIG. 75. Shock particle velocity data for two Au–Ge alloys. (×) 5.8 wt % Ge, (○) 9.3 wt % Ge.

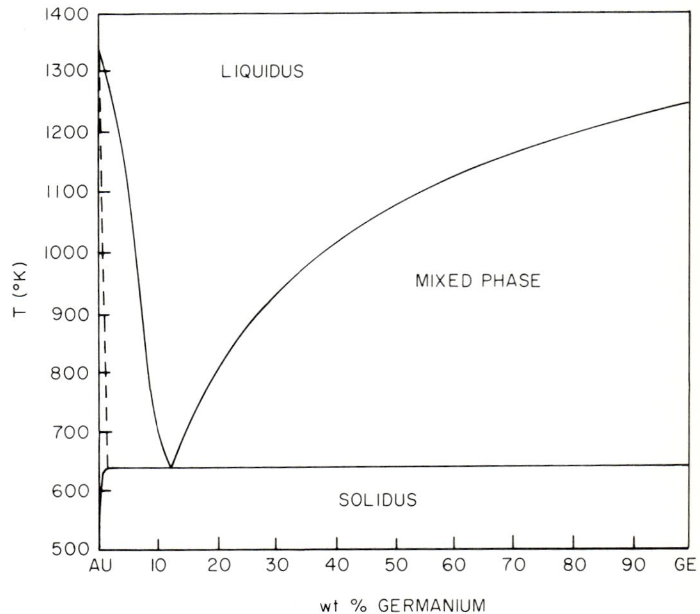

FIG. 76. Phase diagram for the Au–Ge system. As can be seen all the alloys will begin to melt at 356°C. The completion of melting will depend on the composition.

TABLE IX

ZERO-PRESSURE DENSITY AND SOUND VELOCITY FOR TRANSFORMED GE AS A FUNCTION OF SLOPE OF THE LINEAR u_s-u_p HUGONIOT

Slope	c_0 (km/sec)	ρ_0 (g/cm^3)
1.00	5.82	7.29
1.25	5.08	7.10
1.50	4.15	6.79

data in Fig. 77. It is apparent that all three slopes adequately fit the data, giving little reason to choose one slope over the others. However, the calculated volume changes at the transition can be compared with values obtained from static measurements. The slope of 1.5 gives the best agreement.

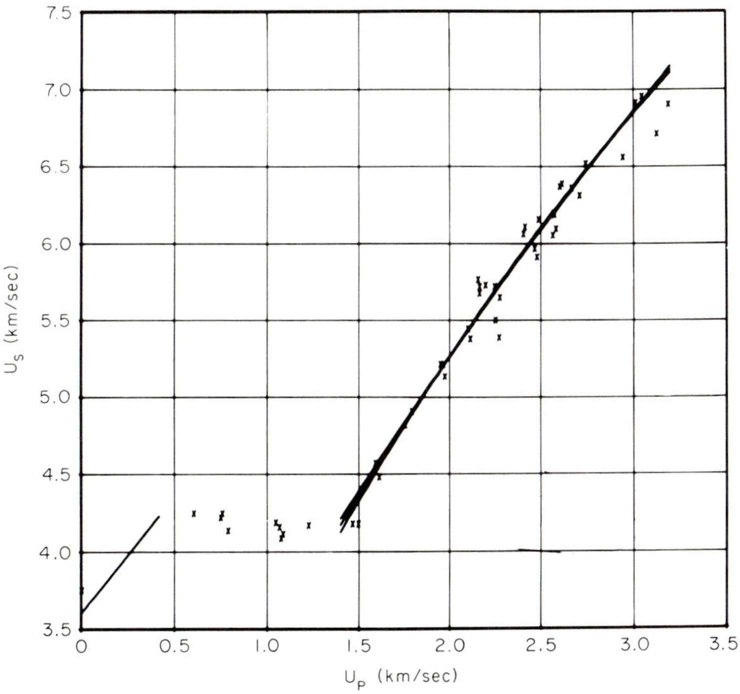

FIG. 77. Shock-velocity particle-velocity Hugoniot and calculations for Ge. (×) impedance match data. The flat region at $u_s = 4.2$ km/sec is the manifestation of a two-wave structure in the shock front due to a transition. The data above $u_s = 4.2$ km/sec have been fitted from calculations using the three Hugoniots listed in Table IX. With so much scatter in the data one fit cannot be preferred to any other.

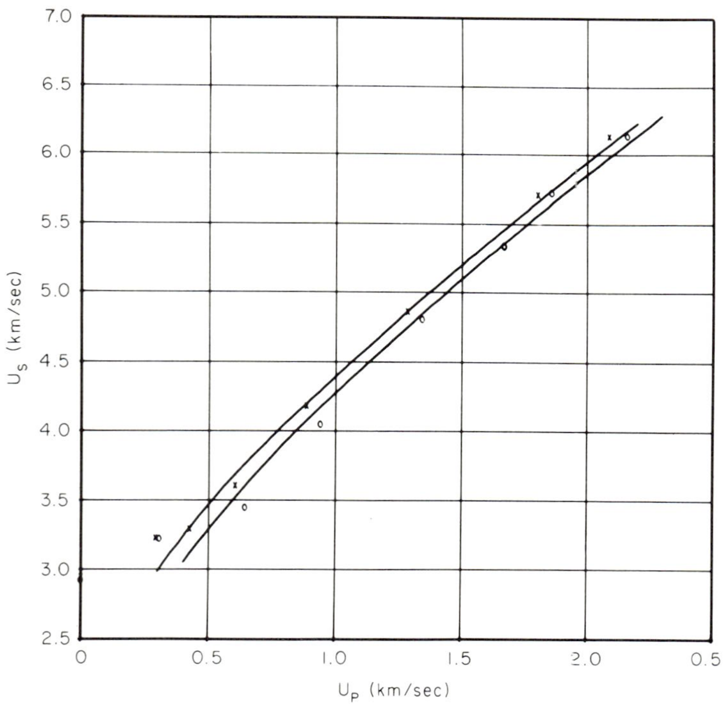

FIG. 78. Calculated shock particle velocity Hugoniots and the experimental data for two Au–Ge alloys. (×) 5.8 wt % Ge, (○) 9.3 wt % Ge.

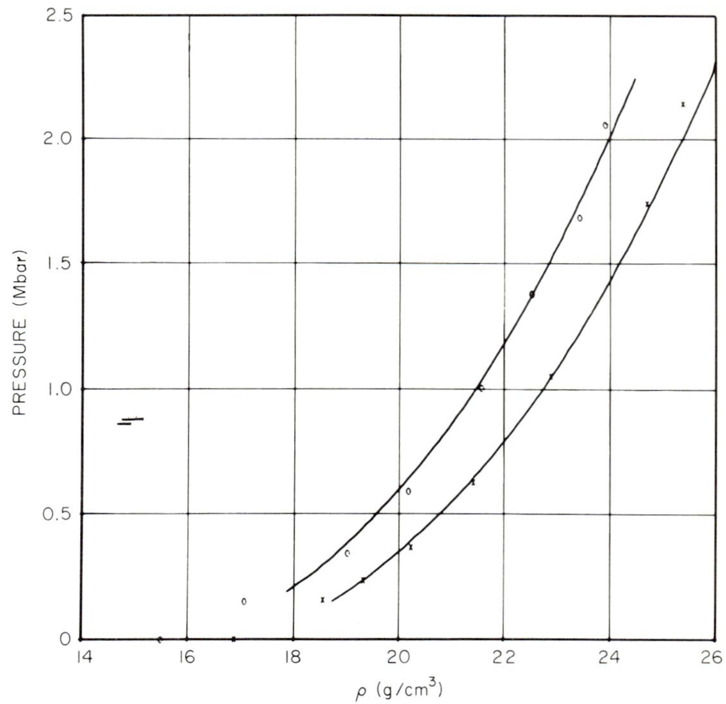

FIG. 79. Calculated pressure–density Hugoniots and the experimental data for two Au–Ge alloys. (×) 5.8 wt % Ge, (○) 9.3 wt % Ge. It should be noted that these calculations represent a considerable extrapolation of the Ge data since the shock wave data of Ge only goes to about 1 Mbar.

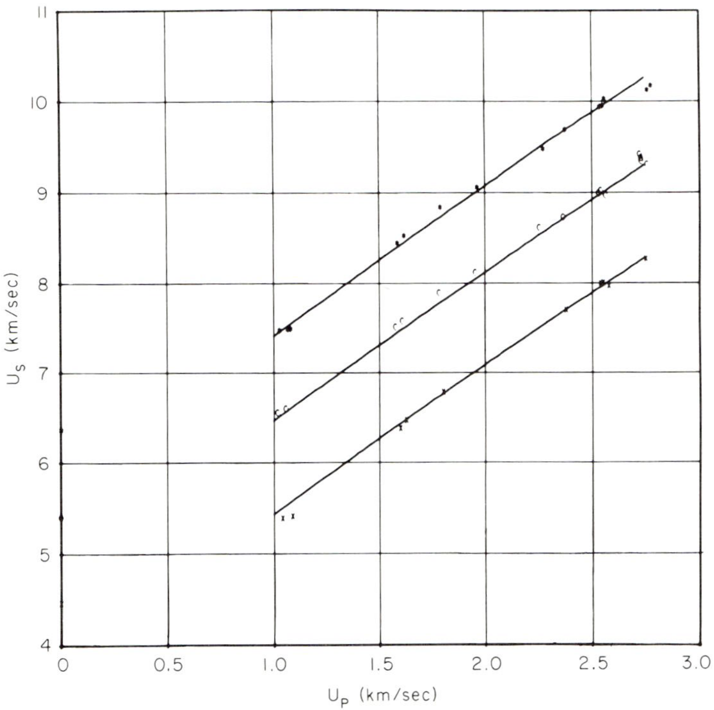

FIG. 80. Comparison of shock particle velocity data for Fe–Ni alloys with Hugoniots calculated by the method of mixtures. (×) Fe–10.0-wt %-Ni (u_s), (○) Fe–17.9-wt %-Ni ($u_s + 1$), (*) Fe–29.2-wt %-Ni ($u_s + 2$). In order to separate the curves and the data 1 and 2 km/sec have been added to 17.9 wt % Ni and 26.2 wt % Ni alloys. Data obtained in the two-wave region is not reported.

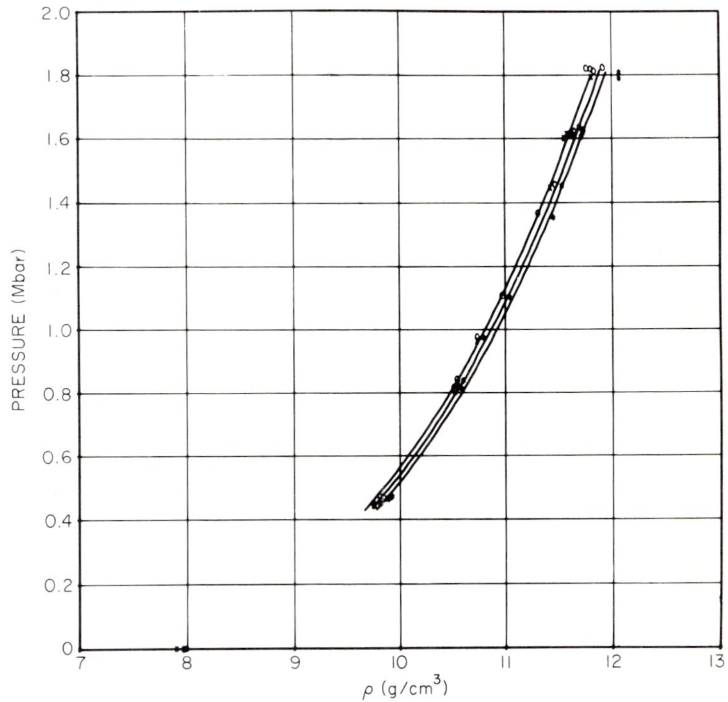

FIG. 81. Pressure–density Hugoniot data and the Hugoniots calculated from Ni and the metastable Fe Hugoniot. (×)Fe–10.0-wt %-Ni, (○) Fe–17.9-wt %-Ni, (*) Fe–26.2-wt %-Ni.

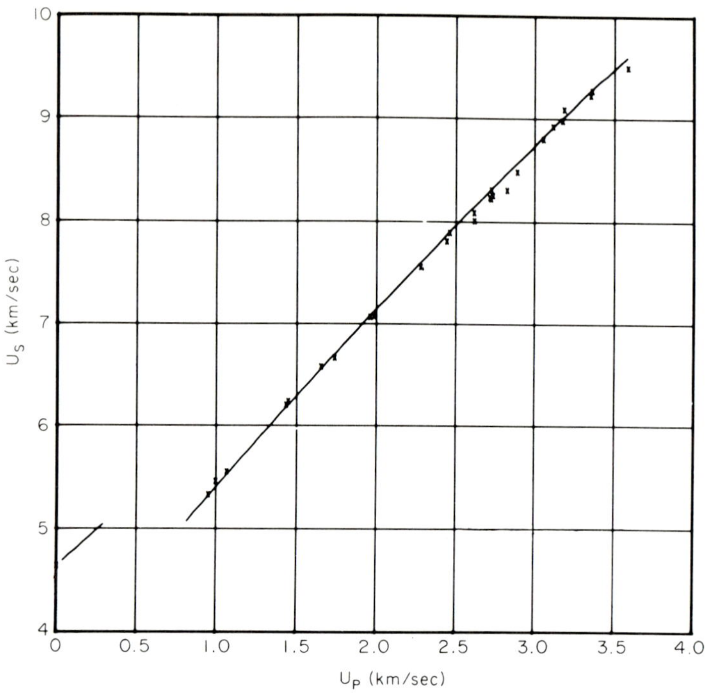

FIG. 82. Shock particle velocity data for iron (×) and the transformed Hugoniot calculated from the derived metastable equation of state.

These equations of state are used in the mixing calculation described earlier. The shock-velocity–particle-velocity and pressure–density results of this calculation are shown in Figs. 78 and 79.

b. The Hugoniot Equation of State of Iron–Nickel Alloys. As a final test of the validity of the method of mixtures, we have chosen three iron–nickel alloys having nickel content ranging from 10 to 26 wt%. Again the standard impedance match was used, with copper as the standard. Data on these materials have been presented before, but in a different context (*25*).

The Hugoniots for the iron–nickel alloys are presented in both the u_s–u_p and P–ρ planes in Figs. 80 and 81. Again it was obvious that a transition has taken place, almost certainly from the zero-pressure bcc phase to an hcp phase or an fcc phase. Above about 30% nickel content the alloy is stabilized in the γ (fcc) phase and will not transform under shock loading. Hence it is necessary to know the metastable Hugoniot for the γ phase of iron in order to perform the mixing calculation.

The metastable Hugoniot for transformed iron has been obtained from the shock particle velocity data of Section III by the same iterative method which was used for germanium. Details of the behavior of iron at and below the 130-kbar transition were obtained from the work of Bancroft *et al.* (*26*) and Hughes and Maurette (*27*). Figure 82 shows the calculated metastable iron Hugoniot, recentered at zero pressure, using the following parameters:

$$u_s = 4.33 + 1.55u_p, \qquad \rho_0 = 8.10 \quad \mathrm{g/cm^3}$$

This Hugoniot was then mixed with the nickel Hugoniot,

$$u_s = 4.581 + 1.463u_p, \qquad \rho_0 = 8.874 \quad \mathrm{g/cm^3}$$

to produce the calculated equations of state of the alloys shown as solid curves in Figs. 80 and 81.

VI. Elastic-Plastic Flow

A. Theoretical Models

Most of the statements and conclusions discussed in the preceding sections of this chapter depend on the condition that a strongly shocked solid be in thermodynamic equilibrium. Aside from considerations of thermal equilibrium, the most obviously possible deviation from thermodynamic equilibrium is an anisotropic stress distribution. Such a condition is known to exist to at least a small degree for virtually all shocked solids and, as mentioned above, can be approximately corrected out of the high-pressure data by assuming a small constant offset between the measured Hugoniot curve and the hydrostat in the pressure–volume plane.

In order to understand the context in which the phrase "elastic-plastic flow" is used in discussions of plane impact loading and plane shock wave propagation, it is necessary to consider explicitly the atomic nature of solids and also the classical theory of elasticity. The assumption that a macroscopically one-dimensional shock wave is truly one dimensional implies that as the shock front passes over any given volume element, the average spacing of the atoms changes only in the direction of motion of the shock front. If this is the case, there are no elastic strains transverse to the shock front.

Let the shock be propagating in the x direction in a material which is presumed to be an isotropic homogeneous linear elastic material. In the infinitesimal strain approximation, the components σ_{ij} of the stress tensor are given by

$$\sigma_{xx} = (\lambda + 2\mu)\,\epsilon_{xx} \tag{50}$$

$$\sigma_{yy} = \lambda\epsilon_{xx} \tag{51}$$

$$\sigma_{zz} = \lambda\epsilon_{xx} \tag{52}$$

$$\sigma_{xy} = \sigma_{yz} = \sigma_{zx} = 0 \tag{53}$$

where λ and μ are the Lamé constants which are related to the bulk modulus K and shear modulus G by the relations

$$G = \mu \tag{54}$$

$$K = \lambda + \tfrac{2}{3}\mu \tag{55}$$

The hydrostatic pressure, which is the mean value of the principal stresses, is given by

$$P = (\lambda + \tfrac{2}{3}\mu)\,\epsilon_{xx} = K\epsilon_{xx} \tag{56}$$

However, since it is really only the component of the stress normal to the shock front which can be deduced from the usual experimental data and the Rankine–Hugoniot equations, shock wave measurements could not really define the thermodynamic equation of state of a material which behaved in a fashion described by this model.

The key to understanding the usefulness of the shock wave technique and to an understanding of the concept how plastic flow under one-dimensional loading conditions results in nearly hydrostatic pressures lies in the realization that a model which requires no transverse microscopic strain also implies that very large shear stresses will develop on planes which lie at significant angles relative to the shock normal. Qualitatively it is known from metallurgical and mineralogical experience that no material can withstand unlimited shear stresses without either flowing plastically or fracturing, whichever comes first. It follows that all materials, under the influence of plane shock waves, will at some level of loading begin to shear plastically in such a manner that the maximum resolved shear stresses are held within the failure limits of the material. Therefore, provided that the maximum tolerable resolved shear stress is small compared with the shock stress under consideration and that the plastic flow can occur sufficiently rapidly, a strongly shocked solid should approximate a fluid behavior. The equation of state deduced from shock wave data will then correspond to a hydrostatic equation of state in the same approximation.

Now reconsider the situation arising in the compression of a solid by a plane shock wave, assert that macroscopically the compression is purely one dimensional, and attempt to discover what this means microscopically. The situation can be treated mathematically within the infinitesimal strain approximation, since it is always possible to consider a real problem in terms of successive differential steps. In any case, the total strain involved in most plane shock wave experiments is no more than 20% and the changes in the elastic moduli are vastly more important than the finiteness of the strains.

Consider a volume element which is sufficiently small to be considered a differential element relative to the physical scale of measurement, but sufficiently large to be considered a representative macroscopic sample of the solid being shock loaded. The shock is propagating in the x direction. Since all neighboring volume elements are being subjected to the same loading conditions, it follows that there can be no net average motion of matter across the boundaries of the volume element and in this coordinate system, which is also the principal axis system of the macroscopic strain tensor, the components of the strain tensor are

$$\epsilon_{xx} = \Delta V / V_0 \tag{57}$$

and all other $\epsilon_{ij} = 0$. However, from a coordinate system represented by planes which make arbitrary angles with the shock normal (the x axis in laboratory coordinates) shear strains appear which have their maximum values on 45° planes. These shear strains, from the macroscopic point of view, are an inevitable permanent feature of one-dimensional compression and can only be prevented from producing large shear stresses if plastic flow occurs in such a manner that the elastic shear strains are removed.

Modern solid-state theory and experiment have shown that plastic flow in crystalline solids proceeds in an essentially heterogeneous manner by the passage of dislocations through the lattice on a few lattice planes. The present theory of dynamic elastic-plastic flow is essentially an application of the present state of the art of dislocation dynamics to the special conditions existing in one-dimensional shock waves. No serious attempt has yet been made to compute wave propagation in a material other than a homogeneous, isotropic solid in which it is assumed that dislocations will move only on planes at 45° with the shock normal, i.e., where the resolved shear stress is a maximum. In this approximation for the one-dimensional shock problem, the total strain in each of the principal directions may be decomposed into an elastic part and a plastic part.

Hence, using symmetry arguments, the principal components of the strain tensor become

$$\epsilon_{xx} = \epsilon^{e}_{xx} + \epsilon^{p}_{xx} \tag{58}$$

$$\epsilon_{yy} = 0 = \epsilon^{e}_{yy} + \epsilon^{p}_{yy} \tag{59}$$

$$\epsilon^{e}_{yy} = \epsilon^{e}_{zz}\,, \qquad \epsilon^{p}_{yy} = \epsilon^{p}_{zz} \tag{60}$$

Also $0 = \epsilon^{p}_{xx} + 2\epsilon^{p}_{yy}$, since the plastic strains merely involve rearrangements of atoms and not compressions.

From dislocation theory the rate of plastic shear strain on any plane is given by

$$d\gamma_p/dt = bn\,\bar{v}(\tau) \tag{61}$$

where n is the dislocation line length per unit volume having a Burger vector b lying along this plane and $\bar{v}$ is the average velocity of these mobile dislocations. It is known from various other studies that $\bar{v}$ is a function, and frequently a strong function, of the resolved shear stress on the glide plane and in the glide direction. Along the planes of maximum resolved shear stress, the 45° planes, γ_p is given by

$$\gamma_p = \tfrac{1}{2}(\epsilon^{p}_{xx} - \epsilon^{p}_{yy}) = \tfrac{3}{4}\epsilon^{p}_{xx} \tag{62}$$

from Eq. (60).

The equation for the stress normal to the shock front now reduces to

$$\sigma_{xx} = (\lambda + 2\mu)\,\epsilon_{xx} - \tfrac{8}{3}\mu\gamma \tag{63}$$

and by taking the time derivative and using Eq. (61) the equation

$$d\sigma_{xx}/dt = (\lambda + 2\mu)(d\epsilon_{xx}/dt) - \tfrac{8}{3}\mu bn\,\bar{v}(\tau) \tag{64}$$

is obtained as the constitutive relation governing the response of a linear relaxing solid. It is of course anticipated that the mobile dislocation density n and the average velocity function $\bar{v}(\tau)$ will be functions of the past history of the volume element, and herein lies the central problem in the use of this constitutive relation in computations. It is also not at present certain at what stress level dislocation loops can actually be created spontaneously by the local resolved shear stresses.

At the present time it appears on the basis of ultrasonic attenuation measurements and extremely careful etch pit data on very pure crystalline solids that the viscous damping force on a moving dislocation is the ultimate factor limiting dislocation velocity. This viscous force has been discussed by Liebfried (*28*), Mason (*29*), Gilman (*30*), and others. The Koehler–Granato–Lücke (*31*) model of a pinned dislocation loop vibrating as a stretched string with a linear viscous damping term has been applied with remarkable success to the interpretation of the attenuation of both shear and longitudinal ultrasonic waves in the 1–1000-MHz frequency range. This frequency range covers essentially the entire spectrum of interest in shock wave measurements between roughly 1 and 100 kbar, but of course the stress and strain amplitudes employed in the ultrasonic damping experiments are very small compared with the amplitudes of interest in most shock wave work. One consequence of this is that, whereas in most ultrasonic damping experiments the dislocation line density is expected to remain constant during the experiment, in shock wave experiments the line density is expected to increase rapidly in the shock front if the sample is initially well annealed and to become significantly redistributed even if the sample is initially well work hardened. As a result, although the viscous dislocation damping appropriate to shock wave conditions may be quite similar in magnitude to that measured in ultrasonic experiments, the effect of the dislocation line tension, or line energy per unit length, as deduced from ultrasonic experiments cannot possibly persist into the strong shock regime because the line tension effect implies an effective relaxed modulus, under one-dimensional loading conditions, which is 10–15% higher than the hydrostatic bulk modulus. This is in disagreement with the observation that, for most metals, the modulus deduced from shock wave Hugoniot extrapolates very nearly exactly to the bulk modulus as measured in static presses.

The explanation undoubtedly lies in the fact that linear work hardening simply does not persist to very large strains under any set of loading conditions.

Within the framework of the linear model the above statements can be formalized by analogy with the current Granato–Lücke theory. Assume that there exists a network of dislocations which intersect with an average free line length l leading to a dislocation line density of $N = l^{-2}/\text{cm}^2$ of dislocation line per cubic centimeter of material. Under the influence of the shock wave the dislocations bow out between pinning points by an amount $\xi(y)$, where y is a coordinate measuring the distance from one pinning point to the next. The plastic strain is given by

$$\gamma = (Nb/l) \int_0^l \xi(y)\, dy \tag{65}$$

where b is the Burger vector of the material. The motion of the pinned dislocation obeys the equation

$$\pi\rho b^2 \frac{\partial^2 \xi}{\partial t^2} + B \frac{\partial \xi}{\partial t} - \frac{2\mu b^2}{\pi(1-\nu)} \frac{\partial^2 \xi}{\partial y^2} = \tau b \tag{66}$$

where τ is the resolved shear stress, ν is Poisson's ratio, B is the Granato–Lücke damping constant, and $\pi\rho b^2$ is the effective mass per unit length of dislocation line. Although the acceleration term is expected to be very small under most circumstances, it can be included in the analysis without complication. If it is assumed that the dislocation density remains constant, the system of Eqs. (63), (65), and (66) can be solved with the boundary condition $\xi(0) = \xi(l) = 0$ and the observation that for an isotropic medium the maximum resolved shear stress is

$$\tau_{\max} = \tfrac{3}{4}[\sigma_{xx} - (\lambda + \tfrac{2}{3}\mu)\,\epsilon_{xx}] \tag{67}$$

By taking Laplace transforms on the time, the complex modulus connecting the transforms of the one-dimensional stress and strain is obtained. The result is

$$\Sigma_{xx} = \left[\frac{(K + \frac{2}{3}\mu)/K + g(z)}{1 + g(z)}\right] KE_{xx} \tag{68}$$

where z is a function of the transform parameter corresponding to frequency and

$$g = \frac{\pi(1 - \nu)}{4z^2}\left[1 - \frac{\tanh}{z}(z)\right] \tag{69}$$

$$z(s) = \frac{l}{2}\left[\frac{\pi\rho b^2 s^2 + Bs}{2\mu b^2/\pi(1 - \nu)}\right]^{1/2} \tag{70}$$

In Eq. (68) Σ and E are the transforms of stress and strain, respectively. It is also assumed that the average loop length is given by $l = N^{-1/2}$. This is in fact an assumption about work hardening. Equations (68), (69), and (70) show that for very high frequencies (large values of s) the effective modulus is $K + \frac{2}{3}\mu$ or the longitudinal modulus. This is therefore the controlling influence in the propagation of the first disturbance. As $s \to 0$, however, corresponding to very low frequencies and long times, the modulus approaches the final value (relaxed value)

$$M_R = K\left[\frac{(K + \frac{4}{3}\mu)/K + \pi(1 - \nu)/12}{1 + \pi(1 - \nu)/12}\right] \tag{71}$$

in disagreement with the experiments at very high pressures.

If, however, the line tension term is deleted from the equation of motion (66), the function g in Eq. (68) becomes

$$g(s) = 2\mu N b^2/(\pi\rho b^2 + Bs) \tag{72}$$

and in the limit of long times the modulus becomes equal to K, the bulk modulus. However, the total plastic strain required to render the stress distribution essentially hydrostatic during a one-dimensional macroscopic compression of the order of 10% requires that for any mobile dislocation density less than approximately 10^{13} cm/cm^3 the average dislocation line element must move a distance equal to several tens or hundreds of times the average dislocation spacing. At the very least, dislocations must move to grain boundaries, but single crystals of cubic metals exhibit essentially the same behavior as polycrystalline samples (*32*). One possible conclusion is that in a shock whose strength is of the order of 100 kbar there must be a very large number of dislocation loops created (and probably annihilated) in a very short time. This would have the same effect as the motion of, say, 10^{10} loops/cm^2 over distances of the order of 10^{-3} cm.

In many impure materials, and at relatively small resolved shear stresses, it has been found experimentally (*33*) that average dislocation velocities obey the equation

$$v(\tau) = v_0 \exp\{-D/\tau\} \tag{73}$$

over many orders of magnitude of dislocation velocity, where D is a material constant called the drag coefficient and depends on impurities such as point defects and the existing dislocation density. The quantity v_0 has previously been assumed to be less than or at most equal to shear wave velocity, except that the Stein–Low (*34*) data on Fe–3%-Si single crystals contradicts this assumption. Chen *et al.* (*35*) have shown that this could be caused by fluctuating internal stress fields. Equation (73) has been used with some success to explain the behavior of the elastic precursor in shocked annealed Armco iron (*36*), and a theoretical thermal activation model has been proposed to justify its validity, particularly in impure metals where point defect and dislocation forest pinning may be expected to be important (*33*). A similar result is obtained for covalent crystals where it is anticipated that strongly localized chemical bonds may be expected to provide an energy barrier to dislocation motion.

It appears that a composite model may well be appropriate to the case of shock-induced dislocation motion, since the external observables in this situation are quite comparable to the etch pit motion experiment. The observed average velocity is given by a distance of motion X_m, either directly measured or inferred. Assume that the average moving dislocation line segment is pinned n times while traversing the distance X_m, with a probability of release per unit time given by

$$p = \nu_0 e^{-E(\tau)/kT} \tag{74}$$

where the activation energy $E(\tau)$ may depend on the resolved shear stress τ in the reciprocal manner predicted by the Fisher–Gilman bowing-out theory. In between pinning point delays, the dislocation moves at a velocity $\tau b/B = v(\tau)$ as long as this velocity is not too near shear wave velocity (in which vicinity the situation is presently unclear). The measured time of motion is then given by

$$t_m = X_m/v(\tau) + ne^{E(\tau)/kT}/\nu_0 \tag{75}$$

where clearly $n = C_P^{1/3}X_m$, C_P being the average concentration of pinning points per unit volume. Hence,

$$1/v_m = t_m/X_m = 1/v(\tau) + C_p^{1/3}e^{E/kT}/\nu_0 \tag{76}$$

or

$$v_m = (\tau b/B)/[1 + (\tau b/B)\, C_p^{1/3}e^{E/kT}/\nu_0] \tag{77}$$

is obtained as the average measured or inferred velocity.

It is now evident that if E is a strong function of the applied stress, there may be expected to be a range of low stresses where the measured relationship will be

$$v_m = (\nu_0/C_P^{1/3})\, e^{-E/kT} \tag{78}$$

and that a plot of $\ln(v_m)$ against stress may well have an apparent extrapolated value at infinite stress which is much greater than the shear wave velocity. As the stress increases, however, the exponential will rapidly and precipitously reach a sufficiently small value that the denominator in Eq. (77) will approach unity and the stress–velocity relationship will become linear, provided that $\nu_0/C_P^{1/3}$ is large enough. Since it may be expected that the number of obstacles sufficiently strong to be classed as pinning points will itself be a function of the applied stress and the temperature, it is probably fortuitous that there are indeed a variety of materials which, in the low-stress range, obey the exponential formula.

No high dislocation velocity measurements ($v \gtrsim 10^3$ cm/sec) have been made by the etch pit technique on any materials other than LiF and NaCl, and in those particular cases the exponential slope which is obtained happens to extrapolate to a terminal velocity near shear wave velocity.

Further discussions of mechanisms are deferred to a later section. The above considerations dictate that for all solids responding to a one-dimensional shock loading the ultimate result of finite shear strength is to produce a "cusp" in the Hugoniot in the stress–volume plane. This effect results from the increased one-dimensional compressibility when the stress deviators begin to be reduced, and has as one consequence the fact that over a wide range of impact loads the shock wave will split up into two waves because of the violation of the Bethe–Weyl conditions. As a result, developers of experimental techniques for studying one-dimensional elastic-plastic flow and its analogs have had to concentrate on obtaining time-resolved stress, strain, or velocity histories.

B. Experimental Techniques

Because the shock wave structures associated with elastic-plastic flow show considerable character, at least up to shock strengths at which the plastic or hydrodynamic wave velocity equals or exceeds the longitudinal velocity of sound in the sample, it has been possible to study these problems by using special techniques. Although in many instances explosively induced shock waves may be preceded by relatively weak elastic precursors, it is generally desirable to start with significantly lower impact stresses than explosives produce. For this reason it has become very common for workers in this field to rely on smooth-bored large-diameter guns firing flat-ended projectiles. Such guns are usually simple one-stage inert gas or propellant fired devices. These guns can accelerate fairly massive projectiles to well-controlled velocities ranging from ten to a few thousand meters per second. Guns also have the virtue that the particle velocity can be maintained constant at the collision interfaces for a relatively long time (a few microseconds). If the projectile has a driver plate of the same material as the target, a relatively simple measurement of the projectile velocity just before impact establishes the shock particle velocity at the collision interface with very high precision. If a thin driver plate is used, it may be backed up by a material of higher or lower impedance to study overtaking waves. Alternatively, ordinary high explosives such as low density nitroguanidine may be used. Continuous writing cameras or electro-optical cameras must be used with guns if photographic records are desired because of the relatively long and imprecise times involved in projectile release and acceleration.

Several useful and largely complementary analysis techniques, both optical and electronic, have been developed to study the detailed structure of the shock wave. Since there are very few ways to "look into" a shocked solid, particularly a metal, virtually all insight into the response of a solid to shock waves has been obtained from measurements made at the free surface opposite the impact surface. For these studies it is highly desirable to monitor the actual free-surface velocity as a function of time during the first few microseconds or more of motion.

Under circumstances where high-speed moving image cameras can be used, two optical monitoring techniques have been employed. The simplest of these (*37*) employs a photographic view of a tilted mirror placed in front of the surface. The streak camera record then yields a time-resolved record of the relative position of the surface as it strikes the mirror. A similar method (*38*) is known as the wire reflection technique in which a time-resolved record of the spatial position of the image of a wire placed just off a highly reflecting free surface is recorded. Direct observation of marks on the free surface also gives x–t data. These techniques have the advantages of simplicity and economy of design, construction, and use, and the obvious disadvantage that the data must be differentiated to obtain the desired velocity–time history. A third purely optical technique, the optical lever arm, measures free surface velocity directly. This method has the disadvantage that, because it employs a plane wave interacting with an oblique surface, the free surface-velocity–time history cannot be specifically determined over long periods of time without considerable analysis. The free surface must also remain a fairly good optically reflecting surface, which puts a severe limitation on the type of materials that can be investigated. As a method of obtaining material behavior, the technique also possesses the dubious advantage of resolving spatial dimensions so well that it differentiates clearly between the response of individual metal grains. Figure 83 shows some typical records obtained with this technique showing the grain-size effect.

Electronic techniques for monitoring free surface behavior are becoming rather manifold. The most straightforward and historically the first method to be applied with real precision was the electrical contactor pin array by Minshall (*26*). This technique employed many pins closely grouped and spaced very accurately at variable distances from the free surface. This technique has the advantages of being conceptually simple, digital rather than analog, and undisturbed by the failure of a few pins. It has the disadvantage of requiring the user to differentiate experimental data, presuming that he possesses the necessary array of oscilloscopes and electronic counters together with the skill and patience in the shot assembly necessary to obtain any data worth differentiating. A second electronic technique which has proven to be valuable in some circumstances is the slanted resistance wire, usually made from platinum foil (*39*). This technique, while it is rather reliable and versatile, still requires numerical differentiation of experimental data. A purely electronic technique which has enjoyed some success is the dc capacitor technique (*40*), wherein the circuit time constants may conveniently be adjusted so that the output voltage is essentially directly

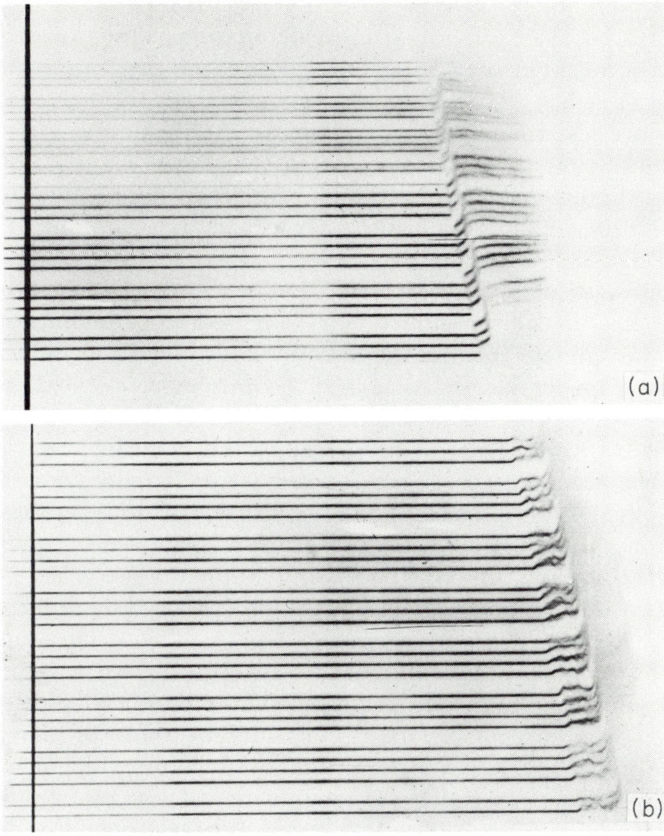

FIG. 83. Two optical lever photographic records. Time is increasing to the right. (a) Obtained on large-grain iron and shows the effect of differential motion between the large individual grains. (b) Obtained on very-fine-grain iron. The vertical displacement is directly proportional to the free-surface velocity.

proportional to the time rate of change of capacitance between a moving plane conducting surface and a polarized parallel plate. This technique has the advantages that it is capable of monitoring a velocity range of more than two orders of magnitude in a single experiment, the experimental data do not have to be differentiated, and timing synchronization, as with the pin and slant resistor wire techniques, is electronically controlled. Furthermore, decreases or even reversals of free surface velocity are readily recorded. Nonmetals may be studied simply by depositing a metal film on the surface. The principal apparent disadvantages are: the device is essentially a very-high-impedance antenna which must be very carefully shielded from such electrical disturbances as the radiation from a detonating explosive charge; the signal is the average over a fairly large area of the free surface; and many metals when sufficiently strongly shocked so that shock front rise times are less than 10^{-8} sec apparently produce an electrical dipole at the shock front (*41*) which makes measurements impossible unless a carefully designed differential measuring technique is employed (*42*). This technique is different from early capacitor techniques that only measured distance and time.

An electrooptical technique which has shown great promise is laser interferometry in two modifications. The earliest modification (*43*), in which one merely records the maxima and minima in light intensity as a surface moves toward a half-silvered mirror, is most useful for measuring very low free-surface velocities (10 m/sec and less). Its principal advantage seems to be the opportunity to look at very small areas of the free surface. This feature virtually eliminates the time smearing effect of nonplanarity in low-velocity collisions, but also means that the measurements do not represent averages over acceptable areas when polycrystalline materials are being studied.

The most recent and exciting development of laser interferometry is the so-called velocity interferometer (*44*). This technique utilizes a beam splitter to produce a difference in optical path length between two components of the same beam, so that a photocell pickup "sees" the position of a moving surface at two different times, or phases, of the laser wave train. The result is amplitude modulation which is a function of the sample's velocity rather than its position. Consequently, velocities can be obtained directly and over a much greater range. The technique requires a good specularly reflecting surface and does not therefore seem to be applicable to the study of "grainy" materials.

The only technique which presents a real hope of actually measuring particle velocities rather than free surface velocities is the magnetic loop method (*45*). This consists of simply measuring the emf produced when a conducting foil imbedded in a shocked insulator moves through a carefully measured transverse magnetic field. Needless to say, although the method shows great promise for studying insulators, it is useless when metals are of interest.

Various electromechanical transducers have been used and several are currently in use. The two most valuable seem to be the imbedded manganin wire or foil (*46*), whose electrical resistance is pressure sensitive, and the quartz transducer. The manganin wire and foil methods have the advantage that they "see" inside the shocked material, and the disadvantage that the response is complicated by thermal effects and the measuring device must be electrically isolated from the sample. The quartz transducer has proved to be very valuable, especially since it can be used over a wide range of temperatures and is the only known way of actually measuring the stress at the impact surface. It suffers from the disadvantages that the gage itself becomes untrustworthy at stress levels above about 30 kbar in the quartz and that the total available observation time is limited to one signal transit time through the quartz. The latter statement is not quite precise, in that it is in fact frequently possible to make differential measurements at later times, but this quickly becomes very complicated.

Although any of the techniques described are available for studying wave profiles, it should be pointed out that the choice of techniques in any given situation depends, in addition to the material under consideration, on the facilities available to the experimenter and on his personal experience and prejudices. Because of its wide range of application, the dc capacitor technique for monitoring the velocity of a free surface has been used rather extensively here.

C. Experimental Wave Profiles and Discussion

As a prelude to discussion of specific wave profiles, it should be noted that in many cases much of the deformation appears to result from twinning. Since this is currently understood to be the result of the motion of particular types of partial dislocations, the dislocation-theoretical approach to the problem is still appropriate.

The question of the applicability of the term elastic-plastic flow to the yielding process in solids which are normally considered brittle has received considerable attention in the past and will certainly receive more in the future. In good single crystals of such materials as germanium, silicon, and alpha quartz the deformation mode in compression cannot be like complete brittle fracture because it has been observed that in some crystallographic orientations the initial jump in free surface velocity associated with the elastic precursor is followed by a period of deceleration (see Fig. 84) before the plastic shock arrives (*47*). The first wave, propagating at approximately longitudinal sound speed (4.94 km/sec), has an amplitude of approximately 50 kbar. It cannot be proven, but it seems reasonable that the second rise in the free surface velocity should be regarded as a second "elastic" disturbance reflected from the advancing high-pressure shock, followed so closely by a third reflection that the two are not resolvable. Of particular interest is the very strong deceleration occurring at approximately 2.75 μsec, indicating that the material is supporting significant tension after

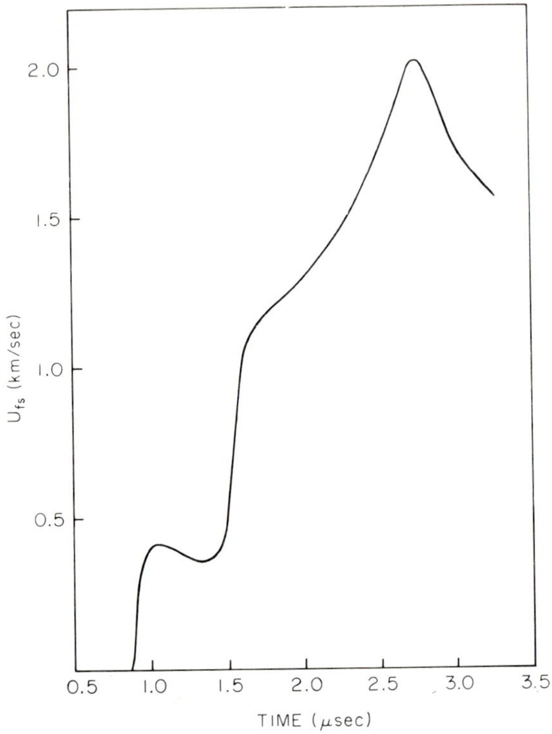

FIG. 84. Free-surface velocity versus time for a (100) single crystal of germanium which was impacted by a thin tungsten carbide driver backed up by aluminum.

yielding in compression. Since the free surface can only be slowed down by tension, and since a material which fails completely in a brittle manner cannot possibly be expected to support tension or to propagate a tension signal, it seems on this basis that such materials yield primarily plastically under the influence of sufficiently strong shocks. It is obvious that there is not enough energy available to allow the material to fail in complete brittle fracture. If the stress relaxation is accompanied by only discrete areas of brittle fracture, there is the possibility that the material could show some dynamic tensile strength. Interpretation of recovered shock-loaded brittle materials is always complicated by secondary interactions, e.g., edge effects.

Gilman (*30*) has noted that the Hugoniot elastic limit of germanium corresponds to the shear stress measured by the Knoop method. In addition, a single crystal of germanium has been observed to propagate a tension signal to the free surface after being shocked far beyond its one-dimensional elastic limit (Fig. 84). Free surface velocity records from single-crystal silicon obtained by two techniques also shows this in Fig. 85.

As an example of a different situation, it has been observed that porous ceramics composed of 85 and 95 wt % alumina (Fig. 86), do not propagate

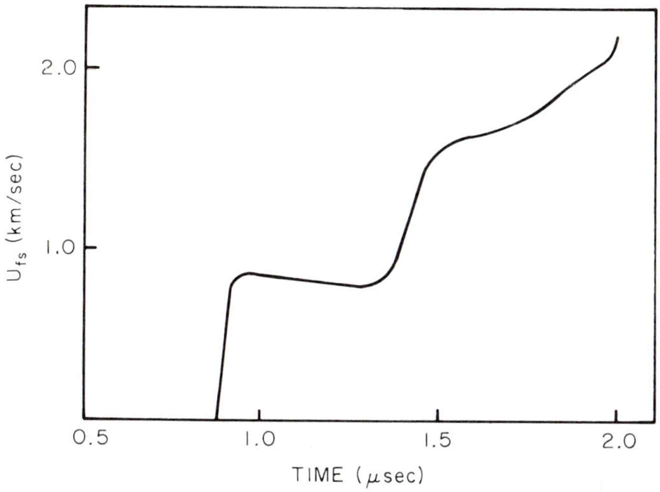

(a)

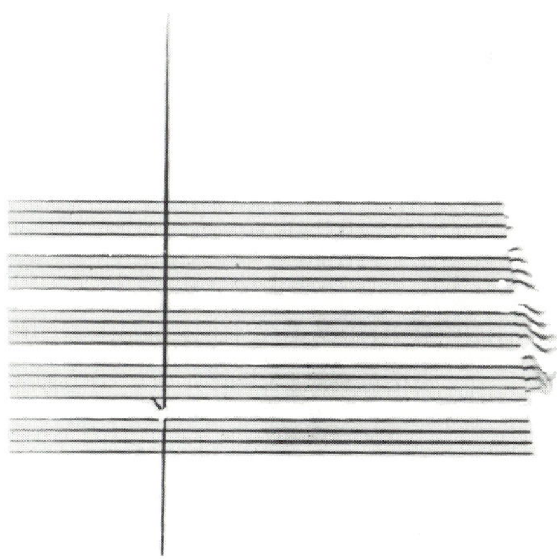

(b)

FIG. 85. Free-surface velocity versus time for Si ⟨100⟩. (a) Reduction of a record obtained by the capacitor technique showing the rapid rise and subsequent decay of the elastic wave. (b) Photographic record obtained under similar conditions by the optical lever technique, clearly showing these same features.

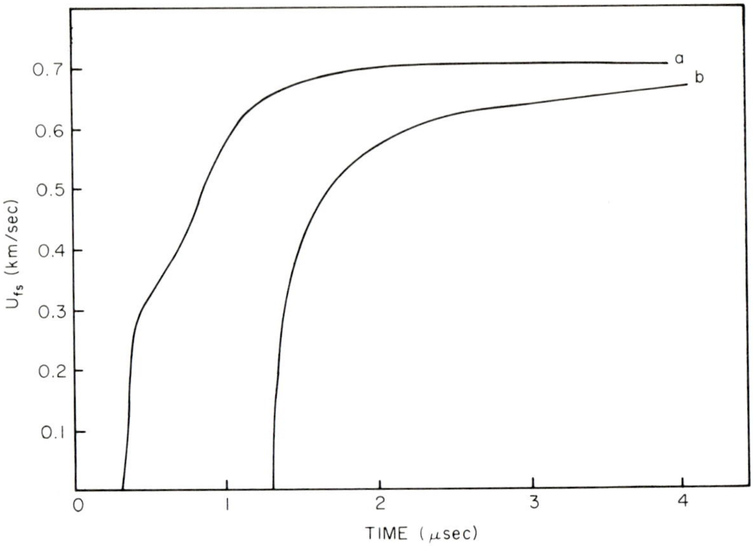

FIG. 86. Free-surface velocity profiles for shocked 85% alumina ceramic (a) and 95% alumina ceramic (b). In both experiments the driver plates were thinner than the targets and some deceleration would be expected during the time motion was observed if the material could support tension after yielding in compression.

any tension after being shocked above their Hugoniot elastic limit. The Hugoniot elastic limits of these materials are much less than the Hugoniot elastic limit of high-density ceramic alumina such as Lucalox (*48*), a fine-grained polycrystalline material of ncarly pure alumina with almost theoretical density. This polycrystalline material is still apparently much weaker than single crystals of sapphire (*49*).

Other hard materials investigated are tungsten carbide, boron carbide, and titanium diboride. Metallwerke Plansee grade 830 tungsten carbide has been found to have a Hugoniot elastic limit of 40 kbar. Typical wave profiles of this material and Rockwell C-62 4150 steel are showh in Fig. 87. The Hugoniot elastic limits are taken to be the stress at whicn observable plastic flow first begins, which is discernable as a "knee" in the free surface velocity profile. For the WC the associated stress is about 40 kbar, and for the steel it is about 35 kbar. The small dips in the free surface velocity are associated with the spall fracture stress of the sample, approximately 8–10 kbar for the WC and 22 kbar for the steel. The reported Hugoniot elastic limit for high-density sintered boron carbide is 150 kbars (*49*). The great difference could well be due to the cobalt binder present in the tungsten carbide. The point is that the carbides themselves are very much stronger than any metallic

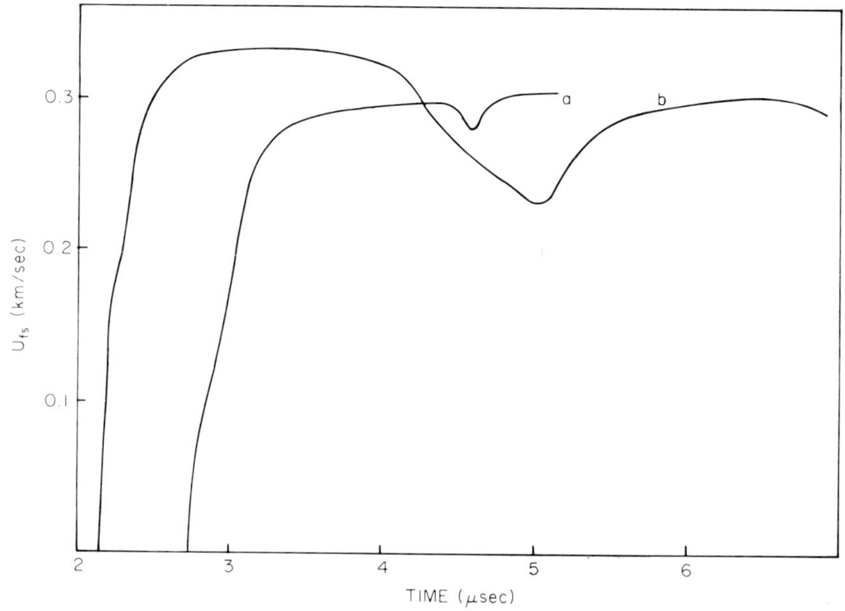

FIG. 87. Free-surface velocity profiles for (a) an 0.5-in.-thick tungsten carbide target impacted by an 0.25-in.-thick tungsten carbide driver, and (b) Rockwell C-4150 steel.

binder, just as the alumina grains in a ceramic are very much stronger than the silicates and aluminosilicates which make up the grain boundaries and the filler material. The porous ceramic has, of course, additional weaknesses due to the stress concentrations at the voids. The free surface velocity profile for a shocked TiB_2 sample is shown in Fig. 88.

The picture which one draws of brittle polycrystalline materials is that their compressive strengths are usually largely determined by free boundary effects. In static-loading experiments these free boundaries may be external macroscopic boundaries, internal grain boundaries, or free surfaces of grains next to voids. Under plane impact loading conditions only the internal boundaries count. Probably only for nearly theoretically dense samples does true plastic flow determine the yield strength.

The effect of porosity on the wave profile of a strong metal is shown in the behavior of a sample of fansteel 77 (Fig. 89). To some extent this might be called a "crush curve." Preshot examination of the sample indicated many fine pits on the ground surface, indicating porosity. The capacitor record reflects this result in the fact that although this

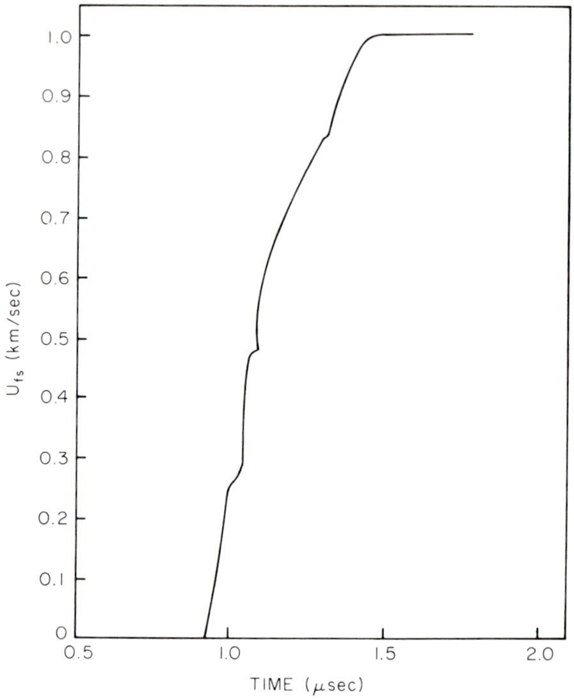

FIG. 88. Free-surface velocity profile for an 0.4-in.-thick sample of TiB_2 with density 4.3 g/cm^3. This sample was explosively loaded with baratol through a 2024 aluminum alloy plate. The Hugoniot elastic limit is about 60 kbar, and several wave reflections are evident.

material is very strong there is no well-defined Hugoniot elastic limit. The spall fracture stress, evidenced by the dip in the velocity, is also small. The words "crush curve" may also apply to the shock wave profiles in Pyrex glass, shown in Fig. 90. This material, not surprisingly, behaves similarly to pure fused silica, the behavior of which was measured in some detail by Wackerle (*10*). The open microstructure of vitreous silica apparently behaves similarly to the voids in a porous solid as far as shock wave response is concerned.

A class of materials whose impact properties are not at present well documented are the plastics, although some attention has been given to Lucite and Plexiglas. There is not yet any evidence that these materials actually undergo an elastic-plastic transition in the sense that metals or even brittle materials do. Shock waves in these materials, so far as is currently known, do not develop a two-wave structure at any stress level investigated (Fig. 91). There is no elastic wave visible; however, the surface of the wedge retains its reflectivity long enough to follow

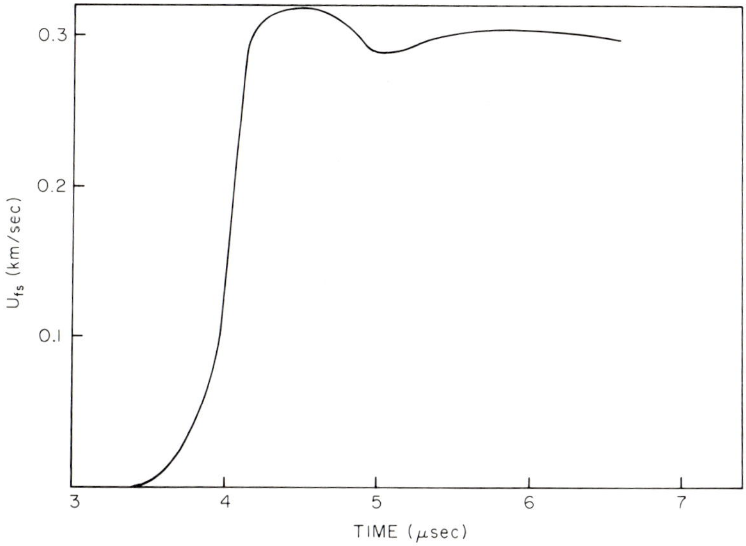

FIG. 89. The wave profile for an 0.5-in.-thick sample of Fansteel 77.

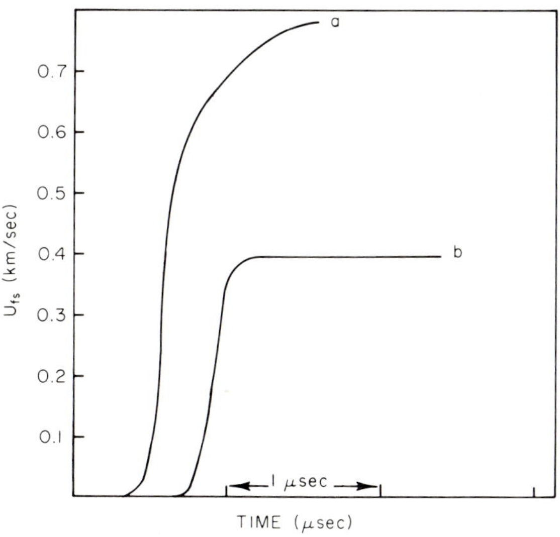

FIG. 90. Free-surface velocity profiles for Pyrex glass impacted at two different velocities. The leading portions of the traces are similar to the Fansteel 77 data of Fig. 89 and to similar results for fused quartz, as discussed in the text.

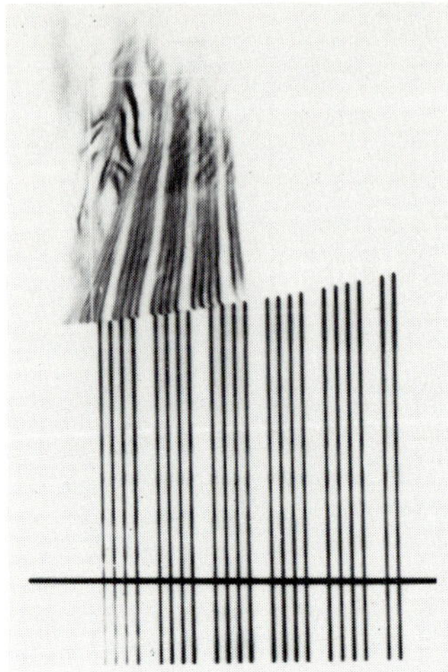

FIG. 91. Photographic record of the wave profile of Plexiglas using the optical lever arm technique.

the plastic wave displacement for several microseconds. Although edge effects are obviously important, the large horizontal displacement of the traces indicates a shock amplitude of 14 kbar for this shot. The fact that the shock particle velocity data for Plexiglas does not extrapolate very well to the bulk sound speed indicates some other type of transition exists. The absence of a two-wave structure is probably due to the fact that plastics are such open structures that individual atoms can move great distances relative to each other without severe interference. It is also possible that chemical bonds can be broken and rearranged so that hydrodynamic equilibrium is easily attained.

An aspect of the subject of elastic-plastic flow which has received increasing attention is the profile of overtaking waves. Both compression and relief waves are of interest. The study of the structure of overtaking relief waves is fundamental to the understanding of the attenuation of short-duration compressive pulses as they propagate through materials. Not surprisingly, it is found that in metals such stress relief waves have an elastic-plastic structure of their own. As discussed in Section III.C, this results from the fact that at any given shock stress level the material can relieve from a state lying above the hydrostat to another state lying at least as far below. This, combined with the fact that such signals move at the high-pressure longitudinal sound speed, results in a more rapid attenuation of short-duration stress pulses than would be anticipated on the basis of hydrodynamic calculations. These effects have been studied by measuring peak stresses as a function of shock run in samples which have been subjected to a short-duration pulse (*50*). In such systems it is possible to follow free surface motion beyond the peak velocity. Here the tension waves resulting from the interaction of the forward- and backward-facing rarefaction waves can be observed, followed by a minimum in the free surface velocity corresponding to the spall tension and subsequent damped oscillations as the spalled sample thermalizes its stored elastic energy. Such data are currently very useful for calibrating computer models which are used to calculate the interactions of more complex pulse shapes. Such calculations are particularly necessary to interpret completely the free-surface velocity records because the free-surface motion, after the first arrival of the elastic precursor, is actually the result of a complicated series of wave interactions. For example, even the apparent arrival of the first plastic shock front is actually the result of an elastic rarefaction signal returning

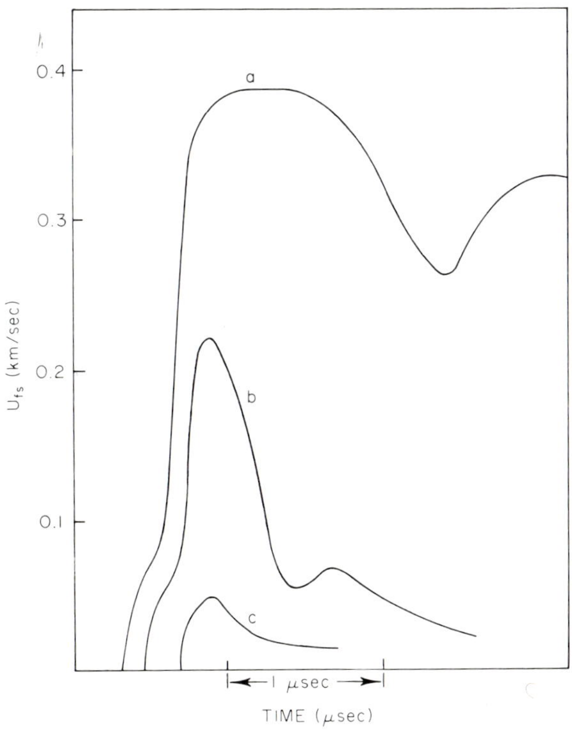

FIG. 92. Free-surface velocity data obtained from three samples of tantalum, 0.378-in.-thick, which were subjected to collisions at 0.4 km/sec by (a) 0.125-in. tantalum, (b) 0.030-in. tantalum, (c) 0.005-in. tantalum. Overtaking waves and spall stresses can be studied by such experiments.

to the advancing plastic shock and a new elastic compression wave returning to the free surface. This process occurs several times while the free surface is accelerating up to its final velocity. Presented in Figs. 92–94 are some overtaking wave data for tantalum, AZ-31B Mg alloy, and 304 Stainless Steel.

One useful feature of the time-resolved free surface velocity records is that the magnitude of the first drop in free surface velocity may be used as a measure of the spall stress of the material under the loading conditions of the experiment. The calculation is correct to the approximation that the acoustic impedance of the material is the same for a tension of a few kilobars as it is for compression. In this approximation

$$\sigma_{sp} = \rho_0 c\, \Delta u_{fs}/2 \tag{79}$$

where the greatest uncertainty is in whether the most appropriate sound speed is the longitudinal or the bulk sound speed.

In addition, shock wave attenuation experiments in metals are providing data from which the longitudinal sound speed, and hence the longitudinal elastic modulus, can be drawn as a function of pressure up to very high pressures. Since it has been shown that the bulk modulus can be ascertained with high confidence from the Hugoniot data, it

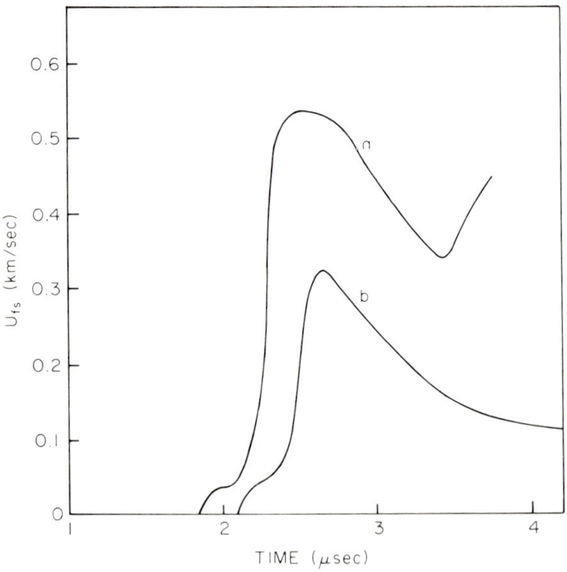

FIG. 93. Free-surface velocity data for 0.500-in.-thick AZ-31B magnesium alloy driven by (a) 0.120-in. of 2024 aluminum at 0.4 km/sec, (b) identical target with 0.032-in. 2024 aluminum driver at 0.4 km/sec.

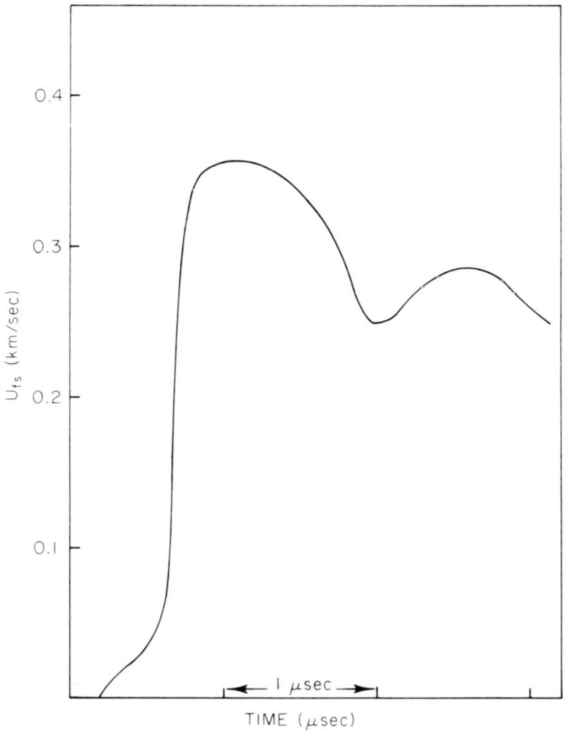

FIG. 94. Free-surface velocity profile for type 301 Stainless Steel 0.500-in.-thick struck by an 0.125-in.-thick 301 Stainless-Steel driver.

has become possible to obtain both the bulk and shear moduli of isotropic materials up to very high pressures.

A general conclusion from these experiments is that materials which are commonly considered ductile exhibit considerably greater spall fracture stresses than materials which are normally considered brittle. There are, however, degrees of brittleness as evidenced by the comparison between the tantalum, magnesium, and stainless-steel data and the tungsten carbide, Fansteel 77, and alumina ceramic data.

A second general conclusion, reasonably well substantiated, is that at least for the relatively pure metals, Poisson's ratio as a function of pressure is approximately constant or slowly decreasing at least up to pressures of the order of 100 kbar. For these weak shocks the temperature rise is usually quite small.

Examples of the range of material types for which free surface velocity data can be obtained and used to help understand shock propagation are given in Figs. 95–98, which show wave profiles in diabase rock,

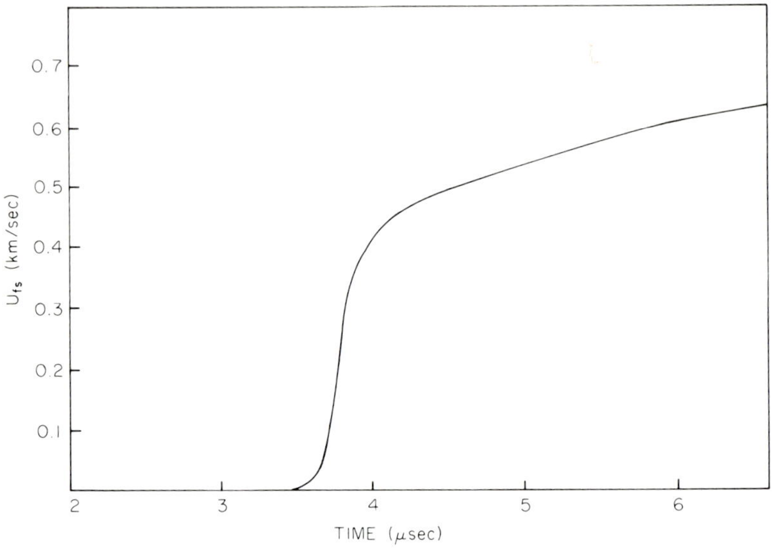

FIG. 95. Time-resolved free surface velocity for a sample of diabase. The curve exhibits to some extent the "compaction" feature, in that only a negligible stress apparently propagates at the measured longitudinal sound speed, but a large disturbance is followed almost immediately by extensive flow.

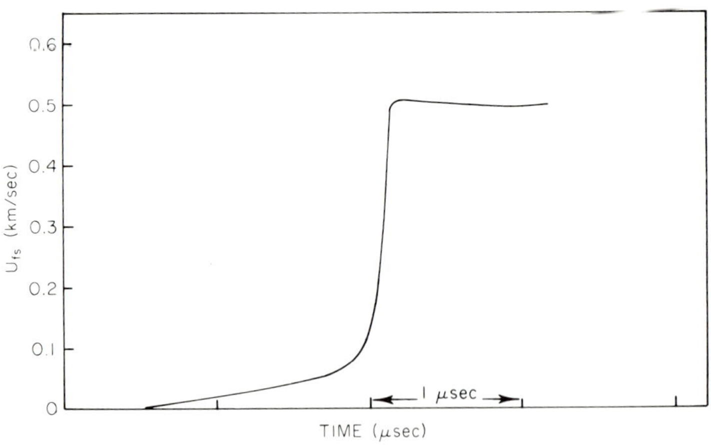

FIG. 96. The free-surface profile for a sample of low-density boron nitride. The elastic precursor has only a very small amplitude, but the plastic wave is very steep, in contrast to the behavior of the PT-0178 graphite discussed earlier.

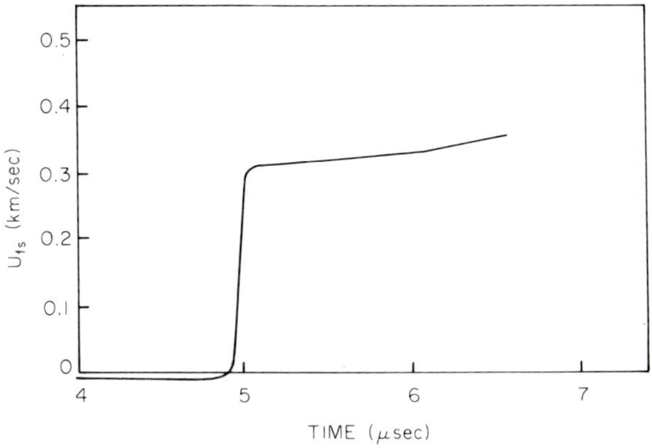

FIG. 97. Time-resolved free-surface velocity for a plastic mockup of a high explosive. The entire disturbance is apparently "elastic."

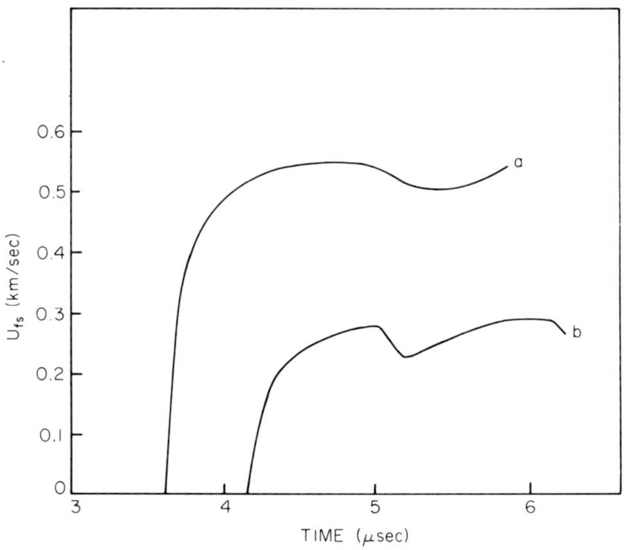

FIG. 98. Time-resolved free-surface velocity data for quartz phenolic Refrasil. The character of the spall tension waves are drastically different in the two shots for reasons not understood.

boron nitride at a density of 2.0 g/cm^3, a plastic mockup of a high explosive, and quartz phenolic Refrasil. Note in particular that the phenolic Refrasil exhibits a very respectable spall stress under plane wave loading conditions.

TABLE X

HUGONIOT ELASTIC LIMITS

Material	Sample thickness (mm)	Elastic limit (kbar)
2024 Al	12	5.4[a]
6061 Al	12	5.4[a]
Be[c]	12	0[a]
Be, *c* axis	6	40[a]
Be, basal plane	6	4[a]
Cr	5	16[b]
Cu, annealed	all	0[a]
Cu, 50% cold worked	12	6.3[a]
Fe		
Fine grain hard	3	15[b]
Large grain hard	3	14[b]
Fine grain soft	3	11[b]
Large grain soft	3	9[b]
Mg (AZ-31B)	12	1.1[a]
Mo	3	16[b]
Ni	12	10[a]
Nb	12	20.7[a]
Pb, annealed	12	0[a]
Ta	10	18.7[a]
Th	12	1.4[a]
Ti	12	18.5[a]
U, cast	all	0[a]
W	5	32[b]
Ge⟨100⟩	8	52[a]; 52[b]
Ge⟨114⟩	10	40[b]
Ge⟨111⟩	10	37[b]
304 stainless steel	12	2.3[a]
Si⟨100⟩	10	82[a]; 68[b]
Si⟨111⟩	10	60[b]
Si⟨110⟩	10	67[b]
WC	12	40[a]
85% alumina ceramic (Porous, $\rho_0 = 3.40$ g/cm^3)	12	45[a]
95% alumina ceramic (Porous, $\rho_0 = 3.75$ g/cm^3)	12	50[a]
BN (Porous, $\rho_0 = 2.06$ g/cm^3)	12	0[a]
Jadeite	10	65[b]
MgO, single crystal	10	40[b]

[a] Capacitor technique.

[b] Optical lever technique.

[c] Yielding begins immediately.

Beyond approximate calculations which enable one to extract rather good values of the Hugoniot elastic limit, estimates of spall strengths and wave velocities, various computer codes exist for analysis and use of the type of data presented in this section. In some of these codes, the yield stress is simply held constant and the calculation of the propagation of shock waves is corrected for small elastic components. In other codes, time-dependent yielding, work hardening, and combinations of these effects are considered. The only such code with which we have worked is STRESS-R, which incorporates these features in one-dimensional flow (*47*). Two-dimensional calculations with rigidity effects have been made by Wilkins with HEMP (*51*) and by Hermann *et al.* with TWODY (*52*). The general conclusion from all the computer work and the experimental work appears to be that yield strengths do not have to be very large to have significant effects in attenuation and propagation of shock pulses, particularly in systems consisting of layers of materials, because of the relatively high velocity of the elastic components of the signals. It has also been learned that one-dimensional experiments give more information on dislocation dynamics than more complicated two-dimensional experiments. This is due to the fact that although the strains are small in the one-dimensional experiments, the high stress rates provide information that is quite difficult to obtain in the more complicated experiments.

Table X is a compilation of Hugoniot elastic limit data obtained by the dc capacitor and optical lever techniques. It is obvious that, except in special cases where there is a well-defined "yield-point" effect, the decision about exact amplitudes is made arbitrarily. A consistent attempt has been made to take the first point where the free surface velocity versus time record shows significant curvature. In no case is the elastic limit determined to within 10%; the error in some cases is probably even greater. It is also well known that for many materials the Hugoniot elastic limit is a decreasing function of the sample thickness.

REFERENCES

1. H. A. Bethe, The theory of shock waves for an arbitrary equation of state. Office of Scientific Res. and Develop. Rept. No. 545, Serial No. 237 (1942).
2. R. Courant and K. O. Friedrichs, "Supersonic Flow and Shock Waves." Wiley (Interscience), New York, 1948.
3. M. H. Rice, R. G. McQueen, and J. M. Walsh, *Solid State Phys.* **6**, 1–63 (1958).
4. S. B. Kormer, M. V. Sinitsyn, G. A. Kirillov, and V. D. Urlin, *J. Expl. Theoret. Phys.* **48**, 1033 (1965); *Soviet Phys.-JETP (English Transl.)* **21**, 689 (1965).
5. W. C. Davis, private communication (1959).

6. J. W. Taylor, *J. Appl. Phys.* **34**, 2727 (1963).
7. R. G. McQueen, E. Zukas, and S. P. Marsh, Residual temperatures of shock-loaded iron, Special Tech. Publ. No. 336, pp. 306–316. Am Soc. for Testing and Materials, Philadelphia (1962).
8. J. M. Ziman, "Electrons and Phonons," p. 56. Oxford Univ. Press (Clarendon), London and New York, 1960.
9. R. G. McQueen and S. P. Marsh, *J. Appl. Phys.* **33**, 654 (1962).
10. J. Wackerle, *J. Appl. Phys.* **33**, 922 (1962).
11. R. G. McQueen and S. P. Marsh, *J. Appl. Phys.* **31**, 1253 (1960).
12. P. W. Bridgman, *Proc. Am. Acad. Arts Sci.* **77**, 189 (1949); **76**, 9 (1945).
13. R. G. McQueen, J. N. Fritz, and S. P. Marsh, *J. Geophys. Res.* **68**, 2319 (1963).
14. R. G. McQueen, S. P. Marsh, and J. N. Fritz, *J. Geophys. Res.* **72**, 4999 (1967).
15. A. H. Jones, W. M. Isbell, and C. J. Maiden, *J. Appl. Phys.* **37**, 3493 (1966).
16. L. V. Al'tshuler, K. K. Krupnikov, B. N. Ledenev, V. I. Zhuchikhin, and M. I. Brazhnik, *J. Exptl. Theoret. Phys.* **34**, 874 (1958); *Soviet Phys. JETP (English Transl.)* **7**, 606 (1958).
17. L. V. Al'tshuler, K. K. Krupnikov, and M. I. Brazhnik, *J. Expl. Theoret. Phys.* **34**, 886 (1958); *Soviet Phys. JETP* (*English Transl.)* **7**, 614 (1958).
18. J. M. Walsh and R. H. Christian, *Phys. Rev.* **97**, 1544 (1955).
19. R. G. Shreffler and W. E. Deal, *J. Appl. Phys.* **24**, 44 (1953).
20. D. E. Grey (ed.), "American Institute of Physics Handbook," pp. 3-80. McGraw-Hill, New York, 1957.
21. O. Jahari, Univ. of California Radiation Lab. Rept. UCRL 10932 (1963).
22. D. I. Balef and J. de Klerk, *Phys. Rev.* **129**, 1063 (1963).
23. S. P. Clark (ed.), "Handbook of Physical Constants," p. 119. The Geological Society of America, New York, 1966.
24. R. I. Jaffe, E. M. Smith, and B. W. Gonser, *Trans. AIME* **161**, 366 (1945); R. I. Jaffe and B. W. Gonser, "Metals Handbook" (T. Lyman, ed.), p. 1172. The American Society for Metals, Cleveland, Ohio, 1948.
25. R. G. McQueen and S. P. Marsh, *J. Geophys. Res.* **71**, 1751 (1966).
26. D. Bancroft, E. L. Peterson, and S. Minshall, *J. Appl. Phys.* **27**, 291 (1956).
27. D. S. Hughes and C. Maurette, *J. Appl. Phys.* **27**, 1184 (1956).
28. G. Liebfried, *Z. Physik* **127**, 344 (1950).
29. W. P. Mason, "Physical Acoustics" (W. P. Mason, ed.), Vol. IV, Part A, p. 299. Academic Press, New York, 1966.
30. J. J. Gilman, Symp. Mechanical Behavior of Materials Under Dynamic Loads, San Antonio, Texas (1967).
31. A. V. Granato and K. Lücke, "Physical Acoustics" (W. P. Mason, ed.), Vol. IV, Part A, p. 225. Academic Press, New York, 1966.
32. J. W. Taylor, *Proc. 1967 Batelle Conf. Dislocation Dynamics* (to be published).
33. J. J. Gilman, *J. Appl. Phys.* **36**, 3195 (1965).
34. D. L. Stein and J. R. Low, *J. Appl. Phys.* **33**, 2499 (1962).
35. H. S. Chen, J. J. Gilman, and A. R. Head, *J. Appl. Phys.* **35**, 2502 (1964).
36. J. W. Taylor, *J. Appl. Phys.* **36**, 3146 (1965).
37. G. R. Fowles, Shock-wave compression of quartz, Doctoral Thesis, Dept. of Geophys., Stanford Univ. (1962).
38. W. C. Davis and B. G. Craig, *Rev. Sci. Instr.* **32**, 579 (1961).
39. L. M. Barker, Measurement of free surface motion by the slanted resistor technique, SCDR 78-61, Sandia Corp., Albuquerque, New Mexico (1961).
40. M. H. Rice, *Rev. Sci. Instr.* **32**, 449 (1961).

41. J. W. Taylor and M. H. Rice, unpublished data.
42. P. Baker and J. W. Taylor, unpublished work.
43. L. M. Barker and R. E. Hollenbach, *Rev. Sci. Instr.* **36**, 1617 (1965).
44. L. M. Barker, Fine structure of compressive and release wave shapes in aluminum measured by the velocity interferometer technique, SC-DC-66-2447 (preprint), Sandia Corp., Albuquerque, New Mexico (1967).
45. A. N. Dremin and K. K. Shvedov, PMTF, No. 2, 154 (1964).
46. D. D. Keough and W. Wilkinson, AFWL-TR-65-170, 1965.
47. R. G. McQueen, "Laboratory Techniques for Very High Pressures and the Behavior of Metals Under Dynamic Loading" (K. A. Gschneidner, Jr., M. T. Hepworth, and N. A. D. Parlee, eds.), Metall. Soc. Conf., Vol. 22. Gordon and Breach, New York, 1964.
48. T. J. Ahrens, W. H. Gust, and E. B. Royce, *J. Appl. Phys.* (to be published).
49. E. B. Royce, private communication (1968).
50. D. R. Curran, *J. Appl. Phys.* **34**, 2677 (1963).
51. M. L. Wilkins, Le calcul de la resistance dynamique d'un matériau, IUTAM, Symposium HDP, Paris, 1967 (to be published).
52. B. J. Thorne and W. Herrmann, SC-RR-66-602, Sandia Corp., Albuquerque, New Mexico (1967).

CHAPTER VIII

METALLURGICAL OBSERVATIONS AND ENERGY PARTITIONING

R. B. POND

THE JOHNS HOPKINS UNIVERSITY
BALTIMORE, MARYLAND

C. M. GLASS

TERMINAL BALLISTICS DIVISION
U.S. ARMY BALLISTICS RESEARCH LABORATORY
ABERDEEN PROVING GROUND
ABERDEEN, MARYLAND

I. Hypervelocity and the Cratering Phenomenon

Hypervelocity deformation is normally considered to be deformation carried out at sufficiently high velocities to cause a cratering effect. Figure 1 is a view of the cross section of a crater in an aluminum target.

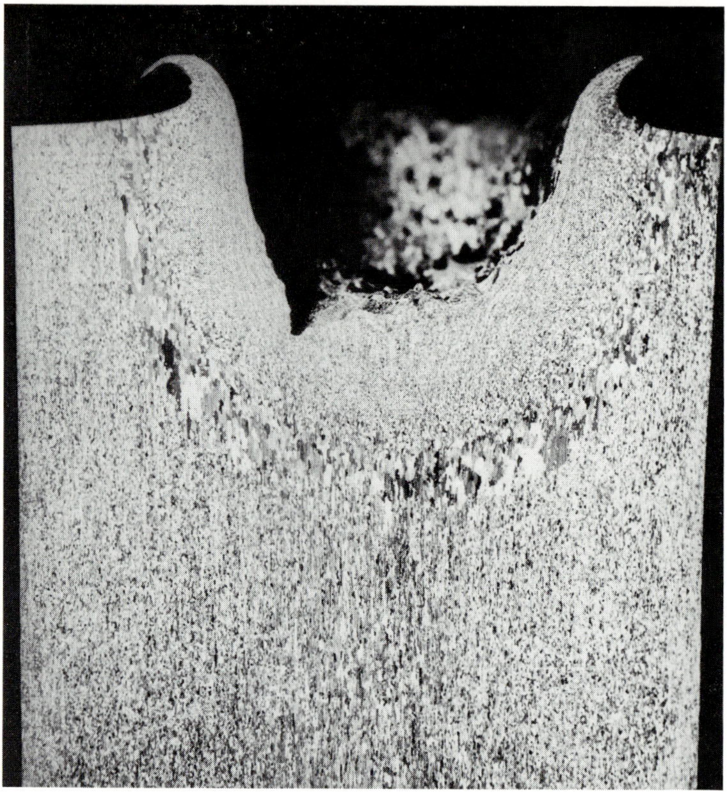

FIG. 1. Delineation of various zones of a crater in 3-in. aluminum target.

This view delineates the various regions associated with the cratering effect. The crater itself is generally hemispherical in shape although, under proper conditions, it can be cylindrical with a hemispherical bottom.

Surrounding the crater at the surface is a lip which gives the appearance of having been "splashed" out of the crater. Surrounding the crater under the surface is a strain-affected region, the constant strain contours of which simulate the geometric pattern of the crater. (This region and its delineation will be discussed later in the chapter.)

Upon inspecting such a crater, many questions suggest themselves. Included in the array of questions are the following:

1. Does the crater volume represent a volume of metal which has been evacuated from the specimen?
2. Where does the material of the lip come from?
3. How much of the input energy is absorbed by the strain-affected region, the evacuated metal, the lip, and the disintegrated projectile?
4. Will the answers to the previous questions vary as material properties vary?

The purpose of this chapter is to respond to the last two questions.

It appears that answers to the first two questions have been available for some time. Davies (*1*), in his discussion of the internal stresses due to contact of a sphere and a semiinfinite solid, described what Thomas and Hoersch discerned experimentally as early as 1930, that when hardened cylinders were forced into steel plates, failure started, not in the area of contact, but at a point below this. If the volume of material under the highest shear stress lies below the surface, then the progression of evacuation should be as illustrated in Fig. 2. This also implies that the material near the surface, having the lowest shear stress when forced out by the ejecta, becomes the lip of the crater.

By placing a three-dimensional nickel net in a copper target, Hsu and Pond (*2*) were able to identify the source of the lip material, as well

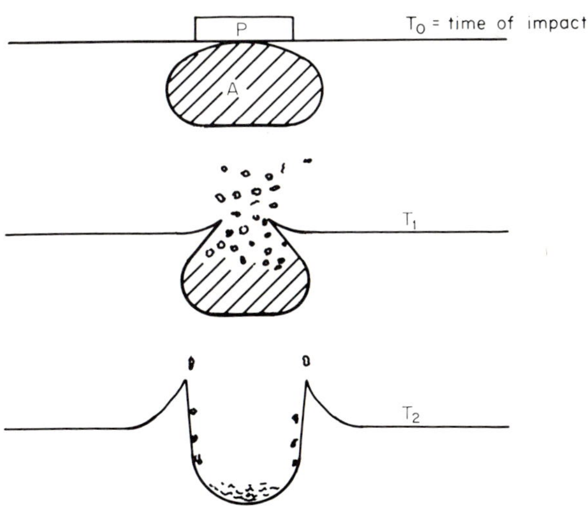

FIG. 2. Schematic representation of highest shear stress and disintegration volume during impact: A, highest shear stress volume; P, projectile.

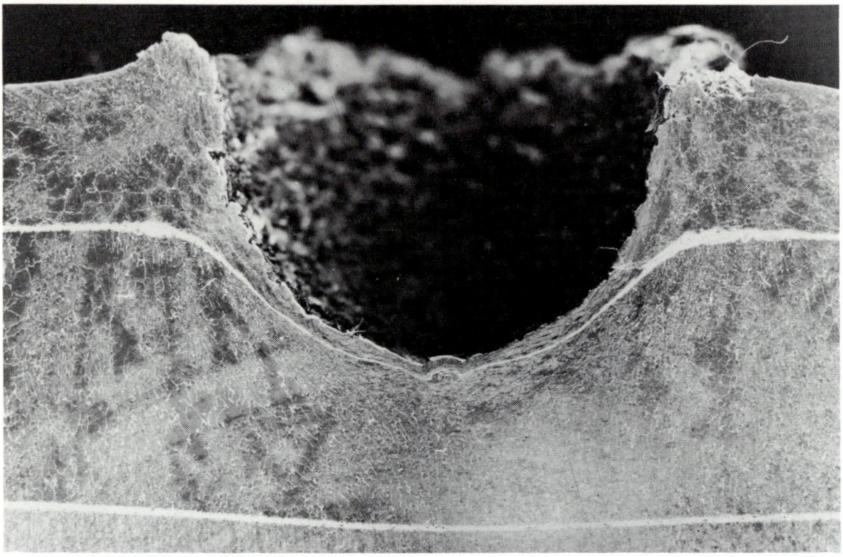

FIG. 3. A $\frac{1}{2}$-in.-diam crater in copper showing nickel marker deflection.

as determine that only a small part (10–30%) of the metal which initially occupied the crater was ejected. Figure 3 is illustrative of this work and shows a nickel wire which originally crossed approximately midway of the crater and has been deformed inwardly rather than ejected.

II. Historical Development

A. Background Experience

Historically, the concept of metallurgical parameter influences on hypervelocity impact phenomena had no significance during the early stages of investigation of this fascinating problem. The process was viewed, originally, from the point of background experience. This background was:

1. The impact of a short, compact, metallic hypervelocity projectile (velocity in the neighborhood of 3–5 km/sec) on a solid metal target produced a large crater. The original concern could be considered an outgrowth of questions related to the encounter of a space vehicle with a meteorite: the velocities in this case could be up to 20–30 km/sec (100,000 ft/sec).

2. Shaped-charge jet penetration theory had been handling similar problems for several years, with the differences that the jet was longer than the usual hypervelocity projectile (meteorite) and the target was very thick. Eichelberger (*3*, *4*) had advanced the original shaped-charge theory of Birkhoff *et al.* (*5*) to the point where it was feasible to make calculations on the penetration of jets into semiinfinite targets using hydrodynamic considerations. In a sense, the high-velocity impact (HVI) problem was made to order for the application of shaped-charge jet penetration work. A number of reports are available concerning shaped-charge penetration (*3–7*). Figure 4 shows the influence of target hardness

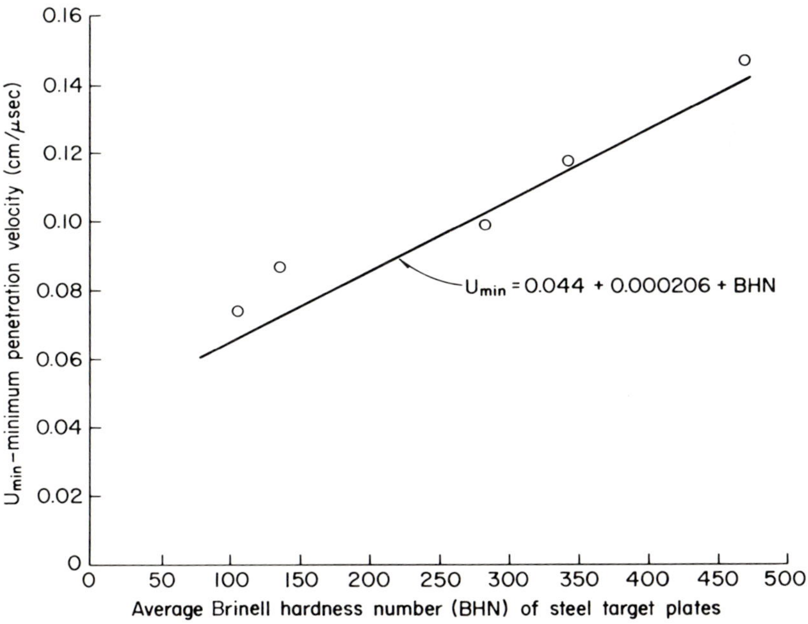

FIG. 4. Influence of increasing hardness on the velocity required for penetration.

on penetration for a series of metal targets. Moss (*8*) showed, however, that microstructure changes the penetration for a given hardness level. This significant study is the only one of its kind and was not carried out until 1966.

3. The high-pressure shocks set up in the penetrator and the target are similar, in many respects, to the shocks set up in a solid by an explosive in contact with the solid. The original analysis of the reaction of a solid to the high-pressure shock loading was done by von Neuman and Richtmeyer (*9*). The equation transforms developed involved an

artificial viscosity term, added to the basic equation of motion. The availability of computers introduced a new tool in solving these equations. Bjork (*10*) was the early leader in developing computer codes to handle the hydrodynamic approach to hypervelocity cratering.

4. The work of Rice *et al.* (*11*) developed the Hugoniot curves and data for equations of state that could be used in the hydrodynamic analysis of HVI.

Early work in the field of HVI concentrated on the knowledge available. However, it is apparent from the above that the only material property of importance in any approach used in the HVI field was the original material density. Thermodynamic equations of state did not, originally, involve thermomechanical relations. Nor could viscosity terms, which are analytical necessities, be considered material properties.

A one-dimensional analysis of shock wave attenuation using plate impact was carried out by Allison and Chou (*12*). Their analysis showed that plane shocks produced by impacts in aluminum are equivalent at late times provided $u_0{}^{\alpha}d$ is constant, where u_0 is the impact velocity, d is the striker thickness, and α is a constant, approximately 1.3. However, the number α takes must be considered a function of not only velocity but material properties, particularly in the lower velocity range.

B. Use of a Material Property Parameter

Recently Walsh and co-workers (*13*) have added strength dependency to the hydrodynamic code (RPM) in order to analyze HVI cratering. They find the crater size can be predicted from basic geometry effects, using this approach. Also, fracture criteria are being added to TOIL code.

However, most computer codes in use develop flow patterns essentially different from those expected by the analysis seen in Fig. 2, and the results seen in Fig. 3.

The majority of studies on the metallurgical property influences on deformation under hypervelocity impact have concentrated on relating crater volume to material hardness. As examples:

1. Eichelberger and Gehring (*14*) proposed that the crater volume V is related to the kinetic energy of the projectile (mv^2), the angle of attack of the projectile, α, and the hardness of the target H (Brinell hardness number) by

$$V = mv^2 \cos \alpha / 2.62H \tag{1}$$

2. Hermann and Jones (*15*) developed the relation

$$h = 1.2(\rho_0/\rho)^{2/3} r \ln(1 + \rho_0^{2/3} \rho^{1/3} v^2 / H) \tag{2}$$

where h is the crater depth, r is the radius of spherical projectile, ρ_0 is the projectile density, ρ is the target density, v is the projectile velocity, and H is the target hardness.

In general, these relations fail to provide coverage of a wide variety of experimental results. The reasons for this are found in the influence of dynamic properties on flow and fracture of a dynamically loaded solid. Attempts to universalize a relation between crater depth and material hardness have resulted in studics of "dynamic hardness." Feldman (*16*) studied the problem but reached no firm conclusion. Fadeyenko (*17*) had the same experience in 1964, deciding that his analysis of dynamic hardness suffered from problems he couldn't take into account.

Several other approaches should be mentioned to show how various investigations have attempted to introduce material properties into the HVI field.

(1) Kineke (*18*) developed a large amount of data on the hole volume created in various materials impacted at hypervelocities, but could add nothing to the material property problem aside from the fact that temperature has an influence. Richards and Glass (*19*) showed that as the temperature increases the energy able to leave the crater area and produce damage in the target decreases. Allison (*20*) also studied the temperature effect.

(2) Wilkens (*21*) developed an elastic-plastic code, using a yield criterion to analyze target and projectile flow under impact conditions. He has extended this work and is currently working on penetration problems.

(3) Rolsten and Hunt (*22*) have analyzed dynamic stress and true tensile strength in their cratering model, relating the crater depth d of a hemispherical crater to the true tensile stress s at failure by

$$d = (mv^2/4\pi s)^{1/3} \tag{3}$$

These are but a few examples of the various approaches available. Refer to the various hypervelocity symposia mentioned in the references for other studies.

III. Metallurgical Property Influences on Metal Deformation as a Function of Test Velocity

There is an obvious devision existing in the analytical approaches taken in the field of high-rate loading of solids—particularly metals. If the imposed dynamic stress is the same order of magnitude as the static ultimate strength of the metal the problem is generally attacked in one way. When the imposed dynamic stress is several orders of magnitude (or more) larger than the static ultimate strength of the material, a different approach, primarily phcnomenological, must be used. Although this division exists, it is quite likely there is no such division of reactions of metals existing in nature.

The inability of the investigator to link these two regimes rests on: (1) a macroscopic–microscopic–submicroscopic problem; (2) an inability to accurately measure theory-sensitive metal reactions to dynamic stresses, when the hydrostatic component of the stress is causing density changes but the deviatoric component of the stress is created by and reacting with the solid to produce flow and fracture. This problem is dominant in any consideration of the strain-rate sensitivity of materials.

A further problem is found in the fact that, even at low dynamic stress, three-dimensional models are difficult to formulate rigorously and more difficult to analyze. The phenomenological—or mechanistic— approach is predominant in both the low and the high dynamic stress regime when real problems are considered—such as the penetration of a solid by a cylinder at velocities up to and including the hypervelocity region.

IV. The Role of Magnification in Modeling

The three basic regions of exploration in materials research that are identifiable are (1) the macroscopic ($>10^0$), (2) the microscopic ($10^{-4} \rightarrow 10^0$), and the submicroscopic region ($<10^{-4}$). While the exact demarcation points may be somewhat arbitrary, their existence is not. Each region has its own peculiarities, its own system of analytics, and its own experimental tools.

The classical approach for several hundred years has been concerned with the *macroscopic*. Stress–strain properties of a given specimen are concerned only with the following quantities: E, Young's modulus; G, shear modulus; ϵ_y the strain at yield of the material (or the strain at an arbitrarily defined yield point); σ_y, the stress at yield; and the ultimate and fracture stress–strain points. Practical design criteria frequently requires a knowledge of Poisson's ratio ν and the impact strength or fracture toughness of the solid. Mechanical engineering, in the applied sense, seldom requires more knowledge than a modulus and a strength parameter.

During the past two decades, emphasis has shifted in the research field from the *macroscopic* to the *submicroscopic*. Dislocations and lattice vacancies have become the subject of extensive experimental effort with each study generally resulting in a relation between some number of dislocations, N, and a stress required to bow, move, generate, annihilate, or otherwise change the makeup of the imperfection structure in the lattice.

The complexities involved in relating macroscopic specimen behavior to submicroscopic specimen character are enormous, and are made all the more so by the absence of an experimentally well-explored and analytically well-defined body of knowledge relating microscopic character to macroscopic behavior.

In the final analysis a specimen reacts according to the imposed stresses. Both the specimen and the stresses are viewed from the macroscopic standpoint. Moduli, yield points, strain hardening coefficients, ultimate and fracture stresses and strains are still the properties required for every day, garden variety, calculations. The influence on these quantities of grain size, subgrain size, micromosaic structure, preferred orientation, and other intrinsic material characteristics found in the microscopic region have not been completely cataloged. Nor are there abundant data available whereby relations may be obtained, for example, between preferred orientation, grain size, and Young's modulus. Popular research has bypassed this prosaic field, leaving it to industrial application groups who develop instantaneously needed data for their particular piece of steel used in a particular missile body.

Hypervelocity impact phenomena are definitely in the *macroscopic* realm of study, at present. So, the first problem encountered when trying to relate metallurgical–mechanical properties of materials to their reaction in hypervelocity impact is the absence of a smoothly connected body of knowledge, however sparse, between the macroscopic, microscopic, and submicroscopic. Two definite areas of knowledge exist, at either end of the spectrum. We are forced to limit the correlation to seeking a connection between the macroscopic and rather simple microscopic parameters.

V. Correlating Model and Test Data

The problem of actual measurement of the influence of metallurgical properties on metal deformation under combined hydrostatic and deviatoric stresses exists at all velocity ranges. Separation of the influences of these two stress wave components is quite difficult.

One approach is to measure geometric changes in the system and relate these changes to intrinsic material parameters. For example, analysis of the amount of plastic deformation developed in a cylindrical rod (length/diameter < 15) penetrating and/or perforating a target may be described by

$$P_l = C_0 e^{-\beta(x/t)} \tag{4}$$

where P_l is the total plastic deformation in the rod, C_0 is the speed of sound in the metal (when like metals are used for the target and penetrator), x is the target thickness, t is the time required to penetrate the target, and β is a constant. The constant β is a function of the dynamic stress–strain properties of the material being studied. However, this function of material parameters has not been determined exactly at present.

Another approach is the development of a theoretical model that may be examined experimentally during the deformation process. However, even at low impact velocities, where the stresses traveling through the specimen are the same order of magnitude as the strength of the specimen, it is difficult to obtain agreement on the influence of loading rate on material properties.

Taylor (*23*), along with White and Griffis (*24*), proposed that the speed of plastic waves in a metal could be determined from the stress–strain relation for the metal by

$$C^2 = (1/\rho)\, d\sigma/d\epsilon \tag{5}$$

where C is the wave speed, $d\sigma/d\epsilon$ is the slope of the applicable stress–strain curve, and ρ is the density of the solid.

The question ever since has been: What stress–strain relation is applicable? Should the nominal low testing rate stress–strain curve be used or is there a new stress–strain curve that applies at high velocity (greater than some critical velocity)?

Entering this question is the effect of grain size, dislocation density, impurity content, crystal structure, preferred orientation, modulus, and all mechanical material properties commonly used in describing a material.

There are several excellent review articles in this field. One of the latest descriptions of new work in the low-velocity field is given by Bell (*25*) in his recent book, primarily from the standpoint of strain-rate insensitivity. Bell references many of the critical papers in the field irrespective of the attitudes expressed on strain-rate sensitivity.

The approach taken by the authors to the problems involved in the influence of metallurgical parameters on HVI metal deformation is basically phenomenological. However, a number of concepts have been borrowed from theoretical and experimental one-dimensional studies. The following presents a number of observations made during the past that have led to the conclusion that intrinsic material properties control the deformation of solids when subjected to dynamic stresses several orders of magnitude larger than their strength.

VI. Supportive Metallurgical Observations

The impact of a projectile on a target at high velocities (in the hypervelocity region) can produce megabar impact pressures. However, if the target is thick enough, the final stages of deformation occur at stress levels just above the static target strength.

At what stress level is it possible to consider intrinsic material properties unimportant?

Under HVI of semiinfinite targets stresses and shock waves are produced that have some resemblance to those generated by explosives in contact with an inert solid. Observations made using explosive loading can assist in answering the question asked above.

The first general treatise in the field of explosive loading was by Rinehart and Pearson (*26*) in 1954. In 1955, Rinehart (*27*) internally loaded a thin-walled aluminum single crystal with explosive and correlated fractured crystal sections to the crystal orientation using stereographic projection. It would appear that Rinehart "proved" that the single crystallinity of his specimen controlled its fracture, even at pressures several orders of magnitude greater than the ultimate strength of the metal. However, Hugoniot studies of polycrystalline materials, extended to single crystals by Walsh *et al.* (*11*), developed data that was interpreted as proving that the Hugoniot for single crystals of a given material were the same as those for polycrystalline specimens of the same material.

In 1956, an extensive program was begun at the Ballistic Research Laboratories, Aberdeen Proving Ground, Maryland, to determine what, if any, influence metal properties have on the response of shock-loaded materials.

Studies by Glass *et al.* (*28*) mainly showed that single crystals began to deform under shock loading according to the crystal orientation to the imposed load, and fractured on crystal planes. Figure 5 shows an aluminum single crystal expanding due to internal explosive loading. Figure 6 compares fractured single crystal and polycrystalline specimens, and Fig. 7 is a summary of data for two single crystals, one tricrystal, one low-angle polycrystal, and one randomly oriented polycrystalline specimen. In the original experimental setup, the crystals were aligned so that framing camera records could be interpreted in terms of the

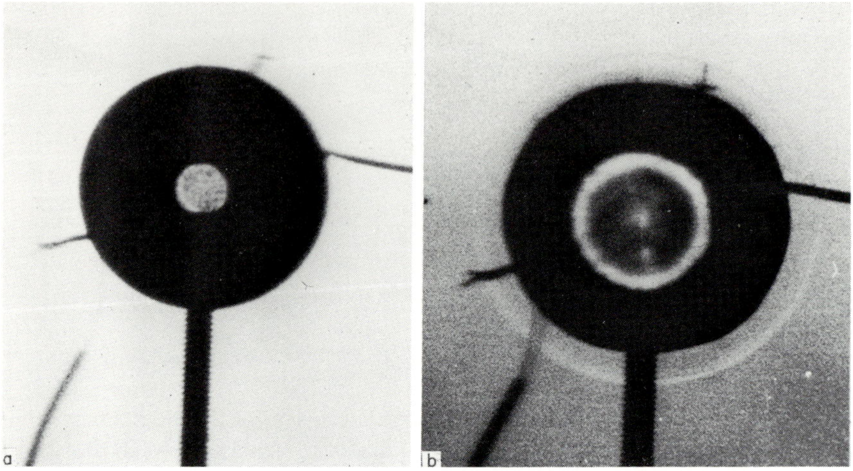

FIG. 5. Framing camera pictures of an expanding aluminum single crystal showing nonsymmetrical deformation. (a) 2 μsec after detonation; (b) 7 μsec after detonation.

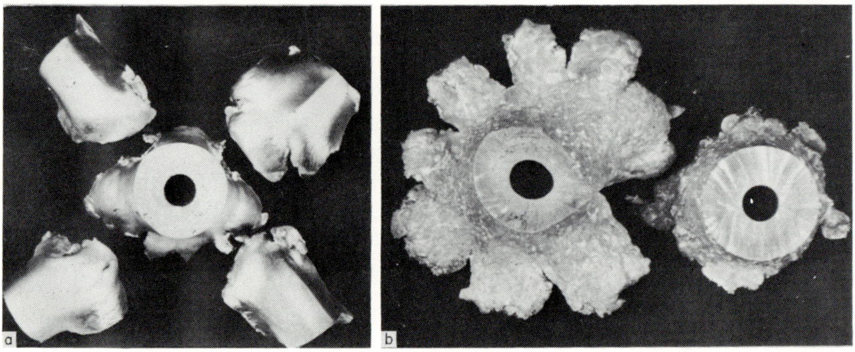

FIG. 6. Aluminum single crystal (a), and a polycrystalline specimen (b) after internal loading.

orientation of the crystals. Two 45° radii were chosen that corresponded, respectively, to [110] and [100] directions in the crystal wall. The free surface velocities on these two directions are shown for the single crystals and one grain of a tricrystal, in Fig. 7. The low-angle polycrystalline specimen corresponded, essentially, to a single crystal possessing a lineage structure. An "average" directionality was defined for the structure, and free surface velocity measurements made as in the other cases. The particle velocity was assumed to be one-half of the free surface velocity.

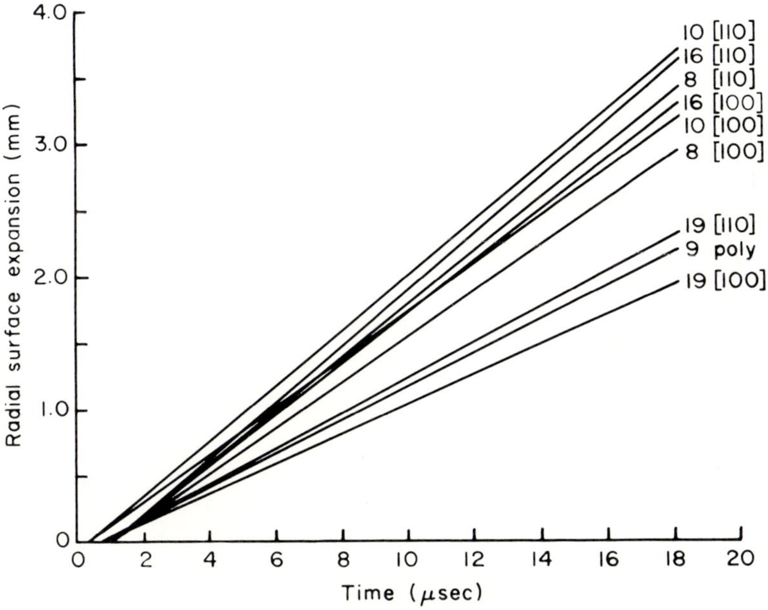

FIG. 7. Surface expansion of aluminum cylinders for various rounds.

Table I lists the particle velocities for these rounds, with average shock velocities and pressure at (just below) the free surface. It is interesting to note the differences between the velocities in the different directions.

TABLE I

ANALYTICAL RESULTS FROM SELECTED ROUNDS OF ALUMINUM CYLINDER EXPERIMENTS

Round number	Specimen type	Average shock velocity (mm/μsec)	Particle velocities referred to crystallographic direction for single crystals (mm/μsec)		Pressure at the free surface (kbar)	
			[110]	[100]	[110]	[100]
10	Single crystal	4.98	0.098	0.084	12.8	11.3
16	Single crystal	5.80	0.089	0.066	13.9	10.3
8	Tricrystal	5.50	0.097	0.083	14.1	11.9
19	Low-angle polycrystal	5.00	0.106	0.093	14.0	12.2
18	Polycrystal	5.07	0.152		20.2	
8	Polycrystal	4.49	0.059		7.13	

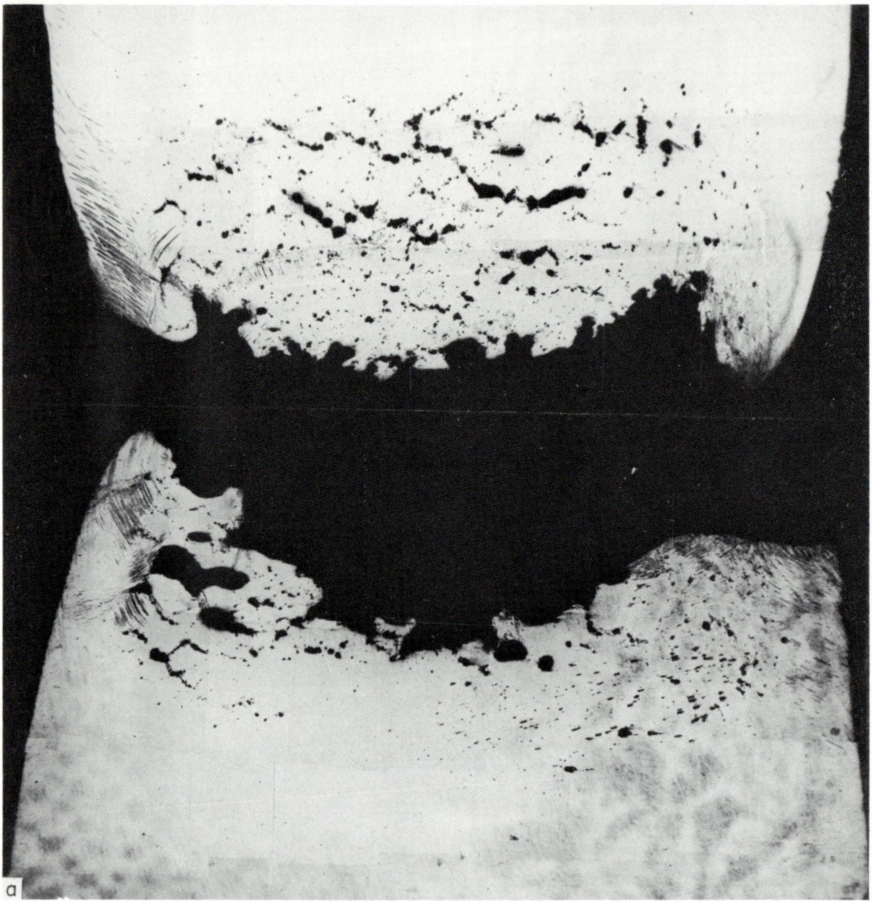

FIG. 8. Composite photomicrographs of a $\frac{1}{2}$-in. copper single crystal (a) and a $\frac{1}{2}$-in. copper polycrystal specimen after shock loading (b).

The fact that the crystals were subjected to stresses 10^2 greater than their ultimate strength, *at the outer surface*, and stresses 10^4 times their strength at the inner surface, and still deformed, from the beginning of deformation, as single crystals, is significant.

Work by Gainer and Glass (*29*) using copper cones having a variable preferred orientation through the cone wall showed that the cone collapse under explosive loading was a direct function of this orientation, and of the nature of the variation in orientation through the wall. Empirical relations were developed to show the influence of the orientation on collapse.

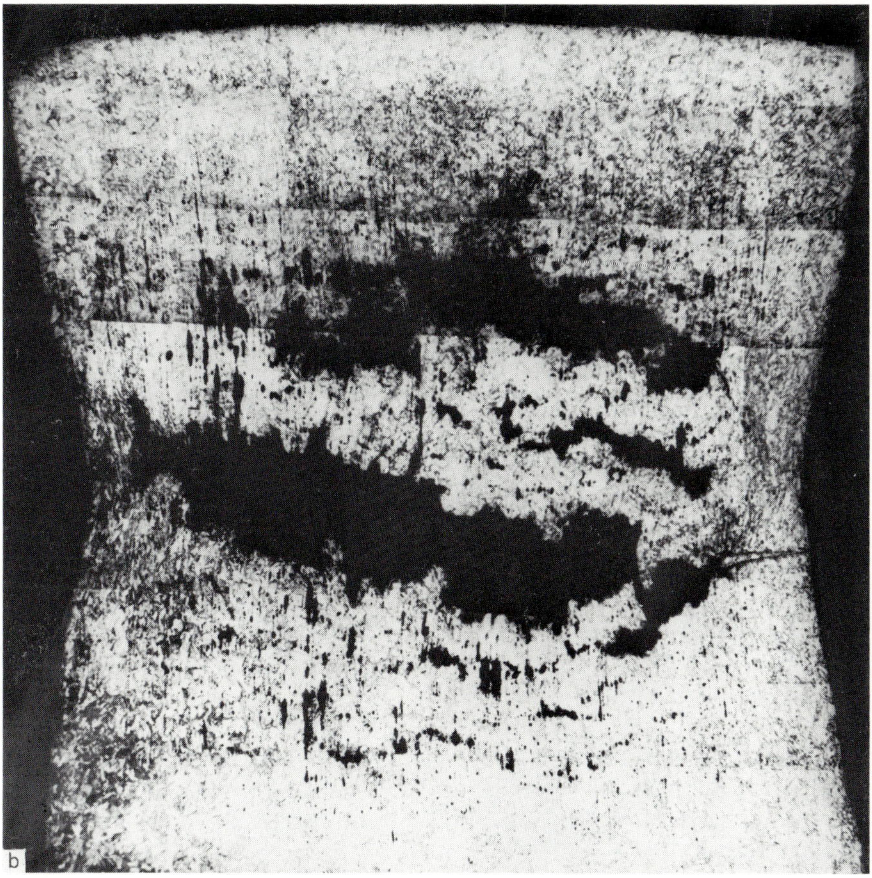

FIG. 8b.

It is interesting to note that recently Green (*30*) has shown that an elastic wave propagating through a single crystal loses energy down crystal planes nonparallel to the original direction of propagation of the wave. That is to say, a longitudinal wave develops shear components (and the converse) due to the crystal orientation. The same appears to occur for shock waves traveling through single crystals. However, attempts to directly measure quantitatively the split in the wave *during loading* have met with little success to date, primarily due to the difficulties in experiment.

To summarize, it has been shown in a semiquantitative manner that the following was true for shock-loaded specimens:

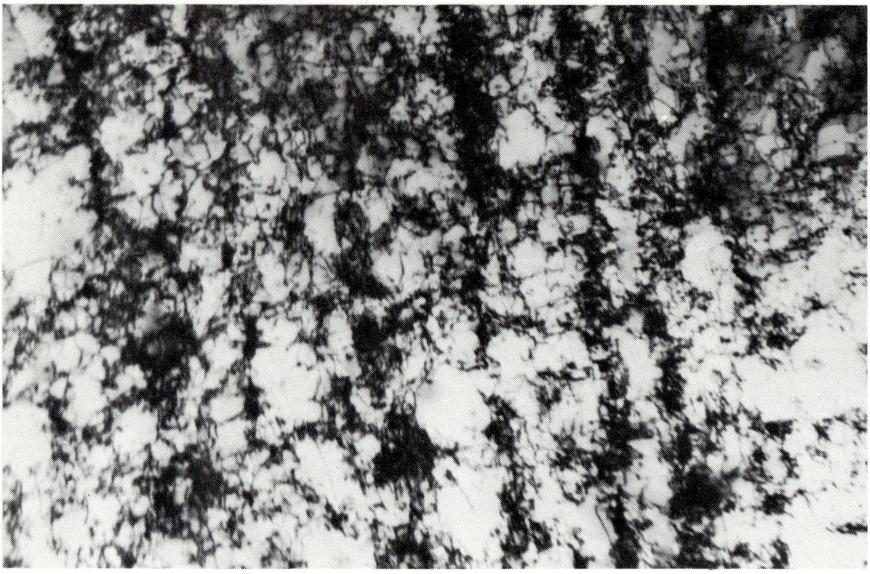

FIG. 9. Electron micrograph of a copper single crystal shock loaded to 90 kbar (40,000×). (N. Brown, private communication.)

1. Crystal structure controls the flow and fracture of shock-loaded specimens even at stress levels 10^2 greater than the strength of the metal.

2. Preferred orientation in polycrystalline materials controls the flow and fracture of shock-loaded solids up to stress levels 10^2 greater than the ultimate strength of the solid.

3. A shock will develop several components, directing energy along paths nonparallel to the original direction of shock propagation. At least two material parameters that influence this phenomenon are (a) crystal structure (orientation of planes to the shock in single crystals, or preferred orientations in polycrystalline specimens) and (b) micromosaic structure of the specimen (i.e., lineage, grain boundaries, etc.).

4. Subsequent studies at Ballistics Research Laboratory (BRL) have shown that the shock loses energy to the material by producing rotation of the microstructure. This rotation occurs on release of the high pressure (it is speculated). As the shock passes through a metal, there is a cyclic occurrence of rotations along the path taken by the shock.

5. Fracture of metals by strong compression shocks reflected from a free surface as tensile shocks is a function of all of the properties mentioned above. Figure 8 shows fractures in a single crystal and a

polycrystalline aggregate. The holes produced during fracture link through microscopic shear to produce the fracture surface.

6. The imperfection structure created by the passage of a compressive shock influences subsequent fracture by reflected waves. X-ray and electron microscopy have been used to show that large numbers of dislocations are produced in a cell-like pattern (See Fig. 9, for example). However, there is a regular pattern of planes on which the imperfection concentration is so high that it has not been resolved. In copper specimens the plane on which this occurs is (110).

7. Fracture under high-velocity (shock) loading is a function of the energy per unit volume imposed on the material and *not* a single-valued function of peak stress, as had been postulated.

While the above conclusions were being arrived at, Pond (*31*) had been studying high-velocity extrusion and had developed methodology for predicting required extrusion velocities based on an elastic–perfectly plastic stress–strain law. The conclusion of these studies was that the material will react elastically to a large impact or dynamic load and then begin to flow plastically at a constant stress (*32*). Interestingly enough, the same curve had been postulated theoretically by Prager (*33*) in 1933, for dynamically loaded materials, in an effort to cope with the yield problem. Drucker (*34*) had proposed the same type curve in studies on high-velocity metal cutting.

The main effort of this work was more than proposing a simple elastic–perfectly plastic stress–strain relation for dynamically loaded metals. The analysis was carried past the stumbling point that had been encountered by previous investigators: viz., how to assign values to the stress and strain points. Pond concluded:

(1) The stress–strain relation applied to high-rate loading problems where metal flow occurred sufficiently rapidly under a constantly applied stress.

(2) The elastic–perfectly plastic relation implied a time delay for strain hardening in a solid metal dynamically loaded.

(3) The end points (σ_y , σ_F , ϵ_y , ϵ_F) of the stress–strain curve could be determined from true-stress–true-strain data obtained at normal (0.05/min) strain rates.

(4) The area under the dynamic curve could be used to compute the energy per unit volume required to flow and fracture a material dynamically, *above some critical velocity*. This critical velocity depends on *the time delay for strain hardening* of a dynamically loaded metal.

The observations made prior to 1960 led to the first serious attempt by any one to introduce metallurgical parameters into the hypervelocity cratering problem. This was reported in a paper by Glass and Pond (*35*) at the Fourth Hypervelocity Impact Symposium. Energy dissipated by the projectile and various target reactions were discussed. Among the new (at that time) concepts introduced in the paper was the idea that the target continued to eject material long after the projectile had been dissipated. This was subsequently shown experimentally by Kineke (*18*) and Kineke and Vitali (*36*).

The following sections develop the concepts in detail.

VII. Theory of Cratering Based on Metal Properties

The fact that the dynamic stress–strain relationships are different from the static is well established, particularly with respect to the nominal stress–strain curve. However, true stress–strain curves have not been explored with respect to dynamic conditions. There are several important considerations of metal flow that may determine the volume of crater formed in the target. These may be listed as follows:

1. If the shear stress exerted on the metal is greater than the theoretical strength, the target flows by block shear, rather than dislocation motion. The theoretical strength of the metal is defined by the relation

$$\sigma_t = G/n \tag{6}$$

where G is the shear modulus, n is a number between 7 and 30, and σ_t is the theoretical shear strength.

A reasonable value of n for most metals is somewhere around 15. The shear modulus is generally in the neighborhood of $0.35E$ to $0.5E$.

The theoretical strength of a metal has a value in the range of explosively developed loads. This implies that the differential between the explovise load and the theoretical strength is about the same as the differential between normal load and the static flow stress.

An interesting result concerning fracture strength of explosively loaded metals is seen in Fig. 10. The plot of specimen diameter versus peak stress at fracture, for seven metals, extrapolates to $E/2\pi$ when the specimen diameter goes to zero. This is tantamount to saying that the material fractures at a stress level equal to the theoretical *tensile strength* when it cannot flow by a shear process (zero diameter).

2. A second parameter of the material of prime importance is the true-stress–true-strain curve. A schematic representation of these may be seen in Fig. 11 where the nominal strain rate and the high strain rate curves are compared.

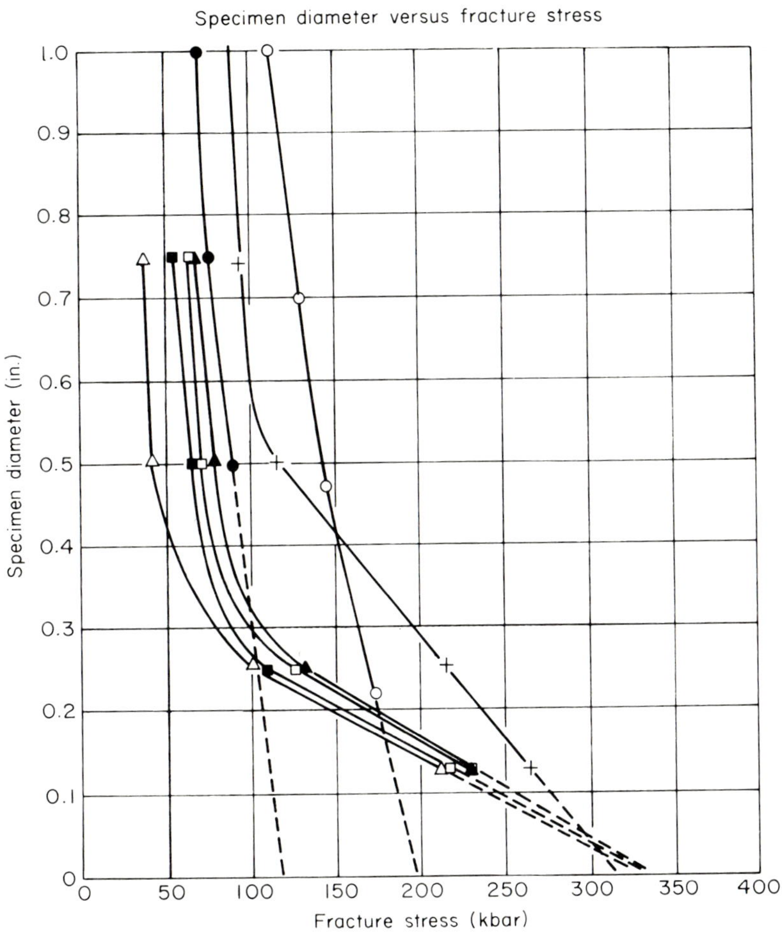

FIG. 10. The effect of specimen diameter on fracture stress: (○) Cu; (+) 304 SS; (●) 1100 Al; (▲) 1045 Steel; (□) 1020 Steel; (■) 4140 Steel; (△) Armco iron. Note: specimens fracture with no neckdown.

The basis of this presentation is as follows:

(a) The pellet delivers its energy instantaneously to the metal. A certain amount of that energy is used up to dissipate the pellet, and the rest goes into the target.

(b) The energy delivered to the target causes the target to start flowing. It can flow out from under the free surfaces, causing the pressure to drop rapidly in the shock front.

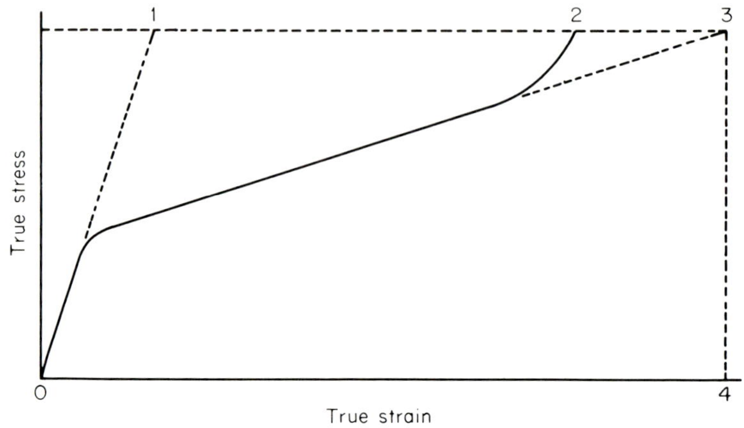

FIG. 11. The relationship between the static and the postulated dynamic true-stress–true-strain curves.

(c) The pressure, while very high at the contact surface, will drop off very rapidly in the metal due to side release flow, and the expansion of the shock from the impact point. The problem is removed from hydrodynamic considerations, except during the time when the pellet is being used up, due to the drop in pressure from the megabar region.

(d) The metal then reacts as if there were a time delay for strain hardening, and this time is longer than the time required for crater formation. This essentially says that the stress–strain curve follows the shape, 0–1–3, seen in Fig. 11. Here, the material is *not* acting hydrodynamically, since it still flows according to crystallographic and strength properties, *but it is flowing without hardening*, so that once it starts it will continue to flow until the energy in the metal has dropped below a certain level.

(e) The time of reaction is determined by a "critical velocity" of metal flow in the target. The material can only get out of the target at velocities determined by this critical velocity.

(f) The process of fracture and flow stops when the energy in the target has dropped below the energy under the true stress–strain curve determined at low (0.05/sec) strain rates.

(g) The energy remaining in the target goes into plastic deformation, recrystallization, and hardening, giving an "affected" zone around the target hole that has changed structure, or has taken on residual stress.

Using the approach presented in the preceding section, the following path is taken.

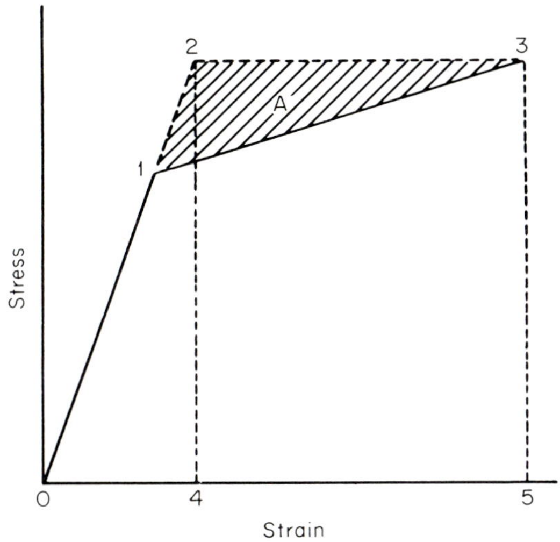

FIG. 12. The postulated dynamic stress–strain curve showing the energy increase over the static curve.

Figure 12 shows a schematic representation of the true stress–strain curve, 0–1–3. If the elastic modulus is extended up to the fracture stress, and a horizontal line drown to the fracture strain, Section A is added to the curve. This is the concept previously referred to (*31*, *32*) which has been used very successfully in analyzing high-velocity metal extrusion problems. The quantity of energy represented by the elastic section of the "new" stress–strain diagram, 0–2–4–0, is the amount that must be put into a metal to cause it to begin flowing without strain hardening. This energy must be put into the metal in less time than the time for strain hardening to begin. Once the metal begins flowing, it will continue to do so until the energy is dissipated, until fracture occurs, or until the time delay for strain hardening is reached.

The energy represented by the area under the total curve, 0–2–3–5–0, may be looked upon as that amount required to flow a metal faster than a critical velocity.

An interesting point concerning this energy–velocity relationship is that once material is moving faster than this velocity the metal no longer requires large amounts of energy to deform. Plastic flow is limited, and the amount of energy required is only that used in fracture of the metal, and in the acceleration of fragments formed (excluding the compression of nonflowing material).

VIII. Strain Energy Distribution under the Crater

A. Mechanical Energy Equivalent

In order to develop the strain energy distribution under the crater it is necessary to develop a mechanical energy equivalent as well as portray the strain distribution around the crater.

Generation of the mechanical energy equivalent is accomplished by developing a static true-stress–true-strain tensile curve for the target material as well as the pellet material. The target material is tested in the dead annealed state since the target itself is provided in this state for penetration. A characteristic tensile curve is illustrated in Fig. 11. The rupture stress (Point 2) is taken as the maximum stress in the proposed dynamic stress–strain diagram. The material is proposed to not strain harden during the high-velocity deformation and therefore the elastic portion of the curve is followed by a plastic portion wherein the strain-hardening index (or the slope of the curve) is zero. However, the rupture strain at high velocity is taken as Point 3 rather than Point 2.

This rupture strain is used for the following reasons. At the beginning of necking during the tensile test, all of the deformation occurs in the necked volume and this continues to be so during the rest of the test. Therefore, in the true-stress–true-strain plot the instantaneous gage length during necking should be taken as the length of the necked region. If this is carried out and the observed strain added to the previously documented strain, the slope of the plastic portion of the static curve is found to remain constant until rupture. This observation was made possible by taking motion pictures of tensile test specimens during the entire tensile test and determining the geometry of the specimen from the picture record at all desired stages of the test. The system used by McGregor (*37*) for determining true strain Q wherein $Q = 2 \log_e D_{\text{original}}/D_{\text{instant}}$ is much less tedious and produces the same curve. This observation was found to be consistent for an array of aluminum alloys, copper, low-carbon steel, and an array of stainless steels.

The mechanical energy equivalent per unit volume is taken as the area under the dynamic curve (0–1–2–3–4–0) in Fig. 11 for portions of the metal which are deformed to the rupture strain limit. The energy equivalent for all other strains is taken as the area of the curve up to the strain limit being considered.

B. Strain Display

The targets to be used were all annealed in an identical manner and were then penetrated with a drill rod pellet. The crater volume was determined by titration. Because it was desired to study the effects of the minor craters on the target surface, the volumes of all the craters were taken. After the volume determination the specimen was again annealed and then sectioned through the center of the crater. The sectioned surface was ground flat, metallographically prepared, and etched. Figure 1 illustrates the characteristic appearance of the crater, the lip, and the strain-affected region after this procedure.

The strain-affected region is delineated because the residual strain energy in a metal determines the grain size for a particular recrystallization schedule. In order that the varying grain-size field of the strain-affected region could be documented in terms of strain, tapered tensile specimens were extended to rupture, the strain measured along the length, and (after annealing) the recrystallized grain size correlated with the strain. Using this correlation, the strain gradient in the strain-affected volume can be determined.

The strain distribution can be delineated in a more complete and efficient manner by the use of hardness measurements instead of grain size. The Vickers microhardness (30-kg load) method was adopted as the standard method for measuring hardness on the target materials.

The Vickers hardness measurements are not only reproducible, they are also extremely sensitive to low strain values (being capable of measuring strains as low as 0.1%). Thus, it is possible to obtain "strain versus distance from crater" graphs more accurately than with the grain-size method and also to measure the strain in any part of the specimens. Development of strain versus distance from crater (r) in 2S aluminum is given in Figs. 13–15.

These plots were developed by first determining the variation of Vickers hardness (cross diameter of pyramidal impressions) with strain level on a tapered tensile of target material, Fig. 13. Then a plot of hardness versus distance from the crater was made (Fig. 14). Finally, by cross plotting the two previous graphs, Fig. 15 is obtained, a plot of strain versus distance from crater. If this plot is made in terms of logarithm of strain versus r, it is possible to read off values of the constants α and c in $\epsilon = e^{-\alpha(r-c)}$ via the slope and intercept method (Fig. 16).

C. Computation of Energy Distribution

The impact energy E_I is determined from measurements of the mass of the projectile (m) and its velocity at impact (v):

$$\text{impact energy} = \tfrac{1}{2}mv^2$$

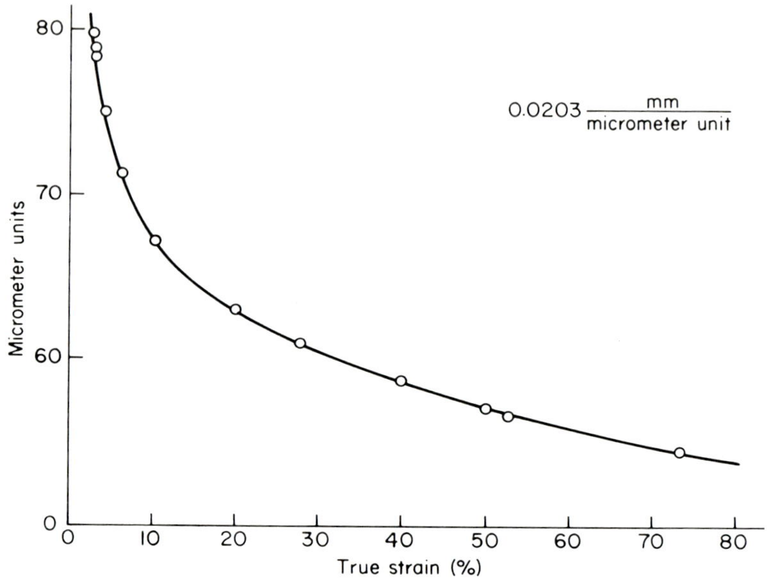

FIG. 13. Average cross-corner Vickers parameter versus percent strain in tapered aluminum tensile specimen.

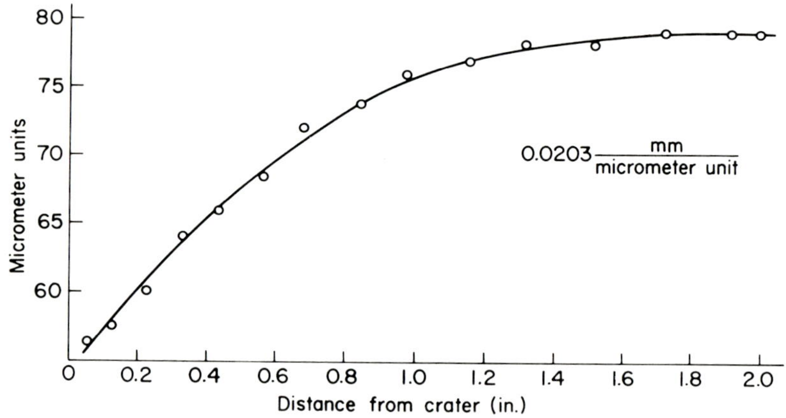

FIG. 14. Cross-corner Vickers parameter versus distance from crater in aluminum target.

Once the mechanical energy equivalent and the strain distribution are determined, an energy balance can be expressed by

$$\begin{aligned}\text{energy of impact} &= \text{crater energy} + \text{strain field energy}\\ &\quad + \text{lip energy} + \text{projectile breakup energy}\end{aligned}$$

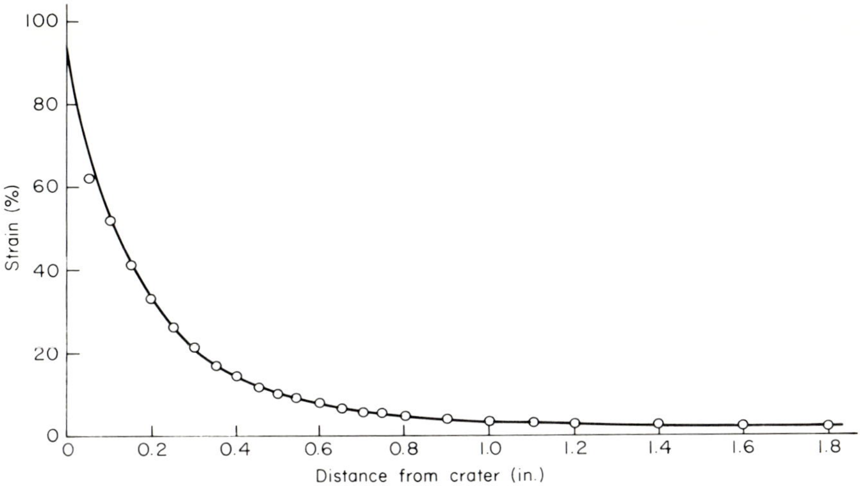

FIG. 15. Percent strain versus distance from crater in aluminum target.

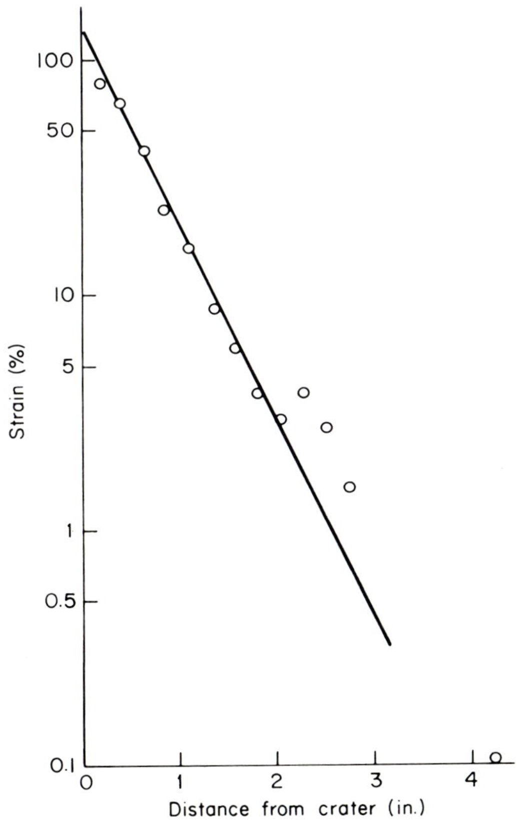

FIG. 16. Log percent strain versus distance from crater in aluminum target.

The crater energy E_c is the crater volume V_c times the mechanical energy equivalent for rupture or the area under the dynamic stress–strain curve of the target material up to the dynamic rupture strain.

The lip energy is found using the same mechanical energy equivalent but multiplying it times the volume of the lip.

The projectile breakup energy is the volume of the projectile times the mechanical energy equivalent for rupture of the projectile material.

The strain field energy is expressed by

$$E_{AR} = (2\pi\sigma_m/\alpha)[\epsilon_m\{(r_0 + 1/\alpha)^2 + 1/\alpha^2\} - \epsilon_0\{(r_m + 1/\alpha)^2 + 1/\alpha^2\}] + (2\pi\sigma_m h/\alpha)[\epsilon_m(r_0 + 1/\alpha) - \epsilon_0(r_m + 1/\alpha)] \quad (7)$$

where r_0 is the crater radius, r_m is the radius of affected region, h is the height of cylindrical portion of crater, σ_m is the rupture stress defined by hypervelocity curve, ϵ_0 is the strain at r_m and ϵ_m is the strain at r_0, both defined by observation of grain growth or hardness measurements, α is a constant in $\epsilon = e^{-\alpha(r-c)}$, c is a constant in $\epsilon = e^{-\alpha(r-c)}$, and E_{AR} is the energy of affected region.

IX. The Relation of Energy Balances to Projectile Orientation

In examining the results of the energy balances in Cu and 2S Al at 2000–3000 J input it was found that some of the balances did not agree with the known impact energy within reasonable limits. Energy balances are intimately dependent upon variables derived from measurements on the targets. Examination of these target variables was necessary to determine whether a systematic change occurred in any of them which was related to the energy balance error. When such a procedure was carried out, the following observations were made:

(1) The ratio of the crater volume of a target to the average of all target crater volumes at a fixed impact energy is low whenever the ratio of the energy balance E_T to the impact energy is low.

(2) The variables r_0 and r_m, defined earlier in the chapter, do not vary much at the tested energy levels in aluminum and copper, but the depth of the crater P_c has a variation an order of magnitude higher than r_0 or r_m (Tables II and III).

(3) Projectile orientation is shown to govern this variation in depth, and hence in volume, of the crater; however, only inferences may be made as to the nature of this effect on energy balances in the strain-affected region.

TABLE II

PROJECTILE ORIENTATION EFFECTS

	Target no.	Penetration (cm)	Diameter crater (cm)	P_c/D_c	E_T/E_I	cos ϕ
2S Al: 2000 J	1	1.26	2.26	0.56	0.748	0.961
	2	1.85	1.99	0.93	—	0.000
	3	1.53	2.18	0.70	—	0.485
	4	1.41	2.32	0.61	—	—
	5	1.77	2.25	0.79	1.045	0.139
	6	1.73	2.14	0.81	1.082	0.000
	7	1.66	2.17	0.76	0.915	0.000
	8	1.46	2.29	0.74	—	0.035
	9	1.36	2.34	0.58	1.020	0.530
	10	1.29	2.28	0.57	0.802	0.643
3000 J	11	1.86	2.38	0.78	—	—
	12	2.35	2.16	1.08	1.029	0.122
	13	2.38	2.04	1.17	—	0.174
	14	2.30	2.15	1.10	1.016	0.139
	15	1.32	2.34	0.56	—	0.857
	16	2.48	2.22	1.12	1.013	0.000
	17	1.64	2.38	0.69	0.728	0.866
	18	2.27	2.04	1.11	1.137	0.358
	19	1.33	2.39	0.56	0.694	0.809
	20	2.28	2.21	1.03	1.039	0.000
9000 J	23	2.12	3.82	0.50	0.770	0.848
Cu: 2000 J	2	0.72	2.39	0.30	0.954	0.755
	3	1.09	1.83	0.60	1.023	0.000
	4	0.83	1.75	0.42	1.06	0.000
	5	0.73	1.63	0.45	1.00	0.485
	6	0.73	1.67	0.44	0.952	0.875

In Table IV the ratios of Volume of crater/average volume of crater ($V_c/\bar{V}_c$) and total computed energy/impact energy (E_T/E_I) are given for the targets discussed. From the plots (Figs. 17 and 18) of crater volume (V_c) versus impact energy (E_I) the mean values of these variables are linear for the velocities tested. Since the projectiles fired into the targets were disk shaped, they could give different effective areas as a source of kinetic energy depending on their angle of tilt relative to the target surface at impingement.

The geometry of the problem is the following: The projectile is a disk. If normals are chosen to the target surface and to the pellet surface,

TABLE III

CRATER PARAMETER VARIATION AS A FUNCTION OF IMPACT ENERGY

Target no.	Impact energy E_I (J)	$\bar{r}_0$ (cm)	$\bar{r}_{\max}$ (cm)	$\bar{V}_c$ (cm³)	P_c (cm)
2S Al:					
1, 5, 6, 7, 9, 10	1948 ± 53.3	1.12 ± 0.023	2.24 ± 0.08	4.27 ± 0.46	1.57 ± 0.226
12, 14, 16, 17, 18, 19, 20	3088 ± 39	1.10 ± 0.05	2.40 ± 0.044	6.38 ± 1.05	2.14 ± 0.326
23	9003	1.85	2.50	17.09	2.20
Tough pitch Cu:					
1–6	2000	0.84 ± 0.05	2.54 ± 0.09	1.18 ± 0.28	0.80 ± 0.09

TABLE IV

COMPARISON OF PROJECTILE ORIENTATION AND ENERGY BALANCE

	2S Al				Cu		
Target	cos ϕ	$V_c/\bar{V}_c$	E_T/E_I	Target	cos ϕ	$V_c/\bar{V}_c$	E_T/E_I
1	0.961	0.798	0.748	1	0.755	0.805	0.733
2	0.000	—	—	2	0.358	0.915	0.954
3	0.485	—	—	3	0.000	1.48	1.023
5	0.139	1.080	1.045	4	0.000	1.18	1.06
6	0.000	1.176	1.082	5	0.485	1.14	1.00
7	0.000	0.948	0.915	6	0.875	0.712	0.952
8	0.035	—	—	7	0.995	—	—
9	0.530	0.972	1.020				
10	0.643	0.842	0.802				
12	0.122	1.171	1.029				
13	0.174	—	—				
14	0.139	1.020	1.016				
15	0.857	—	—				
16	0.000	1.204	1.013				
17	0.866	0.798	0.728				
18	0.358	1.066	1.137				
19	0.809	0.749	0.694				
20	0.000	1.197	1.039				
23	0.848	—	0.770				

then the cosine of the angle between these normals is proportional to the projected area of the pellet on the target surface. That is, if ϕ is the angle in question, then cos ϕ is a minimum where the projected area is a minimum and is a maximum where the projected area is a maximum.

The angle ϕ is measured by stereographic projection using BRL firing-range values for two angles of rotation about independent axes.

From plots of crater volume (V_c) vs. cos ϕ (Fig. 19) at constant energy a linear variation is observed. The minimum volume occurs at $\phi = 0^0$ and the maximum volume at $\phi = 90°$. The dispersion of points on these plots is felt to be due in part to a dispersion of energies about the mean for each charge design.

Another source of error is the uncertainty of the pellet orientation because the documentation of the angle is made at some distance from the target surface. To examine this error the maximum tolerable pellet rotation velocity is estimated and compared with the value calculated from BRL's flash sequence on an array of targets. If a pellet translational

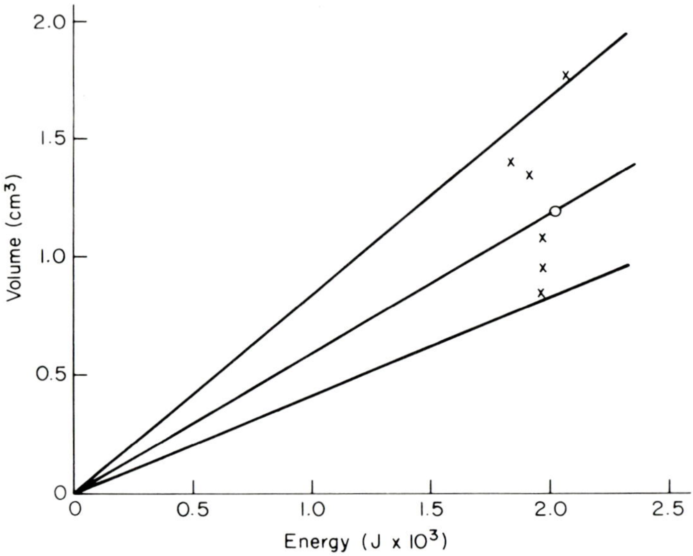

FIG. 17. Crater volume versus impact energy in tough pitch copper. (×) Targets, (○) average.

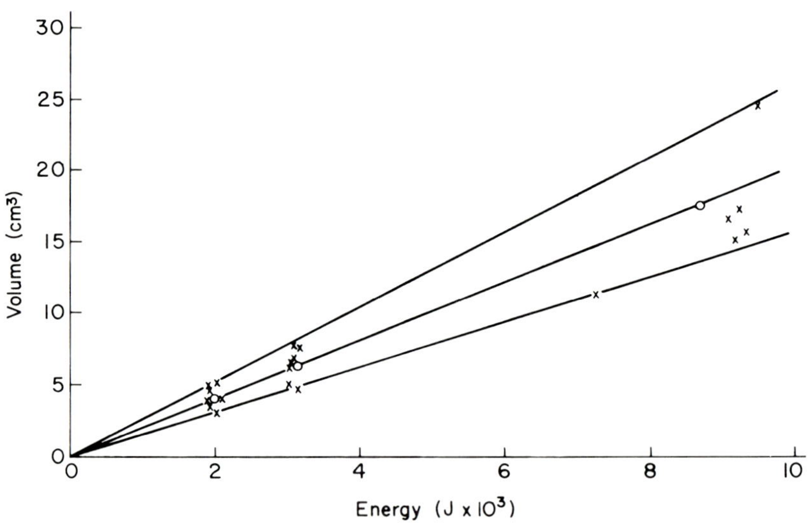

FIG. 18. Crater volume versus impact energy in annealed 2S aluminum. (×) Targets; (○) average.

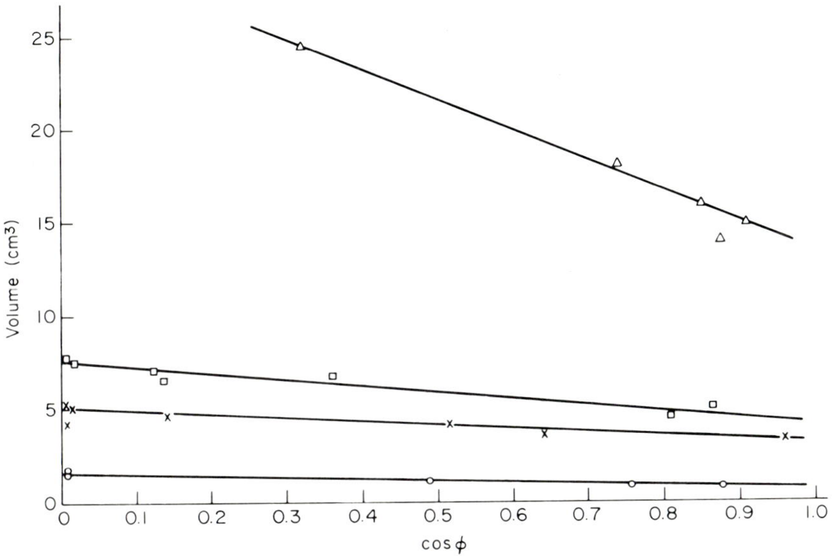

FIG. 19. Crater volume versus cosine of orientation angle ϕ for several impact energies. (×) 2S Al, 2000 J; (□) 2S Al, 3000 J; (△) 2S Al, 9000 J; (○) tough pitch Cu, 2000 J.

velocity of 4–5 km/sec, a distance at radiograph of 5 cm, and a variation in ϕ of 5° are chosen, this corresponds to a maximum tolerable tumble rate of 1100–1400 rps. Perusal of the tumble-rate values for the aforementioned flash sequence shows that all but one of the values at the radiograph closest to the target are below the maximum allowance (on the assumption that the pellet does not turn $180° + \Delta\phi$ or $360° + \Delta\phi$, etc.).

The estimated error in $\cos\phi$ for a 5° variation in ϕ is then

$$(\cos\phi) = \sin\phi\,\Delta\phi$$

or

$$(\cos\phi) = \pm 0.088 \qquad \text{for} \quad \phi = \pm 5° \quad \text{at} \quad \phi = 90°$$

Unfortunately the charge designs were different at different energies and the effect of pellet area on volume may not be isolated from the effect of energy on volume as a function of pellet orientation. Each line on the plot, Fig. 19, for a material and energy gives only relative effects of projected area.

In terms of the energy balances the effect of pellet orientation is not understood. Mathematically, the greatest energy per unit volume is ascribed to the crater region. Thus, when the crater is small and the maximum extent of the strain-affected region has not increased (noting the aforementioned consistency of r_m), the energy balances will obviously be low. Part of the answer in a physical sense may be in the use of the strain gradient function $\epsilon = e^{-\alpha(r-c)}$.

This function contains only a radial variation which might be taken to imply deformation proceeding from a point source. A disk projectile striking on its edge is more closely a point source than is the same projectile striking on its face.

X. Summary of Energy Balances

A summary of values for the targets on which energy balances have been performed is found in Tables V–VII. A plot showing the spread in energy balances for 2S Al, Cu, and mild steel is shown in Fig. 20.

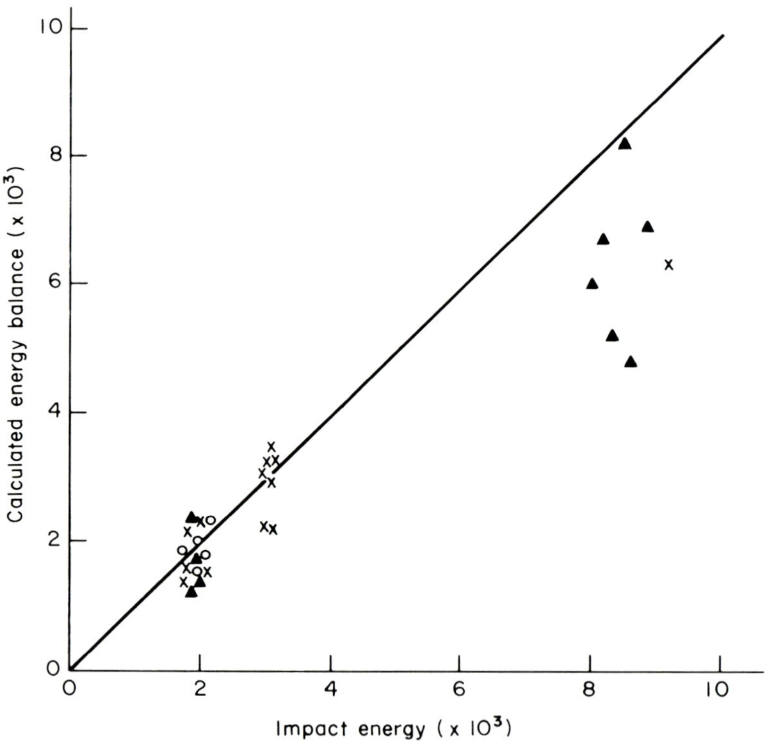

FIG. 20. Comparison of calculated energy balances with impact energy for several metals. (×) 2S Al, (▲) 1014 Steel, (○) tough pitch Cu.

TABLE V

CRATER PARAMETER AND ENERGY BALANCES FOR 2S AL

Target no.	Impact velocity (km/sec)	E_I (J)	r_0 (cm)	r_m (cm)	ϵ_0 (in./in.)	ϵ_m (in./in.)	α (cm^{-1})	h (cm)	V_c (cm^3)	E_c (J)	E_{AR} (J)	E_T (J)	ϕ (deg)
1	4.747	2028	1.10	1.90	0.075	0.735	2.85	0.20	3.41	491	669	1190	16
							2.28				1026	1516	
2	4.660	1954	1.00	2.25	0.075	0.735	1.82	1.00	5.03	708	1511	2219	90
5	4.602	1906	1.10	2.30	0.075	0.735	1.90	0.50	4.61	670	1323	1993	82
6	4.577	1885	1.10	2.25	0.075	0.735	1.98	0.60	5.02	717	1332	2049	90
7	4.768	2046	1.10	2.20	0.075	0.735	2.07	0.30	4.05	584	1085	1669	90
							1.90				1288	1872	
9	4.592	1898	1.10	2.00	0.075	0.735	2.85	0.10	4.15	586	782	1368	58
10	4.604	1908	1.00	2.10	0.075	0.735	2.07	0.50	3.59	515	1025	1540	50
12	3.809	3047	1.05	2.50	0.075	0.735	1.54	1.25	7.18	1001	2118	3150	83
14	3.815	3056	1.10	2.45	0.075	0.735	1.66	1.15	6.51	907	2174	3104	82
16	3.877	3157	1.15	2.45	0.075	0.735	1.76	1.20	7.68	1071	2145	3198	90
17	3.800	3032	1.23	2.30	0.075	0.735	2.07	0.75	5.09	719	1476	2209	30
18	3.834	3087	1.00	2.45	0.075	0.735	1.54	1.25	6.80	948	2550	3511	69
19	3.876	3155	0.95	2.28	0.075	0.735	1.68	0.75	4.78	666	1516	2188	36
20	3.830	3080	1.05	2.50	0.075	0.735	1.54	1.25	7.64	1064	2118	3202	90

TABLE VI

CRATER PARAMETER AND ENERGY BALANCES FOR TOUGH PITCH COPPER[a]

Target no.	Impact velocity (km/sec)	E_I (J)	r_0 (cm)	r_m (cm)	ϵ_0 (in./in.)	ϵ_m (in./in.)	α (cm^{-1})	h (cm)	V_c (cm^3)	E_c (J)	E_{AR} (J)	$E_c + E_{AR}$ (J)	ϕ (deg)
1	4.675	1967	0.78	2.26	0.045	0.556	1.67	0.00	0.95	191	1156	1441	41
2	4.678	1970	0.84	2.54	0.045	0.556	1.47	0.00	1.08	215	1581	1888	69
3	4.827	2097	0.99	2.72	0.045	0.556	1.66	0.00	1.75	349	1707	2146	90
4	4.517	1836	0.84	2.54	0.045	0.556	1.47	0.00	1.40	279	1576	1947	90
5	4.623	1923	0.86	2.54	0.045	0.556	1.48	0.00	1.35	270	1568	1927	61
6	4.694	1983	0.78	2.63	0.045	0.556	1.43	0.00	0.84	168	1628	1887	29

[a] $\sigma_m = 5.195 \times 10^4$ psi for all Cu targets.

TABLE VII

CRATER PARAMETERS AND ENERGY BALANCES FOR 1014 STEEL[a]

Target	Impact velocity (km/sec)	E_I (J)	r_0 (cm)	r_m (cm)	ϵ_0 (in./in.)	ϵ_m (in./in.)	α (cm^{-1})	h (cm)	V_c (cm^3)	E_c (J)	E_{AR} (J)	E_T (J)	ϕ (deg)
3	4.6669	1960	0.68	2.06	0.001	1.37	4.83	0.00	0.341	349	1102	1452	3
4	4.625	1925	0.71	2.76	0.001	1.07	3.70	0.11	0.743	595	1674	2269	72
5	4.646	1943	0.66	2.56	0.001	0.72	3.46	0.19	0.489	263	813	1076	52
6	4.670	1963	0.67	2.17	0.003	0.72	3.77	0.17	0.826	441	976	1417	72
7	4.627	1927	0.74	2.33	0.001	1.09	4.43	0.01	0.710	579	1142	1721	71
10	3.880	8445	1.18	3.39	0.001	1.73	3.52	0.00	2.455	3177	5119	8296	46
11	3.771	7997	1.08	3.68	0.001	0.73	2.74	0.05	3.00	1629	4304	5933	63
12	3.975	8864	1.15	3.73	0.001	1.90	3.23	—0.25	2.337	3321	5119	8439	48
13	3.922	8629	1.14	2.89	0.004	0.72	2.96	—0.25	2.289	1223	2127	3350	42
14	3.975	8864	1.13	2.91	0.001	1.95	4.23	—0.20	2.231	3254	3687	6941	60
16	3.854	8333	1.16	3.54	0.001	0.92	2.87	0.00	3.338	2171	3570	6741	68
17	3.822	8195	1.08	3.19	0.001	1.04	3.33	0.50	3.626	2820	3843	6663	90

[a] $\sigma_m = 1.087 \times 10^5$ psi for all 1014 Steel targets.

XI. Energy Balances on 2S Al Prestrained Targets

In the light of the energy balance technique the effect of prior cold working of 2S Al on its hypervelocity deformation behavior was studied. Since cold working introduces strain energy in a metal, the strained metal should absorb less energy than is absorbed in the annealed state prior to rupturing. This decreased ability to absorb energy would infer a larger crater volume and/or strain-affected region in comparison with hypervelocity penetration of the annealed metal. In effect, it was hypothesized that the energies of two deformation processes raising the strain level above the annealed state would be additive. The dynamic material properties governing the metal's behavior would then be related to an annealed state by consideration of energy addition. Data indicate that this is *not* the case. The crater volume was smaller in the forged targets than that resulting from hypervelocity penetration of annealed 2S Al.

Seven cylindrical 2S Al annealed specimens were forged from approximately 4 in. in height to approximately 3 in. in height. The specimens were then machined on a lathe to dimensions of 3-in. height and 3-in. diam. The specimens were then used as targets for the impact of hypervelocity projectiles.

The targets were sectioned through the center of the crater and polished so that Vichers hardness (30-kg load) measurements could be made. The hardness measurements were analyzed and correlated to their corresponding strain values. Then, a plot of "strain versus distance from crater" was obtained. The "strain versus hardness measurement" curve for annealed 2S Al specimens was used in order to determine the "strain versus distance" curve. The "strain versus distance" curve plotted on semilog paper was quite different from any previous nonforged 2S Al specimens. Generally, there was a short-length exponentially decreasing portion starting at the crater edge and extending approximately 1 cm. Following this portion there was a region of constant strain (generally of a value around 30%) which then decreased to some lower strain value at large distances from the crater (4 cm).

The specimens were analyzed by assuming that the constant strain level observed on the "strain versus distance" curves was the average strain imposed on the specimen during forging. This strain value was subtracted from every strain value on the "strain versus distance from crater" curve in order to obtain the "strain versus distance" graph for the hypervelocity impact. The resulting curve was exponentially decaying.

Energy balances were performed in the same manner as on any of the

TABLE VIII

CRATER PARAMETERS AND ENERGY BALANCES 2S Al PRESTRAINED TARGETS

Target no.	Impact velocity (km/sec)	E_I (J)	r_0 (cm)	r_m (cm)	ϵ_0 (in./in.)	ϵ_m (in./in.)	α (cm^{-1})	h (cm)	V_c (cm^3)	E_c (J)	E_{AR} (J)	E_T (J)	ϕ (deg)
101	4.811	2083	1.06	3.73	0.001	0.508	2.34	0.17	3.271	321	689	1010	—
			1.06	2.35	0.010	0.497	3.33	0.17	3.271	314	424	738	
			1.06	3.12	0.001	0.497	3.03	0.17	3.271	314	450	764	
102	4.557	1869	1.05	3.57	0.001	0.788	2.64	0.04	2.784	424	807	1230	—
103	4.877	2141	1.08	2.05	0.010	1.38	5.12	−0.06	3.121	832	530	1361	48
104	4.280	1649	0.98	2.34	0.001	1.41	5.39	0.65	4.118	1121	783	1904	—
105	4.706	2003	1.043	3.44	0.001	0.651	2.711	0.24	3.666	461	711	1173	52
106	4.839	2107	1.03	2.56	0.001	0.572	4.14	0.17	2.775	307.5	325	632	—
107	4.835	2139	1.10	5.58	0.001	0.380	1.34	0.17	3.685	270	1514	1784	—

previous 2S Al specimens. The energy balances thus obtained are presented in Table VIII.

It can be seen from Tables V and VIII that the crater volumes of forged targets are all less than the mean value of crater volumes in the annealed targets (1–10) penetrated by the same charge design. The energy balances were not in agreement with the impact energy. The energy balance discrepancy is thought to result from the following:

(1) Hardness measurements on the base of the fired targets were extremely variable. This variation was probably introduced in the forging operation which also produced considerable "barreling" of the targets. This inhomogeneity carried over after the firing and made the strain-affected region very difficult to analyze.

(2) The dynamic rupture stress used in the energy balances was the value obtained for annealed aluminum. This had been felt to be justified on the hypothesis that dynamic rupture was dependent upon the metal's ability to absorb energy above some initial state and that energies absorbed would be additive in terms of producing rupture. On the basis of the crater volume data it appears that the material's dynamic strength properties have changed and that a different dynamic rupture stress should be considered.

Energy balances (see Table IX) were also computed using the strain field without correction for the initial deformation in forging. The estimated energy absorbed in forging was subtracted from the energy balance and the remainder compared to the impact energy of hypervelocity deformation. The values listed in Table IX are far too high to be reasonable. Comparison of these values, however, does not test the hypothesis about energy addition because of the uncertainty of the effect of cold working on the dynamic strength properties of the metal. It has only been shown that, using the dynamic strength properties of the annealed metal, neither addition of the strain levels nor addition of the energies imposed in deformation leads to good energy balances.

TABLE IX

Special Analysis of Prestrained Al Targets

Target no.	Impact energy	Estimated energy of forging	Total calculated energy
104	1649	6765	19,862
105	2003	15,200	3180

XII. Calculation of Maximum Impact Damage from Material Properties

The importance of any parameter in a process or reaction can probably be best measured by its use in predicting the results of the reaction. To predict penetration in terms of strength properties a dynamic energy balance is performed:

$$\text{impact energy/volume} = \tfrac{1}{2}\rho V^2_{\text{rupture}} + \text{energy available/volume}$$

The energy associated with crater lip formation and projectile breakup is considered to be small and is ignored. The term energy available/volume is presumed to include the energy required to eject the crater material. The term $\frac{1}{2}\rho V^2_{\text{rupture}}$ is associated with the maximum plastic deformation energy which the material can support. Fitzgerald (*38*) has related this velocity to bond energies of the lattice. This velocity may be expressed

$$V = [(\sigma_m \epsilon_m - \phi)/\rho]^{1/2} \tag{8}$$

where ρ is the density, $\sigma_m \epsilon_m$ is the product of rupture stress and strain, and ϕ is the dynamic elastic energy.

The "energy available" is included because for some spatial interval the impact energy/volume is much greater than the energy of plastic deformation. The crater penetration reaches its final value at the point where a balance occurs between the maximum deformation energy/volume and the impact energy/volume.

When the impact energy has propagated into the target to any radius P, prior to the termination of the crater, this impact energy is considered, for the purpose here, to be distributed uniformly throughout a hemispherical volume defined by the radius P.

Then at the point where the term energy available/volume is equal to zero, the following is true:

$$E_I/\tfrac{2}{3}\pi P^3 = \tfrac{1}{2}\rho V_{\text{rupture}} = \sigma_m \epsilon_m - \phi \tag{9}$$

For spherical attenuation of the energy along the axis of penetration, the penetration is solved for in terms of the impact energy and the dynamic modulus of toughness which represents the maximum energy/volume the metal may sustain without rupturing. Further deformation may occur with a lower absorption of energy.

A plot of P vs. E_I (Fig. 21) has been made for Al and mild steel and compared with BRL data. The experimental points chosen are those for which the disk projectile struck the target with the minimal surface

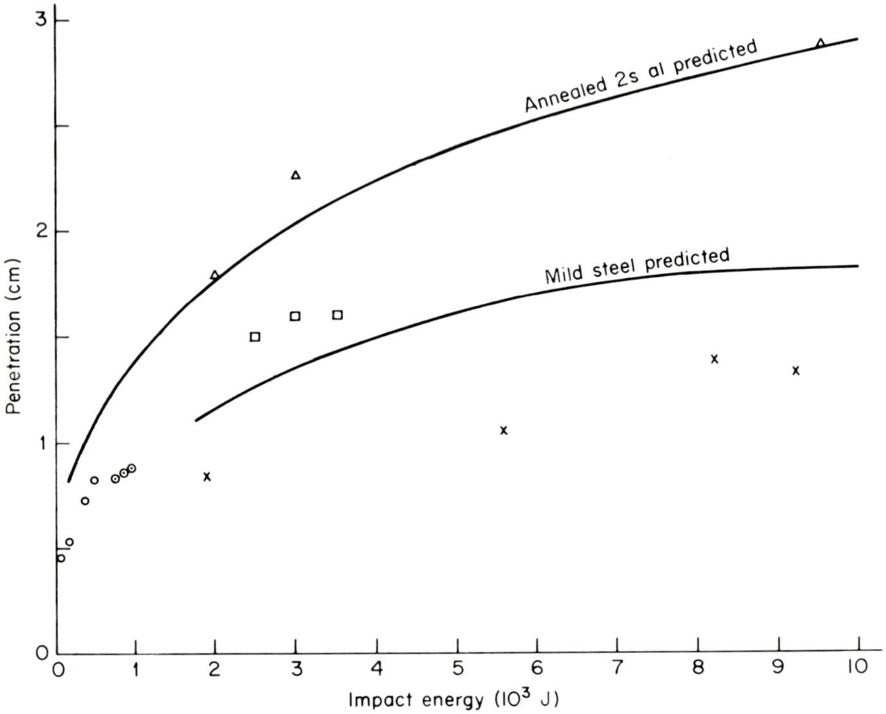

FIG. 21. Computed and experimental comparison of penetration versus impact energy. (△) BRL data (*41*), 2S Al; (○) commercially pure Al projectile, 1.59 mm diam; (⊙) commercially pure Al projectile (*39*), 3.18 mm diam; (□) Stainless Steel projectile, 1.59 mm diam; (×) BRL mild steel targets.

area. The data of Goodman and Liles (*39*) are plotted for commercially pure aluminum, although a dynamic stress–strain curve has not been determined for the test material.

In the case of steel the predicted penetration is consistently too high. This may be due to a phase change observed metallographically at about 130 kbar (*40*) in explosive loading. This phase change is also seen in Hugoniot data of Armco iron (*11*). It has been shown that a plane shock may pass through aluminum and copper prior to major plastic deformation. If this were also the case in steel, then some or all of the gross deformation could be occurring in transformed material. This transformed material would not necessarily exhibit the same modulus of toughness as the mild steel.

The criterion established in this method for predicting hypervelocity penetration is based on an energy consideration and restricted impact geometry, i.e., point impact. The effect of varying projectile material is ignored. The predictions could probably be sharpened by including factors to account for geometry orientation as well as the material of the projectile. However the material parameter $P = [\frac{3}{2}\pi E_I/(\sigma_m \epsilon_m - \phi)]^{1/3}$ seems to be established as one of considerable importance.

XIII. Conclusions

It can be concluded that although originally hypervelocity impact phenomenon was viewed without consideration of the metallurgical or mechanical properties of the target there is a current recognition of the importance of the role of such properties. Perhaps one of the more acute manifestations of the importance of these properties is found in the experiments dealing with the penetration of prestrained aluminum targets. In this case identical impacts on targets which differ only by the previous strain history give entirely different penetrations.

After establishing phenomenological dynamic stress–strain curves, a procedure has been devised to obtain an energy balance from post-impact measurements. The success of the technique seems marked in some of the values obtained and in the subsequently derived equation for predicting maximum impact damage based on material properties.

It is unfortunate that the experiments with prestrained aluminum, which so singularly set forth the importance of material properties, do not provide data to allow good energy balances. Although numerous explanations may be offered for the difficulties encountered, it is possible that the methodology employed does not take into account all properties of importance.

In any event, it appears conclusive that material properties influence the deformation processes occurring during hypervelocity cratering. It is equally obvious that the interrelation is not completely understood and should be studied further.

REFERENCES

1. R. M. Davies, The determination of static and dynamic yield stresses using a steel ball, *Proc. Roy. Soc. (London)* **A197**, 416 (1949).
2. N. Hsu and R. B. Pond, The effect of metal properties on hypervelocity penetration, Progr. Rept. (BRL) No. 5, Contract No. DA 18-001-AMC-745 (X) (June 1966).
3. R. J. Eichelberger, Reexamination of the theories of jet formation and target penetration by lined cavity charges, CIT CEL Rept. No. 1 (June 1954).
4. R. J. Eichelberger, Experimental test of the theory of penetration by multiple jets, *J. Appl. Phys.* **27**, 63 (1965).

5. G. Birkhoff, D. MacDougall, E. Pugh, and G. I. Taylor, Explosives with lined cavities, *J. Appl. Phys.* **19**, 563 (1948).
6. F. E. Allison and R. Vitali, A new method of computing penetration variables for shaped-charge jets, BRL Rept. No. 1184 (Jan. 1963).
7. R. DiPersio, J. Simon, and A. Merendino, Penetration of shaped-charge jets into metallic targets, BRL Rept. No. 1296 (September 1965).
8. G. L. Moss, S. Toms, R. Vitali, and A. Merendino, Effect of target microstructure on penetration by shaped charge jets, BRLM Rept. No. 1739 (April 1966).
9. A. von Neuman and R. D. Richtmeyer, A method for the numerical calculation of hydrodynamic shocks, *J. Appl. Phys.* **21**, 232 (1950).
10. R. L. Bjork, The effect of meteroid impact on steel and aluminum in space, Rand Corp. P-1662 (Dec. 16, 1958). See also *Proc. Symp. Hypervelocity, 6th*, **2**, Part, I (August 1963) for additional references to Bjork's work.
11. J. M. Walsh, M. H. Rice, R. G. McQueen, and F. L. Yarger, Shock-wave compression of twenty-seven metals, *Phys.* Rev. **108** (1957). Also see M. H. Rice, R. G. McQueen, and J. M. Walsh, *Solid State Phys.* **6**, 1–63 (1958).
12. P. C. Chou and F. E. Allison, Strong plane shock produced by hypervelocity impact and late-stage equivalence, *J. Appl. Phys.* **37**, No. 2, 853 (1966).
13. J. K. Dienes, M. W. Evans, L. J. Hagemon, W. E. Johnson, and J. M. Walsh, An Eulerian method of calculating strength dependent deformation, Gulf General Atomic Inc. Rept. GAMD-8497 (Feb. 1968). Also see Chapter III by Dienes and Walsh in this book.
14. R. J. Eichelberger and J. W. Gehring, Jr., The effect of collisions with meteoric bodies on spacecraft, *Rocketry* No. 10 (1962).
15. W. Herrmann and A. H. Jones, Correlation of hypervelocity impact data, *Proc. Symp. Hypervelocity Impact, 5th, Denver, Colorado* (Oct. 1961).
16. J. B. Feldman, Jr., Volume-energy relation from shaped charge jet penetrations, *Proc. Symp. Hypervelocity, 4th, Eglin AFB, Florida* (April 1960).
17. Yu. I. Fadeyenko, Relation of crater size to target hardness, *J. Appl. Mech. Tech. Phys. USSR (English Transl.)* No. 5 (Sept.–Oct. 1964).
18. J. H. Kineke, Jr., Observations of crater formation in ductile materials, *Proc. Hypervelocity Symp., 5th, Denver, Colorado* (Oct. 1961); J. H. Kineke, Jr., and L. G. Richards, Influence of target strength on hypervelocity crater formation in aluminum, *Proc. Symp. Hypervelocity Impact, 6th, Cleveland, Ohio* (May 1963).
19. L. G. Richards and C. M. Glass, unpublished data.
20. F. E. Allison, Effects of target temperature on hypervelocity cratering, *Proc. Symp. Hypervelocity Impact, 4th, Eglin AFB, Florida* (April 1960).
21. M. L. Wilkins, Calculation of elastic-plastic flow, UCRL Rept. No. 7322 (April 1963).
22. R. F. Rolsten and H. H. Hunt, Dynamic stress at hypervelocity impact and true tensile strength, *Can. Aeron. Space J.* **13**, No. 6, 269 (1967).
23. G. I. Taylor, The plastic wave in a wire extended by an impact load, British Ministry Res. Committee Rept. R.C. 329 (1942); "Scientific Papers of Sir Geoffry Ingram Taylor" (G. K. Batchelor, ed.). Cambridge Univ. Press, London and New York, 1958.
24. M. P. White and L. V. Griffis, The propagation on plasticity in uniaxial compression, *Trans. ASME* **70** (1948).
25. J. F. Bell, "The Physics of Large Deformation of Crystalline Solids" (B. D. Coleman, ed.) (Springer Tracts in Natural Philosophy). Springer, Berlin, 1968.
26. J. S. Rinehart and J. Pearson, "Behavior of Metals under Impulsive Loads." Am. Soc. Metals, Cleveland, Ohio, 1954.

27. J. S. Rinehart, Deformation of an explosively loaded aluminum single crystal, *J. Appl. Phys.* **26**, 1315 (1955).
28. C. M. Glass, S. K. Golaski, J. J. Misey, and G. L. Moss, Fracture of single crystals under explosive loading, Symp. ASTM Dynamic Behavior Materials, Tech. Publ. No. 336 (1962); G. L. Moss and C. M. Glass, A study of the relationship between the first compressive stress and gross plastic deformation in explosively loaded copper, BRLM Rept. No. 1236 (1959).
29. M. K. Gainer and C. M. Glass, A study of metallurgical effects in high velocity deformation of copper using rotary extruded liners, BRL Rept. 1167 (May 1962).
30. R. E. Green, Jr., Ultrasonic attenuation study of dislocation motion, Part 1. Theoretical, *Trans. Met. Soc. AIME* **230**, 1383 (1964); R. A. Swanson and R. E. Green, Jr., Part II. Experimental, *ibid.* **230**, 1390 (1964).
31. R. B. Pond, Cold extrude rapidly to produce long, thin-wall aluminum tubes, *Metal Progr.* **89**, No. 6, 77 (1966).
32. R. B. Pond, Unpublished results. The work was used in Ref. 35 and also in R. B. Pond and C. M. Glass, Crystallographic aspects of the high velocity deformation of aluminum single crystals, "Response of Metals to High Velocity Deformation," *Conf. Metal. Soc.* **9**, 145 (1961).
33. W. Prager, On isotropic materials with continuous transition from elastic to plastic, state, *Intern. Congr. Appl. Mech., 5th* (1933).
34. D. C. Drucker, Analogous strain rate effects in jet formation and metal cutting, Tech. Rept. No. 4, BRL Contract No. DA-19-020-ORD-3426 (Aug. 1956).
35. C. M. Glass and R. B. Pond, A metallurgical approach to the hypervelocity problem, *Proc. Symp. Hypervelocity Impact, 4th, Eglin AFB, Florida* (April 1960). Also R. B. Pond, C. Mobley, and C. M. Glass, Energy balances in hypervelocity penetration, *Proc. Symp. Hypervelocity Impact, 6th, Cleveland, Ohio,* (Aug. 1963).
36. J. H. Kineke, Jr. and R. Vitali, Transient observations of crater formation in semi-infinite targets, *Proc. Symp. Hypervelocity Impact, 6th, Cleveland, Ohio* (August 1963).
37. C. W. MacGregor, The true stress–strain tension test—its role in modern materials testing, *J. Franklin Inst.* **238**, 159 (1944).
38. E. R. Fitzgerald, *Phys. Letters* **10**, 42 (1964).
39. E. H. Goodman and C. D. Liles, *Proc. Symp. Hypervelocity Impact, 6th,* **2**, Part II, 563 (1963).
40. C. M. Glass, G. L. Moss, S. K. Golaski, Effects of explosive loading on single crystals and polycrystalline aggregates, "Response of Metals to High Velocity Deformation," *Conf. Metal. Soc.* **9**, 115 (1961).
41. Selected data from *Proc. Symp. Hypervelocity Impact, 5th, Denver, Colorado* (1961).

CHAPTER IX

ENGINEERING CONSIDERATIONS IN HYPERVELOCITY IMPACT

JOHN W. GEHRING, JR.

IMPACT PHYSICS LABORATORY
GENERAL MOTORS CORPORATION
AND
AC ELECTRONICS DEFENSE RESEARCH LABORATORIES
SANTA BARBARA, CALIFORNIA

I. Introduction

Other chapters of this treatise on the phenomena of hypervelocity impact have been concerned with the impact of a simple projectile—usually a sphere or cube impacting against either a semiinfinite metallic target or a simple two-sheet structure. This approach to a simple model is ofttimes inadequate from the viewpoint of the design engineer because of the unique problems associated with reentry vehicles, spacecraft, space radiators, solar panels, etc. These complex structures do not lend themselves to treatment by such a simple model. Therefore, this section will be concerned with the practical aspects, or engineering considerations, that must be understood in order to design an adequate missile or space vehicle.

Impact damage to aerospace vehicles and structures can be broadly related to three classes of targets: semiinfinite, finite, and complex. The semiinfinite target is defined as a homogeneous material of thickness and size such that free-surface effects will have negligible effect on impact damage for a given set of impact conditions. The finite target is one of a thickness such that free-surface effects will influence the impact damage. Included here is the damage resulting in spall off the target, the shock interaction in a laminated target, and the effect on impact damage by using multisheet fillers and honeycombs. Complex targets in the following discussion include materials for unique application that require special consideration. Included here are pressure vessels, laminated space radiator tubes, and space vehicle windows.

As indicated by the previous definitions, simple targets are usually distinguished on the basis of results of impacts, which are related to impact conditions (e.g., projectile mass and velocity), whereas complex targets are usually distinguished on the basis of structure, material, or application. Most military and space vehicle configurations can be classified according to these definitions. For example, the ablative shield or heat sink of a reentry body might be classed as a semiinfinite or thick target for most projectiles; satellites and spacecraft with thin aluminum hulls surrounding control gas tanks, electronic modules, or recoverable instrumentation packages could be considered bumper-protected complex targets; and rocket nozzles could be classed as laminated targets or pressure vessels. Impact damage assessment and the use of test data in establishing critical damage criteria must be related to the mission requirements of the vehicle system or subsystem that is being evaluated.

II. Experimental Observations

A. Semiinfinite Targets

The phenomena of cratering in semiinfinite metal targets has been treated in more detail, both empirically and theoretically, than any other phase of hypervelocity impact. This section will deal with empirical treatments and applications of the phenomena; the analysis of the theoretical aspects of crater formation has been covered in other chapters of this book.

1. *Empirical Relations*

There is no detailed mathematical formulation with which to predict damage to a complex target by a hypervelocity fragment. Although the theoretical treatments have made significant advances in the analytical approach, the treatments do not account in detail for the effects of increased target temperature, variations in the alloy of the target, impacts at angles of obliquity, or for spalling and delaminating in thin and composite targets. A number of empirical relationships which permit at least an approximation of the damage which might occur under limited conditions are offered in the literature. Two of these, given by Eichelberger and Gehring (*1*) and by Charters and Summers (*2*), appear to be valid for a wide variety of impact conditions and do allow for the effects of target temperature and the angle of projectile impact.

Rewriting those two relationships we have the following forms: from Eichelberger and Gehring (*1*),

$$\frac{P}{d_p} = \left(\frac{12 \times 10^{-9}(E/B)\cos^2\theta}{2\pi/d_p}\right)^{1/3} \quad \text{(cgs units)} \tag{1}$$

and, from Charters and Summers (*2*),

$$\frac{P}{d_p} = 2.28\left(\frac{\rho_p}{\rho_t}\right)^{2/3}\left(\frac{V\cos\theta}{C_t}\right)^{2/3} \tag{2}$$

where P is the depth of penetration, E is the energy of projectile, B is the Brinell hardness number (varying with target temperature), θ is the angle of impact (measured from the normal), ρ_p is the projectile density, ρ_t is the target density, V is the projectile velocity, C_t is the velocity of sound in the target, and d_p is the diameter of the projectile.

Eichelberger and Gehring's equation [Eq. (1)] was derived from the relationship $V = 4 \times 10^{-9}E/B$, where V is crater volume in cubic centimeters, E is projectile kinetic energy in ergs, and B is target hardness in standard BHN; this relationship applies to meteoroid impact where velocities can be expected to be in the range of 10–70 km/sec. At lower impact velocities, a factor T, of the primary penetration phase of crater formation, is added to the relationship. This factor, T, is proportional to $(\rho_p/\rho_t)^{1/2}$ and is independent of velocity. Therefore, as velocity increases and the contribution of projectile energy is increased, the contribution of T to the total penetration becomes negligible.

Charters and Summers, in a paper following Ref. 2 (*3*), use a combination of Eq. (2) and a simple model of the impact process to derive another damage formula. This equation, which shifts the emphasis from the speed of sound to the strength of the target material, is given by

$$\frac{P}{d_p} = 1.50\left(\frac{\rho_p}{\rho_t}\right)^{1/3}\left(\frac{\rho_p V^2}{2S_t}\right)^{1/3} \tag{3}$$

where S_t is assumed to be a constant of the deformation strength in the target material. Because the values of S_t are not known for conditions relevant to hypervelocity impact, Charters and Summers suggest that an impact experiment be conducted and, from Eq. (3) and the measured crater depth, a value of S_t be derived. This value can then be used to predict damage in the same target material for a variety of impact conditions.

The importance of considering projectile shape and density and the target strength and hardness will be demonstrated in the following sections.

2. *Projectile Shape and Density Effects*

The great majority of experimental research programs in hypervelocity impact have used solid spheres as the principal projectile. (Most theoretical treatments of hypervelocity impact have considered the impacting projectile to be a sphere, cube, or short cylinder.) There are, however, several other basic projectile configurations that could significantly affect the damage mechanisms that operate in hypervelocity impact, but there is little information available concerning these configurations.

Studies of rod impacts have been carried out by Atkins (*4*), Summers and Niehaus (*5*), Allen and Rogers (*6*), Slattery and Clay (*7*), Nysmith *et al.* (*8*), and Christman and Gehring (*9*). Results have shown that for low-velocity (below the wave velocity in the projectile) end-on impacts, the craters tend to be cone shaped. This can be attributed to deceleration of the back end of the rod during impact of the front end. At high velocities (above the wave velocity of the projectile) the craters tend more toward a cylindrical shape; this is to be expected, since the cratering process involves the addition of energy at a constant rate for a time that is comparable to the toal time of crater formation. Impact at intermediate angles (between "end-on" and "side-on" conditions) does not result in any new phenomena; it merely complicates the prediction and interpretation of results.

On the other hand, with a solid sphere the primary phase of cratering (or the phase during which the projectile is being used up) is short (compared to that of a long and slender rod) and leads to the creation of a hemispherical crater upon impact of an equiaxial projectile, since the projectile energy can be thought of as being deposited at the surface of the target almost instantaneously.

The results of a series of experiments to assess the effects of projectile density and shape are given in Figs. 1 and 2 (*10*). The targets in all cases

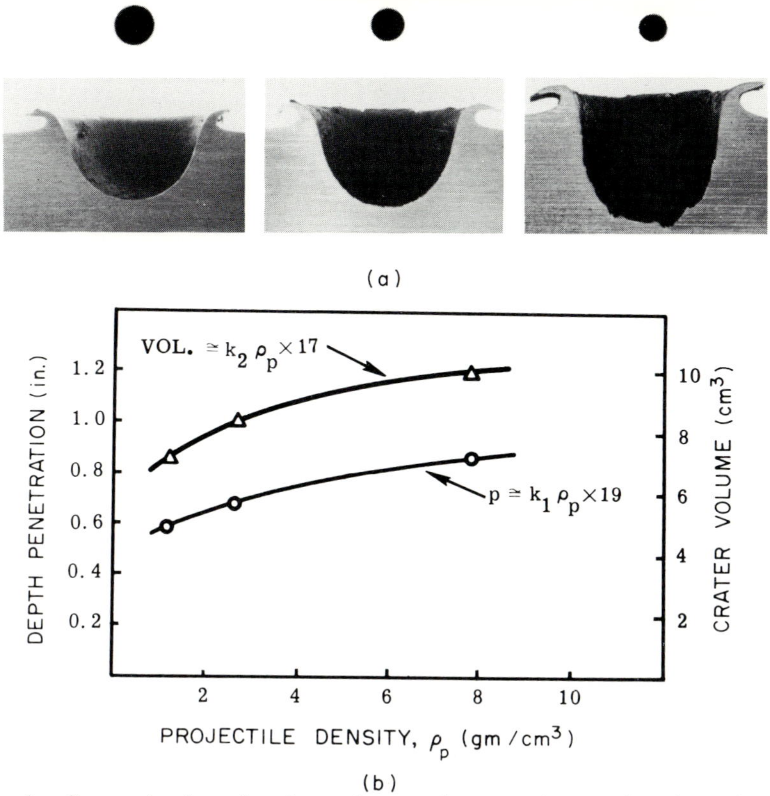

FIG. 1. Crater depth and volume for equal-mass spheres of various densities. (a) Photographic representation; (b) graphical representation: (△) volume versus ρ_p ; (○) penetration versus ρ_p . Target: 110-F Al, semiinfinite. Impact velocity: 6.6 km/sec. Projectile: Zelux-type M (ρ_p = 1.20, diam = 0.313); 2017 Al (ρ_p = 2.70, diam = 0.240); C1020 steel (ρ_p = 7.80 g/cm, diam = 0.169 in.). All projectiles same mass—0.32 g.

were semiinfinite, 1100-F aluminum blocks, and impact velocity was 6.6 km/sec.

In Fig. 1 the penetration depth and crater volume are given as functions of projectile density. The penetration is proportional to $\rho_p^{0.19}$ and volume is proportional to $\rho_p^{0.17}$. Results of a similar series of tests at 2.0 km/sec, reported by Charters and Summers in 1960 (*11*), showed penetration to be proportional to $\rho_p^{0.33}$ and volume to be proportional to $\rho_p^{0.50}$. Note the significant decrease in exponents for an increase in velocity from 2.0 to 6.6 km/sec. Extrapolating these results to meteoroid velocities reveals that the influence of projectile density on penetration and volume may become negligible.

In Fig. 2 the penetration depth and crater volume are given as functions of projectile shape. Note that the penetration and volume are strongly

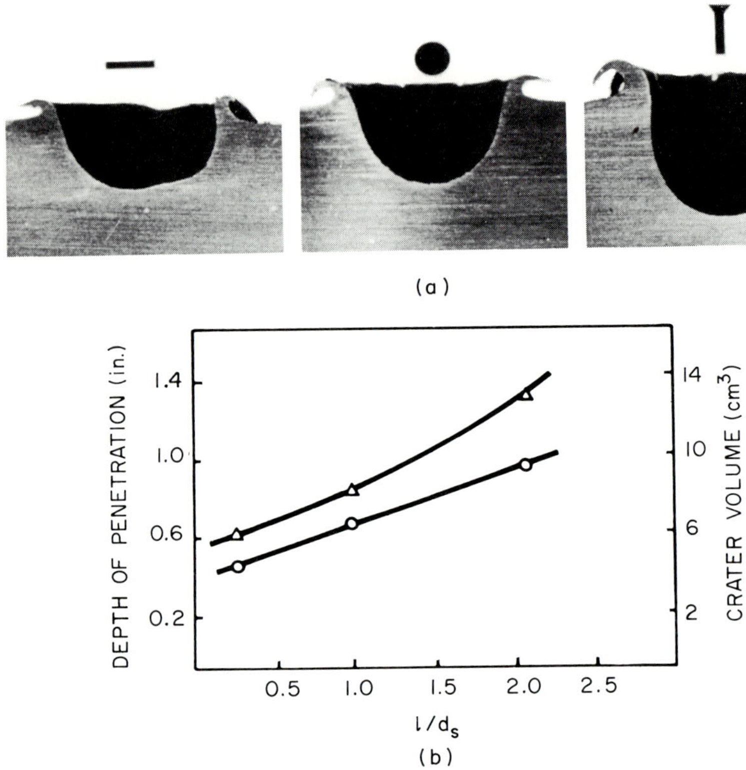

FIG. 2. Crater depth and volume for projectiles of equal mass and various shapes. (a) Photographic representation; (b) graphical representation versus l/d_s ; (○) penetration versus l/d_s ; (△) volume versus l/d_s. Target: 1100-F Al, semiinfinite. Impact velocity: 6.6 km/sec. Projectile: 2017 Al. All projectiles same mass—0.32 g.

influenced by the length of the projectile along its flight axis (normalized with respect to sphere diameter). This result is significant in military applications where impact velocities are below 10 km/sec. However, for meteoroid impacts, it is reasonable to assume that the particles are approximately spherical and, consequently, that shape is not a governing factor in damage.

3. *Target Strength and Temperature*

The effects of variations in the strength of the target material upon the damage sustained as a result of high-speed impact is a matter of concern to a space vehicle designer. The expression "strength of material" is an arbitrary term used to refer to the ability of the material to resist deformation. It is well known that the strength of a material is affected during its metallurgical history by such details as heat treatment, fabrication process, and the effect of the operating environment. The extent to which target strength and temperature are considered in empirical approaches to hypervelocity cratering was discussed earlier.

The variations in mechanical properties of nominally identical target materials has called for a method of measuring resistance that can be used on individual targets. For this purpose, the Brinell hardness number (the ratio of the load to the area of indentation after the load is removed) has been found to provide a surprisingly good criterion (*1*, *12*). In Fig. 3 the correlation [first reported by Feldman (*12*)] between the energy per unit volume $dE/d\tau$ and the Brinell hardness number can be seen. These results were obtained with projectiles ranging in mass from 10^{-11} to 10 gm and in velocity to 15 km/sec; furthermore, pellets of a variety

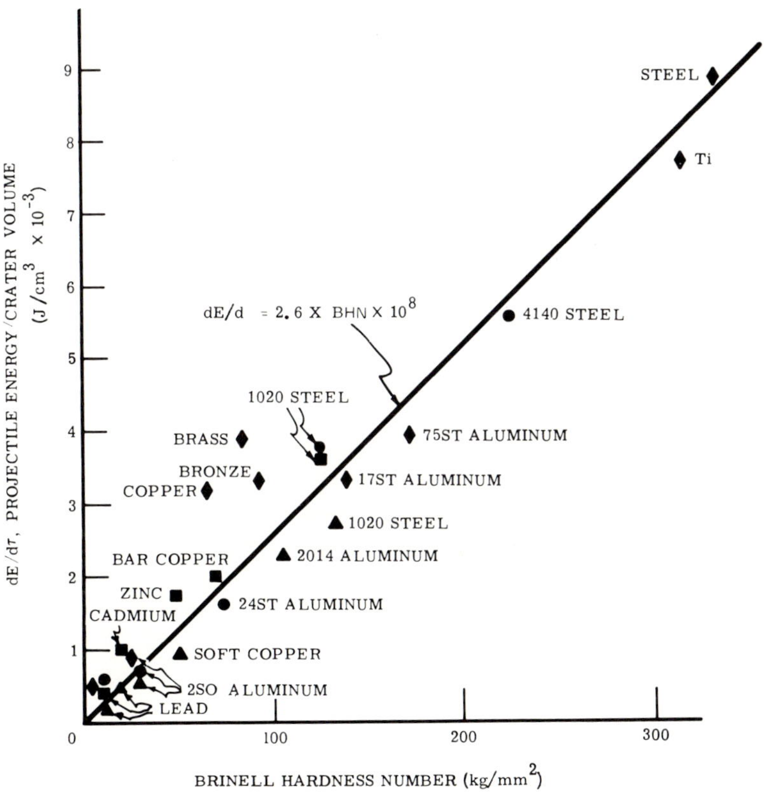

FIG. 3. Cratering efficiency versus target hardness for various materials.

of shapes and of materials ranging from nylon to tungsten carbide have been used. It was reported that, without exception, the points lying farthest from the correlation line represent the cases in which the targets were least well known (handbook values have been used in some cases), and that experiments using targets of known hardness have invariably yielded closer agreement.

As the wealth of experimental data increases and attention is turned to the effects of alloys and their heat treatments, there seems to be little doubt that the plastic behavior of metals in compression will be the important consideration. This is brought into sharp focus by the data reported by Feldman (*12*) and in Fig. 3 for three steels and five aluminum alloys.

Also important is the effect that an increase or decrease in temperature will have on the strength and/or the Brinell hardness number of a target material. Because heated targets are of specific interest in applications such as solar-heated space vehicles, space radiators, and reentry nose cones, they have been studied by several investigators. Rockowitz *et al.* (*13*) performed impact experiments on semiinfinite blocks of heated OFHC copper at temperatures ranging from 24 to 870°C with projectiles of aluminum, steel, and tungsten carbide, all of $\frac{1}{4}$-g mass, at velocities of up to 6.5 km/sec. Allison *et al.* (*14*) studied the impact behavior on lead (25–287°C) and copper (148–466°C) with a 0.18-g steel projectile at velocities of 5 km/sec.

The results from these investigations are illustrated in Figs. 4 and 5. Figure 4 shows the dependence of the crater volume on target temperature, and Fig. 5 compares the dependence of energy per unit volume and tensile strength with temperature. From these it can be concluded that the dimensions of the crater increase with increasing target temperature. However, it has been found (Fig. 5) that an increase in crater volume is not necessarily a linear function of the increasing temperature. The curves in Fig. 5 are related to the changes in the solid-state phase which have taken place at the indicated temperatures.

Although no reference on the behavior of thin targets impacted at high temperatures has been found, some insight can be obtained from Gehring and Lieblein (*15*). Here the investigators studied the meteoroid hazard to space radiators by firing 0.016- and 0.04-g glass spheres against aluminum and columbium targets at temperatures up to 370°C. Qualitatively, two effects were observed:

(1) The size of the hole made by the projectile was larger at higher temperatures.

(2) Less spall was observed at higher temperatures.

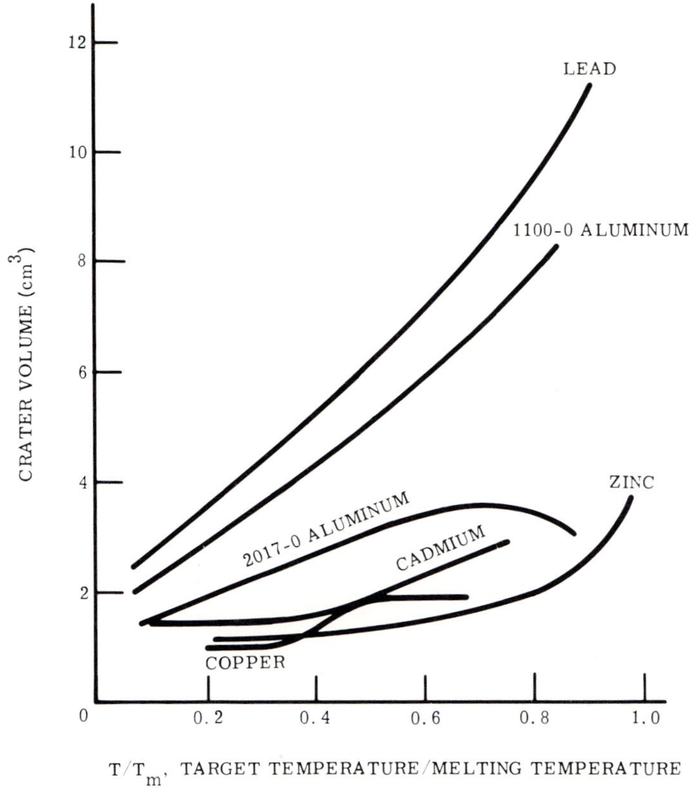

FIG. 4. Crater volume versus target temperature from Eichelberger and Gehring (*1*) and Feldman (*12*): 1095 Steel projectile, 0/18 g; 5.01 km/sec

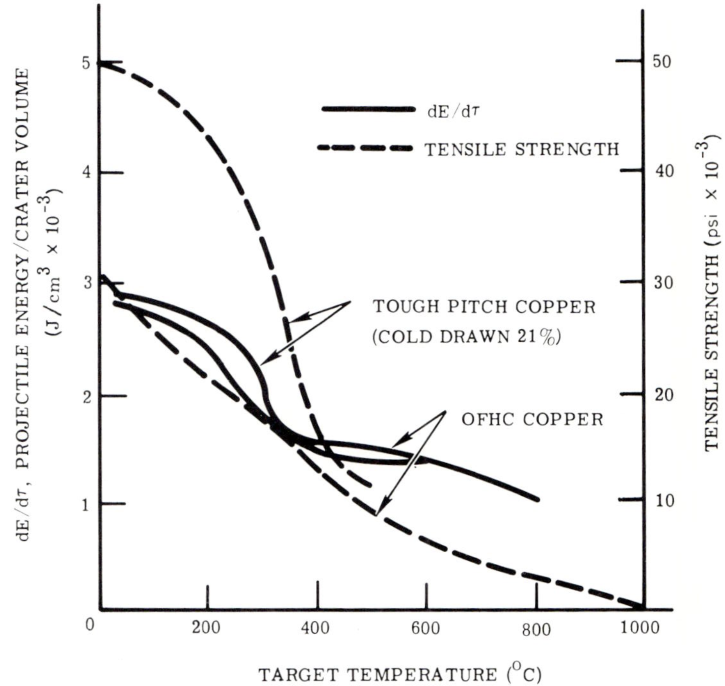

FIG. 5. Cratering efficiency and target strength versus target temperature [from Rockowitz *et al.* (*13*)].

From this, it can be expected that at elevated temperatures the formation of a crater in a semiinfinite target will leave a better defined lip on the surface of the crater. Also, in the case of a thin target less backspall is expected because of increased ductility at higher temperatures.

From these studies, it can be seen that the cratering process is a complex phenomenon in which mechanical and metallurgical properties of the target play an important role.

B. Finite Targets

1. *Effects of Spalling and Use of Laminates*

When a particle strikes a plate at high velocity, it forms a crater in the front surface of the plate and starts a strong compressive shock. This shock, attenuated as it travels through the plate, is generally reflected off the rear surface (or any free surface) as a tension wave. If the plate is thin enough, or if the initial shock is strong enough, the reflected tension wave will be so intense that a portion of the free surface of the plate may be ejected with a momentum sufficient to damage other parts of the structure (*16–21*). This phenomenon, called spalling, can be defined as fracture resulting from reflection of a decaying shock wave from a free surface. Spalling can occur even when the plate is not perforated (Fig. 6). The diameter of the spall is usually two or three times the thickness of the

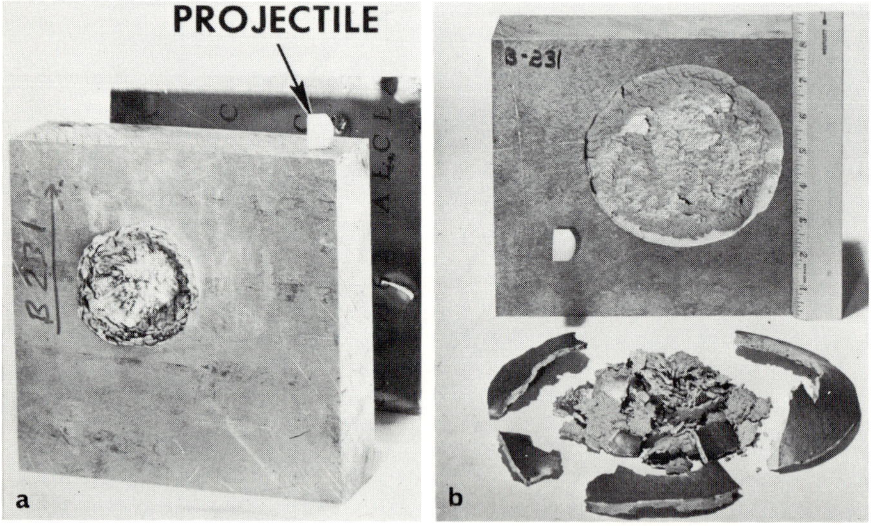

Fig. 6. Back spall from thick target. (a) Front view; (b) rear view. Target: 5.08-cm-thick 2024-T3 Al. Projectile: 4.14-g polyethylene. Velocity: 7.9 km/sec.

plate; the thickness of the spall is usually in the range of $\frac{1}{10}$ to $\frac{1}{2}$ the thickness of the plate. The following formula (*17*) can be used for computing spall thickness:

$$t = (\sigma_c/\sigma_0)(\lambda/2) \tag{4}$$

where t is the spall thickness, σ_c is the critical stress necessary to fracture the material, σ_0 is the maximum stress of the compressive wave, and λ is the length of the pulse. The velocity v of the spall is given by

$$v = (2\sigma_0 - \sigma_c)/\rho_t \tag{5}$$

where ρ_t is the density of the material. Table I (*18*) gives the critical normal fracture stress of several materials determined experimentally.

TABLE I

CRITICAL NORMAL FRACTURE STRESS OF SEVERAL MATERIALS

Metal	Critical normal fracture stress (lb/in.^{2})
4130 Steel	440,000
Copper	415,000
Brass	310,000
1020 Steel	180,000
2024-T4 aluminum	140,000

The calculation of spall thickness and velocity is difficult because neither the magnitude nor the shape of the compressive pulse are well-defined for given conditions of impact.

It is possible to reduce spall by using laminates of properly selected materials (*17*). The pertinent property of a material is its acoustic resistance $G = \rho_t C$, where ρ_t is the density of the target and C is the velocity of shock wave propagation in the target under the conditions encountered. If a compressive stress pulse σ_1 in a material with an acoustic resistance G_1 reaches an interface between this material and one with an acoustic resistance G_2, the pulse σ_r reflected from the interface will have the height of

$$\sigma_r = [(G_2 - G_1)/(G_2 + G_1)]\,\sigma_1 \tag{6}$$

and there will be a pulse σ_t transmitted through the second material with the height of

$$\sigma_t = [2G_2/(G_2 + G_1)]\,\sigma_1 \tag{7}$$

The above two equations contain several implications which aid in understanding why and how the pulse is modified at an interface. Taking the ratio of the stress transmitted to the stress reflected as

$$\sigma_t/\sigma_r = 2G_2/(G_2 - G_1) \tag{8}$$

the following points become evident. When G_2 is much less than G_1, nearly all the stress is reflected; when G_2 is approximately equal to G_1, nearly all the stress is transmitted; and when G_2 is much greater than G_1, about $\frac{2}{3}$ of the stress is transmitted. Thus, the stress pulse may be divided and attenuated to a greater extent in a laminate than in an equal weight of a homogeneous material. It is necessary, however, that the joint efficiency be near unity to avoid spalling at the joint.

The use of laminates can result in significant changes in the nature of the damage to a structure as a result of hypervelocity impact (*22*) as demonstrated in Fig. 7, in which all of the four targets are of equal

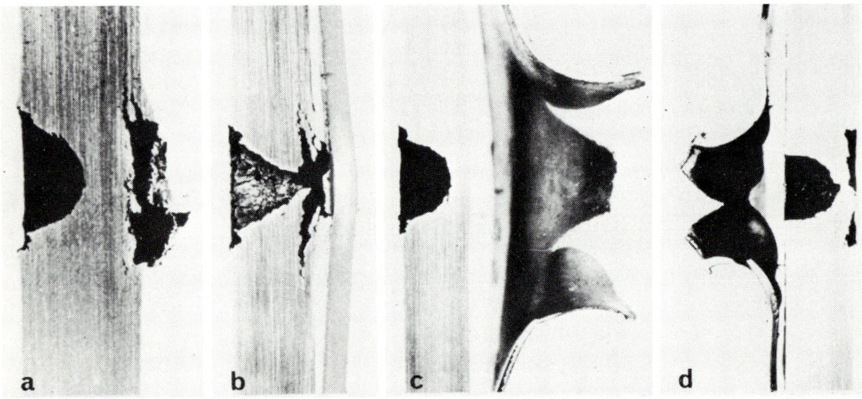

FIG. 7. Effects of laminates on spall and penetration—flat targets. (a) 12-mm Al; (b) Al–polyethylene; (c) Al–Cu; (d) Cu–Al. Projectile: 3-mm Al spheres. Velocity: 7.4 km/sec. All targets equal weight per unit area—3.4 g/cm².

weight per unit area. The plain aluminum target was not perforated but was severely spalled. The addition of a polyethylene laminate behind the aluminum resulted in perforation of the aluminum, but no spall was ejected from the back of the structure. A copper laminate behind the aluminum prevented perforation and spalling of the aluminum, although the copper was severely deformed. The copper–aluminum laminate at the right is shown for comparison only, since this combination does not offer any advantages with regard to penetration or spalling. The main

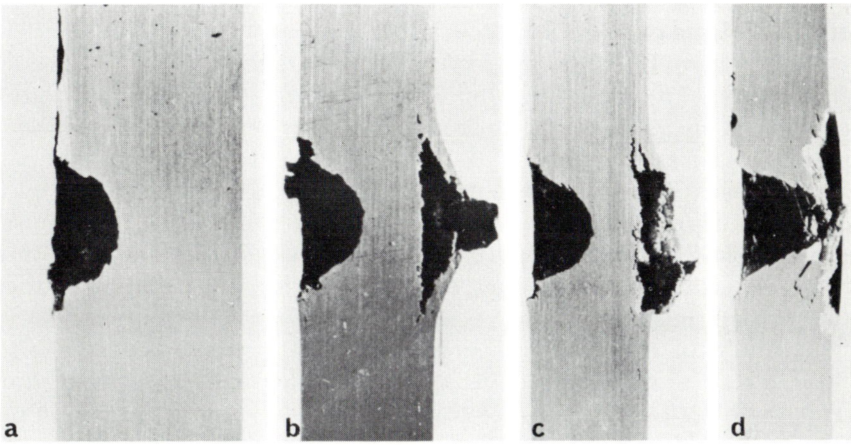

FIG. 8. Depth of penetration in flat targets of various thicknesses. (a) $t = 5.1$ cm, $p = 6.6$ mm; (b) $t = 1.4$ cm, $p = 6.9$ mm; (c) $t = 1.3$ cm, $p = 7.1$ mm; (d) $t = 1.0$ cm, $p =$ hole. Projectile: 0.125-in. Al sphere, 0.047 g. Velocity: 7.4 km/sec. Target: 2024-T3 Al.

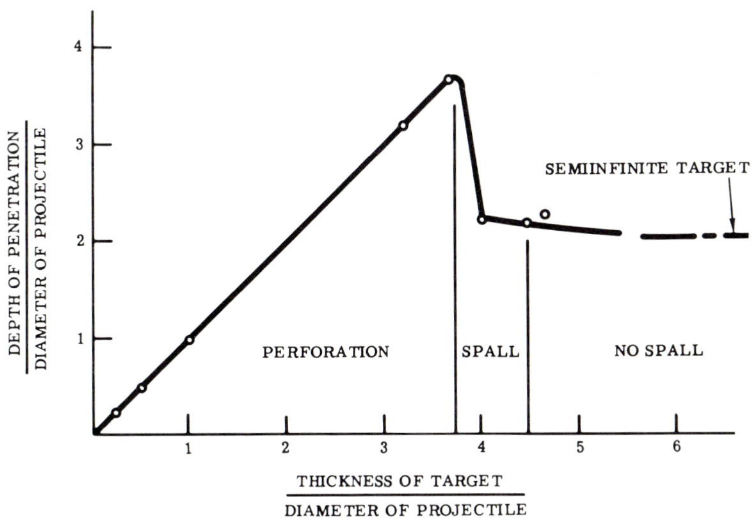

FIG. 9. Depth of penetration versus target thickness—flat targets. Targets: 2024-T3 Al. Projectiles: 3.18-mm 2017 Al spheres, 0.047 g. Velocity: 7.4 km/sec.

point illustrated in Fig. 7 is that the type of protection required (minimum penetration, no perforation, or no spall) will determine the combination of materials and thickness.

Many references state that plate thickness (t) should be 1.5 times the crater depth (p) in a semiinfinite target to prevent either perforation or spall. This value of 1.5 does not have any theoretical basis and is probably a very optimistic (low) value for flat targets. Jaffe and Rittenhouse (*20*) computed that spalling may occur at thicknesses up to 3.0 times crater depth in a semiinfinite target. Loeffler *et al.* (*23*) used 1.75; Bjork (*21*) used 1.5; and Gehring and Lieblein (*19*) found the ratio to vary from 1.1 for a cylindrical tube with a liner, to 1.6 for a cylindrical tube without a liner.

Recent data concerning the onset of spall is illustrated in Fig. 8. Spall is seen to occur at a target thickness of between 0.50 and 0.56 in. This indicates that the ratio of plate thickness to crater depth (0.26 in.) should be a minimum of about 2.1:1 to prevent spall, at least for the conditions cited. Figure 8 also shows perforation to occur at a target thickness of between 0.40 and 0.50 in., indicating that the ratio of plate thickness to crater depth should be a minimum of about 1.7:1 to prevent perforation. These conclusions are shown graphically in Fig. 9, which gives depth of penetration as a function of target thickness. Note the abrupt decrease in penetration as target thickness approaches four projectile diameters and the relatively narrow range over which spall without perforation occurs.

2. *Fillers*

Under some conditions it is desirable to introduce a filler material between the sheets of a two-sheet structure. This filler material can take the form of a honeycomb for structural rigidity or additional sheets or foamed material for thermal insulation. In general the addition of mass to a structure increases its resistance to impact damage, but studies of fillers, honeycombs in particular, have indicated that they may make the structure more vulnerable (*24*).

The first of these fillers considered was additional sheets between the two primary sheets. The strip analysis described in Chapter IV was used to examine the amount of momentum that passed through a sheet when the sheet was less than the fracture thickness. When fracture occurred in any of the six layers in the strip, the momentum of the broken segment was summed. The result of this procedure is shown in Fig. 10. These results indicate that the use of multiple sheets offers little or no advantage over single backup sheets, as there is no significant decrease

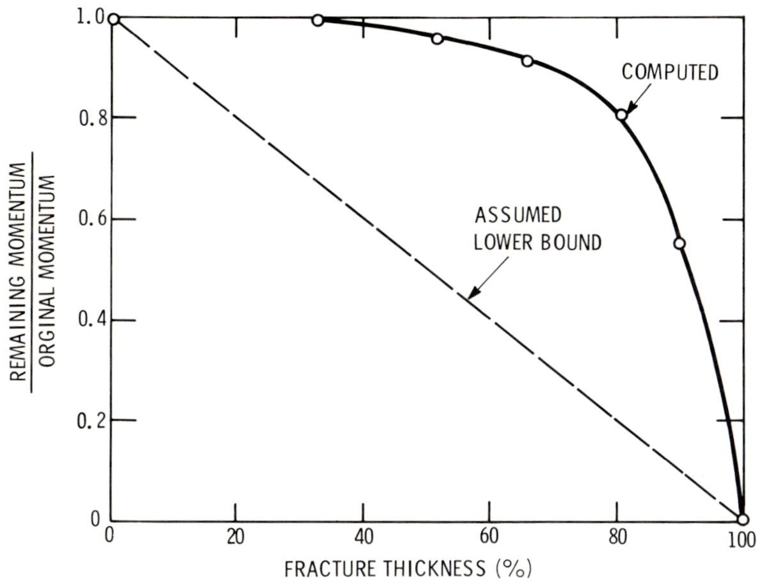

FIG. 10. Remaining momentum versus backup thickness.

of momentum through the second sheet until the second sheet is of approximately 90% of the fracture thickness.

A series of experiments was conducted to test the validity of this approach. The first three experiments utilized the ballistic pendulum behind various thicknesses of the second sheet to attempt to measure the momentum delivered through the second sheet. The results of these tests are summarized in Table II.

TABLE II
MOMENTUM THROUGH BACKUP[a]

Shot number	D-1437	1400	1442
Velocity	8.05 km/sec	7.60	7.50
% of fracture thickness	32	50	100
Momentum through backup (actual)	1.38	0.65	0.47
Momentum through backup (predicted)	0.99	0.95	0

[a] All tests 3.18-mm aluminum spheres, 5.08-cm spacing, 7075-T6 aluminum backups.

The poor agreement between the experiments and the predictions is due in part to spall from the second sheet and also to the momentum multiplication in the catch material on the ballistic pendulum.

On the basis of these experiments a lower bound was set on the momentum coming through the loaded sheet as the momentum fraction remaining being equal to one minus the sheet thickness to true fracture thickness ratio. This lower bound is indicated in Fig. 10 by the dotted line. Using this lower bound it is possible to analyze two cases of multiple sheets. Both cases require that the first sheet or shield breaks up, melts, or vaporizes the impacting projectile and spreads the debris such that the blast-loading analysis is applicable.

First assume that no spread of the debris takes place after the fracture of the second sheet. The situation is illustrated in Fig. 11.

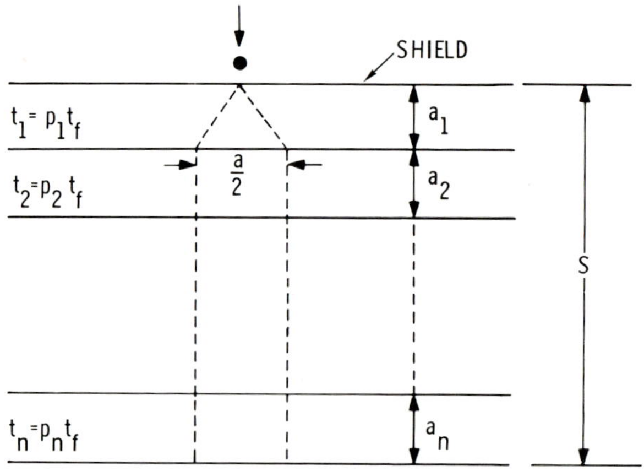

FIG. 11. Schematic of multisheet impact with debris channeled.

In this diagram P_n is arbitrary and t_c is defined as the fracture thickness for a second sheet spaced at S for a two-sheet structure. The true fracture thickness of the nth sheet is T_n. From the scaling results

$$T_1 = (S/S_1)^2 t_F \tag{9}$$

then

$$t_1/T_1 = \frac{P_1 t_F}{(S/S_1)^2 t_F} = P_1(S_1/S)^2 \tag{10}$$

Then the momentum fraction remaining after fracture of this sheet is

$$M_1/M_0 = \frac{\text{momentum to next sheet}}{\text{original momentum}} = 1 - P_1(S_1/S)^2 \tag{11}$$

$$T_2 = [1 - P_1(S_1/S)^2]T_1 = [1 - P_1(S_1/S)^2](S/S_1)^2 t_F \tag{12}$$

$$t_2/T_2 = \frac{P_2 t_F}{[1 - P_1(S_1/S)^2](S/S_1)^2\, t_F} = \frac{P_2(S_1/S)^2}{[1 - P_1(S_1/S)^2]} \tag{13}$$

$$M_2/M_0 = \left[1 - \frac{P_2(S_1/S)^2}{1 - P_1(S_1/S)^2}\right][1 - P_1(S_1/S)^2] = 1 - P_1(S_1/S)^2 - P_2(S_1/S)^2 \tag{14}$$

Generalizing,

$$\begin{aligned} M_{n-1}/M_0 &= 1 - P_1(S_1/S)^2 - P_2(S_1/S)^2 - \cdots - P_{n-1}(S_1/S)^2 \\ &= 1 - (S_1/S)^2(P_1 + P_2 + \cdots + P_{n-1}) \end{aligned} \tag{15}$$

$$\begin{aligned} T_n &= [1 - (S_1/S)^2(P_1 + P_2 + P_3 + \cdots + P_{n-1})]\, T_1 \\ &= [1 - (S_1/S)^2(P_1 + P_2 + P_3 + \cdots + P_{n-1})](S/S_1)^2\, t_F \end{aligned} \tag{16}$$

and

$$t_n/T_n = P_n/[1 - (S_1/S)^2(P_1 + P_2 + P_3 + \cdots + P_{n-1})](S/S_1)^2 \tag{17}$$

If the last sheet is not to fracture $t_n = T_n$ so

$$P_n = [1 - (S_1/S)^2(P_1 + P_2 + P_3 + \cdots + P_{n-1})](S/S_1)^2 \tag{18}$$

Then the total of the p is

$$\begin{aligned} \sum_{i=1}^{n} P_i &= \sum_{i=1}^{n-1} P_i + (S/S_1)^2[1 - (S_1/S)^2(P_1 + P_2 + P_3 + \cdots + P_{n-1})] \\ &= \sum_{i=1}^{n-1} P_i + (S/S_1)^2 - \sum_{i=1}^{n-1} P_i \\ &= (S/S_1)^2 \end{aligned} \tag{19}$$

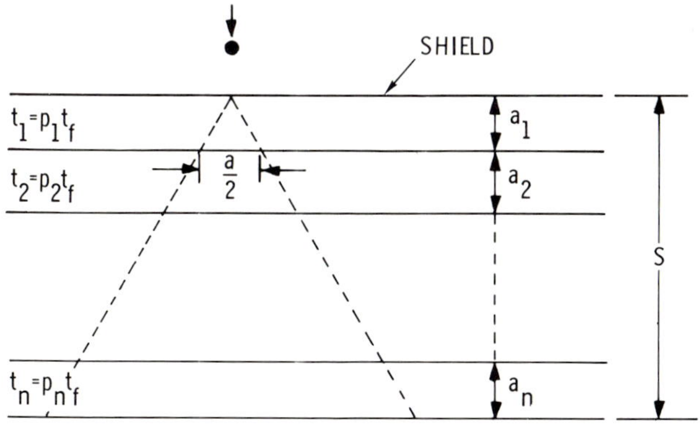

FIG. 12. Schematic of multisheet impact with debris spreading.

If n sheets are better than 2, the sum of the p must be less than 1; so

$$\sum_{i=1}^{n} P_i < 1, \qquad \text{or} \qquad (S/S_1)^2 < 1, \qquad \text{or} \qquad S < S_1$$

which is not so. Therefore two sheets are better than n for the case where the debris is channeled after fracturing the second sheet.

For the second case assume that the debris continues to spread after each sheet fractures and that the sheets are uniformly spaced (Fig. 12). With the notation as above

$$T_1 = (S/S_1)^2 t_F \tag{20}$$

$$t_1/T_1 = P_1(S_1/S)^2 \tag{21}$$

$$M_1/M_0 = 1 - P_1(S_1/S)^2 \tag{22}$$

$$T_2 = [1 - P_1(S_1/S)^2](S/2S_1)^2 t_F \tag{23}$$

$$t_2/T_2 = \frac{P_2(2S_1/S)^2}{[1 - P_1(S_1/S)^2]} \tag{24}$$

$$\begin{aligned} M_2/M_0 &= \left\{1 - \frac{P_2(2S_1/S)^2}{[1 - P_1(S_1/S)^2]}\right\} [1 - P_1(S_1/S)^2] \\ &= 1 - P_1(S_1/S)^2 - P_2(2S_1/S)^2 \end{aligned} \tag{25}$$

Generalizing, the momentum fraction to the last sheet is

$$\begin{aligned} \frac{M_{n-1}}{M_0} &= 1 - P_1(S_1/S)^2 - P_2(2S_1/S)^2 - P_3(3S_1/S)^2 - \cdots - P_{n-1}[(n-1)S_1/S]^2 \\ &= 1 - (S_1/S)^2[P_1 + 2^2P_2 + 3^2P_3 + \cdots + (n-1)^2P_{n-1}] \end{aligned} \tag{26}$$

$$T_n = \{1 - (S_1/S)^2[P_1 + 2^2P_2 + 3^2P_3 + \cdots + (n-1)^2P_{n-1}]\}(S/nS_1)^2 t_F \tag{27}$$

$$t_n/T_n = P_n/\{1 - (S_1/S)^2[P_1 + 2^2P_2 + \cdots + (n-1)^2P_{n-1}]\}(S/nS_1)^2 \tag{28}$$

$$P_n = \{1 - (S_1/S)^2[P_1 + 2^2P_2 + \cdots + (n-1)^2P_{n-1}]\}(S/nS_1)^2 \tag{29}$$

Then the sum of all the p is

$$\sum_{i=1}^{n} P_i = \sum_{i=1}^{n-1} P_i + (S/nS_1)^2 - \frac{1}{n^2}\sum_{i=1}^{n-1} i^2 P_i \tag{30}$$

But $S = nS_1$, so

$$\sum_{i=1}^{n} P_i = 1 + \sum_{i=1}^{n-1} P_i(1 - i^2/n^2) \tag{31}$$

If n sheets are better than 2, the sum of the p must be less than 1 or

$$\sum_{i=1}^{n-1} P_i(1 - i^2/n^2) < 0 \tag{32}$$

which cannot be so. Therefore, two sheets are better than n for the sheets uniformly spaced and with the debris spreading after the fracture of each sheet.

Further generalization of the second case with the sheets placed arbitrarily an S distance from the shield leads to the sum of the p as

$$\sum_{i=1}^{n} P_i = \sum_{i=1}^{n} P_i + (S/S_1)^2 - \sum_{i=1}^{n-1} a_i^2 P_i$$

$$= 1 + \sum_{i=1}^{n-1} P_i(1 - a_i^2) \tag{33}$$

Again for n sheets to be better than 2

$$\sum_{i=1}^{n} P_i < 1$$

or

$$\sum_{i=1}^{n-1} P_i(1 - a_i^2) < 0$$

or at least one of the a_i must be greater than one. This says that one of the intersheet spacings must be greater than S. Therefore for all cases two sheets are better than n within a given spacing, provided the mode of failure is blast loading.

Experiments with aluminum–aluminum and cadmium–cadmium impacts supported the conclusion that two sheets provide more protection than n. In some cases the addition of a third sheet increased the vulnerability of a structure. An impact of a 3.18 mm cadmium projectile at 6.5 km/sec upon a 0.64 mm cadmium shield with 5.08 cm spacing to a 7075-T6 aluminum second sheet 1.02 mm thick did not cause failure of the second sheet. However, when a 7075-T6 aluminum sheet 1.02 mm thick was inserted between the shield and second sheet, all three sheets failed.

Examination of Fig. 13 gives an indication of the reason for this. As can be seen the sheet upon failure seems to cause a restriction in the spread of the debris. This causes a higher load per unit area upon the next sheet than would occur if the debris were allowed to spread.

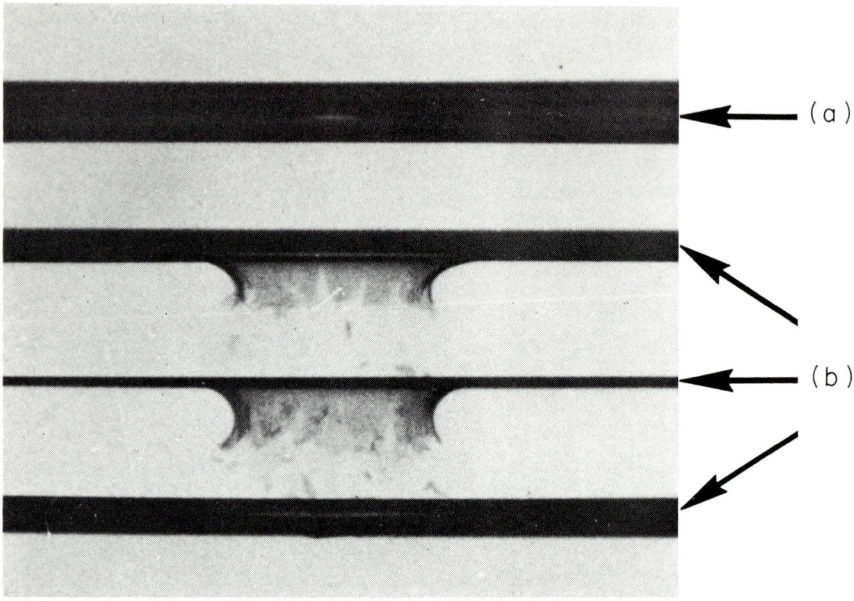

FIG. 13. X-ray of multisheet failure. (a) 0.67-mm 1100-O Al; (b) 0.41-mm 7075-T6 Al. A 3.18-mm Al sphere impacting at 7.74 km/sec, 40 μsec after shield impact; 1.7-cm spacing between each sheet.

Figure 13 also shows that debris from the sheet that has failed will impact the next sheet. This solid debris is not accounted for within the blast-loading treatment. If the primary damage mechanism is the impact of this solid debris, the blast-loading analysis will not apply.

Experimentally, the only advantage that multiple sheets offered was to catch spall fragments from the preceding sheet.

3. *Honeycombs*

The second filler to be considered was the honeycomb added between two sheets, such as on the Apollo Service Module.

To obtain an understanding of the reaction of the honeycomb to the debris passing through a shield, the two-dimensional hydrodynamic computation results were examined. The debris in the cases of interest is in a liquid or vapor form and expands radially at a maximum velocity of about one-half of the axial velocity. This debris would be expected to impulsively load the honeycomb from the inside. Any particulate debris from the late stages of hole formation in the shield would not impact the honeycomb until after the impulsive loading had taken place.

The basic questions to be considered with the honeycomb are whether the honeycomb cell will break and, if it does, will the projectile-shield debris be channeled. To attempt to answer these questions a somewhat simplified analysis was performed. The honeycomb cell was assumed to be a circular cylinder rather than a hexagon. The impact was assumed to take place normal to the shield and in the center of the circular cell. The radial momenta were then taken from the two-dimensional hydrodynamic calculations and used to calculate the impulse delivered to the circular cell. Given the wall thickness of the cell, the maximum radial velocity of the cell wall can be obtained. This velocity can then be compared with the radial velocity of the debris. The velocity of the cell wall computed is not the actual radial velocity but rather the velocity it would have if all the debris radial momentum were instantaneously imparted to it. This is not actually the case, as the debris impacts over some period of time. It is possible that the debris first striking the wall and rebounding interferes with that debris that is trailing. Thus, if the maximum radial velocity of the cell wall is less than the debris radial velocity the debris will strike the wall, rebound, and be channeled. For the debris not to be channeled the computed velocity of the cell wall must be at least equal to the debris radial velocity.

In the case where the debris is channeled, the backup thickness requirements can be obtained from the plate calculations made with the uniform $S/2$ momentum distribution. If the debris is not channeled, the backup thickness requirements will be somewhere between the thicknesses with no filler and the channeled thicknesses. Based upon this analysis a honeycomb will be a detriment to the meteoroid-protection capabilities of any two-sheet structure.

The analysis performed states implicitly that the honeycomb damage mechanism is controlled by the momentum and velocity of the debris. The debris velocity is a function of the impact velocity, thus a velocity-scaled honeycomb experiment is not possible. It was necessary for this reason to perform experiments to test the analysis, and no attempt was made to simulate high-velocity impacts.

The experiments that were performed utilized cadmium–cadmium impacts so that the debris would be in the heated liquid or vapor state to effectively blast load the cell interior. The honeycomb was represented by a thin circular tube of 6061-T6 aluminum. The impacts were in most cases in the center of these cells, and the reaction of the cell is typified by the framing camera sequences shown in Figs. 14 and 15. Figure 16 shows flash x-rays of similar impacts a few microseconds after the first-sheet impact. From these figures it can be seen that the cell wall starts to expand very soon after impact and soon breaks. However, the cell wall has not achieved a high enough velocity to get out of the way of the

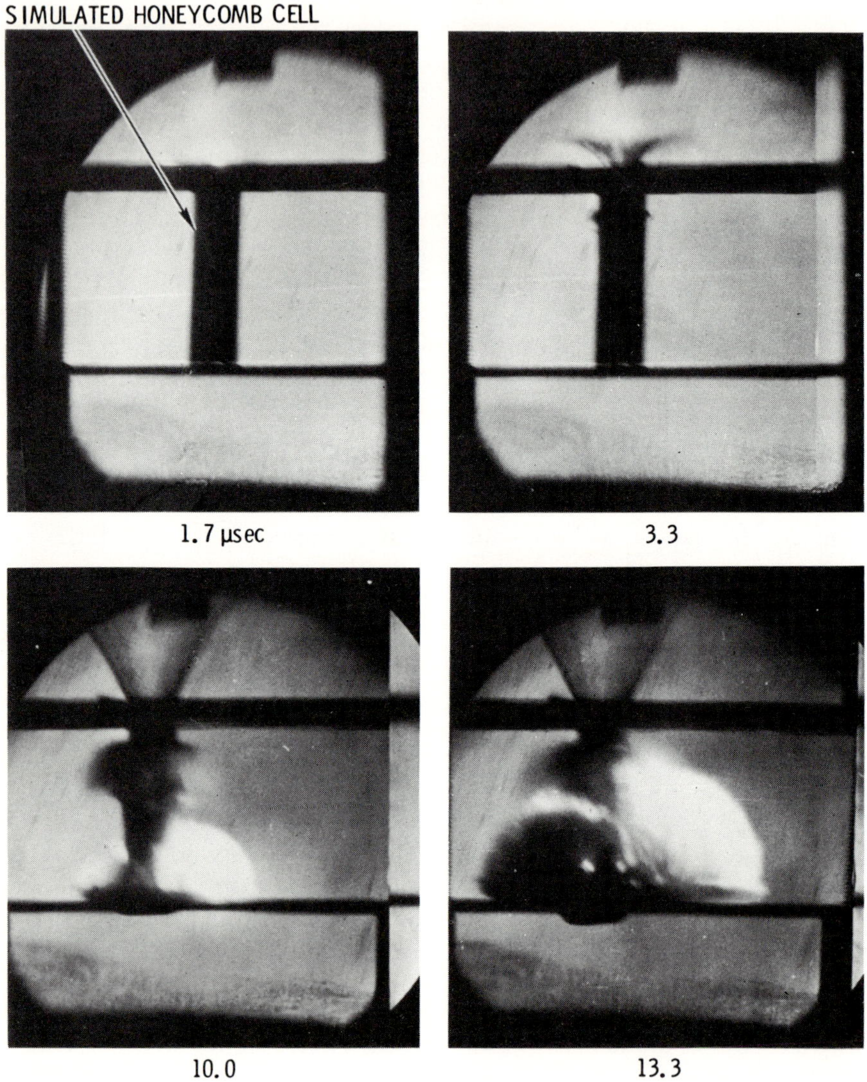

FIG. 14. Framing camera sequence of simulated honeycomb impact. A 14.4-mm-diam 0.25-mm-thick 6061-T6 Al tube (simulated honeycomb cell), 3.18-mm Cd sphere,

debris. As can be seen in Fig. 16, the cell wall has broken but the debris is well channeled down the tube. By channeling the debris the load applied to the second sheet is very much restricted in expanse, producing a very high momentum intensity. This high-intensity loading requires

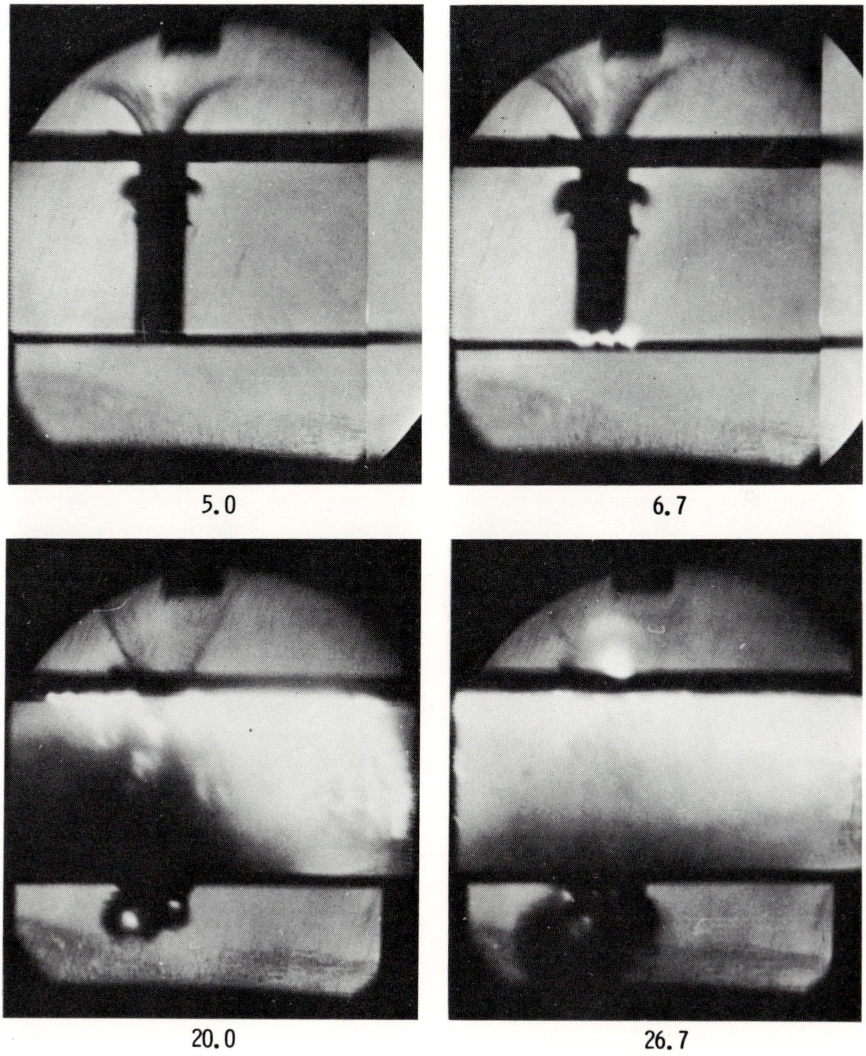

0.33-mm Cd shield, 5.08-cm spacing, 3.18-mm 7075-T6 Al backup, 6.92 km/sec.

that the thickness of the backup sheet be increased to resist any given level of damage.

The radial momenta from hydrocalculations were used to compute the tube wall thicknesses which would be expected to channel or not to channel the debris for cadmium–cadmium impacts. For an impact at 6 km/sec and a shield $t_s/d = 0.5$, a tube 1.27 cm in diameter and

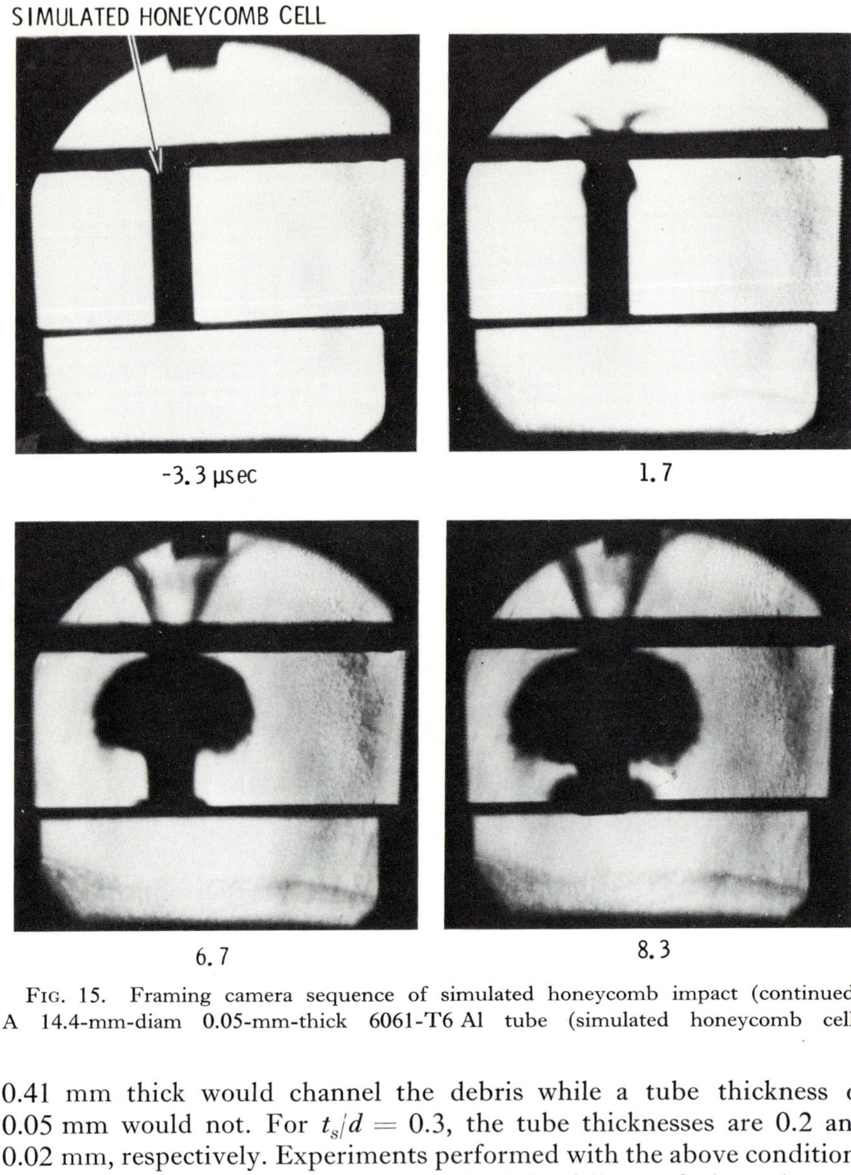

FIG. 15. Framing camera sequence of simulated honeycomb impact (continued). A 14.4-mm-diam 0.05-mm-thick 6061-T6 Al tube (simulated honeycomb cell),

0.41 mm thick would channel the debris while a tube thickness of 0.05 mm would not. For $t_s/d = 0.3$, the tube thicknesses are 0.2 and 0.02 mm, respectively. Experiments performed with the above conditions within the range of tube sizes resulted in failure of the tube and channeling of the debris (see Table III).

As would be expected, larger-diameter tubes and thinner shields also caused channeling. The second-sheet failures are typified by those shown in Fig. 17; the failure in these cases seems to be a shear failure with the

3.3

5.0

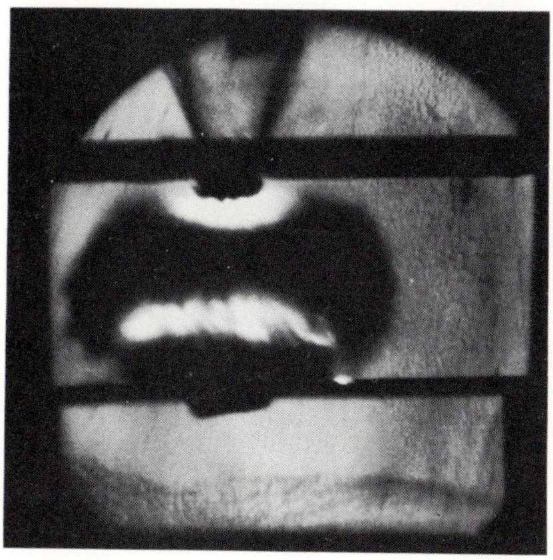

11.7

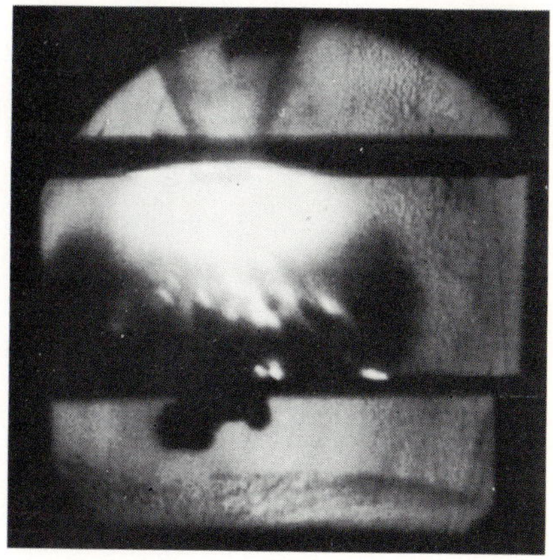

15.0

3.18-mm Cd sphere, 0.48-mm Cd shield, 5.08-cm spacing, 2.54-mm 7075-T6 Al backup, 6.86 km/sec.

portion of the second sheet under the honeycomb all thrown out almost intact. As can be seen by the damage to the third sheets, this fragment can cause considerable further damage.

With a $t_s/d = 0.15$ cadmium shield and a 0.18-mm tube wall thickness, the second sheet had to be between 4.82 and 6.35 mm thick to avoid failure. Using the results of the blast-loaded plate in Chapter IV, the backup thickness to prevent fracture was computed to be 10.8 mm. This is compared with a fracture thickness of between 1.0 and 1.27 mm for

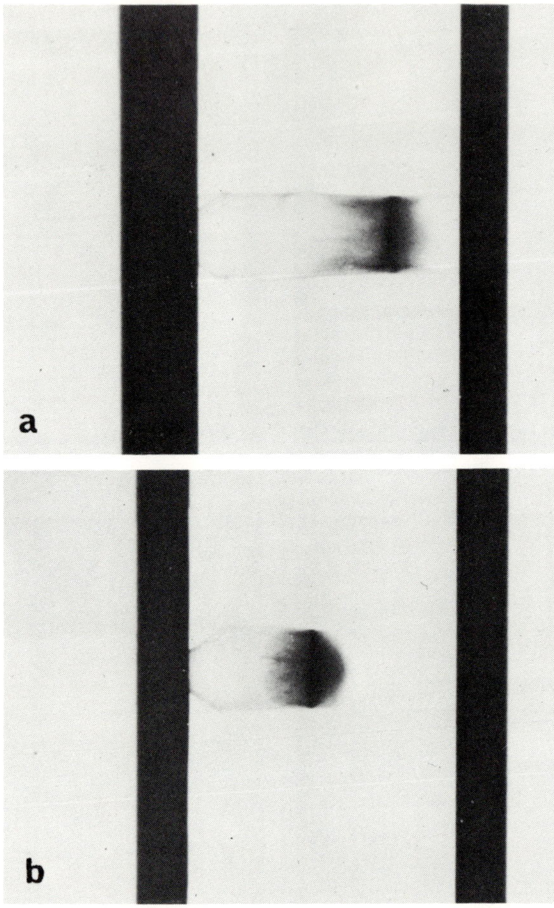

FIG. 16. Flash x-rays of honeycomb cell channeling effect. (a) 0.33-mm Cd shield, 0.25-mm-thick tube, 6.92 km/sec; (b) 0.33-mm Cd shield; 0.05-mm-thick tube, 6.99 km/sec; (c) 0.48-mm Cd shield, 0.18-mm-thick tube, 6.85 km/sec; (d) 0.48-mm Cd shield, 0.05-mm-thick tube, 6.85 km/sec. All projectiles: 3.18-mm Cd spheres. All tubes: 14.4-mm-diam 6061-T6 Al.

the same structure without the tube. This honeycomb then results in an increase of at least a factor of 3.8 in the backup thickness required to prevent failure.

When the simulated honeycomb was subjected to oblique impacts the debris was still channeled to some extent but the main body of the debris impacted the side of the tube, caused it to fail, and then continued to spread.

It is concluded from the analysis and from these tests that a honeycomb

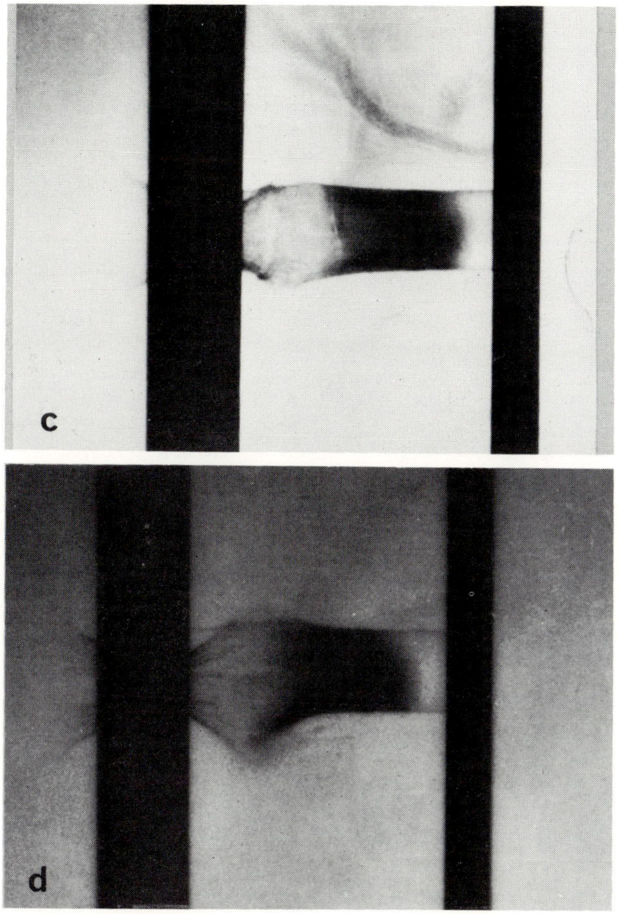

FIG. 16c and d.

will not increase the meteoroid-impact resistance of a structure and can result in a serious degradation of its resistance to failure.

4. *Low-Velocity Impact*

The predictions of second-sheet failure made earlier have assumed that the debris passing through the shield essentially blast-loads the second sheet. This assumption is true only if the impact occurs at velocities sufficient to cause the debris to consist of very small particles or a vapor. For aluminum–aluminum impacts this condition is achieved when the debris contains molten material, say at 7 km/sec. At velocities lower than

TABLE III

3.18-MM CD SPHERES, 5.08-CM SPACING
ALL-CADMIUM SHIELDS, 7075-T6 ALUMINUM SECOND SHEETS

Shot no.	Velocity (km/sec)	Shield thickness (mm)	Second sheet thickness (mm)	Tube Diameter (mm)	Tube Thickness (mm)	Damage
D-1921	6.85	0.48	4.82	14.5	0.18	Perforation
1922	6.98	0.48	6.35	14.5	0.18	Perforation
1912	6.83	0.48	2.54	14.5	0.18	Perforation
1910	6.83	1.22	2.54	14.5	0.18	Perforation
1909	6.50	0.33	3.18	14.5	0.05	Perforation
1913	6.85	0.48	2.54	14.5	0.05	Perforation
1952	6.67	1.60	2.54	14.5	0.05	Perforation
1905	6.95	1.22	1.02	14.5	0.41	Perforation
1906	6.92	1.22	1.02	14.5	0.41	Perforation
1907	6.85	1.22	1.02	14.5	0.05	Perforation
1946	6.67	0.48	2.54	24.6	0.15	Perforation
1947	6.55	0.48	2.54	8.4	0.15	Perforation
1948	6.48	1.22	2.54	8.4	0.15	Perforation
1908	6.92	0.33	3.18	14.5	0.25	Perforation

this incomplete fragmentation of the projectile occurs and the fragments passing through the bumper can inflict substantial damage on the rear sheet.

To investigate this effect, several tests were conducted with 3.18-mm aluminum projectiles at low velocities. Figure 18 shows the penetration results for these experiments along with results at higher velocities. Also shown are the thicknesses to prevent perforation and/or spall over the range of velocities. It can be seen that for this structure (0.64-mm 1100-O aluminum shield with 5.08-cm spacing to a 7075-T6 aluminum backup) a peak in penetration and necessary thickness to prevent second-sheet rear-surface damage occurs at about 2.5 km/sec. At higher velocities the thickness decreases as the projectile is well fragmented, then increases again as the blast loading becomes more severe. The thickness to stop the projectile at 2.5 km/sec is the same as that required to stop the same projectile at 10 km/sec.

The thickness of backup for various shield thicknesses to prevent perforation was determined. These thicknesses are given in Table IV.

All but the first set are for the low-velocity maximum penetration. Although the values are not determined very accurately, there appears to be a trend toward lower values for thicker shields.

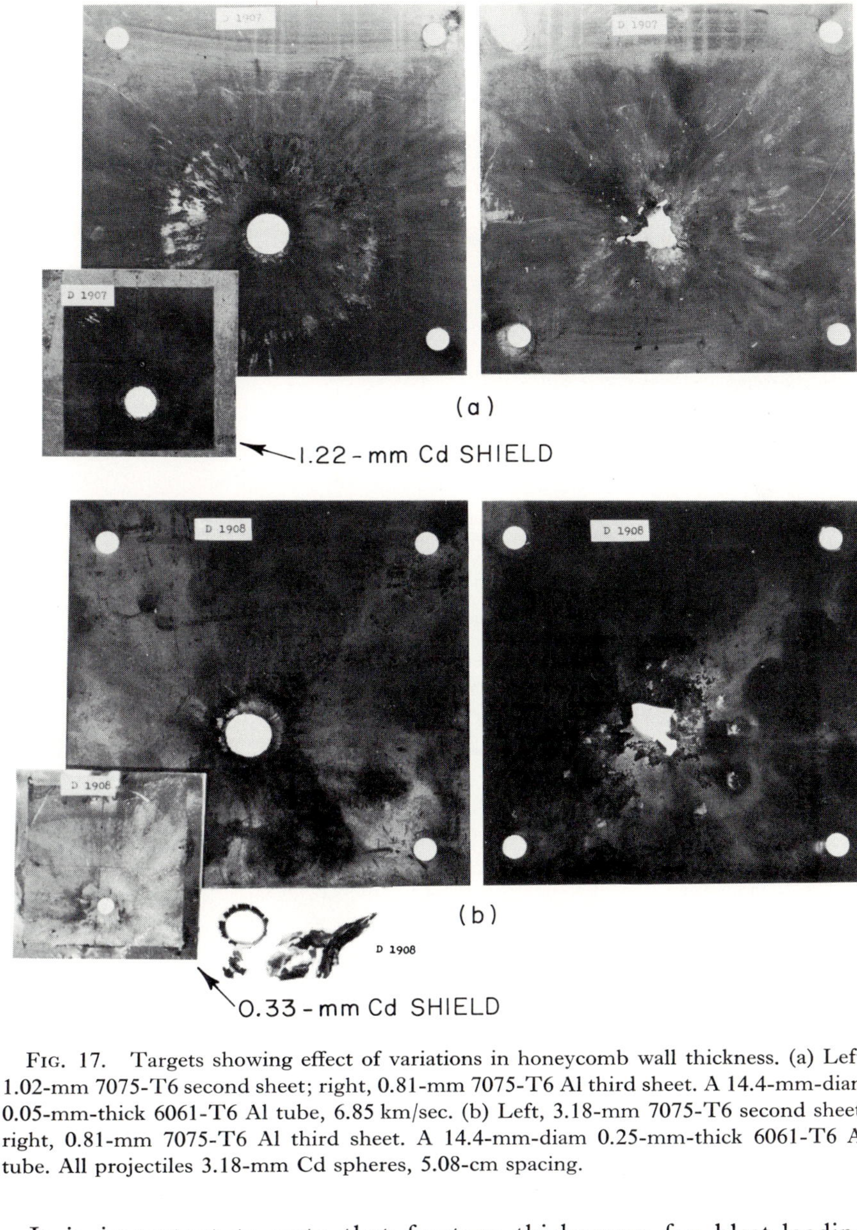

FIG. 17. Targets showing effect of variations in honeycomb wall thickness. (a) Left, 1.02-mm 7075-T6 second sheet; right, 0.81-mm 7075-T6 Al third sheet. A 14.4-mm-diam 0.05-mm-thick 6061-T6 Al tube, 6.85 km/sec. (b) Left, 3.18-mm 7075-T6 second sheet; right, 0.81-mm 7075-T6 Al third sheet. A 14.4-mm-diam 0.25-mm-thick 6061-T6 Al tube. All projectiles 3.18-mm Cd spheres, 5.08-cm spacing.

It is important to note that fracture thicknesses for blast-loading failure scale with projectile mass, while the low-velocity results should scale directly with projectile diameter. Also, the low-velocity results are expected to be independent of intersheet spacing.

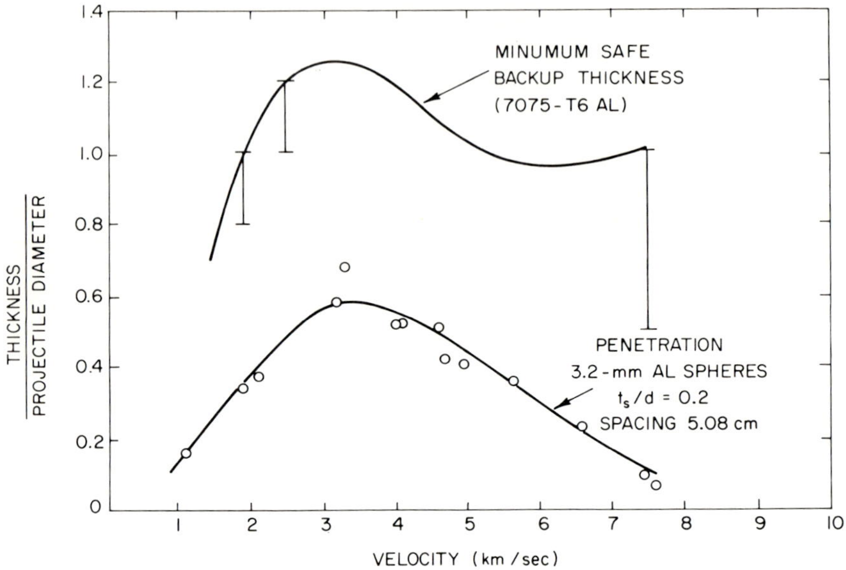

FIG. 18. Backup thickness necessary to prevent perforation.

TABLE IV

SHIELD THICKNESS DATA

t_s/d	$(t_s + t_B)/d$	Velocity (km/sec)
0.20	>1.00 <1.20	1.90
0.20	>1.20 <1.40	2.48
0.32	>1.12 <1.32	2.60
0.50	>1.01 <1.30	2.12

5. *Oblique Impacts*

It would be expected that most meteoroid impacts will not occur normal to the surface of a structure. Consideration of such oblique impacts is not possible with a two-dimensional treatment such as CAMEO because the impact is not axisymmetric. Kreyenhagen *et al.* (*25*) have used a two-dimensional analog of an oblique impact which demonstrates some features of the process. In their computations an infinite-length aluminum cylinder strikes a thin aluminum plate at 6.1 km/sec at an angle of 80.5° from the normal to the plate. The maximum pressure generated is about 330 kbar, whereas a normal impact would result in a maximum pressure of about 800 kbar. The projectile material does not pass through the sheet and the debris to the rear of the sheet is all shield material, leaving the sheet normally at low velocity. For this impact there would be no "in-line" damage to the second sheet; i.e., there would be no damage to the second sheet along an extension of the original flight path. Impacts at smaller angles of obliquity will allow projectile debris to pass through the shield and this material can cause in-line damage.

A cadmium–cadmium impact at an angle of 30° from the normal to the shield is shown a few microseconds after impact in Fig. 19. For the impact conditions here there certainly will be in-line damage. Note that the debris here is concentrated in the front of the debris as opposed to the case of the normal impact where the debris is spread through the

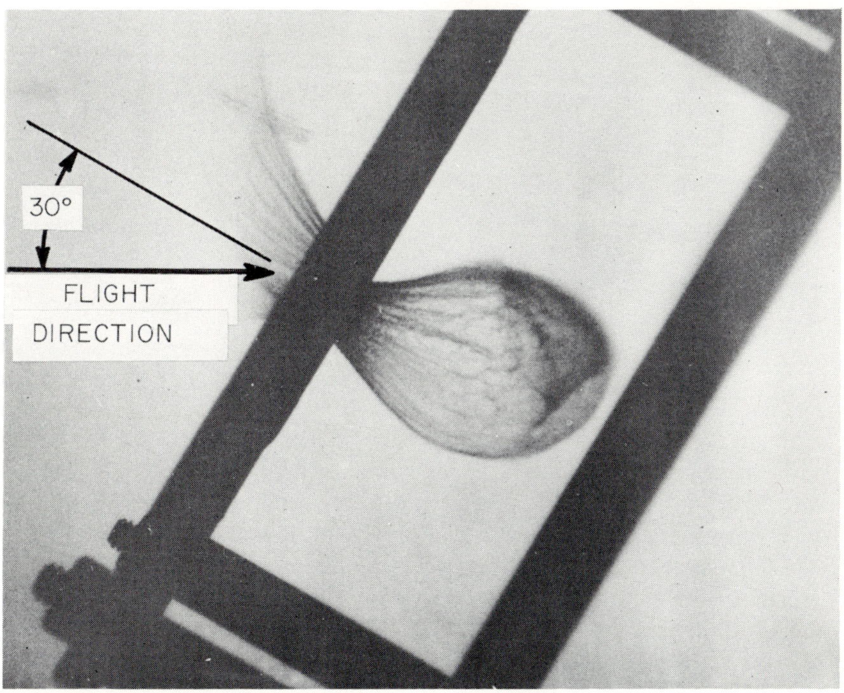

FIG. 19. Flash x-ray showing effect of oblique impact. A 3.18-mm Cd sphere impacting at 6.43 km/sec, 1.22-mm Cd shield, 5.08-cm spacing, 7 μsec after impact.

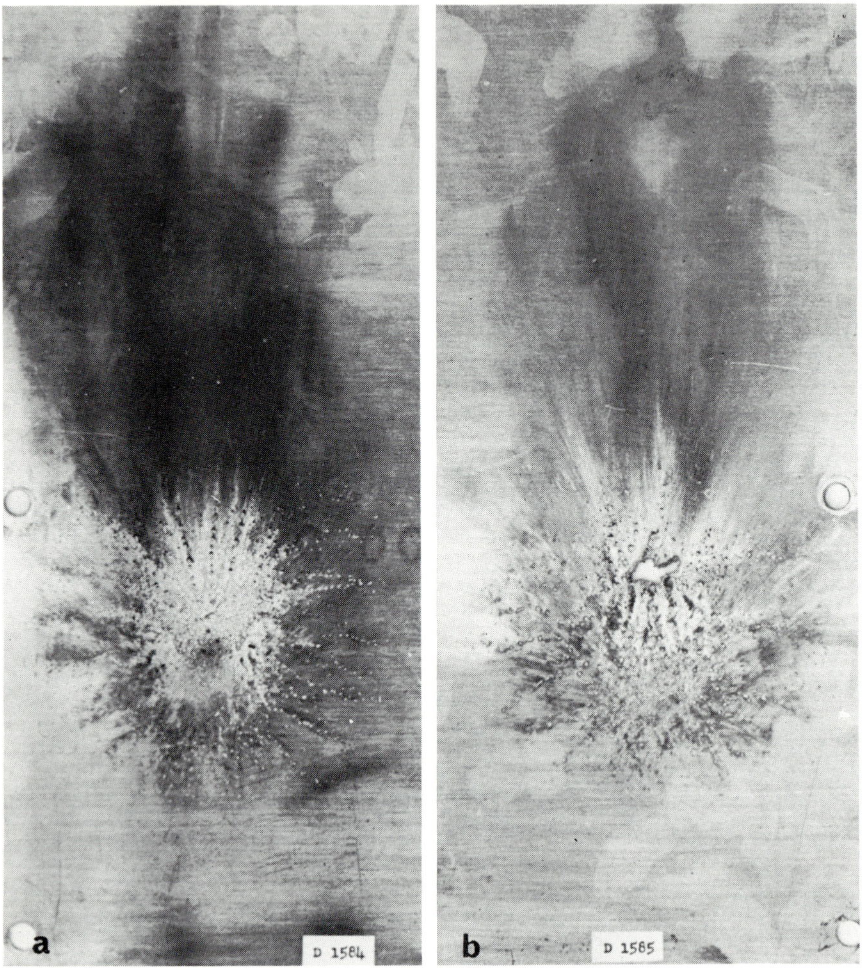

length of the bubble of debris. This concentration of the debris would be expected to deliver a more intense pressure pulse of shorter duration in the second sheet for the oblique impact, and would be more likely to cause rear-surface spall from the second sheet.

Initially experiments were performed with aluminum–aluminum impacts at angles of obliquity of 30, 45, and 60° measured from the normal to the surface of the shield. These experiments, at impact velocities in excess of 7 km/sec, showed that the major source of damage was fragment impact to the second sheet. For normal impacts against these structures fragment damage is very minor due to the melting of the

FIG. 20. Backup targets showing effects of oblique impact: (a) 30°, (b) 45°, (c) 60°. All projectiles 3.18-mm Al spheres impacting at 7.26-7.53 km/sec, 0.64-mm 1100-O Al shield, 5.08-cm spacing, 1.60-mm 7075-T6 Al backups.

debris. The lesser degree of fragmentation is expected because of lower peak pressures associated with oblique impacts. Examination of the second sheets from these experiments, shown in Fig. 20, reveals that the amount of fragmentation of the debris decreases as the angle of obliquity is increased (as evidenced by the crater sizes). Note that the second sheet of the target struck at 45° shows the greatest damage. This would seem to be due to the increase in lateral dispersion as the angle is increased, eventually offsetting the decrease in debris fragmentation.

Further tests were performed in excess of 7 km/sec to determine what the second sheet requirements were to prevent failure. A summary of these experiments is presented in Table V. In all cases of failure the cause was fragment impact. This observation from examination of the targets was supported by flash x-rays of the debris. The term "safe" in Table V means that the second sheet was intact (leakproof) and no material was detached from the rear surface.

TABLE V

OBLIQUE IMPACTS SECOND-SHEET DAMAGE[a]

	Shield thickness 1100-O Al (mm)		
Angle of obliquity	0.64	1.02	1.60
1.60-mm 7075-T6 Al second sheets, 7.26–7.59 km/sec:			
30	Rear spall	Safe	Safe
45	Perforation	Perforation	Perforation
60	Rear spall	Perforation	Perforation
3.18-mm 7075-T6 Al second sheets, 7.50–7.65 km/sec:			
30	Rear spall	—	—
45	Rear spall	Safe	Safe
60	Safe	Safe	Safe
6.35-mm 7075-T6 Al second sheets, 7.60–7.65 km/sec:			
30	Safe	—	—
45	Safe	—	—
60	—	—	—

[a] 3.18-mm Al spheres, 5.08-cm spacing.

Comparison of the total structure $(t_s + t_b/d)$ required to prevent failure from high-velocity oblique impacts with the structures required to prevent failure from low-velocity normal impacts show that, for the aluminum–aluminum impacts, the values all fall in the same range $(1.0 \geqslant t_s + t_b/d \geqslant 1.5)$. It is expected that, as debris heating increases with increasing impact velocity, less fragment damage will be found. This expectation is supported by the results of the cadmium–cadmium impacts partially described in Table VI.

Because of the low melting temperature of the cadmium these represent "higher-velocity" impacts than the aluminum. When the thickness of the rear sheet for the 0.48-mm shields was increased to 6.35 mm the structure was safe. Even though the cadmium projectiles are approximately three times the mass of the aluminum projectiles, the damage was due to second-sheet rear-surface spall and not fragment impact. Also, the second sheets required to resist the cadmium impacts were no more than those required for the aluminum impacts.

TABLE VI

OBLIQUE IMPACTS SECOND-SHEET DAMAGE[a]

Angle of obliquity	Shield thickness Cd (mm)	
	0.48	1.22
30	Rear spall	Safe
45	Rear spall	Safe
60	Safe	Safe

[a] 3.18-mm Cd spheres, 5.08-cm spacing; 3.18-mm 7075-T6 Al second sheets, 6.13–6.43 km/sec.

From these results it is concluded that if a two-sheet structure can resist a low-velocity impact, it can resist the fragment damage due to an oblique impact.

With the exception of one test, no cases of blast-loading failure were found in the tests. The one exception was a very thin backup sheet with the impact at 30°. Even in this case the second sheet was perforated in several places (see Fig. 21). Momentum transfer experiments were performed with aluminum–aluminum impacts, as described previously. The results are shown in Table VII.

TABLE VII

3.18-MM AL SPHERES IMPACTING AT 7.6 KM/SEC, 5.08-CM SPACING, MAXIMUM MEASURED MOMENTUM INTENSITY

t_s/d	Angle of impact			
	0°	30°	45°	60°
0.2	10,000	—	3000	700
0.5	2100	1000	1000	1000 g cm/sec-cm^2

The maximum momentum intensity measured for the oblique impacts is less in both cases than that for comparable normal impacts. This is to be expected, as the debris must travel a greater distance from the shield to the second sheet as the angle is increased. Also, as the angle is increased, the amount of projectile debris which passes through the shield decreases. From the momentum intensity measurements it may be concluded that the debris which strikes the second sheet is widely dispersed in spray angle. The gross-deformation tensile failure mode is

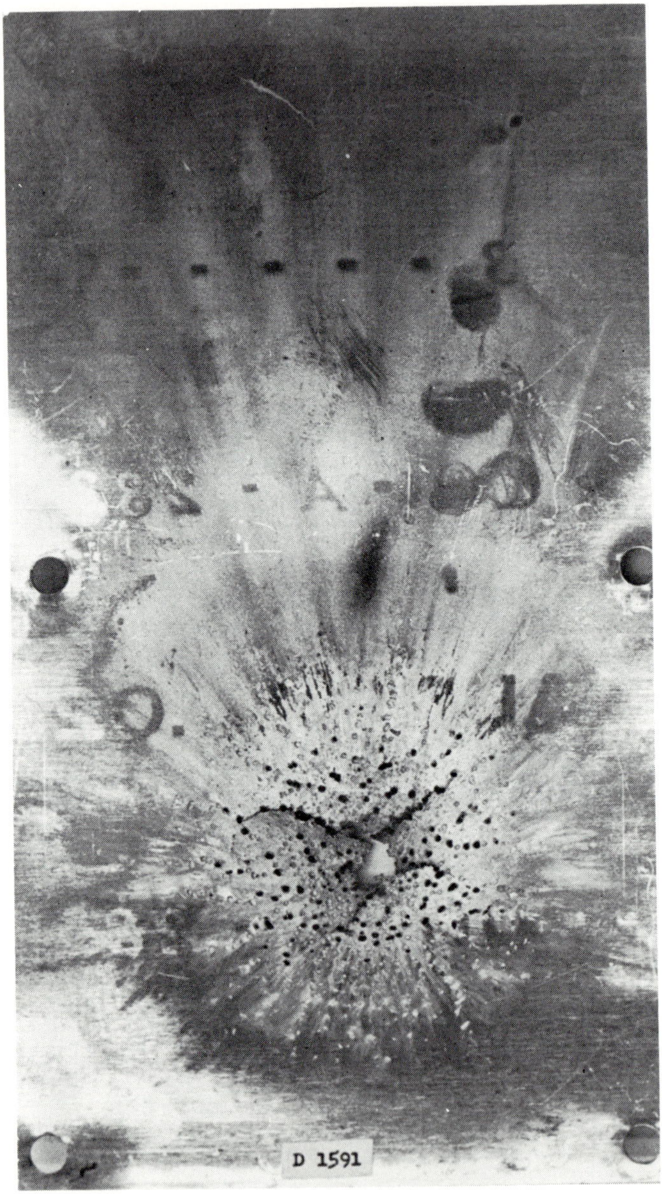

FIG. 21. Perforation of very thin backup sheet. A 3.18-mm Al sphere, 7.47 km/sec, 30° from normal, 1.60-mm 1100-O Al shield, 5.08-cm spacing, 0.64-mm 7075-T6 Al backup.

more strongly influenced by momentum intensity than by total momentum. Therefore, it is concluded that any structure which can resist the gross-deformation tensile failure for a normal impact will not fail through this mechanism for an oblique impact.

6. *Nonoptimum Shields*

Both Maiden and McMillan (*26*) and Sandorff (*27*) designated what was termed an optimum shield. This was defined as a shield that produced a debris cloud in a particular form. The combination of the hydrodynamic CAMEO calculations coupled with the gross-deformation failure analysis and the spall calculations show that an optimum shield is rather hard to define. On the basis of the analysis and the experiments it is easily seen that if the shield is too thin a severe weight penalty is imposed. However, there is a very wide range of minimum-weight structures available (see Chapter IV), where the weight can be split between the shield and backup. Because of the rapid increase in backup thickness required to prevent fracture when the t_s/d ratio becomes small and the severe slope of the strain versus thickness for large strains, it seems advisable to keep the t_s/d ratio as large as possible and to design the backup thickness to accept low maximum strain. As was seen previously, the overriding factor in determining the spall or gross-deformation failure is the intersheet spacing and the fact that the failure thicknesses are proportional to the reciprocal of the spacing squared.

It is very likely that meteoroid impacts will occur where t_s/d is greater than 1. Impacts of this type will result in debris clouds with fragments.

The greater-than-optimum shield situation was experimentally investigated using 3.2-mm aluminum projectiles, 5.08-cm spacing, a velocity of 7.5 km/sec, and various-thickness shields and backups of 7075-T6 aluminum. The damage inflicted upon 1.6-mm-thick backup targets in these tests is shown in Figs. 22 and 23. The results of all the tests can be conveniently summarized in the solid-line construction of Fig. 24. Straight lines have been drawn between known points A, B, and C, where t_A (thickness at Point A) corresponds to the thickness of the rear sheet necessary to prevent spall detachment with no shield, t_B (thickness at Point B) is the total thickness of the structure to give 4% maximum strain at optimum shield thickness of $t_s/d = 0.15$, and $t_C = t_A$ is the thickness of shield necessary to prevent spall detachment and thus require no backup sheet. The experimental results indicate that, for design purposes, if the structure is designed above ABC, then no perforation of the rear sheet will occur. Note that, at a given velocity, thickness $t_A = t_C$ scales approximately as particle diameter d, whereas t_B scales as d^3.

7. *Two-Sheet Scaling*

The results of the analysis and experiments indicate that shield thicknesses scale with projectile diameter and backup thicknesses scale with projectile mass or diameter cubed. The total structure would then scale as between the diameter to the first and third power. It is generally agreed that the failure thickness of a single-sheet structure is proportional to the projectile diameter (*28–30*). These two statements would indicate

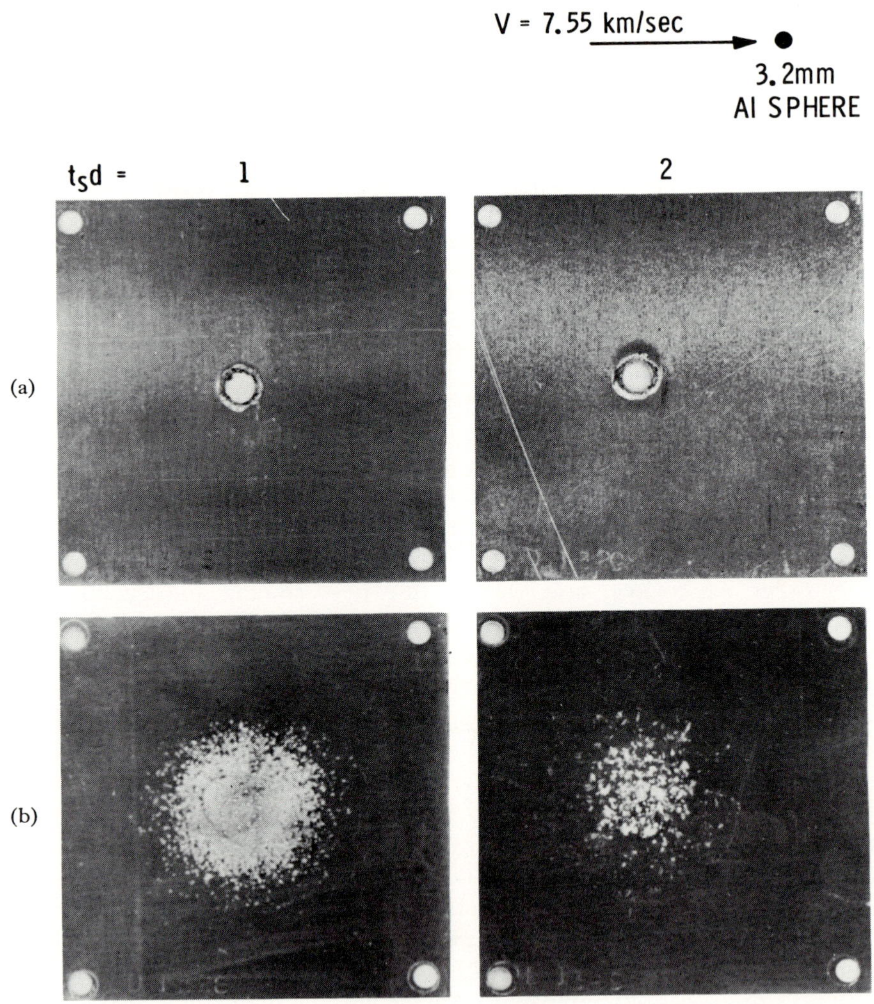

FIG. 22. Targets having greater than optimum shields (front). (a) Left and right,

that if the projectile mass were very large, a single-sheet structure would be better than a two-sheet structure. However the two-sheet results are qualified as applying only to structures in which the spacing is greater than eight projectile diameters. With this restriction two sheets are better than one. For spacings less than this the projectile–shield debris is still quite dense and results, upon impact on the second sheet, in

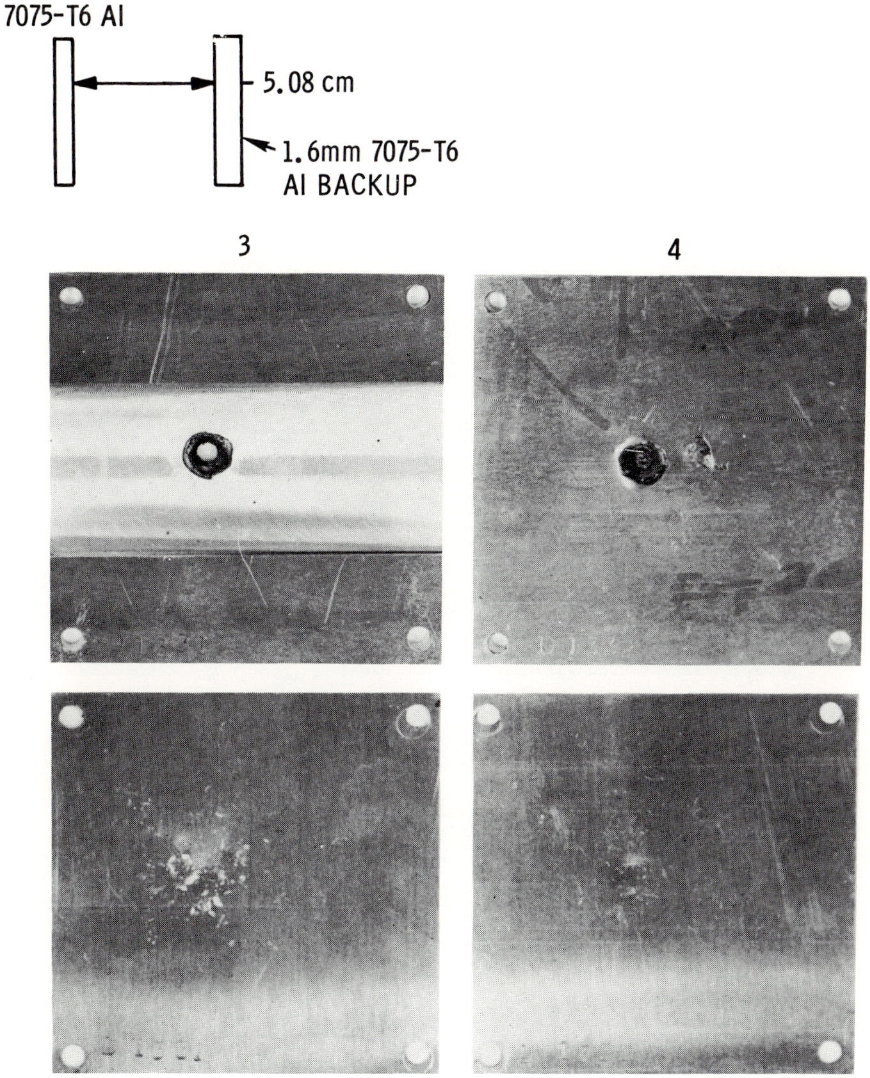

front of shield; (b) left and right, front of backup.

cratering of the second sheet. The analysis utilized herein does not apply to this case, but it is reasoned that a two-sheet structure is always better than one because any high-velocity impact upon a shield results in the spread of the projectile–shield debris, the loss of energy, and, at most, a slight increase in the total momentum. The impact of the second sheet is then less severe than if there were no shield because the energy per unit area and momentum per unit area are less. This same reasoning also applies to velocity scaling; thus it is concluded that two sheets are always better than one.

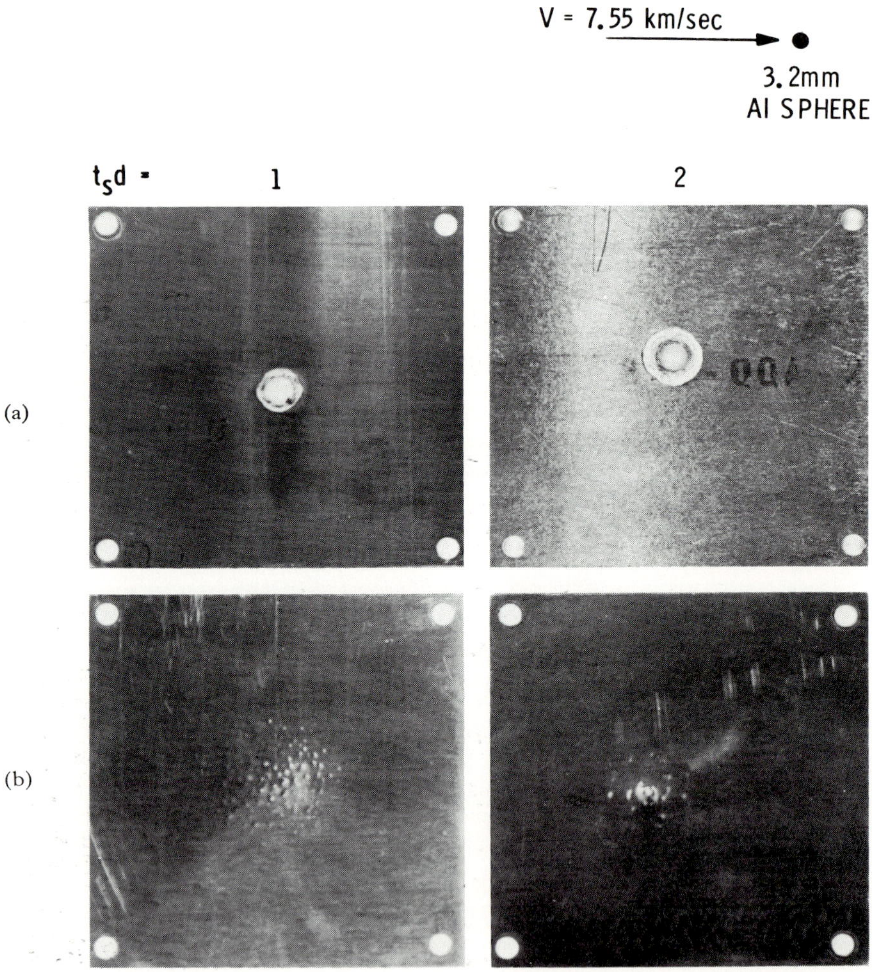

FIG. 23. Targets having greater than optimum shields (rear). (a) Left and right, rear

8. *Evaluation of Two-Sheet Structures*

The impact resistance of two-sheet structures can be evaluated utilizing the information given previously. The basic two-sheet structure must be evaluated to cover the probable impact conditions. Evaluations of three structures have been performed and are presented as examples of the method.

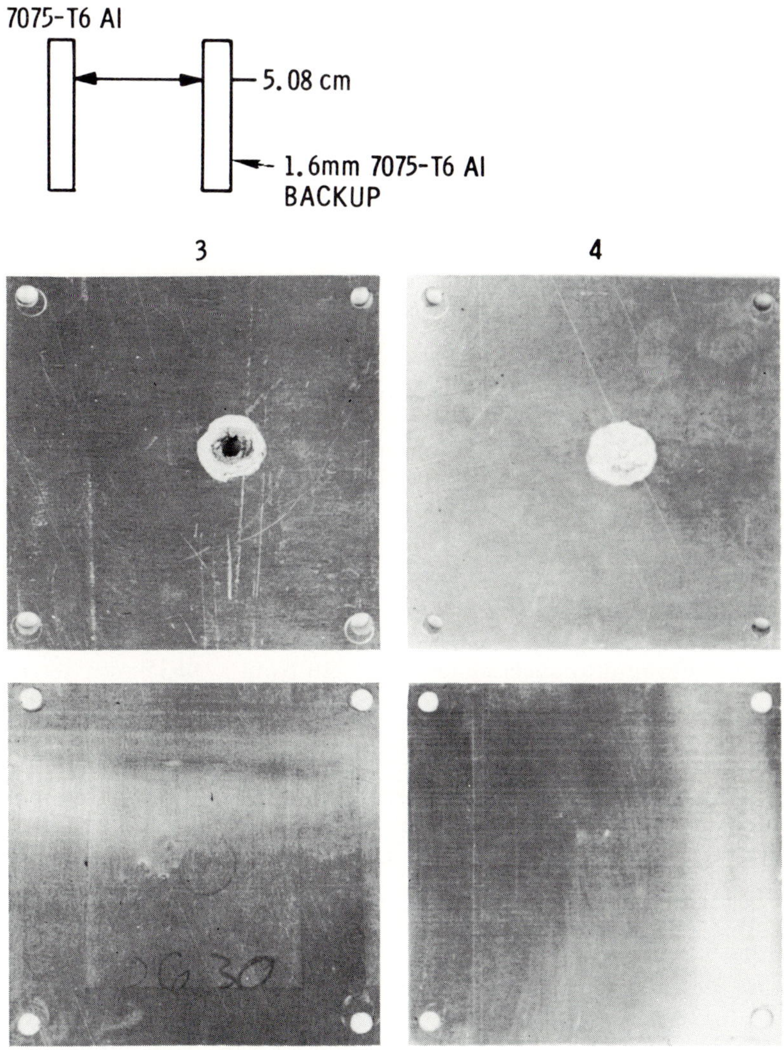

of shield; (b) left and right, rear of backup.

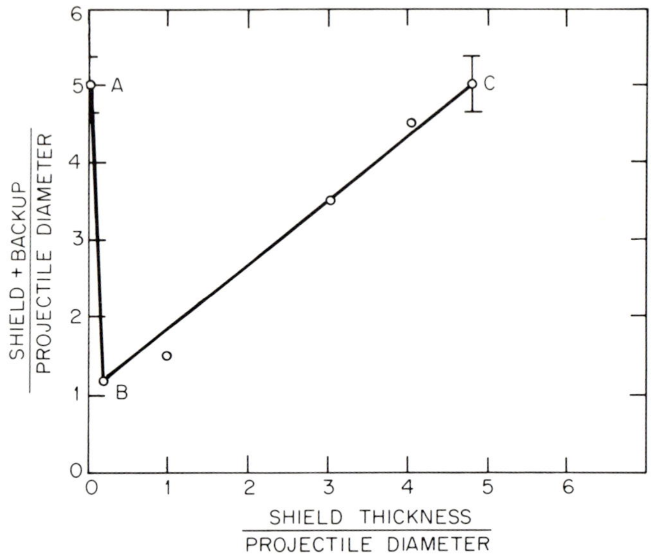

FIG. 24. Safe thickness versus shield thickness. (○) Safe points; 3.2-mm Al spheres, 7.4 km/sec, 5.08-cm spacing, 7075-T6 Al targets.

The three structures are the Apollo Service Module Block 1 and 2 and the Lunar Excursion Module. The Service Module structure consists of two thin aluminum sheets 0.405 mm thick spaced 2.54 cm apart. The Block 1 structure has a honeycomb between the two sheets with a cell size of 6.35 mm and a wall thickness of 0.0254 mm. The Block 2 structure has a honeycomb with a 4.75-mm cell size and a wall thickness of 0.0178-mm. The Lunar Excursion Module structure consists of a 0.102-mm aluminum sheet spaced 5.08 cm from a 0.635-mm aluminum second sheet. All three structures were assumed to consist of 7075-T6 aluminum.

The structures are evaluated on the basis of their vulnerability to solid aluminum impacts. This is conservative, as an underdense material would cause less damage. By using Eq. (10) in Chapter IV the critical masses can be determined for impacts at 30 km/sec. These masses are 2.64×10^{-3} g for the Service Module (the honeycomb has initially been ignored) and 2.1×10^{-3} g for the Lunar Excursion Module.

Next the critical masses are determined for the honeycomb by scaling the uniform plate gross-deformation results given in Figs. 28 and 29 of Chapter IV. The critical mass for the Block 1 Service Module structure is 1.78×10^{-4} g, and is 10^{-4} g for the Block 2 structure.

Finally the low-velocity impact critical masses are determined from the results given in Table IV. Here the honeycomb can be ignored, as the particle does not break up upon impact. The critical masses are 3.4×10^{-4} g for the Service Module structure and 2.54×10^{-4} for the Lunar Excursion Module structure.

These critical masses can then be employed to determine fluxes and probabilities of no failure from the NASA environment (*31*). For a probability of no failure of 0.99 the Service Module Block 1 structure has an area time of exposure of 91,500 ft^2-days ignoring the honeycomb, 2510 ft^2-days for all impacts normal and in the center of the honeycomb cell, and 5900 ft^2-days for the low-velocity impact. For the Block 2 structure the area-times are the same as for the Block 1 structure except for the honeycomb, which is 1160 ft^2-days.

For the Lunar Excursion Module structure the area-times are 66,200 ft^2-days for normal impacts at 30 km/sec and 2050 ft^2-days for low-velocity impacts. The low-velocity evaluation also provides for oblique impact fragment damage.

The exposure times given should be taken as lower bounds as all considerations have been conservative. The honeycomb analysis considers only the worst case, low-velocity (2.5 km/sec) impacts seem unlikely, and oblique impact fragment damage is not expected to be a problem for high-velocity impacts. The true exposure times would therefore be expected to tend toward the higher values.

C. Complex Targets

Complex targets are defined as those which represent functional vehicle structures and are of great concern to the design engineer. These targets are not amenable to theoretical analysis and relatively few experimental tests have been carried out. Particular attention is directed here to tubular targets, pressure vessels, and space vehicle windows. This cursory look at complex targets is intended only to yield an insight into the multitude of novel target shapes, materials, and structural configurations that may be subjected to a meteoroid impact in space flight.

1. *Tubular Targets*

Typical of the unique and complex targets is the tubular-shaped space radiator configuration that in a real situation offers a meteoroid a target of hundreds of square feet of exposed area. To consider the damage to a space radiator or tabular target McMillan (*22*) studied the effect on the damage to thin targets by changing the target radius, as shown in Fig. 25. For a wall thickness of 0.446 in., the depth of pene-

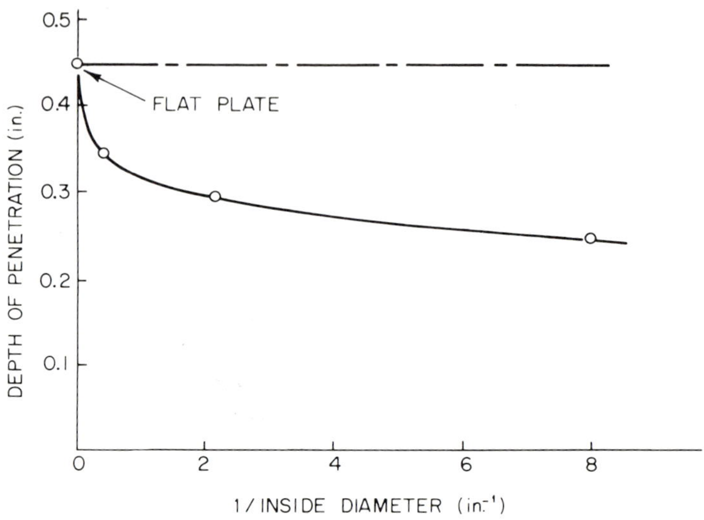

FIG. 25. Depth of penetration versus target diameter-constant wall thickness. Projectile: 0.094-in.-o.d. 0.016-g Pyrex sphere. Target: 356 Al tubes, 700°F. Velocity: 7.5 km/sec. Constant wall thickness: 0.446 in.

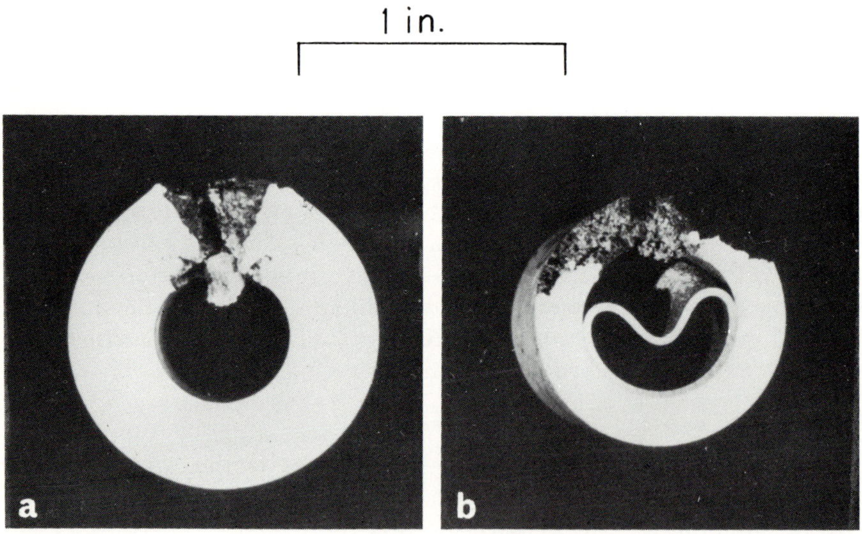

FIG. 26. Effect of liner on spall-cylindrical targets. (a) 2024-T6 Al, no liner; (b) 356-T51 Al, 316 Stainless-Steel liner. Projectile: 0.094-in. Pyrex spheres. Velocity: 7.3 km/sec.

tration is substantially decreased as the inside diameter is decreased from infinitity (flat target) to 0.50 in. Below 0.50 in., little change in penetration is observed. The ability of the circular target shape to sustain impact damage stems from the fact that shock propagation through the circular section is affected by more free surface. Consequently, the rarefaction waves which act to diminish the intensity of the transient pulse react more quickly, thus weakening the shock and diminishing its ability to perforate the target or to cause spall.

In order to prevent spalling on the inside of a tubular target the use of a ductile liner was proposed for a space radiator (*19*), where a perforation would cause a fluid leak, while spalling without perforation could cause damage or failure of the liquid-circulating pumps by injecting metal particles into the fluid. The beneficial effects of having an inner liner can be seen in Fig. 26. One target was lined with steel; the second target had no liner, but the aluminum armor was thicker, thus keeping the weight per unit length constant. Although severely dimpled, the inner liner prevented spalling of metal into the coolant tube.

There was no damage to the inner tube of the lined target, even though the depth of penetration exceeded the depth predicted to cause spalling.

2. *Space Vehicle Windows*

Hypervelocity meteoroid impacts against windows can seriously affect guidance and control systems or docking maneuvers, especially in manned spacecraft, by impairing the transparency of the windows. Two examples of this are given in Fig. 27. The acrylic window has been severely damaged in the area of impact but still permits outside observation. The laminated glass window, however, has been rendered almost useless as a result of the impact. Also of concern, of course, would be any through cracks in the windows that might result in loss of pressure inside the vehicle.

3. *Pressurized Vessels and Stressed Structures*

The impact of meteoroid particles on those space vehicles that have components which are highly pressurized with fluid or gas of varying temperatures have reportedly (*32–34*) caused catastrophic ruptures. Typical pressurized vessels in space vehicles which may encounter meteoroids are power plants, space radiators (*19*), fuel tanks, and storage bottles. Since it is impossible to control meteoroids in space, the designer must choose a safeguard that can protect the system from impact by the largest meteoroid it is likely to encounter during its time of exposure.

When a high-speed particle strikes the surface of a pressurized vessel, one of two things will usually happen: First, the projectile may com-

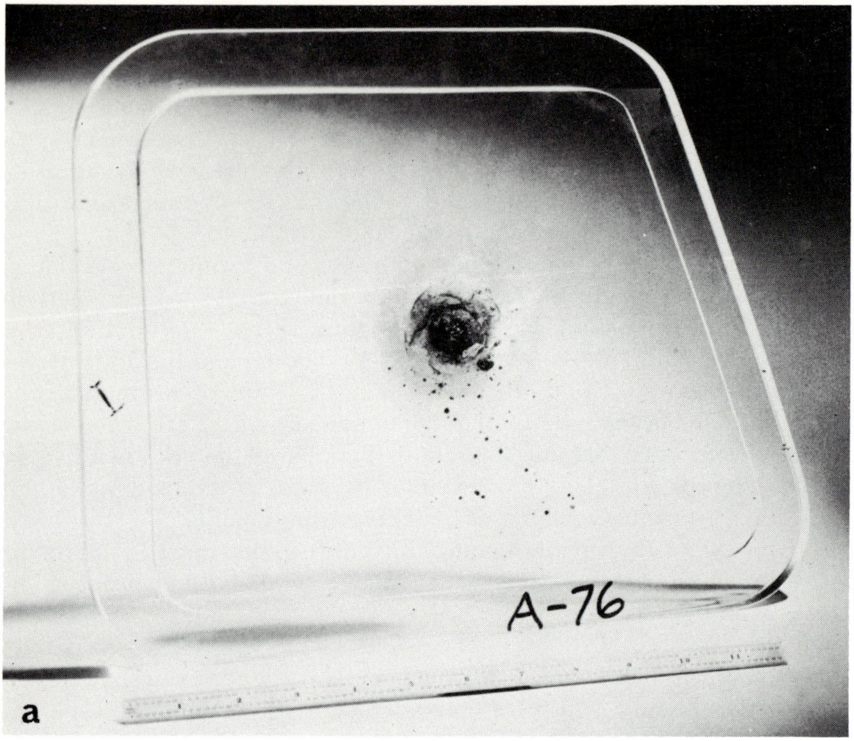

FIG. 27. Impact damage to windows. (a) 2.54-cm stressed acrylic plastic window; (b) 1.27-cm-thick laminated tempered safety glass window. Projectile: 0.125-in. 0.04-g glass sphere. Velocity: 7.7 km/sec.

pletely perforate the vessel, spewing particle fragments and spall into the inner material and increasing the internal pressure, or (in the case of a very thin-walled vessel) the projectile may perforate both the front and rear surfaces of the container; second, the projectile may not perforate but may generate a shock wave into the internal fluid, setting up a compression wave in the fluid. In both cases, the internal pressure will be increased; when the internal pressure exceeds the hoop tensile strength of the container, the pressure vessel will fail.

For example, in the case of a structure which has a very thin wall, such as a pressurized missile-propellant tank, it was observed by Rolsten *et al.* (*33*) and Stepka and Morse (*34*) that the chances of catastrophic failure are high. Rolsten *et al.* cite the case of impact of a liquid-pressurized, thin-walled vessel by a projectile, wherein there was catastrophic failure of the vessel wall, attributed to the stored energy from the hoop

FIG. 27b.

tension stresses. On the other hand, in the case of a liquid-pressurized, thick-walled vessel, the probability that catastrophic failure will occur is lower.

General Motors Defense Research Laboratories conducted a test where a $\frac{1}{8}$-in.-diam glass sphere was impacted against a sphere of 2024-T3 aluminum (16 in. in diameter and $\frac{1}{4}$ in. thick) at 7.0 km/sec. The aluminum sphere contained water at a pressure of 800 psi. Figure 28 shows that there was no catastrophic failure—instead, the result was a clean perforation. However, if the projectile had been larger, or the thickness of the structure reduced, the chances of a catastrophic failure would have been greater. Also, the failure mechanisms under conditions of gas pressurization, where shock propagation to the rest of the vessel would be significantly different, could have been different in each case. Some results of experimental work on pressure vessels can be found in the classified literature (*35*).

Another problem is that the high temperature associated with an

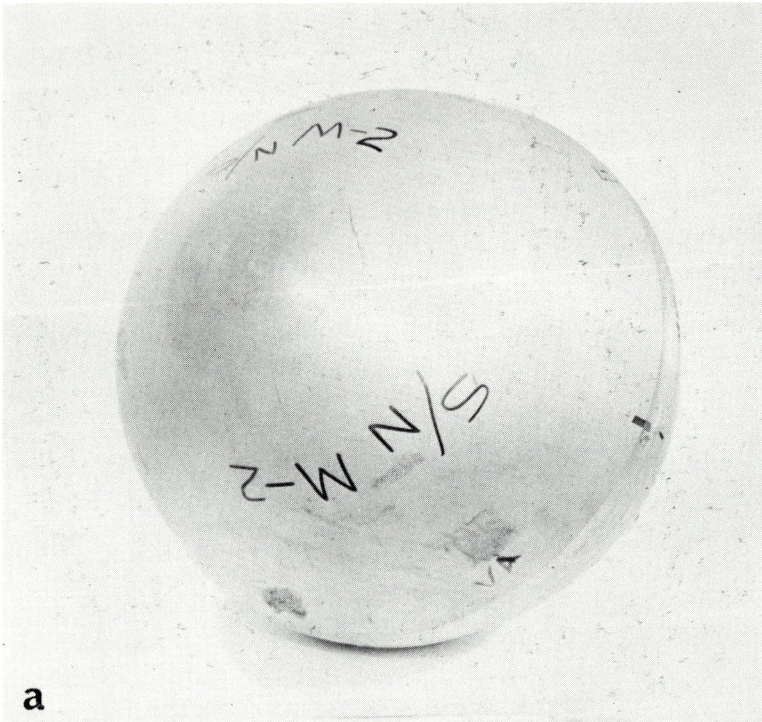

FIG. 28. Impact damage to thick-walled, liquid-pressurized vessel. (a) 39-cm-diam sphere, 6.35-mm-thick Al; (b) arrow pointing to projectile, a 3.18-mm 0.04-g glass sphere. Sphere contained water at 800 psi at time of impact. Impact velocity: 7.0 km/sec.

impact flash created by the impact of a high-speed particle against a fluid-rich pressurized space vehicle may cause it to burn or explode. An astronaut may experience shock, concussion, or disastrous decompression (*36*, *37*).

The frictional removal or change of temperatures of structural material as the result of repeated impacts can contribute to failure by either (1) reducing the net area of the structure, thus increasing the stress level, or (2) reducing the strength of the material by uncontrolled heat treating, so that failure of the structure could result from a shorter critical crack length.

In conclusion, catastrophic failure depends upon the structure of the material, stress level, internal pressure of the medium, type of internal fluid, area of hole, compression wave behavior in the material, and reactivity between the medium and structural material.

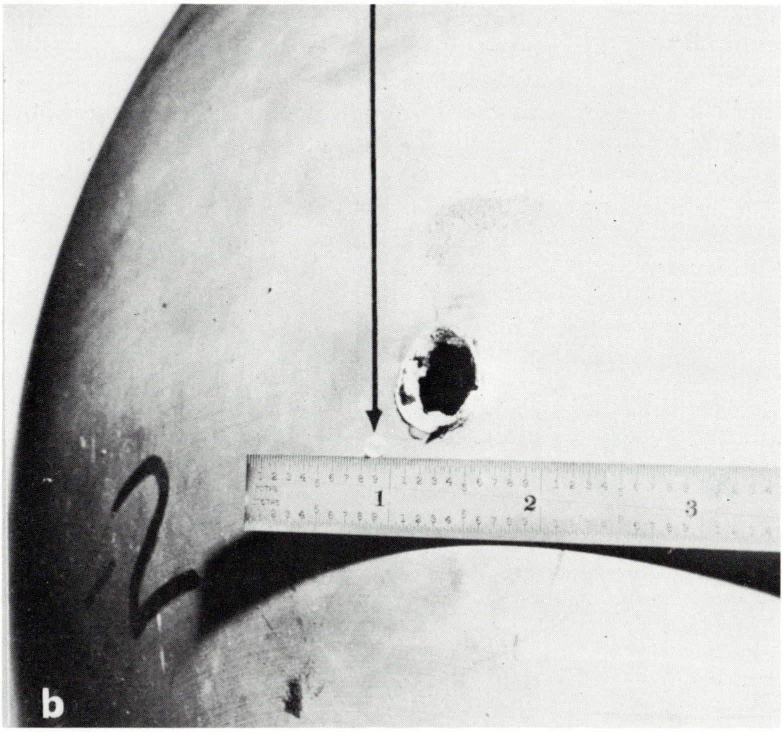

FIG. 28b.

III. Discussion

An attempt has been made in this chapter to present a brief look at the many engineering aspects of the phenomena of hypervelocity impact. Seldom in the real world does a meteoroid or fragment impact a target under conditions of a normal angle of attack against a homogeneous isotropic target material. Hence, it is necessary to critically examine the cases of impact upon complex targets under a variety of impact conditions. Such an attempt has been made here. Unfortunately it is impossible to cover any and all structural materials in configurations that the engineer might encounter in his design of a space vehicle. A cross section of targets has been looked at ranging from the most simple infinite target to the most complex target currently envisioned. Empirical relations were presented to allow the calculation and approximation of the damage which would be sustained to a simple target, and a design approach has been offered to the problem of the two-sheet vehicle hull structure. In addition, a brief survey has been made of the cases of impact against laminated targets, space vehicle windows, and pressurized vessels. An attempt has been made to check through the many references which are available and to pick those which demonstrate a particular phenomena. Space will not allow for the coverage of many references that are excellent in their detailed examination of specific structures and materials, and an apology is offered to the many authors whose work could not be covered herein.

Acknowledgments

The author wishes to express his gratitude to his many co-workers at the General Motors Defense Research Laboratories who have, over the recent years, contributed to the research programs summarized in this chapter. Particular appreciation is extended to D. R. Christman, R. L. Warnica, A. R. McMillan, and B. L. Lathrop. Much of the work leading to the manuscript was supported by NASA, both at the Manned Spacecraft Center in Houston, Texas, under the direction of Mr. Burton Cour-Palais, and at the Lewis Research Center, under the direction of Mr. Seymour Lieblein. Their cooperation and expert advice during the conduct of the program is sincerely appreciated.

REFERENCES

1. R. J. Eichelberger and J. W. Gehring, Effects of meteoroid impacts on space vehicles, *Am. Rocket Soc. J.* **32**, No. **10**, **1583–1951** (**1962**),
2. A. C. Charters and J. L. Summers, High speed impact of metal projectiles in target of various materials, *Proc. Symp. Hypervelocity Impact, 3rd, Armor Res. Foundation, Chicago, Illinois* (1958).
3. A. C. Charters and J. L. Summers, Some comments on the phenomena of high speed impacts, *Proc. Decennial Symp., White Oak U.S. Naval Ord. Lab.* (1959).
4. W. W. Atkins, Hypervelocity penetration studies, *Proc. Symp. Hypervelocity Impact, 3rd, Armour Res. Foundation, Chicago, Illinois* (Feb. 1959).
5. J. L. Summers and W. R. Niehaus, A Preliminary Investigation of the Penetration of Slender Metal Rods in Thick Metal Targets, NASA TN-D-137 (Dec. 1959).
6. W. Allen and J. W. Rogers, Penetration of a rod into a thick aluminum target, *J. Franklin Inst.* **272**, No. 4, 275–284 (1961).
7. R. E. Slattery and W. G. Clay, The penetration of thin rods into aluminum, *Proc. Symp. Hypervelocity Impact, 4th, Eglin AFB, Florida* **3**, APGC-TR-60-39 (Sept. 1960).
8. C. R. Nysmith, J. L. Summers, and B. P. Denardo, Investigation of the impact of copper filaments into aluminum targets at velocities to 16,000 ft/sec, paper presented at Symp. Hypervelocity Impact, 6th, Cleveland, Ohio (April 30–May 2, 1963).
9. D. R. Christman and J. W. Gehring, Analysis of high-velocity projectile penetration mechanics, *J. Appl. Phys.* **37**, No. 4, 1579–1587 (1966).
10. D. R. Christman, J. W. Gehring, C. J. Maiden, and A. B. Wenzel, Study of the phenomena of hypervelocity impact, GM TR63-216 (June 1963).
11. A. C. Charters and J. L. Summers, High-speed impact of metal projectiles in targets

of various materials, *Proc. Symp. Hypervelocity Impact, 3rd, Armour Res. Foundation, Chicago, Illinois* (Feb. 1959).

12. J. G. Feldman, Volume-energy relations from shape charge jet penetrations, *Proc. Symp. Hypervelocity Impact, 4th, Eglin AFB, Florida* **2,** APGC TR-60-39 (Sept. 1960).
13. M. Rockowitz, C. Carey, and J. Dignam, Hypervelocity impact of heated copper, *Proc. Symp. Hypervelocity Impact, 5th* **1,** Part 2, *Colorado School of Mines, Colorado* Nonr-(G)-0020-62(X) (April 1962).
14. F. E. Allison, K. R. Becker, and R. Vitali, Effects of target temperature on hypervelocity cratering, *Proc. Symp. Hypervelocity Impact, 4th, Eglin AFB, Florida* **1,** APGC TR-60-39 (Sept. 1960).
15. J. W. Gehring and S. Lieblein, Preliminary results on effects of hypervelocity impact on space radiator tubes, ARS Paper No. 2544-62, presented at Conf. ARS Space Power Systems, Santa Monica, California (Sept. 1962).
16. D. V. Keller and J. G. Trulis, Mechanism of spall in lucite, *J. Appl. Phys.* **34,** No. 1, 172–175 (1963).
17. J. S. Rinehart, Pratical countermeasures for the prevention of spallation, Colorado School of Mines Res. Foundation, Rept. AF-SWC-TR-60-7 to Air Force Special Weapons Center, Alburquerque, New Mexico (1960).
18. J. S. Rinehart, Some Quantitative data bearing on the scabbing of metals under explosive attack, *J. Appl. Phys.* **22,** No. 5, 555–560 (1951).
19. J. W. Gehring and S. Lieblein, Preliminary results on effects of hypervelocity impact on space radiator tubes, ARS Paper No. 2544-62, presented at Conf., ARS Space Power Systems, Santa Monica, California (Sept. 1962).
20. L. D. Jaffe and J. B. Rittenhouse, Behavior of materials in space environments, *Am. Rocket Soc. J.* **32,** No. 3, 320–346 (1962).
21. R. L. Bjork, Meteoroids vs. space vehicles, *Am. Rocket Soc. J.* **31,** No. 6, 803–807 (1961).
22. A. R. McMillan, An investigation of the penetration of hypervelocity projectiles into composite laminates, *Proc. Symp. Hypervelocity Impact, 6th, Firestone Tire and Rubber Co.* **3,** DA-31-124-ARO(D)-16 (Aug. 1963).
23. I. J. Loeffler, S. Lieblein, and N. Clough, Meteoroid protection for space radiators, ARS Paper No. 2543-62, presented at Conf. ARS Space Power Systems, Santa Monica, California (Sept. 1962).
24. J. L. Summers and C. R. Nysmith, The resistance of a variety of composite space structures to hypervelocity impact, *Proc. Structures Materials Conf., 5th, Palm Springs, California,* 386–393 AIAA (April 1964).
25. K. N. Kreyenhagen, R. L. Bjork, and N. B. Brooks, Numerical solution of oblique impacts, *Proc. Symp. Hypervelocity Impact, 7th* **3,** 55–74 AD 463229 (Feb. 1965).
26. C. J. Maiden and A. R. McMillan, An investigation of the protection afforded a spacecraft by a thin shield, *AIAA J.* **2,** No. 11, 1992–1998 (1964).
27. P. Sandorff, A meteoroid bumper design criterion, *Proc. Symp. Hypervelocity Impact, 6th* **3,** 41–67, AD 423063 (Aug. 1963).
28. R. L. Bjork, Review of physical processes in hypervelocity impact and penetration, *Proc. Symp. Hypervelocity Impact, 6th* **2,** 1–58, AD 423063 (Aug. 1963).
29. T. D. Riney and J. F. Heyda, Hypervelocity impact calculations, *Proc. Symp. Hypervelocity Impact, 7th* **2,** 77–186, AD 463228 (Feb. 1965).
30. J. M. Walsh and W. E. Johnson, On the theory of hypervelocity impact, *Proc. Symp. Hypervelocity Impact, 7th* **2,** 1–76, AD 463228 (Feb. 1965).
31. P. B. Burbank, B. G. Cour-Palais, and W. E. McAllum, A meteoroid environment for near-earth, cislunar, and near-lunar operations, NASA TN D-2747 (April 1965).

32. R. H. Kemp, Flow and fracture problems in aerospace vehicles, *Proc. NASA-Univ. Conf. Sci. Technol. Space Exploration, Chicago, Illinois, Nov. 1–3* **2**, NASA SP-11 (1962).
33. R. F. Rolsten, H. H. Hunt, and J. N. Wellnitz, Study of principles of meteoroid protection, General Dynamics/Astronautics Rept. No. AE 62-0413 (April 1962).
34. F. S. Stepka and C. R. Morse, Preliminary investigations of catastrophic fracture of liquid-filled tanks impacted by high-velocity particles, NASA TN-D-1537 (May 1963).
35. K. N. Kreyenhagen, R. B. Mortensen, and H. S. Zimney, Impact effects against propulsion systems (U), *Proc. Symp. Hypervelocity Impact, 5th, Colorado School of Mines, Colorado* **2**, Nonr-(G)-0020-62(X) (April 1962).
36. M. Kornhauser, Satellite pressure losses caused by meteoroid impacts, *Am. Rocket Soc. J.* **30**, No. 5, 475–479 (1960).
37. J. W. Gehring and B. L. Lathrop, Spacecraft interior hazards from hypervelocity impact, GM TR66-13 (March 1966).

APPENDIX A

SHOCK WAVE DATA FOR STANDARD MATERIALS

2024 ALUMINUM

ρ_0 (g/cm^3)	u_s (km/sec)	u_p (km/sec)	P (Mb)	ρ (g/cm^3)
2.785	6.02	0.50	0.08	3.04
2.785	6.10	0.50	0.09	3.03
2.785	6.05	0.51	0.09	3.04
2.782	6.16	0.67	0.12	3.12
2.785	6.35	0.77	0.14	3.17
2.782	6.49	0.86	0.16	3.21
2.785	6.52	0.87	0.16	3.22
2.782	6.61	0.97	0.18	3.26
2.789	6.86	1.15	0.22	3.35
2.789	6.89	1.16	0.22	3.35
2.780	6.91	1.16	0.22	3.34
2.789	6.86	1.21	0.23	3.38
2.789	7.27	1.43	0.29	3.47
2.789	7.33	1.58	0.32	3.55
2.779	7.51	1.62	0.34	3.54
2.789	7.68	1.72	0.37	3.60
2.789	7.60	1.73	0.37	3.61
2.789	7.66	1.77	0.38	3.63
2.789	7.77	1.81	0.39	3.64
2.789	7.69	1.85	0.40	3.67
2.779	7.97	1.95	0.43	3.68
2.789	8.13	2.13	0.48	3.78
2.789	8.33	2.16	0.50	3.76
2.789	8.23	2.21	0.51	3.81
2.779	8.40	2.31	0.54	3.83
2.789	8.53	2.45	0.58	3.91
2.789	8.83	2.59	0.64	3.95
2.789	8.76	2.60	0.64	3.97
2.789	8.80	2.65	0.65	3.99
2.789	9.14	2.82	0.72	4.03
2.789	9.40	2.99	0.78	4.09
2.789	9.67	3.22	0.87	4.19
2.789	9.64	3.29	0.88	4.23
2.778	9.87	3.42	0.94	4.25
2.778	10.19	3.72	1.05	4.37
2.789	10.37	3.75	1.08	4.37

921T ALUMINUM

ρ_0 (g/cm^3)	u_s (km/sec)	u_p (km/sec)	P (Mb)	ρ (g/cm^3)
2.810	5.72	0.50	0.08	3.08
2.810	5.74	0.53	0.08	3.09
2.810	5.98	0.63	0.11	3.14
2.811	5.88	0.65	0.11	3.16
2.828	6.72	1.19	0.23	3.43
2.828	7.04	1.40	0.28	3.53
2.828	7.18	1.49	0.30	3.57
2.828	7.21	1.57	0.32	3.61
2.828	7.42	1.63	0.34	3.63
2.828	7.42	1.63	0.34	3.63
2.828	7.73	1.86	0.41	3.73
2.828	8.06	2.13	0.49	3.84
2.828	8.09	2.13	0.49	3.84
2.828	8.35	2.28	0.54	3.89
2.828	8.42	2.40	0.57	3.96
2.828	8.51	2.44	0.59	3.96
2.828	9.07	2.87	0.74	4.13
2.828	9.38	3.09	0.82	4.22
2.828	9.40	3.10	0.82	4.22

COPPER

ρ_0 (g/cm^3)	u_s (km/sec)	u_p (km/sec)	P (Mb)	ρ (g/cm^3)
8.930	4.48	0.41	0.16	9.82
8.931	4.57	0.41	0.17	9.82
8.928	4.89	0.62	0.27	10.23

COPPER *(continued)*

ρ_0 (g/cm^3)	u_s (km/sec)	u_p (km/sec)	P (Mb)	ρ (g/cm^3)
8.930	4.91	0.64	0.28	10.26
8.930	4.91	0.73	0.32	10.49
8.930	5.07	0.74	0.33	10.45
8.930	4.89	0.74	0.32	10.53
8.930	5.18	0.85	0.39	10.67
8.930	5.36	0.95	0.46	10.86
8.930	5.84	1.25	0.65	11.37
8.930	6.19	1.48	0.82	11.74
8.930	6.51	1.71	0.99	12.10
8.928	6.55	1.74	1.02	12.15
8.930	6.91	1.99	1.23	12.54
8.930	6.88	2.04	1.25	12.68
8.930	7.03	2.10	1.32	12.73
8.928	7.46	2.37	1.58	13.08
8.930	7.47	2.37	1.58	13.08

IRON

ρ_0 (g/cm^3)	u_s (km/sec)	u_p (km/sec)	P (Mb)	ρ (g/cm^3)
7.870	4.59	0.	0.	7.87
7.850	5.33	0.95	0.40	9.55
7.843	5.46	1.00	0.43	9.60
7.850	5.56	1.06	0.46	9.71
7.847	6.20	1.43	0.70	10.20
7.850	6.24	1.44	0.71	10.22
7.850	6.58	1.65	0.85	10.48
7.850	6.66	1.73	0.91	10.61
7.847	7.07	1.96	1.09	10.85
7.850	7.08	1.98	1.10	10.90
7.831	7.56	2.27	1.35	11.20
7.847	7.81	2.44	1.50	11.42
7.832	7.90	2.46	1.52	11.37
7.809	8.09	2.61	1.65	11.54
7.850	8.01	2.62	1.65	11.66
7.847	8.24	2.72	1.76	11.70
7.850	8.23	2.72	1.76	11.73
7.850	8.32	2.73	1.78	11.68
7.850	8.26	2.73	1.77	11.73
7.868	8.31	2.82	1.84	11.91
7.840	8.49	2.89	1.92	11.88
7.842	8.81	3.06	2.11	12.01
7.850	8.94	3.12	2.19	12.05
7.850	8.99	3.18	2.24	12.14
7.850	9.10	3.19	2.28	12.08
7.860	9.24	3.35	2.43	12.34
7.860	9.28	3.36	2.45	12.32
7.855	9.51	3.59	2.68	12.61

URANIUM- 3 WT PER CENT MOLYBDENUM

ρ_0 (g/cm^3)	u_s (km/sec)	u_p (km/sec)	P (Mb)	ρ (g/cm^3)
18.450	3.17	0.41	0.24	21.15
18.450	3.33	0.49	0.30	21.63
18.450	3.67	0.72	0.49	22.97
18.450	3.73	0.75	0.52	23.08
18.450	4.24	1.11	0.87	24.97
18.450	4.54	1.29	1.08	25.80
18.450	4.53	1.32	1.10	26.01
18.480	5.19	1.76	1.68	27.95
18.450	5.31	1.80	1.77	27.93
18.450	5.36	1.84	1.82	28.08
18.450	5.83	2.11	2.27	28.90
18.450	6.06	2.36	2.65	30.25
18.450	6.27	2.38	2.75	29.75
18.450	6.51	2.54	3.05	30.24
18.450	6.76	2.75	3.43	31.07

APPENDIX B

SHOCK WAVE DATA FOR POROUS MATERIALS

2024 ALUMINUM (0.91 CRYSTAL DENSITY)

ρ_0 (g/cm³)	u_s (km/sec)	u_p (km/sec)	P (Mb)	ρ (g/cm³)	ID	u_s(ID) (km/sec)
2.571	4.78	0.72	0.09	3.03	1	6.17
2.647	5.36	1.00	0.14	3.26	1	6.53
2.524	5.16	1.03	0.13	3.15	1	6.52
2.575	5.98	1.33	0.20	3.31	1	6.94
2.566	6.85	1.76	0.31	3.45	1	7.52
2.532	6.00	1.92	0.29	3.73	1	7.58
2.472	6.93	1.99	0.34	3.47	1	7.75
2.554	7.55	2.23	0.43	3.62	1	8.12
2.465	7.91	2.58	0.50	3.66	1	8.52
2.554	8.32	2.70	0.57	3.78	1	8.74
2.605	8.37	2.79	0.61	3.91	1	8.87
2.583	8.81	3.03	0.69	3.94	1	9.19
2.506	8.72	3.05	0.67	3.86	1	9.17
2.603	9.09	3.23	0.77	4.04	1	9.47
2.535	9.15	3.30	0.77	3.97	1	9.52
2.585	9.59	3.55	0.88	4.11	1	9.89
2.543	9.58	3.59	0.87	4.07	1	9.90
2.522	9.54	3.62	0.87	4.06	1	9.92
2.510	9.63	3.64	0.88	4.04	1	9.95
2.631	9.84	3.73	0.97	4.24	1	10.15
2.618	9.84	3.80	0.98	4.27	1	10.22
2.617	9.90	3.82	0.99	4.26	1	10.25

2024 ALUMINUM (0.59 CRYSTAL DENSITY)

ρ_0 (g/cm³)	u_s (km/sec)	u_p (km/sec)	P (Mb)	ρ (g/cm³)	ID	u_s(ID) (km/sec)
1.659	2.47	0.98	0.04	2.75	1	6.16
1.622	2.31	1.01	0.04	2.89	1	6.17
1.699	3.33	1.34	0.08	2.84	1	6.53
1.618	2.98	1.38	0.07	3.01	1	6.52
1.648	4.05	1.75	0.12	2.90	1	6.94
1.663	5.15	2.25	0.19	2.95	1	7.52
1.687	5.13	2.31	0.20	3.07	1	7.58
1.626	5.64	2.45	0.22	2.88	1	7.75
1.700	6.26	2.74	0.29	3.02	1	8.12
1.661	6.68	3.12	0.35	3.12	1	8.52
1.671	7.09	3.30	0.39	3.13	1	8.74
1.687	7.41	3.38	0.42	3.11	1	8.87
1.685	7.60	3.67	0.47	3.25	1	9.17
1.691	7.74	3.67	0.48	3.22	1	9.19
1.657	8.18	3.92	0.53	3.18	1	9.47
1.656	8.20	3.97	0.54	3.21	1	9.52
1.707	8.76	4.24	0.63	3.31	1	9.89
1.596	8.71	4.35	0.60	3.19	1	9.90
1.603	8.65	4.37	0.61	3.24	1	9.92
1.544	8.64	4.45	0.59	3.18	1	9.95
1.673	9.00	4.50	0.68	3.35	1	10.15
1.713	9.02	4.54	0.70	3.45	1	10.22
1.694	9.12	4.57	0.71	3.39	1	10.25

2024 ALUMINUM (0.80 CRYSTAL DENSITY)

ρ_0 (g/cm³)	u_s (km/sec)	u_p (km/sec)	P (Mb)	ρ (g/cm³)	ID	u_s(ID) (km/sec)
2.241	3.47	0.85	0.07	2.97	1	6.17
2.261	4.30	1.14	0.11	3.08	1	6.52
2.183	4.10	1.18	0.11	3.07	1	6.53
2.062	4.89	1.55	0.16	3.02	1	6.94
2.271	6.12	1.93	0.27	3.31	1	7.52
2.136	6.54	1.98	0.28	3.06	1	7.58
2.247	6.40	2.12	0.31	3.36	1	7.75
2.193	6.90	2.44	0.37	3.39	1	8.12
2.260	7.63	2.70	0.47	3.50	1	8.52
2.266	7.91	2.88	0.52	3.56	1	8.74
2.186	7.88	3.04	0.52	3.56	1	8.87
2.301	8.47	3.21	0.63	3.71	1	9.19
2.203	8.30	3.26	0.60	3.63	1	9.17
2.280	8.53	3.48	0.68	3.85	1	9.47
2.198	8.80	3.53	0.68	3.67	1	9.52
2.287	9.27	3.77	0.80	3.85	1	9.89
2.208	9.27	3.83	0.78	3.76	1	9.90
2.217	9.26	3.84	0.79	3.79	1	9.92
2.229	9.28	3.86	0.80	3.82	1	9.95
2.211	9.64	4.01	0.86	3.79	1	10.15
2.265	9.65	4.04	0.88	3.90	1	10.22
2.289	9.57	4.06	0.89	3.98	1	10.25

COPPER (0.88 CRYSTAL DENSITY)

ρ_0 (g/cm³)	u_s (km/sec)	u_p (km/sec)	P (Mb)	ρ (g/cm³)	ID	u_s(ID) (km/sec)
7.877	3.38	0.63	0.17	9.67	10	4.71
7.885	3.32	0.65	0.17	9.79	10	4.73
7.790	3.71	0.78	0.23	9.87	10	4.92
7.890	3.77	0.82	0.24	10.09	10	4.98
7.897	4.50	1.11	0.39	10.48	10	5.40
7.908	4.55	1.12	0.40	10.50	10	5.42
7.876	4.41	1.16	0.40	10.71	10	5.45
7.914	4.75	1.18	0.44	10.54	10	5.51
7.880	5.00	1.39	0.55	10.92	10	5.80
7.910	5.78	1.76	0.81	11.38	10	6.35
7.932	5.69	1.79	0.81	11.58	10	6.37
7.881	6.65	2.31	1.21	12.09	10	7.13
7.927	6.64	2.33	1.22	12.20	10	7.15
7.894	6.75	2.33	1.24	12.06	10	7.17
7.915	7.01	2.47	1.37	12.22	10	7.38
7.968	7.23	2.59	1.49	12.42	10	7.57
7.995	7.84	2.93	1.84	12.76	10	8.07
7.866	7.76	2.97	1.81	12.74	10	8.09

2024 ALUMINUM (0.70 CRYSTAL DENSITY)

ρ_0 (g/cm³)	u_s (km/sec)	u_p (km/sec)	P (Mb)	ρ (g/cm³)	ID	u_s(ID) (km/sec)
1.993	2.91	0.92	0.05	2.91	1	6.17
1.957	2.77	0.93	0.05	2.94	1	6.16
1.966	3.62	1.25	0.09	3.01	1	6.52
1.929	3.62	1.27	0.09	2.97	1	6.53
1.908	4.40	1.64	0.14	3.05	1	6.94
2.010	5.62	2.07	0.23	3.18	1	7.52
1.970	5.49	2.16	0.23	3.24	1	7.58
1.995	5.99	2.26	0.27	3.21	1	7.75
1.981	6.61	2.57	0.34	3.24	1	8.12
1.916	7.05	2.94	0.40	3.28	1	8.52
1.959	7.47	3.09	0.45	3.34	1	8.74
1.880	7.56	3.25	0.46	3.30	1	8.87
2.019	8.00	3.40	0.55	3.52	1	9.17
2.037	8.04	3.42	0.56	3.54	1	9.19
1.933	8.48	3.69	0.61	3.43	1	9.47
1.888	8.38	3.78	0.60	3.44	1	9.52
2.065	8.92	3.99	0.73	3.73	1	9.92
1.996	9.02	4.01	0.72	3.59	1	9.90
1.983	9.03	4.06	0.73	3.60	1	9.95
1.825	8.76	4.15	0.66	3.47	1	9.89
1.916	9.16	4.29	0.75	3.60	1	10.15
1.951	9.32	4.33	0.79	3.65	1	10.25
1.899	9.25	4.36	0.77	3.59	1	10.22

COPPER (0.82 CRYSTAL DENSITY)

ρ_0 (g/cm³)	u_s (km/sec)	u_p (km/sec)	P (Mb)	ρ (g/cm³)	ID	u_s(ID) (km/sec)
7.368	3.00	0.67	0.15	9.49	10	4.71
7.384	2.79	0.70	0.14	9.86	10	4.73
7.182	3.32	0.84	0.20	9.61	10	4.92
7.364	3.32	0.88	0.21	10.02	10	4.98
7.208	4.09	1.19	0.35	10.16	10	5.40
7.201	4.12	1.21	0.36	10.19	10	5.42
7.364	4.15	1.22	0.37	10.43	10	5.45
7.193	4.33	1.27	0.39	10.17	10	5.51
7.354	4.72	1.46	0.51	10.65	10	5.80
7.216	5.44	1.86	0.73	10.98	10	6.35
7.372	5.47	1.87	0.75	11.19	10	6.37
7.489	6.52	2.39	1.17	11.82	10	7.15
7.366	6.42	2.40	1.14	11.76	10	7.13
7.219	6.53	2.44	1.15	11.53	10	7.17
7.201	6.62	2.61	1.24	11.88	10	7.38
7.384	6.98	2.70	1.39	12.05	10	7.57
7.376	7.51	3.06	1.70	12.46	10	8.07
7.374	7.59	3.06	1.71	12.36	10	8.09

COPPER (0.71 CRYSTAL DENSITY)

ρ_0 (g/cm^3)	u_s (km/sec)	u_p (km/sec)	P (Mb)	ρ (g/cm^3)	ID	u_s(ID) (km/sec)
6.372	2.41	0.74	0.11	9.22	10	4.71
6.322	2.28	0.78	0.11	9.58	10	4.73
6.270	2.80	0.92	0.16	9.34	10	4.92
6.325	2.79	0.97	0.17	9.70	10	4.98
6.290	3.56	1.30	0.29	9.89	10	5.40
6.309	3.64	1.31	0.30	9.85	10	5.42
6.395	3.69	1.33	0.31	9.99	10	5.45
6.279	3.89	1.37	0.34	9.71	10	5.51
6.300	4.27	1.59	0.43	10.04	10	5.80
6.278	5.08	2.00	0.64	10.35	10	6.35
6.332	5.10	2.01	0.65	10.46	10	6.37
6.392	6.17	2.56	1.01	10.92	10	7.13
6.350	6.13	2.58	1.01	10.98	10	7.15
6.278	6.14	2.61	1.01	10.92	10	7.17
6.264	6.39	2.77	1.11	11.05	10	7.38
6.327	6.70	2.89	1.22	11.12	10	7.57
6.342	7.37	3.26	1.52	11.37	10	8.09
6.343	7.21	3.27	1.50	11.60	10	8.07

COPPER (0.64 CRYSTAL DENSITY)

ρ_0 (g/cm^3)	u_s (km/sec)	u_p (km/sec)	P (Mb)	ρ (g/cm^3)	ID	u_s(ID) (km/sec)
5.983	2.26	0.77	0.10	9.06	10	4.71
5.833	2.22	0.80	0.10	9.08	10	4.73
5.567	2.51	0.98	0.14	9.11	10	4.92
5.718	2.61	1.01	0.15	9.35	10	4.98
5.564	3.35	1.37	0.26	9.40	10	5.40
5.830	3.52	1.38	0.28	9.61	10	5.45
5.605	3.38	1.39	0.26	9.50	10	5.42
5.605	3.65	1.45	0.30	9.28	10	5.51
5.876	4.17	1.64	0.40	9.67	10	5.80
5.712	4.95	2.10	0.59	9.91	10	6.37
5.557	4.87	2.10	0.57	9.78	10	6.35
5.978	6.10	2.62	0.96	10.48	10	7.13
5.785	5.93	2.69	0.92	10.58	10	7.15
5.627	5.99	2.72	0.92	10.32	10	7.17
5.598	6.20	2.89	1.00	10.50	10	7.38
5.892	6.72	2.95	1.17	10.51	10	7.57
5.972	7.33	3.33	1.46	10.93	10	8.09
5.801	7.20	3.36	1.41	10.89	10	8.07

COPPER (0.51 CRYSTAL DENSITY)

ρ_0 (g/cm^3)	u_s (km/sec)	u_p (km/sec)	P (Mb)	ρ (g/cm^3)	ID	u_s(ID) (km/sec)
4.453	2.83	1.50	0.19	9.51	10	5.40
4.467	2.92	1.52	0.20	9.30	10	5.42
4.533	3.21	1.58	0.23	8.90	10	5.51
4.577	5.63	2.94	0.76	9.60	10	7.17
4.513	5.86	3.13	0.83	9.68	10	7.38

IRON (0.89 CRYSTAL DENSITY)

ρ_0 (g/cm^3)	u_s (km/sec)	u_p (km/sec)	P (Mb)	ρ (g/cm^3)	ID	u_s(ID) (km/sec)
7.002	3.21	0.57	0.13	8.51	10	4.60
6.992	3.26	0.70	0.16	8.90	10	4.75
7.001	3.37	0.70	0.17	8.85	10	4.76
7.061	3.47	0.79	0.19	9.13	10	4.87
6.965	3.57	0.85	0.21	9.14	10	4.95
6.960	4.42	1.15	0.36	9.41	10	5.38
6.945	5.29	1.43	0.52	9.51	10	5.79
6.940	5.14	1.45	0.52	9.68	10	5.80
6.988	5.42	1.57	0.59	9.82	10	5.98
6.954	5.51	1.65	0.63	9.92	10	6.08
7.025	6.14	1.77	0.76	9.88	10	6.31
6.968	5.90	1.84	0.76	10.12	10	6.35
6.986	5.92	1.86	0.77	10.19	10	6.39
7.070	6.24	1.95	0.86	10.29	10	6.55
6.913	6.29	2.01	0.87	10.16	10	6.60
6.963	6.30	2.05	0.90	10.32	10	6.65
6.979	6.43	2.09	0.94	10.34	10	6.72
6.933	6.94	2.44	1.17	10.69	10	7.19
6.959	7.60	2.80	1.48	11.02	10	7.71
7.046	7.93	3.01	1.68	11.35	10	8.02
6.945	8.07	3.05	1.71	11.17	10	8.08
6.910	7.99	3.13	1.73	11.36	10	8.15

IRON (0.76 CRYSTAL DENSITY)

ρ_0 (g/cm^3)	u_s (km/sec)	u_p (km/sec)	P (Mb)	ρ (g/cm^3)	ID	u_s(ID) (km/sec)
5.973	2.24	0.54	0.07	7.86	10	4.48
5.993	2.44	0.65	0.09	8.15	10	4.60
5.884	2.74	0.79	0.13	8.26	10	4.76
5.913	3.03	0.94	0.17	8.58	10	4.95
6.054	3.62	1.12	0.25	8.76	10	5.20
6.041	4.02	1.25	0.30	8.76	10	5.38
6.000	4.68	1.56	0.44	9.01	10	5.79
6.050	4.97	1.70	0.51	9.18	10	5.98
6.104	5.50	2.00	0.67	9.60	10	6.39
5.920	5.73	2.19	0.74	9.58	10	6.60
5.919	5.91	2.28	0.80	9.62	10	6.72
5.820	6.62	2.63	1.01	9.66	10	7.19
6.080	7.22	2.98	1.31	10.37	10	7.71
6.055	7.58	3.33	1.53	10.80	10	8.15

IRON (0.60 CRYSTAL DENSITY)

ρ_0 (g/cm^3)	u_s (km/sec)	u_p (km/sec)	P (Mb)	ρ (g/cm^3)	ID	u_s(ID) (km/sec)
4.808	1.77	0.59	0.05	7.22	10	4.48
4.629	1.93	0.72	0.06	7.39	10	4.60
4.697	2.35	0.86	0.10	7.41	10	4.76
4.761	2.48	0.96	0.11	7.77	10	4.87
4.532	2.44	1.06	0.12	8.00	10	4.95
4.662	2.76	1.13	0.15	7.91	10	5.05
4.744	2.93	1.26	0.18	8.32	10	5.20
4.651	3.20	1.43	0.21	8.40	10	5.38
4.650	4.23	1.74	0.34	7.90	10	5.79
4.720	3.99	1.77	0.33	8.48	10	5.80
4.680	4.29	1.91	0.38	8.43	10	5.98
4.827	4.57	1.96	0.43	8.47	10	6.08
4.729	5.05	2.19	0.52	8.34	10	6.35
4.908	5.08	2.19	0.55	8.64	10	6.39
4.901	5.27	2.38	0.61	8.92	10	6.60
4.734	5.33	2.44	0.62	8.74	10	6.65
4.619	5.43	2.51	0.63	8.60	10	6.72
4.880	6.07	2.85	0.85	9.20	10	7.19
4.770	7.16	3.53	1.20	9.39	10	8.02
4.958	7.35	3.58	1.31	9.66	10	8.15

IRON (0.43 CRYSTAL DENSITY)

ρ_0 (g/cm^3)	u_s (km/sec)	u_p (km/sec)	P (Mb)	ρ (g/cm^3)	ID	u_s(ID) (km/sec)
3.313	1.41	0.64	0.03	6.11	10	4.48
3.222	1.50	0.94	0.05	8.58	10	4.73
3.472	2.14	1.05	0.08	6.79	10	4.87
3.351	1.97	1.07	0.07	7.37	10	4.88
3.388	1.97	1.18	0.08	8.52	10	4.98
3.380	2.30	1.24	0.10	7.38	10	5.05
3.323	2.83	1.65	0.16	7.96	10	5.45
3.538	3.46	1.95	0.24	8.12	10	5.80
3.219	3.59	1.96	0.23	7.10	10	5.79
3.351	3.49	2.01	0.24	7.92	10	5.85
3.716	4.05	2.16	0.32	7.97	10	6.08
3.494	4.65	2.37	0.38	7.12	10	6.31
3.310	4.45	2.48	0.37	7.49	10	6.37
3.133	4.24	2.51	0.33	7.71	10	6.35
3.260	4.73	2.70	0.42	7.59	10	6.60
3.317	4.87	2.73	0.44	7.54	10	6.65
3.260	5.62	3.16	0.58	7.44	10	7.13
3.323	5.63	3.16	0.59	7.59	10	7.15
3.486	6.34	3.47	0.77	7.71	10	7.57
3.424	6.29	3.65	0.79	8.14	10	7.71
3.443	6.82	3.95	0.93	8.17	10	8.07

APPENDIX C

SHOCK WAVE DATA FOR HUGONIOT CROSS CHECKS

921T ALUMINUM ON 2024 ALUMINUM BASE PLATES

ρ_0 (g/cm^3)	u_s (km/sec)	u_p (km/sec)	P (Mb)	ρ (g/cm^3)	ID	u_s(ID) (km/sec)
2.812	5.46	0.35	0.05	3.01	1	5.79
2.812	5.51	0.35	0.05	3.01	1	5.79
2.838	5.69	0.46	0.07	3.09	1	5.94
2.812	5.78	0.46	0.08	3.06	1	5.94
2.812	5.75	0.47	0.08	3.06	1	5.95
2.812	5.71	0.48	0.08	3.07	1	5.96
2.814	5.75	0.49	0.08	3.08	1	5.98
2.812	5.83	0.49	0.08	3.07	1	5.98
2.812	6.03	0.64	0.11	3.15	1	6.18
2.812	6.33	0.86	0.15	3.26	1	6.48
2.842	6.37	0.90	0.16	3.31	1	6.53
2.814	6.42	0.90	0.16	3.27	1	6.53
2.842	6.39	0.91	0.17	3.32	1	6.55
2.842	6.37	0.91	0.17	3.32	1	6.55
2.842	6.76	1.15	0.22	3.43	1	6.87
2.842	6.74	1.15	0.22	3.43	1	6.87
2.814	6.57	1.20	0.22	3.44	1	6.91
2.842	7.35	1.61	0.34	3.64	1	7.48
2.842	7.35	1.61	0.34	3.64	1	7.48
2.814	7.47	1.66	0.35	3.62	1	7.55
2.842	8.45	2.34	0.56	3.93	1	8.48
2.842	8.42	2.34	0.56	3.94	1	8.48
2.803	8.82	2.65	0.65	4.00	1	8.87
2.842	9.34	3.02	0.80	4.20	1	9.40
2.842	9.33	3.03	0.80	4.21	1	9.40
2.842	9.68	3.28	0.90	4.30	1	9.74
2.842	9.67	3.28	0.90	4.30	1	9.74
2.838	9.71	3.38	0.93	4.35	1	9.86
2.838	10.21	3.68	1.07	4.43	1	10.28

COPPER ON 2024 ALUMINUM BASE PLATES

ρ_0	u_s	u_p	P	ρ	ID	u_s(ID)
8.920	4.31	0.21	0.08	9.38	1	5.79
8.925	4.21	0.21	0.08	9.40	1	5.79
8.900	4.22	0.22	0.08	9.40	1	5.81
8.928	4.34	0.28	0.11	9.55	1	5.94
8.925	4.35	0.28	0.11	9.54	1	5.94
8.920	4.32	0.29	0.11	9.55	1	5.95
8.920	4.35	0.29	0.11	9.55	1	5.96
8.925	4.38	0.30	0.12	9.58	1	5.98
8.930	4.30	0.30	0.12	9.60	1	5.98
8.925	4.51	0.39	0.16	9.78	1	6.18
8.928	4.50	0.40	0.16	9.79	1	6.18
8.933	4.47	0.43	0.17	9.89	1	6.25
8.933	4.50	0.44	0.18	9.90	1	6.26
8.900	4.71	0.49	0.20	9.93	1	6.38
8.900	4.69	0.49	0.20	9.94	1	6.38
8.899	4.71	0.52	0.22	10.01	1	6.44
8.924	4.73	0.52	0.22	10.03	1	6.45
8.925	4.80	0.53	0.23	10.03	1	6.47
8.920	4.75	0.53	0.23	10.05	1	6.48
8.925	4.70	0.54	0.23	10.08	1	6.47
8.900	4.71	0.55	0.23	10.07	1	6.49
8.930	4.77	0.56	0.24	10.12	1	6.53
8.918	4.71	0.58	0.24	10.16	1	6.55
8.933	4.85	0.66	0.29	10.34	1	6.72
8.900	5.01	0.69	0.31	10.32	1	6.79
8.900	4.95	0.71	0.31	10.39	1	6.82
8.918	5.04	0.73	0.33	10.42	1	6.87
8.924	5.06	0.75	0.34	10.47	1	6.91
8.930	4.94	0.76	0.33	10.55	1	6.91
8.933	5.07	0.79	0.36	10.57	1	6.98
8.933	5.19	0.80	0.37	10.56	1	7.03
8.933	5.18	0.81	0.37	10.58	1	7.04
8.933	5.22	0.83	0.39	10.63	1	7.09
8.933	5.15	0.87	0.40	10.74	1	7.14
8.933	5.33	0.94	0.45	10.85	1	7.30
8.933	5.33	0.94	0.45	10.86	1	7.31
8.899	5.44	0.99	0.48	10.89	1	7.41
8.900	5.44	1.01	0.49	10.92	1	7.43
8.918	5.51	1.03	0.50	10.96	1	7.48
8.924	5.50	1.06	0.52	11.05	1	7.54
8.930	5.53	1.06	0.52	11.05	1	7.55
8.930	6.03	1.39	0.75	11.62	1	8.21
8.930	6.02	1.40	0.75	11.63	1	8.21
8.924	5.98	1.40	0.75	11.66	1	8.21
8.918	6.92	2.00	1.24	12.56	1	9.40
8.918	7.23	2.17	1.40	12.74	1	9.74
8.929	7.17	2.21	1.41	12.90	1	9.79
8.924	7.65	2.41	1.65	13.04	1	10.23
8.924	7.60	2.46	1.67	13.20	1	10.29
8.924	7.73	2.56	1.77	13.34	1	10.47

IRON ON 2024 ALUMINUM BASE PLATES

ρ_0 (g/cm^3)	u_s (km/sec)	u_p (km/sec)	P (Mb)	ρ (g/cm^3)	ID	u_s(ID) (km/sec)
7.861	4.84	0.57	0.22	8.92	1	6.48
7.840	5.39	0.97	0.41	9.56	1	7.23
7.840	5.37	0.97	0.41	9.57	1	7.23
7.840	5.41	0.99	0.42	9.59	1	7.27
7.840	5.58	1.06	0.46	9.68	1	7.41
7.865	5.52	1.06	0.46	9.74	1	7.41
7.857	5.48	1.07	0.46	9.76	1	7.41
7.861	5.62	1.10	0.49	9.78	1	7.49
7.862	5.97	1.30	0.61	10.05	1	7.88
7.861	6.24	1.46	0.72	10.26	1	8.19
7.861	6.93	1.85	1.01	10.72	1	8.95
7.862	7.15	1.99	1.12	10.89	1	9.22
7.861	7.47	2.07	1.21	10.87	1	9.41
7.861	7.45	2.15	1.26	11.06	1	9.54
7.861	7.80	2.34	1.43	11.23	1	9.90
7.861	7.75	2.35	1.43	11.27	1	9.90
7.861	8.17	2.58	1.66	11.50	1	10.36
7.861	8.11	2.59	1.65	11.56	1	10.36

U- 3 WT PERCENT MO ON 2024 ALUMINUM BASE PLATES

ρ_0 (g/cm^3)	u_s (km/sec)	u_p (km/sec)	P (Mb)	ρ (g/cm^3)	ID	u_s(ID) (km/sec)
18.320	2.96	0.24	0.13	19.94	1	5.98
18.330	3.07	0.31	0.18	20.41	1	6.18
18.480	3.25	0.41	0.25	21.17	1	6.45
18.480	3.20	0.42	0.25	21.24	1	6.45
18.330	3.24	0.43	0.25	21.11	1	6.48
18.450	3.26	0.45	0.27	21.43	1	6.55
18.450	3.44	0.58	0.37	22.18	1	6.87
18.480	3.49	0.59	0.38	22.24	1	6.91
18.480	3.46	0.59	0.38	22.31	1	6.91
18.450	3.80	0.82	0.57	23.49	1	7.48
18.480	3.85	0.83	0.59	23.60	1	7.54
18.480	4.18	1.11	0.86	25.18	1	8.21
18.480	4.68	1.40	1.21	26.34	1	8.95
18.480	5.10	1.55	1.47	26.58	1	9.41
18.480	5.10	1.62	1.53	27.12	1	9.54
18.450	5.30	1.69	1.65	27.10	1	9.74
18.480	5.31	1.75	1.72	27.56	1	9.86
18.480	5.64	1.88	1.96	27.73	1	10.23
18.480	5.60	1.91	1.98	28.05	1	10.28

2024 ALUMINUM ON 921T ALUMINUM BASE PLATES

ρ_0 (g/cm^3)	u_s (km/sec)	u_p (km/sec)	P (Mb)	ρ (g/cm^3)	ID	u_s(ID) (km/sec)
2.784	6.00	0.50	0.08	3.04	2	5.76
2.783	5.95	0.51	0.08	3.04	2	5.77
2.784	5.95	0.51	0.08	3.04	2	5.77
2.785	6.12	0.61	0.10	3.09	2	5.91
2.785	6.10	0.61	0.10	3.09	2	5.91
2.785	6.53	0.91	0.17	3.24	2	6.34
2.785	6.53	0.91	0.17	3.24	2	6.34
2.784	6.56	0.98	0.18	3.27	2	6.43
2.784	6.51	0.99	0.18	3.28	2	6.44
2.780	6.49	0.99	0.18	3.28	2	6.44
2.785	6.95	1.26	0.24	3.40	2	6.83
2.777	6.94	1.26	0.24	3.40	2	6.83
2.784	6.94	1.28	0.25	3.41	2	6.86
2.781	7.61	1.73	0.37	3.60	2	7.49
2.781	7.61	1.73	0.37	3.60	2	7.49
2.784	7.62	1.74	0.37	3.61	2	7.52
2.783	8.70	2.47	0.60	3.88	2	8.54
2.783	8.62	2.48	0.59	3.91	2	8.54
2.781	9.60	3.18	0.85	4.16	2	9.53
2.781	9.55	3.19	0.85	4.17	2	9.53
2.779	9.62	3.27	0.87	4.21	2	9.65
2.781	9.92	3.40	0.94	4.23	2	9.84
2.781	9.87	3.41	0.94	4.25	2	9.84

COPPER ON 921T ALUMINUM BASE PLATES

ρ_0 (g/cm^3)	u_s (km/sec)	u_p (km/sec)	P (Mb)	ρ (g/cm^3)	ID	u_s(ID) (km/sec)
8.930	4.32	0.31	0.12	9.63	2	5.76
8.925	4.53	0.41	0.16	9.81	2	5.98
8.925	4.49	0.41	0.16	9.82	2	5.98
8.930	4.76	0.62	0.26	10.27	2	6.43
8.925	5.08	0.79	0.36	10.56	2	6.80
8.925	5.07	0.79	0.36	10.57	2	6.80
8.930	5.05	0.82	0.37	10.65	2	6.86
8.930	5.05	0.82	0.37	10.65	2	6.86
8.925	5.58	1.11	0.55	11.14	2	7.48
8.925	5.53	1.12	0.55	11.18	2	7.48
8.930	5.53	1.13	0.56	11.23	2	7.52
8.928	6.12	1.46	0.80	11.74	2	8.21
8.928	6.08	1.47	0.80	11.77	2	8.21
8.928	6.56	1.78	1.04	12.25	2	8.84
8.928	6.56	1.78	1.04	12.24	2	8.84
8.928	7.24	2.16	1.40	12.73	2	9.65
8.928	7.14	2.18	1.39	12.84	2	9.65
8.928	7.67	2.48	1.69	13.19	2	10.26
8.928	7.65	2.48	1.69	13.20	2	10.26

IRON ON 921T ALUMINUM BASE PLATES

ρ_0 (g/cm^3)	u_s (km/sec)	u_p (km/sec)	P (Mb)	ρ (g/cm^3)	ID	u_s(ID) (km/sec)
7.861	5.67	1.18	0.53	9.93	2	7.48
7.861	5.66	1.18	0.53	9.94	2	7.48
7.861	6.35	1.54	0.77	10.38	2	8.21
7.861	6.33	1.54	0.77	10.39	2	8.21
7.861	6.86	1.86	1.00	10.78	2	8.84
7.861	6.84	1.86	1.00	10.80	2	8.84
7.861	7.58	2.26	1.35	11.19	2	9.65
7.861	7.55	2.26	1.34	11.22	2	9.65
7.861	8.17	2.56	1.64	11.45	2	10.26
7.861	8.09	2.57	1.64	11.53	2	10.26

U- 3 WT PERCENT MO ON 921T ALUMINUM BASE PLATES

ρ_0 (g/cm^3)	u_s (km/sec)	u_p (km/sec)	P (Mb)	ρ (g/cm^3)	ID	u_s(ID) (km/sec)
18.480	3.05	0.32	0.18	20.68	2	5.98
18.480	3.50	0.62	0.40	22.49	2	6.80
18.480	3.87	0.88	0.63	23.95	2	7.48
18.480	4.29	1.17	0.92	25.38	2	8.21
18.480	4.66	1.41	1.21	26.52	2	8.84
18.480	5.20	1.71	1.65	27.58	2	9.65
18.480	5.60	1.95	2.02	28.37	2	10.26

2024 ALUMINUM ON COPPER BASE PLATES

ρ_0 (g/cm^3)	u_s (km/sec)	u_p (km/sec)	P (Mb)	ρ (g/cm^3)	ID	u_s(ID) (km/sec)
2.784	5.81	0.28	0.04	2.92	10	4.24
2.784	5.78	0.28	0.04	2.93	10	4.24
2.784	6.02	0.44	0.07	3.00	10	4.41
2.782	6.05	0.47	0.08	3.02	10	4.45
2.784	6.23	0.65	0.11	3.11	10	4.64
2.782	6.28	0.68	0.12	3.12	10	4.67
2.785	6.37	0.72	0.13	3.14	10	4.72
2.784	6.32	0.73	0.13	3.15	10	4.72
2.784	6.31	0.73	0.13	3.15	10	4.72
2.783	6.39	0.78	0.14	3.17	10	4.78
2.784	6.52	0.86	0.16	3.21	10	4.87
2.782	6.59	0.89	0.16	3.22	10	4.90
2.784	6.59	0.90	0.16	3.22	10	4.91
2.784	6.58	0.90	0.16	3.22	10	4.91
2.782	6.66	0.97	0.18	3.25	10	4.98
2.782	7.06	1.32	0.26	3.42	10	5.37
2.781	7.09	1.35	0.27	3.44	10	5.40
2.782	7.22	1.38	0.28	3.44	10	5.44
2.785	7.16	1.44	0.29	3.48	10	5.49
2.782	7.21	1.45	0.29	3.48	10	5.51
2.782	7.85	1.86	0.41	3.64	10	5.97
2.782	8.11	2.09	0.47	3.75	10	6.23
2.781	8.08	2.10	0.47	3.76	10	6.22
2.782	8.36	2.33	0.54	3.86	10	6.48
2.785	8.42	2.33	0.55	3.85	10	6.49
2.784	8.44	2.37	0.56	3.87	10	6.53
2.781	8.57	2.45	0.58	3.89	10	6.62
2.782	8.75	2.60	0.63	3.96	10	6.79
2.782	8.91	2.73	0.68	4.01	10	6.93
2.782	8.92	2.74	0.68	4.01	10	6.94
2.782	9.23	2.93	0.75	4.08	10	7.16
2.782	9.77	3.35	0.91	4.23	10	7.62
2.780	9.75	3.36	0.91	4.24	10	7.63
2.780	9.80	3.38	0.92	4.24	10	7.65
2.782	9.75	3.38	0.92	4.26	10	7.65
2.782	9.82	3.39	0.93	4.25	10	7.67
2.780	10.04	3.62	1.01	4.35	10	7.92
2.784	10.24	3.63	1.03	4.31	10	7.95

921T ALUMINUM ON COPPER BASE PLATES

ρ_0 (g/cm^3)	u_s (km/sec)	u_p (km/sec)	P (Mb)	ρ (g/cm^3)	ID	u_s(ID) (km/sec)
2.838	5.52	0.28	0.04	2.99	10	4.24
2.842	5.86	0.47	0.08	3.09	10	4.45
2.814	6.19	0.72	0.13	3.19	10	4.72
2.838	6.17	0.73	0.13	3.22	10	4.72
2.814	6.17	0.73	0.13	3.19	10	4.72
2.838	6.19	0.74	0.13	3.22	10	4.73
2.812	6.29	0.78	0.14	3.21	10	4.78
2.842	6.40	0.89	0.16	3.30	10	4.90
2.838	6.40	0.90	0.16	3.30	10	4.91
2.838	6.40	0.90	0.16	3.30	10	4.91
2.822	6.53	0.97	0.18	3.31	10	4.98
2.842	6.95	1.32	0.26	3.51	10	5.37
2.842	7.05	1.38	0.28	3.54	10	5.44
2.814	7.08	1.44	0.29	3.53	10	5.49
2.822	7.22	1.44	0.29	3.52	10	5.51
2.842	7.70	1.86	0.41	3.75	10	5.97
2.822	8.06	2.09	0.48	3.81	10	6.23
2.842	8.27	2.32	0.55	3.95	10	6.48
2.814	8.36	2.33	0.55	3.90	10	6.49
2.812	8.28	2.51	0.58	4.04	10	6.66
2.842	8.69	2.59	0.64	4.05	10	6.79
2.842	8.82	2.73	0.68	4.11	10	6.93
2.842	8.81	2.73	0.68	4.12	10	6.94
2.822	9.15	2.93	0.76	4.15	10	7.16
2.842	9.75	3.33	0.92	4.32	10	7.62
2.822	9.66	3.37	0.92	4.33	10	7.65
2.842	9.76	3.38	0.94	4.35	10	7.67

IRON ON COPPER BASE PLATES

ρ_0 (g/cm^3)	u_s (km/sec)	u_p (km/sec)	P (Mb)	ρ (g/cm^3)	ID	u_s(ID) (km/sec)
7.864	5.25	0.99	0.41	9.69	10	5.33
7.856	5.41	1.01	0.43	9.65	10	5.37
7.859	5.49	1.06	0.46	9.74	10	5.44
7.861	5.48	1.07	0.46	9.77	10	5.45
7.864	5.52	1.10	0.48	9.81	10	5.49
7.864	5.49	1.10	0.47	9.83	10	5.49
7.861	5.95	1.30	0.61	10.06	10	5.80
7.856	6.20	1.41	0.69	10.17	10	5.97
7.861	6.58	1.70	0.88	10.60	10	6.37
7.856	6.74	1.77	0.94	10.65	10	6.48
7.861	6.74	1.78	0.94	10.68	10	6.49
7.857	7.29	2.07	1.19	10.98	10	6.93
7.856	7.26	2.08	1.19	11.01	10	6.94
7.861	7.49	2.23	1.31	11.19	10	7.15
7.867	7.56	2.25	1.34	11.20	10	7.19
7.867	7.46	2.26	1.33	11.29	10	7.19
7.867	7.62	2.28	1.37	11.23	10	7.24
7.867	7.51	2.30	1.36	11.33	10	7.24
7.859	8.03	2.55	1.61	11.51	10	7.62
7.861	8.05	2.59	1.64	11.59	10	7.67
7.852	8.55	2.83	1.90	11.74	10	8.05

U- 3 WT PERCENT MO ON COPPER BASE PLATES

ρ_0 (g/cm^3)	u_s (km/sec)	u_p (km/sec)	P (Mb)	ρ (g/cm^3)	ID	u_s(ID) (km/sec)
18.480	2.86	0.19	0.10	19.78	10	4.27
18.480	3.03	0.26	0.15	20.25	10	4.41
18.480	3.17	0.39	0.23	21.10	10	4.64
18.480	3.23	0.41	0.25	21.20	10	4.68
18.330	3.29	0.47	0.29	21.40	10	4.78
18.480	3.35	0.53	0.33	21.92	10	4.87
18.450	3.42	0.55	0.35	22.02	10	4.93
18.490	3.79	0.80	0.56	23.41	10	5.37
18.480	3.80	0.82	0.58	23.58	10	5.40
18.450	3.77	0.82	0.57	23.60	10	5.40
18.450	3.78	0.84	0.58	23.70	10	5.43
18.480	3.82	0.84	0.60	23.72	10	5.44
18.480	3.83	0.86	0.61	23.86	10	5.47
18.450	3.91	0.88	0.64	23.82	10	5.51
18.480	4.25	1.14	0.89	25.24	10	5.97
18.480	4.55	1.30	1.09	25.88	10	6.27
18.450	4.57	1.35	1.14	26.19	10	6.35
18.480	4.68	1.42	1.23	26.54	10	6.48
18.480	5.10	1.67	1.57	27.43	10	6.93
18.480	5.09	1.67	1.57	27.53	10	6.94
18.450	5.31	1.80	1.76	27.92	10	7.18
18.480	5.48	1.91	1.93	28.38	10	7.37
18.450	5.47	1.91	1.93	28.38	10	7.38
18.480	5.73	2.00	2.12	28.40	10	7.57
18.480	5.75	2.04	2.16	28.60	10	7.62
18.480	5.72	2.04	2.15	28.71	10	7.61
18.480	5.78	2.07	2.21	28.76	10	7.67
18.480	5.95	2.21	2.42	29.38	10	7.91
18.480	5.88	2.24	2.43	29.84	10	7.94
18.480	6.10	2.29	2.58	29.61	10	8.07
18.320	6.12	2.31	2.59	29.39	10	8.08

2024 ALUMINUM ON IRON BASE PLATES

ρ_0 (g/cm^3)	u_s (km/sec)	u_p (km/sec)	P (Mb)	ρ (g/cm^3)	ID	u_s(ID) (km/sec)
2.784	6.62	0.95	0.18	3.25	12	4.93
2.784	6.62	0.95	0.18	3.25	12	4.93
2.784	7.14	1.36	0.27	3.44	12	5.49
2.784	7.14	1.36	0.27	3.44	12	5.49
2.784	8.15	2.15	0.49	3.78	12	6.53
2.784	8.15	2.15	0.49	3.78	12	6.53
2.784	8.85	2.64	0.65	3.97	12	7.15
2.784	8.80	2.64	0.65	3.98	12	7.15
2.784	9.76	3.24	0.88	4.17	12	7.90
2.784	9.43	3.27	0.86	4.26	12	7.90
2.784	9.76	3.29	0.89	4.20	12	7.96
2.784	9.72	3.30	0.89	4.21	12	7.96
2.784	9.67	3.38	0.91	4.28	12	8.05
2.784	9.61	3.39	0.91	4.30	12	8.05
2.784	10.46	3.77	1.10	4.35	12	8.54
2.784	10.41	3.78	1.09	4.37	12	8.54
2.784	10.43	3.78	1.10	4.36	12	8.54
2.784	10.34	3.79	1.09	4.39	12	8.54

921T ALUMINUM ON IRON BASE PLATES

ρ_0 (g/cm^3)	u_s (km/sec)	u_p (km/sec)	P (Mb)	ρ (g/cm^3)	ID	u_s(ID) (km/sec)
2.838	6.49	0.95	0.18	3.33	12	4.93
2.838	7.08	1.36	0.27	3.51	12	5.49
2.838	8.16	2.14	0.50	3.85	12	6.53
2.838	8.73	2.64	0.65	4.07	12	7.15
2.838	9.59	3.24	0.88	4.28	12	7.90
2.838	9.73	3.28	0.91	4.28	12	7.96
2.838	9.75	3.36	0.93	4.33	12	8.05
2.838	10.35	3.76	1.11	4.46	12	8.54
2.838	10.30	3.77	1.10	4.48	12	8.54

COPPER ON IRON BASE PLATES

ρ_0 (g/cm^3)	u_s (km/sec)	u_p (km/sec)	P (Mb)	ρ (g/cm^3)	ID	u_s(ID) (km/sec)
8.930	5.01	0.68	0.30	10.33	12	4.93
8.930	4.99	0.68	0.30	10.34	12	4.93
8.930	5.47	0.98	0.48	10.88	12	5.49
8.930	5.47	0.98	0.48	10.88	12	5.49
8.930	6.31	1.57	0.89	11.90	12	6.53
8.930	6.87	1.94	1.19	12.44	12	7.15
8.930	6.84	1.94	1.19	12.47	12	7.15
8.929	7.59	2.39	1.62	13.04	12	7.90
8.929	7.65	2.42	1.66	13.07	12	7.96
8.929	7.68	2.48	1.70	13.20	12	8.05
8.929	7.65	2.49	1.70	13.24	12	8.05
8.929	8.23	2.78	2.05	13.50	12	8.54
8.929	8.17	2.79	2.04	13.56	12	8.54
8.929	8.16	2.79	2.03	13.57	12	8.54
8.929	8.14	2.81	2.04	13.64	12	8.56
8.929	8.10	2.82	2.04	13.69	12	8.56
8.929	8.09	2.82	2.04	13.70	12	8.56
8.929	8.08	2.82	2.03	13.72	12	8.56
8.929	8.41	2.92	2.20	13.68	12	8.75
8.929	8.35	2.93	2.18	13.76	12	8.75
8.929	8.32	2.93	2.18	13.79	12	8.75
8.929	8.28	2.94	2.17	13.85	12	8.75

U- 3 WT PERCENT MO ON IRON BASE PLATES

ρ_0 (g/cm^3)	u_s (km/sec)	u_p (km/sec)	P (Mb)	ρ (g/cm^3)	ID	u_s(ID) (km/sec)
18.330	3.49	0.57	0.36	21.89	12	4.93
18.330	3.85	0.82	0.58	23.29	12	5.49
18.330	4.56	1.31	1.10	25.74	12	6.53
18.330	5.04	1.61	1.49	26.95	12	7.15
18.320	5.77	2.05	2.17	28.45	12	8.05
18.320	6.13	2.31	2.59	29.38	12	8.54

2024 AL ON U- 3 WT PERCENT MO BASE PLATES

ρ_0 (g/cm^3)	u_s (km/sec)	u_p (km/sec)	P (Mb)	ρ (g/cm^3)	ID	u_s(ID) (km/sec)
2.782	6.26	0.63	0.11	3.09	6	3.20
2.782	6.23	0.63	0.11	3.09	6	3.20
2.784	6.36	0.79	0.14	3.18	6	3.37
2.782	6.46	0.80	0.14	3.18	6	3.38
2.782	6.39	0.80	0.14	3.18	6	3.38
2.782	6.82	1.08	0.21	3.31	6	3.66
2.784	6.84	1.12	0.21	3.33	6	3.70
2.784	7.01	1.22	0.24	3.37	6	3.81
2.784	6.98	1.22	0.24	3.37	6	3.81
2.785	7.69	1.74	0.37	3.60	6	4.34
2.784	7.76	1.78	0.38	3.61	6	4.38
2.785	7.77	1.94	0.42	3.71	6	4.53
2.782	8.05	1.95	0.44	3.67	6	4.56
2.782	8.01	1.96	0.44	3.68	6	4.56
2.784	9.24	2.97	0.76	4.10	6	5.59
2.784	10.39	3.71	1.07	4.33	6	6.35
2.789	10.14	3.73	1.05	4.41	6	6.35

921T ALUMINUM ON U- 3 WT PERCENT MO BASE PLATES

ρ_0 (g/cm^3)	u_s (km/sec)	u_p (km/sec)	P (Mb)	ρ (g/cm^3)	ID	u_s(ID) (km/sec)
2.812	6.29	0.79	0.14	3.22	6	3.37
2.812	6.30	0.79	0.14	3.22	6	3.37
2.812	6.76	1.12	0.21	3.37	6	3.70
2.812	6.87	1.22	0.24	3.42	6	3.81
2.812	6.88	1.22	0.24	3.42	6	3.81
2.814	7.62	1.74	0.37	3.65	6	4.34
2.814	7.60	1.74	0.37	3.65	6	4.34
2.838	7.79	1.77	0.39	3.67	6	4.38
2.838	7.70	1.77	0.39	3.69	6	4.38
2.814	7.94	1.92	0.43	3.71	6	4.53
2.814	7.88	1.93	0.43	3.72	6	4.53
2.838	9.12	2.97	0.77	4.21	6	5.59
2.838	9.10	2.97	0.77	4.21	6	5.59
2.838	10.07	3.72	1.06	4.50	6	6.35

COPPER ON U- 3 WT PERCENT BASE PLATES

ρ_0 (g/cm^3)	u_s (km/sec)	u_p (km/sec)	P (Mb)	ρ (g/cm^3)	ID	u_s(ID) (km/sec)
8.923	4.59	0.48	0.20	9.97	6	3.20
8.923	4.60	0.48	0.20	9.97	6	3.20
8.930	4.80	0.61	0.26	10.23	6	3.37
8.930	4.79	0.61	0.26	10.23	6	3.37
8.923	4.81	0.62	0.27	10.24	6	3.38
8.923	4.80	0.62	0.26	10.24	6	3.38
8.918	5.24	0.83	0.39	10.60	6	3.66
8.921	5.24	0.83	0.39	10.61	6	3.66
8.930	5.22	0.86	0.40	10.70	6	3.70
8.930	5.21	0.86	0.40	10.71	6	3.70
8.930	5.33	0.94	0.45	10.85	6	3.81
8.930	6.01	1.34	0.72	11.50	6	4.34
8.929	5.98	1.38	0.74	11.61	6	4.38
8.930	6.18	1.50	0.83	11.78	6	4.53
8.920	6.25	1.52	0.85	11.78	6	4.56
8.929	7.36	2.31	1.52	13.02	6	5.59
8.931	7.72	2.46	1.70	13.11	6	5.79
8.930	8.47	2.87	2.17	13.52	6	6.35

IRON ON U- 3 WT PERCENT MO BASE PLATES

ρ_0 (g/cm^3)	u_s (km/sec)	u_p (km/sec)	P (Mb)	ρ (g/cm^3)	ID	u_s(ID) (km/sec)
7.882	5.22	0.90	0.37	9.53	6	3.70
7.882	5.17	0.91	0.37	9.56	6	3.70
7.882	5.34	0.98	0.41	9.66	6	3.81
7.861	6.18	1.39	0.68	10.14	6	4.34
7.861	6.17	1.39	0.68	10.15	6	4.34
7.861	6.25	1.42	0.70	10.17	6	4.38
7.861	6.40	1.54	0.78	10.36	6	4.53
7.861	6.39	1.54	0.78	10.37	6	4.53
7.861	7.74	2.37	1.44	11.33	6	5.59
7.861	7.74	2.37	1.44	11.33	6	5.59
7.861	8.08	2.49	1.59	11.37	6	5.76
7.861	8.66	2.97	2.02	11.96	6	6.35

APPENDIX D

SHOCK WAVE DATA FOR SOLIDS

BERYLLIUM

ρ_0 (g/cm^3)	u_s (km/sec)	u_p (km/sec)	P (Mb)	ρ (g/cm^3)	ID	u_s(ID) (km/sec)
1.850	8.14	0.	0.	1.85	40	0.
1.850	7.93	0.	0.	1.85	30	0.
1.850	8.99	0.34	0.06	1.92	10	4.31
1.850	8.58	0.48	0.08	1.96	1	5.96
1.850	8.67	0.49	0.08	1.96	1	5.98
1.850	8.60	0.50	0.08	1.96	1	5.98
1.850	8.59	0.50	0.08	1.96	1	5.98
1.850	8.78	0.68	0.11	2.01	1	6.22
1.851	9.04	0.88	0.15	2.05	1	6.47
1.851	9.01	0.88	0.15	2.05	1	6.47
1.850	9.05	0.93	0.16	2.06	1	6.53
1.850	9.02	0.93	0.16	2.06	1	6.53
1.850	9.05	0.94	0.16	2.06	1	6.54
1.850	9.05	1.02	0.17	2.08	2	6.43
1.850	9.03	1.02	0.17	2.08	2	6.43
1.850	9.42	1.25	0.22	2.13	1	6.92
1.850	9.06	1.25	0.21	2.15	1	6.91
1.850	9.02	1.26	0.21	2.15	1	6.91
1.850	9.39	1.34	0.23	2.16	2	6.86
1.850	9.36	1.34	0.23	2.16	2	6.86
1.850	9.96	1.75	0.32	2.25	1	7.55
1.850	9.94	1.75	0.32	2.25	1	7.55
1.851	10.00	1.76	0.33	2.25	1	7.56
1.851	9.98	1.76	0.33	2.25	1	7.56
1.851	9.97	1.76	0.33	2.25	1	7.56
1.850	10.01	1.77	0.33	2.25	2	7.43
1.850	9.97	1.78	0.33	2.25	2	7.43
1.850	9.97	1.80	0.33	2.26	1	7.60
1.850	10.01	1.84	0.34	2.27	2	7.52
1.850	9.99	1.85	0.34	2.27	2	7.52
1.850	10.58	2.21	0.43	2.34	1	8.12
1.851	10.60	2.27	0.45	2.36	1	8.19
1.851	10.98	2.62	0.53	2.43	1	8.61
1.851	10.96	2.63	0.53	2.43	1	8.61
1.853	11.07	2.78	0.57	2.47	2	8.72
1.850	11.41	2.92	0.62	2.49	1	8.98
1.851	11.36	2.93	0.62	2.49	1	8.99
1.851	11.29	2.94	0.61	2.50	1	8.99
1.855	11.42	3.16	0.67	2.57	1	9.24
1.851	11.84	3.38	0.74	2.59	2	9.47
1.851	11.78	3.39	0.74	2.60	2	9.47
1.835	11.86	3.44	0.75	2.58	2	9.54
1.851	11.98	3.54	0.79	2.63	1	9.72
1.851	11.99	3.54	0.79	2.63	1	9.72
1.851	12.08	3.72	0.83	2.67	2	9.88
1.851	12.27	3.72	0.84	2.66	1	9.94
1.851	12.31	3.74	0.85	2.66	1	9.96
1.851	12.35	3.87	0.89	2.70	1	10.11

MAGNESIUM

ρ_0 (g/cm^3)	u_s (km/sec)	u_p (km/sec)	P (Mb)	ρ (g/cm^3)	ID	u_s(ID) (km/sec)
1.740	4.44	0.	0.	1.74	40	0.
1.740	4.45	0.	0.	1.74	30	0.
1.745	5.66	0.88	0.09	2.07	1	6.26
1.745	5.65	0.88	0.09	2.07	1	6.26
1.745	5.87	1.08	0.11	2.14	1	6.48
1.735	5.91	1.08	0.11	2.12	1	6.48
1.745	5.86	1.08	0.11	2.14	1	6.48
1.734	6.12	1.20	0.13	2.16	1	6.61
1.734	6.28	1.34	0.15	2.21	1	6.77
1.734	6.24	1.37	0.15	2.22	1	6.80
1.745	6.32	1.47	0.16	2.28	1	6.91
1.770	6.15	1.48	0.16	2.33	1	6.91
1.745	6.27	1.48	0.16	2.28	1	6.91
1.741	6.38	1.58	0.18	2.32	2	6.86
1.745	6.98	1.88	0.23	2.39	1	7.38
1.745	6.89	1.89	0.23	2.40	1	7.38
1.734	7.01	1.93	0.23	2.39	1	7.43
1.734	6.76	1.93	0.23	2.43	1	7.41
1.734	6.87	1.95	0.23	2.42	1	7.43
1.735	7.00	2.03	0.25	2.44	1	7.53
1.745	7.17	2.12	0.27	2.48	1	7.64
1.745	7.13	2.13	0.26	2.49	1	7.64
1.740	7.11	2.14	0.26	2.49	2	7.52
1.734	7.25	2.15	0.27	2.46	1	7.67
1.745	7.27	2.21	0.28	2.51	1	7.74
1.745	7.27	2.21	0.28	2.51	1	7.74
1.745	7.45	2.35	0.31	2.55	1	7.90
1.745	7.52	2.41	0.32	2.57	1	7.97
1.745	7.45	2.42	0.31	2.58	1	7.97

MAGNESIUM *(continued)*

ρ_0 (g/cm^3)	u_s (km/sec)	u_p (km/sec)	P (Mb)	ρ (g/cm^3)	ID	u_s(ID) (km/sec)
1.745	8.24	3.01	0.43	2.75	1	8.64
1.745	8.22	3.01	0.43	2.75	1	8.64
1.734	8.83	3.36	0.51	2.80	1	9.06
1.734	8.85	3.36	0.52	2.79	1	9.06
1.734	8.79	3.37	0.51	2.81	1	9.06
1.734	9.03	3.50	0.55	2.83	1	9.23
1.734	8.93	3.51	0.54	2.86	1	9.22
1.735	9.04	3.72	0.58	2.95	1	9.45
1.745	9.28	3.73	0.60	2.92	1	9.50
1.745	9.15	3.75	0.60	2.96	1	9.50
1.734	9.75	4.02	0.68	2.95	1	9.84
1.734	9.65	4.03	0.68	2.98	1	9.83
1.734	9.54	4.05	0.67	3.01	1	9.83
1.745	9.85	4.25	0.73	3.07	1	10.08
1.745	9.74	4.27	0.73	3.10	1	10.08

AZ-31B MAGNESIUM

ρ_0 (g/cm^3)	u_s (km/sec)	u_p (km/sec)	P (Mb)	ρ (g/cm^3)	ID	u_s(ID) (km/sec)
1.780	4.52	0.	0.	1.78	40	0.
1.770	5.53	0.39	0.04	1.90	10	4.31
1.780	5.30	0.60	0.06	2.01	1	5.96
1.770	5.35	0.60	0.06	1.99	1	5.96
1.770	5.33	0.61	0.06	2.00	1	5.98
1.770	5.36	0.64	0.06	2.01	2	5.76
1.780	5.52	0.82	0.08	2.09	1	6.20
1.770	5.60	0.83	0.08	2.08	1	6.22
1.770	5.95	1.12	0.12	2.18	1	6.53
1.780	5.91	1.12	0.12	2.20	1	6.53
1.770	5.93	1.13	0.12	2.18	1	6.54
1.770	5.96	1.22	0.13	2.22	2	6.43
1.780	6.28	1.42	0.16	2.30	1	6.87
1.770	6.39	1.47	0.17	2.30	1	6.92
1.770	7.00	2.03	0.25	2.49	1	7.55
1.780	7.04	2.04	0.26	2.51	1	7.56
1.770	7.14	2.04	0.26	2.48	2	7.43
1.770	7.09	2.07	0.26	2.50	1	7.60
1.780	7.12	2.12	0.27	2.54	1	7.65
1.770	7.66	2.53	0.34	2.65	1	8.12
1.780	7.70	2.60	0.36	2.69	1	8.19
1.780	8.23	3.02	0.44	2.81	1	8.67
1.780	8.54	3.29	0.50	2.90	1	8.98
1.780	8.72	3.36	0.52	2.90	1	9.08
1.780	9.13	3.71	0.60	3.00	1	9.48
1.780	9.63	4.05	0.69	3.07	1	9.88
1.780	10.00	4.14	0.74	3.04	1	10.03

TITANIUM

ρ_0 (g/cm^3)	u_s (km/sec)	u_p (km/sec)	P (Mb)	ρ (g/cm^3)	ID	u_s(ID) (km/sec)
4.510	5.22	0.	0.	4.51	30	0.
4.517	5.54	0.40	0.10	4.87	1	5.98
4.517	5.58	0.52	0.13	4.98	1	6.18
4.517	5.71	0.55	0.14	5.00	10	4.62
4.539	5.68	0.57	0.15	5.05	1	6.26
4.517	5.71	0.63	0.16	5.08	10	4.71
4.539	5.75	0.71	0.19	5.18	1	6.48
4.539	5.73	0.71	0.19	5.19	1	6.48
4.517	5.69	0.75	0.19	5.21	1	6.53
4.517	5.73	0.78	0.20	5.23	10	4.88
4.539	5.95	1.00	0.27	5.46	1	6.91
4.539	5.89	1.01	0.27	5.48	1	6.91
4.517	5.87	1.03	0.27	5.48	2	6.76
4.517	6.09	1.16	0.32	5.58	10	5.33
4.539	6.29	1.31	0.37	5.74	1	7.38
4.539	6.23	1.32	0.37	5.76	1	7.38
4.517	6.36	1.44	0.41	5.84	2	7.41
4.539	6.45	1.49	0.44	5.90	1	7.64
4.539	6.44	1.49	0.44	5.91	1	7.64
4.539	6.53	1.56	0.46	5.96	1	7.74
4.539	6.50	1.56	0.46	5.97	1	7.74
4.539	6.68	1.66	0.50	6.04	1	7.90
4.539	6.65	1.66	0.50	6.05	1	7.90
4.539	6.69	1.71	0.52	6.09	1	7.97
4.539	6.68	1.71	0.52	6.10	1	7.97
4.517	6.99	1.94	0.61	6.26	2	8.20
4.539	7.14	2.17	0.70	6.52	1	8.64
4.539	7.12	2.17	0.70	6.53	1	8.64

TITANIUM *(continued)*

ρ_0 (g/cm^3)	u_s (km/sec)	u_p (km/sec)	P (Mb)	ρ (g/cm^3)	ID	u_s(ID) (km/sec)
4.510	7.31	2.22	0.73	6.48	11	6.42
4.510	7.25	2.23	0.73	6.51	11	6.42
4.510	7.26	2.23	0.73	6.51	11	6.43
4.517	7.27	2.28	0.75	6.58	2	8.72
4.510	7.51	2.60	0.88	6.89	1	9.24
4.539	7.77	2.75	0.97	7.03	1	9.50
4.539	7.68	2.76	0.96	7.09	1	9.50
4.517	7.83	2.85	1.01	7.11	10	7.35
4.510	7.84	2.89	1.02	7.15	11	7.18
4.510	7.82	2.89	1.02	7.16	11	7.18
4.539	8.29	3.01	1.14	7.13	2	9.88
4.539	8.19	3.15	1.17	7.38	1	10.08
4.539	8.12	3.16	1.17	7.44	1	10.08

ZIRCONIUM

ρ_0	u_s	u_p	P	ρ	ID	u_s(ID)
6.500	3.74	0.	0.	6.50	30	0.
6.510	4.34	0.53	0.15	7.41	10	4.61
6.510	4.28	0.54	0.15	7.44	10	4.62
6.506	4.38	0.60	0.17	7.54	10	4.70
6.510	4.38	0.61	0.17	7.56	10	4.71
6.488	4.44	0.68	0.20	7.67	1	6.49
6.510	4.51	0.74	0.22	7.78	10	4.87
6.510	4.52	0.74	0.22	7.79	10	4.88
6.506	4.54	0.80	0.24	7.90	10	4.95
6.509	4.59	0.81	0.24	7.91	10	4.97
6.488	4.62	0.93	0.28	8.12	1	6.88
6.510	4.73	1.11	0.34	8.52	10	5.33
6.503	4.73	1.13	0.35	8.53	10	5.35
6.510	4.86	1.21	0.38	8.66	10	5.45
6.505	4.87	1.22	0.39	8.69	10	5.47
6.488	4.86	1.24	0.39	8.72	1	7.38
6.505	5.64	1.86	0.68	9.72	10	6.29
6.512	5.67	1.87	0.69	9.71	10	6.31
6.505	5.67	1.91	0.70	9.80	10	6.35
6.509	5.97	2.10	0.82	10.05	10	6.62
6.510	6.11	2.23	0.89	10.25	10	6.77
6.506	6.37	2.42	1.01	10.50	10	7.03
6.510	6.52	2.57	1.09	10.75	10	7.22
6.510	6.73	2.66	1.17	10.77	10	7.35
6.506	6.77	2.68	1.18	10.77	10	7.37
6.505	7.11	3.01	1.39	11.28	10	7.80

HAFNIUM

ρ_0	u_s	u_p	P	ρ	ID	u_s(ID)
12.890	2.91	0.	0.	12.89	30	0.
12.890	2.98	0.	0.	12.89	40	0.
12.870	3.34	0.32	0.14	14.24	10	4.44
12.900	3.43	0.44	0.19	14.79	10	4.62
12.890	3.50	0.50	0.23	15.03	10	4.71
12.900	3.53	0.50	0.23	15.05	10	4.72
12.890	3.62	0.61	0.29	15.51	10	4.88
12.880	3.86	0.92	0.46	16.89	10	5.33
12.900	3.90	0.98	0.49	17.23	10	5.42
12.900	3.90	1.00	0.50	17.37	10	5.45
12.890	3.85	1.02	0.51	17.54	10	5.47
12.900	3.96	1.02	0.52	17.41	10	5.49
12.890	4.25	1.40	0.77	19.20	10	6.04
12.830	4.46	1.55	0.89	19.66	10	6.27
12.890	4.51	1.61	0.93	20.01	10	6.36
12.910	4.65	1.69	1.01	20.27	10	6.49
12.900	4.74	1.77	1.08	20.60	10	6.62
12.840	4.88	1.88	1.18	20.87	10	6.77
12.920	5.21	2.17	1.46	22.13	10	7.22
12.890	5.44	2.23	1.56	21.86	10	7.35
12.910	5.42	2.25	1.58	22.08	10	7.37
12.890	5.52	2.42	1.72	22.96	10	7.61

VANADIUM

ρ_0	u_s	u_p	P	ρ	ID	u_s(ID)
6.100	5.18	0.	0.	6.10	30	0.
6.100	5.71	0.56	0.19	6.76	1	6.39
6.100	5.66	0.56	0.19	6.77	1	6.39
6.100	6.08	0.78	0.29	6.99	1	6.81
6.100	6.00	0.79	0.29	7.02	1	6.81
6.100	6.00	0.79	0.29	7.02	1	6.81
6.100	6.00	0.79	0.29	7.02	1	6.81
6.100	5.98	0.79	0.29	7.03	1	6.81
6.100	6.42	1.09	0.43	7.34	1	7.35
6.100	6.41	1.09	0.43	7.35	1	7.35
6.100	6.38	1.09	0.43	7.36	1	7.35
6.100	7.25	1.80	0.79	8.11	11	6.18
6.100	7.23	1.80	0.79	8.12	11	6.18
6.100	7.20	1.80	0.79	8.13	11	6.18
6.100	7.19	1.80	0.79	8.14	11	6.18
6.100	8.07	2.43	1.19	8.72	11	7.01
6.100	8.10	2.52	1.25	8.86	11	7.12

NIOBIUM

ρ_0 (g/cm^3)	u_s (km/sec)	u_p (km/sec)	P (Mb)	ρ (g/cm^3)	ID	u_s(ID) (km/sec)
8.580	4.43	0.	0.	8.58	30	0.
8.580	4.39	0.	0.	8.58	40	0.
8.583	5.12	0.50	0.22	9.50	10	4.70
8.604	5.11	0.53	0.23	9.59	1	6.49
8.586	5.26	0.67	0.30	9.85	10	4.95
8.604	5.25	0.72	0.33	9.97	1	6.86
8.584	5.60	0.94	0.45	10.32	10	5.35
8.604	5.57	0.97	0.46	10.41	1	7.36
8.583	5.71	1.02	0.50	10.46	10	5.47
8.584	6.36	1.60	0.87	11.48	10	6.29
8.580	6.34	1.61	0.88	11.50	10	6.30
8.578	6.38	1.61	0.88	11.47	10	6.31
8.584	7.01	2.11	1.27	12.28	10	7.03
8.571	7.31	2.36	1.48	12.66	10	7.39
8.585	7.41	2.46	1.56	12.85	10	7.52
8.584	7.62	2.65	1.74	13.17	10	7.80

TANTALUM

ρ_0	u_s	u_p	P	ρ	ID	u_s(ID)
16.660	3.41	0.	0.	16.66	30	0.
16.660	3.39	0.	0.	16.66	40	0.
16.660	3.92	0.40	0.26	18.57	10	4.70
16.640	4.07	0.55	0.37	19.23	10	4.95
16.650	4.33	0.77	0.56	20.28	10	5.35
16.660	4.43	0.84	0.62	20.57	10	5.47
16.650	4.99	1.32	1.10	22.67	10	6.29
16.660	5.02	1.32	1.11	22.63	10	6.30
16.660	5.54	1.75	1.62	24.37	10	7.03
16.640	5.80	1.96	1.89	25.14	10	7.39
16.660	5.88	2.04	2.00	25.54	10	7.52
16.650	6.01	2.22	2.22	26.38	10	7.80

CHROMIUM

ρ_0	u_s	u_p	P	ρ	ID	u_s(ID)
7.100	5.15	0.	0.	7.10	30	0.
7.130	5.85	0.52	0.22	7.82	1	6.43
7.130	5.97	0.52	0.22	7.82	1	6.46
7.130	6.28	0.72	0.32	8.05	1	6.85
7.130	6.28	0.72	0.32	8.05	1	6.85
7.130	6.30	0.72	0.32	8.05	1	6.86
7.130	6.29	0.72	0.32	8.06	1	6.86
7.130	6.58	0.98	0.46	8.38	1	7.36
7.130	6.59	0.98	0.46	8.38	1	7.36
7.123	7.20	1.44	0.74	8.90	10	6.04
7.100	7.54	1.65	0.89	9.09	11	6.17
7.100	7.50	1.66	0.88	9.11	11	6.18
7.100	8.34	2.19	1.30	9.63	11	6.93
7.100	8.53	2.19	1.33	9.56	11	6.96
7.100	8.47	2.21	1.33	9.61	11	6.97

MOLYBDENUM

ρ_0	u_s	u_p	P	ρ	ID	u_s(ID)
10.200	5.19	0.	0.	10.20	30	0.
10.200	5.04	0.	0.	10.20	40	0.
10.200	5.63	0.42	0.24	11.02	1	6.44
10.200	5.58	0.43	0.24	11.05	1	6.45
10.220	5.66	0.45	0.26	11.10	10	4.72
10.210	5.67	0.46	0.27	11.11	10	4.74
10.200	5.88	0.57	0.34	11.30	1	6.81
10.200	5.79	0.59	0.35	11.35	1	6.83
10.210	5.88	0.59	0.36	11.36	10	4.96
10.200	6.05	0.77	0.48	11.69	1	7.25
10.200	6.14	0.83	0.52	11.80	1	7.38
10.210	6.22	0.86	0.54	11.84	10	5.39
10.220	6.23	0.87	0.55	11.87	10	5.41
10.210	6.23	0.87	0.55	11.86	10	5.41
10.210	6.24	0.89	0.57	11.90	10	5.44
10.210	6.70	1.26	0.86	12.58	10	6.04
10.210	6.90	1.48	1.04	12.99	10	6.36
10.210	7.05	1.56	1.12	13.11	10	6.49
10.200	7.20	1.63	1.20	13.19	11	6.45
10.200	7.21	1.64	1.20	13.20	11	6.46
10.200	7.11	1.65	1.20	13.27	11	6.46
10.210	7.25	1.69	1.25	13.32	10	6.70
10.210	7.28	1.75	1.30	13.43	10	6.78
10.210	7.57	1.99	1.54	13.85	10	7.15
10.200	7.56	2.00	1.54	13.88	11	7.00
10.200	7.62	2.00	1.56	13.84	11	7.01
10.200	7.66	2.01	1.57	13.83	11	7.02
10.210	8.01	2.39	1.96	14.56	10	7.75
10.210	8.05	2.44	2.01	14.66	10	7.82
10.210	8.22	2.49	2.09	14.64	10	7.91

TUNGSTEN

ρ_0 (g/cm^3)	u_s (km/sec)	u_p (km/sec)	P (Mb)	ρ (g/cm^3)	ID	u_s(ID) (km/sec)
19.300	4.05	0.	0.	19.30	30	0.
19.250	4.43	0.32	0.27	20.73	10	4.64
19.250	4.45	0.36	0.31	20.93	10	4.72
19.250	4.52	0.36	0.32	20.94	10	4.74
19.270	4.52	0.37	0.32	20.97	10	4.75
19.170	4.51	0.44	0.38	21.23	1	6.80
19.170	4.50	0.44	0.38	21.23	1	6.80
19.250	4.65	0.47	0.43	21.44	10	4.96
19.170	4.77	0.62	0.57	22.05	1	7.34
19.170	4.73	0.62	0.57	22.09	1	7.34
19.250	4.92	0.69	0.66	22.40	10	5.39
19.250	4.92	0.70	0.66	22.44	10	5.41
19.250	4.92	0.70	0.66	22.44	10	5.41
19.250	4.95	0.72	0.68	22.51	10	5.44
19.170	5.42	1.13	1.17	24.21	11	6.07
19.170	5.40	1.13	1.17	24.25	11	6.07
19.250	5.80	1.38	1.55	25.28	10	6.70
19.250	5.84	1.43	1.61	25.49	10	6.78
19.250	6.05	1.64	1.91	26.39	10	7.15
19.250	6.05	1.64	1.91	26.42	10	7.15
19.170	6.12	1.68	1.98	26.45	11	7.08
19.170	6.14	1.69	1.98	26.43	11	7.09
19.170	6.17	1.69	2.00	26.39	11	7.10
19.250	6.42	1.98	2.45	27.84	10	7.75
19.250	6.45	2.02	2.51	28.05	10	7.82
19.250	6.56	2.06	2.61	28.09	10	7.91
19.250	6.65	2.07	2.65	27.95	10	7.94

RHENIUM

ρ_0 (g/cm^3)	u_s (km/sec)	u_p (km/sec)	P (Mb)	ρ (g/cm^3)	ID	u_s(ID) (km/sec)
21.020	4.16	0.	0.	21.02	35	0.
20.990	4.69	0.33	0.32	22.56	10	4.72
20.960	4.74	0.43	0.43	23.07	10	4.94
20.980	5.04	0.66	0.70	24.17	10	5.42
20.990	5.19	0.69	0.75	24.21	10	5.49
20.990	5.51	0.96	1.11	25.42	10	6.04
20.990	5.71	1.13	1.35	26.15	10	6.36

RHENIUM (0.98 CRYSTAL DENSITY)

ρ_0 (g/cm^3)	u_s (km/sec)	u_p (km/sec)	P (Mb)	ρ (g/cm^3)	ID	u_s(ID) (km/sec)
20.530	4.10	0.	0.	20.53	40	0.
20.530	4.15	0.21	0.18	21.60	10	4.41
20.530	4.08	0.37	0.31	22.59	10	4.74
20.530	4.29	0.48	0.42	23.11	10	4.96
20.530	4.77	0.68	0.66	23.94	10	5.39
20.530	4.79	0.72	0.70	24.14	10	5.46
20.530	5.74	1.35	1.59	26.82	10	6.70
20.530	5.87	1.38	1.66	26.82	10	6.78
20.530	5.81	1.44	1.72	27.30	10	6.87
20.530	6.04	1.58	1.97	27.83	10	7.15
20.530	6.15	1.72	2.18	28.53	10	7.39
20.530	6.69	2.03	2.79	29.45	10	7.99

RHODIUM

ρ_0 (g/cm^3)	u_s (km/sec)	u_p (km/sec)	P (Mb)	ρ (g/cm^3)	ID	u_s(ID) (km/sec)
12.420	5.40	0.37	0.25	13.33	1	5.68
12.430	5.36	0.43	0.28	13.50	10	4.74
12.430	5.47	0.54	0.37	13.80	10	4.94
12.420	5.79	0.71	0.51	14.17	1	6.02
12.430	5.86	0.80	0.58	14.39	10	5.39
12.430	5.88	0.84	0.62	14.51	10	5.46
12.430	6.26	1.06	0.82	14.95	10	5.84
12.420	6.04	1.21	0.91	15.53	10	6.04
12.430	7.03	1.57	1.37	15.99	10	6.70
12.430	7.11	1.61	1.42	16.07	10	6.78
12.430	7.08	1.68	1.48	16.29	10	6.87
12.430	7.39	1.84	1.69	16.55	10	7.15
12.430	7.47	2.00	1.86	16.99	10	7.39
12.430	7.80	2.19	2.12	17.28	10	7.71

PALLADIUM

ρ_0 (g/cm^3)	u_s (km/sec)	u_p (km/sec)	P (Mb)	ρ (g/cm^3)	ID	u_s(ID) (km/sec)
12.000	3.80	0.	0.	12.00	30	0.
12.000	3.90	0.	0.	12.00	40	0.
12.000	4.67	0.44	0.25	13.25	10	4.70
11.880	4.62	0.45	0.25	13.18	1	6.48
12.000	4.92	0.59	0.35	13.64	10	4.95
11.950	4.94	0.60	0.36	13.61	1	6.86
12.020	5.31	0.80	0.51	14.16	1	7.35
12.000	5.32	0.82	0.53	14.20	10	5.35
12.010	5.44	0.90	0.59	14.38	10	5.47
12.000	6.18	1.40	1.04	15.51	10	6.29
12.000	6.21	1.40	1.04	15.49	10	6.30
12.000	6.21	1.40	1.05	15.50	10	6.31
12.000	6.88	1.84	1.52	16.40	10	7.03
12.000	7.19	2.07	1.78	16.84	10	7.39
12.000	7.29	2.15	1.88	17.03	10	7.52
12.000	7.58	2.32	2.11	17.28	10	7.80

IRIDIUM

ρ_0 (g/cm^3)	u_s (km/sec)	u_p (km/sec)	P (Mb)	ρ (g/cm^3)	ID	u_s(ID) (km/sec)
22.420	4.18	0.19	0.18	23.50	10	4.41
22.500	4.90	0.67	0.74	26.05	10	5.46
22.420	5.44	0.93	1.14	27.06	10	6.04
22.500	5.91	1.36	1.81	29.21	10	6.87
22.500	6.26	1.63	2.29	30.41	10	7.39
22.500	6.73	1.93	2.92	31.54	10	7.99

PLATINUM

ρ_0 (g/cm^3)	u_s (km/sec)	u_p (km/sec)	P (Mb)	ρ (g/cm^3)	ID	u_s(ID) (km/sec)
21.430	3.64	0.	0.	21.43	30	0.
21.430	3.51	0.	0.	21.43	40	0.
21.370	4.15	0.32	0.28	23.13	1	6.48
21.430	4.12	0.34	0.30	23.35	10	4.70
21.370	4.25	0.44	0.40	23.84	1	6.86
21.430	4.30	0.46	0.42	24.00	10	4.95
21.430	4.34	0.47	0.43	24.01	10	4.97
21.370	4.44	0.60	0.57	24.68	1	7.32
21.430	4.61	0.65	0.64	24.94	10	5.35
21.430	4.71	0.71	0.72	25.25	10	5.48
21.430	5.34	1.12	1.28	27.11	10	6.29
21.430	5.39	1.12	1.30	27.07	10	6.31
21.430	5.36	1.15	1.32	27.28	10	6.35
21.430	5.93	1.49	1.90	28.64	10	7.03
21.430	6.22	1.66	2.22	29.25	10	7.37
21.430	6.28	1.75	2.36	29.72	10	7.52
21.430	6.45	1.90	2.63	30.36	10	7.80

SILICON CARBIDE (RHO 0= APPROX 3.1 G/CM**3

ρ_0 (g/cm^3)	u_s (km/sec)	u_p (km/sec)	P (Mb)	ρ (g/cm^3)	ID	u_s(ID) (km/sec)
3.120	8.00	0.	0.	3.12	40	0.
3.124	10.77	0.46	0.16	3.26	1	6.22
3.114	10.05	0.67	0.21	3.34	1	6.52
3.127	9.99	0.92	0.29	3.44	1	6.90
3.120	9.88	1.37	0.42	3.62	1	7.53
3.110	9.56	1.56	0.47	3.72	1	7.76
3.128	9.83	1.67	0.51	3.77	1	7.93
3.116	9.78	1.93	0.59	3.88	1	8.26
3.126	10.03	2.16	0.68	3.98	1	8.59
3.123	10.42	2.47	0.80	4.09	1	9.03
3.115	10.65	2.79	0.93	4.22	1	9.45
3.124	10.83	2.83	0.96	4.23	1	9.53
3.113	10.70	3.00	1.00	4.32	1	9.71
3.129	10.59	3.20	1.06	4.49	1	9.95
3.129	10.59	3.20	1.06	4.49	1	9.95
3.124	10.84	3.24	1.10	4.45	1	10.03
3.113	10.56	3.56	1.17	4.69	10	8.02

SILICON CARBIDE (RHO 0= APPROX 2.3 G/CM**3

ρ_0 (g/cm^3)	u_s (km/sec)	u_p (km/sec)	P (Mb)	ρ (g/cm^3)	ID	u_s(ID) (km/sec)
2.368	5.93	2.05	0.29	3.62	1	7.65
2.295	6.34	2.24	0.33	3.55	1	7.88
2.364	6.91	2.44	0.40	3.66	1	8.19
2.423	7.19	2.48	0.43	3.70	1	8.30
2.307	6.86	2.52	0.40	3.65	1	8.24
2.311	7.42	2.86	0.49	3.76	1	8.67
2.292	8.14	3.15	0.59	3.74	1	9.08
2.308	8.37	3.19	0.62	3.73	1	9.16
2.302	8.21	3.27	0.62	3.83	1	9.22
2.357	8.51	3.44	0.69	3.96	1	9.48
2.302	8.82	3.82	0.78	4.06	1	9.88

BORON CARBIDE

ρ_0 (g/cm^3)	u_s (km/sec)	u_p (km/sec)	P (Mb)	ρ (g/cm^3)	ID	u_s(ID) (km/sec)
1.980	5.46	1.47	0.16	2.71	1	6.90
1.948	6.72	2.10	0.27	2.83	1	7.65
1.924	7.03	2.30	0.31	2.86	1	7.88
1.970	7.77	2.50	0.38	2.90	1	8.19
1.889	7.50	2.62	0.37	2.90	1	8.24
1.944	8.43	2.90	0.48	2.97	1	8.67
1.937	9.02	3.22	0.56	3.01	1	9.08
1.909	9.01	3.38	0.58	3.06	1	9.22
1.940	9.45	3.56	0.65	3.11	1	9.48
1.903	9.78	3.94	0.73	3.19	1	9.88

TUNGSTEN CARBIDE

ρ_0 (g/cm^3)	u_s (km/sec)	u_p (km/sec)	P (Mb)	ρ (g/cm^3)	ID	u_s(ID) (km/sec)
15.020	4.92	0.	0.	15.02	40	0.
15.050	5.57	0.22	0.18	15.67	10	4.44
15.010	5.71	0.35	0.30	15.99	10	4.71
15.010	5.67	0.37	0.31	16.05	10	4.74
15.000	5.73	0.44	0.38	16.24	10	4.87
15.060	5.72	0.44	0.38	16.32	10	4.88
15.020	5.96	0.68	0.61	16.95	10	5.33
14.990	6.01	0.71	0.64	17.01	10	5.39
14.990	5.97	0.75	0.67	17.14	10	5.45
14.990	6.86	1.44	1.49	18.99	10	6.70
15.030	6.93	1.48	1.55	19.13	10	6.77
15.020	6.91	1.49	1.55	19.14	10	6.78
15.000	7.11	1.71	1.83	19.76	10	7.15
15.010	7.17	1.75	1.89	19.86	10	7.22
15.010	7.33	1.82	2.00	19.96	10	7.35

STAINLESS STEEL (TYPE 304)

ρ_0 (g/cm^3)	u_s (km/sec)	u_p (km/sec)	P (Mb)	ρ (g/cm^3)	ID	u_s(ID) (km/sec)
7.890	4.51	0.	0.	7.89	40	0.
7.890	4.92	0.23	0.09	8.28	10	4.29
7.890	5.06	0.34	0.14	8.46	10	4.45
7.890	5.35	0.53	0.22	8.75	10	4.73
7.890	5.58	0.66	0.29	8.95	10	4.92
7.890	5.65	0.74	0.33	9.09	10	5.04
7.890	5.89	0.89	0.41	9.29	10	5.26
7.890	6.01	0.97	0.46	9.41	10	5.37
7.890	6.08	1.01	0.48	9.46	10	5.44
7.890	6.15	1.06	0.51	9.53	10	5.51
7.890	6.64	1.38	0.72	9.97	10	5.98
7.890	6.73	1.41	0.75	9.98	10	6.03
7.890	6.73	1.41	0.75	9.98	10	6.03
7.890	7.01	1.65	0.91	10.33	10	6.37
7.890	7.46	1.91	1.13	10.61	10	6.76
7.890	8.05	2.33	1.48	11.11	10	7.37
7.903	8.56	2.63	1.78	11.40	10	7.81
7.890	8.60	2.71	1.84	11.52	10	7.91
7.890	8.67	2.77	1.90	11.60	10	8.00
7.903	8.62	2.79	1.90	11.70	10	8.13

STAINLESS STEEL (TYPE 304L)

ρ_0 (g/cm^3)	u_s (km/sec)	u_p (km/sec)	P (Mb)	ρ (g/cm^3)	ID	u_s(ID) (km/sec)
7.903	4.49	0.	0.	7.90	40	0.
7.903	4.90	0.23	0.09	8.30	10	4.29
7.903	5.10	0.34	0.14	8.46	10	4.45
7.903	5.35	0.53	0.22	8.77	10	4.73
7.903	5.55	0.66	0.29	8.97	10	4.92
7.903	5.65	0.74	0.33	9.10	10	5.04
7.903	5.86	0.89	0.41	9.32	10	5.26
7.903	6.02	0.97	0.46	9.42	10	5.37
7.903	6.06	1.01	0.48	9.49	10	5.44
7.903	6.26	1.14	0.56	9.66	10	5.62
7.903	6.60	1.38	0.72	10.00	10	5.98
7.903	6.68	1.41	0.75	10.02	10	6.03
7.903	6.98	1.65	0.91	10.36	10	6.37
7.903	7.46	1.91	1.13	10.63	10	6.76
7.903	8.05	2.33	1.48	11.13	10	7.37
7.903	8.57	2.71	1.84	11.56	10	7.91

STAINLESS STEEL (TYPE 347)

ρ_0 (g/cm^3)	u_s (km/sec)	u_p (km/sec)	P (Mb)	ρ (g/cm^3)	ID	u_s(ID) (km/sec)
7.920	4.55	0.	0.	7.92	40	0.
7.920	5.60	0.72	0.32	9.08	1	6.84
7.920	6.03	0.96	0.46	9.43	1	7.34

ADIPRENE, BULK

ρ_0 (g/cm^3)	u_s (km/sec)	u_p (km/sec)	P (Mb)	ρ (g/cm^3)	ID	u_s(ID) (km/sec)
1.096	4.56	1.42	0.07	1.59	1	6.56
1.095	4.99	1.74	0.09	1.68	1	6.86
1.091	6.24	2.61	0.18	1.87	1	7.71
1.094	7.93	3.61	0.31	2.01	1	8.75

NEOPRENE

ρ_0 (g/cm^3)	u_s (km/sec)	u_p (km/sec)	P (Mb)	ρ (g/cm^3)	ID	u_s(ID) (km/sec)
1.439	4.37	1.14	0.07	1.95	1	6.38
1.438	5.01	1.53	0.11	2.07	1	6.78
1.439	5.72	2.08	0.17	2.26	1	7.34
1.440	5.76	2.09	0.17	2.26	1	7.36
1.439	5.74	2.10	0.17	2.27	1	7.36

SILASTIC RUBBER (RTV-521)

ρ_0 (g/cm^3)	u_s (km/sec)	u_p (km/sec)	P (Mb)	ρ (g/cm^3)	ID	u_s(ID) (km/sec)
1.370	3.69	1.40	0.07	2.21	1	6.55
1.380	4.17	1.77	0.10	2.40	1	6.91
1.380	5.48	2.37	0.18	2.44	1	7.56
1.370	7.03	3.50	0.34	2.73	1	8.74
1.370	7.46	3.73	0.38	2.74	1	9.00
1.360	7.42	3.78	0.38	2.77	1	9.03
1.380	7.54	3.88	0.40	2.84	1	9.16
1.370	8.31	4.56	0.52	3.03	1	9.86
1.370	8.10	4.66	0.52	3.23	1	9.93

POLYRUBBER (RHO 0= 1.01 G/CM**3)

ρ_0 (g/cm^3)	u_s (km/sec)	u_p (km/sec)	P (Mb)	ρ (g/cm^3)	ID	u_s(ID) (km/sec)
1.012	3.23	1.28	0.04	1.68	1	6.36
1.014	3.33	1.30	0.04	1.66	1	6.38
1.015	4.12	1.73	0.07	1.75	1	6.78
1.009	4.06	1.78	0.07	1.80	1	6.81
1.019	5.19	2.30	0.12	1.83	1	7.33

POLYRUBBER (RHO 0= 0.77 G/CM**3)

ρ_0 (g/cm^3)	u_s (km/sec)	u_p (km/sec)	P (Mb)	ρ (g/cm^3)	ID	u_s(ID) (km/sec)
0.784	2.78	1.37	0.03	1.55	1	6.37
0.771	2.80	1.38	0.03	1.53	1	6.38
0.778	3.66	1.84	0.05	1.57	1	6.78
0.775	3.64	1.90	0.05	1.62	1	6.82
0.775	4.34	2.51	0.08	1.83	1	7.33
0.775	4.56	2.52	0.09	1.73	1	7.36
0.769	4.46	2.54	0.09	1.78	1	7.37

POLYURETHANE (RHO 0= 1.265 G/CM**3)

ρ_0 (g/cm^3)	u_s (km/sec)	u_p (km/sec)	P (Mb)	ρ (g/cm^3)	ID	u_s(ID) (km/sec)
1.265	2.07	0.	0.	1.27	40	0.
1.265	4.15	1.05	0.06	1.69	1	6.26
1.265	4.14	1.05	0.06	1.70	1	6.26
1.265	4.50	1.28	0.07	1.77	1	6.48
1.265	4.48	1.28	0.07	1.77	1	6.48
1.265	5.20	1.71	0.11	1.89	1	6.91
1.265	5.20	1.71	0.11	1.89	1	6.91
1.265	5.95	2.17	0.16	1.99	1	7.38
1.265	5.90	2.18	0.16	2.00	1	7.38
1.265	6.34	2.43	0.19	2.05	1	7.64
1.265	6.32	2.43	0.19	2.06	1	7.64
1.265	6.46	2.53	0.21	2.08	1	7.74
1.265	6.44	2.53	0.21	2.08	1	7.74
1.265	6.68	2.68	0.23	2.11	1	7.90
1.265	6.65	2.68	0.23	2.12	1	7.90
1.265	6.75	2.74	0.23	2.13	1	7.97
1.265	7.61	3.39	0.33	2.28	1	8.64
1.265	7.59	3.40	0.33	2.29	1	8.64
1.265	8.42	4.15	0.44	2.49	1	9.42
1.265	8.42	4.15	0.44	2.50	1	9.42
1.265	9.28	4.76	0.56	2.60	1	10.08
1.265	9.20	4.78	0.56	2.63	1	10.08

POLYURETHANE (RHO 0= 0.35 TO 0.48 G/CM**3)

ρ_0 (g/cm^3)	u_s (km/sec)	u_p (km/sec)	P (Mb)	ρ (g/cm^3)	ID	u_s(ID) (km/sec)
0.476	2.34	1.62	0.02	1.54	1	6.48
0.381	2.55	1.62	0.02	1.05	1	6.48
0.355	3.13	2.15	0.02	1.13	1	6.86

POLYURETHANE (RHO 0= 0.31 TO 0.34 G/CM**3)

ρ_0 (g/cm^3)	u_s (km/sec)	u_p (km/sec)	P (Mb)	ρ (g/cm^3)	ID	u_s(ID) (km/sec)
0.310	2.52	1.68	0.01	0.93	1	6.51
0.339	3.12	2.15	0.02	1.09	1	6.86
0.308	3.21	2.16	0.02	0.95	1	6.86
0.333	3.25	2.16	0.02	1.00	1	6.87
0.319	3.88	2.72	0.03	1.07	1	7.29
0.320	3.68	2.74	0.03	1.25	1	7.29
0.315	3.75	2.74	0.03	1.17	1	7.29

POLYURETHANE (RHO 0= 0.25 TO 0.30 G/CM**3)

ρ_0 (g/cm^3)	u_s (km/sec)	u_p (km/sec)	P (Mb)	ρ (g/cm^3)	ID	u_s(ID) (km/sec)
0.289	3.33	2.19	0.02	0.84	1	6.88
0.267	3.08	2.21	0.02	0.94	1	6.88
0.250	3.05	2.22	0.02	0.92	1	6.88
0.296	3.79	2.75	0.03	1.07	1	7.29
0.286	3.75	2.85	0.03	1.19	1	7.36
0.285	3.78	2.86	0.03	1.17	1	7.36
0.282	3.78	2.86	0.03	1.16	1	7.36
0.273	3.68	2.87	0.03	1.24	1	7.36
0.285	4.00	2.87	0.03	1.01	1	7.38
0.289	4.00	2.87	0.03	1.03	1	7.38

POLYURETHANE (RHO 0= 0.14 TO 0.21 G/CM**3)

ρ_0 (g/cm^3)	u_s (km/sec)	u_p (km/sec)	P (Mb)	ρ (g/cm^3)	ID	u_s(ID) (km/sec)
0.145	2.23	1.73	0.01	0.65	1	6.51
0.150	2.26	1.74	0.01	0.65	1	6.51
0.149	2.26	1.74	0.01	0.64	1	6.51
0.147	2.23	1.74	0.01	0.67	1	6.52
0.145	2.24	1.75	0.01	0.66	1	6.52
0.159	2.95	2.21	0.01	0.64	1	6.85
0.159	3.06	2.22	0.01	0.58	1	6.86
0.157	3.01	2.22	0.01	0.60	1	6.86
0.149	2.94	2.23	0.01	0.62	1	6.86
0.202	3.06	2.23	0.01	0.75	1	6.88
0.178	3.75	2.88	0.02	0.77	1	7.33
0.158	3.74	2.88	0.02	0.68	1	7.32
0.160	3.78	2.98	0.02	0.75	1	7.39
0.175	3.66	2.98	0.02	0.95	1	7.39
0.156	3.63	2.98	0.02	0.87	1	7.39
0.159	3.68	3.00	0.02	0.85	1	7.40
0.159	3.61	3.01	0.02	0.95	1	7.40

POLYURETHANE (RHO 0= 0.07 TO 0.11 G/CM**3)

ρ_0 (g/cm^3)	u_s (km/sec)	u_p (km/sec)	P (Mb)	ρ (g/cm^3)	ID	u_s(ID) (km/sec)
0.107	2.30	1.73	0.00	0.44	1	6.50
0.107	2.32	1.77	0.00	0.46	1	6.53
0.093	2.97	2.26	0.01	0.39	1	6.86
0.098	3.00	2.26	0.01	0.40	1	6.87
0.095	2.87	2.27	0.01	0.46	1	6.87
0.091	2.92	2.29	0.01	0.42	1	6.88
0.099	3.41	3.05	0.01	0.94	1	7.40
0.094	3.37	3.05	0.01	1.01	1	7.40
0.077	3.36	3.07	0.01	0.90	1	7.40
0.081	3.49	3.07	0.01	0.68	1	7.41
0.077	3.39	3.08	0.01	0.82	1	7.41

POLYETHYLENE

ρ_0 (g/cm^3)	u_s (km/sec)	u_p (km/sec)	P (Mb)	ρ (g/cm^3)	ID	u_s(ID) (km/sec)
0.922	1.89	0.	0.	0.92	36	0.
0.967	2.22	0.	0.	0.97	36	0.
0.913	4.70	1.17	0.05	1.22	1	6.32
0.920	4.98	1.40	0.06	1.28	1	6.53
0.916	5.01	1.48	0.07	1.30	2	6.40
0.916	5.01	1.49	0.07	1.30	2	6.40
0.913	5.43	1.72	0.09	1.34	1	6.81
0.916	5.60	1.77	0.09	1.34	1	6.87
0.916	5.63	1.83	0.09	1.36	2	6.74
0.916	5.57	1.83	0.09	1.37	2	6.74
0.913	6.17	2.23	0.13	1.43	1	7.29
0.908	6.15	2.35	0.13	1.47	1	7.40
0.913	6.49	2.41	0.14	1.45	1	7.47
0.916	6.54	2.44	0.15	1.46	1	7.50
0.916	6.58	2.46	0.15	1.46	2	7.38
0.916	6.58	2.46	0.15	1.46	2	7.38
0.916	6.67	2.49	0.15	1.46	2	7.41
0.916	6.64	2.50	0.15	1.47	2	7.41
0.916	6.63	2.52	0.15	1.48	2	7.44
0.916	7.47	3.04	0.21	1.54	1	8.09
0.916	7.56	3.08	0.21	1.55	2	8.02
0.916	7.51	3.09	0.21	1.56	2	8.02
0.916	8.42	3.66	0.28	1.62	1	8.71
0.916	8.41	3.77	0.29	1.66	1	8.79
0.916	9.07	4.11	0.34	1.68	2	9.11
0.916	9.04	4.12	0.34	1.68	2	9.11
0.916	9.03	4.14	0.34	1.69	1	9.17
0.916	8.98	4.15	0.34	1.70	1	9.18
0.916	9.55	4.51	0.39	1.74	1	9.55
0.916	9.57	4.53	0.40	1.74	1	9.56
0.916	9.97	4.83	0.44	1.78	1	9.86

POLYSTYRENE

ρ_0 (g/cm^3)	u_s (km/sec)	u_p (km/sec)	P (Mb)	ρ (g/cm^3)	ID	u_s(ID) (km/sec)
1.044	4.35	1.27	0.06	1.47	1	6.41
1.044	4.51	1.29	0.06	1.46	1	6.44
1.044	5.66	2.21	0.13	1.71	1	7.30

STYROFOAM (RHO 0= 0.174)

ρ_0 (g/cm^3)	u_s (km/sec)	u_p (km/sec)	P (Mb)	ρ (g/cm^3)	ID	u_s(ID) (km/sec)
0.174	1.81	1.59	0.00	1.42	1	6.41
0.174	1.91	1.76	0.01	2.21	1	6.53
0.174	2.11	1.85	0.01	1.38	1	6.59
0.174	2.51	2.27	0.01	1.81	1	6.88
0.174	2.49	2.29	0.01	2.16	1	6.90
0.174	3.48	2.97	0.02	1.19	1	7.38
0.174	3.44	3.03	0.02	1.47	1	7.42
0.174	3.33	3.04	0.02	2.00	1	7.43

STYROFOAM (RHO 0= 0.063)

ρ_0 (g/cm^3)	u_s (km/sec)	u_p (km/sec)	P (Mb)	ρ (g/cm^3)	ID	u_s(ID) (km/sec)
0.063	3.33	3.10	0.01	0.92	1	7.42
0.063	3.81	3.28	0.01	0.45	1	7.54
0.063	4.37	3.63	0.01	0.37	1	7.78
0.063	5.66	4.84	0.02	0.43	1	8.59
0.063	6.57	5.52	0.02	0.39	1	9.05
0.063	6.69	5.73	0.02	0.44	1	9.19
0.063	7.52	6.21	0.03	0.36	1	9.51
0.063	9.51	6.62	0.04	0.21	1	9.81
0.063	9.22	6.84	0.04	0.24	1	9.95
0.063	8.94	7.29	0.04	0.34	1	10.23

STYROFOAM (RHO 0= 0.032)

ρ_0 (g/cm^3)	u_s (km/sec)	u_p (km/sec)	P (Mb)	ρ (g/cm^3)	ID	u_s(ID) (km/sec)
0.032	4.04	3.30	0.00	0.17	1	7.54
0.032	4.52	3.66	0.01	0.17	1	7.78
0.032	6.36	4.89	0.01	0.14	1	8.59
0.032	7.05	5.59	0.01	0.15	1	9.05
0.032	6.46	5.81	0.01	0.32	1	9.19
0.032	8.21	6.30	0.02	0.14	1	9.51
0.031	7.78	6.78	0.02	0.24	1	9.81
0.032	10.72	6.95	0.02	0.09	1	9.95
0.032	9.67	7.42	0.02	0.14	1	10.23

LUCITE

ρ_0 (g/cm^3)	u_s (km/sec)	u_p (km/sec)	P (Mb)	ρ (g/cm^3)	ID	u_s(ID) (km/sec)
1.184	2.17	0.	0.	1.18	40	0.
1.180	4.66	1.25	0.07	1.61	1	6.44
1.180	4.65	1.28	0.07	1.63	1	6.47
1.180	4.64	1.28	0.07	1.63	1	6.47
1.180	5.20	1.64	0.10	1.72	1	6.81
1.180	6.00	2.10	0.15	1.81	1	7.28
1.180	6.18	2.19	0.16	1.83	1	7.38
1.186	9.56	4.66	0.53	2.32	1	9.95
1.186	9.74	5.09	0.59	2.48	1	10.35

PLEXIGLAS

ρ_0 (g/cm^3)	u_s (km/sec)	u_p (km/sec)	P (Mb)	ρ (g/cm^3)	ID	u_s(ID) (km/sec)
1.186	2.27	0.	0.	1.19	40	0.
1.190	2.18	0.	0.	1.19	36	0.
1.186	2.19	0.	0.	1.19	40	0.
1.186	3.45	0.54	0.02	1.41	1	5.79
1.186	3.70	0.72	0.03	1.47	1	5.94
1.189	3.76	0.76	0.03	1.49	1	5.98
1.186	3.72	0.76	0.03	1.49	1	5.98
1.189	3.74	0.79	0.04	1.51	2	5.76
1.186	4.04	0.97	0.05	1.56	1	6.17
1.186	4.05	0.98	0.05	1.56	1	6.18
1.180	4.26	1.13	0.06	1.61	1	6.32
1.182	4.28	1.15	0.06	1.61	1	6.33
1.180	4.24	1.15	0.06	1.62	1	6.33
1.186	4.78	1.34	0.08	1.65	1	6.53
1.186	4.78	1.34	0.08	1.65	1	6.53
1.189	4.67	1.34	0.07	1.67	1	6.53
1.186	4.66	1.36	0.07	1.67	1	6.54
1.184	4.59	1.37	0.07	1.69	1	6.55
1.189	4.70	1.46	0.08	1.73	2	6.43
1.180	5.04	1.62	0.10	1.74	1	6.79
1.180	5.05	1.63	0.10	1.74	1	6.80
1.186	5.33	1.69	0.11	1.74	1	6.88
1.186	5.33	1.69	0.11	1.74	1	6.88
1.189	5.14	1.74	0.11	1.80	1	6.91
1.186	5.30	1.78	0.11	1.79	1	6.95

PLEXIGLAS *(continued)*

ρ_0 (g/cm^3)	u_s (km/sec)	u_p (km/sec)	P (Mb)	ρ (g/cm^3)	ID	u_s(ID) (km/sec)
1.189	5.30	1.87	0.12	1.83	2	6.86
1.189	5.30	1.87	0.12	1.83	2	6.86
1.186	5.51	1.89	0.12	1.81	1	7.07
1.180	5.81	2.17	0.15	1.88	1	7.33
1.189	6.24	2.37	0.18	1.92	1	7.55
1.186	6.24	2.38	0.18	1.92	1	7.55
1.189	6.32	2.39	0.18	1.91	2	7.43
1.186	6.32	2.40	0.18	1.91	1	7.58
1.186	6.19	2.42	0.18	1.94	1	7.58
1.186	6.44	2.47	0.19	1.93	1	7.66
1.189	6.32	2.48	0.19	1.95	2	7.52
1.186	6.69	2.64	0.21	1.96	1	7.83
1.186	6.69	2.64	0.21	1.96	1	7.83
1.186	6.84	2.79	0.23	2.00	1	7.97
1.186	6.83	2.79	0.23	2.00	1	7.97
1.186	6.83	2.83	0.23	2.03	1	8.01
1.186	6.82	2.83	0.23	2.03	1	8.01
1.186	8.77	4.35	0.45	2.35	1	9.57
1.186	8.76	4.35	0.45	2.35	1	9.57
1.189	9.56	4.66	0.53	2.32	1	9.95
1.189	9.81	5.07	0.59	2.46	1	10.35

PARAFFIN

ρ_0 (g/cm^3)	u_s (km/sec)	u_p (km/sec)	P (Mb)	ρ (g/cm^3)	ID	u_s(ID) (km/sec)
0.919	4.47	1.06	0.04	1.20	1	6.22
0.904	5.00	1.32	0.06	1.23	1	6.46
0.918	5.15	1.39	0.07	1.26	1	6.53
0.918	4.97	1.41	0.06	1.28	1	6.53
0.919	5.41	1.46	0.07	1.26	1	6.60
0.918	5.75	1.81	0.10	1.34	1	6.91
0.919	5.88	1.81	0.10	1.33	1	6.92
0.904	6.49	2.43	0.14	1.44	1	7.48
0.918	6.75	2.47	0.15	1.45	1	7.54
0.919	6.67	2.51	0.15	1.47	1	7.57
0.919	7.05	2.53	0.16	1.43	1	7.61
0.918	7.25	2.90	0.19	1.53	1	7.95
0.918	7.81	3.18	0.23	1.55	1	8.24
0.918	8.13	3.31	0.25	1.55	1	8.38
0.919	8.55	3.68	0.29	1.62	1	8.74
0.918	8.58	3.70	0.29	1.62	1	8.76
0.919	9.02	3.86	0.32	1.61	10	7.30
0.918	9.43	4.08	0.35	1.62	1	9.16
0.919	9.13	4.10	0.34	1.67	1	9.16
0.919	9.71	4.31	0.38	1.65	1	9.39
0.919	10.39	4.60	0.44	1.65	10	8.00
0.919	10.17	4.63	0.43	1.69	1	9.72
0.919	10.09	4.74	0.44	1.73	10	8.13
0.918	10.93	5.28	0.53	1.77	1	10.35
0.919	10.76	5.30	0.52	1.81	1	10.36
0.919	10.83	5.33	0.53	1.81	1	10.39

EPOXY

ρ_0 (g/cm^3)	u_s (km/sec)	u_p (km/sec)	P (Mb)	ρ (g/cm^3)	ID	u_s(ID) (km/sec)
1.199	2.26	0.	0.	1.20	40	0.
1.184	2.19	0.	0.	1.18	40	0.
1.184	3.26	0.37	0.01	1.34	10	4.24
1.184	3.35	0.41	0.02	1.35	1	5.67
1.184	3.42	0.49	0.02	1.38	1	5.75
1.184	3.80	0.64	0.03	1.42	10	4.47
1.184	3.71	0.70	0.03	1.46	1	5.93
1.184	3.80	0.72	0.03	1.46	1	5.94
1.184	3.70	0.74	0.03	1.48	1	5.96
1.184	3.72	0.78	0.03	1.50	1	6.00
1.184	4.31	0.92	0.05	1.51	10	4.72
1.184	4.17	0.93	0.05	1.52	1	6.14
1.184	4.22	0.96	0.05	1.53	10	4.75
1.184	4.19	0.96	0.05	1.54	10	4.75
1.184	4.09	0.99	0.05	1.56	1	6.19
1.184	4.48	1.15	0.06	1.59	10	4.92
1.184	4.56	1.19	0.06	1.60	10	4.96
1.184	4.57	1.19	0.06	1.60	10	4.96
1.198	4.54	1.21	0.07	1.64	1	6.41
1.198	4.53	1.22	0.07	1.64	1	6.41
1.184	4.74	1.34	0.08	1.65	1	6.53
1.184	4.74	1.36	0.08	1.66	1	6.55
1.184	4.68	1.42	0.08	1.70	1	6.60
1.198	5.24	1.70	0.11	1.77	1	6.88
1.198	5.24	1.70	0.11	1.77	1	6.88
1.198	5.24	1.70	0.11	1.77	1	6.88
1.184	5.30	1.70	0.11	1.74	10	5.41
1.184	5.37	1.72	0.11	1.74	10	5.42

EPOXY *(continued)*

ρ_0 (g/cm^3)	u_s (km/sec)	u_p (km/sec)	P (Mb)	ρ (g/cm^3)	ID	u_s(ID) (km/sec)
1.184	5.36	1.73	0.11	1.75	1	6.91
1.184	5.33	1.75	0.11	1.76	1	6.92
1.184	5.35	1.80	0.11	1.78	1	6.97
1.184	5.33	1.81	0.11	1.79	10	5.50
1.184	5.63	1.89	0.13	1.78	10	5.58
1.198	6.12	2.25	0.16	1.89	1	7.43
1.198	6.09	2.25	0.16	1.90	1	7.43
1.184	6.34	2.36	0.18	1.88	1	7.54
1.184	6.28	2.39	0.18	1.91	1	7.57
1.184	6.38	2.40	0.18	1.90	1	7.58
1.184	6.63	2.41	0.19	1.86	1	7.61
1.184	6.43	2.52	0.19	1.95	10	6.15
1.184	6.60	2.58	0.20	1.94	1	7.76
1.184	6.88	2.76	0.22	1.98	1	7.95
1.184	7.17	2.84	0.24	1.96	10	6.46
1.184	7.21	3.18	0.27	2.12	10	6.75
1.184	7.34	3.19	0.28	2.09	1	8.38
1.184	7.61	3.56	0.32	2.22	1	8.74
1.184	7.65	3.58	0.32	2.22	1	8.76
1.184	7.62	3.61	0.33	2.25	10	7.13
1.184	7.95	3.78	0.36	2.26	10	7.30
1.184	8.39	3.89	0.39	2.21	10	7.42
1.184	8.24	3.95	0.39	2.27	1	9.16
1.184	8.06	3.95	0.38	2.32	10	7.45
1.181	8.26	4.12	0.40	2.36	10	7.60
1.181	8.48	4.14	0.42	2.31	10	7.64
1.184	8.73	4.15	0.43	2.26	1	9.39
1.184	8.90	4.49	0.47	2.39	1	9.72
1.184	8.82	4.65	0.49	2.51	10	8.13
1.184	9.16	4.68	0.51	2.42	10	8.14
1.198	9.25	4.86	0.54	2.52	1	10.10
1.198	9.23	4.86	0.54	2.53	1	10.10
1.184	9.74	5.10	0.59	2.48	1	10.36
1.184	9.76	5.13	0.59	2.50	1	10.39

EXON

ρ_0 (g/cm^3)	u_s (km/sec)	u_p (km/sec)	P (Mb)	ρ (g/cm^3)	ID	u_s(ID) (km/sec)
1.682	3.99	1.19	0.08	2.40	1	6.44
1.681	3.94	1.22	0.08	2.43	1	6.47
1.681	4.48	1.55	0.12	2.57	1	6.82
1.681	5.33	2.02	0.18	2.71	1	7.42

DURITE

ρ_0 (g/cm^3)	u_s (km/sec)	u_p (km/sec)	P (Mb)	ρ (g/cm^3)	ID	u_s(ID) (km/sec)
1.379	4.62	1.30	0.08	1.92	1	6.53
1.378	7.02	2.94	0.28	2.37	1	8.24
1.382	8.27	3.79	0.43	2.55	1	9.16
1.382	9.63	4.89	0.65	2.81	1	10.35

PHENOXY

ρ_0 (g/cm^3)	u_s (km/sec)	u_p (km/sec)	P (Mb)	ρ (g/cm^3)	ID	u_s(ID) (km/sec)
1.178	2.18	0.	0.	1.18	40	0.
1.170	4.68	1.36	0.07	1.65	1	6.54
1.178	4.69	1.37	0.08	1.66	1	6.55
1.179	5.24	1.75	0.11	1.77	1	6.92
1.178	5.27	1.77	0.11	1.77	1	6.93
1.178	6.19	2.34	0.17	1.89	1	7.51
1.179	6.17	2.35	0.17	1.91	1	7.52
1.179	6.64	2.56	0.20	1.92	1	7.75
1.178	7.03	3.05	0.25	2.08	1	8.22
1.178	7.00	3.06	0.25	2.10	1	8.23
1.178	7.46	3.59	0.32	2.27	1	8.75
1.179	7.45	3.61	0.32	2.29	1	8.76
1.178	8.88	4.62	0.48	2.45	1	9.82
1.179	8.83	4.72	0.49	2.53	1	9.91

PHENOLIC REFRASIL

ρ_0 (g/cm^3)	u_s (km/sec)	u_p (km/sec)	P (Mb)	ρ (g/cm^3)	ID	u_s(ID) (km/sec)
1.667	3.74	0.66	0.04	2.03	1	5.95
1.641	3.72	0.68	0.04	2.01	1	5.96
1.647	3.90	0.93	0.06	2.16	1	6.20
1.699	4.24	1.20	0.09	2.37	1	6.48
1.642	4.24	1.27	0.09	2.34	1	6.53
1.645	4.60	1.60	0.12	2.53	1	6.87
1.637	5.34	2.28	0.20	2.85	1	7.56
1.646	5.36	2.37	0.21	2.96	1	7.65
1.646	5.32	2.63	0.23	3.25	1	7.88
1.645	5.77	2.91	0.28	3.32	1	8.19
1.643	6.42	3.34	0.35	3.42	1	8.67
1.650	7.13	3.66	0.43	3.39	1	9.08
1.651	7.18	3.81	0.45	3.52	1	9.22
1.643	7.53	4.04	0.50	3.54	1	9.48
1.630	8.17	4.38	0.58	3.51	1	9.88

LITHIUM TETRABORATE- EPOXY

ρ_0 (g/cm^3)	u_s (km/sec)	u_p (km/sec)	P (Mb)	ρ (g/cm^3)	ID	u_s(ID) (km/sec)
2.180	3.14	0.	0.	2.18	40	0.
2.174	5.54	1.07	0.13	2.70	1	6.54
2.181	5.56	1.08	0.13	2.71	1	6.55
2.177	6.09	1.39	0.18	2.82	1	6.92
2.175	6.09	1.40	0.19	2.83	1	6.93
2.188	6.81	1.87	0.28	3.02	1	7.51
2.181	6.83	1.88	0.28	3.01	1	7.52
2.177	7.09	2.07	0.32	3.08	1	7.75
2.179	7.42	2.48	0.40	3.27	1	8.22
2.176	7.45	2.49	0.40	3.27	1	8.23
2.178	8.04	2.91	0.51	3.41	1	8.75
2.176	8.23	2.92	0.52	3.37	1	8.79
2.178	8.00	2.93	0.51	3.43	1	8.76
2.173	9.22	3.78	0.76	3.68	1	9.82
2.172	9.32	3.85	0.78	3.70	1	9.91

COPPER OXIDE- EPOXY

ρ_0 (g/cm^3)	u_s (km/sec)	u_p (km/sec)	P (Mb)	ρ (g/cm^3)	ID	u_s(ID) (km/sec)
1.761	4.25	1.28	0.10	2.52	1	6.56
1.684	4.85	1.65	0.13	2.55	1	6.94
1.717	5.73	2.18	0.21	2.77	1	7.54
1.682	7.74	3.73	0.49	3.24	1	9.24
1.651	8.20	3.99	0.54	3.22	1	9.53
1.766	8.76	4.44	0.69	3.58	1	10.12

ZERIFAC- EPOXY

ρ_0 (g/cm^3)	u_s (km/sec)	u_p (km/sec)	P (Mb)	ρ (g/cm^3)	ID	u_s(ID) (km/sec)
1.777	4.41	1.27	0.10	2.49	1	6.56
1.776	5.42	2.19	0.21	2.98	1	7.54
1.776	6.82	3.15	0.38	3.30	1	8.62
1.774	6.82	3.44	0.42	3.58	1	8.89
1.775	7.83	3.96	0.55	3.59	1	9.53
1.775	8.35	4.48	0.66	3.83	1	10.10

ZERIFAC- POLYURETHANE FOAM

ρ_0 (g/cm^3)	u_s (km/sec)	u_p (km/sec)	P (Mb)	ρ (g/cm^3)	ID	u_s(ID) (km/sec)
0.702	2.58	1.62	0.03	1.87	1	6.53
0.699	2.71	1.63	0.03	1.76	1	6.55
0.705	3.30	2.06	0.05	1.88	1	6.91
0.712	3.72	2.39	0.06	2.00	1	7.18
0.703	4.35	2.79	0.09	1.96	1	7.53
0.708	4.43	2.82	0.09	1.95	1	7.56
0.721	4.33	2.86	0.09	2.13	1	7.59
0.709	4.53	3.01	0.10	2.11	1	7.71
0.707	5.69	3.67	0.15	1.99	1	8.32
0.706	5.52	3.67	0.14	2.11	1	8.30
0.709	5.71	3.98	0.16	2.33	1	8.56
0.705	6.19	4.13	0.18	2.12	1	8.72
0.708	6.33	4.45	0.20	2.38	1	8.98
0.701	6.48	4.47	0.20	2.26	1	9.01
0.704	6.43	4.51	0.20	2.36	1	9.03
0.707	6.39	4.52	0.20	2.41	1	9.04
0.704	6.77	4.62	0.22	2.21	1	9.16
0.706	7.79	5.38	0.30	2.28	1	9.86
0.703	7.91	5.45	0.30	2.27	1	9.93
0.717	8.06	5.79	0.33	2.54	1	10.23

PYREX

ρ_0 (g/cm^3)	u_s (km/sec)	u_p (km/sec)	P (Mb)	ρ (g/cm^3)	ID	u_s(ID) (km/sec)
2.230	3.87	0.	0.	2.23	40	0.
2.230	4.93	0.56	0.06	2.52	1	5.96
2.230	4.91	0.78	0.09	2.65	1	6.20
2.230	4.82	1.11	0.12	2.90	1	6.53
2.230	4.79	1.44	0.15	3.19	1	6.87
2.230	5.11	2.10	0.24	3.79	1	7.56
2.230	5.18	2.19	0.25	3.86	1	7.65
2.230	5.35	2.40	0.29	4.04	1	7.88
2.230	5.77	2.65	0.34	4.13	1	8.19
2.230	6.30	3.06	0.43	4.34	1	8.67
2.230	7.03	3.35	0.53	4.26	1	9.08
2.230	7.04	3.49	0.55	4.42	1	9.22
2.230	7.41	3.69	0.61	4.44	1	9.48
2.230	8.02	3.99	0.71	4.44	1	9.88

MELMAC

ρ_0 (g/cm^3)	u_s (km/sec)	u_p (km/sec)	P (Mb)	ρ (g/cm^3)	ID	u_s(ID) (km/sec)
1.494	4.80	1.03	0.07	1.90	1	6.32
1.411	4.86	1.04	0.07	1.80	1	6.32
1.494	5.40	1.50	0.12	2.07	1	6.80
1.494	5.31	1.52	0.12	2.09	1	6.81
1.411	5.33	1.54	0.12	1.98	1	6.80
1.411	5.13	1.56	0.11	2.03	1	6.81
1.494	6.09	2.01	0.18	2.23	1	7.33
1.411	6.00	2.05	0.17	2.14	1	7.33

TEFLON

ρ_0 (g/cm^3)	u_s (km/sec)	u_p (km/sec)	P (Mb)	ρ (g/cm^3)	ID	u_s(ID) (km/sec)
2.190	1.27	0.	0.	2.19	36	0.
2.150	1.14	0.	0.	2.15	40	0.
2.152	2.89	0.66	0.04	2.79	1	5.94
2.152	3.25	0.85	0.06	2.91	1	6.14
2.152	3.71	1.09	0.09	3.04	10	4.92
2.147	3.75	1.13	0.09	3.07	1	6.45
2.152	3.91	1.20	0.10	3.11	1	6.53
2.147	4.32	1.47	0.14	3.25	1	6.82
2.152	4.51	1.54	0.15	3.27	1	6.91
2.152	4.64	1.69	0.17	3.38	10	5.50
2.147	5.15	1.89	0.21	3.39	1	7.33
2.152	5.43	2.07	0.24	3.48	1	7.54
2.152	5.77	2.32	0.29	3.60	10	6.15
2.152	5.99	2.41	0.31	3.60	1	7.95
2.152	6.38	2.61	0.36	3.65	10	6.46
2.152	6.65	2.75	0.39	3.66	1	8.38
2.152	6.72	2.90	0.42	3.78	10	6.75
2.152	7.08	3.06	0.47	3.79	1	8.76
2.152	7.38	3.26	0.52	3.86	10	7.13
2.152	7.62	3.38	0.55	3.86	1	9.16
2.152	8.13	3.53	0.62	3.81	1	9.39
2.152	7.85	3.56	0.60	3.94	10	7.45
2.152	8.00	3.71	0.64	4.01	10	7.60
2.151	8.04	3.75	0.65	4.03	10	7.64
2.152	8.39	3.82	0.69	3.95	1	9.72
2.152	8.95	4.39	0.85	4.23	1	10.39

MICARTA

ρ_0 (g/cm^3)	u_s (km/sec)	u_p (km/sec)	P (Mb)	ρ (g/cm^3)	ID	u_s(ID) (km/sec)
1.394	4.73	1.22	0.08	1.88	1	6.47
1.394	4.72	1.23	0.08	1.88	1	6.47
1.394	4.90	1.28	0.09	1.89	1	6.53
1.394	5.20	1.55	0.11	1.99	1	6.80
1.394	6.15	2.08	0.18	2.11	1	7.37
1.394	6.18	2.18	0.19	2.15	1	7.46
1.394	7.15	2.92	0.29	2.36	1	8.24
1.394	8.51	3.75	0.44	2.49	1	9.16
1.394	9.84	4.85	0.67	2.75	1	10.35

SELECTRON

ρ_0 (g/cm^3)	u_s (km/sec)	u_p (km/sec)	P (Mb)	ρ (g/cm^3)	ID	u_s(ID) (km/sec)
1.219	4.23	1.14	0.06	1.67	1	6.33
1.215	4.28	1.26	0.07	1.72	1	6.44
1.215	4.61	1.34	0.08	1.71	1	6.53
1.215	5.02	1.64	0.10	1.81	1	6.82
1.219	5.60	2.18	0.15	2.00	1	7.34
1.216	6.68	3.08	0.25	2.26	1	8.24
1.215	8.16	3.94	0.39	2.35	1	9.16
1.215	9.45	5.09	0.59	2.63	1	10.35

MIN-K-2000

ρ_0 (g/cm^3)	u_s (km/sec)	u_p (km/sec)	P (Mb)	ρ (g/cm^3)	ID	u_s(ID) (km/sec)
0.270	2.19	1.76	0.01	1.38	1	6.55
0.280	2.67	2.26	0.02	1.81	1	6.91
0.280	5.28	4.12	0.06	1.27	1	8.30
0.280	5.25	4.14	0.06	1.32	1	8.32
0.270	5.82	4.68	0.07	1.38	1	8.72
0.290	6.39	4.99	0.09	1.32	1	8.98
0.290	6.32	5.02	0.09	1.41	1	9.00
0.280	6.77	5.06	0.10	1.11	1	9.04
0.280	6.39	5.08	0.09	1.37	1	9.03
0.270	6.81	5.24	0.10	1.17	1	9.16
0.290	7.64	6.13	0.14	1.46	1	9.86
0.280	8.29	6.19	0.14	1.11	1	9.9[illegible]

ELKONITE (TC-10)

ρ_0 (g/cm^3)	u_s (km/sec)	u_p (km/sec)	P (Mb)	ρ (g/cm^3)	ID	u_s(ID) (km/sec)
11.670	4.25	0.	0.	11.67	40	0.
11.670	4.13	0.20	0.10	12.26	10	4.27
11.660	4.30	0.22	0.11	12.28	10	4.31
11.670	4.36	0.28	0.14	12.47	10	4.41
11.460	4.33	0.30	0.15	12.33	10	4.45
11.690	4.65	0.43	0.23	12.88	10	4.67
11.670	4.93	0.62	0.36	13.34	10	4.98
11.680	5.42	0.94	0.59	14.12	10	5.51
11.670	6.09	1.38	0.98	15.08	10	6.23
11.680	6.61	1.72	1.33	15.80	10	6.79
11.660	6.96	1.95	1.59	16.21	10	7.16
11.660	7.38	2.26	1.95	16.81	10	7.65

ELKONITE (2125C)

ρ_0 (g/cm^3)	u_s (km/sec)	u_p (km/sec)	P (Mb)	ρ (g/cm^3)	ID	u_s(ID) (km/sec)
9.725	3.20	0.	0.	9.73	40	0.
9.734	2.87	0.29	0.08	10.81	10	4.31
9.724	3.28	0.40	0.13	11.08	10	4.48
9.722	3.44	0.51	0.17	11.42	10	4.63
9.642	3.76	0.59	0.21	11.45	10	4.76
9.611	3.98	0.67	0.25	11.54	10	4.88
9.547	4.14	0.79	0.31	11.80	10	5.05
9.551	4.57	1.01	0.44	12.25	10	5.38
9.767	5.00	1.17	0.57	12.77	10	5.66
9.638	5.15	1.27	0.63	12.80	10	5.80
9.762	5.56	1.45	0.79	13.20	10	6.10
9.672	5.47	1.45	0.77	13.17	10	6.08
9.796	5.79	1.62	0.92	13.60	10	6.35
9.669	5.85	1.69	0.96	13.61	10	6.45
9.776	5.92	1.76	1.02	13.90	10	6.55
9.654	6.66	2.19	1.41	14.37	10	7.20
9.775	7.10	2.44	1.70	14.91	10	7.61
9.758	7.50	2.67	1.95	15.14	10	7.96
9.578	7.61	2.77	2.02	15.05	10	8.08

ELKONITE (1W3)

ρ_0 (g/cm^3)	u_s (km/sec)	u_p (km/sec)	P (Mb)	ρ (g/cm^3)	ID	u_s(ID) (km/sec)
12.380	3.11	0.	0.	12.38	40	0.
12.440	3.67	0.34	0.16	13.72	10	4.48
12.410	3.83	0.44	0.21	14.00	10	4.63
12.170	4.09	0.51	0.26	13.92	10	4.76
12.430	4.21	0.58	0.30	14.40	10	4.88
12.450	4.42	0.68	0.37	14.72	10	5.05
11.950	4.44	0.80	0.43	14.59	10	5.21
12.460	4.73	0.88	0.52	15.30	10	5.38
12.500	5.03	1.05	0.66	15.79	10	5.66
12.460	5.18	1.16	0.75	16.04	10	5.84
12.450	5.43	1.30	0.88	16.38	10	6.08
12.240	5.31	1.34	0.87	16.37	10	6.10
12.360	5.65	1.48	1.03	16.74	10	6.35
12.320	5.78	1.53	1.09	16.76	10	6.45
12.360	5.90	1.59	1.16	16.92	10	6.55
11.790	6.50	2.03	1.56	17.15	10	7.20
12.300	6.79	2.26	1.89	18.45	10	7.61
12.020	7.22	2.48	2.16	18.33	10	7.96
12.460	7.31	2.56	2.33	19.20	10	8.13

ELKONITE (3W3)

ρ_0 (g/cm^3)	u_s (km/sec)	u_p (km/sec)	P (Mb)	ρ (g/cm^3)	ID	u_s(ID) (km/sec)
13.860	4.35	0.53	0.32	15.79	10	4.87
13.870	5.18	1.07	0.77	17.50	10	5.80
13.820	6.02	1.57	1.30	18.69	10	6.65
13.710	7.20	2.44	2.41	20.72	10	8.08

ELKONITE (10W3)

ρ_0 (g/cm^3)	u_s (km/sec)	u_p (km/sec)	P (Mb)	ρ (g/cm^3)	ID	u_s(ID) (km/sec)
14.880	3.76	0.	0.	14.88	40	0.
14.890	3.53	0.19	0.10	15.73	10	4.27
14.830	3.62	0.21	0.11	15.74	10	4.31
14.840	3.78	0.26	0.15	15.96	10	4.41
14.830	3.73	0.29	0.16	16.07	10	4.45
14.830	4.02	0.41	0.24	16.50	10	4.67
14.810	4.39	0.58	0.38	17.07	10	4.98
14.900	4.79	0.82	0.58	17.97	10	5.41
14.860	4.72	0.82	0.58	18.00	10	5.41
14.880	4.81	0.84	0.60	18.01	10	5.44
14.840	4.88	0.87	0.63	18.08	10	5.51
14.830	5.53	1.28	1.05	19.31	10	6.23
14.820	5.97	1.62	1.43	20.33	10	6.79
14.860	6.26	1.84	1.71	21.03	10	7.16
14.820	6.65	2.13	2.10	21.80	10	7.65
14.880	6.72	2.19	2.19	22.08	10	7.75
14.890	6.74	2.24	2.24	22.29	10	7.82

GOLD- 5.8 WT PERCENT GERMANIUM

ρ_0 (g/cm^3)	u_s (km/sec)	u_p (km/sec)	P (Mb)	ρ (g/cm^3)	ID	u_s(ID) (km/sec)
16.880	2.95	0.	0.	16.88	40	0.
16.880	3.23	0.29	0.16	18.55	10	4.45
16.820	3.29	0.42	0.24	19.31	10	4.67
16.850	3.61	0.60	0.37	20.22	10	4.98
16.880	4.19	0.88	0.63	21.40	10	5.51
16.830	4.86	1.29	1.05	22.88	10	6.23
16.910	5.71	1.80	1.74	24.70	10	7.16
16.760	6.14	2.09	2.15	25.38	10	7.65

GOLD- 9.3 WT PERCENT GERMANIUM

ρ_0 (g/cm^3)	u_s (km/sec)	u_p (km/sec)	P (Mb)	ρ (g/cm^3)	ID	u_s(ID) (km/sec)
15.490	2.93	0.	0.	15.49	40	0.
15.460	3.23	0.30	0.15	17.07	10	4.45
15.480	3.45	0.64	0.34	19.03	10	4.98
15.510	4.05	0.94	0.59	20.19	10	5.51
15.550	4.81	1.34	1.01	21.57	10	6.23
15.490	5.34	1.67	1.38	22.52	10	6.79
15.480	5.34	1.67	1.38	22.52	10	6.79
15.840	5.73	1.86	1.69	23.43	10	7.16
15.520	6.15	2.16	2.06	23.92	10	7.65

GERMANIUM

ρ_0 (g/cm^3)	u_s (km/sec)	u_p (km/sec)	P (Mb)	ρ (g/cm^3)	ID	u_s(ID) (km/sec)
5.330	3.83	0.	0.	5.33	35	0.
5.330	3.75	0.	0.	5.33	35	0.
5.330	4.25	0.60	0.14	6.21	1	6.24
5.330	4.22	0.75	0.17	6.48	1	6.45
5.311	4.25	0.75	0.17	6.46	1	6.46
5.311	4.14	0.79	0.17	6.56	1	6.49
5.330	4.19	1.05	0.23	7.11	1	6.84
5.330	4.16	1.07	0.24	7.17	1	6.86
5.330	4.09	1.07	0.23	7.23	1	6.86
5.311	4.12	1.09	0.24	7.22	1	6.87
5.330	4.17	1.23	0.27	7.55	1	7.06
5.311	4.18	1.46	0.32	8.18	1	7.36
5.311	4.30	1.47	0.33	8.06	1	7.38
5.311	4.18	1.50	0.33	8.28	1	7.40
5.330	4.57	1.59	0.39	8.18	1	7.60
5.330	4.55	1.60	0.39	8.21	1	7.60
5.330	4.49	1.61	0.39	8.32	1	7.61
5.330	4.82	1.75	0.45	8.37	1	7.85
5.330	4.91	1.79	0.47	8.39	1	7.92
5.330	5.22	1.95	0.54	8.50	1	8.17
5.330	5.22	1.96	0.55	8.54	1	8.19
5.330	5.14	1.97	0.54	8.65	1	8.19
5.330	5.45	2.09	0.61	8.66	1	8.41
5.330	5.38	2.11	0.60	8.76	1	8.41
5.330	5.77	2.15	0.66	8.50	1	8.55
5.330	5.68	2.16	0.65	8.60	1	8.55
5.330	5.72	2.16	0.66	8.57	1	8.55
5.330	5.73	2.19	0.67	8.63	1	8.59
5.330	5.72	2.24	0.68	8.76	1	8.66
5.330	5.50	2.25	0.66	9.01	1	8.61
5.330	5.39	2.27	0.65	9.19	1	8.61
5.330	5.65	2.27	0.68	8.91	1	8.68
5.330	6.07	2.40	0.78	8.82	1	8.93
5.330	6.12	2.41	0.79	8.80	1	8.96
5.330	5.97	2.46	0.78	9.07	1	8.99
5.330	5.91	2.47	0.78	9.17	1	8.99

GERMANIUM *(continued)*

ρ_0 (g/cm^3)	u_s (km/sec)	u_p (km/sec)	P (Mb)	ρ (g/cm^3)	ID	u_s(ID) (km/sec)
5.330	6.16	2.48	0.81	8.92	1	9.05
5.330	6.06	2.56	0.83	9.23	1	9.13
5.330	6.20	2.56	0.85	9.09	1	9.17
5.330	6.19	2.57	0.85	9.11	1	9.17
5.330	6.10	2.58	0.84	9.23	1	9.16
5.330	6.37	2.60	0.88	9.00	1	9.25
5.330	6.39	2.61	0.89	9.01	1	9.27
5.330	6.36	2.66	0.90	9.16	1	9.32
5.330	6.32	2.71	0.91	9.33	1	9.37
5.330	6.52	2.74	0.95	9.19	1	9.46
5.330	6.56	2.94	1.03	9.65	1	9.72
5.330	6.92	3.01	1.11	9.43	1	9.89
5.330	6.96	3.04	1.13	9.47	1	9.94
5.330	6.71	3.12	1.12	9.96	1	9.98
5.330	6.91	3.19	1.17	9.90	1	10.11

FE- 10.0 WT PERCENT NI

ρ_0 (g/cm^3)	u_s (km/sec)	u_p (km/sec)	P (Mb)	ρ (g/cm^3)	ID	u_s(ID) (km/sec)
7.890	4.46	0.	0.	7.89	40	0.
7.883	5.39	1.04	0.44	9.76	10	5.40
7.885	5.41	1.09	0.47	9.87	10	5.47
7.883	6.38	1.60	0.80	10.52	10	6.22
7.895	6.47	1.63	0.83	10.54	10	6.27
7.883	6.79	1.80	0.96	10.73	10	6.53
7.896	7.71	2.38	1.45	11.42	10	7.37
7.885	8.00	2.54	1.60	11.55	10	7.61
7.896	8.02	2.55	1.62	11.59	10	7.63
7.870	7.98	2.58	1.62	11.63	10	7.65
7.883	8.27	2.75	1.80	11.82	10	7.92

FE- 17.9 WT PERCENT NI

ρ_0 (g/cm^3)	u_s (km/sec)	u_p (km/sec)	P (Mb)	ρ (g/cm^3)	ID	u_s(ID) (km/sec)
7.960	4.40	0.	0.	7.96	40	0.
7.962	5.56	1.02	0.45	9.75	10	5.40
7.962	5.61	1.06	0.47	9.81	10	5.45
7.962	6.52	1.58	0.82	10.51	10	6.22
7.962	6.58	1.61	0.84	10.54	10	6.27
7.962	6.90	1.78	0.98	10.74	10	6.53
7.962	7.13	1.95	1.11	10.97	10	6.77
7.962	7.62	2.26	1.37	11.31	10	7.22
7.962	7.74	2.37	1.46	11.48	10	7.37
7.962	8.00	2.53	1.61	11.65	10	7.61
7.962	8.03	2.54	1.63	11.65	10	7.63
7.962	7.98	2.56	1.63	11.74	10	7.65
7.962	8.42	2.72	1.83	11.77	10	7.91
7.962	8.38	2.73	1.82	11.81	10	7.92
7.962	8.35	2.73	1.82	11.83	10	7.91
7.962	8.32	2.76	1.83	11.92	10	7.94

FE- 26.2 WT PERCENT NI

ρ_0 (g/cm^3)	u_s (km/sec)	u_p (km/sec)	P (Mb)	ρ (g/cm^3)	ID	u_s(ID) (km/sec)
7.974	4.37	0.	0.	7.97	40	0.
7.974	5.48	1.03	0.45	9.81	10	5.40
7.974	5.49	1.06	0.47	9.89	10	5.45
7.974	5.49	1.08	0.47	9.92	10	5.47
7.974	6.44	1.58	0.81	10.58	10	6.22
7.974	6.53	1.61	0.84	10.59	10	6.27
7.974	6.84	1.79	0.97	10.79	10	6.53
7.974	7.07	1.96	1.10	11.03	10	6.77
7.974	7.48	2.27	1.36	11.45	10	7.22
7.974	7.69	2.37	1.46	11.54	10	7.37
7.974	7.94	2.54	1.61	11.72	10	7.61
7.974	7.96	2.55	1.62	11.73	10	7.63
7.974	8.03	2.56	1.64	11.70	10	7.65
7.974	8.13	2.76	1.79	12.07	10	7.92
7.974	8.18	2.78	1.81	12.08	10	7.94

APPENDIX E

CALCULATED THERMODYNAMIC RESULTS FOR SOLIDS

2024 ALUMINUM

93.4/4.5/1.5/0.6 WT PER CENT AL/CU/MG/MN

COEFFICIENTS IN US=CO+S*UP+S2*UP**2+... 5.328 1.338
DE/DP)V= .180 CM**3/G GAMMA=2.000
DEBYE PARAMETERS RHOO= 2.785 T= 293 THETA= 350 3NK= 8.990E+06 CV= 8.380E+06

**10^{10} ergs/g
*10^{6} ergs/g-deg K

P	PARAMETERS ON THE HUGONIOT							RELEASE ISENTROPE(P=0)			ZERO KELVIN		T 0 ISOTHERM	
MB	RHO G/CM**3	T DEG-K	S *	E **	C KM/SEC	US KM/SEC	UP KM/SEC	RHO G/CM**3	T DEG-K	LFS KM/SEC	RHO G/CM**3	E **	RHO G/CM**3	C KM/SEC
0.00	2.785	293	10.70	.166	5.33	5.33	0.00	2.785	293	0.00	2.817	0.000	2.78	5.33
.05	2.945	330	10.78	.215	5.82	5.75	.31	2.785	296	.63	2.976	.045	2.95	5.83
.10	3.081	374	11.14	.338	6.22	6.11	.59	2.782	308	1.18	3.112	.154	3.09	6.23
.15	3.200	433	11.81	.515	6.56	6.45	.84	2.778	332	1.67	3.233	.304	3.22	6.57
.20	3.306	508	12.70	.732	6.85	6.75	1.06	2.772	367	2.13	3.343	.482	3.33	6.87
.25	3.402	599	13.72	.980	7.11	7.04	1.28	2.764	411	2.56	3.445	.679	3.43	7.14
.30	3.490	707	14.80	1.254	7.35	7.30	1.48	2.755	462	2.97	3.540	.892	3.53	7.38
.35	3.572	830	15.89	1.550	7.57	7.55	1.66	2.744	519	3.36	3.628	1.116	3.62	7.61
.40	3.647	968	16.96	1.864	7.77	7.79	1.84	2.732	581	3.72	3.712	1.349	3.70	7.82
.45	3.718	1120	17.99	2.194	7.97	8.02	2.01	2.720	646	4.08	3.791	1.589	3.78	8.02
.50	3.785	1286	18.98	2.538	8.15	8.24	2.18	2.706	715	4.42	3.867	1.834	3.86	8.21
.55	3.848	1464	19.92	2.894	8.32	8.45	2.34	2.692	786	4.75	3.939	2.083	3.93	8.40
.60	3.908	1654	20.81	3.262	8.48	8.66	2.49	2.676	858	5.07	4.009	2.335	4.00	8.57
.65	3.965	1855	21.65	3.640	8.64	8.85	2.64	2.661	932	5.38	4.075	2.590	4.07	8.74
.70	4.020	2067	22.45	4.026	8.79	9.05	2.78	2.645	1006	5.68	4.139	2.847	4.13	8.90
.75	4.072	2290	23.21	4.421	8.93	9.23	2.92	2.628	1080	5.98	4.201	3.104	4.19	9.06
.80	4.122	2522	23.93	4.823	9.07	9.41	3.05	2.611	1154	6.27	4.261	3.363	4.25	9.22
.85	4.169	2764	24.61	5.233	9.21	9.59	3.18	2.594	1227	6.55	4.319	3.622	4.31	9.37
.90	4.215	3014	25.25	5.649	9.34	9.76	3.31	2.576	1300	6.83	4.375	3.882	4.37	9.51
.95	4.260	3273	25.87	6.071	9.47	9.93	3.44	2.558	1371	7.10	4.429	4.141	4.42	9.66
1.00	4.303	3540	26.46	6.498	9.59	10.09	3.56	2.540	1442	7.37	4.482	4.400	4.48	9.80
1.05	4.344	3815	27.02	6.931	9.71	10.25	3.68	2.521	1510	7.63	4.533	4.659	4.53	9.93
1.10	4.384	4098	27.55	7.369	9.83	10.41	3.80	2.503	1577	7.89	4.583	4.918	4.58	10.07
1.15	4.423	4387	28.07	7.811	9.95	10.56	3.91	2.484	1642	8.14	4.630	5.176	4.63	10.19
1.20	4.460	4684	28.56	8.258	10.06	10.71	4.02	2.466	1705	8.40	4.675	5.434	4.67	10.32

921-T ALUMINUM

92.0/3.5/2.0/1.5(MAX) WT PER CENT AL/CU/SI/OTHER ELEMENTS (PIONEER)

COEFFICIENTS IN US=C0+S*UP+S2*UP**2+... 5.041 1.420
DE/DP)V= .168 CM**3/G GAMMA=2.100
DEBYE PARAMETERS RHO0= 2.833 T= 293 THETA= 359 3NK= 9.050E+06 CV= 8.405E+06

**10^{10} ergs/g
*10^{6} ergs/g-deg K

P	PARAMETERS ON THE HUGONIOT							RELEASE ISENTROPE(P=0)			ZERO KELVIN		T 0 ISOTHERM	
MB	RHO G/CM**3	T DEG-K	S *	E **	C KM/SEC	US KM/SEC	UP KM/SEC	RHO G/CM**3	T DEG-K	UFS KM/SEC	RHO G/CM**3	E **	RHO G/CM**3	C KM/SEC
0.00	2.833	293	10.56	.165	5.04	5.04	0.00	2.833	293	0.00	2.871	0.000	2.83	5.04
.05	3.009	335	10.65	.217	5.60	5.50	.32	2.832	296	.64	3.044	.047	3.02	5.61
.10	3.154	385	11.06	.345	6.04	5.89	.60	2.829	310	1.20	3.190	.158	3.17	6.05
.15	3.278	451	11.80	.525	6.41	6.24	.85	2.824	337	1.70	3.317	.307	3.30	6.43
.20	3.387	534	12.74	.743	6.73	6.57	1.07	2.816	374	2.16	3.431	.482	3.41	6.76
.25	3.485	634	13.81	.991	7.02	6.87	1.29	2.806	419	2.59	3.536	.675	3.52	7.06
.30	3.574	750	14.92	1.263	7.28	7.15	1.48	2.794	471	2.99	3.632	.880	3.62	7.33
.35	3.656	882	16.02	1.555	7.52	7.41	1.67	2.781	529	3.37	3.721	1.095	3.71	7.58
.40	3.731	1028	17.09	1.865	7.75	7.66	1.84	2.766	590	3.73	3.805	1.317	3.79	7.82
.45	3.801	1188	18.12	2.188	7.96	7.90	2.01	2.751	654	4.08	3.884	1.545	3.87	8.05
.50	3.867	1362	19.10	2.524	8.16	8.13	2.17	2.734	720	4.42	3.959	1.776	3.95	8.27
.55	3.928	1547	20.03	2.872	8.35	8.34	2.33	2.717	787	4.75	4.030	2.010	4.02	8.47
.60	3.986	1744	20.90	3.229	8.54	8.56	2.48	2.699	856	5.06	4.098	2.246	4.09	8.67
.65	4.041	1951	21.73	3.595	8.71	8.76	2.62	2.680	924	5.37	4.163	2.483	4.16	8.87
.70	4.093	2169	22.52	3.969	8.88	8.96	2.76	2.661	992	5.67	4.225	2.722	4.22	9.05
.75	4.143	2397	23.26	4.351	9.04	9.15	2.89	2.641	1059	5.96	4.285	2.960	4.28	9.24
.80	4.191	2634	23.96	4.740	9.20	9.34	3.02	2.621	1126	6.24	4.342	3.199	4.34	9.41
.85	4.236	2879	24.63	5.134	9.35	9.52	3.15	2.601	1191	6.52	4.397	3.438	4.39	9.59
.90	4.280	3133	25.26	5.535	9.50	9.69	3.28	2.580	1254	6.80	4.451	3.676	4.45	9.76
.95	4.321	3395	25.86	5.940	9.64	9.87	3.40	2.559	1315	7.06	4.502	3.914	4.50	9.93
1.00	4.362	3664	26.44	6.351	9.78	10.04	3.52	2.538	1375	7.33	4.552	4.151	4.55	10.09
1.05	4.400	3940	26.99	6.766	9.92	10.20	3.63	2.516	1432	7.59	4.600	4.387	4.60	10.25
1.10	4.438	4223	27.51	7.186	10.05	10.36	3.75	2.495	1487	7.84	4.646	4.622	4.64	10.41
1.15	4.474	4513	28.01	7.610	10.18	10.52	3.86	2.474	1540	8.10	4.690	4.857	4.69	10.56
1.20	4.509	4809	28.49	8.037	10.31	10.68	3.97	2.452	1590	8.34	4.731	5.090	4.73	10.69

COPPER

DEOXIDIZED COPPER (DHP) AND OXYGEN-FREE COPPER (OF)

COEFFICIENTS IN US=C0+S*UP+S2*UP**2+... 3.940 1.489
DE/DP)V= .056 CM**3/G GAMMA=1.990
DEBYE PARAMETERS RHO0= 8.930 T= 293 THETA= 306 3NK= 3.925E+06 CV= 3.718E+06

**10^{10} ergs/g
*10^{6} ergs/g-deg K

P	PARAMETERS ON THE HUGONIOT							RELEASE ISENTROPE(P=0)			ZERO KELVIN		T 0 ISOTHERM	
MB	RHO G/CM**3	T DEG-K	S *	E **	C KM/SEC	US KM/SEC	UP KM/SEC	RHO G/CM**3	T DEG-K	UFS KM/SEC	RHO G/CM**3	E **	RHO G/CM**3	C KM/SEC
0.00	8.930	293	5.16	.077	3.94	3.94	0.00	8.930	293	0.00	9.017	0.000	8.93	3.94
.10	9.499	336	5.22	.110	4.42	4.33	.26	8.928	298	.52	9.578	.031	9.51	4.43
.20	9.959	395	5.51	.193	4.81	4.66	.48	8.918	320	.96	10.041	.102	9.99	4.82
.30	10.349	479	5.98	.307	5.13	4.95	.68	8.899	362	1.36	10.440	.196	10.40	5.15
.40	10.688	589	6.57	.445	5.41	5.22	.86	8.873	419	1.72	10.794	.306	10.76	5.44
.50	10.990	726	7.19	.602	5.67	5.47	1.02	8.840	490	2.06	11.114	.426	11.08	5.71
.60	11.262	888	7.82	.772	5.90	5.70	1.18	8.801	571	2.38	11.407	.552	11.38	5.95
.70	11.510	1072	8.42	.955	6.12	5.91	1.33	8.757	660	2.68	11.678	.684	11.65	6.18
.80	11.738	1279	8.99	1.148	6.32	6.12	1.46	8.709	755	2.96	11.930	.820	11.91	6.40
.90	11.949	1505	9.52	1.350	6.51	6.32	1.60	8.658	855	3.24	12.166	.958	12.14	6.60
1.00	12.145	1751	10.01	1.559	6.69	6.50	1.72	8.603	958	3.50	12.388	1.098	12.37	6.80
1.10	12.329	2014	10.48	1.775	6.86	6.68	1.84	8.545	1064	3.76	12.598	1.239	12.58	6.99
1.20	12.502	2294	10.91	1.997	7.03	6.86	1.96	8.485	1170	4.01	12.797	1.381	12.78	7.17
1.30	12.666	2589	11.31	2.224	7.19	7.03	2.07	8.422	1277	4.25	12.986	1.523	12.97	7.35
1.40	12.820	2900	11.69	2.456	7.34	7.19	2.18	8.357	1383	4.48	13.167	1.665	13.15	7.52
1.50	12.967	3224	12.04	2.692	7.49	7.35	2.29	8.291	1488	4.71	13.339	1.808	13.32	7.69
1.60	13.107	3561	12.37	2.932	7.63	7.50	2.39	8.223	1592	4.94	13.504	1.950	13.49	7.86
1.70	13.241	3910	12.69	3.176	7.77	7.65	2.49	8.154	1693	5.15	13.662	2.091	13.65	8.02
1.80	13.368	4271	12.98	3.423	7.91	7.79	2.59	8.084	1791	5.37	13.815	2.232	13.80	8.18
1.90	13.491	4643	13.26	3.673	8.04	7.93	2.68	8.012	1886	5.58	13.961	2.372	13.95	8.34
2.00	13.608	5026	13.53	3.926	8.17	8.07	2.77	7.940	1978	5.79	14.102	2.512	14.09	8.49
2.10	13.721	5419	13.78	4.182	8.30	8.21	2.87	7.866	2067	5.99	14.238	2.651	14.23	8.64
2.20	13.829	5821	14.02	4.441	8.42	8.34	2.95	7.793	2151	6.20	14.367	2.789	14.36	8.79
2.30	13.934	6232	14.25	4.702	8.54	8.47	3.04	7.718	2231	6.39	14.491	2.927	14.48	8.93
2.40	14.035	6653	14.47	4.965	8.66	8.60	3.13	7.643	2307	6.59	14.608	3.063	14.60	9.06

IRON

1018 STEEL

COEFFICIENTS IN US=CO+S*UP+S2*UP**2+... 3.574 1.920 -.068
DE/DP)V= .075 CM**3/G GAMMA=1.690
DEBYE PARAMETERS RHOO= 7.850 T= 293 THETA= 175 3NK= 4.466E+06 CV= 4.387E+06

** 10^{10} ergs/g
* 10^{6} ergs/g-deg K

	PARAMETERS ON THE HUGONIOT							RELEASE ISENTROPE(P=0)			ZERO KELVIN		T O ISOTHERM	
P MB	RHO G/CM**3	T DEG-K	S *	E **	C KM/SEC	US KM/SEC	UP KM/SEC	RHO G/CM**3	T DEG-K	UFS KM/SEC	RHO G/CM**3	E **	RHO G/CM**3	C KM/SEC
0.00	7.850	293	8.30	.105	3.57	3.57	0.00	7.850	293	0.00	7.956	0.000	7.85	3.57
.10	8.475	343	8.43	.152	4.39	4.16	.31	7.846	302	.61	8.558	.040	8.49	4.41
.20	8.917	416	8.95	.258	4.99	4.61	.55	7.829	338	1.11	8.998	.124	8.95	5.02
.30	9.266	521	9.70	.397	5.47	5.00	.76	7.799	398	1.53	9.352	.229	9.31	5.52
.40	9.556	657	10.53	.560	5.88	5.34	.95	7.760	476	1.92	9.653	.344	9.62	5.94
.50	9.806	820	11.36	.740	6.24	5.65	1.13	7.713	568	2.27	9.916	.468	9.88	6.32
.60	10.027	1007	12.14	.935	6.56	5.93	1.29	7.658	669	2.60	10.151	.596	10.12	6.66
.70	10.225	1215	12.86	1.141	6.85	6.20	1.44	7.598	776	2.92	10.365	.728	10.34	6.96
.80	10.406	1444	13.53	1.357	7.11	6.44	1.58	7.531	888	3.22	10.563	.863	10.54	7.23
.90	10.573	1691	14.14	1.582	7.35	6.67	1.72	7.459	1001	3.51	10.746	1.000	10.72	7.49
1.00	10.729	1953	14.71	1.814	7.57	6.89	1.85	7.382	1115	3.78	10.918	1.140	10.90	7.72
1.10	10.874	2231	15.23	2.054	7.77	7.10	1.97	7.299	1228	4.06	11.081	1.281	11.06	7.93
1.20	11.012	2523	15.71	2.300	7.95	7.30	2.10	7.212	1338	4.32	11.236	1.424	11.22	8.12
1.30	11.142	2828	16.15	2.551	8.13	7.49	2.21	7.120	1444	4.58	11.383	1.568	11.37	8.30
1.40	11.266	3145	16.57	2.809	8.29	7.67	2.33	7.024	1544	4.83	11.525	1.714	11.51	8.47
1.50	11.384	3473	16.96	3.071	8.44	7.85	2.44	6.922	1639	5.08	11.662	1.861	11.65	8.62
1.60	11.498	3812	17.32	3.339	8.58	8.01	2.54	6.816	1726	5.33	11.794	2.010	11.78	8.76
1.70	11.608	4161	17.66	3.610	8.71	8.18	2.65	6.704	1804	5.57	11.922	2.160	11.91	8.89
1.80	11.714	4520	17.99	3.887	8.83	8.34	2.75	6.590	1873	5.81	12.047	2.312	12.03	9.01
1.90	11.816	4887	18.29	4.167	8.94	8.49	2.85				12.168	2.465	12.15	9.12
2.00	11.916	5264	18.58	4.452	9.05	8.64	2.95				12.287	2.620	12.27	9.22
2.10	12.012	5648	18.86	4.740	9.15	8.79	3.04				12.403	2.777	12.39	9.31
2.20	12.107	6040	19.12	5.032	9.24	8.93	3.14				12.518	2.935	12.51	9.39
2.30	12.199	6440	19.37	5.328	9.33	9.07	3.23				12.630	3.095	12.62	9.46
2.40	12.289	6846	19.60	5.627	9.41	9.20	3.32				12.741	3.257	12.73	9.53
2.50	12.377	7260	19.83	5.929	9.49	9.33	3.41				12.849	3.421	12.84	9.59
2.60	12.463	7680	20.05	6.235	9.56	9.46	3.50				12.955	3.586	12.94	9.64
2.70	12.548	8107	20.26	6.544	9.62	9.58	3.59				13.059	3.753	13.05	9.69

URANIUM-MOLYBDENUM ALLOY
97/3 WEIGHT PERCENT U/MO

COEFFICIENTS IN US=C0+S*UP+S2*UP**2+... 2.565 1.531
DE/DP)V= .027 CM**3/G GAMMA=2.030
DEBYE PARAMETERS RHO0=18.450 T= 293 THETA= 275 3NK= 1.200E+06 CV= 1.149E+06

**10^{10} ergs/g
*10^{6} ergs/g-deg K

P	PARAMETERS ON THE HUGONIOT							RELEASE ISENTROPE(P=0)			ZERO KELVIN		T 0 ISOTHERM	
MB	RHO G/CM**3	T DEG-K	S *	E **	C KM/SEC	US KM/SEC	UP KM/SEC	RHO G/CM**3	T DEG-K	UFS KM/SEC	RHO G/CM**3	E **	RHO G/CM**3	C KM/SEC
0.00	18.450	293	1.70	.025	2.56	2.56	0.00	18.450	293	0.00	18.588	0.000	18.45	2.56
.10	19.764	347	1.74	.043	2.93	2.86	.19	18.444	303	.38	19.890	.016	19.79	2.94
.20	20.795	436	1.90	.086	3.22	3.10	.35	18.415	345	.70	20.934	.053	20.85	3.23
.30	21.652	573	2.13	.145	3.46	3.32	.49	18.363	420	.98	21.820	.102	21.75	3.47
.40	22.389	759	2.40	.215	3.67	3.51	.62	18.291	522	1.24	22.597	.156	22.54	3.69
.50	23.036	992	2.66	.294	3.85	3.69	.73	18.203	644	1.48	23.292	.216	23.24	3.88
.60	23.614	1267	2.90	.380	4.02	3.86	.84	18.100	782	1.70	23.923	.278	23.88	4.06
.70	24.138	1582	3.13	.472	4.18	4.01	.95	17.986	931	1.92	24.503	.342	24.46	4.23
.80	24.616	1934	3.33	.568	4.33	4.16	1.04	17.862	1088	2.12	25.039	.407	25.00	4.39
.90	25.056	2320	3.52	.668	4.47	4.30	1.13	17.729	1251	2.31	25.538	.474	25.51	4.55
1.00	25.465	2737	3.69	.771	4.60	4.44	1.22	17.589	1417	2.50	26.005	.541	25.98	4.69
1.10	25.845	3183	3.84	.877	4.73	4.56	1.31	17.442	1584	2.68	26.445	.608	26.42	4.84
1.20	26.202	3656	3.98	.987	4.85	4.69	1.39	17.289	1750	2.85	26.860	.675	26.83	4.97
1.30	26.537	4154	4.12	1.098	4.97	4.81	1.47	17.131	1914	3.02	27.252	.742	27.23	5.11
1.40	26.853	4676	4.24	1.212	5.08	4.92	1.54	16.969	2075	3.19	27.625	.809	27.60	5.24
1.50	27.153	5219	4.35	1.327	5.19	5.04	1.61	16.804	2232	3.35	27.981	.875	27.96	5.37
1.60	27.437	5784	4.46	1.445	5.30	5.15	1.69	16.635	2384	3.50	28.319	.941	28.30	5.49
1.70	27.708	6367	4.56	1.564	5.40	5.25	1.75	16.463	2529	3.66	28.643	1.007	28.62	5.61
1.80	27.966	6969	4.65	1.684	5.50	5.35	1.82	16.289	2668	3.81	28.953	1.073	28.94	5.73
1.90	28.213	7589	4.74	1.806	5.60	5.46	1.89	16.112	2800	3.96	29.251	1.138	29.23	5.85
2.00	28.450	8225	4.82	1.930	5.69	5.55	1.95	15.934	2925	4.11	29.536	1.202	29.52	5.97
2.10	28.676	8876	4.90	2.054	5.79	5.65	2.01	15.755	3041	4.25	29.811	1.266	29.80	6.09
2.20	28.894	9542	4.98	2.180	5.88	5.74	2.08	15.575	3149	4.39	30.076	1.330	30.06	6.20
2.30	29.104	10222	5.05	2.306	5.97	5.84	2.14	15.395	3250	4.54	30.331	1.392	30.32	6.31
2.40	29.306	10916	5.12	2.434	6.06	5.93	2.20				30.577	1.455	30.56	6.42
2.50	29.501	11622	5.18	2.562	6.14	6.01	2.25				30.815	1.517	30.80	6.53
2.60	29.690	12341	5.24	2.692	6.23	6.10	2.31				31.045	1.578	31.03	6.64
2.70	29.872	13071	5.30	2.822	6.31	6.19	2.37				31.268	1.639	31.26	6.75
2.80	30.048	13813	5.36	2.953	6.39	6.27	2.42				31.483	1.699	31.47	6.86
2.90	30.219	14565	5.41	3.085	6.47	6.35	2.47				31.691	1.759	31.68	6.96
3.00	30.384	15327	5.47	3.218	6.55	6.43	2.53				31.890	1.818	31.88	7.06
3.10	30.545	16100	5.52	3.351	6.63	6.51	2.58				32.080	1.877	32.07	7.16
3.20	30.701	16882	5.57	3.485	6.71	6.59	2.63				32.262	1.935	32.25	7.25
3.30	30.853	17674	5.62	3.620	6.79	6.67	2.68				32.434	1.992	32.43	7.34
3.40	31.000	18474	5.66	3.755	6.86	6.75	2.73				32.597	2.049	32.59	7.43
3.50	31.144	19282	5.71	3.891	6.94	6.82	2.78				32.750	2.106	32.74	7.51

BERYLLIUM

COEFFICIENTS IN US=C0+S*UP+S2*UP**2+... 7.998 1.124
DE/DP)V= .466 CM**3/G GAMMA=1.160
DEBYE PARAMETERS RHO0= 1.851 T= 293 THETA= 930 3NK= 2.769E+07 CV= 1.755E+07

**10^{10} ergs/g
*10^{6} ergs/g-deg K

P	PARAMETERS ON THE HUGONIOT							RELEASE ISENTROPE(P=0)			ZERO KELVIN		T 0 ISOTHERM	
MB	RHO G/CM**3	T DEG-K	S *	E **	C KM/SEC	US KM/SEC	UP KM/SEC	RHO G/CM**3	T DEG-K	UFS KM/SEC	RHO G/CM**3	E **	RHO G/CM**3	C KM/SEC
0.00	1.851	293	10.87	.213	8.00	8.00	0.00	1.851	293	0.00	1.858	0.000	1.85	8.00
.10	1.993	324	11.21	.406	8.72	8.70	.62	1.851	299	1.24	2.000	.181	1.99	8.73
.20	2.115	376	12.77	.888	9.29	9.30	1.16	1.849	325	2.32	2.123	.608	2.12	9.30
.30	2.222	453	15.49	1.568	9.77	9.85	1.65	1.846	372	3.30	2.232	1.182	2.23	9.78
.40	2.319	553	18.93	2.395	10.17	10.35	2.09	1.841	435	4.19	2.332	1.852	2.33	10.20
.50	2.408	673	22.68	3.337	10.54	10.81	2.50	1.835	510	5.01	2.425	2.588	2.42	10.56
.60	2.490	813	26.52	4.372	10.86	11.24	2.88	1.828	595	5.79	2.512	3.369	2.51	10.90
.70	2.566	969	30.29	5.484	11.16	11.65	3.25	1.820	688	6.53	2.594	4.183	2.59	11.21
.80	2.638	1142	33.92	6.662	11.44	12.03	3.59	1.810	787	7.23	2.671	5.022	2.67	11.50
.90	2.706	1330	37.37	7.896	11.69	12.40	3.92	1.800	892	7.90	2.745	5.878	2.74	11.76

MAGNESIUM

COEFFICIENTS IN US=C0+S*UP+S2*UP**2+... 4.492 1.263
DE/DP)V= .403 CM**3/G GAMMA=1.425
DEBYE PARAMETERS RHO0= 1.740 T= 293 THETA= 237 3NK= 1.026E+07 CV= 9.933E+06

**10^{10} ergs/g
*10^{6} ergs/g-deg K

P	PARAMETERS ON THE HUGONIOT							RELEASE ISENTROPE(P=0)			ZERO KELVIN		T 0 ISOTHERM	
MB	RHO G/CM**3	T DEG-K	S *	E **	C KM/SEC	US KM/SEC	UP KM/SEC	RHO G/CM**3	T DEG-K	UFS KM/SEC	RHO G/CM**3	E **	RHO G/CM**3	C KM/SEC
0.00	1.740	293	16.02	.222	4.49	4.49	0.00	1.740	293	0.00	1.767	0.000	1.74	4.49
.10	2.106	457	17.99	.721	5.81	5.75	1.00	1.732	355	2.00	2.132	.415	2.12	5.83
.20	2.346	803	22.47	1.707	6.62	6.67	1.72	1.707	541	3.47	2.386	1.143	2.38	6.67
.30	2.532	1314	26.70	2.918	7.24	7.42	2.32	1.671	794	4.71	2.589	1.954	2.58	7.33
.40	2.684	1958	30.22	4.266	7.77	8.08	2.84	1.628	1075	5.81	2.760	2.789	2.75	7.91
.50	2.815	2711	33.11	5.708	8.23	8.68	3.31	1.578	1360	6.82	2.910	3.623	2.90	8.42
.60	2.929	3552	35.53	7.220	8.65	9.22	3.74	1.525	1630	7.77	3.043	4.445	3.04	8.90
.70	3.031	4469	37.59	8.788	9.04	9.72	4.14	1.469	1873	8.66	3.162	5.253	3.16	9.35
.80	3.122	5451	39.38	10.400	9.39	10.19	4.51	1.411	2081	9.52	3.269	6.043	3.26	9.76

AZ31B ALLCY

95.8/3/1/0.2 WT PER CENT MG/AL/ZN/MN

COEFFICIENTS IN US=C0+S*UP+S2*UP**2+... 4.522 1.242
DE/DP)V= .392 CM**3/G GAMMA=1.434
DEBYE PARAMETERS RHO0= 1.780 T= 293 THETA= 322 3NK= 1.023E+07 CV= 9.638E+06

**10^{10} ergs/g
*10^{6} ergs/g-deg K

P	PARAMETERS ON THE HUGONIOT							RELEASE ISENTRCPE(P=0)			ZERO KELVIN		T 0 ISOTHERM	
MB	RHO G/CM**3	T DEG-K	S *	E **	C KM/SEC	US KM/SEC	UP KM/SEC	RHC G/CM**3	T DEG-K	UFS KM/SEC	RHO G/CM**3	E **	RHO G/CM**3	C KM/SEC
0.00	1.780	293	12.98	.196	4.52	4.52	C.00	1.780	293	0.00	1.804	0.000	1.78	4.52
.10	2.146	452	14.82	.676	5.77	5.74	.98	1.773	352	1.96	2.171	.403	2.16	5.79
.20	2.392	786	19.10	1.633	6.54	6.63	1.70	1.749	531	3.41	2.430	1.120	2.42	6.59
.30	2.583	1279	23.23	2.815	7.14	7.36	2.29	1.714	775	4.64	2.639	1.925	2.63	7.22
.40	2.740	1903	26.69	4.134	7.63	8.01	2.81	1.672	1049	5.73	2.817	2.757	2.81	7.75
.50	2.876	2633	29.56	5.548	8.07	8.59	3.27	1.624	1329	6.73	2.973	3.592	2.97	8.24
.60	2.995	3451	31.96	7.034	8.46	9.11	3.70	1.573	1597	7.66	3.112	4.418	3.11	8.68
.70	3.102	4345	34.01	8.575	8.82	9.61	4.09	1.518	1842	8.55	3.236	5.231	3.23	9.09

TITANIUM

COEFFICIENTS IN US=C0+S*UP+S2*UP**2+... 5.220 .767
DE/DP)V= .203 CM**3/G GAMMA=1.090
DEBYE PARAMETERS RHO0= 4.528 T= 293 THETA= 9C 3NK= 5.210E+06 CV= 5.186E+06

**10^{10} ergs/g
*10^{6} ergs/g-deg K

P	PARAMETERS ON THE HUGONIOT							RELEASE ISENTRCPE(P=0)			ZERO KELVIN		T 0 ISOTHERM	
MB	RHO G/CM**3	T DEG-K	S *	E **	C KM/SEC	US KM/SEC	UP KM/SEC	RHC G/CM**3	T DEG-K	UFS KM/SEC	RHO G/CM**3	E **	RHO G/CM**3	C KM/SEC
0.00	4.528	293	13.11	.136	5.22	5.22	C.C0	4.528	293	0.00	4.553	0.000	4.53	5.22
.05	4.708	306	13.12	.157	5.32	5.38	.21	4.528	294	.41	4.733	.020	4.71	5.32
.10	4.881	323	13.20	.216	5.41	5.53	.40	4.527	298	.80	4.906	.076	4.88	5.41
.15	5.049	346	13.39	.307	5.48	5.67	.58	4.526	309	1.17	5.075	.161	5.05	5.48
.20	5.211	378	13.69	.426	5.54	5.80	.76	4.525	327	1.52	5.240	.269	5.22	5.54
.25	5.370	419	14.08	.569	5.59	5.93	.93	4.522	353	1.86	5.402	.397	5.38	5.59

ZIRCONIUM

COEFFICIENTS IN US=CO+S*UP+S2*UP**2+... 3.757 1.018
DE/DP)V= .141 CM**3/G GAMMA=1.090
DEBYE PARAMETERS RHOO= 6.505 T= 293 THETA= 90 3NK= 2.734E+06 CV= 2.721E+06

**10^{10} ergs/g
*10^{6} ergs/g-deg K

	PARAMETERS ON THE HUGONIOT							RELEASE ISENTROPE(P=0)			ZERO KELVIN		T 0 ISOTHERM	
P MB	RHO G/CM**3	T DEG-K	S *	E **	C KM/SEC	US KM/SEC	UP KM/SEC	RHO G/CM**3	T DEG-K	UFS KM/SEC	RHO G/CM**3	E **	RHO G/CM**3	C KM/SEC
0.00	6.505	293	6.88	.071	3.76	3.76	0.00	6.505	293	0.00	6.541	0.000	6.50	3.76
.05	6.841	311	6.90	.090	3.95	3.95	.19	6.505	295	.39	6.876	.018	6.84	3.95
.10	7.147	339	7.01	.141	4.11	4.14	.37	6.503	308	.74	7.182	.064	7.15	4.11
.15	7.431	383	7.24	.215	4.25	4.30	.54	6.499	334	1.07	7.468	.130	7.44	4.25
.20	7.695	446	7.56	.309	4.37	4.46	.69	6.494	376	1.38	7.736	.211	7.71	4.37
.25	7.944	529	7.95	.420	4.48	4.61	.83	6.486	433	1.67	7.991	.304	7.97	4.48

HAFNIUM

COEFFICIENTS IN US=CO+S*UP+S2*UP**2+... 2.954 1.121
DE/DP)V= .079 CM**3/G GAMMA= .980
DEBYE PARAMETERS RHOO=12.885 T= 293 THETA= 90 3NK= 1.500E+06 CV= 1.493E+06

**10^{10} ergs/g
*10^{10} ergs/g-deg K

	PARAMETERS ON THE HUGONIOT							RELEASE ISENTROPE(P=0)			ZERO KELVIN		T 0 ISOTHERM	
P MB	RHO G/CM**3	T DEG-K	S *	E **	C KM/SEC	US KM/SEC	UP KM/SEC	RHO G/CM**3	T DEG-K	UFS KM/SEC	RHO G/CM**3	E **	RHO G/CM**3	C KM/SEC
0.00	12.885	293	3.77	.039	2.95	2.95	0.00	12.885	293	0.00	12.942	0.000	12.89	2.95
.05	13.429	306	3.78	.047	3.10	3.09	.13	12.885	295	.25	13.483	.008	13.43	3.10
.10	13.925	327	3.83	.068	3.24	3.22	.24	12.883	303	.48	13.978	.027	13.93	3.24
.15	14.382	358	3.92	.100	3.35	3.34	.35	12.878	323	.70	14.436	.055	14.39	3.35
.20	14.808	403	4.06	.140	3.46	3.46	.45	12.872	354	.90	14.866	.090	14.83	3.46
.25	15.208	463	4.23	.187	3.55	3.56	.54	12.862	398	1.09	15.271	.130	15.23	3.55
.30	15.585	538	4.43	.241	3.64	3.67	.64	12.850	453	1.27	15.656	.175	15.62	3.64
.35	15.943	629	4.64	.300	3.72	3.76	.72	12.835	520	1.44	16.024	.222	15.99	3.73
.40	16.284	736	4.85	.363	3.80	3.86	.80	12.818	597	1.61	16.375	.272	16.34	3.80
.45	16.610	858	5.05	.431	3.87	3.95	.89	12.799	684	1.77	16.714	.325	16.68	3.88
.50	16.923	994	5.25	.502	3.94	4.03	.96	12.778	780	1.93	17.039	.379	17.01	3.95
.55	17.224	1145	5.45	.577	4.00	4.12	1.04	12.754	886	2.08	17.354	.435	17.32	4.02
.60	17.514	1310	5.63	.655	4.07	4.20	1.11	12.729	999	2.22	17.658	.492	17.63	4.08

VANADIUM

COEFFICIENTS IN US=C0+S*UP+S2*UP**2+... 5.077 1.201
DE/DP)V= .127 CM**3/G GAMMA=1.290
DEBYE PARAMETERS RHO0= 6.100 T= 293 THETA= 196 3NK= 4.899E+06 CV= 4.791E+06

** 10^{10} ergs/g
* 10^{6} ergs/g-deg K

	PARAMETERS ON THE HUGONIOT							RELEASE ISENTROPE(P=0)			ZERO KELVIN		T 0 ISOTHERM	
P MB	RHO G/CM**3	T DEG-K	S *	E **	C KM/SEC	US KM/SEC	UP KM/SEC	RHO G/CM**3	T DEG-K	UFS KM/SEC	RHO G/CM**3	E **	RHO G/CM**3	C KM/SEC
0.00	6.100	293	8.56	.111	5.08	5.08	0.00	6.100	293	0.00	6.134	0.000	6.10	5.08
.10	6.458	319	8.62	.157	5.48	5.44	.30	6.099	297	.60	6.490	.043	6.46	5.48
.20	6.769	361	8.95	.273	5.81	5.76	.57	6.096	318	1.14	6.801	.147	6.78	5.81
.30	7.046	428	9.55	.441	6.09	6.05	.81	6.090	360	1.63	7.082	.292	7.06	6.10
.40	7.297	524	10.34	.649	6.34	6.32	1.04	6.081	422	2.08	7.339	.465	7.32	6.35
.50	7.528	648	11.21	.888	6.56	6.57	1.25	6.068	504	2.50	7.578	.657	7.56	6.58
.60	7.741	801	12.10	1.154	6.77	6.81	1.44	6.053	603	2.90	7.801	.865	7.78	6.79
.70	7.941	981	12.96	1.441	6.96	7.04	1.63	6.036	717	3.28	8.011	1.083	8.00	6.98
.80	8.128	1187	13.78	1.747	7.13	7.25	1.81	6.016	845	3.64	8.211	1.310	8.20	7.16
.90	8.305	1418	14.55	2.070	7.30	7.45	1.98	5.994	984	3.98	8.401	1.544	8.39	7.34
1.00	8.473	1673	15.27	2.407	7.45	7.65	2.14	5.970	1133	4.32	8.582	1.783	8.57	7.50
1.10	8.633	1950	15.94	2.757	7.60	7.84	2.30	5.944	1291	4.64	8.756	2.026	8.74	7.65
1.20	8.786	2249	16.56	3.118	7.74	8.02	2.45	5.917	1457	4.95	8.923	2.272	8.91	7.80
1.30	8.932	2568	17.13	3.489	7.88	8.20	2.60	5.889	1629	5.26	9.084	2.520	9.07	7.95

NIOBIUM

COEFFICIENTS IN US=C0+S*UP+S2*UP**2+... 4.438 1.207
DE/DP)V= .079 CM**3/G GAMMA=1.470
DEBYE PARAMETERS RHO0= 8.586 T= 293 THETA= 143 3NK= 2.685E+06 CV= 2.653E+06

**10^{10} ergs/g
*10^{6} ergs/g-deg K

	PARAMETERS ON THE HUGONIOT							RELEASE ISENTROPE(P=0)			ZERO KELVIN		T 0 ISOTHERM	
P MB	RHO G/CM**3	T DEG-K	S *	E **	C KM/SEC	US KM/SEC	UP KM/SEC	RHO G/CM**3	T DEG-K	UFS KM/SEC	RHO G/CM**3	E **	RHO G/CM**3	C KM/SEC
0.00	8.586	293	5.52	.065	4.44	4.44	C.CO	8.586	293	0.00	8.628	0.000	8.59	4.44
.10	9.056	321	5.56	.096	4.77	4.73	.25	8.585	298	.49	9.096	.029	9.06	4.77
.20	9.468	369	5.77	.174	5.04	5.00	.47	8.581	321	.93	9.510	.099	9.48	5.04
.30	9.837	446	6.14	.288	5.27	5.24	.67	8.573	369	1.33	9.883	.198	9.86	5.28
.40	10.172	558	6.63	.429	5.48	5.47	.85	8.560	442	1.71	10.227	.317	10.20	5.49
.50	10.480	705	7.16	.592	5.66	5.68	1.03	8.544	537	2.06	10.548	.450	10.52	5.68
.60	10.767	888	7.69	.773	5.83	5.87	1.19	8.523	652	2.39	10.848	.595	10.83	5.85
.70	11.035	1104	8.19	.970	5.99	6.06	1.35	8.500	785	2.70	11.132	.747	11.11	6.01
.80	11.287	1354	8.67	1.180	6.13	6.24	1.49	8.473	934	3.00	11.402	.906	11.38	6.16
.90	11.525	1635	9.12	1.402	6.27	6.41	1.63	8.444	1096	3.29	11.659	1.071	11.64	6.30
1.00	11.751	1947	9.53	1.634	6.40	6.58	1.77	8.412	1271	3.57	11.906	1.240	11.89	6.43
1.10	11.967	2287	9.91	1.875	6.52	6.73	1.90	8.377	1455	3.84	12.143	1.411	12.13	6.56
1.20	12.172	2655	10.26	2.124	6.63	6.89	2.03	8.341	1649	4.10	12.371	1.586	12.35	6.68
1.30	12.370	3048	10.59	2.381	6.74	7.04	2.15	8.303	1849	4.36	12.591	1.762	12.57	6.80
1.40	12.559	3467	10.89	2.645	6.85	7.18	2.27	8.263	2056	4.61	12.804	1.940	12.79	6.91
1.50	12.741	3910	11.18	2.914	6.95	7.32	2.39	8.222	2268	4.85	13.010	2.120	12.99	7.02
1.60	12.917	4375	11.44	3.190	7.05	7.46	2.50	8.179	2484	5.08	13.209	2.300	13.20	7.12
1.70	13.087	4862	11.69	3.470	7.15	7.59	2.61	8.134	2703	5.32	13.403	2.481	13.39	7.23
1.80	13.251	5369	11.93	3.756	7.24	7.72	2.72	8.089	2924	5.54	13.591	2.662	13.58	7.32

TANTALUM

COEFFICIENTS IN US=C0+S*UP+S2*UP**2+... 3.414 1.200
DE/DP)V= .038 CM**3/G GAMMA=1.600
DEBYE PARAMETERS RHO0=16.654 T= 293 THETA= 180 3NK= 1.400E+06 CV= 1.374E+06

** 10^{10} ergs/g
* 10^{6} ergs/g-deg K

P	PARAMETERS ON THE HUGONIOT							RELEASE ISENTROPE(P=0)			ZERO KELVIN		T 0 ISOTHERM	
MB	RHO G/CM**3	T DEG-K	S *	E **	C KM/SEC	US KM/SEC	UP KM/SEC	RHO G/CM**3	T DEG-K	UFS KM/SEC	RHO G/CM**3	E **	RHO G/CM**3	C KM/SEC
0.00	16.654	293	2.56	.032	3.41	3.41	0.00	16.654	293	0.00	16.728	0.000	16.65	3.41
.10	17.457	319	2.58	.046	3.64	3.61	.17	16.653	297	.33	17.528	.013	17.46	3.64
.20	18.170	361	2.66	.083	3.82	3.79	.32	16.647	315	.63	18.244	.046	18.19	3.82
.30	18.815	427	2.83	.136	3.98	3.96	.45	16.634	354	.91	18.898	.094	18.85	3.98
.40	19.407	522	3.05	.203	4.12	4.11	.58	16.615	414	1.17	19.505	.151	19.46	4.13
.50	19.956	648	3.29	.281	4.25	4.26	.70	16.589	494	1.41	20.072	.216	20.03	4.26
.60	20.468	804	3.55	.368	4.37	4.40	.82	16.558	592	1.64	20.608	.287	20.57	4.38
.70	20.949	992	3.80	.463	4.48	4.53	.93	16.520	705	1.86	21.117	.363	21.08	4.49
.80	21.403	1209	4.04	.565	4.58	4.65	1.03	16.478	833	2.08	21.602	.443	21.57	4.59
.90	21.834	1455	4.26	.673	4.67	4.77	1.13	16.431	974	2.28	22.067	.526	22.03	4.68
1.00	22.244	1728	4.47	.787	4.76	4.89	1.23	16.380	1126	2.48	22.513	.611	22.48	4.78
1.10	22.636	2029	4.67	.905	4.84	5.00	1.32	16.325	1287	2.67	22.944	.698	22.91	4.86
1.20	23.011	2355	4.85	1.028	4.92	5.11	1.41	16.267	1457	2.85	23.359	.787	23.33	4.94
1.30	23.371	2706	5.02	1.154	5.00	5.21	1.50	16.205	1634	3.03	23.761	.878	23.73	5.02
1.40	23.718	3081	5.18	1.284	5.07	5.31	1.58	16.140	1818	3.20	24.151	.970	24.12	5.10
1.50	24.052	3478	5.32	1.418	5.14	5.41	1.66	16.073	2007	3.38	24.529	1.062	24.50	5.17
1.60	24.375	3897	5.46	1.554	5.21	5.51	1.74	16.003	2200	3.54	24.898	1.156	24.87	5.25
1.70	24.687	4337	5.59	1.693	5.27	5.60	1.82	15.932	2397	3.71	25.256	1.250	25.23	5.32
1.80	24.989	4797	5.72	1.835	5.34	5.69	1.90	15.857	2596	3.87	25.605	1.344	25.58	5.38
1.90	25.282	5277	5.83	1.979	5.40	5.78	1.97	15.782	2798	4.02	25.946	1.439	25.92	5.45
2.00	25.567	5775	5.94	2.126	5.46	5.87	2.05	15.704	3001	4.18	26.278	1.534	26.26	5.51
2.10	25.843	6291	6.05	2.274	5.51	5.95	2.12	15.624	3205	4.33	26.603	1.629	26.58	5.57
2.20	26.113	6824	6.14	2.425	5.57	6.04	2.19	15.544	3410	4.48	26.920	1.725	26.90	5.64
2.30	26.375	7374	6.24	2.577	5.63	6.12	2.26	15.461	3614	4.63	27.228	1.821	27.21	5.69

CHROMIUM

COEFFICIENTS IN US=C0+S*UP+S2*UP**2+... 5.173 1.473
DE/DP)V= .118 CM**3/G GAMMA=1.190
DEBYE PARAMETERS RHOC= 7.117 T= 293 THETA= 365 3NK= 4.797E+06 CV= 4.444E+06

** 10^{10} ergs/g
* 10^{6} ergs/g-deg K

	PARAMETERS ON THE HUGONIOT							RELEASE ISENTROPE(P=0)			ZERO KELVIN		T 0 ISOTHERM	
P MB	RHO G/CM**3	T DEG-K	S *	E **	C KM/SEC	US KM/SEC	LP KM/SEC	RHO G/CM**3	T DEG-K	LFS KM/SEC	RHO G/CM**3	E **	RHO G/CM**3	C KM/SEC
0.00	7.117	293	5.52	.086	5.17	5.17	0.00	7.117	293	0.00	7.144	0.000	7.12	5.17
.10	7.458	313	5.57	.118	5.65	5.55	.25	7.117	296	.51	7.482	.030	7.46	5.65
.20	7.747	344	5.81	.200	6.06	5.88	.48	7.114	312	.96	7.771	.104	7.75	6.06
.30	8.001	395	6.27	.319	6.42	6.18	.68	7.109	346	1.36	8.026	.206	8.01	6.43
.40	8.226	465	6.89	.465	6.74	6.46	.87	7.102	395	1.74	8.255	.326	8.24	6.75
.50	8.431	556	7.60	.633	7.04	6.71	1.05	7.093	460	2.09	8.464	.460	8.45	7.06
.60	8.618	666	8.34	.820	7.31	6.96	1.21	7.081	538	2.43	8.657	.604	8.64	7.34
.70	8.790	796	9.09	1.022	7.57	7.19	1.37	7.067	629	2.74	8.835	.756	8.82	7.61
.80	8.950	943	9.81	1.237	7.81	7.41	1.52	7.051	731	3.04	9.002	.913	8.99	7.86
.90	9.100	1108	10.50	1.464	8.04	7.62	1.66	7.033	843	3.33	9.159	1.074	9.15	8.10
1.00	9.240	1289	11.15	1.700	8.27	7.82	1.80	7.014	964	3.61	9.306	1.239	9.30	8.33
1.10	9.373	1485	11.77	1.946	8.48	8.01	1.93	6.993	1092	3.87	9.446	1.406	9.44	8.56
1.20	9.498	1696	12.35	2.199	8.68	8.20	2.06	6.971	1228	4.13	9.579	1.575	9.57	8.77
1.30	9.617	1922	12.89	2.460	8.88	8.38	2.18	6.947	1370	4.39	9.706	1.745	9.70	8.98
1.40	9.730	2160	13.40	2.727	9.07	8.56	2.30	6.922	1517	4.63	9.827	1.916	9.82	9.19

MOLYBDENUM

COEFFICIENTS IN US=C0+S*UP+S2*UP**2+... 5.124 1.233
DE/DP)V= .064 CM**3/G GAMMA=1.520
DEBYE PARAMETERS RHO0=10.206 T= 293 THETA= 314 3NK= 2.600E+06 CV= 2.457E+06

**10^{10} ergs/g
*10^{6} ergs/g-deg K

	PARAMETERS ON THE HUGONIOT							RELEASE ISENTROPE(P=0)			ZERO KELVIN		T 0 ISOTHERM	
P MB	RHO G/CM**3	T DEG-K	S *	E **	C KM/SEC	US KM/SEC	UP KM/SEC	RHO G/CM**3	T DEG-K	UFS KM/SEC	RHO G/CM**3	E **	RHO G/CM**3	C KM/SEC
0.00	10.206	293	3.36	.050	5.12	5.12	0.00	10.206	293	0.00	10.235	0.000	10.21	5.12
.10	10.568	311	3.38	.067	5.38	5.35	.18	10.206	295	.37	10.596	.016	10.57	5.38
.20	10.897	336	3.46	.112	5.61	5.56	.35	10.204	305	.71	10.926	.059	10.90	5.61
.30	11.200	376	3.64	.180	5.81	5.75	.51	10.201	329	1.02	11.231	.121	11.21	5.82
.40	11.482	433	3.91	.268	6.00	5.94	.66	10.195	365	1.32	11.517	.198	11.50	6.00
.50	11.746	509	4.24	.371	6.16	6.11	.80	10.188	416	1.60	11.787	.287	11.77	6.17
.60	11.995	604	4.61	.488	6.32	6.28	.94	10.178	480	1.88	12.042	.386	12.03	6.33
.70	12.230	719	4.99	.618	6.47	6.44	1.07	10.166	556	2.14	12.286	.493	12.27	6.48
.80	12.453	853	5.37	.757	6.60	6.59	1.19	10.153	643	2.39	12.519	.606	12.50	6.61
.90	12.666	1007	5.74	.906	6.73	6.74	1.31	10.137	742	2.63	12.743	.725	12.73	6.75
1.00	12.870	1179	6.10	1.064	6.85	6.88	1.42	10.121	850	2.86	12.958	.849	12.94	6.87
1.10	13.065	1370	6.44	1.229	6.97	7.02	1.54	10.102	967	3.09	13.166	.977	13.15	6.99
1.20	13.253	1579	6.76	1.402	7.08	7.15	1.64	10.082	1092	3.31	13.367	1.108	13.35	7.11
1.30	13.434	1805	7.07	1.580	7.19	7.28	1.75	10.061	1226	3.52	13.562	1.243	13.55	7.22
1.40	13.608	2049	7.36	1.765	7.29	7.41	1.85	10.038	1366	3.73	13.751	1.379	13.74	7.32
1.50	13.777	2309	7.63	1.955	7.39	7.53	1.95	10.014	1512	3.93	13.935	1.518	13.92	7.43
1.60	13.940	2585	7.89	2.149	7.48	7.65	2.05	9.989	1665	4.13	14.113	1.659	14.10	7.53
1.70	14.098	2877	8.14	2.349	7.58	7.77	2.14	9.963	1822	4.33	14.288	1.802	14.28	7.62
1.80	14.251	3184	8.37	2.553	7.67	7.88	2.24	9.936	1985	4.52	14.457	1.946	14.45	7.72
1.90	14.400	3506	8.59	2.761	7.75	8.00	2.33	9.908	2151	4.71	14.623	2.091	14.61	7.81
2.00	14.545	3841	8.80	2.973	7.84	8.11	2.42	9.879	2321	4.90	14.785	2.237	14.78	7.90
2.10	14.686	4191	9.00	3.188	7.92	8.21	2.51	9.849	2495	5.08	14.943	2.384	14.93	7.99

TUNGSTEN

COEFFICIENTS IN US=C0+S*UP+S2*UP**2+... 4.029 1.237
DE/DP)V= .034 CM**3/G GAMMA=1.540
DEBYE PARAMETERS RHO0=19.224 T= 293 THETA= 247 3NK= 1.357E+06 CV= 1.310E+06

**10^{10} ergs/g
*10^{6} ergs/g-deg K

	PARAMETERS ON THE HUGONIOT							RELEASE ISENTROPE(P=0)			ZERO KELVIN		T 0 ISOTHERM	
P MB	RHO G/CM**3	T DEG-K	S *	E **	C KM/SEC	US KM/SEC	UP KM/SEC	RHO G/CM**3	T DEG-K	UFS KM/SEC	RHO G/CM**3	E **	RHO G/CM**3	C KM/SEC
0.00	19.224	293	2.07	.029	4.03	4.03	0.00	19.224	293	0.00	19.276	0.000	19.22	4.03
.10	19.813	308	2.07	.036	4.21	4.18	.12	19.224	294	.25	19.863	.007	19.82	4.21
.20	20.355	330	2.11	.058	4.37	4.33	.24	19.222	303	.48	20.406	.027	20.36	4.37
.30	20.859	362	2.19	.090	4.51	4.46	.35	19.217	321	.70	20.912	.057	20.87	4.51
.40	21.331	409	2.30	.131	4.64	4.59	.45	19.210	351	.91	21.389	.094	21.35	4.64
.50	21.775	472	2.46	.181	4.76	4.71	.55	19.200	393	1.10	21.841	.138	21.81	4.76
.60	22.195	551	2.63	.238	4.87	4.83	.65	19.186	447	1.29	22.271	.186	22.24	4.87
.70	22.594	647	2.81	.300	4.97	4.94	.74	19.170	512	1.48	22.682	.239	22.65	4.98
.80	22.975	760	3.00	.368	5.07	5.05	.82	19.152	588	1.65	23.076	.295	23.05	5.08
.90	23.339	891	3.19	.441	5.17	5.15	.91	19.130	674	1.82	23.456	.355	23.43	5.18
1.00	23.688	1038	3.37	.519	5.26	5.25	.99	19.106	769	1.99	23.822	.417	23.79	5.27
1.10	24.024	1202	3.54	.600	5.34	5.35	1.07	19.079	874	2.15	24.176	.482	24.15	5.35
1.20	24.347	1383	3.71	.685	5.42	5.45	1.15	19.051	986	2.30	24.519	.548	24.49	5.44
1.30	24.659	1580	3.87	.774	5.50	5.54	1.22	19.020	1107	2.45	24.852	.617	24.83	5.52
1.40	24.960	1793	4.02	.865	5.58	5.63	1.29	18.987	1235	2.60	25.176	.686	25.15	5.60
1.50	25.252	2021	4.17	.960	5.65	5.72	1.36	18.952	1369	2.75	25.490	.757	25.47	5.67
1.60	25.535	2264	4.30	1.057	5.72	5.80	1.43	18.916	1509	2.89	25.797	.830	25.78	5.74
1.70	25.809	2522	4.43	1.157	5.79	5.89	1.50	18.878	1655	3.03	26.096	.903	26.08	5.81
1.80	26.076	2795	4.55	1.259	5.85	5.97	1.57	18.838	1807	3.16	26.388	.977	26.37	5.88
1.90	26.336	3081	4.67	1.363	5.92	6.05	1.63	18.796	1963	3.30	26.674	1.052	26.65	5.95
2.00	26.588	3380	4.78	1.469	5.98	6.13	1.70	18.753	2123	3.43	26.953	1.128	26.93	6.02
2.10	26.834	3693	4.89	1.578	6.04	6.21	1.76	18.709	2287	3.56	27.226	1.204	27.21	6.08
2.20	27.074	4019	4.99	1.688	6.10	6.28	1.82	18.663	2455	3.69	27.494	1.281	27.48	6.14
2.30	27.309	4357	5.09	1.800	6.16	6.36	1.88	18.617	2627	3.81	27.756	1.358	27.74	6.20
2.40	27.538	4707	5.18	1.913	6.21	6.43	1.94	18.569	2801	3.93	28.013	1.436	28.00	6.26
2.50	27.762	5070	5.27	2.028	6.27	6.50	2.00	18.519	2977	4.06	28.266	1.514	28.25	6.32
2.60	27.981	5443	5.35	2.145	6.32	6.57	2.06	18.469	3157	4.18	28.513	1.593	28.50	6.38
2.70	28.195	5828	5.44	2.263	6.37	6.64	2.11	18.418	3338	4.29	28.756	1.671	28.74	6.44

RHENIUM

COEFFICIENTS IN US=CO+S*UP+S2*UP**2+... 4.184 1.367
DE/DP)V= .019 CM**3/G GAMMA=2.440
DEBYE PARAMETERS RHOO=21.020 T= 293 THETA= 236 3NK= 1.400E+06 CV= 1.356E+06

**10^{10} ergs/g
*10^{6} ergs/g-deg K

	PARAMETERS ON THE HUGONIOT							RELEASE ISENTROPE(P=0)			ZERO KELVIN		T 0 ISOTHERM	
P MB	RHO G/CM**3	T DEG-K	S *	E **	C KM/SEC	US KM/SEC	UP KM/SEC	RHO G/CM**3	T DEG-K	UFS KM/SEC	RHO G/CM**3	E **	RHO G/CM**3	C KM/SEC
0.00	21.020	293	2.19	.030	4.18	4.18	0.00	21.020	293	0.00	21.108	0.000	21.02	4.18
.10	21.566	313	2.20	.036	4.37	4.33	.11	21.020	294	.22	21.652	.006	21.57	4.37
.20	22.069	337	2.22	.053	4.53	4.47	.21	21.017	300	.43	22.155	.021	22.08	4.53
.30	22.535	368	2.28	.078	4.68	4.61	.31	21.012	312	.62	22.625	.045	22.56	4.68
.40	22.971	410	2.37	.111	4.81	4.73	.40	21.004	333	.80	23.068	.074	23.01	4.82
.50	23.381	464	2.48	.150	4.94	4.85	.49	20.992	361	.98	23.487	.109	23.43	4.94
.60	23.767	530	2.62	.195	5.05	4.97	.57	20.977	398	1.15	23.886	.148	23.83	5.06
.70	24.134	609	2.76	.245	5.16	5.08	.66	20.959	442	1.32	24.269	.191	24.22	5.17
.80	24.483	702	2.92	.299	5.26	5.19	.73	20.938	492	1.47	24.636	.237	24.59	5.27
.90	24.816	808	3.07	.358	5.36	5.29	.81	20.914	549	1.63	24.989	.286	24.94	5.36
1.00	25.135	927	3.23	.420	5.45	5.39	.88	20.888	612	1.77	25.330	.337	25.29	5.46
1.10	25.441	1060	3.38	.485	5.54	5.49	.95	20.859	681	1.92	25.661	.390	25.62	5.54
1.20	25.735	1205	3.53	.553	5.62	5.58	1.02	20.828	754	2.06	25.981	.446	25.94	5.63
1.30	26.018	1364	3.67	.624	5.70	5.67	1.09	20.794	831	2.20	26.292	.502	26.25	5.71
1.40	26.291	1535	3.81	.698	5.77	5.76	1.16	20.759	913	2.33	26.594	.561	26.56	5.79
1.50	26.555	1719	3.94	.774	5.85	5.85	1.22	20.722	998	2.46	26.889	.621	26.85	5.86
1.60	26.810	1915	4.06	.852	5.92	5.94	1.28	20.683	1087	2.59	27.177	.682	27.14	5.93
1.70	27.057	2123	4.18	.932	5.99	6.02	1.34	20.643	1178	2.72	27.457	.744	27.42	6.00
1.80	27.297	2342	4.30	1.015	6.05	6.10	1.40	20.601	1272	2.84	27.732	.807	27.70	6.07
1.90	27.530	2573	4.40	1.099	6.12	6.18	1.46	20.558	1368	2.96	28.000	.871	27.97	6.14
2.00	27.756	2815	4.51	1.185	6.18	6.26	1.52	20.514	1466	3.08	28.263	.935	28.23	6.20
2.10	27.977	3068	4.61	1.272	6.24	6.34	1.58	20.468	1566	3.20	28.521	1.001	28.49	6.26
2.20	28.191	3332	4.70	1.361	6.30	6.41	1.63	20.421	1668	3.32	28.773	1.067	28.74	6.32
2.30	28.400	3606	4.80	1.452	6.36	6.49	1.69	20.374	1770	3.43	29.021	1.134	28.99	6.38
2.40	28.604	3890	4.88	1.544	6.41	6.56	1.74	20.325	1874	3.54	29.264	1.201	29.24	6.44
2.50	28.802	4183	4.97	1.637	6.47	6.63	1.79	20.276	1978	3.65	29.503	1.269	29.48	6.50
2.60	28.997	4487	5.05	1.731	6.52	6.71	1.84	20.226	2084	3.76	29.738	1.337	29.71	6.55
2.70	29.186	4800	5.13	1.827	6.58	6.78	1.90	20.175	2189	3.87	29.967	1.406	29.94	6.61
2.80	29.372	5121	5.20	1.924	6.63	6.84	1.95	20.123	2295	3.98	30.192	1.475	30.17	6.66

RHODIUM

COEFFICIENTS IN US=CO+S*UP+S2*UP**2+... 4.807 1.376
DE/DP)V= .043 CM**3/G GAMMA=1.880
DEBYE PARAMETERS RHOO=12.428 T= 293 THETA= 161 3NK= 2.424E+06 CV= 2.388E+06

**10^{10} ergs/g
*10^{6} ergs/g-deg K

	PARAMETERS ON THE HUGONIOT							RELEASE ISENTROPE (P=0)			ZERO KELVIN		T 0 ISOTHERM	
P MB	RHO G/CM**3	T DEG-K	S *	E **	C KM/SEC	US KM/SEC	UP KM/SEC	RHO G/CM**3	T DEG-K	UFS KM/SEC	RHO G/CM**3	E **	RHO G/CM**3	C KM/SEC
0.00	12.428	293	4.70	.058	4.81	4.81	0.00	12.428	293	0.00	12.486	0.000	12.43	4.81
.10	12.837	313	4.71	.071	5.08	5.03	.16	12.428	294	.32	12.892	.012	12.84	5.08
.20	13.205	338	4.78	.105	5.32	5.23	.31	12.426	303	.62	13.260	.044	13.21	5.32
.30	13.541	375	4.92	.157	5.53	5.42	.45	12.421	321	.89	13.598	.091	13.56	5.53
.40	13.850	426	5.14	.223	5.72	5.60	.57	12.414	350	1.15	13.912	.149	13.87	5.73
.50	14.138	492	5.40	.301	5.90	5.77	.70	12.404	390	1.40	14.207	.216	14.17	5.91
.60	14.408	574	5.69	.389	6.06	5.93	.81	12.392	441	1.63	14.485	.290	14.45	6.07
.70	14.661	672	6.01	.487	6.22	6.08	.93	12.377	501	1.86	14.748	.370	14.72	6.23
.80	14.900	786	6.32	.592	6.36	6.23	1.03	12.359	569	2.07	14.999	.455	14.97	6.38
.90	15.127	916	6.64	.704	6.50	6.37	1.14	12.340	646	2.28	15.239	.544	15.21	6.52
1.00	15.343	1061	6.94	.822	6.63	6.51	1.24	12.319	730	2.49	15.469	.637	15.44	6.65
1.10	15.548	1222	7.23	.946	6.75	6.64	1.33	12.295	821	2.68	15.690	.732	15.66	6.78
1.20	15.745	1397	7.51	1.075	6.87	6.77	1.43	12.270	918	2.87	15.903	.830	15.88	6.91
1.30	15.934	1587	7.78	1.209	6.99	6.89	1.52	12.244	1020	3.06	16.108	.931	16.08	7.03
1.40	16.116	1791	8.03	1.347	7.10	7.02	1.61	12.216	1127	3.24	16.307	1.033	16.28	7.14
1.50	16.291	2009	8.27	1.489	7.21	7.13	1.69	12.186	1239	3.42	16.500	1.137	16.48	7.26
1.60	16.459	2239	8.50	1.634	7.31	7.25	1.78	12.155	1355	3.59	16.687	1.242	16.67	7.36
1.70	16.622	2482	8.71	1.783	7.41	7.36	1.86	12.123	1474	3.76	16.868	1.348	16.85	7.47
1.80	16.779	2738	8.92	1.936	7.51	7.47	1.94	12.090	1595	3.93	17.045	1.456	17.02	7.57
1.90	16.931	3006	9.12	2.091	7.60	7.58	2.02	12.056	1720	4.09	17.216	1.564	17.20	7.67

PALLADIUM

COEFFICIENTS IN US=CO+S*UP+S2*UP**2+... 3.948 1.588
DE/DP)V= .037 CM**3/G GAMMA=2.260
DEBYE PARAMETERS RHOO=11.991 T= 293 THETA= 220 3NK= 2.400E+06 CV= 2.334E+06

** 10^{10} ergs/g
* 10^{6} ergs/g-deg K

P	PARAMETERS ON THE HUGONIOT							RELEASE ISENTROPE(P=0)			ZERO KELVIN		T 0 ISOTHERM	
MB	RHO G/CM**3	T DEG-K	S *	E **	C KM/SEC	US KM/SEC	UP KM/SEC	RHO G/CM**3	T DEG-K	LFS KM/SEC	RHO G/CM**3	E **	RHO G/CM**3	C KM/SEC
0.00	11.991	293	3.92	.053	3.95	3.95	0.00	11.991	293	0.00	12.083	0.000	11.99	3.95
.10	12.569	329	3.95	.072	4.36	4.26	.20	11.989	297	.39	12.654	.018	12.58	4.36
.20	13.051	379	4.09	.121	4.69	4.53	.37	11.982	315	.74	13.136	.061	13.08	4.70
.30	13.466	450	4.35	.190	4.98	4.78	.52	11.967	350	1.05	13.559	.120	13.51	4.99
.40	13.832	547	4.68	.275	5.24	5.01	.67	11.945	401	1.34	13.938	.190	13.89	5.25
.50	14.160	668	5.05	.373	5.47	5.22	.80	11.917	466	1.61	14.282	.267	14.24	5.49
.60	14.458	813	5.42	.480	5.68	5.42	.92	11.883	542	1.86	14.599	.351	14.56	5.71
.70	14.731	982	5.79	.596	5.88	5.60	1.04	11.845	627	2.10	14.893	.439	14.86	5.92
.80	14.982	1173	6.14	.719	6.07	5.78	1.15	11.803	721	2.33	15.168	.530	15.14	6.12
.90	15.216	1386	6.47	.849	6.24	5.95	1.26	11.757	821	2.55	15.426	.624	15.40	6.31
1.00	15.434	1618	6.78	.983	6.41	6.11	1.36	11.708	926	2.76	15.670	.719	15.64	6.49
1.10	15.639	1870	7.07	1.123	6.57	6.27	1.46	11.656	1034	2.97	15.900	.816	15.87	6.66
1.20	15.832	2139	7.35	1.267	6.72	6.42	1.56	11.601	1146	3.17	16.120	.915	16.10	6.82
1.30	16.015	2426	7.60	1.415	6.87	6.57	1.65	11.543	1260	3.37	16.329	1.014	16.31	6.99
1.40	16.188	2729	7.84	1.567	7.01	6.71	1.74	11.483	1375	3.55	16.529	1.114	16.51	7.14
1.50	16.352	3047	8.06	1.721	7.15	6.85	1.83	11.422	1490	3.74	16.720	1.214	16.70	7.30
1.60	16.509	3380	8.27	1.879	7.28	6.98	1.91	11.359	1606	3.92	16.904	1.315	16.88	7.45
1.70	16.659	3726	8.47	2.039	7.41	7.11	1.99	11.296	1721	4.10	17.080	1.416	17.06	7.59
1.80	16.802	4086	8.66	2.202	7.54	7.24	2.07	11.230	1836	4.27	17.250	1.517	17.23	7.74
1.90	16.940	4459	8.84	2.368	7.66	7.36	2.15	11.162	1948	4.44	17.413	1.617	17.40	7.88
2.00	17.072	4843	9.01	2.535	7.78	7.49	2.23	11.093	2059	4.61	17.568	1.718	17.55	8.01
2.10	17.199	5239	9.17	2.705	7.90	7.60	2.30	11.023	2168	4.77	17.716	1.819	17.70	8.14

IRIDIUM

COEFFICIENTS IN US=C0+S*UP+S2*UP**2+... 3.916 1.457
DE/DP)V= .023 CM**3/G GAMMA=1.970
DEBYE PARAMETERS RHO0=22.484 T= 293 THETA= 168 3NK= 1.350E+06 CV= 1.328E+06

**10^{10} ergs/g
*10^{6} ergs/g-deg K

	PARAMETERS ON THE HUGONIOT							RELEASE ISENTROPE(P=0)			ZERO KELVIN		T 0 ISOTHERM	
P MB	RHO G/CM**3	T DEG-K	S *	E **	C KM/SEC	US KM/SEC	UP KM/SEC	RHO G/CM**3	T DEG-K	UFS KM/SEC	RHO G/CM**3	E **	RHO G/CM**3	C KM/SEC
0.00	22.484	293	2.56	.032	3.92	3.92	0.00	22.484	293	0.00	22.576	0.000	22.48	3.92
.10	23.103	310	2.57	.038	4.12	4.08	.11	22.484	294	.22	23.190	.006	23.11	4.12
.20	23.664	332	2.60	.054	4.30	4.22	.21	22.481	301	.42	23.751	.021	23.68	4.30
.30	24.180	362	2.66	.079	4.46	4.36	.31	22.475	315	.61	24.268	.043	24.20	4.47
.40	24.657	403	2.75	.110	4.62	4.49	.40	22.466	338	.79	24.751	.071	24.69	4.62
.50	25.102	456	2.88	.148	4.75	4.62	.48	22.454	370	.96	25.203	.104	25.14	4.76
.60	25.519	522	3.02	.191	4.89	4.74	.56	22.438	411	1.13	25.631	.140	25.58	4.89
.70	25.912	602	3.17	.238	5.01	4.85	.64	22.418	461	1.29	26.037	.180	25.99	5.02
.80	26.284	694	3.33	.289	5.12	4.96	.72	22.395	518	1.44	26.424	.222	26.37	5.14
.90	26.637	800	3.50	.344	5.23	5.07	.79	22.370	583	1.59	26.794	.266	26.75	5.25
1.00	26.973	920	3.65	.402	5.34	5.17	.86	22.341	654	1.73	27.148	.313	27.10	5.36
1.10	27.294	1052	3.81	.463	5.44	5.27	.93	22.310	732	1.87	27.489	.360	27.45	5.46
1.20	27.601	1197	3.96	.527	5.54	5.37	.99	22.276	815	2.00	27.818	.410	27.78	5.56
1.30	27.896	1354	4.10	.593	5.63	5.46	1.06	22.239	904	2.13	28.135	.460	28.10	5.66
1.40	28.179	1523	4.24	.661	5.72	5.55	1.12	22.201	998	2.26	28.441	.512	28.40	5.75
1.50	28.452	1704	4.37	.732	5.81	5.64	1.18	22.160	1095	2.39	28.738	.565	28.70	5.84
1.60	28.715	1897	4.50	.804	5.89	5.73	1.24	22.118	1197	2.51	29.026	.618	28.99	5.93
1.70	28.969	2101	4.62	.878	5.97	5.81	1.30	22.073	1303	2.63	29.305	.672	29.27	6.02
1.80	29.215	2315	4.73	.954	6.05	5.89	1.36	22.027	1412	2.75	29.577	.727	29.54	6.11
1.90	29.453	2540	4.84	1.032	6.13	5.98	1.41	21.979	1523	2.86	29.841	.783	29.81	6.19
2.00	29.683	2776	4.94	1.111	6.21	6.06	1.47	21.930	1638	2.97	30.098	.838	30.07	6.27
2.10	29.907	3021	5.04	1.191	6.28	6.13	1.52	21.879	1754	3.09	30.349	.895	30.32	6.35
2.20	30.124	3276	5.14	1.273	6.36	6.21	1.58	21.826	1873	3.20	30.594	.951	30.56	6.43
2.30	30.335	3541	5.23	1.356	6.43	6.29	1.63	21.772	1994	3.30	30.833	1.008	30.80	6.50
2.40	30.541	3814	5.31	1.440	6.50	6.36	1.68	21.718	2116	3.41	31.066	1.066	31.04	6.58
2.50	30.741	4097	5.40	1.525	6.56	6.43	1.73	21.661	2239	3.52	31.294	1.123	31.27	6.65
2.60	30.936	4388	5.48	1.612	6.63	6.51	1.78	21.604	2364	3.62	31.518	1.181	31.49	6.73
2.70	31.126	4688	5.56	1.699	6.70	6.58	1.83	21.546	2490	3.72	31.736	1.239	31.71	6.80
2.80	31.312	4995	5.63	1.787	6.76	6.65	1.87	21.486	2616	3.82	31.950	1.297	31.93	6.87
2.90	31.493	5311	5.70	1.877	6.82	6.71	1.92	21.426	2743	3.92	32.159	1.355	32.13	6.94
3.00	31.670	5635	5.77	1.967	6.89	6.78	1.97	21.364	2870	4.02	32.362	1.413	32.34	7.01

PLATINUM

COEFFICIENTS IN US=C0+S*UP+S2*UP**2+... 3.598 1.544
DE/DP)V= .019 CM**3/G GAMMA=2.400
DEBYE PARAMETERS RHO0=21.419 T= 293 THETA= 200 3NK= 1.300E+06 CV= 1.270E+06

	PARAMETERS ON THE HUGONIOT							RELEASE ISENTROPE(P=0)			ZERO KELVIN		T 0 ISOTHERM	
P MB	RHO G/CM**3	T DEG-K	S *	E **	C KM/SEC	US KM/SEC	UP KM/SEC	RHO G/CM**3	T DEG-K	UFS KM/SEC	RHO G/CM**3	E **	RHO G/CM**3	C KM/SEC
0.00	21.419	293	2.24	.029	3.60	3.60	0.00	21.419	293	0.00	21.535	0.000	21.42	3.60
.10	22.139	319	2.25	.037	3.85	3.79	.12	21.418	295	.25	22.250	.007	22.15	3.85
.20	22.774	352	2.30	.057	4.06	3.96	.24	21.413	305	.47	22.885	.026	22.80	4.06
.30	23.343	399	2.39	.087	4.25	4.12	.34	21.402	327	.68	23.459	.052	23.38	4.26
.40	23.860	462	2.51	.125	4.42	4.27	.44	21.385	360	.88	23.987	.085	23.92	4.43
.50	24.334	542	2.66	.169	4.58	4.41	.53	21.361	404	1.06	24.477	.123	24.41	4.59
.60	24.773	640	2.83	.219	4.72	4.55	.62	21.333	458	1.24	24.936	.164	24.88	4.74
.70	25.181	757	3.00	.274	4.86	4.68	.70	21.299	522	1.40	25.367	.208	25.31	4.88
.80	25.564	891	3.17	.332	4.99	4.80	.78	21.261	593	1.57	25.776	.255	25.72	5.01
.90	25.924	1043	3.34	.394	5.11	4.92	.85	21.218	672	1.72	26.164	.304	26.11	5.13
1.00	26.264	1212	3.50	.460	5.22	5.03	.93	21.171	757	1.87	26.534	.354	26.49	5.25
1.10	26.586	1398	3.66	.528	5.33	5.14	1.00	21.121	847	2.02	26.888	.407	26.84	5.36
1.20	26.892	1599	3.80	.600	5.44	5.25	1.07	21.068	943	2.16	27.228	.460	27.19	5.47
1.30	27.184	1816	3.94	.673	5.54	5.35	1.13	21.012	1042	2.30	27.555	.514	27.51	5.58
1.40	27.463	2047	4.07	.749	5.63	5.45	1.20	20.953	1144	2.43	27.869	.570	27.83	5.68
1.50	27.730	2293	4.20	.826	5.73	5.55	1.26	20.892	1250	2.56	28.173	.626	28.14	5.78
1.60	27.986	2553	4.31	.906	5.82	5.64	1.32	20.828	1358	2.69	28.466	.682	28.43	5.88
1.70	28.232	2826	4.42	.987	5.91	5.73	1.38	20.763	1468	2.82	28.750	.740	28.72	5.98
1.80	28.469	3111	4.53	1.070	5.99	5.83	1.44	20.695	1579	2.94	29.025	.797	28.99	6.07
1.90	28.698	3410	4.63	1.154	6.08	5.91	1.50	20.626	1691	3.06	29.293	.855	29.26	6.16
2.00	28.918	3720	4.72	1.240	6.16	6.00	1.56	20.555	1805	3.18	29.552	.914	29.52	6.25
2.10	29.131	4041	4.82	1.327	6.24	6.09	1.61	20.483	1918	3.30	29.804	.972	29.78	6.34
2.20	29.338	4374	4.90	1.416	6.32	6.17	1.67	20.409	2032	3.41	30.049	1.031	30.02	6.42
2.30	29.537	4718	4.99	1.505	6.39	6.25	1.72	20.334	2146	3.52	30.288	1.090	30.26	6.51
2.40	29.731	5072	5.06	1.596	6.47	6.33	1.77	20.258	2260	3.64	30.521	1.150	30.50	6.59
2.50	29.919	5436	5.14	1.687	6.54	6.41	1.82	20.181	2373	3.75	30.748	1.209	30.72	6.67
2.60	30.102	5810	5.21	1.780	6.61	6.49	1.87	20.103	2485	3.85	30.967	1.268	30.94	6.75
2.70	30.279	6193	5.28	1.874	6.68	6.56	1.92	20.024	2596	3.96	31.179	1.328	31.16	6.83

** 10^{10} ergs/g
* 10^{6} ergs/g-deg K

SILICON CARBIDE

COEFFICIENTS IN US=C0+S*UP+S2*UP**2+... 8.000 .950
DE/DP)V= .256 CM**3/G GAMMA=1.250
DEBYE PARAMETERS RHO0= 3.120 T= 293 THETA=1140 3NK= 1.244E+07 CV= 6.463E+06

** 10^{10} ergs/g
* 10^{6} ergs/g-deg K

	PARAMETERS ON THE HUGONIOT							RELEASE ISENTROPE(P=0)			ZERO KELVIN		T 0 ISOTHERM	
P MB	RHO G/CM**3	T DEG-K	S *	E **	C KM/SEC	US KM/SEC	UP KM/SEC	RHC G/CM**3	T DEG-K	UFS KM/SEC	RHO G/CM**3	E **	RHO G/CM**3	C KM/SEC
0.00	3.120	293	3.42	.070	8.00	8.0C	C.C0	3.120	293	0.00	3.124	0.000	3.12	8.00
.10	3.270	314	3.49	.143	8.33	8.36	.38	3.120	296	.77	3.274	.071	3.27	8.33
.20	3.409	348	3.87	.341	8.60	8.70	.74	3.119	313	1.47	3.414	.257	3.41	8.60
.30	3.539	404	4.64	.639	8.83	9.01	1.07	3.118	348	2.13	3.545	.527	3.54	8.83
.40	3.662	479	5.74	1.018	9.03	9.31	1.38	3.115	398	2.76	3.670	.864	3.67	9.03
.50	3.779	574	7.05	1.466	9.21	9.59	1.67	3.112	460	3.35	3.791	1.251	3.79	9.21
.60	3.891	685	8.48	1.974	9.37	9.85	1.95	3.107	532	3.91	3.906	1.681	3.90	9.37
.70	3.998	813	9.96	2.533	9.52	10.11	2.22	3.102	613	4.45	4.018	2.144	4.02	9.52
.80	4.101	957	11.45	3.137	9.65	10.35	2.48	3.096	703	4.97	4.127	2.636	4.13	9.65
.90	4.201	1117	12.92	3.781	9.77	10.59	2.72	3.089	800	5.47	4.233	3.152	4.23	9.77
1.00	4.298	1292	14.35	4.461	9.88	10.82	2.96	3.082	903	5.96	4.337	3.687	4.34	9.88
1.10	4.391	1483	15.73	5.173	9.98	11.04	3.19	3.074	1014	6.43	4.438	4.240	4.44	9.98

TUNGSTEN CARBIDE

COEFFICIENTS IN US=C0+S*UP+S2*UP**2+... 4.920 1.339
DE/DP)V= .044 CM**3/G GAMMA=1.500
DEBYE PARAMETERS RHO0=15.020 T= 293 THETA= 605 3NK= 2.547E+06 CV= 2.077E+06

**10^{10} ergs/g
*10^{6} ergs/g-deg K

P	PARAMETERS ON THE HUGONIOT							RELEASE ISENTROPE(P=0)			ZERO KELVIN		T 0 ISOTHERM	
MB	RHO G/CM**3	T DEG-K	S *	E **	C KM/SEC	US KM/SEC	UP KM/SEC	RHO G/CM**3	T DEG-K	UFS KM/SEC	RHO G/CM**3	E **	RHO G/CM**3	C KM/SEC
0.00	15.020	293	1.80	.032	4.92	4.92	0.00	15.020	293	0.00	15.050	0.000	15.02	4.92
.10	15.415	305	1.81	.041	5.13	5.09	.13	15.020	294	.26	15.444	.008	15.42	5.14
.20	15.780	322	1.85	.064	5.33	5.26	.25	15.019	299	.51	15.808	.031	15.78	5.33
.30	16.118	345	1.93	.100	5.50	5.41	.37	15.016	312	.74	16.148	.064	16.13	5.51
.40	16.435	377	2.07	.147	5.67	5.56	.48	15.012	331	.96	16.467	.106	16.45	5.67
.50	16.733	419	2.24	.202	5.82	5.70	.58	15.007	359	1.17	16.769	.155	16.75	5.83
.60	17.015	471	2.45	.266	5.96	5.84	.68	14.999	394	1.37	17.056	.210	17.04	5.97
.70	17.283	533	2.69	.337	6.10	5.97	.78	14.990	436	1.56	17.330	.270	17.31	6.11
.80	17.538	604	2.94	.414	6.23	6.09	.87	14.979	485	1.75	17.592	.334	17.58	6.24
.90	17.781	685	3.20	.497	6.35	6.21	.96	14.967	540	1.93	17.844	.403	17.83	6.36
1.00	18.015	776	3.46	.585	6.47	6.33	1.05	14.953	601	2.11	18.086	.474	18.07	6.48
1.10	18.238	876	3.72	.678	6.58	6.44	1.14	14.938	667	2.28	18.320	.548	18.31	6.60
1.20	18.454	986	3.97	.775	6.68	6.55	1.22	14.921	738	2.45	18.546	.624	18.53	6.71
1.30	18.661	1104	4.22	.876	6.79	6.66	1.30	14.903	814	2.61	18.764	.703	18.75	6.81
1.40	18.862	1232	4.46	.981	6.89	6.76	1.38	14.883	895	2.77	18.976	.783	18.97	6.92
1.50	19.055	1368	4.69	1.089	6.99	6.87	1.45	14.863	980	2.92	19.182	.865	19.17	7.02
1.60	19.243	1513	4.91	1.201	7.08	6.97	1.53	14.841	1069	3.07	19.382	.948	19.37	7.12
1.70	19.424	1667	5.13	1.315	7.17	7.07	1.60	14.818	1162	3.22	19.577	1.033	19.57	7.21
1.80	19.601	1828	5.33	1.432	7.26	7.16	1.67	14.794	1259	3.37	19.767	1.119	19.76	7.31
1.90	19.772	1998	5.53	1.552	7.35	7.25	1.74	14.769	1358	3.51	19.952	1.206	19.94	7.40
2.00	19.939	2176	5.72	1.674	7.43	7.35	1.81	14.744	1461	3.65	20.132	1.293	20.12	7.49

STAINLESS STEEL
TYPE 304
68/19/10/2(MAX)/1(MAX)/0.08(MAX)
FE/CR/NI/MN /SI /C

COEFFICIENTS IN US=C0+S*UP+S2*UP**2+... 4.569 1.490
DE/DP)V= .058 CM**3/G GAMMA=2.170
DEBYE PARAMETERS RHO0= 7.896 T= 293 THETA= 230 3NK= 4.550E+06 CV= 4.413E+06

**10^{10} ergs/g
*10^{6} ergs/g-deg K

P	PARAMETERS ON THE HUGONIOT							RELEASE ISENTROPE(P=0)			ZERO KELVIN		T 0 ISOTHERM	
MB	RHO G/CM**3	T DEG-K	S *	E **	C KM/SEC	US KM/SEC	UP KM/SEC	RHO G/CM**3	T DEG-K	UFS KM/SEC	RHO G/CM**3	E **	RHO G/CM**3	C KM/SEC
0.00	7.896	293	7.24	.099	4.57	4.57	0.00	7.896	293	0.00	7.977	0.000	7.90	4.57
.10	8.326	332	7.29	.132	5.05	4.95	.26	7.895	296	.51	8.401	.030	8.34	5.06
.20	8.684	382	7.54	.214	5.44	5.28	.48	7.888	313	.96	8.760	.102	8.71	5.45
.30	8.992	452	7.99	.331	5.77	5.58	.68	7.876	345	1.36	9.074	.200	9.03	5.79
.40	9.264	544	8.57	.473	6.06	5.86	.86	7.859	391	1.74	9.356	.316	9.32	6.08
.50	9.507	659	9.21	.636	6.32	6.11	1.04	7.836	448	2.08	9.613	.445	9.58	6.35
.60	9.728	795	9.87	.815	6.56	6.35	1.20	7.810	515	2.41	9.850	.582	9.82	6.60
.70	9.931	951	10.52	1.008	6.79	6.58	1.35	7.780	590	2.72	10.071	.726	10.04	6.84
.80	10.118	1127	11.14	1.212	6.99	6.79	1.49	7.747	671	3.02	10.277	.876	10.25	7.05
.90	10.293	1322	11.73	1.426	7.19	7.00	1.63	7.711	757	3.30	10.472	1.029	10.45	7.26
1.00	10.456	1534	12.29	1.649	7.37	7.19	1.76	7.672	847	3.57	10.656	1.186	10.63	7.46
1.10	10.609	1763	12.81	1.880	7.55	7.38	1.89	7.632	939	3.84	10.830	1.344	10.81	7.65
1.20	10.753	2007	13.31	2.118	7.72	7.56	2.01	7.589	1033	4.10	10.996	1.505	10.98	7.84
1.30	10.890	2266	13.77	2.362	7.88	7.74	2.13	7.545	1128	4.35	11.155	1.666	11.14	8.02
1.40	11.019	2540	14.20	2.612	8.04	7.91	2.24	7.500	1224	4.59	11.307	1.829	11.29	8.19
1.50	11.143	2826	14.61	2.867	8.19	8.07	2.35	7.453	1320	4.83	11.453	1.992	11.43	8.36
1.60	11.260	3125	14.99	3.126	8.34	8.24	2.46	7.405	1415	5.06	11.592	2.155	11.58	8.52
1.70	11.373	3436	15.35	3.390	8.48	8.39	2.57	7.357	1509	5.29	11.727	2.319	11.71	8.69
1.80	11.481	3758	15.70	3.658	8.62	8.54	2.67	7.307	1602	5.51	11.856	2.482	11.84	8.84
1.90	11.585	4091	16.02	3.930	8.75	8.69	2.77	7.256	1693	5.73	11.978	2.646	11.96	8.99

NEOPRENE

COMMERCIAL

COEFFICIENTS IN US=CO+S*UP+S2*UP**2+... 2.785 1.419
DE/DP)V= .500 CM**3/G GAMMA=1.390
DEBYE PARAMETERS RHOO= 1.439 T= 293 THETA= 322 3NK= 1.750E+07 CV= 1.649E+07

$**10^{10}$ ergs/g
$*10^{6}$ ergs/g-deg K

	PARAMETERS ON THE HUGONIOT							RELEASE ISENTRCPE(P=0)			ZERO KELVIN		T 0 ISOTHERM	
P MB	RHO G/CM**3	T DEG-K	S *	E **	C KM/SEC	US KM/SEC	UP KM/SEC	RHC G/CM**3	T DEG-K	UFS KM/SEC	RHO G/CM**3	E **	RHO G/CM**3	C KM/SEC
0.00	1.439	293	22.20	.345	2.78	2.78	0.00	1.439	293	0.00	1.522	0.000	1.44	2.78
.02	1.640	354	22.50	.430	3.48	3.37	.41	1.437	298	.83	1.705	.063	1.66	3.52
.04	1.778	415	23.57	.610	3.96	3.82	.73	1.428	315	1.46	1.839	.188	1.80	4.02
.06	1.886	486	25.12	.839	4.35	4.20	.99	1.415	341	2.00	1.947	.338	1.92	4.43
.08	1.975	566	26.91	1.099	4.68	4.53	1.23	1.397	373	2.48	2.039	.498	2.02	4.79
.10	2.050	656	28.78	1.381	4.98	4.83	1.44	1.377	407	2.92	2.118	.663	2.10	5.12
.12	2.117	755	30.65	1.680	5.26	5.10	1.63	1.354	443	3.33	2.189	.830	2.17	5.42
.14	2.176	861	32.46	1.992	5.51	5.36	1.81	1.329	479	3.72	2.252	.997	2.24	5.70
.16	2.229	973	34.19	2.315	5.75	5.60	1.98	1.301	513	4.09	2.310	1.163	2.30	5.97
.18	2.277	1091	35.84	2.647	5.97	5.83	2.15	1.273	546	4.44	2.361	1.328	2.35	6.22

ADIPRENE

L-167/MOCA (LASL)

COEFFICIENTS IN US=CO+S*UP+S2*UP**2+... 2.332 1.536
DE/DP)V= .729 CM**3/G GAMMA=1.480
DEBYE PARAMETERS RHOO= .927 T= 293 THETA= 25C 3NK= 2.240E+07 CV= 2.161E+07

$**10^{10}$ ergs/g
$*10^{6}$ ergs/g-deg K

	PARAMETERS ON THE HUGONIOT							RELEASE ISENTRCPE(P=0)			ZERO KELVIN		T 0 ISOTHERM	
P MB	RHO G/CM**3	T DEG-K	S *	E **	C KM/SEC	US KM/SEC	UP KM/SEC	RHC G/CM**3	T DEG-K	UFS KM/SEC	RHO G/CM**3	E **	RHO G/CM**3	C KM/SEC
0.00	.927	293	33.82	.511	2.33	2.33	C.00	.927	293	0.00	1.037	0.000	.93	2.33
.02	1.151	413	35.03	.722	3.53	3.33	.65	.918	305	1.30	1.223	.123	1.18	3.64
.04	1.271	528	37.92	1.096	4.26	3.99	1.08	.892	334	2.19	1.336	.325	1.31	4.44
.06	1.355	658	41.28	1.533	4.84	4.53	1.43	.858	366	2.93	1.419	.542	1.40	5.10
.08	1.419	801	44.60	2.008	5.35	4.99	1.73	.816	393	3.59	1.485	.759	1.47	5.69
.10	1.471	955	47.74	2.507	5.81	5.40	2.00	.776	412	4.20	1.540	.972	1.53	6.23
.12	1.515	1119	50.65	3.024	6.23	5.78	2.24				1.586	1.179	1.58	6.74
.14	1.553	1290	53.34	3.555	6.62	6.12	2.47				1.626	1.381	1.62	7.22
.16	1.586	1469	55.81	4.097	6.99	6.45	2.68				1.661	1.576	1.65	7.68
.18	1.615	1653	58.10	4.649	7.34	6.75	2.88				1.693	1.765	1.69	8.12
.20	1.642	1843	60.22	5.208	7.68	7.04	3.06				1.721	1.949	1.71	8.56
.22	1.666	2037	62.19	5.774	8.01	7.32	3.24				1.746	2.127	1.74	8.98
.24	1.688	2236	64.03	6.346	8.33	7.58	3.42				1.770	2.30C	1.76	9.38
.26	1.708	2439	65.75	6.923	8.63	7.83	3.58				1.791	2.469	1.79	9.78
.28	1.726	2644	67.37	7.505	8.93	8.08	3.74				1.810	2.633	1.81	10.14
.30	1.744	2854	68.90	8.091	9.22	8.31	3.89				1.826	2.792	1.82	10.46
.32	1.760	3066	70.34	8.680	9.50	8.54	4.C4				1.840	2.947	1.84	10.74

SILASTIC RUBBER

RTV-521 (DOW CORNING) P. 284 PLASTICS ENCYC. (1967)

COEFFICIENTS IN US=C0+S*UP+S2*UP**2+... .218 2.694 -.208
DE/DP)V= .521 CM**3/G GAMMA=1.400
DEBYE PARAMETERS RHO0= 1.372 T= 293 THETA= 258 3NK= 1.960E+07 CV= 1.886E+07

** 10^{10} ergs/g
* 10^{6} ergs/g-deg K

P MB	PARAMETERS ON THE HUGONIOT RHO G/CM**3	T DEG-K	S *	E **	C KM/SEC	US KM/SEC	LP KM/SEC	RELEASE ISENTRCPE(P=0) RHC G/CM**3	T DEG-K	LFS KM/SEC	ZERO KELVIN RHO G/CM**3	E **	T 0 ISOTHERM RHO G/CM**3	C KM/SEC
0.00	1.372	293	29.00	.621	.22	.22	0.00	1.372	293	0.00	2.090	0.000	1.37	.22
.05	2.211	809	38.25	1.313	5.61	3.10	1.18				2.281	.079	2.27	6.17
.10	2.310	1178	44.83	2.101	6.57	4.24	1.72				2.397	.236	2.38	6.73
.15	2.392	1559	49.74	2.952	6.85	5.06	2.16				2.508	.466	2.50	6.68
.20	2.470	1951	53.62	3.862	6.86	5.73	2.55				2.627	.782	2.61	6.36
.25	2.550	2353	56.80	4.830	6.72	6.28	2.90				2.762	1.202	2.75	5.91
.30	2.634	2765	59.49	5.860	6.50	6.76	3.24				2.924	1.751	2.91	5.36
.35	2.726	3186	61.79	6.956	6.21	7.17	3.56				3.125	2.470	3.10	4.75
.40	2.827	3617	63.77	8.123	5.89	7.53	3.87				3.392	3.415	3.36	4.07
.45	2.941	4056	65.50	9.370	5.53	7.84	4.18				3.759	4.614	3.72	3.35
.50	3.073	4505	67.01	10.707	5.15	8.11	4.49				4.280	6.069	4.23	2.62
.55	3.229	4962	68.31	12.148	4.75	8.35	4.80				5.058	7.779	4.98	1.93

POLYRUBBER

DENSITY 1.01 G/CM**3

COEFFICIENTS IN US=C0+S*UP+S2*UP**2+... .852 1.865
DE/DP)V= .660 CM**3/G GAMMA=1.500
DEBYE PARAMETERS RHO0= 1.010 T= 293 THETA= 93 3NK= 2.000E+07 CV= 1.990E+07

** 10^{10} ergs/g
* 10^{6} ergs/g-deg K

P MB	PARAMETERS ON THE HUGONIOT RHO G/CM**3	T DEG-K	S *	E **	C KM/SEC	US KM/SEC	LP KM/SEC	RELEASE ISENTRCPE(P=0) RHC G/CM**3	T DEG-K	LFS KM/SEC	ZERO KELVIN RHO G/CM**3	E **	T 0 ISOTHERM RHO G/CM**3	C KM/SEC
0.00	1.010	293	49.67	.729	.85	.85	0.00	1.010	293	0.00	1.487	0.000	1.01	.85
.02	1.543	590	53.28	1.071	3.01	2.39	.83				1.674	.058	1.63	3.74
.04	1.662	789	57.67	1.506	4.25	3.18	1.25				1.755	.138	1.73	5.27
.06	1.728	992	61.56	1.963	5.29	3.78	1.57				1.805	.215	1.79	6.59
.08	1.771	1200	64.94	2.431	6.23	4.29	1.85				1.840	.288	1.83	7.81
.10	1.803	1413	67.90	2.906	7.09	4.74	2.09				1.866	.356	1.86	8.94
.12	1.828	1630	70.52	3.387	7.90	5.15	2.31				1.883	.420	1.88	9.88
.14	1.848	1849	72.87	3.872	8.67	5.53	2.51				1.893	.480	1.89	10.47

POLYRUBBER

DENSITY 0.77 G/CM**3

COEFFICIENTS IN US=C0+S*UP+S2*UP**2+... .875 1.432
DE/DP)V= .866 CM**3/G GAMMA=1.500
DEBYE PARAMETERS RHO0= .770 T= 293 THETA= 93 3NK= 2.000E+07 CV= 1.990E+07

**10^{10} ergs/g
*10^{6} ergs/g-deg K

	PARAMETERS ON THE HUGONIOT							RELEASE ISENTROPE(P=0)			ZERO KELVIN		T 0 ISOTHERM	
P MB	RHO G/CM**3	T DEG-K	S *	E **	C KM/SEC	US KM/SEC	UP KM/SEC	RHO G/CM**3	T DEG-K	UFS KM/SEC	RHO G/CM**3	E **	RHO G/CM**3	C KM/SEC
0.00	.770	293	49.67	.779	.87	.87	0.00	.770	293	0.00	1.282	0.000	.77	.87
.02	1.388	739	54.78	1.358	2.45	2.42	1.08				1.594	.114	1.54	3.01
.04	1.563	1076	60.43	2.097	3.27	3.20	1.62				1.734	.260	1.71	4.10
.06	1.666	1424	65.12	2.875	3.93	3.81	2.05				1.822	.397	1.80	5.03
.08	1.738	1783	69.04	3.672	4.52	4.32	2.41				1.872	.523	1.86	5.72
.10	1.792	2150	72.38	4.483	5.06	4.77	2.72				1.879	.639	1.88	5.94

POLYETHYLENE

LOW DENSITY

COEFFICIENTS IN US=C0+S*UP+S2*UP**2+... 2.901 1.481
DE/DP)V= .665 CM**3/G GAMMA=1.644
DEBYE PARAMETERS RHO0= .915 T= 293 THETA= 320 3NK= 2.440E+07 CV= 2.300E+07

**10^{10} ergs/g
*10^{6} ergs/g-deg K

	PARAMETERS ON THE HUGONIOT							RELEASE ISENTROPE(P=0)			ZERO KELVIN		T 0 ISOTHERM	
P MB	RHO G/CM**3	T DEG-K	S *	E **	C KM/SEC	US KM/SEC	UP KM/SEC	RHO G/CM**3	T DEG-K	UFS KM/SEC	RHO G/CM**3	E **	RHO G/CM**3	C KM/SEC
0.00	.915	293	31.09	.499	2.90	2.90	0.00	.915	293	0.00	.996	0.000	.92	2.90
.05	1.226	535	35.39	1.191	4.84	4.64	1.18	.889	336	2.38	1.283	.433	1.26	5.00
.10	1.372	832	42.76	2.319	5.93	5.73	1.91	.835	411	3.94	1.437	1.038	1.42	6.23
.15	1.471	1187	49.51	3.596	6.80	6.59	2.49	.779	473	5.23	1.545	1.635	1.54	7.26
.20	1.545	1582	55.27	4.954	7.54	7.32	2.99				1.627	2.205	1.62	8.19
.25	1.604	2008	60.18	6.367	8.22	7.97	3.43				1.694	2.745	1.69	9.05
.30	1.653	2459	64.41	7.820	8.84	8.57	3.83				1.749	3.256	1.74	9.87
.35	1.695	2929	68.12	9.303	9.42	9.12	4.20				1.796	3.742	1.79	10.65
.40	1.732	3416	71.41	10.810	9.96	9.63	4.54				1.836	4.204	1.83	11.38
.45	1.764	3916	74.35	12.337	10.49	10.11	4.87				1.869	4.643	1.87	12.00
.50	1.793	4429	77.02	13.880	10.99	10.56	5.17				1.893	5.059	1.89	12.47

POLYURETHANE

CPR 1009-78 (CPR INTERNATIONAL)

COEFFICIENTS IN US=C0+S*UP+S2*UP**2+... 2.486 1.577
DE/DP)V= .510 CM**3/G GAMMA=1.550
DEBYE PARAMETERS RHO0= 1.265 T= 293 THETA= 281 3NK= 1.890E+07 CV= 1.806E+07

**10^{10} ergs/g
*10^{6} ergs/g-deg K

P	PARAMETERS ON THE HUGONIOT							RELEASE ISENTROPE(P=0)			ZERO KELVIN		T 0 ISOTHERM	
MB	RHO G/CM**3	T DEG-K	S *	E **	C KM/SEC	US KM/SEC	UP KM/SEC	RHO G/CM**3	T DEG-K	UFS KM/SEC	RHO G/CM**3	E **	RHO G/CM**3	C KM/SEC
0.00	1.265	293	26.42	.407	2.49	2.49	0.00	1.265	293	0.00	1.379	0.000	1.27	2.49
.05	1.671	508	29.58	.888	4.33	4.03	.98	1.231	334	1.98	1.742	.290	1.71	4.48
.10	1.855	770	35.10	1.664	5.39	4.99	1.59	1.158	407	3.26	1.929	.692	1.91	5.68
.15	1.975	1081	40.22	2.539	6.25	5.74	2.06				2.056	1.088	2.04	6.69
.20	2.065	1427	44.63	3.469	7.00	6.39	2.47				2.152	1.464	2.14	7.60
.25	2.136	1800	48.40	4.436	7.67	6.96	2.84				2.229	1.821	2.22	8.44
.30	2.194	2195	51.67	5.428	8.30	7.48	3.17				2.292	2.159	2.29	9.24
.35	2.244	2606	54.54	6.441	8.89	7.96	3.47				2.345	2.481	2.34	10.00
.40	2.286	3032	57.09	7.470	9.44	8.41	3.76				2.391	2.787	2.39	10.74
.45	2.324	3470	59.37	8.512	9.98	8.84	4.03				2.431	3.081	2.43	11.45
.50	2.358	3919	61.44	9.566	10.49	9.24	4.28				2.466	3.362	2.46	12.11
.55	2.388	4377	63.33	10.628	10.98	9.62	4.52				2.495	3.631	2.49	12.68
.60	2.415	4843	65.06	11.700	11.46	9.98	4.75				2.518	3.887	2.52	13.14

POLYSTYRENE

FREE FLOWING, GENERAL PURPOSE, UNFILLED MOLDING COMPOUND

COEFFICIENTS IN US=C0+S*UP+S2*UP**2+... 2.746 1.319
DE/DP)V= .812 CM**3/G GAMMA=1.180
DEBYE PARAMETERS RHO0= 1.044 T= 293 THETA= 264 3NK= 1.400E+07 CV= 1.345E+07

**10^{10} ergs/g
*10^{6} ergs/g-deg K

P	PARAMETERS ON THE HUGONIOT							RELEASE ISENTROPE(P=0)			ZERO KELVIN		T 0 ISOTHERM	
MB	RHO G/CM**3	T DEG-K	S *	E **	C KM/SEC	US KM/SEC	UP KM/SEC	RHO G/CM**3	T DEG-K	UFS KM/SEC	RHO G/CM**3	E **	RHO G/CM**3	C KM/SEC
0.00	1.044	293	20.41	.296	2.75	2.75	0.00	1.044	293	0.00	1.090	0.000	1.04	2.75
.02	1.241	371	21.05	.448	3.55	3.47	.55	1.041	306	1.10	1.277	.117	1.25	3.58
.04	1.371	470	23.00	.753	4.08	4.01	.96	1.030	349	1.92	1.407	.329	1.39	4.13
.06	1.471	599	25.49	1.130	4.50	4.45	1.29	1.014	411	2.60	1.511	.569	1.50	4.57
.08	1.553	754	28.04	1.551	4.85	4.84	1.58	.994	483	3.21	1.597	.819	1.59	4.96
.10	1.622	930	30.47	2.003	5.17	5.18	1.85	.970	559	3.77	1.672	1.071	1.66	5.30
.12	1.683	1125	32.73	2.478	5.46	5.50	2.09	.944	634	4.29	1.738	1.320	1.73	5.62
.14	1.737	1334	34.79	2.971	5.72	5.80	2.31	.915	706	4.78	1.797	1.567	1.79	5.92

LUCITE

ACRYLIC RESIN

COEFFICIENTS IN US=CO+S*UP+S2*UP**2+... 2.263 1.816
DE/DP)V=1.129 CM**3/G GAMMA= .750
DEBYE PARAMETERS RHOO= 1.181 T= 293 THETA= 233 3NK= 1.510E+07 CV= 1.463E+07

**10^{10} ergs/g
*10^{6} ergs/g-deg K

P	PARAMETERS ON THE HUGONIOT							RELEASE ISENTROPE(P=0)			ZERO KELVIN		T 0 ISOTHERM	
MB	RHO G/CM**3	T DEG-K	S *	E **	C KM/SEC	US KM/SEC	UP KM/SEC	RHO G/CM**3	T DEG-K	UFS KM/SEC	RHO G/CM**3	E **	RHO G/CM**3	C KM/SEC
0.00	1.181	293	23.83	.332	2.26	2.26	0.00	1.181	293	0.00	1.244	0.000	1.18	2.26
.05	1.572	480	28.36	.858	4.89	4.12	1.03	1.156	392	2.05	1.593	.319	1.58	4.98
.10	1.716	784	34.97	1.652	6.52	5.21	1.62	1.104	588	3.26	1.737	.692	1.73	6.71
.15	1.804	1156	40.44	2.525	7.87	6.06	2.09				1.826	1.038	1.82	8.17
.20	1.867	1573	44.83	3.442	9.07	6.79	2.49				1.889	1.358	1.89	9.47
.25	1.915	2022	48.43	4.388	10.17	7.43	2.85				1.938	1.656	1.94	10.68

PLEXIGLASS

ACRYLIC RESIN

COEFFICIENTS IN US=CO+S*UP+S2*UP**2+... 2.572 1.536
DE/DP)V= .870 CM**3/G GAMMA= .970
DEBYE PARAMETERS RHOO= 1.185 T= 293 THETA= 266 3NK= 1.530E+07 CV= 1.469E+07

**10^{10} ergs/g
*10^{6} ergs/g-deg K

P	PARAMETERS ON THE HUGONIOT							RELEASE ISENTROPE(P=0)			ZERO KELVIN		T 0 ISOTHERM	
MB	RHO G/CM**3	T DEG-K	S *	E **	C KM/SEC	US KM/SEC	UP KM/SEC	RHO G/CM**3	T DEG-K	UFS KM/SEC	RHO G/CM**3	E **	RHO G/CM**3	C KM/SEC
0.00	1.185	293	22.19	.322	2.57	2.57	0.00	1.185	293	0.00	1.238	0.000	1.19	2.57
.05	1.572	475	25.82	.842	4.49	4.14	1.02	1.163	368	2.04	1.603	.347	1.59	4.56
.10	1.751	770	31.96	1.686	5.64	5.11	1.65	1.109	527	3.35	1.786	.807	1.78	5.80
.15	1.870	1144	37.33	2.640	6.58	5.88	2.15	1.042	701	4.42	1.909	1.253	1.90	6.83
.20	1.958	1572	41.74	3.655	7.39	6.54	2.58	.974	860	5.37	2.002	1.676	2.00	7.74
.25	2.029	2039	45.40	4.709	8.13	7.12	2.96				2.077	2.076	2.07	8.59
.30	2.087	2538	48.50	5.793	8.82	7.65	3.31				2.138	2.455	2.13	9.38
.35	2.137	3062	51.18	6.900	9.47	8.14	3.63				2.190	2.817	2.19	10.13
.40	2.180	3606	53.52	8.024	10.08	8.60	3.92				2.235	3.162	2.23	10.85
.45	2.218	4169	55.60	9.163	10.67	9.03	4.20				2.275	3.493	2.27	11.54
.50	2.251	4746	57.46	10.314	11.23	9.44	4.47				2.310	3.811	2.31	12.20
.55	2.282	5338	59.15	11.476	11.78	9.83	4.72				2.342	4.118	2.34	12.84
.60	2.309	5940	60.70	12.647	12.31	10.20	4.96				2.369	4.414	2.37	13.42

PARAFFIN

COEFFICIENTS IN US=C0+S*UP+S2*UP**2+... 2.960 1.531
DE/DP)V= .942 CM**3/G GAMMA=1.180
DEBYE PARAMETERS RHO0= .900 T= 293 THETA= 350 3NK= 3.100E+07 CV= 2.890E+07

** 10^{10} ergs/g
* 10^{6} ergs/g-deg K

P	PARAMETERS ON THE HUGONIOT							RELEASE ISENTROPE(P=0)			ZERO KELVIN		T 0 ISOTHERM	
MB	RHO G/CM**3	T DEG-K	S *	E **	C KM/SEC	US KM/SEC	UP KM/SEC	RHO G/CM**3	T DEG-K	UFS KM/SEC	RHO G/CM**3	E **	RHO G/CM**3	C KM/SEC
0.00	.900	293	36.90	.590	2.96	2.96	0.00	.900	293	0.00	.965	0.000	.90	2.96
.05	1.194	461	41.66	1.274	5.11	4.75	1.17	.880	336	2.35	1.232	.430	1.21	5.24
.10	1.330	677	50.41	2.386	6.38	5.86	1.90	.835	422	3.86	1.369	1.020	1.36	6.62
.15	1.421	938	58.74	3.644	7.40	6.74	2.47	.782	507	5.12	1.464	1.599	1.46	7.78
.20	1.488	1230	65.99	4.982	8.30	7.50	2.96				1.535	2.151	1.53	8.81
.25	1.542	1546	72.26	6.374	9.11	8.17	3.40				1.592	2.674	1.59	9.77
.30	1.587	1881	77.70	7.804	9.86	8.78	3.80				1.639	3.171	1.64	10.66
.35	1.625	2231	82.50	9.264	10.57	9.34	4.17				1.680	3.645	1.68	11.51
.40	1.658	2595	86.76	10.747	11.24	9.86	4.51				1.715	4.098	1.71	12.33
.45	1.687	2969	90.58	12.250	11.88	10.35	4.83				1.745	4.533	1.74	13.11
.50	1.713	3353	94.05	13.770	12.49	10.82	5.13				1.772	4.950	1.77	13.85
.55	1.736	3745	97.22	15.304	13.08	11.27	5.42				1.794	5.349	1.79	14.49

EPOXY

100 PARTS BY WT EPON 828 - 14 PARTS BY WT DIETHANOLAMINE

COEFFICIENTS IN US=C0+S*UP+S2*UP**2+... 2.678 1.520
DE/DP)V= .739 CM**3/G GAMMA=1.130
DEBYE PARAMETERS RHO0= 1.198 T= 293 THETA= 248 3NK= 1.080E+07 CV= 1.042E+07

** 10^{10} ergs/g
* 10^{6} ergs/g-deg K

P	PARAMETERS ON THE HUGONIOT							RELEASE ISENTROPE(P=0)			ZERO KELVIN		T 0 ISOTHERM	
MB	RHO G/CM**3	T DEG-K	S *	E **	C KM/SEC	US KM/SEC	UP KM/SEC	RHO G/CM**3	T DEG-K	UFS KM/SEC	RHO G/CM**3	E **	RHO G/CM**3	C KM/SEC
0.00	1.198	293	16.39	.232	2.68	2.68	0.00	1.198	293	0.00	1.240	0.000	1.20	2.68
.02	1.397	362	16.93	.351	3.60	3.42	.49	1.195	308	.98	1.427	.091	1.41	3.63
.04	1.522	457	18.53	.587	4.23	3.96	.84	1.185	355	1.69	1.553	.254	1.54	4.28
.06	1.615	582	20.55	.878	4.74	4.41	1.14	1.169	423	2.29	1.644	.437	1.63	4.82
.08	1.688	733	22.62	1.202	5.18	4.79	1.39	1.149	503	2.81	1.721	.625	1.71	5.30
.10	1.750	905	24.58	1.548	5.58	5.14	1.62	1.126	590	3.29	1.786	.814	1.78	5.74
.12	1.803	1096	26.38	1.912	5.95	5.46	1.83	1.099	678	3.74	1.841	1.000	1.83	6.14
.14	1.849	1302	28.03	2.289	6.30	5.76	2.03	1.069	763	4.17	1.891	1.183	1.88	6.53
.16	1.890	1521	29.54	2.677	6.62	6.04	2.21	1.036	843	4.58	1.935	1.363	1.93	6.89
.18	1.927	1752	30.91	3.074	6.93	6.30	2.38	1.000	914	4.97	1.974	1.539	1.97	7.24
.20	1.961	1994	32.17	3.479	7.22	6.55	2.55	.960	971	5.36	2.010	1.712	2.01	7.57

EXON

POLYVINYL CHLORIDE (FIRESTONE PLASTICS)

COEFFICIENTS IN US=C0+S*UP+S2*UP**2+... 1.952 1.660
DE/DP)V= .684 CM**3/G GAMMA= .870
DEBYE PARAMETERS RHO0= 1.681 T= 293 THETA= 255 3NK= 2.170E+07 CV= 2.090E+07

**10^{10} ergs/g
*10^{6} ergs/g-deg K

P MB	RHO G/CM**3	T DEG-K	S *	E **	C KM/SEC	US KM/SEC	UP KM/SEC	RHO G/CM**3	T DEG-K	UFS KM/SEC	RHO G/CM**3	E **	RHO G/CM**3	C KM/SEC
	PARAMETERS ON THE HUGONIOT							RELEASE ISENTROPE(P=0)			ZERO KELVIN		T 0 ISOTHERM	
0.00	1.681	293	32.35	.470	1.95	1.95	0.00	1.681	293	0.00	1.833	0.000	1.68	1.95
.02	2.013	348	32.93	.568	2.93	2.69	.44	1.674	300	.88	2.099	.059	2.03	2.98
.04	2.194	399	34.56	.748	3.58	3.19	.75	1.654	321	1.49	2.263	.159	2.22	3.66
.06	2.321	458	36.69	.962	4.11	3.60	.99	1.624	349	1.99	2.383	.269	2.35	4.23
.08	2.418	523	38.99	1.195	4.58	3.95	1.20	1.585	381	2.43	2.477	.380	2.45	4.73
.10	2.497	595	41.31	1.442	5.00	4.27	1.39	1.539	413	2.83	2.554	.490	2.53	5.19
.12	2.563	671	43.57	1.699	5.40	4.55	1.57	1.484	443	3.20	2.620	.597	2.60	5.62
.14	2.620	752	45.74	1.963	5.77	4.82	1.73	1.425	470	3.56	2.677	.703	2.66	6.03
.16	2.670	836	47.80	2.233	6.12	5.07	1.88				2.727	.805	2.71	6.42
.18	2.714	924	49.76	2.508	6.46	5.30	2.02				2.771	.905	2.76	6.79
.20	2.754	1014	51.60	2.788	6.78	5.53	2.15				2.811	1.003	2.80	7.15

DURITE

FURFURAL FILLED PHENOLIC (DURITE PLASTICS)

COEFFICIENTS IN US=C0+S*UP+S2*UP**2+... 2.847 1.404
DE/DP)V=1.460 CM**3/G GAMMA= .500
DEBYE PARAMETERS RHO0= 1.370 T= 293 THETA= 309 3NK= 1.550E+07 CV= 1.467E+07

**10^{10} ergs/g
*10^{6} ergs/g-deg K

P MB	RHO G/CM**3	T DEG-K	S *	E **	C KM/SEC	US KM/SEC	UP KM/SEC	RHO G/CM**3	T DEG-K	UFS KM/SEC	RHO G/CM**3	E **	RHO G/CM**3	C KM/SEC
	PARAMETERS ON THE HUGONIOT							RELEASE ISENTROPE(P=0)			ZERO KELVIN		T 0 ISOTHERM	
0.00	1.370	293	20.27	.301	2.85	2.85	0.00	1.370	293	0.00	1.395	0.000	1.37	2.85
.05	1.751	387	22.79	.698	4.34	4.10	.89	1.363	347	1.77	1.766	.298	1.76	4.36
.10	1.959	580	28.23	1.399	5.28	4.93	1.48	1.343	494	2.95	1.976	.735	1.97	5.33
.15	2.107	843	33.58	2.216	6.04	5.59	1.96	1.314	692	3.90	2.127	1.178	2.12	6.12
.20	2.222	1156	38.18	3.101	6.71	6.17	2.37	1.276	920	4.73	2.246	1.610	2.24	6.82
.25	2.317	1508	42.10	4.031	7.30	6.68	2.73	1.231	1162	5.48	2.344	2.026	2.34	7.45
.30	2.397	1891	45.44	4.994	7.85	7.15	3.06	1.182	1409	6.18	2.427	2.426	2.42	8.03
.35	2.467	2300	48.34	5.982	8.37	7.58	3.37	1.130	1654	6.84	2.499	2.811	2.50	8.58
.40	2.529	2730	50.89	6.991	8.86	7.98	3.66				2.563	3.183	2.56	9.11
.45	2.584	3179	53.16	8.016	9.32	8.36	3.93				2.620	3.543	2.62	9.61
.50	2.633	3643	55.20	9.056	9.77	8.72	4.18				2.671	3.892	2.67	10.08
.55	2.679	4122	57.04	10.108	10.19	9.06	4.43				2.718	4.230	2.72	10.55
.60	2.720	4613	58.72	11.171	10.61	9.39	4.66				2.761	4.559	2.76	11.00
.65	2.759	5116	60.27	12.243	11.01	9.71	4.89				2.800	4.880	2.80	11.43

PHENOXY

COEFFICIENTS IN US=C0+S*UP+S2*UP**2+... 2.266 1.698
DE/DP)V=1.543 CM**3/G GAMMA= .550
DEBYE PARAMETERS RHO0= 1.178 T= 293 THETA= 245 3NK= 1.730E+07 CV= 1.671E+07

**10^{10} ergs/g
*10^{6} ergs/g-deg K

P	PARAMETERS ON THE HUGONIOT							RELEASE ISENTROPE(P=0)			ZERO KELVIN		T0 ISOTHERM	
MB	RHO G/CM**3	T DEG-K	S *	E **	C KM/SEC	US KM/SEC	UP KM/SEC	RHO G/CM**3	T DEG-K	UFS KM/SEC	RHO G/CM**3	E **	RHO G/CM**3	C KM/SEC
0.00	1.178	293	26.46	.369	2.27	2.27	0.00	1.178	293	0.00	1.221	0.000	1.18	2.27
.02	1.417	339	27.37	.512	3.51	3.17	.53	1.174	309	1.07	1.439	.101	1.42	3.53
.04	1.543	407	29.81	.771	4.34	3.79	.90	1.164	355	1.78	1.561	.260	1.55	4.39
.06	1.630	496	32.78	1.076	5.04	4.28	1.19	1.148	420	2.36	1.647	.425	1.64	5.10
.08	1.697	601	35.79	1.408	5.65	4.71	1.44	1.128	496	2.87	1.714	.589	1.71	5.74
.10	1.750	719	38.65	1.757	6.20	5.10	1.67	1.103	579	3.32	1.768	.748	1.76	6.33
.12	1.795	847	41.30	2.120	6.72	5.44	1.87	1.075	665	3.75	1.813	.902	1.81	6.87
.14	1.833	983	43.73	2.494	7.21	5.77	2.06	1.040	750	4.15	1.852	1.052	1.85	7.39
.16	1.867	1126	45.97	2.875	7.67	6.07	2.24	1.001	834	4.53	1.886	1.197	1.88	7.88
.18	1.896	1275	48.03	3.263	8.11	6.35	2.41	.961	914	4.90	1.916	1.338	1.91	8.36
.20	1.923	1430	49.93	3.658	8.54	6.62	2.56				1.942	1.475	1.94	8.81
.22	1.947	1590	51.68	4.057	8.95	6.88	2.72				1.967	1.608	1.96	9.26
.24	1.968	1754	53.31	4.460	9.35	7.12	2.86				1.989	1.738	1.99	9.68
.26	1.988	1922	54.84	4.867	9.74	7.36	3.00				2.009	1.865	2.01	10.10

PHENOLIC REFRASIL

OBLIQUE, TAPE-WOUND SILICA IMBEDDED IN PHENOLIC

COEFFICIENTS IN US=C0+S*UP+S2*UP**2+... 3.016 1.005
DE/DP)V= .373 CM**3/G GAMMA=1.630
DEBYE PARAMETERS RHO0= 1.644 T= 293 THETA= 330 3NK= 1.340E+07 CV= 1.259E+07

**10^{10} ergs/g
*10^{6} ergs/g-deg K

P	PARAMETERS ON THE HUGONIOT							RELEASE ISENTROPE(P=0)			ZERO KELVIN		T0 ISOTHERM	
MB	RHO G/CM**3	T DEG-K	S *	E **	C KM/SEC	US KM/SEC	UP KM/SEC	RHO G/CM**3	T DEG-K	UFS KM/SEC	RHO G/CM**3	E **	RHO G/CM**3	C KM/SEC
0.00	1.644	293	16.69	.262	3.02	3.02	0.00	1.644	293	0.00	1.721	0.000	1.64	3.02
.10	2.395	679	20.91	1.216	3.92	4.40	1.38	1.601	390	2.79	2.514	.733	2.48	3.91
.20	2.887	1346	27.38	2.881	4.27	5.32	2.29	1.507	576	4.71	3.116	1.847	3.09	4.23
.30	3.280	2223	32.57	4.814	4.49	6.05	3.02	1.401	744	6.31	3.650	3.005	3.63	4.42
.40	3.618	3252	36.63	6.900	4.65	6.68	3.64	1.309	873	7.72	4.145	4.141	4.13	4.57
.50	3.918	4395	39.89	9.089	4.78	7.24	4.20				4.612	5.235	4.60	4.69
.60	4.191	5628	42.61	11.353	4.89	7.75	4.71				5.057	6.280	5.05	4.79
.70	4.443	6934	44.92	13.674	4.98	8.22	5.18				5.483	7.277	5.47	4.89
.80	4.678	8301	46.92	16.042	5.06	8.66	5.62				5.890	8.227	5.88	4.98
.90	4.899	9719	48.69	18.449	5.14	9.08	6.03				6.255	9.131	6.25	5.05
1.00	5.108	11181	50.26	20.887	5.20	9.47	6.42				6.553	9.989	6.55	5.09
1.10	5.307	12681	51.69	23.353	5.27	9.85	6.80				6.757	10.800	6.75	5.12
1.20	5.497	14216	52.98	25.844	5.32	10.20	7.15				6.847	11.565	6.84	5.13

LITHIUM TETRABORATE- EPOXY
90 WT PER CENT LITHIUM TETRABORATE - 10 WT PER CENT EPOXY (100 PARTS BY WT EPON 815 TO 11 PARTS BY WT DIETHYLENE TRIAMINE)

COEFFICIENTS IN US=C0+S*UP+S2*UP**2+... 4.052 1.486 -.033
DE/DP)V= .459 CM**3/G GAMMA=1.000
DEBYE PARAMETERS RHO0= 2.178 T= 293 THETA= 250 3NK= 1.000E+07 CV= 9.645E+06

$^{**}10^{10}$ ergs/g
$^{*}10^{6}$ ergs/g-deg K

P	PARAMETERS ON THE HUGONIOT							RELEASE ISENTROPE(P=0)			ZERO KELVIN		T 0 ISOTHERM	
MB	RHO G/CM**3	T DEG-K	S *	E **	C KM/SEC	US KM/SEC	UP KM/SEC	RHO G/CM**3	T DEG-K	UFS KM/SEC	RHO G/CM**3	E **	RHO G/CM**3	C KM/SEC
0.00	2.178	293	15.10	.211	4.05	4.05	0.00	2.178	293	0.00	2.206	0.000	2.18	4.05
.05	2.423	337	15.47	.327	4.93	4.76	.48	2.177	304	.96	2.445	.100	2.43	4.94
.10	2.601	411	16.81	.585	5.55	5.31	.86	2.171	348	1.73	2.623	.304	2.61	5.57
.15	2.744	522	18.73	.922	6.04	5.78	1.19	2.161	422	2.39	2.769	.553	2.76	6.07
.20	2.866	667	20.80	1.313	6.44	6.19	1.48	2.148	517	2.98	2.895	.826	2.88	6.49
.25	2.972	840	22.81	1.745	6.79	6.55	1.75	2.131	629	3.52	3.006	1.113	3.00	6.85
.30	3.068	1039	24.70	2.209	7.10	6.89	2.00	2.113	753	4.02	3.107	1.410	3.10	7.17
.35	3.155	1260	26.43	2.700	7.37	7.20	2.23	2.092	887	4.50	3.200	1.714	3.19	7.45
.40	3.236	1502	28.01	3.213	7.61	7.50	2.45	2.069	1028	4.96	3.287	2.022	3.28	7.71
.45	3.311	1763	29.45	3.746	7.83	7.77	2.66	2.044	1173	5.39	3.368	2.334	3.36	7.94
.50	3.382	2040	30.78	4.297	8.03	8.03	2.86	2.017	1320	5.82	3.446	2.650	3.44	8.14
.55	3.449	2333	31.99	4.863	8.21	8.28	3.05	1.988	1467	6.23	3.519	2.968	3.51	8.33
.60	3.513	2640	33.11	5.444	8.38	8.52	3.24	1.958	1614	6.63	3.590	3.289	3.58	8.50
.65	3.574	2960	34.15	6.039	8.53	8.74	3.41	1.925	1757	7.02	3.658	3.612	3.65	8.65
.70	3.632	3293	35.11	6.645	8.67	8.96	3.59	1.891	1895	7.40	3.723	3.937	3.72	8.79
.75	3.689	3636	36.01	7.264	8.79	9.17	3.76	1.854	2027	7.78	3.786	4.265	3.78	8.91

COPPER OXICE- EPOXY

44 WT PER CENT CUO - 56 WT PER CENT EPOXY (100 PARTS BY WT EPON 815 TO 11 PARTS BY WT DIETHYLENE TRIAMINE)

COEFFICIENTS IN US=CO+S*UP+S2*UP**2+... 2.110 1.767 -.062
DE/DP)V= .585 CM**3/G GAMMA=1.000
DEBYE PARAMETERS RHOO= 1.710 T= 293 THETA= 25C 3NK= 1.000E+07 CV= 9.645E+06

** 10^{10} ergs/g
* 10^{6} ergs/g-deg K

P	PARAMETERS ON THE HUGONIOT							RELEASE ISENTRCPE(P=0)			ZERO KELVIN		T 0 ISOTHERM	
MB	RHO G/CM**3	T DEG-K	S *	E **	C KM/SEC	US KM/SEC	UP KM/SEC	RHC G/CM**3	T DEG-K	LFS KM/SEC	RHO G/CM**3	E **	RHO G/CM**3	C KM/SEC
0.00	1.710	293	15.10	.216	2.11	2.11	0.00	1.710	293	0.00	1.788	0.000	1.71	2.11
.05	2.234	480	17.62	.559	3.95	3.53	.83	1.674	372	1.66	2.274	.224	2.25	4.01
.10	2.467	776	21.64	1.113	4.91	4.37	1.34	1.592	530	2.72	2.512	.526	2.50	5.03
.15	2.627	1147	25.10	1.747	5.60	5.01	1.75	1.491	697	3.60	2.681	.835	2.67	5.76
.20	2.753	1571	27.93	2.432	6.13	5.56	2.11				2.817	1.148	2.81	6.31
.25	2.860	2033	30.28	3.155	6.55	6.03	2.42				2.934	1.464	2.93	6.74
.30	2.954	2527	32.26	3.909	6.88	6.46	2.72				3.038	1.785	3.03	7.07
.35	3.039	3046	33.97	4.690	7.15	6.84	2.99				3.134	2.112	3.13	7.33
.40	3.117	3588	35.46	5.496	7.37	7.20	3.25				3.224	2.446	3.22	7.53
.45	3.191	4148	36.78	6.323	7.55	7.53	3.49				3.310	2.790	3.31	7.67
.50	3.262	4726	37.97	7.171	7.69	7.84	3.73				3.394	3.143	3.39	7.78
.55	3.330	5319	39.05	8.039	7.79	8.13	3.96				3.476	3.507	3.47	7.84
.60	3.396	5926	40.03	8.926	7.87	8.41	4.17				3.557	3.883	3.55	7.88
.65	3.461	6546	40.93	9.831	7.93	8.67	4.39				3.637	4.272	3.63	7.88
.70	3.524	7177	41.76	10.753	7.97	8.92	4.59				3.716	4.673	3.71	7.87

ZERIFAC- EPOXY

71 WT PER CENT ZERIFAC (COMMERCIALLY MINED LITHIUM ALUMINUM SILICATE) -
29 WT PER CENT EPOXY (100 PARTS BY WT EPON 815 TO 11 PARTS BY WT
DIETHYLENE TRIAMINE)

COEFFICIENTS IN US=C0+S*UP+S2*UP**2+... 2.765 1.247
DE/DP)V= .563 CM**3/G GAMMA=1.000
DEBYE PARAMETERS RHO0= 1.775 T= 293 THETA= 25C 3NK= 1.000E+07 CV= 9.645E+06

** 10^{10} ergs/g
* 10^{6} ergs/g-deg K

P	PARAMETERS ON THE HUGONIOT							RELEASE ISENTRCPE(P=0)			ZERO KELVIN		T 0 ISOTHERM	
MB	RHO G/CM**3	T DEG-K	S *	E **	C KM/SEC	US KM/SEC	UP KM/SEC	RHC G/CM**3	T DEG-K	UFS KM/SEC	RHO G/CM**3	E **	RHO G/CM**3	C KM/SEC
0.00	1.775	293	15.10	.213	2.76	2.76	0.00	1.775	293	0.00	1.823	0.000	1.77	2.76
.05	2.231	416	16.52	.501	3.75	3.71	.76	1.764	337	1.52	2.270	.225	2.25	3.77
.10	2.517	638	19.83	1.043	4.35	4.37	1.29	1.733	464	2.59	2.565	.593	2.55	4.40
.15	2.732	948	23.19	1.693	4.82	4.91	1.72	1.687	634	3.48	2.795	.988	2.78	4.89
.20	2.908	1324	26.13	2.408	5.21	5.38	2.10	1.631	821	4.27	2.984	1.384	2.97	5.33
.25	3.057	1754	28.63	3.166	5.56	5.80	2.43	1.565	1008	4.99	3.148	1.772	3.14	5.71
.30	3.186	2227	30.78	3.956	5.88	6.18	2.74	1.491	1183	5.67	3.291	2.152	3.28	6.07
.35	3.302	2736	32.64	4.772	6.18	6.53	3.02	1.412	1333	6.33	3.419	2.520	3.41	6.40
.40	3.405	3276	34.28	5.607	6.45	6.86	3.28				3.534	2.878	3.53	6.72
.45	3.500	3843	35.73	6.459	6.71	7.17	3.53				3.640	3.226	3.63	7.02
.50	3.586	4433	37.04	7.326	6.96	7.47	3.77				3.737	3.564	3.73	7.31
.55	3.666	5043	38.22	8.205	7.19	7.75	4.00				3.826	3.894	3.82	7.59
.60	3.741	5673	39.30	9.094	7.42	8.02	4.21				3.910	4.214	3.91	7.86
.65	3.810	6319	40.29	9.993	7.64	8.28	4.42				3.987	4.527	3.98	8.12
.70	3.875	6980	41.21	10.900	7.85	8.53	4.62				4.057	4.831	4.05	8.35

ZERIFAC- PCLYURETHANE FOAM
50 WT PER CENT ZERIFAC (COMMERCIALLY MINED LITHIUM ALUMINUM SILICATE) -
50 WT PER CENT POLYURETHANE FOAM

COEFFICIENTS IN US=C0+S*UP+S2*UP**2+... .327 1.479 -.021
DE/DP)V=1.414 CM**3/G GAMMA=1.000
DEBYE PARAMETERS RHO0= .707 T= 293 THETA= 25C 3NK= 1.000E+07 CV= 9.645E+06

** 10^{10} ergs/g
* 10^{6} ergs/g deg K

P	PARAMETERS ON THE HUGONIOT							RELEASE ISENTRCPE(P=0)			ZERO KELVIN		T 0 ISOTHERM	
MB	RHO G/CM**3	T DEG-K	S *	E **	C KM/SEC	US KM/SEC	UP KM/SEC	RHC G/CM**3	T DEG-K	UFS KM/SEC	RHO G/CM**3	E **	RHO G/CM**3	C KM/SEC
0.00	.707	293	15.10	.302	.33	.33	0.00	.707	293	0.00	1.405	0.000	.71	.33
.02	1.710	1200	23.19	1.132	2.59	2.20	1.29				1.800	.139	1.79	3.07
.04	1.860	2045	28.16	2.056	3.63	3.02	1.87				1.945	.257	1.94	4.25
.06	1.948	2918	31.54	3.006	4.36	3.65	2.33				2.036	.370	2.03	5.05
.08	2.013	3811	34.09	3.973	4.93	4.18	2.71				2.105	.483	2.10	5.63
.10	2.065	4717	36.13	4.953	5.38	4.64	3.05				2.163	.597	2.16	6.07
.12	2.109	5634	37.84	5.945	5.74	5.05	3.36				2.214	.713	2.21	6.42
.14	2.149	6560	39.30	6.946	6.05	5.43	3.65				2.260	.834	2.26	6.69
.16	2.185	7494	40.57	7.956	6.31	5.78	3.91				2.303	.958	2.30	6.91
.18	2.218	8434	41.71	8.974	6.52	6.11	4.16				2.344	1.086	2.34	7.08
.20	2.249	9381	42.73	10.000	6.71	6.42	4.40				2.383	1.219	2.38	7.22
.22	2.278	10334	43.65	11.033	6.87	6.72	4.63				2.421	1.356	2.42	7.33
.24	2.306	11292	44.50	12.072	7.01	7.00	4.85				2.457	1.498	2.46	7.42
.26	2.333	12254	45.29	13.118	7.12	7.26	5.06				2.492	1.644	2.49	7.48
.28	2.359	13221	46.01	14.171	7.23	7.52	5.27				2.526	1.796	2.52	7.54
.30	2.385	14192	46.69	15.229	7.32	7.77	5.46				2.558	1.952	2.56	7.58
.32	2.410	15167	47.32	16.293	7.39	8.00	5.66				2.589	2.112	2.59	7.60

MELMAC

MELAMINE FORMALDEHYDE MOLDING COMPOUND (AMERICAN CYANAMID)

COEFFICIENTS IN US=C0+S*UP+S2*UP**2+... 3.505 1.213
DE/DP)V= .774 CM**3/G GAMMA= .890
DEBYE PARAMETERS RHOO= 1.452 T= 293 THETA= 405 3NK= 1.840E+07 CV= 1.676E+07

**10^{10} ergs/g
*10^{6} ergs/g-deg K

P	PARAMETERS ON THE HUGONIOT							RELEASE ISENTROPE(P=0)			ZERO KELVIN		T 0 ISOTHERM	
MB	RHO G/CM**3	T DEG-K	S *	E **	C KM/SEC	US KM/SEC	UP KM/SEC	RHO G/CM**3	T DEG-K	UFS KM/SEC	RHO G/CM**3	E **	RHO G/CM**3	C KM/SEC
0.00	1.452	293	19.43	.313	3.50	3.50	0.00	1.452	293	0.00	1.484	0.000	1.45	3.50
.02	1.594	320	19.57	.375	3.97	3.93	.35	1.452	295	.70	1.622	.053	1.60	3.98
.04	1.708	351	20.22	.520	4.33	4.28	.64	1.449	307	1.29	1.734	.171	1.72	4.35
.06	1.805	391	21.35	.717	4.63	4.60	.90	1.446	327	1.80	1.832	.323	1.82	4.65
.08	1.890	439	22.82	.952	4.89	4.88	1.13	1.440	355	2.26	1.918	.494	1.90	4.91
.10	1.966	497	24.49	1.213	5.12	5.13	1.34	1.434	389	2.69	1.996	.677	1.98	5.15
.12	2.035	562	26.28	1.497	5.32	5.37	1.54	1.426	429	3.09	2.067	.867	2.06	5.37
.14	2.098	635	28.09	1.798	5.52	5.60	1.72	1.417	472	3.46	2.134	1.062	2.12	5.57
.16	2.157	715	29.90	2.114	5.70	5.81	1.90	1.407	519	3.82	2.195	1.260	2.19	5.76
.18	2.212	800	31.66	2.442	5.86	6.01	2.06	1.396	569	4.16	2.253	1.459	2.24	5.94

TEFLON

POLYTETRAFLUOROETHYLENE (DU PONT)

COEFFICIENTS IN US=C0+S*UP+S2*UP**2+... 1.682 1.819
DE/DP)V= .789 CM**3/G GAMMA= .590
DEBYE PARAMETERS RHOO= 2.147 T= 293 THETA= 190 3NK= 1.070E+07 CV= 1.048E+07

**10^{10} ergs/g
*10^{6} ergs/g-deg K

P	PARAMETERS ON THE HUGONIOT							RELEASE ISENTROPE(P=0)			ZERO KELVIN		T 0 ISOTHERM	
MB	RHO G/CM**3	T DEG-K	S *	E **	C KM/SEC	US KM/SEC	UP KM/SEC	RHO G/CM**3	T DEG-K	UFS KM/SEC	RHO G/CM**3	E **	RHO G/CM**3	C KM/SEC
0.00	2.147	293	19.01	.247	1.68	1.68	0.00	2.147	293	0.00	2.247	0.000	2.15	1.68
.02	2.565	338	19.52	.323	2.69	2.39	.39	2.140	307	.78	2.613	.051	2.57	2.71
.04	2.777	399	20.85	.459	3.37	2.86	.65	2.120	346	1.29	2.815	.131	2.79	3.41
.06	2.921	476	22.49	.618	3.93	3.25	.86	2.090	401	1.71	2.955	.214	2.94	4.00
.08	3.029	567	24.17	.790	4.43	3.58	1.04	2.051	464	2.07	3.062	.296	3.05	4.52
.10	3.115	668	25.78	.971	4.89	3.87	1.20	2.004	533	2.40	3.148	.376	3.14	5.00
.12	3.187	777	27.29	1.159	5.32	4.14	1.35	1.947	603	2.71	3.219	.453	3.21	5.45
.14	3.248	892	28.70	1.352	5.72	4.39	1.49	1.877	671	3.00	3.280	.528	3.27	5.88
.16	3.301	1014	30.00	1.550	6.10	4.62	1.61				3.333	.601	3.33	6.29
.18	3.347	1141	31.20	1.750	6.47	4.84	1.73				3.380	.671	3.37	6.68
.20	3.388	1272	32.31	1.954	6.83	5.04	1.85				3.422	.740	3.42	7.06
.22	3.426	1407	33.34	2.160	7.17	5.24	1.96				3.459	.806	3.45	7.43

MICARTA

PULP FILLED PHENOL FORMALDEHYDE (WESTINGHOUSE ELECTRIC)

COEFFICIENTS IN US=CO+S*UP+S2*UP**2+... 3.048 1.422
DE/DP)V=1.259 CM**3/G GAMMA= .570
DEBYE PARAMETERS RHOO= 1.394 T= 293 THETA= 333 3NK= 1.560E+07 CV= 1.464E+07

** 10^{10} ergs/g
* 10^{6} ergs/g-deg K

P	PARAMETERS ON THE HUGONIOT							RELEASE ISENTROPE(P=0)			ZERO KELVIN		T 0 ISOTHERM	
MB	RHO G/CM**3	T DEG-K	S *	E **	C KM/SEC	US KM/SEC	UP KM/SEC	RHO G/CM**3	T DEG-K	UFS KM/SEC	RHO G/CM**3	E **	RHO G/CM**3	C KM/SEC
0.00	1.394	293	19.30	.293	3.05	3.05	C.CO	1.394	293	0.00	1.419	0.000	1.39	3.C5
.05	1.740	378	21.40	.650	4.49	4.25	.84	1.388	337	1.68	1.755	.274	1.74	4.51
.10	1.936	549	26.31	1.297	5.43	5.06	1.42	1.371	464	2.82	1.953	.692	1.95	5.47
.15	2.076	785	31.35	2.061	6.18	5.72	1.88	1.345	637	3.75	2.097	1.125	2.09	6.26
.20	2.187	1068	35.81	2.894	6.84	6.29	2.28	1.313	838	4.56	2.211	1.552	2.21	6.95
.25	2.278	1388	39.66	3.772	7.43	6.80	2.64	1.274	1054	5.30	2.305	1.966	2.30	7.58
.30	2.355	1738	42.97	4.684	7.98	7.26	2.96	1.226	1275	5.99	2.385	2.366	2.38	8.16
.35	2.422	2114	45.87	5.623	8.49	7.69	3.26	1.173	1491	6.64	2.455	2.753	2.45	8.71
.40	2.482	2510	48.42	6.582	8.97	8.09	3.55	1.118	1697	7.26	2.517	3.128	2.51	9.23
.45	2.535	2924	50.70	7.558	9.43	8.47	3.81				2.573	3.492	2.57	9.73
.50	2.583	3354	52.75	8.550	9.87	8.83	4.06				2.623	3.844	2.62	10.20
.55	2.627	3799	54.61	9.554	10.29	9.17	4.30				2.668	4.188	2.67	10.67
.60	2.668	4256	56.30	10.569	10.70	9.49	4.53				2.711	4.521	2.71	11.11
.65	2.705	4724	57.87	11.594	11.10	9.81	4.75				2.749	4.847	2.75	11.55
.70	2.740	5202	59.31	12.627	11.49	10.11	4.97				2.785	5.165	2.78	11.96

SELECTRON

POLYESTER (PITTSBURGH PLATE GLASS)

COEFFICIENTS IN US=CO+S*UP+S2*UP**2+... 2.722 1.333
DE/DP)V= .775 CM**3/G GAMMA=1.060
DEBYE PARAMETERS RHOO= 1.217 T= 293 THETA= 254 3NK= 1.090E+07 CV= 1.050E+07

** 10^{10} ergs/g
* 10^{6} ergs/g-deg K

P	PARAMETERS ON THE HUGONIOT							RELEASE ISENTROPE(P=0)			ZERO KELVIN		T 0 ISOTHERM	
MB	RHO G/CM**3	T DEG-K	S *	E **	C KM/SEC	US KM/SEC	UP KM/SEC	RHO G/CM**3	T DEG-K	UFS KM/SEC	RHO G/CM**3	E **	RHO G/CM**3	C KM/SEC
0.00	1.217	293	16.29	.232	2.72	2.72	C.CO	1.217	293	0.00	1.257	0.000	1.22	2.72
.05	1.619	504	19.25	.742	4.17	4.07	1.01	1.199	381	2.03	1.650	.371	1.64	4.21
.10	1.835	890	24.36	1.616	5.01	4.94	1.66	1.153	587	3.37	1.877	.901	1.87	5.11
.15	1.988	1402	28.69	2.623	5.67	5.64	2.19	1.C88	820	4.49	2.042	1.432	2.04	5.84
.20	2.108	2004	32.17	3.705	6.24	6.24	2.64	1.C16	1037	5.48	2.173	1.944	2.17	6.48
.25	2.207	2672	35.01	4.839	6.75	6.77	3.04				2.282	2.434	2.28	7.C6
.30	2.290	3394	37.38	6.008	7.21	7.25	3.40				2.374	2.902	2.37	7.61
.35	2.363	4159	39.41	7.206	7.65	7.70	3.73				2.455	3.350	2.45	8.12
.40	2.428	4960	41.17	8.427	8.06	8.12	4.C5				2.526	3.779	2.52	8.61
.45	2.485	5791	42.72	9.667	8.44	8.51	4.34				2.589	4.193	2.59	9.08
.50	2.537	6649	44.11	10.922	8.82	8.89	4.62				2.647	4.591	2.64	9.53
.55	2.585	7530	45.36	12.191	9.17	9.24	4.89				2.699	4.975	2.70	9.96
.60	2.629	8431	46.50	13.471	9.52	9.58	5.15				2.744	5.345	2.74	10.34

MIN-K-2000

INSULATING MATERIAL (JOHNS-MANVILLE)

COEFFICIENTS IN US=CO+S*UP+S2*UP**2+... 2.581 1.334
DE/DP)V=3.571 CM**3/G GAMMA=1.000
DEBYE PARAMETERS RHOO= .280 T= 293 THETA= 143 3NK= 9.800E+06 CV= 9.684E+06

**10^{10} ergs/g
*10^{6} ergs/g-deg K

P	PARAMETERS ON THE HUGONIOT							RELEASE ISENTROPE(P=0)			ZERO KELVIN		T 0 ISOTHERM	
MB	RHO G/CM**3	T DEG-K	S *	E **	C KM/SEC	US KM/SEC	UP KM/SEC	RHO G/CM**3	T DEG-K	UFS KM/SEC	RHO G/CM**3	E **	RHO G/CM**3	C KM/SEC
0.00	.280	293	20.15	.236	2.58	2.58	0.00	.280	293	0.00	.288	0.000	.28	2.58
.02	.419	845	27.23	1.422	4.72	4.64	1.54	.269	580	3.11	.429	.764	.43	4.81
.04	.481	1893	34.28	3.225	5.88	5.84	2.45	.247	1081	5.01	.495	1.667	.49	6.09
.06	.523	3210	39.00	5.215	6.80	6.79	3.16				.541	2.500	.54	7.15
.08	.554	4697	42.44	7.307	7.60	7.60	3.76				.575	3.266	.57	8.08
.10	.580	6305	45.11	9.465	8.32	8.31	4.30				.603	3.977	.60	8.94
.12	.600	8004	47.28	11.672	8.99	8.96	4.78				.625	4.640	.62	9.74
.14	.618	9775	49.11	13.916	9.61	9.56	5.23				.645	5.265	.64	10.50
.16	.634	11605	50.68	16.189	10.20	10.12	5.65				.661	5.854	.66	11.15

ELKONITE (2125C)

COPPER-INFILTRATED TUNGSTEN - 75/25 WT PER CENT CU/W (P. R. MALLORY)

COEFFICIENTS IN US=CO+S*UP+S2*UP**2+... 2.285 2.510 -.216
DE/DP)V= .054 CM**3/G GAMMA=1.920
DEBYE PARAMETERS RHOO= 9.690 T= 293 THETA= 300 3NK= 3.272E+06 CV= 3.107E+06

**10^{10} ergs/g
*10^{6} ergs/g-deg K

P	PARAMETERS ON THE HUGONIOT							RELEASE ISENTROPE(P=0)			ZERO KELVIN		T 0 ISOTHERM	
MB	RHO G/CM**3	T DEG-K	S *	E **	C KM/SEC	US KM/SEC	UP KM/SEC	RHO G/CM**3	T DEG-K	UFS KM/SEC	RHO G/CM**3	E **	RHO G/CM**3	C KM/SEC
0.00	9.690	293	4.37	.065	2.28	2.28	0.00	9.690	293	0.00	9.908	0.000	9.69	2.29
.20	11.479	541	5.36	.226	4.27	3.64	.57	9.564	391	1.14	11.612	.113	11.56	4.34
.40	12.265	959	6.87	.499	5.28	4.43	.93	9.255	586	1.90	12.437	.279	12.40	5.40
.60	12.811	1497	8.10	.820	5.94	5.04	1.23				13.036	.462	13.01	6.08
.80	13.249	2119	9.08	1.174	6.40	5.54	1.49				13.535	.659	13.51	6.53
1.00	13.630	2808	9.87	1.557	6.72	5.98	1.73				13.984	.871	13.97	6.79
1.20	13.975	3552	10.52	1.964	6.93	6.36	1.95				14.408	1.102	14.39	6.93
1.40	14.299	4342	11.08	2.394	7.07	6.69	2.16				14.822	1.354	14.81	6.96
1.60	14.611	5172	11.56	2.846	7.14	7.00	2.36				15.238	1.631	15.22	6.90
1.80	14.916	6039	11.99	3.320	7.16	7.28	2.55				15.664	1.934	15.65	6.78
2.00	15.220	6938	12.36	3.815	7.13	7.54	2.74				16.100	2.266	16.08	6.59

ELKONITE (1W3)

COPPER-INFILTRATED TUNGSTEN - 45/55 WT PER CENT CU/W (P. R. MALLORY)

COEFFICIENTS IN US=C0+S*UP+S2*UP**2+... 3.087 1.909 -.106
DE/DP)V= .045 CM**3/G GAMMA=1.800
DEBYE PARAMETERS RHO0=12.400 T= 293 THETA= 300 3NK= 2.488E+06 CV= 2.362E+06

** 10^{10} ergs/g
* 10^{6} ergs/g-deg K

	PARAMETERS ON THE HUGONIOT							RELEASE ISENTROPE(P=0)			ZERO KELVIN		T 0 ISOTHERM	
P MB	RHO G/CM**3	T DEG-K	S *	E **	C KM/SEC	US KM/SEC	UP KM/SEC	RHO G/CM**3	T DEG-K	UFS KM/SEC	RHO G/CM**3	E **	RHO G/CM**3	C KM/SEC
0.00	12.400	293	3.32	.049	3.09	3.09	0.00	12.400	293	0.00	12.513	0.000	12.40	3.09
.20	13.901	409	3.66	.136	4.14	3.87	.42	12.375	336	.84	13.994	.073	13.94	4.16
.40	14.849	646	4.51	.315	4.81	4.42	.73	12.293	473	1.47	14.970	.210	14.93	4.84
.60	15.568	997	5.41	.541	5.30	4.88	.99	12.169	668	2.00	15.734	.370	15.70	5.36
.80	16.159	1443	6.19	.800	5.69	5.27	1.23	12.012	895	2.48	16.377	.544	16.35	5.76
1.00	16.670	1968	6.85	1.082	6.01	5.61	1.44	11.829	1137	2.93	16.945	.728	16.92	6.09
1.20	17.125	2561	7.42	1.384	6.27	5.92	1.63	11.623	1382	3.35	17.460	.919	17.44	6.35
1.40	17.539	3213	7.91	1.703	6.49	6.21	1.82	11.396	1618	3.76	17.939	1.117	17.92	6.56
1.60	17.923	3917	8.33	2.037	6.67	6.47	1.99	11.150	1838	4.15	18.391	1.322	18.37	6.73
1.80	18.284	4669	8.71	2.385	6.82	6.72	2.16	10.884	2036	4.52	18.823	1.535	18.81	6.86
2.00	18.626	5463	9.04	2.745	6.95	6.95	2.32	10.600	2205	4.89	19.241	1.754	19.23	6.96
2.20	18.954	6295	9.34	3.116	7.05	7.16	2.48	10.302	2341	5.26	19.649	1.981	19.63	7.03
2.40	19.270	7164	9.62	3.499	7.13	7.37	2.63	10.009	2447	5.61	20.045	2.215	20.03	7.07

ELKONITE (3W3)

COPPER-INFILTRATED TUNGSTEN - 31/69 WT PER CENT CU/W (P. R. MALLORY)

COEFFICIENTS IN US=C0+S*UP+S2*UP**2+... 3.418 1.776 -.090
DE/DP)V= .042 CM**3/G GAMMA=1.740
DEBYE PARAMETERS RHO0=13.800 T= 293 THETA= 300 3NK= 2.180E+06 CV= 2.070E+06

** 10^{10} ergs/g
* 10^{6} ergs/g-deg K

	PARAMETERS ON THE HUGONIOT							RELEASE ISENTROPE(P=0)			ZERO KELVIN		T 0 ISOTHERM	
P MB	RHO G/CM**3	T DEG-K	S *	E **	C KM/SEC	US KM/SEC	UP KM/SEC	RHO G/CM**3	T DEG-K	UFS KM/SEC	RHO G/CM**3	E **	RHO G/CM**3	C KM/SEC
0.00	13.800	293	2.91	.043	3.42	3.42	0.00	13.800	293	0.00	13.887	0.000	13.80	3.42
.20	15.143	376	3.11	.107	4.25	4.04	.36	13.788	322	.72	15.218	.057	15.17	4.26
.40	16.082	551	3.72	.249	4.81	4.52	.64	13.740	427	1.29	16.177	.171	16.14	4.83
.60	16.825	825	4.44	.434	5.25	4.92	.88	13.663	593	1.78	16.954	.312	16.92	5.28
.80	17.450	1187	5.12	.649	5.59	5.26	1.10	13.563	800	2.22	17.623	.467	17.60	5.64
1.00	17.997	1625	5.71	.888	5.88	5.57	1.30	13.443	1034	2.63	18.218	.634	18.19	5.94
1.20	18.488	2132	6.22	1.145	6.13	5.86	1.48	13.308	1286	3.02	18.761	.808	18.74	6.19
1.40	18.936	2698	6.67	1.419	6.34	6.12	1.66	13.159	1546	3.38	19.265	.989	19.25	6.40
1.60	19.352	3318	7.06	1.706	6.51	6.36	1.82	12.998	1809	3.73	19.740	1.176	19.72	6.57
1.80	19.742	3988	7.41	2.006	6.67	6.58	1.98	12.826	2070	4.07	20.192	1.369	20.18	6.72
2.00	20.112	4702	7.72	2.317	6.80	6.80	2.13	12.643	2322	4.40	20.627	1.567	20.61	6.84
2.20	20.465	5457	8.00	2.639	6.91	7.00	2.28	12.450	2564	4.73	21.048	1.771	21.03	6.94
2.40	20.804	6250	8.25	2.970	7.01	7.19	2.42	12.248	2791	5.04	21.455	1.980	21.44	7.01

ELKONITE (10W3)

COPPER-INFILTRATED TUNGSTEN - 24/76 WT PER CENT CU/W (P. R. MALLORY)

COEFFICIENTS IN US=C0+S*UP+S2*UP**2+... 3.183 2.133 -.241
DE/DP)V= .040 CM**3/G GAMMA=1.700
DEBYE PARAMETERS RHO0=14.850 T= 293 THETA= 300 3NK= 1.979E+06 CV= 1.879E+06

**10^{10} ergs/g
*10^{6} ergs/g-deg K

P	PARAMETERS ON THE HUGONIOT							RELEASE ISENTROPE(P=0)			ZERO KELVIN		T 0 ISOTHERM	
MB	RHO G/CM**3	T DEG-K	S *	E **	C KM/SEC	US KM/SEC	UP KM/SEC	RHO G/CM**3	T DEG-K	UFS KM/SEC	RHO G/CM**3	E **	RHO G/CM**3	C KM/SEC
0.00	14.850	293	2.64	.039	3.18	3.18	0.00	14.850	293	0.00	14.946	0.000	14.85	3.18
.20	16.299	382	2.86	.099	4.18	3.89	.35	14.833	328	.69	16.375	.052	16.32	4.19
.40	17.252	569	3.46	.226	4.80	4.40	.61	14.775	445	1.23	17.347	.152	17.31	4.82
.60	17.999	851	4.12	.392	5.23	4.81	.84	14.684	620	1.69	18.128	.275	18.10	5.26
.80	18.635	1213	4.73	.586	5.53	5.15	1.05	14.567	831	2.11	18.808	.414	18.78	5.56
1.00	19.204	1644	5.25	.802	5.75	5.45	1.24	14.430	1064	2.50	19.428	.567	19.40	5.77
1.20	19.729	2133	5.69	1.038	5.91	5.72	1.41	14.276	1308	2.87	20.013	.732	19.99	5.91
1.40	20.227	2674	6.07	1.292	6.00	5.96	1.58	14.105	1555	3.23	20.579	.910	20.56	5.98
1.60	20.708	3259	6.41	1.563	6.06	6.17	1.75	13.920	1799	3.58	21.138	1.103	21.12	5.99
1.80	21.180	3885	6.70	1.850	6.07	6.37	1.90	13.721	2032	3.92	21.698	1.310	21.68	5.96
2.00	21.648	4547	6.96	2.154	6.06	6.55	2.06	13.508	2252	4.26	22.268	1.534	22.25	5.89
2.20	22.118	5241	7.19	2.473	6.01	6.71	2.21	13.283	2453	4.59	22.854	1.776	22.83	5.78

ELKONITE (TC-10)

COPPER-INFILTRATED TUNGSTEN CARBIDE - 44/56 WT PER CENT CU/WC
(P. R. MALLORY)

COEFFICIENTS IN US=C0+S*UP+S2*UP**2+... 3.865 1.742 -.083
DE/DP)V= .049 CM**3/G GAMMA=1.740
DEBYE PARAMETERS RHO0=11.650 T= 293 THETA= 300 3NK= 2.389E+06 CV= 2.268E+06

**10^{10} ergs/g
*10^{6} ergs/g-deg K

P	PARAMETERS ON THE HUGONIOT							RELEASE ISENTROPE(P=0)			ZERO KELVIN		T 0 ISOTHERM	
MB	RHO G/CM**3	T DEG-K	S *	E **	C KM/SEC	US KM/SEC	UP KM/SEC	RHO G/CM**3	T DEG-K	UFS KM/SEC	RHO G/CM**3	E **	RHO G/CM**3	C KM/SEC
0.00	11.650	293	3.19	.047	3.86	3.86	0.00	11.650	293	0.00	11.713	0.000	11.65	3.86
.20	12.721	372	3.40	.119	4.72	4.52	.38	11.641	321	.76	12.777	.064	12.74	4.73
.40	13.489	543	4.07	.281	5.31	5.02	.68	11.608	426	1.37	13.560	.197	13.53	5.33
.60	14.105	816	4.87	.495	5.76	5.44	.95	11.551	594	1.90	14.203	.363	14.18	5.79
.80	14.627	1180	5.62	.746	6.13	5.81	1.18	11.478	807	2.38	14.760	.548	14.74	6.17
1.00	15.087	1626	6.28	1.025	6.43	6.14	1.40	11.390	1051	2.83	15.259	.747	15.24	6.48
1.20	15.501	2144	6.85	1.327	6.68	6.44	1.60	11.290	1317	3.24	15.716	.956	15.70	6.74
1.40	15.882	2727	7.35	1.648	6.90	6.72	1.79	11.179	1594	3.64	16.142	1.174	16.13	6.95
1.60	16.236	3369	7.79	1.986	7.08	6.97	1.97	11.059	1878	4.02	16.545	1.400	16.53	7.13
1.80	16.569	4063	8.18	2.340	7.24	7.21	2.14	10.932	2162	4.39	16.930	1.633	16.92	7.28
2.00	16.885	4806	8.52	2.708	7.37	7.44	2.31	10.796	2442	4.75	17.299	1.874	17.29	7.40

AUTHOR INDEX

Numbers in parentheses are reference numbers and indicate that an author's work is referred to although his name is not cited in the text. Numbers in italics show the page on which the complete reference is listed.

SUBJECT INDEX

V

W